pe	NMR Frequency in MHz at a 7.0463 T Field	Natural Abundance %	Relative Sensitivity for Equal Numbers of Nuclei at constant field	Magnetic Moment, μ_N,[a] in multiples of nuclear magneton ($e\eta/4\pi\, mc$)	in multiples of $h/2\pi$	of 10^{-24} cm^2	in MHz[b]	in MHz
Rb	98.163	27.2	0.177	2.7415	3/2	0.15	—	3,417*
Sr	13.001	7.02	2.69×10^{-3}	−1.0893	9/2	—	—	—
Y	14.697	100	1.17×10^{-4}	−0.1368	1/2	—	—	—
Zr	27.991	11.23	9.4×10^{-3}	−1.3	5/2	—	—	—
Nb	73.328	100	0.482	6.1435	9/2	-0.4 ± 0.3	—	—
Mo	19.554	15.78	3.22×10^{-3}	−0.9099	5/2	—	—	−3,528
Mo	19.557	9.60	3.42×10^{-3}	−0.9290	5/2	—	—	−3,601
Ru	10.169	12.81	—	—	6/2	—	—	—
Ru	14.824	16.98	—	—	5/2	—	—	—
Rh	9.442	100	3.12×10^{-5}	−0.0879	1/2	—	—	—
Pd	13.728	22.23	7.79×10^{-4}	−0.57	5/2	—	—	—
Ag	12.139	51.35	6.69×10^{-5}	−0.1130	1/2	—	—	−3,520
Ag	13.956	48.65	1.01×10^{-4}	−0.1299	1/2	—	—	−4,044
Cd	63.616	12.86	9.54×10^{-3}	−0.5922	1/2	—	—	—
Cd	66.548	12.34	1.09×10^{-2}	−0.6195	1/2	—	—	—
In	65.742	95.84	0.348	5.5072	9/2	1.161	—	—
Sn	106.875	7.67	4.53×10^{-2}	−0.9949	1/2	—	—	—
Sn	111.817	8.68	5.18×10^{-2}	−1.0409	1/2	—	—	—
Sb	71.791	57.25	0.160	3.3417	5/2	−0.8	—	—
Sb	38.878	42.75	4.57×10^{-2}	2.5334	7/2	−1.0	—	—
Te	94.790	7.03	3.16×10^{-2}	−0.8824	1/2	—	—	—
I	60.021	100	9.35×10^{-2}	2.7939	5/2	−0.75	—	—
Xe	80.981	26.44	2.12×10^{-2}	—	1/2	—	—	—
Hf	9.361	18.39	6.38×10^{-4}	—	7/2	—	—	—
Hf	5.609	13.78	2.16×10^{-4}	—	9/2	—	—	—
Ta	35.910	100	3.60×10^{-3}	2.1	7/2	6.5	—	—
W	12.48	14.28	7.20×10^{-4}	0.115	1/2	—	—	—
Re	67.541	37.07	0.133	3.1437	5/2	2.8	—	—
Re	68.233	62.93	0.137	3.1760	5/2	2.6	—	—
Os	23.276	16.1	2.24×10^{-3}	0.6507	3/2	2.0	—	—
Ir	5.156	38.5	3.5×10^{-5}	0.16	3/2	∼1.2	—	—
Ir	5.614	61.5	4.2×10^{-5}	0.17	3/2	∼1.0	—	—
Pt	64.447	33.7	9.94×10^{-3}	0.6004	1/2	—	—	—
Au	5.138	100	2.14×10^{-5}	0.136	3/2	0.56	—	—
Hg	53.481	16.86	5.72×10^{-3}	0.4993	1/2	—	—	—
Hg	19.799	13.24	1.90×10^{-3}	−0.607	3/2	0.5	—	—
Ti	171.448	29.52	0.187	1.5960	1/2	—	—	—
Ti	173.124	70.48	0.192	1.6114	1/2	—	—	—
Pb	62.765	21.11	9.13×10^{-3}	0.5837	1/2	—	—	—
Bi	48.208	100	0.137	4.0389	9/2	−0.4	—	—
ectron	197.246	—	2.85×10^8	−1836	1/2	—	—	—

Physical Methods for Chemists

SECOND EDITION

RUSSELL S. DRAGO
University of Florida, Gainesville

SURFSIDE
SCIENTIFIC PUBLISHERS

P.O. Box 13413
Gainesville, FL 32604

Copyright © 1992, 1977
All rights reserved. No part of this publication may be reproduced or transmitted in any form or by any means, electronic or mechanical, including photocopy, recording or any information storage and retrieval system, without permission in writing from the publisher.

Requests for permission to make copies of any part of the work should be mailed to: Surfside Scientific Publishers, P.O. Box 13413, Gainesville, FL 32604-1413.

Text Typeface: Times Roman

Library of Congress Cataloging-in-Publication Data

Drago, Russell S.
    Physical methods for chemists / Russell S. Drago. — 2nd ed. p. cm.
    Rev. ed. of: Physical methods in chemistry. 1977.
    Includes bibliographical references (p.   ) and index.
    ISBN 0-03-075176-4
    1. Chemistry, Physical and theoretical. 2. Spectrum analysis.
I. Drago, Russell S. Physical methods in chemistry. II. Title.
QD453.2.D7 1992
543—dc20   91-41550
                                                          CIP

OTHER SURFSIDE TITLES:

SOLUTION MANUAL - PHYSICAL METHODS FOR CHEMISTS by Ngai M. Wong and Russell S. Drago

APPLICATIONS OF ELECTROSTATIC-COVALENT MODELS IN CHEMISTRY by Russell S. Drago

Send orders to Surfside Scientific Publishers, P.O. Box 13413, Gainesville, FL 32604-1413

To RUTH
with love

Preface

The revision of my earlier textbook, "Physical Methods in Chemistry," was motivated by a sense of gratitude to those who continued to use the book even though several parts were out of date. Those parts of the last edition that remain current have not been changed. As in the earlier edition, topics have been subdivided into numbered paragraphs so some can be dropped, if desired, permitting use of the text in either a theoretical or more applications-oriented course.

The popularity of the earlier edition is based on the recognition that there is more to the application of physical methods in chemistry than memorizing where functional groups appear. Core material in the area of spectroscopy is presented that is of importance to both organic and inorganic chemists. This text is designed to reflect this unity. A fundamental subdivision of chemistry involves dividing it into systems that do and do not contain unpaired d-electrons. The material in this book has been so subdivided. In using this book, a chemist with an interest in organic or organometallic sysnthesis can omit Chapters 10, 11, 12, and 13.

This book contains more material than can be covered reasonably in a one-semester course. Accordingly, several methods in less common usage (Mossbauer, NQR, surface techniques) are presented in a succinct, basic manner. In this way, the reader is at least made aware of their existence.

The book continues in the philosophy of the earlier edition, namely, that chemists without an advanced mathematical background learn to use spectroscopic methods by reading about how problems have been solved with them. Thus, the text relies heavily on examples which illustrate the kinds of information that can be obtained by application of the methods. Whenever a choice exists regarding the selection of an application, an example from my own research has been chosen, for I am most familiar with the details of this work. I recommend that instructors supplement this material with examples reflecting their particular interests. Hopefully, more widespread knowledge of the core material in this text will upgrade or at least make the reader more critical of conclusions from spectroscopic measurements. I feel this is one of the important contributions of the earlier text and the original motivation for writing this book.

This text has evolved from the author's presentation of courses in physical methods. He gratefully acknowledges the contributions to the development of this material by the faculty, teaching assistants, and students involved, particularly those mentioned in earlier editions. I appreciate very much the fine contributions to this edition by Ivano Bertini, Claudio Luchinat, Arnie Rheingold, and Joe Ziller, who found the time to help a friend. Thanks go to Kitty Williams for Figure 8–25, Dave Powell for suggestions on the mass spectroscopy section, and to Roy King for many valuable comments and a very thorough review of Chapters 7 and 8. Thanks go to Steve Showalter, Steve Petrosius, and Mike Robbins for hours of proofreading. This revision would not have occurred if it were not for the help of "super secretary" Maribel Lisk and her able assistant April Kirch. I appreciate their dedication, the pride they take in a job well done, and their tolerance of a hyper, "I want it yesterday," professor. I also thank Dyana Drago for help in preparing the manuscript, but most of all, for collaborating with son Steve and giving us (Ruth and I) granddaughters Stacie and Olivia. John Vondeling's faith in the value of this effort is greatly appreciated, with the fond hope that, for this edition, he will provide a cover that does not self-destruct in a year. Finally, I would like to apologize to my wife, Ruth, for the many hours of my time spent on chemistry instead of with her and thank her for encouragement and unselfishness.

RUSSELL S. DRAGO

Contents

1

Symmetry and the Point Groups — 1

Introduction — 1

1–1 Definition of Symmetry — 1
1–2 Symmetry Elements — 2
1–3 Point Groups — 8
1–4 Space Symmetry — 11
1–5 Some Definitions and Applications of Symmetry Considerations — 12

References Cited — 14
Additional Reading — 14
Exercises — 14

2

Group Theory and the Character Tables — 18

2–1 Introduction — 18
2–2 Rules for Elements that Constitute a Group — 19
2–3 Group Multiplication Tables — 19
2–4 Summary of the Properties of Vectors and Matrices — 25
2–5 Representations; Geometric Transformations — 30
2–6 Irreducible Representations — 34
2–7 Character Tables — 35
2–8 Non-Diagonal Representations — 36
2–9 More on Character Tables — 41
2–10 More on Representations — 43
2–11 Simplified Procedures for Generating and Factoring Total Representations: The Decomposition Formula — 44
2–12 Direct Products — 46

Additional Reading — 47
Exercises — 47

3

Molecular Orbital Theory and Its Symmetry Aspects — 52

Introduction — 52

- 3–1 Operators — 52
- 3–2 A Matrix Formulation of Molecular Orbital Calculations — 56
- 3–3 Perturbation Theory — 57

Symmetry in Quantum Mechanics — 59

- 3–4 Wave Functions as a Basis for Irreducible Representations — 59
- 3–5 Projecting Molecular Orbitals — 60

Molecular Orbital Calculations — 68

- 3–6 Hückel Procedure — 68
- 3–7 Properties Derived from Wave Functions — 72
- 3–8 Extended Hückel Procedure — 74
- 3–9 SCF-INDO (Intermediate Neglect of Differential Overlap) — 78
- 3–10 Some Predictions from M.O. Theory on Alternately Double Bonded Hydrocarbons — 83
- 3–11 More on Product Ground State Wave Functions — 84

References Cited — 85
Additional Reading — 85
Compilations — 86
Exercises — 86

4

General Introduction to Spectroscopy — 90

- 4–1 Nature of Radiation — 90
- 4–2 Energies Corresponding to Various Kinds of Radiation — 91
- 4–3 Atomic and Molecular Transitions — 92
- 4–4 Selection Rules — 95
- 4–5 Relaxation and Chemical Exchange Influences on Spectral Line Width — 95

General Applications — 98

- 4–6 Determination of Concentration — 99
- 4–7 Isosbestic Points — 103
- 4–8 Job's Method of Isomolar Solutions — 106
- 4–9 "Fingerprinting" — 106

References Cited — 107
Exercises — 107

5

Electronic Absorption Spectroscopy — 109

Introduction — 109

- 5–1 Vibrational and Electronic Energy Levels in a Diatomic Molecule — 109

5–2	Relationship of Potential Energy Curves to Electronic Spectra	111
5–3	Nomenclature	113

Assignment of Transitions   117

5–4	Spin-Orbit Coupling	117
5–5	Configuration Interaction	117
5–6	Criteria to Aid in Band Assignment	118

The Intensity of Electronic Transitions   120

5–7	Oscillator Strengths	120
5–8	Transition Moment Integral	120
5–9	Derivation of Some Selection Rules	123
5–10	Spectrum of Formaldehyde	123
5–11	Spin-Orbit and Vibronic Coupling Contributions to Intensity	124
5–12	Mixing of d and p Orbitals in Certain Symmetries	126
5–13	Magnetic Dipole and Electric Quadrupole Contributions to Intensity	127
5–14	Charge Transfer Transitions	127
5–15	Polarized Absorption Spectra	128

Applications   130

5–16	Fingerprinting	130
5–17	Molecular Addition Compounds of Iodine	133
5–18	Effect of Solvent Polarity on Charge-Transfer Spectra	135
5–19	Structures of Excited States	137

Optical Rotary Dispersion, Circular Dichroism, and Magnetocircular Dichroism   137

5–20	Introduction	137
5–21	Selection Rules	139
5–22	Applications	140
5–23	Magnetocircular Dichroism	141

References Cited   143
Additional References   144
Exercises   145

6

Vibration and Rotation Spectroscopy: Infrared, Raman, and Microwave   149

Introduction   149

6–1	Harmonic and Anharmonic Vibrations	149
6–2	Absorption of Radiation by Molecular Vibrations—Selection Rules	150
6–3	Force Constant	151

Vibrations in a Polyatomic Molecule   153

6–4	The $3N - 6(5)$ Rule	153
6–5	Effects Giving Rise to Absorption Bands	154

6–6	Normal Coordinate Analyses and Band Assignments	156
6–7	Group Vibrations and the Limitations of This Idea	160

Raman Spectroscopy — 162

6–8	Introduction	162
6–9	Raman Selection Rules	164
6–10	Polarized and Depolarized Raman Lines	168
6–11	Resonance Raman Spectroscopy	170

Symmetry Aspects of Molecular Vibrations — 172

6–12	Significance of the Nomenclature Used to Describe Various Vibrations	172
6–13	Use of Symmetry Considerations to Determine the Number of Active Infrared and Raman Lines	172
6–14	Symmetry Requirements for Coupling, Combination Bands, and Fermi Resonance	176
6–15	Microwave Spectroscopy	177
6–16	Rotational Raman Spectra	179

Applications of Infrared and Raman Spectroscopy — 179

6–17	Procedures	179
6–18	Fingerprinting	184
6–19	Spectra of Gases	186
6–20	Application of Raman and Infrared Selection Rules to the Determination of Inorganic Structures	192

Bond Strength–Frequency Shift Relations — 194

6–21	Changes in the Spectra of Donor Molecules upon Coordination	196
6–22	Change in Spectra Accompanying Change in Symmetry upon Coordination	198

References Cited — 202
Additional References — 205
Exercises — 206

7

Nuclear Magnetic Resonance Spectroscopy—Elementary Aspects — 211

Introduction — 211

Classical Description of the NMR Experiment—The Bloch Equations — 212

7–1	Some Definitions	212
7–2	Behavior of a Bar Magnet in a Magnetic Field	213
7–3	Rotating Axis Systems	214
7–4	Magnetization Vectors and Relaxation	215
7–5	The NMR Transition	217
7–6	The Bloch Equations	218
7–7	The NMR Experiment	220

The Quantum Mechanical Description of the NMR Experiment   224

7–8	Properties of $\hat{I}$	224
7–9	Transition Probabilities	225

Relaxation Effects and Mechanisms   227

7–10	Measuring the Chemical Shift	229
7–11	Interpretation of the Chemical Shift	232
7–12	Interatomic Ring Currents	241
7–13	Examples of Chemical Shift Interpretation	241

Spin-Spin Splitting   243

7–14	Effect of Spin-Spin Splitting on the Spectrum	243
7–15	Discovering Non-Equivalent Protons	246
7–16	Effect of the Number and Nature of the Bonds on Spin-Spin Coupling	247
7–17	Scalar Spin-Spin Coupling Mechanisms	249
7–18	Application of Spin-Spin Coupling to Structure Determination	252

Factors Influencing the Appearance of the NMR Spectrum   257

7–19	Effect of Fast Chemical Reactions on the Spectrum	257
7–20	Quantum Mechanical Description of Coupling	259
7–21	Effects of the Relative Magnitudes of J and Δ on the Spectrum of an AB Molecule	263
7–22	More Complicated Second-Order Systems	265
7–23	Double Resonance and Spin-Tickling Experiments	267
7–24	Determining Signs of Coupling Constants	268
7–25	Effects on the Spectrum of Nuclei with Quadrupole Moments	269

References Cited   271
Compilations of Chemical Shifts   272
Exercises   273

8

Dynamic and Fourier Transform NMR   290

Introduction   290
Evaluation of Theormodynamic Data with NMR   290
NMR Kinetics   291

8–1	Rate Constants and Activation Enthalpies from NMR	291
8–2	Determination of Reaction Orders by NMR	295
8–3	Some Applications of NMR Kinetic Studies	297
8–4	Intramolecular Rearrangements Studied by NMR—Fluxional Behavior	300
8–5	Spin Saturation Labeling	305
8–6	The Nuclear Overhauser Effect	306

Fourier Transform NMR — 309

8–7	Principles	309
8–8	Optimizing the FTNMR Experiment	314
8–9	The Measurement of T_1 by FTNMR	315
8–10	Use of T_1 for Peak Assignments	317
8–11	NMR of Quadrupolar Nuclei	319

Applications and Strategies in FTNMR — 319

8–12	^{13}C	319
8–13	Other Nuclei	323

More on Relaxation Processes — 326

8–14	Spectral Density	326

Multipulse Methods — 328

8–15	Introduction	328
8–16	Spin Echoes	329
8–17	Sensitivity-Enhancement Methods	331
8–18	Selective Excitation and Suppression	332
8–19	Two-Dimensional NMR	334

NMR in Solids and Liquid Crystals — 340

8–20	Direct Dipolar Coupling	340
8–21	NMR Studies of Solids	341
8–22	NMR Studies in Liquid Crystal Solvents	342
8–23	High Resolution NMR of Solids	347

References Cited — 348
Additional References — 351
Exercises — 352

9

Electron Paramagnetic Resonance Spectroscopy — 360

Introduction — 360

9–1	Principles	360

Nuclear Hyperfine Splitting — 363

9–2	The Hydrogen Atom	363
9–3	Presentation of the Spectrum	368
9–4	Hyperfine Splittings in Isotropic Systems Involving More Than One Nucleus	370
9–5	Contributions to the Hyperfine Coupling Constant in Isotropic Systems	374

Anisotropic Effects — 380

9–6	Anisotropy in the g Value	380
9–7	Anisotropy in the Hyperfine Coupling	383
9–8	The EPR of Triplet States	390

9–9	Nuclear Quadrupole Interaction	392
9–10	Line Widths in EPR	394
9–11	The Spin Hamiltonian	396
9–12	Miscellaneous Applications	397

References Cited — 400
Additional References — 401
Exercises — 401

10

The Electronic Structure and Spectra of Transition Metal Ions — 409

Introduction — 409

Free Ion Electronic States — 409

10–1	Electron-Electron Interactions and Term Symbols	409
10–2	Spin-Orbit Coupling in Free Ions	413

Crystal Fields — 416

10–3	Effects of Ligands on the d Orbital Energies	416
10–4	Symmetry Aspects of the d Orbital Splitting by Ligands	419
10–5	Double Groups	427
10–6	The Jahn-Teller Effect	429
10–7	Magnetic Coupling in Metal Ion Clusters	430

Applications — 433

10–8	Survey of the Electronic Spectra of O_h Complexes	433
10–9	Calculation of Dq and β for O_h Ni(II) Complexes	437
10–10	Effect of Distortions on the d Orbital Energy Levels	443
10–11	Structural Evidence from the Electronic Spectrum	447

Bonding Parameters from Spectra — 450

10–12	σ and π Bonding Parameters from the Spectra of Tetragonal Complexes	450
10–13	The Angular Overlap Model	452

Miscellaneous Topics Involving Electronic Ttransitions — 458

10–14	Electronic Spectra of Oxo-Bridged Dinuclear Iron Centers	458
10–15	Intervalence Electron Transfer Bands	459
10–16	Photoreactions	462

References Cited — 462
Exercises — 463

11

Magnetism — 469

11–1	Introduction	469
11–2	Types of Magnetic Behavior	471

11–3	Van Vleck's Equation	476
11–4	Applications of Susceptibility Measurements	483
11–5	Intramolecular Effects	486
11–6	High Spin–Low Spin Equilibria	489
11–7	Measurement of Magnetic Susceptibilities	490
11–8	Superparamagnetism	491

References Cited   494
Exercises   496

12

Nuclear Magnetic Resonance of Paramagnetic Substances in Solution   500
by Ivano Bertini and C. Luchinat

12–1	Introduction	500
12–2	Properties of Paramagnetic Compounds	501
12–3	Considerations Concerning Electron Spin	503
12–4	The Contact Shift	507
12–5	The Pseudocontact Shift	508
12–6	Lanthanides	510
12–7	Factoring the Contact and Pseudocontact Shifts	512
12–8	The Contact Shift and Spin Density	514
12–9	Factors Affecting Nuclear Relaxation in Paramagnetic System	519
12–10	Relaxometry	524
12–11	Electronic Relaxation Times	527
12–12	Contrast Agents	530
12–13	Trends in the Development of Paramagnetic NMR	531
12–14	Some Applications	534
12–15	The Investigation of Bimetallic Systems	543

References Cited   552
Exercises   556

13

Electron Paramagnetic Resonance Spectra of Transition Metal Ion Complexes   559

13–1	Introduction	559
13–2	Interpretation of the g-Values	564
13–3	Hyperfine Couplings and Zero Field Splittings	571
13–4	Ligand Hyperfine Couplings	576
13–5	Survey of the EPR Spectra of First-Row Transition Metal Ion Complexes	578
13–6	The EPR of Metal Clusters	591
13–7	Double Resonance and Fourier Transform EPR Techniques	594

References Cited   594
Additional References   596
Exercises   597

14

Nuclear Quadrupole Resonance Spectroscopy, NQR   604

    14–1   Introduction   604
    14–2   Energies of the Quadrupole Transitions   607
    14–3   Effect of a Magnetic Field on the Spectra   611
    14–4   Relationship Between Electric Field Gradient and Molecular Structure   612
    14–5   Applications   616
    14–6   Double Resonance Techniques   620

References Cited   622
Additional References   623
Exercises   624

15

Mössbauer Spectroscopy   626

    15–1   Introduction   626
    15–2   Interpretation of Isomer Shifts   630
    15–3   Quadrupole Interactions   631
    15–4   Paramagnetic Mössbauer Spectra   633
    15–5   Mössbauer Emission Spectroscopy   635
    15–6   Applications   635

References Cited   646
Series   646
Exercises   647

16

Ionization Methods: Mass Spectrometry, Ion Cyclotron Resonance, Photoelectron Spectroscopy   650

    Mass Spectrometry   650

    16–1   Instrument Operation and Presentation of Spectra   650
    16–2   Processes That Can Occur When a Molecule and a High Energy Electron Combine   655
    16–3   Fingerprint Application   657
    16–4   Interpretation of Mass Spectra   659
    16–5   Effect of Isotopes on the Appearance of a Mass Spectrum   660
    16–6   Molecular Weight Determinations; Field Ionization Techniques   662
    16–7   Evaluation of Heats of Sublimation and Species in the Vapor Over High Melting Solids   663
    16–8   Appearance Potentials and Ionization Potentials   664

    FTICR/MS   665

    16–9   The Fourier Transform Ion Cyclotron Resonance Technique   665

Surface Science Techniques 667

    16–10   Introduction 667
    16–11   Photoelectron Spectroscopy 667
    16–12   SIMS (Secondary Ion Mass Spectrometry) 677
    16–13   LEED, AES, and HREELS Spectroscopy 678
    16–14   STM (Scanning Tunneling Microscopy) and AFM
              (Atomic Force Microscopy) 680

EXAFS and XANES 681

    16–15   Introduction 681
    16–16   Applications 682

References Cited 682
Exercises 685

17

X-Ray Crystallography 689
by Joseph W. Ziller and Arnold L. Rheingold

    17–1   Introduction 689

Principles 690

    17-2   Diffraction of X-Rays 690
    17-3   Reflection and Reciprocal Space 691
    17-4   The Diffraction Pattern 693
    17-5   X-Ray Scattering by Atoms and Structures 693

Crystals 695

    17-6   Crystal Growth 695
    17-7   Selection of Crystals 696
    17-8   Mounting Crystals 697

Methodology 698

    17-9   Diffraction Equipment 698
    17-10   Diffractometer Data Collection 699
    17-11   Computers 700

Some Future Developments 700

    17-12   Area Detectors 700
    17-13   X-Ray versus Neutron Diffraction 701
    17-14   Synchrotron Radiation 701

Symmetry and Related Concerns 701

    17-15   Crystal Classes 701
    17-16   Space Groups 703
    17-17   Space-Group Determination 704
    17-18   Avoiding Crystallographic Mistakes 705
    17-19   Molecular versus Crystallographic Symmetry 707
    17-20   Quality Assessment 707
    17-21   Crystallographic Data 710

References Cited 711
Exercises 711

Appendix A

Character Tables for Chemically Important Symmetry Groups 713

Appendix B

Character Tables for Some Double Groups 724

Appendix C

Normal Vibration Modes for Common Structures 727

Appendix D

Tanabe and Sugano Diagrams for O_h Fields 732

Appendix E

Calculation of Δ ($10Dq$) and β for O_h Ni^{II} and T_d Co^{II} Complexes 735

Appendix F

Conversion of Chemical Shift Data 739

Appendix G

Solution of the Secular Determinant for the NMR Coupling of the AB Spin System 740

Index 745

Symmetry and the Point Groups

1

Introduction

1-1 DEFINITION OF SYMMETRY

Symmetry considerations are fundamental to many areas of chemical reactivity, electronic structure, and spectroscopy. It is customary to describe the structures of molecules in terms of the symmetry that the molecules possess. Spectroscopists have described molecular vibrations in terms of symmetry for many years. Modern applications of spectroscopic methods to the problem of structure determination require a knowledge of symmetry properties. We shall be concerned mainly with the description of the symmetry of an isolated molecule, the so-called point symmetry. Point symmetry refers to the set of operations transforming a system about a common point, which usually turns out to be the center of gravity of a molecule.

If a molecule has two or more orientations in space that are indistinguishable, the molecule possesses *symmetry*. Two possible orientations for the hydrogen molecule can be illustrated in Fig. 1–1 only by labeling the two equivalent hydrogen atoms in this figure with prime and double prime marks. Actually, the two hydrogens are indistinguishable, the two orientations are equivalent, and the molecule has symmetry. The two orientations in Fig. 1–1 can be obtained by rotation of the molecule through 180° about an axis through the center of and perpendicular to the hydrogen-hydrogen bond axis. This rotation is referred to as a *symmetry operation*, and the rotation axis is called a *symmetry element*. The terms symmetry element and symmetry operation should not be confused or used interchangeably. The symmetry element is the line, point, or plane about which the symmetry operation is carried out. The operation can be defined only with respect to the element, and the existence of the element can be shown only by carrying out the operation.

A simple test can be performed to verify the presence of symmetry. If you were to glance at a structure, turn your back and have someone perform a symmetry operation, on once again examining the molecule, you would not be able to determine that a change had been made. The symmetry operations of a molecule form a group and, as a result, are amenable to group theoretical procedures. Mastery of the material in this chapter is essential to the understanding of most of the material in this book.

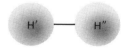

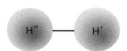

Figure 1–1 Equivalent orientations for H_2.

1

1-2 SYMMETRY ELEMENTS

Five types of symmetry elements will be considered for point symmetry: (1) the center of symmetry (inversion center), (2) the identity, (3) the rotation axis, (4) the mirror plane, and (5) the rotation-reflection axis. The symmetry operations corresponding to these elements will be defined in the course of a further discussion of each element.

The Center of Symmetry, or Inversion Center

A molecule is said to possess a center of symmetry or inversion center when every atom in the molecule, if moved in a straight line through this center and an equal distance on the other side of the center, encounters a like atom. If oxygen atom A of the hyponitrite ion (Fig. 1-2) is moved through the inversion center, it comes into coincidence with another oxygen atom, B. The same arguments must apply to atom B and also to both nitrogen atoms if the molecule is to possess a center of symmetry. This operation corresponds to placing this center at the center of a coordinate system, and taking every atom with coordinates (x, y, z) and changing its coordinates to $(-x, -y, -z)$. *At most* one atom can be at the center, and all other atoms in the molecule must exist in pairs. Neither of the two structures indicated in Fig. 1-3 nor any other tetrahedral molecule possesses an inversion center. Other molecules or ions that should be examined for the presence of an inversion center are 1,4-dioxane, tetracyanonickelate(II), *trans*-dichloroethylene, and *trans*-dichlorotetraammine cobalt(III). The molecule HCCl=CHBr (*cis* or *trans*) does not possess a center of symmetry because the symmetry operation on bromine does not result in coincidence. All points in the molecule must be inverted in this operation, if the molecule as a whole is to possess this symmetry element. The symbol used to indicate an inversion center (center of symmetry) is i.

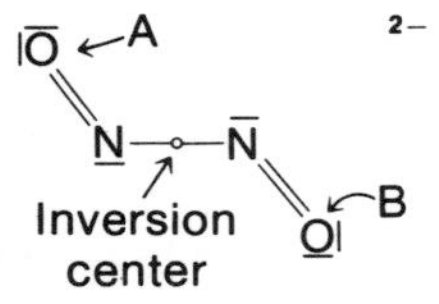

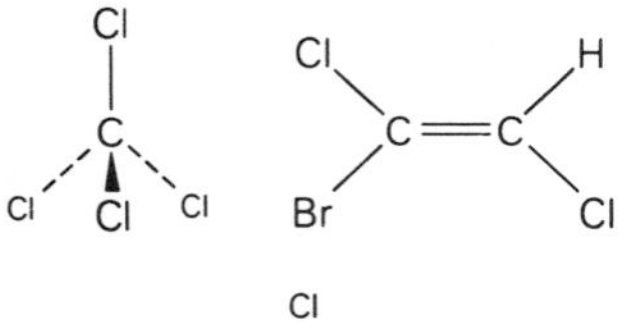

FIGURE 1-2 The center of symmetry in the hyponitrite ion.

FIGURE 1-3 Absence of a center of symmetry in CCl$_4$ and HCl=CBrCl.

The Identity

In the identity operation, no change is made in the molecule. Obviously, the result of this operation is not only to produce an equivalent orientation but an identical one; *i.e.*, even if similar atoms were labeled with prime or double prime marks, or any other notation, no change would be detected. All molecules possess this symmetry element, and it is indicated by the symbol E. This operation probably seems trivial at present, but, as will be seen in Chapter 2, this operation is required so that symmetry elements can be treated by group theory.

The Rotation Axis

If an imaginary axis can be constructed in a molecule, around which the molecule can be rotated to produce an *equivalent* (*i.e.*, indistinguishable from the original) orientation, this molecule is said to possess a rotation axis. The symmetry element previously discussed for the hydrogen molecule is a rotation axis. This element is usually referred to as a *proper rotation axis*. It may be possible to carry out several symmetry operations around a single rotation axis. If the molecule can occupy n different equivalent positions about the axis, the axis is said to be of order n. For example, consider the axis through the center of the boron atom in

BCl_3 perpendicular to the plane of the molecule. Rotation, which by convention is in a clockwise direction, about this axis two times through an angle of 120° each time produces two equivalent orientations. Taken with the initial orientation, we have the three different equivalent orientations, illustrated in Fig. 1–4. The order, n, of this axis is three, for three rotations are needed to return to the original position. The molecule is said to possess a threefold rotation axis, indicated by the symbol C_3. Rotation of the molecule through $2\pi/n$ (*i.e.*, 120° when $n = 3$) produces equivalent orientations, and n operations produce the starting configuration referred to as the *identity*. The symbol C_3^2 is employed to indicate a rotation of 240° around a C_3 axis. The C_3^2 operation is identical to a counterclockwise rotation of 120°, which is indicated as C_3^-. It should be clear that an axis through the center and perpendicular to the plane of a benzene ring is a sixfold axis, C_6. Since $n = 6$, rotation by 60° ($= 360°/6$) six times produces the six equivalent orientations. Further examination of the BCl_3 molecule indicates the lack of a center of symmetry and the presence of the three additional twofold rotation axes, C_2, illustrated in Fig. 1–5. One twofold axis, arbitrarily selected, is labeled C_2; the other two are labeled C_2' and C_2''. If $n = 1$, the molecule must be rotated 360° to produce an equivalent (and in this case identical) orientation. A molecule is said to possess no symmetry if no elements other than the identity are present.

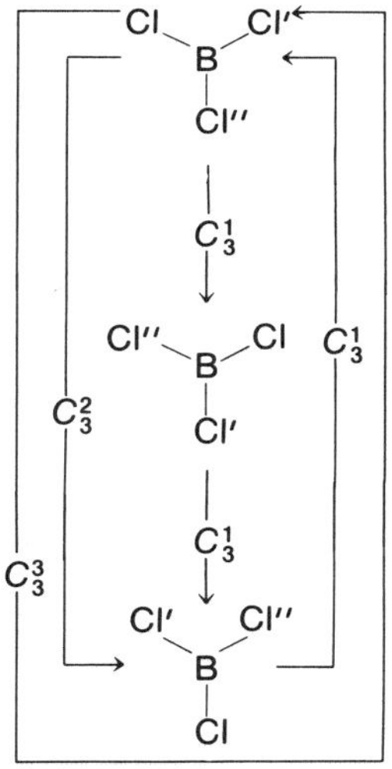

FIGURE 1–4 The results of clockwise rotations about the threefold rotation axis of BCl_3.

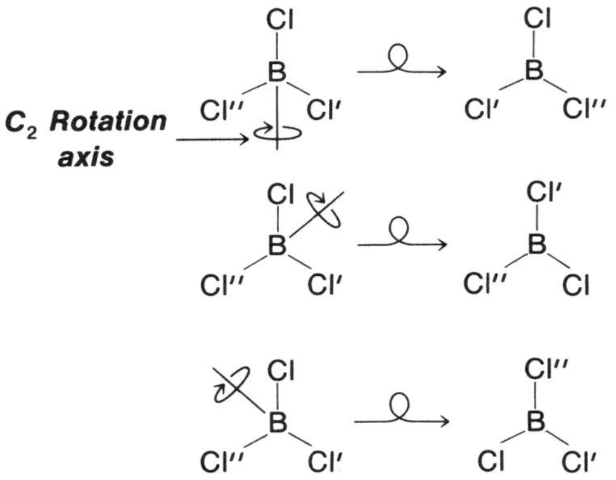

FIGURE 1–5 The results of rotations about the three twofold rotation axes in BCl_2. The 180° rotation on the left produces the result on the right.

A rotation axis of order n generates n operations: $C_n, C_n^2, C_n^3, \ldots, C_n^{n-1}, C_n^n$. Furthermore, the operation C_4^2 is equivalent to C_2, the operation C_6^2 is equivalent to C_3, and C_n^n is the identity. *The highest-fold rotation axis is referred to as the principal axis in a molecule.* If all of the C_n axes are equivalent, any one may be chosen as the principal axis.

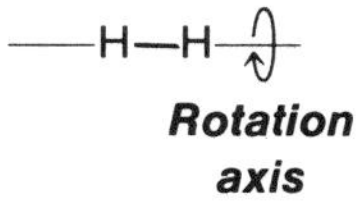

In Fig. 1–6, a rotation axis is illustrated for the H_2 molecule, for which $n = \infty$. The C_2 rotation axes are not illustrated but are perpendicular to the bond axis and centered between the two hydrogen atoms. There are an infinite number of these twofold axes. It should also be obvious that benzene possesses six twofold axes that lie in the plane of the molecule. Three pass through pairs of opposite carbon atoms, and the other three pass through the centers of C—C bonds.

The molecule ClF_3 is illustrated in Fig. 1–7. The geometry is basically a trigonal bipyramid, with two lone pairs of electrons in the equatorial plane and with two fluorines bent down toward the equatorial fluorine. This molecule has only one rotation axis, the C_2 axis shown in Fig. 1–7. The reader should verify the absence of symmetry along the axes indicated by dashed lines.

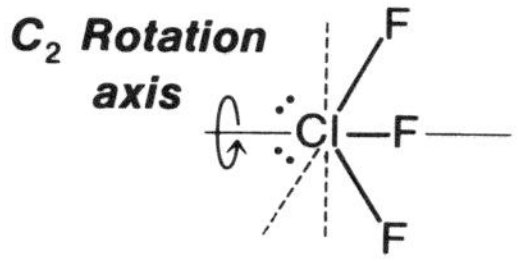

FIGURE 1–6 The C_∞ axis in H_2.

FIGURE 1–7 The symmetry axis in ClF_3.

For many purposes, it is convenient to locate a molecule and its symmetry elements in a Cartesian coordinate system. A right-handed coordinate system will be used. Movement of y to x is a positive rotation. The center of gravity of the molecule is located at the center of the coordinate system. If there is only one symmetry axis in the molecule, it is selected as the z-axis. If there is more than one symmetry axis, the principal rotation axis in the molecule is selected as the z-axis. If there is more than one highest-fold axis, the one connecting the most atoms is selected. If the molecule is planar and if the z-axis lies in this plane (as in the water molecule), the x-axis is chosen perpendicular to this plane. If the molecule is planar and the z-axis is perpendicular to this plane, e.g., in

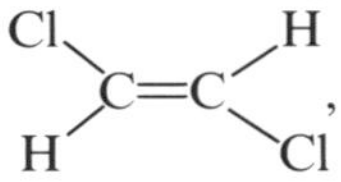

$$\begin{array}{c}ClH\\ \diagdown\diagup\\ C{=}C\\ \diagup\diagdown\\ HCl\end{array}$$, the x-axis is chosen to pass through the largest number of atoms.

In *trans*-dichloroethylene the x-axis passes through the two carbons.

The Mirror Plane or Plane of Symmetry

If a plane exists in a molecule that separates the molecule into two halves that are mirror images of each other, the molecule possesses the symmetry element of a *mirror plane*. This plane cannot lie outside the molecule but must pass through it. Another way of describing this operation involves selecting a plane, dropping a perpendicular from every atom in the molecule to the plane, and placing the atom at the end of the line an equal distance to the opposite side of the plane. If an equivalent configuration is obtained after this is done to all the atoms, the plane selected is a mirror plane. Reflection through the mirror plane is indicated by Fig. 1–8. A linear molecule possesses an infinite number of mirror planes.

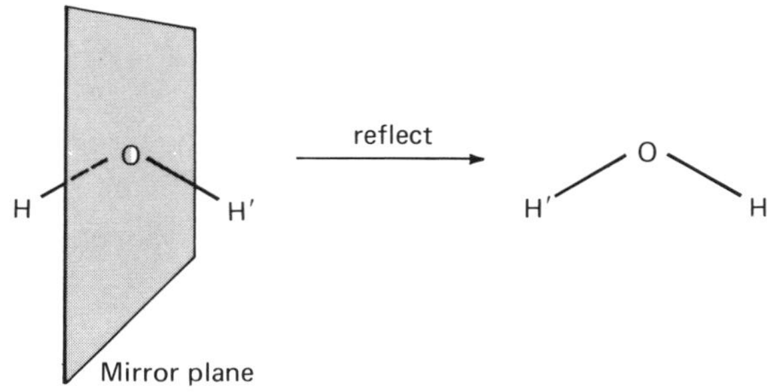

FIGURE 1–8 Operation of a mirror plane (σ) reflection on H_2O.

Often rotation axes lie in a mirror plane (see Fig. 1–9, for example), but there are examples in which this is not the case. The tetrahedral molecule $POBr_2Cl$, illustrated in Fig. 1–10, is an example of a molecule that contains no rotation axis but does contain a mirror plane. The atoms P, Cl, and O lie in the mirror plane. In general, the presence of a mirror plane is denoted by the symbol σ. In those molecules that contain more than one mirror plane (e.g., BCl_3), the horizontal plane σ_h is taken as the one perpendicular to the principal (highest-fold rotation) axis. In Fig. 1–9, the plane of the paper is σ_h and there are then three vertical planes σ_v (two others, similar to the one illustrated, each containing the boron atom and one chlorine atom) perpendicular to σ_h.

When a molecule is located in a coordinate system, the z-axis always lies in the vertical plane(s).

Some molecules have mirror planes containing the principal axis but none of the perpendicular C_2 axes. These planes bisect the angle between two of the C_2 axes (in the xy-plane); they are referred to as *dihedral planes* and are abbreviated as σ_d. Two σ_d planes are illustrated in Fig. 1–14, *vide infra*.* In some molecules there is more than one set of mirror planes containing the highest-fold axes; *e.g.*, in $PtCl_4^{2-}$, one set consists of the xz- and yz-planes (z being the fourfold axis, and the Cl—Pt—Cl bond axes being x and y), and the other set bisects the angle between the x- and y-axes. The former set, including the chlorine atoms, is called σ_v by convention, and the latter set is called σ_d. The diagonal plane is always taken as one that bisects x, y, z, or twofold axes in the molecule.

The Rotation-Reflection Axis; Improper Rotations

This operation involves rotation about an axis followed by reflection through a mirror plane that is perpendicular to the rotation axis, or vice versa (*i.e.*, rotation-reflection is equivalent to reflection-rotation). When the result of the two operations produces an equivalent structure, the molecule is said to possess a rotation-reflection axis. This operation is referred to as an improper rotation, and the rotation-reflection axis is often called an alternating axis. The symbol S is used to indicate this symmetry element. The subscript n in S_n indicates rotation (clockwise by our convention) through $2\pi/n$.

Obviously, if an axis C_n exists and there is a σ perpendicular to it, C_n will also be an S_n. Now, we shall consider a case in which S_n exists when neither C_n or the mirror plane perpendicular to it exists separately. In the staggered form of ethane, Fig. 1–11, the C—C bond defines a C_3 axis, but there is no perpendicular mirror plane. However, if we rotate the molecule $60°$ and then reflect it through a plane perpendicular to the C—C bond, we have an equivalent configuration. Consequently, an S_6 axis exists and, clearly, there is no C_6.

The dotted line in Fig. 1–12 indicates the S_2 element in the molecule *trans*-dichloroethylene. The subscript two indicates clockwise rotation through $180°$. S_2 is equivalent to i and, by convention, is usually called i.

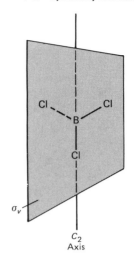

FIGURE 1–9 A mirror plane in BCl_3.

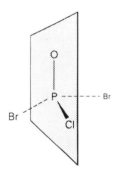

FIGURE 1–10 The mirror plane in $POBr_2Cl$.

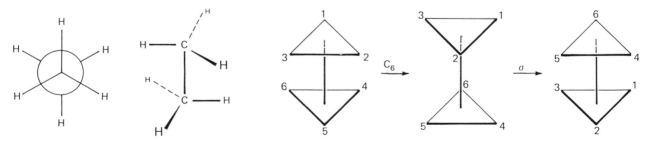

FIGURE 1–11 Improper rotation axis in the staggered form of ethane.

A particular orientation of the molecule methane is illustrated in Fig. 1–13. Open circles or squares represent hydrogen atoms in a plane parallel to but above the plane of the paper, and the solid squares or circles are those below the plane

---

*We will use this term to indicate that a statement will be treated more fully later in the text.

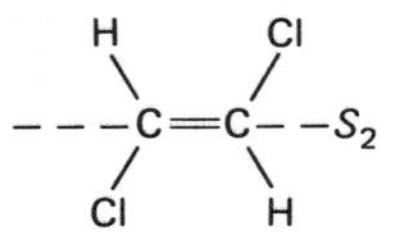

FIGURE 1–12 Rotation-reflection axis of symmetry.

of the paper. The plane of the paper is the reflection plane, and it contains the carbon atom. The C_4 operation is straightforward. The operation of reflection, σ, moves the hydrogens below the plane to above the plane and vice versa. This is indicated by changing the solid squares to open squares and the open circles to solid circles. However, since all four hydrogens are identical, the initial and final orientations are equivalent. The molecule contains a fourfold rotation-reflection axis, abbreviated S_4. This operation can be repeated three more times, or four times in all. These four operations are indicated by the symbols S_4^1, S_4^2, S_4^3, and S_4^4. The reader should convince himself or herself that S_4^2 is equivalent to a C_2 operation on this axis. It should also be mentioned that the molecule possesses two other rotation-reflection axes, and these symmetry elements are abbreviated by the symbols S' and S''. The operation S_4^3 is equivalent to a counterclockwise rotation of 90° followed by reflection. This is often indicated as S_4^-.

FIGURE 1–13 Effect of the operations C_4 and σ perpendicular to C_4 on the hydrogens of CH_4. (Open symbols above the page; solid symbols below.)

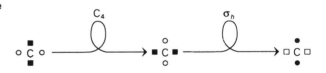

Next we shall consider some differences in improper rotation axes of even and odd order. With n even, S_n^m generates the set $S_n^1, S_n^2, S_n^3, \ldots, S_n^n$. This is equivalent to $C_n^1\sigma^1, C_n^2\sigma^2, C_n^3\sigma^3, \ldots, C_n^n\sigma^n$. (Note: In carrying out $C_n^n\sigma^n$, one reflects first and then rotates, while in carrying out $\sigma^nC_n^n$, one rotates and then reflects.) We have the relation $\sigma^m = \sigma$ when m is odd, and $\sigma^m = E$ when m is even. The latter leads to the identities $S_n^n = C_n^n E = E$ and $S_n^m = C_n^m$.

Notice that the existence of an S_n axis of even order always requires a $C_{n/2}$, since $S_n^2 = C_n^2\sigma^2 = C_{n/2}$.

Consider the set $S_6, S_6^2, S_6^3, S_6^4, S_6^5, S_6^6$ by carrying out the operations described below on the staggered ethane example.

S_6 cannot be written any other way.

$S_6^2 = C_6^2\sigma^2 = C_3$

$S_6^3 = S_2 = i$ (An S_2 axis always equals i.)

$S_6^4 = C_3^2$

S_6^5 cannot be written any other way.

$S_6^6 = E$

The complete set would normally be written:

$$S_6 \quad C_3 \quad i \quad C_3^2 \quad S_6^5 \quad E$$

When n is odd, S_n^m generates the set $S_n, S_n^2, S_n^3, S_n^4, \ldots, S_n^{2n}$. An odd-order S_n requires that C_n and a σ perpendicular to it exist. Note that the operation S_n^n when n is odd is equivalent to $C_n^n\sigma^n = C_n^n\sigma = \sigma$. (Compare this with S_n^n when n is even.)

For example, consider S_3. First one would reflect, producing a configuration that is equivalent to the starting configuration (because of the existence of σ in the molecule). Then one rotates by C_3 to give the configuration corresponding to S_3^1. Since S_3^1 is a symmetry operation for the molecule, the configuration after the C_3 rotation must be equivalent to the starting one, and therefore C_3 exists. In general, for any S_n (with n odd), C_n is also a symmetry operation. For practice, consider the S_5^m operations:

$$\mathbf{S_5 = C_5 \text{ and then } \sigma}$$
$$S_5^2 = C_5^2$$
$$\mathbf{S_5^3 = C_5^3 \text{ and then } \sigma}$$
$$S_5^4 = C_5^4$$
$$\mathbf{S_5^5 = \sigma}$$
$$S_5^6 = C_5$$
$$\mathbf{S_5^7 = C_5^2 \text{ and then } \sigma}$$
$$S_5^8 = C_5^3$$
$$\mathbf{S_5^9 = C_5^4 \text{ and then } \sigma}$$
$$S_5^{10} = E$$
$$S_5^{11} = S_5, \text{ etc.}$$

The boldface indicates S_5 operations that cannot be written as a single operation. Thus, S_n with n odd generates $2n$ operations. Table 1–1 summarizes the key aspects of our discussion of symmetry operations and elements.

TABLE 1–1. A Summary of Symmetry Elements and Symmetry Operations

Symmetry Operation	Symmetry Element	Symbol	Examples
identity		E	all molecules
reflection	plane	σ	H_2O, BF_3 (planar)
inversion	point (center of symmetry)	i	Cl_2B-BCl_2
proper rotation	axis	C_n (n-order)	NH_3, H_2O
improper[a] rotation (rotation by $2\pi/n$ followed by reflection in plane perpendicular to axis)	axis *and* plane	S_n (n-order)	ethane, ferrocene (staggered structures)

[a] Ferrocene is staggered and possesses an S_{10} improper rotation axis.

1-3 POINT GROUPS

A point group is a collection of all the symmetry operations that can be carried out on a molecule belonging to this group. It is possible to classify any given molecule into one of the point groups. We shall first examine molecules belonging to some of the very common, simple point groups for the symmetry elements they possess. This will be followed by a general set of rules for assigning molecules to the appropriate point groups.

A molecule in the point group C_n has only one symmetry element, an n-fold rotation axis. (Of course, all molecules possess the element E, so this will be assumed in subsequent discussion.) A molecule such as *trans*-dichloroethylene, which has a horizontal mirror plane perpendicular to its C_n rotation axis, belongs to the point group C_{nh}. The point group C_{nv} includes molecules like water and sulfuryl chloride, which have n vertical mirror planes containing the rotation axis, but no horizontal mirror plane (the horizontal plane by definition must be perpendicular to the highest-fold rotation axis). The C_{2v} molecules, H_2O and SO_2Cl_2, contain only one rotation axis and two σ_v planes. The molecule is assigned to the point group that contains all of the symmetry elements in the molecule. For example, H_2O, which has a C_2 axis and two vertical planes, is assigned to the higher-order (more symmetry elements) point group C_{2v} rather than to C_2.

The symbol D_n is used for point groups that have, in addition to a C_n axis, n C_2 axes perpendicular to it. Therefore, the D_n point group has greater symmetry (*i.e.*, more symmetry operations) than the C_n group. A D_n molecule that also has a horizontal mirror plane perpendicular to the C_n axis belongs to the point group D_{nh}, and as a consequence will also have n vertical mirror planes. The addition of a horizontal mirror plane to the point group C_{nv} necessarily implies the presence of n C_2 axes in the horizontal plane, and the result is the point group D_{nh}. BCl_3 is an example of a molecule belonging to the D_{3h} point group. In the D_{nh} point groups, the σ_h planes are perpendicular to the principal axis and contain all the C_2 axes. Each σ_v plane contains the principal axis and one of the C_2 axes.

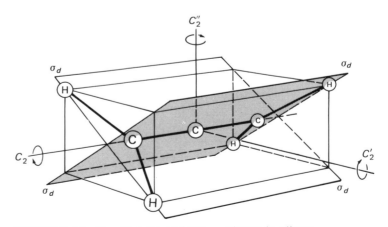

FIGURE 1-14 The C_2-axes and two σ_d planes in allene.

TABLE 1–2. Symmetry Elements in Some Common Point Groups

Point Group	Symmetry Elements[a]	Examples
C_1	no symmetry	SiBrClFI
C_2	one C_2 axis	H_2O_2
C_{nh}	one n-fold axis and a horizontal plane σ_h which must be perpendicular to the n-fold axis	trans-$C_2H_2Cl_2$ (C_{2h})
C_{2v}	one C_2 axis and two σ_v planes	H_2O, SO_2Cl_2, $SiCl_2Br_2$
C_{3v}	one C_3 axis and three σ_v planes	NH_3, CH_3Cl, $POCl_3$
D_{2h}	three C_2 axes all $\perp$, two σ_v planes, one σ_h plane, and a center of symmetry	N_2O_4 (planar)
D_{3h}	one C_3, three C_2 axes $\perp$ to C_3, three σ_v planes, and one σ_h	BCl_3
D_{2d}	three C_2 axes, two σ_d planes, and one S_4 (coincident with one C_2)	$H_2C{=}C{=}CH_2$
T_d	three C_2 axes $\perp$ to each other, four C_3, six σ, and three S_4 containing C_2	CH_4, $SiCl_4$

[a] All point groups possess the identity element, E.

D_n molecules may also have σ_d planes that contain the principal (C_n) axis but none of the perpendicular C_2 axes. As mentioned previously, these dihedral planes, σ_d, bisect the angle between two of the C_2 axes. The notation for a D_n molecule containing this symmetry element is D_{nd}. Such a molecule will contain an n-fold axis, n twofold axes perpendicular to C_n, and in addition n (vertical) planes of symmetry bisecting the angles between two twofold axes and containing the n-fold axis. Allene, $H_2C{=}C{=}CH_2$, is an example of a molecule belonging to the D_{2d} point group. Some of its symmetry elements are illustrated in Fig. 1–14. The two hydrogens on one carbon are in a σ_d plane perpendicular to the σ_d plane containing the two hydrogens on the other carbon. The C_2 axes (the principal axis being labeled C_2, the others C_2' and C_2'') and the dihedral planes containing the principal axis are indicated. The two other C_2 axes, C_2' and C_2'', which are perpendicular to the principal C_2 axis, form 45° angles with the two dihedral planes. (Study the figure carefully to see this. Constructing a model may help.)

Table 1–2 contains the symmetry elements and examples of some of the more common point groups. The reader should examine the examples for the presence of the symmetry elements required for each point group and the absence of others. Remember that the symbol C indicates that the molecule has only one rotation axis. The symbol D_n indicates n C_2 axes in addition to the n-fold axis; e.g., the point group D_4 contains a C_4 axis and four C_2 axes.

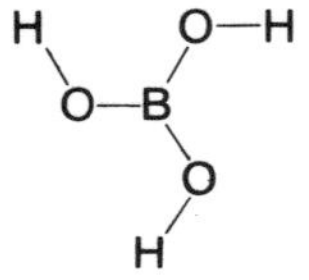

FIGURE 1–15 A structure with C_{3h} symmetry.

A more exhaustive compilation of the important point groups is contained in Herzberg.[1] In addition to the point groups listed in Table 1–2, the O_h point group which includes molecules whose structures are perfect octahedra (SF_6, PCl_6^-) is very common.

It is important to realize that although $CHCl_3$ has a tetrahedral geometry, it does not have tetrahedral symmetry, so it belongs to the point group C_{3v} and not T_d. A tetragonal complex, *trans*-dichlorotetraammine cobalt(III) ion, belongs to the point group D_{4h} (ignoring the hydrogens) and not O_h. Phosphorus pentachloride belongs to the point group D_{3h}, and not to C_{3h}, for it has three C_2 axes perpendicular to the C_3 axis. The structure of monomeric boric acid (assumed to be rigid), illustrated in Fig. 1–15, is an example of C_{3h} symmetry. It has a threefold axis and a σ_h plane but does not have the three C_2 axes or σ_v planes necessary for the D_{3h} point group.

It is helpful but not always possible to classify additional structures by memorizing the above examples and using analogy as the basis for classification instead of searching for all the possible symmetry elements. The following sequence of steps has been proposed[2] for classifying molecules into point groups and is a more reliable procedure than analogy.

1. Determine whether or not the molecule belongs to one of the special point groups, $C_{\infty v}$, $D_{\infty h}$, I_h, O_h, or T_d. The group I_h contains the regular dodecahedron and regular icosahedron. Only linear molecules belong to $C_{\infty v}$ or $D_{\infty h}$.

2. If the molecule does not belong to any of the special groups, look for a proper rotation axis. If any are found, proceed to step (3); if not, look for a center of symmetry, i, or a mirror plane, σ. If the element i is present, the molecule belongs to the point group C_i; if a mirror is present, the molecule belongs to the point group C_s. If no symmetry other than E is present, the molecule belongs to C_1.

3. Locate the principal axis, C_n. See if a rotation-reflection axis S_{2n} exists that is coincident with the principal axis. If this element exists and there are no other elements except possibly i, the molecule belongs to one of the S_n point groups (where n is even). If other elements are present or if the S_{2n} element is absent, proceed to step (4).

4. Look for a set of n twofold axes lying in the plane perpendicular to C_n. If this set is found, the molecule belongs to one of the groups D_n, D_{nh}, or D_{nd}. Proceed to step (5). If not, the molecule must belong to either C_n, C_{nh}, or C_{nv}. Proceed to step (6) and skip (5).

5. By virtue of having arrived at this step, the molecule must be assigned to D_n, D_{nh}, or D_{nd}. If the molecule contains the symmetry element σ_h, it belongs to D_{nh}. If this element is not present, look for a set of n σ_d's, the presence of which enables assignment of the molecule to D_{nd}. If σ_d and σ_h are both absent, the molecule belongs to D_n.

6. By virtue of having arrived at this step, the molecule must be assigned to C_n, C_{nh}, or C_{nv}. If the molecule contains σ_h, the point group is C_{nh}. If σ_h is absent, look for a set of n σ_v's, which places the molecule in C_{nv}. If neither σ_v nor σ_h is present, the molecule belongs to the point group C_n.

The following flow chart provides a systematic way to approach the classification of molecules by point groups:

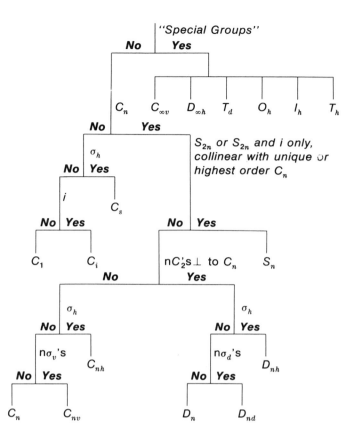

1-4 SPACE SYMMETRY

In describing the symmetry of an arrangement of atoms in a crystal (essential to x-ray crystallography), two additional symmetry elements are required: (1) the glide plane and (2) the screw axis, corresponding to mixtures of point group operations and translation. A space group is a collection of symmetry operations referring to three-dimensional space. A glide plane is illustrated in Fig. 1–16. The unit (represented by a comma) is translated a fraction (in this case $1/2$) of a unit cell dimension and is reflected. (A unit cell is the smallest repeating unit of the crystal.) The comma must lie in a plane perpendicular to the plane of the paper and all of the comma tails must point into the paper. A screw axis is illustrated in Fig. 1–17. Each position, as one proceeds from top to bottom of the axis, represents rotation by $120°$ along with translation parallel to the axis. This is a threefold screw axis. Rotations by $90°$, $60°$, and other angles are found in other structures. There are 230 possible space groups. These are of importance in x-ray crystallography, and will be covered in more detail in that chapter.

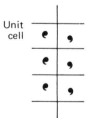

FIGURE 1–16 A glide plane. (The tails of these commas point into the page.)

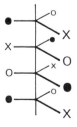

FIGURE 1–17 A screw axis.

1-5 SOME DEFINITIONS AND APPLICATIONS OF SYMMETRY CONSIDERATIONS

Products or Combinations of Symmetry Operations

The product of any two symmetry operations, defined as their consecutive application, must be a symmetry operation. Hence, the product of the C_2 and σ_v' operations on a molecule with C_{2v} symmetry (*e.g.*, H_2O) is σ_v. (Recall that the C_2 axis is the z-axis, the plane of the molecule is yz, and σ_v' is the yz-plane.) This can be written as:

$$C_2 \times \sigma_v' = \sigma_v \quad \text{or}$$
$$C_2 \sigma_v' = \sigma_v$$

Instead of the term *product*, the term *combination* is a better description of the above operations. The order in which the operations are written (*i.e.*, from left to right) is the reverse of the order in which they are applied. For the above example, the σ_v' operation is carried out first and is followed by C_2 to produce the same result as σ_v. The final result often depends on the order in which the operations are carried out, but, in some cases, it does not. When the result is independent of the order (*e.g.*, $C_2\sigma_v' = \sigma_v'C_2$), the two symmetry elements C_2 and σ_v' are said to *commute*. In a D_{3h} molecule (see, *e.g.*, Fig. 1–4), the two operations C_3 and σ_v do not commute; *i.e.*,

$$C_3\sigma_v \neq \sigma_v C_3$$

Equivalent Symmetry Elements and Equivalent Atoms

If a symmetry *element A* can be moved into the *element B* by an *operation* corresponding to the element X, then A and B are said to be *equivalent*. If we define X^{-1} as the reverse operation of X (*e.g.*, a counterclockwise rotation instead of a clockwise one), then X^{-1} will take B back into A. Furthermore, if A can be carried into C, then there must be a symmetry operation (or a sequence of symmetry operations) that carries B into C, since B can be carried into A. The elements A, B, and C are said to form an *equivalent set*.

Any set of symmetry elements chosen so that any member can be transformed into each and every other member of the set by application of some symmetry operation is said to be a set of *equivalent* symmetry elements. This collection of elements is said to constitute a *class*. To make this discussion clearer, consider the planar molecule $PtCl_4^{2-}$ illustrated in Fig. 1–18. The C_2' and C_2 axes form an equivalent set, as do the C_2'' and C_2'''. However, C_2' is not equivalent to C_2'', because the molecule possesses no symmetry operation that takes one into the other.

Equivalent atoms in a molecule are defined as those that may be interchanged with one another by a symmetry operation that the molecule possesses. Accordingly, all of the chlorines in $PtCl_4^{2-}$ (Fig. 1–18) and all of the hydrogens in methane, benzene, cyclopropane, or ethane are equivalent. The fluorines in gaseous PF_5 (trigonal bipyramidal structure) are not all equivalent, but form two sets of equivalent atoms, one containing the three equatorial fluorines and one containing the two axial fluorine atoms. These considerations are very important

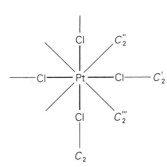

FIGURE 1–18 Equivalent symmetry elements and atoms in the molecule $PtCl_4^{2-}$.

Optical Activity

If the mirror image of a molecule cannot be superimposed on the original, the molecule is optically active; if it can be superimposed, the molecule is optically inactive. In using this criterion, the mirror is understood to be external to the whole molecule, and reflection through the mirror gives an image of the whole molecule. With complicated molecules, the visualization of superimposability is difficult. Accordingly, it is to our advantage to have a symmetry basis for establishing the existence of optically active isomers. *Any molecule that has no improper rotation axis is said to be dissymmetric, and optically active molecules must be dissymmetric.* One often hears the incomplete statement that in order for optical isomerism to exist, the molecule must lack a plane or center of symmetry. Since $S_1 = \sigma$ (S_1 is a rotation by 360° followed by a reflection) and $S_2 = i$, if the molecule lacks an improper rotation axis, both i and σ must be lacking. To show the incompleteness of the earlier statement, we need to find a molecule that has neither σ nor i, but does contain an S_n axis and is not optically active. Such a molecule is 1,3,5,7-tetramethylcyclooctatetraene, shown in Fig. 1–19. This molecule does not have a plane or center of symmetry. However, since it has an S_4 axis, it is *not* optically active.

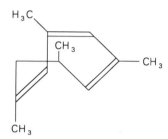

FIGURE 1–19 Structure of 1,3,5,7-tetramethylcyclo-octatetraene.

A dissymmetric molecule differs from an asymmetric one, for the latter type is completely lacking in symmetry. The molecule *trans*-1,2-dichlorocyclopropane, shown in Fig. 1–20, is dissymmetric (there is no S_n axis) and hence optically active, but it is not asymmetric, for it possesses a C_2 axis.

Many molecules can exist in some conformation that is optically active. However, if rotation of the molecule about a bond produces a conformation with an improper axis, the molecule will not be optically active. If the conformation is frozen in a form that does not possess an improper axis, optical activity could result.

In summary, then, we can state that if a molecule possesses only C_n, it is dissymmetric and optically active. If $n = 1$, the molecule is asymmetric as well as dissymmetric; and if $n > 1$, the molecule is dissymmetric. If a molecule possesses S_n with any n, it cannot be optically active.

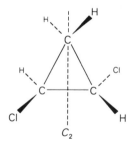

FIGURE 1–20 *Trans*-1,2-dichlorocyclopropane.

Dipole Moments

Molecules possess a center of gravity of positive charge, which is determined by the nuclear positions. When the center of gravity of the negative charge from the electrons is at some other point, the molecule has a dipole moment, which is related to the magnitude of the charge times the distance between the centers. The dipole moment is a vector property; that is, it has both a magnitude and a direction. For a fixed geometry (*i.e.*, a non-vibrating molecule) the dipole moment is a non-fluctuating property of the molecule; as a result it, like the total energy, must remain unchanged by the operation of every symmetry element of the molecule. In order for this to occur, the dipole moment vector must be coincident with *each* of the symmetry elements.

There are several obvious consequences of these principles. Molecules that have a center of symmetry cannot possess a dipole moment, because the vector cannot be a point, and any vector would be changed by inversion through the center. Molecules with more than one C_n axis cannot have a dipole moment, for a dipole moment vector could not be coincident with more than one axis. Thus, only the following types of molecules may have dipole moments: those with one C_n ($n > 1$), those with one σ and no C_n, those with a C_n and symmetry planes that include C_n, and those that have no symmetry. In all cases, where the molecule has symmetry, the direction of the dipole vector is determined; for it must lie in all of the symmetry elements that the molecule possesses.

REFERENCES CITED

1. G. Herzberg, "Infrared and Raman Spectra," p. 11, Van Nostrand, New York, 1945.
2. F. A. Cotton, "Chemical Applications of Group Theory," p. 54, Wiley-Interscience, New York, 3rd ed., 1990.

ADDITIONAL READING

M. Tinkham, "Group Theory and Quantum Mechanics," McGraw-Hill, New York, 1964.
F. A. Cotton, "Chemical Applications of Group Theory," 3rd ed., Wiley-Interscience, New York, 1990.
M. Orchin and H. H. Jaffé, J. Chem. Educ., *47*, 246, 372, 510 (1970).
M. Orchin and H. H. Jaffé, "Symmetry, Orbitals, and Spectra," Wiley-Interscience, 1971.
C. D. H. Chisholm, "Group Theoretical Techniques in Quantum Chemistry," Academic Press, New York, 1976.
A. Vincent, "Molecular Symmetry and Group Theory," Wiley, New York, 1977.
For space groups, see: J. D. Donaldson and S. D. Ross, "Symmetry and Stereochemistry," Wiley, 1972.

EXERCISES

1. Classify the following molecules in the appropriate point groups and, for (c) through (k), indicate all symmetry elements except E.

   a. $CoCl_4^{2-}$

   b. $Ni(CN)_4^{2-}$

   c. *cis*-$CoCl_4(NH_3)_2^-$ (ignore the hydrogen atoms)

   d. C_6H_{12} (chair form)

   e. $Si(CH_3)_3 \cdot A \cdot B$ (with A and B *trans* in a trigonal bipyramid)

   f. PF_3

   g. $(CH_3)_2B\begin{smallmatrix}H\\ \\H\end{smallmatrix}B(CH_3)_2$

   h. $Cl-I-Cl^-$

   i. planar *cis*-$PdCl_2B_2$ (B = base)

j. planar *trans*-PdCl$_2$B$_2$ (B = base)

k. staggered configuration for C$_2$H$_6$

l. ferrocene (staggered)

2. a. How does a D_{2d} complex for formula MCl$_4{}^{2-}$ differ from a T_d complex (*i.e.*, which symmetry elements differ)?

b. What important symmetry element is absent in the PF$_3$ molecule that is present in the D_{3h} point group?

c. If each Ni—C≡N bond in planar Ni(CN)$_4{}^{2-}$ was not linear but was bent

(Ni⟋C⫽⟍N), to what point group would this ion belong? What essential symmetry element present in D_{4h} is missing in C_{4h}?

d. Indicate which operation is equivalent to the following products for the T_d point group
   (1) $S_4 \times S_4 = ?$
   (2) $C_3 \times C_2 = ?$
   (3) $\sigma_d \times C_2 = ?$

3. Does POCl$_2$Br possess a rotation-reflection axis coincident with the P—O bond axis?

4. a. Rotate the molecule in Fig. 1–14, 180° around the S_4 axis, and draw the resulting structure.

b. Locate the mirror plane of the S_4 operation and indicate on the structure resulting from part (a) where all the atoms are located after reflection through the plane. Use script letters to indicate the final location of each atom.

c. Is S_2 a symmetry operation for the molecule?

5. To which point group does the molecule triethylenediamine (also called 1,4-diazabicyclo[2.2.2]octane, or dabco) belong?

6. Assign PtCl$_4{}^{2-}$ to a point group and locate one symmetry element in each class of this point group.

7. Give the point group for each of the following and indicate whether the entity could have a dipole moment:

a. three-bladed propeller

b. hexahelicene

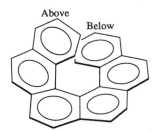

c. 2,2-cyclophane

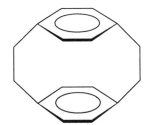

d. hexadentate ligand on Fe^{2+}

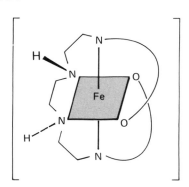

e. $Co(C_5H_5NO)_6^{2+}$ (C_5H_5NO is pyridine N-oxide)

○ = Co
● = O

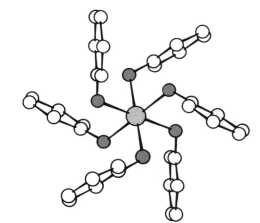

8. Is the following compound optically active?

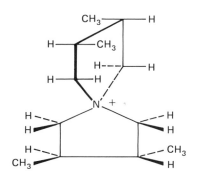

9. $Cr(en)_3^{3+}$ (where en = ethylenediamine) belongs to the D_3 point group. Is it optically active? Why?

10. Determine the point groups of the following molecules, and find the requested equivalent atoms within these molecules.

   a. [structure with three Cl and H₁, H₂, H₃]   Which protons, if any, are equivalent?

   b. [ferrocene with Cl substituents, H₅–H₈ on top ring, H₁–H₄ on bottom ring]   Which protons are equivalent?

   c. [cube with H₁–H₆ and Cl at corners]   Which protons are equivalent?

   d. 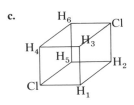   Which chlorides are equivalent?

11. Locate the σ_d planes in C_6H_6. Are these equivalent to σ_v? Why?

2 Group Theory and the Character Tables

2–1 INTRODUCTION

Group theory is a topic in abstract algebra that can be applied to certain systems if they meet specific requirements. There are many systems of interest to chemists that can be treated by these techniques, and by the end of this chapter you will only have begun to gain an appreciation for the power of this method. We shall use these concepts frequently throughout the rest of the book. We shall be involved in some abstract and seemingly irrelevant concepts in the course of our development of this topic. The reader is encouraged to persevere, for the rewards are many.

There are *two common major types* of applications of group theory.

    a. It can be used in the *generation of symmetry combinations*. If you are given some basis set, that is, a set of orbitals or mathematical functions pertaining to a molecule, you can use group theory to *construct linear combinations* of the things in the basis set that *reflect* the symmetry of the molecule. For example, if you are given a set of atomic orbitals for a molecule, you can use group theory to assist you in mapping out the shape of the molecular orbitals for the given molecule. There are many areas where such an application can be made, and a list of a few include our ability to:

1. determine *hybrid orbitals* used in bonding the atoms in the molecule.
2. determine which *atomic orbitals* can contribute to the various *molecular orbitals* in a molecule
3. determine the number and symmetries of *molecular vibrations*
4. predict how the *degeneracies of the d-orbitals* are removed by *crystal fields* of various symmetries
5. predict and generate the *spin functions* to be used in the Hamiltonian for epr, nqr, and Mössbauer spectroscopy
6. construct symmetry adapted combinations of nuclear spin functions to work out nmr equations.

    b. It can be used to *ascertain which quantum mechanical integrals are zero;* i.e., is $\int \psi^* \text{ op } \psi \, d\tau$ equal to zero? Such integrals are important to:

1. determine the allowedness of electronic transitions
2. determine the activity of infrared and Raman vibrations
3. determine the allowedness of any given transition in nmr, esr, etc.

2–2 RULES FOR ELEMENTS THAT CONSTITUTE A GROUP

The symmetry operations can be treated with group theory, and we shall use them to illustrate the principles. In order for any set of elements to form a mathematical group the following conditions must be satisfied.

    1. The combination (Chapter 1; often referred to as the "product") of any two elements, and the square of each element, must produce an element of the group.

    This combination is not the same thing as a product in arithmetic. We are referring to the consecutive operation of elements. It is a simple matter to square and take all possible combinations of the elements of the C_{2v} point group to illustrate the point that another element is obtained. The reader is encouraged to do so, using H_2O as an example. Unlike multiplication in ordinary arithmetic, we must worry about the order of the combination, i.e., $C_2 \times \sigma_v$ or $\sigma_v \times C_2$.

    2. *One operation of the group must commute with all others and leave them unchanged. This is the identity element.*

$$E\sigma_2 = \sigma_2 E = \sigma_2$$

    3. The associative law of multiplication must hold, *i.e.*,

$$(XY)Z = X(YZ)$$

The result of combining three elements must be the same if the first element is combined with the product of the second two (*i.e.*, $\sigma_v(C_2\sigma_v')$) or if the product of the first two is combined with the last element (*i.e.*, $(\sigma_v C_2)\sigma_v'$).

    4. Every element must have a reciprocal that is also an element of the group. This requires that for every symmetry operation there be another operation that will undo what the first operation did. For any mirror plane, the inverse is the identical mirror plane, *e.g.*, $\sigma \times \sigma = E$. For a proper rotation C_n^m, the inverse is C_n^{n-m}; $C_n^m \times C_n^{n-m} = E$. If A has the reciprocal element B, then $AB = BA = E$. If B is the reciprocal of A, then A is the reciprocal of B. In general, the reciprocal of A can be written as A^{-1}.

C_3 has the reciprocal element C_3^2:

$$C_3 C_3^2 = C_3^2 C_3 = E$$

We shall illustrate these rules in an abstract way in the next section.

2–3 GROUP MULTIPLICATION TABLES

Properties of the Multiplication Tables

If we have a complete and non-redundant list of all the elements in a finite group, and we know all the products, then the group is completely and uniquely defined. This information can be summarized with a *group multiplication table*. This table is simply an array in which each column and row is headed by an element. The matrix format is employed here because it is a convenient way to insure that we have taken all possible permutations of products of the elements. The number of elements in the group is referred to as the *order of the group, h*. The group multiplication table will consist of h rows and h columns. Certain groups have an infinite number of elements, *e.g.*, H_2 belongs to $D_{\infty h}$.

In order to illustrate simply the rules of group theory discussed earlier, we will consider a group of order three containing the abstract elements A and B as well as the identity E. We shall determine what requirements these elements must satisfy (*i.e.*, how their combinations must be defined) in order for these elements to constitute a group. Each matrix element is the result of the product of the column element times the row element:

	E	A	B
E			
A			
B			

Since multiplication is generally not commutative, we must *adhere to a consistent order for multiplication*. The convention is to carry out the operations in the order, *column element times row element*; *i.e.*, when we write a product RC, where C is column and R is row, we take the column element first, followed by the row.

Each of the original h elements in the group must appear once and only once in each row and each column of the group multiplication table, and no two rows or columns may be alike. The following argument proves this. For a group of h different elements $E, A_1, \ldots A_n, \ldots A_{h-1}$, the elements of the nth row (A_n) are $A_n E, A_n A_1, \ldots A_n A_{h-1}$. If any two products $A_n A_i$ and $A_n A_j$ were the same, left multiplication by the inverse of A_n would give $A_n^{-1} A_n A_i = A_i$ and $A_n^{-1} A_n A_j = A_j$ respectively, and thus require A_i and A_j to be the same. This conclusion is contrary to our original assumption that all elements of the group were different. Thus, all the products and all the h elements of the row must be different. Each of the original h elements in the group must thus appear once and only once in the row. A similar argument can be used for the columns. We can illustrate the requirements for the general, unspecified set of elements E, A, and B described above to constitute a group, and show how these rules define the multiplication table.

Since E is the identity, it is a simple matter to indicate the matrix elements for products involving E and write (REMEMBER column first, then row multiplication):

	E	A	B
E	E	A	B
A	A		
B	B		

There are only two ways to complete this table; these involve defining either $AA = E$ or $AA = B$. If $AA = E$, then according to rule 4, A is its own reciprocal, so $BB = E$. (A cannot be reciprocal to both B and A unless $A = B$, so B must be the reciprocal of B.) Then, if $BA = A$ or B, we would not be able to complete the table without repeating an element in a row or column; *i.e.*, $BA = A$ makes

A repeat in a column, while $BA = B$ repeats B in the row. However, if we define $AA = B$, the following table results:

	E	A	B
E	E	A	B
A	A	B	E
B	B	E	A

If $AA = B$, we know that A must have a reciprocal, so $AB = BA = E$.

We can consider this entire group to be generated by taking an element and all its powers, *e.g.*, A, $A^2 (= B)$, and $A^3 (= E)$. Such a group is a *cyclic group*. For such a group, all multiplications must commute.

Thus, by adhering to the conditions specified for a collection of elements to constitute a group, we have been able to define our elements so as to construct a group multiplication table that specifies the result of all possible combinations of elements.

The very same procedures and rules can be used to construct group multiplication tables of higher order. When a table of order four is made, two possibilities result, which we shall label G_4^1 and G_4^2:

G_4^1	E	A	B	C		G_4^2	E	A	B	C
E	E	A	B	C		E	E	A	B	C
A	A	B	C	E	and	A	A	E	C	B
B	B	C	E	A		B	B	C	E	A
C	C	E	A	B		C	C	B	A	E

The superscripts arbitrarily number the two possibilities.

We can make these procedures more specific by considering a system in which the elements are symmetry operations. In so doing, our definition of the products will not seem so arbitrary, for we know physically what the results of the products of the symmetry operations are. We shall proceed by considering the C_{3v} ammonia molecule, whose point group contains the symmetry operations E, $2C_3$, and $3\sigma_v$. The three σ_v planes are labeled as in Fig. 2-1, and the group multiplication table is given as Table 2-1. The reader is encouraged to carry out

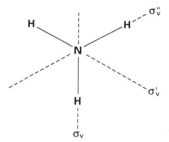

FIGURE 2-1 The σ_v planes ($\perp$ to page) in the C_{3v} ammonia molecule.

TABLE 2-1. Group Multiplication Table for the C_{3v} Point Group

	E	C_3	C_3^2	σ_v	σ_v'	σ_v''
E	E	C_3	C_3^2	σ_v	σ_v'	σ_v''
C_3	C_3	C_3^2	E	σ_v''	σ_v	σ_v'
C_3^2	C_3^2	E	C_3	σ_v'	σ_v''	σ_v
σ_v	σ_v	σ_v'	σ_v''	E	C_3	C_3^2
σ_v'	σ_v'	σ_v''	σ_v	C_3^2	E	C_3
σ_v''	σ_v''	σ_v	σ_v'	C_3	C_3^2	E

and verify the combinations given in this table. All of the points made in conjunction with the discussion of the group A, B, E should also be verified.

The group multiplication table (Table 2–1) contains sub-groups, which are smaller groups of elements in the point group that satisfy all of the requirements for constituting a group. E is such a sub-group, as is the collection of operations E, C_3, C_3^2. The order of any sub-group must be an integral divisor of the order, h, of the full group.

Similarity Transforms

A second way in which the symmetry operations of a group may be subdivided into smaller sets is by application of similarity transforms. If A and X are two operations of a group, then $X^{-1}AX$ will be equal to some operation of the group, say B:

$$B = X^{-1}AX$$

The operation B is said to be the similarity transform of A by X. In the general definition of a similarity tranform, X does not necessarily represent an operation of the group. However, when taking the similarity transform of A in order for the result B to be some other operation of the group, *X must be an operation of the group. We then say that A and B are conjugate.* There are three important properties of conjugate operations:

1. Every operation is conjugate with itself. If we select one operation, say A, it must be possible to find at least one operation X such that $A = X^{-1}AX$.

Proof:
Left multiply by A^{-1}:

$$A^{-1}A = E = A^{-1}X^{-1}AX = (XA)^{-1}AX$$

(The reciprocal of the product of two or more operations is equal to the product of the reciprocals in reverse order; see Cotton,[2] "*Chemical Applications of Group Theory,*" if you desire to see the proof.) By definition, E must also equal $(AX)^{-1}(AX)$.

Both equations can be true only if A and X commute. Thus, the operation X may always be E, but it may also be any other operation that commutes with the selected operation A.

2. If A is conjugate with B, then B is conjugate with A; i.e., if $A = X^{-1}BX$, then there must be some operation in the group, Y, such that $B = Y^{-1}AY$. Note that X equals Y^{-1} and vice versa.

3. If A is conjugate with both B and C, then B and C are conjugate with each other.

The use and meaning of the similarity transform will be made more concrete as we employ it in subsequent discussion in this chapter. For the moment, it may help to point out that a similarity transform can be used to change the coordinate system selected to describe a problem. A right-handed coordinate system is given by:

$$\begin{array}{c} Z \\ \swarrow Y \\ X \end{array}$$

or any rotation of these axes. A left-handed coordinate system is given by:

$$\begin{array}{c} Z \\ \swarrow X \\ Y \end{array}$$

The two coordinate systems are seen to be related to each other by a mirror plane σ_d that contains the Z-axis and bisects the X- and Y-axes.

Now, consider a C_4^1 rotation about the Z-axis. A point with coordinates (a, b) (the X coordinate is always given first) in a right-handed system goes to $(b, -a)$ upon a 90° clockwise rotation, but to $(-b, a)$ with a 270° rotation. In a left-handed system C_4^1 takes (a, b) to $(-b, a)$, while C_4^3 yields $(b, -a)$. (Construct a figure if needed.) Thus, the roles of C_4 and C_4^3 are interchanged by changing the coordinate system. We can say that there is a similarity transform of C_4^3 to C_4 by σ_d; i.e.,

$$C_4 = \sigma_d^{-1} C_4^3 \sigma_d$$

or, since $\sigma_d^{-1} = \sigma_d$ (note that $\sigma_d \sigma_d^{-1} = E$ but $\sigma_d \sigma_d = E$ so $\sigma_d = \sigma_d^{-1}$)

$$C_4 = \sigma_d C_4^3 \sigma_d$$

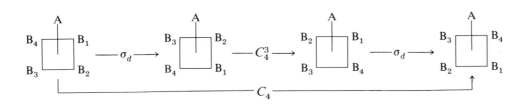

Classes of Elements

A complete set of elements that are conjugate to one another is called a *class* of the group. The order of each class must be an integral divisor of the order of the group. To determine the classes, begin with one element, say A, and work out all of its similarity transforms with all the elements of the group, including itself. Then find an element that is not one of those conjugate with A, and determine all of its transforms. Repeat this procedure until all of the elements in the group have been divided into classes. E will always constitute a class by itself of order 1, $E^{-1}EE = E$.

We shall demonstrate this procedure by working out all of the classes of elements in the group:

G_6^1	E	A	B	C	D	F
E	E	A	B	C	D	F
A	A	E	D	F	B	C
B	B	F	E	D	C	A
C	C	D	F	E	A	B
D	D	C	A	B	F	E
F	F	B	C	A	E	D

First of all, E is a class by itself. Next, work out all of the similarity transforms of A. These would be EAE, $A^{-1}AA$, $B^{-1}AB$, $C^{-1}AC$, $D^{-1}AD$, and $F^{-1}AF$. We already know that $EAE = A$. To carry out the combination $A^{-1}AA$, we first note from our table that $AA = E$. Thus, $A^{-1}AA = A^{-1}E$, so we have to find out what element is equal to A^{-1}. We know that $A^{-1}A = E$; accordingly, we go to the A column and determine which element, when combined with A, produces the result (matrix element) E. This element is A, and therefore $A^{-1} = A$. As a result, $A^{-1}AA = A^{-1}E = AE = A$. In a similar fashion, we can show that $B^{-1} = B$. To determine $B^{-1}AB$, we see that $AB = D$ and $B^{-1}D = BD = C$. Proceeding in a similar fashion, all of the similarity transforms of A can be determined:

$$E^{-1}AE = A$$
$$A^{-1}AA = A$$
$$B^{-1}AB = C$$
$$C^{-1}AC = B$$
$$D^{-1}AD = C$$
$$F^{-1}AF = B$$

Thus, we have determined that A, B, and C are all conjugate and members of the same class. We now know that all transforms of B and C are also A, B, or C (see rule 3, p. 19), so these do not have to be worked out.

Next, we find an element that is not conjugate with A and determine its transforms. Such an element is D, and its transforms are:

$$E^{-1}DE = D$$
$$A^{-1}DA = F$$
$$B^{-1}DB = F$$
$$C^{-1}DC = F$$
$$D^{-1}DD = D$$
$$F^{-1}DF = D$$

All transforms of D are either F or D, so these two elements constitute a class. Note that we have now placed every element in a class.

We can again make this procedure less abstract by working out the classes for the symmetry group of the ammonia molecule from Table 2–1. (You might try closing the book, working this out yourself, and then using the ensuing discussion to check your result.)

$$EEE = E$$
$$C_3{}^2 E C_3 = E \text{ (note: } C_3{}^{-1} = C_3{}^2\text{)}$$
$$\vdots$$
$$\sigma_v'' E \sigma_v'' = E$$

The first result is, of course, that E is always in a class by itself. Next, it can be shown that C_3 and $C_3{}^2$ form a class of the group.

$$E C_3 E = C_3$$
$$C_3{}^2 C_3 C_3 = C_3$$
$$C_3 C_3 C_3{}^2 = C_3$$
$$\sigma_v C_3 \sigma_v = C_3{}^2$$
$$\sigma_v' C_3 \sigma_v' = C_3{}^2$$
$$\sigma_v'' C_3 \sigma_v'' = C_3{}^2$$

Similarly, σ_v, σ_v', and σ_v'' form a class of the group:

$$E \sigma_v E = \sigma_v$$
$$C_3{}^2 \sigma_v C_3 = \sigma_v''$$
$$C_3 \sigma_v C_3{}^2 = \sigma_v'$$
$$\sigma_v \sigma_v \sigma_v = \sigma_v$$
$$\sigma_v' \sigma_v \sigma_v' = \sigma_v''$$
$$\sigma_v'' \sigma_v \sigma_v'' = \sigma_v'$$

In general, equivalent symmetry operations belong to the same class. (Recall that equivalent elements are ones that can be taken into one another by symmetry operations of the group.) For example, in the planar ion $PtCl_4{}^{2-}$ (D_{4h}), shown in Fig. 2–2, σ_v and σ_v' can never be taken into σ_d and σ_d'. Therefore, σ_v and σ_v' form one class of the group, while σ_d and σ_d' form another.

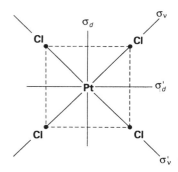

FIGURE 2–2 The σ_v and σ_d planes in the D_{4h} $PtCl_4{}^{2-}$

2–4 SUMMARY OF THE PROPERTIES OF VECTORS AND MATRICES

Vectors

A vector in three-dimensional space has a magnitude and a direction that can be specified by the lengths of its projections on the three orthogonal axes of a Cartesian coordinate system. Vector properties may be more than three-

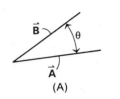

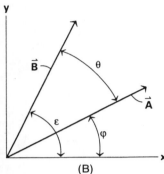

FIGURE 2–3 a. Two vectors, **A** and **B**, separated by an angle θ. b. The vectors placed in a two-dimensional xy-coordinate system.

dimensional, so the above statement can be extended to p-dimensional space and p orthogonal axes in p-space.

It is often necessary to take the product of vectors. One type of product that produces a number (i.e., a scalar) is called the scalar or dot product. This is given by:

$$\vec{\mathbf{A}} \cdot \vec{\mathbf{B}} = AB \cos \theta \qquad (2\text{–}1)$$

Here the boldface $\vec{\mathbf{A}}$ and $\vec{\mathbf{B}}$ refer to the two vectors, and the dot refers to their dot product. On the right-hand side of the equation, A and B refer to the lengths of the $\vec{\mathbf{A}}$ and $\vec{\mathbf{B}}$ vectors, and θ is the angle between them as shown in Fig. 2–3a. Accordingly, if the angle θ between vectors is 90°, the dot product is zero ($\cos 90° = 0$) and the vectors are orthogonal.

It is necessary to reference our vectors to a coordinate system. This is done for a two-dimensional example (i.e., the xy-plane) in Fig. 2–3b. The angle θ is now seen to be $\varepsilon - \varphi$, so the dot product becomes

$$\vec{\mathbf{A}} \cdot \vec{\mathbf{B}} = AB \cos(\varepsilon - \varphi) \qquad (2\text{–}2)$$

Simple trigonometry now gives us the projections of the $\vec{\mathbf{A}}$ and $\vec{\mathbf{B}}$ vectors on the x and y axes as:

$$A_x = A \cos \varphi \qquad (2\text{–}3)$$

$$A_y = A \sin \varphi \qquad (2\text{–}4)$$

$$B_x = B \cos \varepsilon \qquad (2\text{–}5)$$

$$B_y = B \sin \varepsilon \qquad (2\text{–}6)$$

Using a trigonometric identity, equation (2–2) can be rewritten as

$$\vec{\mathbf{A}} \cdot \vec{\mathbf{B}} = AB(\cos \varphi \cos \varepsilon + \sin \varphi \sin \varepsilon)$$
$$= A \cos \varphi \, B \cos \varepsilon + A \sin \varphi \, B \sin \varepsilon \qquad (2\text{–}7)$$

Substituting equations (2–3) through (2–6) into (2–7) produces

$$\vec{\mathbf{A}} \cdot \vec{\mathbf{B}} = A_x B_x + A_y B_y$$

Thus, the dot product of two vectors in two-dimensional space is the product of the components with all cross terms ($A_x B_y$ and so forth) absent. In p-space, the result obtained is

$$\vec{\mathbf{A}} \cdot \vec{\mathbf{B}} = \sum_{i=1}^{p} A_i B_i$$

where i ranges over the p orthogonal axes in p-space. Accordingly, the scalar square of a vector is given by

$$\vec{\mathbf{A}}^2 = \sum_{i=1}^{p} A_i^2$$

Matrices

A matrix is a rectangular array of numbers or symbols that has the following general form:

$$\begin{bmatrix} a_{11} & a_{12} & a_{13} \\ a_{21} & a_{22} & a_{23} \\ a_{31} & a_{32} & a_{33} \end{bmatrix}$$

The square brackets indicate that this is a matrix, as opposed to a determinant. The entire matrix can be abbreviated with a script letter or by the symbol $[a_{ij}]$. The symbol a_{ij} refers to the matrix element in the ith row and jth column. When the number of rows equals the number of columns, the matrix is called a *square matrix*. The elements a_{ij} of a square matrix for which $i = j$ (i.e., a_{11}, a_{22}, a_{33}, etc.) are called the *diagonal elements*, and the other elements are called *off-diagonal*. When all of the off-diagonal elements of a matrix are zero, the matrix is said to be *diagonalized* or to be a diagonal matrix. When each of the diagonal elements of a square matrix equals 1 and all off-diagonal elements are zero, the matrix is called a *unit matrix*. The unit matrix is often abbreviated by the Kronecker delta, δ.

$$\begin{bmatrix} 1 & 0 & 0 & 0 \\ 0 & 1 & 0 & 0 \\ 0 & 0 & 1 & 0 \\ 0 & 0 & 0 & 1 \end{bmatrix} = \delta$$

Unless $i = j$, the matrix element ij of a diagonal matrix has a value of zero. The *trace or character* of a square matrix, an important property (*vide infra*), is simply the sum of the diagonal elements. A one-row matrix can be conveniently written on a single line. In order to write a one-column matrix on a single line, it is enclosed in braces, { }.

A vector is conveniently represented by a one-column matrix. In a three-dimensional orthogonal coordinate system, a vector initiating at the origin of the coordinate system is completely specified by the x, y, and z coordinates of the other end. Thus, the matrix $\{x, y, z\}$ is a one-column matrix that represents the vector. In p-space, a p by 1 column vector is needed. In both instances, the elements of the matrix give the projections of the vector on the orthogonal coordinates.

Matrices may be added, subtracted, multiplied, or divided by using the appropriate rules of matrix algebra. In order to add or subtract two matrices $\mathfrak{A}$ and $\mathfrak{B}$ to give a matrix $\mathfrak{C}$, the matrices must all be of the same *dimension*; i.e., they must contain the same number of rows and columns. The elements of the $\mathfrak{C}$ matrix are given by:

$$c_{ij} = a_{ij} \pm b_{ij}$$

A matrix can be multiplied by a scalar (a single number). When multiplying by a scalar, each matrix element is multiplied by this scalar

$$k[a_{ij}] = [ka_{ij}]$$

The ijth matrix element of a product matrix is obtained by multiplying the ith row of the first matrix by the jth column of the second matrix, i.e., row by column to produce the product matrix. That is, the matrix elements of $\mathfrak{C}$, the product of $\mathfrak{A}$ and $\mathfrak{B}$, are given by

$$c_{ik} = \sum_{j=1}^{n} a_{ij} b_{jk}$$

where n is the number of elements in the ith row and in the jth column. This matrix multiplication is equivalent to taking the dot product of two vectors.

It should be clear that, in order to multiply a matrix by another matrix, the two matrices must be conformable; i.e., if we wish to multiply $\mathfrak{A}$ by $\mathfrak{B}$ to give $\mathfrak{C}$, the number of columns in $\mathfrak{A}$ must equal the number of rows in $\mathfrak{B}$. If the dimensionality of matrix $\mathfrak{A}$ is i by j and that of $\mathfrak{B}$ is j by k, then $\mathfrak{C}$ will have a dimensionality of i by k. This can be seen by carrying out the following matrix multiplication:

$$\begin{bmatrix} a_{11} & a_{12} \\ a_{21} & a_{22} \\ a_{31} & a_{32} \end{bmatrix} \begin{bmatrix} b_{11} & b_{12} & b_{13} \\ b_{21} & b_{22} & b_{23} \end{bmatrix} = \begin{bmatrix} c_{11} & c_{12} & c_{13} \\ c_{21} & c_{22} & c_{23} \\ c_{31} & c_{32} & c_{33} \end{bmatrix}$$

$$\text{3 by 2} \qquad \text{2 by 3} \qquad \text{3 by 3}$$

The number of columns (two) in $\mathfrak{A}$ equals the number of rows (two) in $\mathfrak{B}$. The matrix elements of $\mathfrak{C}$ are obtained by a row by column multiplication; i.e.,

$$c_{11} = a_{11}b_{11} + a_{12}b_{21}$$
$$c_{12} = a_{11}b_{12} + a_{12}b_{22}$$
$$c_{13} = a_{11}b_{13} + a_{12}b_{23}$$
$$c_{21} = a_{21}b_{11} + a_{22}b_{21}$$
$$c_{22} = a_{21}b_{12} + a_{22}b_{22}$$
$$c_{23} = a_{21}b_{13} + a_{22}b_{23}$$
$$c_{31} = a_{31}b_{11} + a_{32}b_{21}$$
$$c_{32} = a_{31}b_{12} + a_{32}b_{22}$$
$$c_{33} = a_{31}b_{13} + a_{32}b_{23}$$

Matrix multiplication always obeys the associative law, but is not necessarily commutative. Conformable matrices in the order $\mathfrak{AB}$ may not be conformable in the order $\mathfrak{BA}$.

Division of matrices is based on the fact that $\mathfrak{A}$ divided by $\mathfrak{B}$ equals $\mathfrak{AB}^{-1}$, where $\mathfrak{B}^{-1}$ is defined as that matrix such that $\mathfrak{BB}^{-1} = \delta_{ij}$. Thus, the only new problem associated with division is the finding of an inverse. Only square matrices can have inverses. The procedure for obtaining the inverse is described in matrix algebra books, which should be consulted should the need arise.

Conjugate matrices deserve special mention. If two matrices $\mathfrak{A}$ and $\mathfrak{B}$ are conjugate, they are related by a similarity transform just as conjugate elements of a group are; *i.e.*, there is a matrix $\mathfrak{R}$ such that

$$\mathfrak{A} = \mathfrak{R}^{-1}\mathfrak{B}\mathfrak{R} \tag{2-8a}$$

One advantage of matrices that will be of significance to us is that they can be used in describing the transformations of points, vectors, functions, and other entities in space. The transformation of points will be discussed in the next section, where we deal with a very important concept in group theory: how symmetry operations (elements) can be represented by matrices and what advantages are thus obtained.

We can use matrices and vectors to gain a further appreciation of similarity transforms. Imagine three distinct fixed points O, P_1, and P_2 in three-dimensional space. Let $\vec{X}$ be the vector from O to P_1, and let $\vec{Y}$ be the vector from O to P_2. A set of three-dimensional Cartesian coordinates centered at the point O will be referred to as frame I. Suppose that an operator A, *also thought of as being associated with frame I*, transforms the vector $\vec{X}$ in frame I into the vector $\vec{Y}$ in frame I according to the equation

$$\vec{Y} = A\vec{X}$$

Next rotate frame I into a new position, keeping the origin of the coordinates fixed at O and leaving the points P_1 and P_2 fixed in space. Frame I in its new position is called frame II. Notice that the vectors $\vec{X}$ and $\vec{Y}$ have been unaffected by this procedure, since the three points O, P_1, and P_2 have remained fixed in space. However, the projections of $\vec{X}$ and $\vec{Y}$ on the coordinate axes of frame II will give numerical values different from those obtained when the vectors were projected on the coordinate axes of frame I. Thus, in frame II we write the vectors as $\vec{X}'$ and $\vec{Y}'$. We now want to find an operator A' *associated with frame II* that transforms the vector $\vec{X}'$ in frame II into the vector $\vec{Y}'$ in frame II according to

$$\vec{Y}' = A'\vec{X}'$$

The operator A' in frame II is said to be *similar* to the operator A in frame I. Thus, A' sends $\vec{X}'$ (expressed in frame II) into $\vec{Y}'$ (also expressed in frame II), while A does the same thing in frame I. In order to evaluate the operator A', we need to know the relationship between the components of $\vec{X}$ and $\vec{Y}$ in frame I and the components of $\vec{X}'$ and $\vec{Y}'$ in frame II. This information is given in the form of a transformation of coordinates expressed by a matrix S. Thus,

$$\vec{X} = S\vec{X}'$$
$$\vec{Y} = S\vec{Y}'$$

so that

$$\vec{Y} = A\vec{X}$$

becomes

$$S\vec{Y}' = AS\vec{X}'$$

Since $|S| \neq 0$, we know that S^{-1} exists, and

$$\vec{Y}' = S^{-1}AS\vec{X}'$$

Thus,

$$\vec{Y}' = A'\vec{X}'$$

if

$$A' = S^{-1}AS \tag{2-8b}$$

The similar matrices A' and A are connected by equation (2–8b), which is known as a *similarity transformation*. *The trace of a matrix is invariant under a similarity transformation.*

2-5 REPRESENTATIONS; GEOMETRIC TRANSFORMATIONS

As discussed previously, E, σ, i, C_n, and S_n describe the symmetry of objects. *Each of these operations can be described by a matrix.* Consider the point P in Fig. 2–4 with x, y, and z coordinates of 1, 1, and 1 corresponding to the projections of the point on these axes. The identity operation on this point corresponds to giving rise to a new set of coordinates that are the same as the old ones. The following matrix does this.

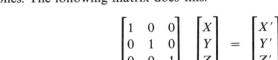

FIGURE 2–4 A point in the Cartesian coordinate system.

$$\begin{bmatrix} 1 & 0 & 0 \\ 0 & 1 & 0 \\ 0 & 0 & 1 \end{bmatrix} \begin{bmatrix} X \\ Y \\ Z \end{bmatrix} = \begin{bmatrix} X' \\ Y' \\ Z' \end{bmatrix}$$

Matrix multiplication yields:

$$X = X'$$
$$Y = Y'$$
$$Z = Z'$$

This unit matrix is said to be a representation of the identity operation.

A reflection in the xy-plane, σ_{xy}, changes the sign of the z-coordinate, but leaves x and y unchanged. The following matrix does this.

$$\begin{bmatrix} 1 & 0 & 0 \\ 0 & 1 & 0 \\ 0 & 0 & -1 \end{bmatrix} \begin{bmatrix} X \\ Y \\ Z \end{bmatrix} = \begin{bmatrix} X' \\ Y' \\ Z' \end{bmatrix}$$

Matrix multiplication gives $X = X'$, $Y = Y'$, and $Z = -Z'$, which is the result of reflection in the xy-plane on the xyz-coordinates of the point P. Similarly, the result of the operation σ_{yz} is given by

$$\begin{bmatrix} -1 & 0 & 0 \\ 0 & 1 & 0 \\ 0 & 0 & 1 \end{bmatrix} \begin{bmatrix} X \\ Y \\ Z \end{bmatrix} = \begin{bmatrix} X' \\ Y' \\ Z' \end{bmatrix}$$

and σ_{xz} is represented by

$$\begin{bmatrix} 1 & 0 & 0 \\ 0 & -1 & 0 \\ 0 & 0 & 1 \end{bmatrix} \begin{bmatrix} X \\ Y \\ Z \end{bmatrix} = \begin{bmatrix} X' \\ Y' \\ Z' \end{bmatrix}$$

The inversion operation is given by

$$\begin{bmatrix} -1 & 0 & 0 \\ 0 & -1 & 0 \\ 0 & 0 & -1 \end{bmatrix} \begin{bmatrix} X \\ Y \\ Z \end{bmatrix} = \begin{bmatrix} X' \\ Y' \\ Z' \end{bmatrix}$$

Call the z-axis the rotation axis; we shall derive the matrix for a clockwise (as viewed down the positive z-axis) rotation of the *point* by an angle φ. This rotation is illustrated in Fig. 2–5. The point (x_1, y_1) defines a vector $\mathbf{r}_1$ that connects it to the origin, and rotation leads to the vector $\mathbf{r}_2$. For any rotation of $\mathbf{r}_1$ about the z-axis by φ, the z component is unchanged, leading to:

$$\begin{bmatrix} & & 0 \\ & & 0 \\ 0 & 0 & 1 \end{bmatrix} \begin{bmatrix} x \\ y \\ z \end{bmatrix} = \begin{bmatrix} x' \\ y' \\ z' \end{bmatrix}$$

In Fig. 2–6, the rotation of $\mathbf{r}_1$ to $\mathbf{r}_2$ is shown. The problem is to determine the new coordinates for $\mathbf{r}_2$. The dots show rotation of the y_1 coordinate of $\mathbf{r}_1$ by φ,

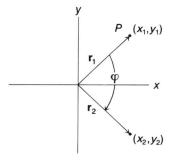

FIGURE 2–5 Clockwise rotation of the point (x_1, y_1) by the angle φ.

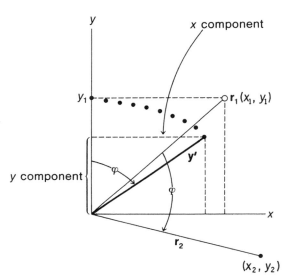

FIGURE 2–6 Rotation of the vector $\mathbf{r}_1$ and its y-coordinate through the angle φ.

producing the vector $\mathbf{y}'$. The vector $\mathbf{y}'$ now has both x- and y-components, which are given by:

$$\mathbf{y}' = y_1 \sin \varphi + y_1 \cos \varphi$$

This result is obtained by the following trigonometric arguments: $\mathbf{y}'$ still has a length of y_1. Using trigonometry on the triangle defined by φ, $\mathbf{y}'$, and the y-component of $\mathbf{y}'$, we see that

$$\cos \varphi = \frac{y \text{ component}}{y'} = \frac{y \text{ component}}{y_1}$$

or y component $= y_1 \cos \varphi$. Similarly, since $\sin \varphi = x$ component$/y_1$, we have x component $= y_1 \sin \varphi$. The vector $\mathbf{y}'$ then has components given by:

$$\mathbf{y}' = x \text{ component} + y \text{ component} = y_1 \sin \varphi + y_1 \cos \varphi$$

Figure 2–7 illustrates the rotation of the x-coordinate of the vector $\mathbf{r}_1$, producing a new vector $\mathbf{x}'$ with both x- and y-components. As before, we can show that the y-component is $-x_1 \sin \varphi$ and the x-component is $x_1 \cos \varphi$. This leads to:

$$\mathbf{x}' = x_1 \cos \varphi - x_1 \sin \varphi$$

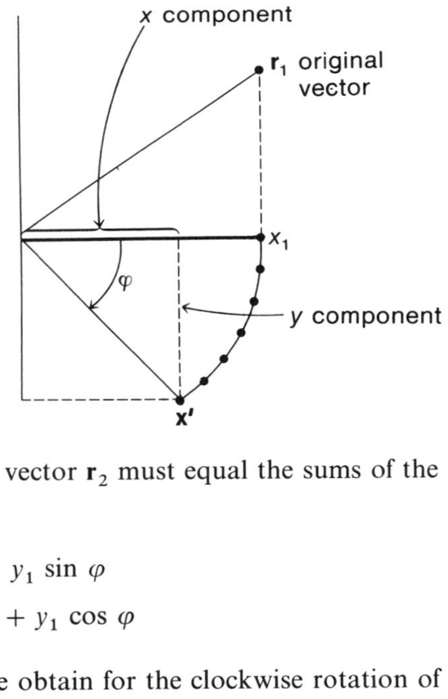

FIGURE 2–7 Rotation of the x-coordinate of the vector $\mathbf{r}_1$ by the angle φ.

The x_2- and y_2-components of the rotated vector $\mathbf{r}_2$ must equal the sums of the x- and y-components of $\mathbf{x}'$ and $\mathbf{y}'$, so:

$$x_2 = x_1 \cos \varphi + y_1 \sin \varphi$$
$$y_2 = -x_1 \sin \varphi + y_1 \cos \varphi$$

Writing these equations in matrix form, we obtain for the clockwise rotation of a point in a fixed axis system:

$$\begin{bmatrix} \cos \varphi & \sin \varphi \\ -\sin \varphi & \cos \varphi \end{bmatrix} \begin{bmatrix} x_1 \\ y_1 \end{bmatrix} = \begin{bmatrix} x_2 \\ y_2 \end{bmatrix}$$

For the reverse transformation (*i.e.*, counterclockwise rotation of a point in a fixed axis system) the transformation matrix is given by*:

$$\begin{bmatrix} \cos\varphi & -\sin\varphi \\ \sin\varphi & \cos\varphi \end{bmatrix}$$

The total transformation matrix for a clockwise proper rotation is:

$$\begin{bmatrix} \cos\varphi & \sin\varphi & 0 \\ -\sin\varphi & \cos\varphi & 0 \\ 0 & 0 & 1 \end{bmatrix}$$

For a C_2 rotation we substitute $\varphi = -180°$ into the clockwise proper rotation matrix. A negative angle corresponds to a clockwise rotation according to the trigonometric convention.

$$\begin{bmatrix} -1 & 0 & 0 \\ 0 & -1 & 0 \\ 0 & 0 & 1 \end{bmatrix}$$

A C_3 rotation matrix can be written by substituting $\varphi = -120°$ in the clockwise proper rotation matrix.

For an *improper* rotation around the z-axis, we rotate in the xy-plane and then reflect through it, leading to the matrix representation

$$\begin{bmatrix} \cos\varphi & \sin\varphi & 0 \\ -\sin\varphi & \cos\varphi & 0 \\ 0 & 0 & -1 \end{bmatrix}$$

The multiplication of these matrix representations of the symmetry operations produces the same result as the product of the symmetry operations, as they must if they are indeed the correct representations.

$$\sigma_{xz} \times \sigma_{xz} = E$$

$$\begin{bmatrix} 1 & 0 & 0 \\ 0 & -1 & 0 \\ 0 & 0 & 1 \end{bmatrix} \begin{bmatrix} 1 & 0 & 0 \\ 0 & -1 & 0 \\ 0 & 0 & 1 \end{bmatrix} = \begin{bmatrix} 1 & 0 & 0 \\ 0 & 1 & 0 \\ 0 & 0 & 1 \end{bmatrix}$$

All matrices that describe the tranformations of a set of orthogonal coordinates* by the symmetry operations of a group are *orthogonal* matrices. Their inverses can be obtained by transposing rows and columns. For example,

---

*This is equivalent to a clockwise rotation of the axis system, which would be written as:

$$\begin{bmatrix} \cos\varphi & -\sin\varphi \\ \sin\varphi & \cos\varphi \end{bmatrix} \begin{bmatrix} \vec{X} \\ \vec{Y} \end{bmatrix} = \begin{bmatrix} \vec{X}' \\ \vec{Y}' \end{bmatrix}$$

*Orthogonal coordinates are those whose dot product is zero.

the inverse of the matrix that corresponds to a 30° rotation about the z-axis [$\cos(-30°) = \sqrt{3}/2$]

$$\begin{bmatrix} \frac{\sqrt{3}}{2} & -\frac{1}{2} & 0 \\ \frac{1}{2} & \frac{\sqrt{3}}{2} & 0 \\ 0 & 0 & 1 \end{bmatrix} \text{ is } \begin{bmatrix} \frac{\sqrt{3}}{2} & \frac{1}{2} & 0 \\ -\frac{1}{2} & \frac{\sqrt{3}}{2} & 0 \\ 0 & 0 & 1 \end{bmatrix}$$

2–6 IRREDUCIBLE REPRESENTATIONS

The total representation describing the effect of all of the symmetry operations in the C_{2v} point group on the point with coordinates x, y, and z is

$$\overset{E}{\begin{bmatrix} 1 & & \\ & 1 & \\ & & 1 \end{bmatrix}} \overset{C_2}{\begin{bmatrix} -1 & & \\ & -1 & \\ & & 1 \end{bmatrix}} \overset{\sigma_{xx}}{\begin{bmatrix} 1 & & \\ & -1 & \\ & & 1 \end{bmatrix}} \overset{\sigma_{yz}}{\begin{bmatrix} -1 & & \\ & 1 & \\ & & 1 \end{bmatrix}}$$

Notice each of these matrices is *block diagonalized*; i.e., the total matrix can be broken up into blocks of smaller matrices with no off-diagonal elements between the blocks. The fact that it is block diagonalized indicates that the total representation must consist of a set of so-called one-dimensional representations. We can see this as follows. If we were to be concerned only with operations on a point that had only an x-coordinate (i.e., the column matrix $\{x, 0, 0\}$), then only the first row of the total representation would be required (i.e., 1, −1, 1, −1). This is an irreducible representation, which in this case is a set of one-dimensional matrices describing the symmetry properties of the one-dimensional x-vector in the specified point group. The symbol B_1 will be used to symbolize this irreducible representation. Do not concern yourself with the meaning of the B and 1 for now, but just think of this as a label. The irreducible representation for y is:

$$1 \quad -1 \quad -1 \quad 1$$

which is labeled B_2; and that for z is:

$$1 \quad 1 \quad 1 \quad 1$$

which is labeled A_1.

The total representation—the four 3 × 3 matrices—is a reducible representation. It is reducible, as discussed above, into three sets of 1 by 1 matrices, each set a representation by itself. The trace or character of each of the total representation matrices is the sum of the characters of each of the component irreducible representations; in order, these are 3, −1, 1, and 1.

We will show shortly that the block diagonalized 3 × 3 matrices for the effect of symmetry operations on our point resulted because the x-, y-, and z-axes of our coordinate system were selected as the *basis set of vectors* for the symmetry operations, *i.e.*, z for rotation, xy for reflection. We shall subsequently consider a representation that is not block diagonalized.

There is one additional irreducible representation in the C_{2v} point group. We can illustrate it by carrying out the operations of the group on a rotation R_z described by the sketch on the right. The identity and C_2 rotation do not change the direction in which the arrow points, but reflection by σ_{xz} and σ_{yx} do. None of the irreducible representations we now have, nor any combination of these irreducible representations, describes the effect of the C_{2v} symmetry operations on this curved arrow representing rotation around the z-axis. The result is another irreducible representation:

$$1 \quad 1 \quad -1 \quad -1$$

which is labeled A_2.

2–7 CHARACTER TABLES

We now have empirically derived all of the irreducible representations of the C_{2v} point group, but we cannot be sure we have all of them with the procedures employed. There are rigorous procedures for deriving all of the irreducible representations of the various point groups, which are covered in many treatments of group theory. These procedures will not be covered here, for we will be more concerned with using the irreducible representations and will not have to generate them; they are readily available. The results of the preceding discussion for C_{2v} symmetry are summarized by *the character table* of the C_{2v} point group shown in Table 2–2.

TABLE 2–2. Character Table for the C_{2v} Point Group

	E	C_2	σ_{xz}	σ'_{yz}	
A_1	+1	+1	+1	+1	z
A_2	+1	+1	−1	−1	R_z
B_1	+1	−1	+1	−1	x, R_y
B_2	+1	−1	−1	+1	y, R_x

Each entry in the character table is the character of the matrix for that operation in that representation. For 1×1 matrices, the character and the matrix are the same. There is a row for each irreducible representation.

The labels have the following general meaning:

1. The symbol A indicates a singly degenerate state (*i.e.*, it consists of only one representation) that is symmetric about the principal axis: *i.e.*, the character table contains values of $+1$ under the column for the principal axis for all A species.

2. The symbol B indicates a singly degenerate state that is antisymmetric about the principal axis; *i.e.*, the value -1 appears under the column for the principal axis for all B species.

3. The subscripts 1 and 2 indicate symmetry or antisymmetry relative to a rotation axis other than the principal axis. If there is no second axis, the subscripts refer to the symmetry about a σ_v plane (*e.g.*, in the C_{2v} group, subscript 1 is symmetric about the *xz*-plane and 2 is antisymmetric).

All properties of a molecule with C_{2v} symmetry can be expressed in mathematical terms that will, for a basis set, form a general representation that consists of one (or a combination of more than one) of these irreducible representations. As we shall see, appropriate combinations of the irreducible representations summarize the results of our matrices operating on anything belonging to the C_{2v} point group. Character tables are available for all of the point group symmetries, and are listed in Appendix A.

2–8 NON-DIAGONAL REPRESENTATIONS

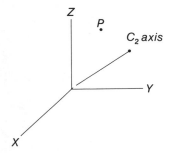

FIGURE 2–8 A basis set for geometric transformations on *p* that are *not* coincident with the Cartesian coordinates. (*P* has positive *x, y, z* coordinates.)

If, instead of selecting the C_2 and σ elements to be coincident with the axes of our coordinate system, we selected as our basis the C_2 axis shown in Fig. 2–8 (and the other elements appropriately), then our matrix representation would not be block diagonalized. For example, the C_2 rotation would not simply change the sign of the *x*-coordinate as we saw before, but instead would have to keep the sign the same and change the magnitude of the coordinates. Instead of

$$\begin{bmatrix} -1 & 0 & 0 \\ 0 & -1 & 0 \\ 0 & 0 & 1 \end{bmatrix} \begin{bmatrix} X \\ Y \\ Z \end{bmatrix}$$

the new *x*-coordinate would be dependent on what the *y*- and *z*-coordinates of the original point were; *i.e.*, there would be off-diagonal matrix elements. We shall not work this all out, for it is not worth the effort. However, the point should be made that *the trace of the matrices for all of the symmetry operations would be the same for any Cartesian basis set*, *i.e.*, 3, −1, 1, 1. The matrix representation for the axis selection we have made here can be converted into a block diagonal matrix by using an appropriate matrix, Q, for the similarity transform; *i.e.*,

$$Q^{-1}C_2Q, \ Q^{-1}\sigma Q, \text{ etc.}$$

In this case, the similarity transform is a rotation to the new basis set (for the symmetry operation with the rotation matrix) from the basis set of the original *X, Y, Z* axis system. The basis set is rotated, not the points or the molecule. In practice, when off-diagonal elements result, we realize that a poor selection of a coordinate system was made. We choose a new orthogonal coordinate system, each basis vector of which transforms as one of the irreducible representations, so that a block diagonal representation will result. This will be made clear by picking a simpler example that we can work through completely. Such a representation, which is not block diagonalized, can be obtained by considering the operations of the C_{2v} point group on two vectors **a** and **b**, using *these vectors as the basis set* to describe the representation. The problem is illustrated in Fig.

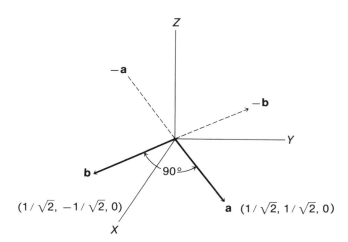

FIGURE 2–9 Geometric transformations in the **a,b** basis.

2-9. The coordinates were selected as $1/\sqrt{2}$ to produce normalized (*i.e.*, unit) vectors:

$$\left(\frac{1}{\sqrt{2}}\right)^2 + \left(\frac{1}{\sqrt{2}}\right)^2 = 1$$

The angle between **a** and **b** is 90°. We are going to set up our matrices to operate on the **a, b** basis set, so these matrices will not work for the *x, y, z* basis set of the *X, Y, Z*-coordinate system. They will tell us what **a** and **b** are changed to, in terms of **a** and **b**. Since we wish to illustrate with this example the dependence of our matrices on the location of our basis set in the coordinate system, we shall carry out all of the symmetry operations about axes and planes in the *X*-, *Y*-, *Z*-coordinate system. In terms of the **a, b** basis, we have

$$\begin{array}{ll} E\mathbf{a} = \mathbf{a} \\ E\mathbf{b} = \mathbf{b} \end{array} \quad \begin{bmatrix} 1 & 0 \\ 0 & 1 \end{bmatrix}$$

Now, the C_2 operation changes **a** to $-\mathbf{a}$ and **b** to $-\mathbf{b}$

$$\begin{array}{ll} C_2\mathbf{a} = -\mathbf{a} \\ C_2\mathbf{b} = -\mathbf{b} \end{array} \quad \begin{bmatrix} -1 & 0 \\ 0 & -1 \end{bmatrix}$$

Reflection in the *xz*-plane moves **a** into **b** and **b** into **a**:

$$\begin{array}{ll} \sigma_{xz}\mathbf{a} = \mathbf{b} \\ \sigma_{xz}\mathbf{b} = \mathbf{a} \end{array} \quad \begin{bmatrix} 0 & 1 \\ 1 & 0 \end{bmatrix}$$

Furthermore,

$$\begin{array}{ll} \sigma_{yz}\mathbf{a} = -\mathbf{b} \\ \sigma_{yz}\mathbf{b} = -\mathbf{a} \end{array} \quad \begin{bmatrix} 0 & -1 \\ -1 & 0 \end{bmatrix}$$

Note that these reflections have moved **a** into **b** and **b** into **a** with off-diagonal elements, and that the trace is zero. Further note that the trace of their representation matrices is the same as those for the x, y parts of the matrices generated earlier in the x, y basis for the C_{2v} point group. The off-diagonal elements immediately tell us that we made a bad choice of axes for this basis set. We know from earlier discussion that x and y form the basis of irreducible representations in C_{2v}; thus, if we locate our vectors along x and y, and use matrices that work on the x- and y-coordinates of the point defined by the vector (*i.e.*, an x, y basis set), then diagonal matrices will result. We can do this with a similarity transform.

For our first example of this procedure, we shall consider the operation σ_{yz}. The similarity transform is $Q^{-1}\sigma_{yz}Q$, where Q is a clockwise rotation matrix of the xy-axis system by $45°$ (this operation makes x and y coincide with **a** and **b**).

$$\begin{bmatrix} \cos\varphi & -\sin\varphi \\ \sin\varphi & \cos\varphi \end{bmatrix} = Q \text{ when } \varphi = 45°$$

so

$$Q = \begin{bmatrix} \frac{1}{\sqrt{2}} & -\frac{1}{\sqrt{2}} \\ \frac{1}{\sqrt{2}} & \frac{1}{\sqrt{2}} \end{bmatrix} \text{ and } Q^{-1} = \begin{bmatrix} \frac{1}{\sqrt{2}} & \frac{1}{\sqrt{2}} \\ -\frac{1}{\sqrt{2}} & \frac{1}{\sqrt{2}} \end{bmatrix}$$

$$Q^{-1}\sigma_{yz}^{(ab)}Q = \begin{bmatrix} \frac{1}{\sqrt{2}} & \frac{1}{\sqrt{2}} \\ -\frac{1}{\sqrt{2}} & \frac{1}{\sqrt{2}} \end{bmatrix} \begin{bmatrix} 0 & -1 \\ -1 & 0 \end{bmatrix} \begin{bmatrix} \frac{1}{\sqrt{2}} & -\frac{1}{\sqrt{2}} \\ \frac{1}{\sqrt{2}} & \frac{1}{\sqrt{2}} \end{bmatrix} = \begin{bmatrix} -1 & 0 \\ 0 & 1 \end{bmatrix} = \sigma_{yz}$$

Note that the trace of the resulting matrix is the same as that of the starting one. The trace of a matrix is always invariant under a similarity transform. When we operate on all of the matrices for the various operations in C_{2v} with this similarity transform, we will have

$$\begin{array}{cccc} E & C_{2v} & \sigma_{vxz} & \sigma_{vyz} \\ \begin{bmatrix} 1 & 0 \\ 0 & 1 \end{bmatrix} & \begin{bmatrix} -1 & 0 \\ 0 & -1 \end{bmatrix} & \begin{bmatrix} 1 & 0 \\ 0 & -1 \end{bmatrix} & \begin{bmatrix} -1 & 0 \\ 0 & 1 \end{bmatrix} \end{array}$$

corresponding to the B_1 (upper left) and B_2 (lower right) representations for x and y, respectively; *i.e.*, our reducible representation in the **a**, **b** basis is decomposed into B_1 and B_2.

Degenerate Representations

For further practice, consider the ammonia molecule, represented in Fig. 2–10 with an x, y, z-coordinate axis basis set. The symmetry operations are E, $2C_3$,

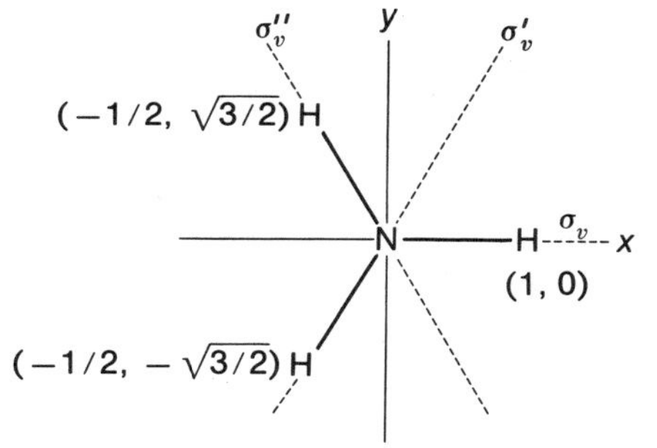

FIGURE 2–10 The ammonia molecule in a Cartesian coordinate system. (N is above the *xy*-plane, and the projection of the N–H bond on the *x*-axis is taken as 1.)

and $3\sigma_v$. Consider the *x*-, *y*-, and *z*-axes all together:

$$E = \begin{bmatrix} 1 & 0 & 0 \\ 0 & 1 & 0 \\ 0 & 0 & 1 \end{bmatrix} \quad C_3 = \begin{bmatrix} -\frac{1}{2} & -\frac{\sqrt{3}}{2} & 0 \\ \frac{\sqrt{3}}{2} & -\frac{1}{2} & 0 \\ 0 & 0 & 1 \end{bmatrix}$$

The C_3 rotation matrix is obtained by substituting $\varphi = 2\pi/3$ ($= 120°$) into the general proper rotation matrix defined earlier (p. 33).

In order to determine the matrices for σ_v, we must work out the reflection matrix for each plane. The matrix for σ_v is quite simple:

$$\begin{bmatrix} 1 & 0 & 0 \\ 0 & -1 & 0 \\ 0 & 0 & 1 \end{bmatrix} \begin{bmatrix} X \\ Y \\ Z \end{bmatrix} = \begin{bmatrix} X' \\ Y' \\ Z' \end{bmatrix}$$

We see that reflection in σ_v just changes the sign of *y*. The following matrices give the new *x*- and *y*-coordinates after reflection by σ_v' and σ_v'':

$$\sigma_v' = \begin{bmatrix} -\frac{1}{2} & \frac{\sqrt{3}}{2} & 0 \\ \frac{\sqrt{3}}{2} & \frac{1}{2} & 0 \\ 0 & 0 & 1 \end{bmatrix} \quad \sigma_v'' = \begin{bmatrix} -\frac{1}{2} & -\frac{\sqrt{3}}{2} & 0 \\ -\frac{\sqrt{3}}{2} & \frac{1}{2} & 0 \\ 0 & 0 & 1 \end{bmatrix}$$

To employ these matrices, substitute the *x*, *y*, *z*-coordinates of one of the hydrogens for *X*, *Y*, and *Z* to get the new coordinates x', y', and z' after the operation.*

---

*For example, to determine the effect of σ_v' on $(-1/2, \sqrt{3}/2)$, we see that the *z*-coordinate will not change, and *x* and *y* are given by

$$\begin{bmatrix} -\frac{1}{2} & \frac{\sqrt{3}}{2} \\ \frac{\sqrt{3}}{2} & \frac{1}{2} \end{bmatrix} \begin{bmatrix} -\frac{1}{2} \\ \frac{\sqrt{3}}{2} \end{bmatrix} = \begin{bmatrix} \frac{1}{4} + \frac{3}{4} \\ -\frac{\sqrt{3}}{4} + \frac{\sqrt{3}}{4} \end{bmatrix} = \begin{bmatrix} 1 \\ 0 \end{bmatrix}$$

Note that $\sigma_v{'}$ does not change $(-1/2, -\sqrt{3}/2)$ but does interchange the other two hydrogens. For this symmetry we have off-diagonal elements that simultaneously change the old X-coordinate into a new one with x- and y-components.

Next, we shall work out the irreducible representations for the x, y, and z axes in the C_{3v} point group. The z-axis, which is the rotation axis, transforms unchanged and can be thought of as being described by the three 1×1 matrices [1].

$$\begin{array}{ccc} E & 2C_3 & 3\sigma_v \\ [1] & [1] & [1] \end{array}$$

A 120° rotation of a vector on the x-axis produces a vector with both x- and y-components (Fig. 2–11). The same thing occurs for rotation of y. Substituting $\varphi = -120°$ into the rotation matrix and multiplying by $\{x, y\}$ gives x' and y' as:

$$x' = -\left(\frac{x}{2}\right) - \left(\frac{y\sqrt{3}}{2}\right)$$

$$y' = +\left(\frac{x\sqrt{3}}{2}\right) - \left(\frac{y}{2}\right)$$

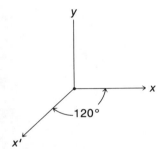

FIGURE 2–11 A rotation of the x-axis by 120° about a z-axis represented by the dot.

As we have just shown, the two vectors, x and y, are related and cannot be transformed independently. The x- and y-vectors are represented in the character table by the symbol E, used to indicate double degeneracy. The trace of the matrix (obtained by summing the diagonal elements, and which has been shown to be independent of the basis selected) is the character reported in the character table. The character is invariant under any similarity transformation carried out on the matrix. The significance of this will be made clearer when we consider more complicated transformations.

In summary, the characters of all non-degenerate point groups (*i.e.*, those having a C_2 as the highest-fold rotation axis) are 1×1 matrices with characters of $+1$ or -1, which indicate symmetric or antisymmetric behavior of an appropriate basis function under the symmetry operation. A degenerate point group contains a character that is the trace of the transformation matrix and that summarizes how the degenerate basis set transforms together.

Generally speaking, then, we shall end up with an $n \times n$ matrix for our representation (where n is the size of our basis set) for each operation in the point group; *e.g.*, in C_{2v}

$$\overset{E}{[\quad]} \overset{C_{2v}}{[\quad]} \overset{\sigma_{xz}}{[\quad]} \overset{\sigma_{yz}}{[\quad]}$$

If each of these is block diagonal, they correspond to irreducible representations. If not, we want a similarity transform to block diagonalize them.

2-9 MORE ON CHARACTER TABLES

We earlier defined a class as a complete set of elements that are conjugate to each other. We also mentioned that conjugate matrices have identical characters. (You can confirm this by checking the trace of the σ_v, σ_v', and σ_v'' matrices just given for the degenerate x and y vectors.) Thus, classes are grouped into one entry in the character tables; see, for example, $3\sigma_v$ of the C_{3v} character table.

The different irreducible representations may be thought of as a series of orthonormal vectors in k-dimensional space, where k is the number of classes in the point group. Since vectors that form the basis for two different irreducible representations, χ_i and χ_j, are orthogonal, we have

$$\sum_R g\chi_i^*(R)\chi_j(R) = 0 \qquad \text{when } i \neq j \qquad (2\text{-}9)$$

The sum is taken over all h symmetry operations (each denoted by R) in the point group, and g represents the number of elements in the class. The sum of the squares of the characters of any irreducible representation equals h, the order of the point group.

$$\sum_R g[\chi_i(R)]^2 = h \qquad (2\text{-}10)$$

where R is a sum over all operations, so $3\sigma_v$ *is counted as three times* σ_v in this summation for $g = 3$.

A character table (Table 2-3) is divided into four main areas, as explained in the following paragraphs.

TABLE 2-3. Character Table for the C_{3v} Point Group

C_{3v}	E	$2C_3$	$3\sigma_v$		
A_1	1	1	1	z	$x^2 + y^2, z^2$
A_2	1	1	-1	R_z	
E	2	-1	0	$(x, y)\, R_x R_y$	$(x^2 - y^2, xy)(xz, yz)$
I	II		III		IV

Area I: The meanings of A, B, E, and the 1 and 2 subscripts have been discussed. Other symbols are encountered in the character tables; see Appendix A for examples. The symbols E and T represent doubly and triply degenerate states, respectively. (F is often employed instead of T). If the molecule has a center of symmetry (see the C_{2h}, D_{2h}, D_{4h}, and O_h character tables in Appendix A), the subscript g is used to indicate symmetry ($+1$) with respect to this center, while u indicates antisymmetry (-1). Prime and double prime marks are employed to indicate symmetry and antisymmetry, respectively, relative to a σ_h plane of symmetry.

Examples of the application of these rules can be obtained by referring to the character tables in Appendix A. All A species in the D_{3h} group have values of $+1$ in the column for the principal axis, $2C_3$. The species A_1 are symmetric (i.e., have $+1$ values) to $3C_2$, and A_2 species are antisymmetric (-1) to these C_2 axes. In D_{4h}, all A species are symmetric with respect to the perpendicular C_2 axes. The prime species are symmetric ($+$) to the horizontal plane σ_h, while the double prime species are antisymmetric ($-$) to this plane. The u species are antisymmetric ($-$) to the center of symmetry, while the g species are symmetric ($+$).

Area II: This has already been discussed, but there are a few additional points to be made. Some character tables contain imaginary or complex characters. Whenever complex characters occur, they occur in pairs such that one is the complex conjugate of the other. While they are mathematically distinct irreducible representations, in all physical problems no difference is observed and the results are as if they had been a doubly degenerate irreducible representation. In the T_d point group, two triply degenerate species exist, T_1 and T_2. For T_2, x, y, and z form a basis; for T_1, all three of the rotations about these axes form a basis.

Area III: This area lists the transformation properties of vectors along the x, y, and z axes and rotations R_x, R_y, and R_z (represented as curved arrows) about the x, y, and z axes.

Area IV: This area indicates the transformation properties of the squares and binary products of the coordinates. Although there are six possible square and binary products (x^2, y^2, z^2, xy, xz, and yz), only five are ever indicated because $x^2 + y^2 + z^2 = r^2$ and, as a result, one of the linear combinations is redundant (r^2 is the square of the radius of the $x^2 + y^2 + z^2$ sphere). The direct product of two vectors is obtained by multiplying the species for each, e.g., $XY = B_1 \times B_2 = A_2$. To perform this multiplication, the following procedure is used:

$$E(B_1) \times E(B_2) \quad C_2(B_1) \times C_2(B_2) \quad \sigma_v(B_1) \times \sigma_v(B_2) \quad \sigma_v'(B_1) \times \sigma_v'(B_2)$$

or

$$(1)(1) \quad\quad (-1)(-1) \quad\quad (1)(-1) \quad\quad (-1)(1)$$

giving the result
$$+1 \quad +1 \quad -1 \quad -1$$

The result $+1$, $+1$, -1, -1 is identical to the irreducible representation A_2; hence, $B_1 \times B_2 = A_2$. In subsequent chapters we shall have occasion to take the direct product of irreducible representations, and this procedure will be employed. Many of the combinations in this area have symmetry properties identical to those of the d-orbitals (e.g., xy and d_{xy} are identical). Thus, these products can be used to indicate which orbitals will remain degenerate in transition metal ion complexes of various symmetries. The pair $x^2 - y^2$ and xy are degenerate in a C_{3v} molecule as are xz and yz, but xz and yz belong to a different class than do $x^2 - y^2$ and xy. The different classes can have different energies in complexes with different symmetries.

The wave functions of a molecule are an example of a basis for a representation; that is, they can serve as the basis set for the representation matrices of a group. These wave functions must possess the same transformation properties under the operations of the group as the irreducible representations.

2–10 MORE ON REPRESENTATIONS

If, instead of vectors, we choose to obtain a representation using a set of atom coordinates, the resulting matrix could have dimensions of $3N \times 3N$ associated with the three Cartesian coordinates for each of the N atoms in the molecule (Fig. 2–12).

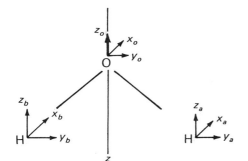

FIGURE 2–12 The $3N$ coordinates for H_2O in a Cartesian coordinate system.

For the E matrix, we have a 9×9 representation

$$\begin{bmatrix} 1 & & & & & & & & \\ & 1 & & & & & & & \\ & & 1 & & & & & & \\ & & & 1 & & & & & \\ & & & & 1 & & & & \\ & & & & & 1 & & & \\ & & & & & & 1 & & \\ & & & & & & & 1 & \\ & & & & & & & & 1 \end{bmatrix} \begin{bmatrix} x_o \\ y_o \\ z_o \\ x_a \\ y_a \\ z_a \\ x_b \\ y_b \\ z_b \end{bmatrix} = \begin{bmatrix} x_o' \\ y_o' \\ z_o' \\ x_a' \\ y_a' \\ z_a' \\ x_b' \\ y_b' \\ z_b' \end{bmatrix}$$

Any 1×1 diagonal matrix element in this large matrix corresponds to a non-degenerate irreducible representation in the final block diagonalized matrix. Accordingly, we shall need nine irreducible representations to describe this system.

For the C_2 symmetry operation on our nine dimensional basis set, we have

$$\begin{bmatrix} -1 & & & & & & & & \\ & -1 & & & & & & & \\ & & 1 & & & & & & \\ & & & 0 & 0 & 0 & -1 & 0 & 0 \\ & & & 0 & 0 & 0 & 0 & -1 & 0 \\ & & & 0 & 0 & 0 & 0 & 0 & 1 \\ & & & -1 & 0 & 0 & 0 & 0 & 0 \\ & & & 0 & -1 & 0 & 0 & 0 & 0 \\ & & & 0 & 0 & 1 & 0 & 0 & 0 \end{bmatrix} \begin{bmatrix} x_o \\ y_o \\ z_o \\ x_a \\ y_a \\ z_a \\ x_b \\ y_b \\ z_b \end{bmatrix} = \begin{bmatrix} x_o' \\ y_o' \\ z_o' \\ x_a' \\ y_a' \\ z_a' \\ x_b' \\ y_b' \\ z_b' \end{bmatrix}$$

Similar matrices can be constructed for $\sigma_{v_{xz}}(\sigma_v)$ and $\sigma_{v_{yz}}(\sigma_v')$.

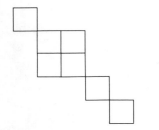

FIGURE 2–13 Form of a matrix after diagonalization.

These larger representations can be broken up, *i.e.*, are reducible into the sum of many of the smaller irreducible ones given earlier for the C_{2v} point group. We could proceed by looking for a similarity transform Q that block diagonalizes the 9 × 9 E, C_2, σ_v, and σ_v' matrices. Then the blocks in the diagonalized matrix would correspond to the irreducible representations for this molecule and in this basis set. Although this would not occur in the C_{2v} point group (or any point group with a lower than threefold axis), matrices that cannot be reduced to a 1 × 1 matrix would correspond to degenerate representations. After diagonalization, the matrix would have the form shown in Fig. 2–13, and each block can be treated separately as a representation, provided that the reducible representations of the other symmetry operations are similarly block diagonalized. There is an easier way to solve this problem, and we shall describe it next.

2–11 SIMPLIFIED PROCEDURES FOR GENERATING AND FACTORING TOTAL REPRESENTATIONS; THE DECOMPOSITION FORMULA

The trace of the matrices for E, C_2, σ_v, and σ_v' operations for the problem described above is

$$\begin{array}{ccccc} & E & C_2 & \sigma_{v_{xz}} & \sigma'_{v_{yz}} \\ \chi_T = & 9 & -1 & +1 & 3 \end{array}$$

In actual practice, similarity transforms are not used to find the irreducible representations that constitute this representation. Even the total representation is found by simpler procedures, and this is then factored into the individual irreducible representations by using the decomposition formula, *vide infra*.

With a little reflection, it can be seen that the following rules summarize what we have already said about generating the character of the total representation.

1. Any vector (or part of the basis set) that is unchanged by a symmetry operation is assigned $+1$. (This diagonal matrix element left the coordinates unchanged in our earlier discussion.)

2. Any vector or part of the basis set that is changed into the opposite direction is assigned -1. (The diagonal matrix element reflects this: in the C_{2v} example discussed earlier, all changes were either symmetric ($+1$) or unsymmetric (-1)).

3. A vector or part of the basis set that is moved onto another vector by a symmetry operation is counted as zero. Recall that in our discussion of the **a** and **b** vectors, the vectors were interchanged in our earlier example of H_2O with off-diagonal elements, and the trace of the matrix dealing with H_a and H_b is zero. These rules give us directly the value of the trace of the matrices we generated in our earlier discussion of the nine coordinates ($3N$) of the water molecule.

Applying these rules to these nine coordinates, we see that E gives a trace for the total representation, χ_T, of 9. The C_2 operation moves all H_a coordinates into H_b and vice versa for a result of zero; while for oxygen, x goes to $-x$ (-1), y goes to $-y$ (-1), and z remains unchanged ($+1$). Accordingly, χ_T for the C_2 operation is -1. With x perpendicular to the plane of the atoms, reflection in the xz-plane moves all H_a and H_b vectors for a result of zero. The x and z vectors remain unchanged, while y goes to $-y$ for a net result of $+1$. Reflection in the yz-plane does not change any of the six y or z vectors ($+6$), while the three x vectors go to $-x$ (-3) for a total ($+6-3$) of $+3$. Thus, $\chi_T = 9, -1, 1, 3$ as described before.

When an operation does not interchange atoms but moves them into a new position that is some combination of the old coordinates (*e.g.*, C_3 on $\hat{x}$), the trace of the matrix must be determined by working out the matrix for the geometric transformation as was done earlier. For rotations, the rotation matrix can be employed.

Next, we shall describe how to determine all of the different irreducible representations that make up the total representation. The following formula shows how to do this; books on group theory should be consulted by those interested in the derivation. The formula summarizing how to factor a total representation, the so-called *decomposition formula*, is:

$$a_i = \frac{1}{h} \sum_R g\chi_i(R)\chi_T(R) \qquad (2\text{--}11)$$

Here a_i is the number of contributions from the ith irreducible representation, R refers to a particular symmetry operation (or, if there is more than one element in the class, to the whole class); h, the order of the group, is given by the total number of symmetry operations in the point group (in determining this, be sure to count $3\sigma_v$'s as three); g is the number of elements in the class; $\chi_i(R)$ is the character of the irreducible representation; and $\chi_T(R)$ is the character for the analogous operation (R) in the total representation. The use of this formula is best demonstrated by decomposing the total character for the $3N$ coordinates of water,

$$\chi_T = 9 \quad -1 \quad 1 \quad 3$$

using the C_{2v} character table. The value of h is 4, and $g = 1$ for all symmetry operations. We sum over the R (in this case, 4) symmetry operations as follows:

$$a_{A_1} = \frac{1}{4}[g\chi_{A_i}(E)\chi_T(E) + g\chi_{A_i}(C_2)\chi_T(C_2) + g\chi_{A_i}(\sigma_v)\chi_T(\sigma_v) + g\chi_{A_i}(\sigma_v')\chi_T(\sigma_v')]$$

$$= \frac{1}{4}[1 \times 1 \times 9 + 1 \times 1 \times (-1) + 1 \times 1 \times 1 + 1 \times 1 \times 3] = 3$$

$$a_{A_2} = \frac{1}{4}[1 \times 1 \times 9 + 1 \times 1 \times (-1) + 1 \times (-1) \times 1 + 1 \times (-1) \times 3] = 1$$

$$a_{B_1} = \frac{1}{4}[1 \times 1 \times 9 + 1 \times (-1) \times (-1) + 1 \times 1 \times 1 + 1 \times (-1) \times 3] = 2$$

$$a_{B_2} = \frac{1}{4}[1 \times 1 \times 9 + 1 \times (-1) \times (-1) + 1 \times (-1) \times 1 + 1 \times 1 \times 3] = 3$$

Thus, we find that the total representation consists of the sum of the irreducible representations $3A_1, 1A_2, 2B_1,$ and $3B_2$. The total representation above describes all of the degrees of freedom of the molecule. In infrared spectroscopy, we are concerned with all of the possible vibrations that a molecule can undergo. If one subtracts the irreducible representations for the three basic translational modes and three rotational modes from the total representation, the remaining irreducible representations describe the vibrations.

A physical view of the decomposition formula can be had by considering the irreducible representations as vectors in a given point group. The general representation is also taken as a vector. The dot product of the general representation vector with an irreducible representation vector that does not contribute will be zero. The group order is used as a normalization factor to insure that the dot product of a contributing irreducible representation vector will give the number of contributions from the vector.

Another basis set, used to give a more transparent view of certain molecular vibrations, employs the bonds to be stretched during a vibration as vectors. Consider the two vectors for stretching the OH bonds of water, illustrated in Fig. 2–14. The matrices using these bond vectors (S_1 and S_2) as the basis set are:

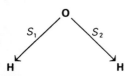

FIGURE 2–14 Vectors representing the two bonds in H_2O.

$$E = \begin{bmatrix} 1 & 0 \\ 0 & 1 \end{bmatrix} \quad \sigma_{v(xz)} = \begin{bmatrix} 1 & 0 \\ 0 & 1 \end{bmatrix}$$

$$C_2 = \begin{bmatrix} 0 & 1 \\ 1 & 0 \end{bmatrix} \quad \sigma'_{v(yz)} = \begin{bmatrix} 0 & 1 \\ 1 & 0 \end{bmatrix}$$

The total representation χ_T is given by:

	E	C_2	$\sigma_{v(xz)}$	$\sigma'_{v(yz)}$
$\chi_T =$	2	0	2	0

Factoring the total representation for these bond stretches leads to the symmetry of the two stretching modes as A_1 and B_2, which correspond to the symmetric and asymmetric stretch, respectively.

2–12 DIRECT PRODUCTS

We have already mentioned that the characters of the representation of a direct product are obtained by taking the products of the characters of the individual sets of functions. In C_{2v},

$$B_1 \times B_1 = \underbrace{\chi_E(B_1) \times \chi_E(B_1)}_{1} \underbrace{\chi_{C_2}(B_1) \times \chi_{C_2}(B_1)}_{1} \underbrace{\chi_{\sigma_v}(B_1) \times \chi_{\sigma_v}(B_1)}_{1} \underbrace{\chi_{\sigma_{v'}}(B_1) \times \chi_{\sigma_{v'}}(B_1)}_{1}$$

$$= A_1$$

That is, $B_1 \times B_1 = A_1$. When dealing with degenerate irreducible representations, remember that the degenerate bases are transformed in pairs, with the character corresponding to the trace of the transformation matrix for both. This can be illustrated by the product of $E \times E$ in the C_{4v} point group. In terms of the x, y basis for C_{4v}, we have

$$E = \begin{bmatrix} 1 & 0 \\ 0 & 1 \end{bmatrix} \qquad C_4 = \begin{bmatrix} 0 & -1 \\ 1 & 0 \end{bmatrix} \qquad C_2 = \begin{bmatrix} -1 & 0 \\ 0 & -1 \end{bmatrix}$$

$$E \times E = \begin{bmatrix} 1 & 0 \\ 0 & 1 \end{bmatrix} \times \begin{bmatrix} 1 & 0 \\ 0 & 1 \end{bmatrix} = \begin{bmatrix} 1 & 0 & 0 & 0 \\ 0 & 1 & 0 & 0 \\ 0 & 0 & 1 & 0 \\ 0 & 0 & 0 & 1 \end{bmatrix}$$

By analogy to taking the cross product, the identity gives a total representation of four because the cross product of two matrices increases the dimensionality of the matrix. The formula for taking the cross product of two matrices involves taking the a_{11} element (where a is the left matrix and b is the right) times the b matrix to give the upper left 2×2 part of the resulting 4×4 c matrix; that is, $a_{11}b_{11} = c_{11}, a_{11}b_{12} = c_{12}, a_{11}b_{21} = c_{21}$, and $a_{11}b_{22} = c_{22}$. The a_{12} element times the b matrix gives the upper right quarter of the c matrix, the a_{21} element times the b matrix gives the lower left quarter, and the a_{22} element times the b matrix gives the lower right quarter. The character of the direct product matrix is the product of the characters of the original matrices. In general, the direct products of irreducible representation matrices may be reducible.

ADDITIONAL READING

M. Tinkham, "Group Theory and Quantum Mechanics," McGraw-Hill, New York, 1964.
F. A. Cotton, "Chemical Applications of Group Theory," 3rd ed., Wiley-Interscience, New York, 1990.
M. Orchin and H. H. Jaffé, J. Chem. Educ., 47, 246, 372, 510 (1970).
M. Orchin and H. H. Jaffé, "Symmetry, Orbitals, and Spectra," Wiley-Interscience, 1971.
C. D. H. Chisholm, "Group Theoretical Techniques in Quantum Chemistry," Academic Press, New York, 1976.
A. Vincent, "Molecular Symmetry and Group Theory," Wiley, New York, 1977.
For space groups, see: J. D. Donaldson and S. D. Ross, "Symmetry and Stereochemistry," Wiley, 1972.

EXERCISES

1. Work out all the classes (of symmetry operations) in a two-dimensional equilateral triangle. (Only x and y dimensions apply, so there is only a threefold rotation axis and $3\sigma_v$ planes.)

2. Work out the group multiplication table for NH_3. Determine all classes.

3. Identify the subgroups in $G_4{}^1$, $G_4{}^2$, and $G_6{}^1$. (The groups were given in this chapter.)

4. Demonstrate that the E_g and B_{2g} irreducible representations of the D_{4h} point group are orthogonal.

5. In the C_{4v} point group, indicate the operations that comprise

   a. the $2C_4$ given in the character table; that is, which of C_4^1, C_4^2, C_4^3, or C_4^4 comprise $2C_4$.

   b. the $2\sigma_v$.

   c. the $2\sigma_d$.

   d. Construct a group multiplication table for the symmetry operations in C_{4v}.

   e. Can a molecule with C_{4v} symmetry have a dipole moment? If it can, where is it located?

6. In addition to vectors and coordinates of a point, the orbitals of an atom can be assigned to irreducible representations. Using the character table for the D_{4h} point group, indicate the representations for the p-orbitals and d-orbitals. Take the center of the orbitals as the point about which all operations are carried out. Why do the p-orbital representations have a u subscript?

7. For which irreducible representation do the following vibrational modes form a basis?

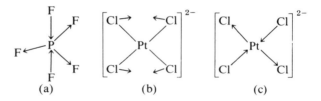

(a)   (b)   (c)

8. Recall the problem of a point in the Cartesian coordinate system with an x, y, z basis set.

   a. Write the matrix for an inversion operation.

   b. Perform the matrix multiplication that shows

   $$i \times \sigma_h = C_2.$$

   c. Complete the multiplication table for the C_{2h} point group.

9. Consider the C_{2v} point group in the xy plane. Using unit vectors along the x- and y-axes as the basis, one finds the following representation:

$$\begin{array}{cccc} E & C_2 & \sigma_{xz} & \sigma_{yz} \\ \begin{bmatrix} 1 & 0 \\ 0 & 1 \end{bmatrix} & \begin{bmatrix} -1 & 0 \\ 0 & -1 \end{bmatrix} & \begin{bmatrix} 1 & 0 \\ 0 & -1 \end{bmatrix} & \begin{bmatrix} -1 & 0 \\ 0 & 1 \end{bmatrix} \end{array}$$

where C_2 is along the z-axis.

Consider the axes x' and y', which are at an angle of $\alpha = 30°$ counterclockwise with respect to the x- and y-axes and in the same plane. Keeping C_2 along the z-axis and the σ planes in the xz- and yz-planes, give the representation for which unit vectors along x' and y' serve as a basis set. Show that the traces are the same in the new representation for each operation as those in the representation above.

10. Generate the matrices describing how the sets of functions (x, y, z), (xy, xz, yz), and $(2z^2 - x^2 - y^2, x^2 - y^2)$ are transformed by the various operations of the group O. Show how the trace transforms as irreducible matrix representations designated T_1, T_2, and E, respectively.

11. Write the matrices describing the effect on a point (x, y, z) of reflections in vertical planes that lie halfway between the xz- and yz-planes. By matrix methods, determine what operation results when each of these reflections is followed by reflection in the xy-plane.

12. Consider $AuCl_4^-$ with D_{4h} symmetry. Use the x, y, and z Cartesian coordinate vectors at each atom as a basis set to form a total representation. Determine the irreducible representations.

13. In the groups specified, determine the direct product representations and reduce them when possible.

   a. $A_2 \times B_1$ in C_{2v}.

   b. $E \times E$ in C_{4v}.

   c. $E \times E$ in C_{3v}.

14. Consider the two complex ions cis-$[CoF_4Cl_2]^{3-}$ and $trans$-$[CoF_4Cl_2]^{3-}$. For each of these ions, find the symmetry designations of the d orbitals.

15. In problem 10, we worked to find out that in the O point group the sets of functions (x, y, z), (xy, yz, xy), and $(x^2 - y^2, 2z^2 - x^2 - y^2)$ transformed as T_1, T_2, and E, respectively. The character table contains this information, but is truncated after giving information on product functions $(xy, xz,$ etc.) and does not show how the triple product functions (e.g., the xyz) transform. The triple products correspond to the f-orbital functions $(x^3, y^3, z^3, xyz, x(y^2 - z^2), y(z^2 - x^2), z(x^2 - y^2))$. Take all of these combinations to ascertain how the seven f-orbitals transform in the C_{2v} point group.

16. Use this square in the following problems. You are to assume that the square is like a table; i.e., it cannot be turned upside down and still be the same. The two coordinate systems share a common z-axis perpendicular to the plane of the figure.

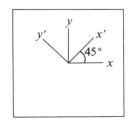

Consider a reflection, σ_{xz}. (Since the z-coordinate is always constant, work the following problems as two-dimensional cases.)

a. Derive the matrix corresponding to this reflection in the x, y-axis system.

b. Derive the matrix corresponding to this reflection in the x', y'-axis system.

c. Show that σ_{xz}' in the new axis system is related to σ_{xz} by a similarity transformation of the type

$$\sigma_{xz}' = \mathbf{T}^{-1}\sigma_{xz}\mathbf{T}$$

Derive the **T** matrix and verify this equation. The bold symbols represent matrices.

17. a. We discussed the problem of using the OH bonds in water to form a basis for 2×2 representation matrices in C_{2v} symmetry. Consider the H$_2$S molecule and assume a 90° bond angle. Write the four matrices for the symmetry operations in C_{2v} using the S—H bonds as a basis set.

b. Actually, the same information is obtained from a set of axes $\bar{z}$ and $\bar{y}$ set up at the S nucleus as shown, with $\bar{z}$ along the C_2-axis and $\bar{y}$ in the molecular plane. Write out the matrices using this basis of $\bar{z}$ and $\bar{y}$.

c. Suppose that R is the matrix for some operation in the z, y frame and that R' is the corresponding matrix in the S—H bond frame. Then a 135° counterclockwise rotation of the points in the plane will convert R to R' by a similarity transformation, i.e.,

$$R' = \pi^{-1}R\pi$$

Write out the matrix π.

18. The following is excerpted from Strommen and Lippincott *J. Chem. Ed.*, **49**, 341 (1972):

In a recent issue it was pointed out by Schäfer and Cyvin that the well-known group theoretical equation

$$n(\gamma) = \frac{1}{g}\sum_R \chi_R \chi_R^{(\gamma)} \qquad (1)$$

fails to work when applied to linear groups. Equation (1) is used to determine the number of times, $n(\gamma)$, that a given symmetry species γ will occur in the reducible representation of a particular molecule. χ_R is the character of the reducible representation Γ; $\chi_R^{(\gamma)}$ is the character of the irreducible representation; and R is the index used to denote each of the symmetry operations of the group.

Outline of Method
1. Assume a lower molecule symmetry which corresponds to a subgroup G of the molecular group G^0.
2. Place a set of cartesian coordinate vectors on each atom. (The z-axis must be placed along the maximum symmetry axis of the parent group.)
3. Using standard methods obtain the characters of $\Gamma_{\text{reducible}}$.

4. Calculate the values of each $n(\gamma)$ by application of equation (1).
5. Finally compare the basis vectors of the irreducible representations under G^0 with those obtained for the molecule assuming G.

(By "basis vectors," rule 5 means the functions x, y, z, $x^2 - y^2$, etc., found in the right-hand columns of the character table.)

a. To what symmetry group does the molecule acetylene (C_2H_2) belong?

b. Why does equation (1) fail to work for groups to which linear molecules belong?

c. Use the method outlined (with $G = D_{2h}$) to obtain the irreducible representations (in the group to which C_2H_2 belongs) for a basis set consisting of x, y, and z vectors located at each atom.

3 Molecular Orbital Theory and Its Symmetry Aspects

Introduction

3–1 OPERATORS

As is true for many of the topics in this text, whole books have been written on the subject of this chapter. This chapter will be a very brief presentation of those fundamentals of molecular orbital theory that should be a part of the background of any modern chemist. We shall concentrate on those aspects that will be needed for our considerations of spectroscopy. One can well appreciate that a complete understanding of the spectroscopy, reactivity, and physical properties of substances will not be possible until we can interpret these phenomena in terms of the electron distribution in the molecule.

In principle, all the information about the properties of a system of N particles is contained in a wave function, ψ, which is a function of only the coordinates of the N particles and time. If time is included explicitly, ψ is called a time dependent wave function; if not, the system is said to be in a stationary state. The quantity $\psi^*\psi$ (where ψ^* is the complex conjugate of ψ) is proportional to the probability of finding the electron at a particular point.

For every observable property of a system, there exists a linear Hermitian operator, $\hat{\alpha}$, and the observable can be inferred from the mathematical properties of its associated operator. A Hermitian operation is defined by

$$\int \psi_i^* \hat{\alpha} \psi_j \, d\tau = \int \psi_j (\hat{\alpha} \psi_i)^* \, d\tau$$

We shall abbreviate the above integrals by the so-called bra $\langle \, | $ and ket $ | \, \rangle$ notation as:

$$\langle \psi_i | \hat{\alpha} | \psi_j \rangle$$

To construct an operator, $\hat{\alpha}$, for a given observable, one first writes the classical expression for the observable of interest in terms of coordinates, q, momenta and time. Then the time and coordinates are left as they are and for Cartesian coordinates the momenta, p_q, are replaced by the differential operator, *i.e.*,

$$p_q = -i\hbar(\partial/\partial_q)$$

where q is the coordinate that is conjugate to p_q. For example, the quantum mechanical operator for kinetic energy, $\hat{T}$, can be constructed by first writing the classical expression for the kinetic energy:

$$T = \frac{1}{2}mv_x^2 + \frac{1}{2}mv_y^2 + \frac{1}{2}mv_z^2$$

Momenta $= mv = p$, so,

$$T = \frac{1}{2m}(p_x^2 + p_y^2 + p_z^2)$$

$$\hat{T} = \frac{1}{2m}\left[\left(-i\hbar \frac{\partial}{\partial x}\right)^2 + \left(-i\hbar \frac{\partial}{\partial y}\right)^2 + \left(-i\hbar \frac{\partial}{\partial z}\right)^2\right]$$

$$= -\frac{\hbar^2}{2m}\left(\frac{\partial^2}{\partial x^2} + \frac{\partial^2}{\partial y^2} + \frac{\partial^2}{\partial z^2}\right) = -\frac{\hbar^2}{2m}\nabla^2$$

where

$$\nabla^2 = \frac{\partial^2}{\partial x^2} + \frac{\partial^2}{\partial y^2} + \frac{\partial^2}{\partial z^2}.$$

If an operator $\hat{\alpha}$ corresponds to an observable, and *if ψ_s, the wave function for the state, is an eigenfunction of the operator $\hat{\alpha}$, then $\hat{\alpha}\psi_s = a_s\psi_s$ where a_s is a number.* An experimentalist making a series of measurements of the quantity corresponding to $\hat{\alpha}$ will always get a_s. The wave functions are eigenfunctions of the Hamiltonian operator which yields energies as *eigenvalues*, i.e., numbers;

$$\hat{H}\psi_n = E\psi_n \tag{3-1}$$

The operator for angular momentum about the Z-axis is also such an operator:

$$\hat{L}_z\psi_n = m\hbar\psi_n \tag{3-2}$$

where m, the azimuthal quantum number, has values of $+l \ldots 0 \ldots -l$.

Some properties of a system are not characterized by an eigenfunction for the appropriate operator which describes that property. Then a series of measurements of this property will not give the same result, but instead a distribution of results. The *average value* or expectation value, $\langle a_s \rangle$, is given by:

$$\langle a_s \rangle = \frac{\langle \psi_s | \hat{\alpha} | \psi_s \rangle}{\langle \psi_s | \psi_s \rangle} \tag{3-3}$$

ψ is generally not an eigenfunction of the operator $\hat{\alpha}$, but using equation (3-3), one can obtain the average value.

In the course of this book, we shall have occasion to consider many operators. We shall discuss them as the need arises. Our immediate concern will be with the Hamiltonian operator and the Schrödinger equation:

$$\hat{H}\psi = E\psi$$

This equation cannot be solved directly for the energy, E, on molecular systems because one cannot find a function ψ such that $\hat{H}\psi/\psi$ is a constant and not a variable function of the position of the electron. Consequently, we average $\hat{H}\psi$ over all space by multiplying by ψ^* and integrating over all space:

$$\int \psi^* \hat{H} \psi \, d\tau = \int \psi^* E \psi \, d\tau = E \int \psi^* \psi \, d\tau$$

so

$$E = \int \frac{\psi^* \hat{H} \psi \, d\tau}{\psi^* \psi \, d\tau} = \frac{\langle \psi | \hat{H} | \psi \rangle}{\langle \psi | \psi \rangle} \qquad (3\text{--}4)$$

The Hamiltonian operator for most systems is easily written as

$$\hat{H} = -(h^2/8\pi^2 m)\nabla^2 + V,$$

where V is the potential energy of the system. Other operators are just as easily written by writing the classical expression for the property of interest and replacing the momentum by $-i\hbar \partial/\partial_x$. Thus, if we had accurate wave functions and could solve the resulting integrals, we could calculate all of the properties of a molecule. Unfortunately, such calculations are rarely practical.

In the LCAO–MO method (linear combination of atomic orbitals–molecular obital) of solving this problem, the wave functions, ψ_j, of the resulting j molecular orbitals are approximated as linear combinations of atomic orbitals:

$$\psi_j = \sum_r C_{jr} \varphi_r \qquad (3\text{--}5)$$

where C_{jr} is the coefficient describing the contribution of φ_r to the jth molecular orbital. The total electronic ground state wave function Ψ is a product of the individual molecular orbital wave functions. As we shall show shortly, the ψ_j's form the basis for irreducible representations of the molecular point group. Therefore, the symmetry of the total electronic ground state wave function is obtained by taking the direct product of the irreducible representations of the occupied ψ_j's.

The various ways of carrying out molecular orbital calculations correspond to different approximations that are used in the solution of the problem. One such approximation ignores the correlation of the motion of the electrons in the molecule. If there are two electrons in a molecule, we know that at any instant one of the electrons will tend to avoid the region occupied by the other. Over a longer time period, the motion of one electron will be correlated with the motion of the second electron. When this correlation is ignored, the resulting ψ_j's describe the system in terms of the coordinates of a single electron. Such wave functions are labeled *one electron molecular orbitals*. In describing the molecule, two electrons are added to each of the one electron m.o.'s.

For clarity, we can proceed with the problem of determining our wave functions by using a two-term basis set φ_1 and φ_2 (i.e., two atomic orbitals on two different atoms) without any loss of generality, i.e., equation (3–5) for our wave function is

$$\psi = C_1 \varphi_1 + C_2 \varphi_2$$

Substituting this equation for ψ into equation (3–4) produces

$$E = \frac{\int (C_1\varphi_1 + C_2\varphi_2)\hat{H}(C_1\varphi_1 + C_2\varphi_2)\, d\tau}{\int (C_1\varphi_1 + C_2\varphi_2)(C_1\varphi_1 + C_2\varphi_2)\, d\tau}$$

$$= \frac{\int (C_1\varphi_1\hat{H}C_1\varphi_1 + C_1\varphi_1\hat{H}C_2\varphi_2 + C_2\varphi_2\hat{H}C_1\varphi_1 + C_2\varphi_2\hat{H}C_2\varphi_2]\, d\tau}{\int [C_1\varphi_1 C_1\varphi_1 + C_2\varphi_2 C_1\varphi_1 + C_1\varphi_1 C_2\varphi_2 + C_2\varphi_2 C_2\varphi_2]\, d\tau}$$

Since C_1 and C_2 are constants, they can be taken out of the integrals. We can abbreviate $\int \varphi_i \hat{H} \varphi_j\, d\tau$ as H_{ij} and $\int \varphi_i \varphi_j\, d\tau$ as S_{ij}. Furthermore, $H_{21} = H_{12}$ and $S_{12} = S_{21}$, leading to equation (3–6).

$$E = \frac{C_1{}^2 H_{11} + 2C_1 C_2 H_{12} + C_2{}^2 H_{22}}{C_1{}^2 S_{11} + 2C_1 C_2 S_{12} + C_2{}^2 S_{22}} \tag{3–6}$$

The various types of integrals are given different names:

H_{ii} integrals are referred to as *Coulomb integrals*;
H_{ij} as *resonance integrals*; and
S_{ij} as *overlap integrals*. The value of $S_{ii} = 1$ for normalized a.o.'s.

The *variational principle* is employed to find the set of coefficients C_1 and C_2 that minimizes the energy, *i.e.*, gives the most stable system. The energy we get, called the *variation energy*, will depend upon how good our wave function is.

$$E_{\text{var}} > E_{0\text{(ground state energy)}} \tag{3–7}$$

This minimization (*i.e.*, the variation procedure) is carried out by setting the partial derivative of E with respect to each of the coefficients (C's) equal to zero. Taking the partial derivative of equation (3–6) with respect to each of the coefficients, setting $\partial E/\partial C_1$ and $\partial E/\partial C_2$ both equal to zero, and collecting terms, we get

$$\begin{aligned} C_1(H_{11} - ES_{11}) + C_2(H_{12} - ES_{12}) &= 0 \\ C_1(H_{21} - ES_{21}) + C_2(H_{22} - ES_{22}) &= 0 \end{aligned} \tag{3–8}$$

This gives us two homogeneous linear equations in two unknowns. The trivial solution is $C_1 = C_2 = 0$, but this would require $\psi = 0$ (where $\psi = C_1\varphi_1 + C_2\varphi_2$). Other solutions exist only for certain values of E, and these values of E can be found by solving the secular determinant:

$$\begin{vmatrix} H_{11} - ES_{11} & H_{12} - ES_{12} \\ H_{21} - ES_{21} & H_{22} - ES_{22} \end{vmatrix} = 0 \tag{3–9}$$

The determinant is solved by standard mathematical procedures. Essentially, we are solving two linear, homogeneous equations in two unknowns. An $n \times n$ determinant gives an nth-order polynomial equation in energy with n roots. Since the size of the determinant, n, equals the number of atomic orbitals, we get n

energies or n molecular orbitals from n atomic orbitals. In the case where the two atoms are the same, we get

$$(H_{11} - E)^2 - (H_{12} - S_{12}E)^2 = 0$$

(with S_{11} and $S_{22} = 1$). The two roots for this equation are:

$$E_1 = \frac{H_{11} + H_{12}}{1 + S_{12}} \quad \text{and} \quad E_2 = \frac{H_{11} - H_{12}}{1 - S_{12}} \tag{3-10}$$

Solving or approximating the integrals enables one to solve for the energies. The energies so obtained can be substituted one at a time back into equation (3–8), and the coefficients are then obtained by using a third equation: $C_1^2 + C_2^2 = 1$. The two equations that result are not independent (the determinant equals 0), so we need the third equation.

The evaluation of the S_{12} integrals is straightforward when the structure of the molecule is known. The integrals H_{11} and H_{12} cannot be determined accurately for complex molecules.

For the hydrogen molecule, the Hamiltonian operator is

$$\sum_{1,2} (-h^2/8\pi^2 m)\nabla_{1,2}^2 - e^2/r_{A,1} - e^2/r_{B,1} - e^2/r_{A,2} - e^2/r_{B,2} + e^2/r_{A,B} + e^2/r_{12})$$

The two hydrogen nuclei are labeled A and B and the two electrons 1 and 2. The charge on the electron is e, the distance from nucleus A to electron 1 is $r_{A,1}$, and so forth. We solve the integrals by using the various individual terms in the Hamiltonian separately. Accordingly, for one $\int \varphi_i \hat{H} \varphi_j \, d\tau$, there are many integrals to be evaluated. The electron-electron repulsion integrals $\langle \varphi_i | e^2/r_{12} | \varphi_j \rangle$ are particularly difficult. You can well imagine how complicated the Hamiltonian is and how many integrals would be needed even for PCl_5. *Ab initio* calculations do all the integrals and thus are practical only for small molecules or wealthy chemists.

The various ways to "guesstimate" values for these integrals correspond to various ways to do molecular orbital calculations. We shall return to a more detailed description of some of these procedures subsequently.

3–2 A MATRIX FORMULATION OF MOLECULAR ORBITAL CALCULATIONS

Modern approaches to the solution of eigenvalue problems employ a matrix formulation. Since the quantum mechanical description of many of the spectroscopic methods treated in this text is discussed in matrix terms, the previous section will be reformulated in matrix terms to introduce the general approach. The solutions of the secular equations are eigenvalues of the Hamiltonian operator $\hat{H}$, so $\hat{H}\psi = E\psi$ can be written in the LCAO approximation in matrix form as

$$[\hat{H}_{ij}][C] = [S][C][E] \tag{3-11}$$

where $[C]$ is the coefficient matrix, $[\hat{H}_{ij}]$ is the energy matrix of elements $\langle \varphi_i|\hat{H}|\varphi_j \rangle$, $[S]$ is the overlap matrix, and $[E]$ is the diagonal matrix of orbital energies, one for each m.o. The $[E]$ and $[C]$ matrices are to be determined.

Dividing both sides by $[C]$, we obtain

$$[C]^{-1}[\hat{H}_{ij}][C] = [S][E] \tag{3-12}$$

The coefficient matrix is, in effect, a similarity transform that diagonalizes the $[\hat{H}_{ij}]$ matrix to produce a unit $[S]$ matrix. The energies result directly from the diagonalized $[\hat{H}]$ matrix, as do the LCAO coefficients of the corresponding molecular orbital.

3-3 PERTURBATION THEORY

Perturbation theory is a useful technique in quantum mechanics when one is trying to ascertain the effect of a small perturbation on a system that has been solved in the absence of this perturbation. Basically, we have the solution to

$$\hat{H}^\circ \psi_i^\circ = E_i^\circ \psi_i^\circ$$

We wish to determine the effect of some perturbation on this ith molecular orbital, i.e.,

$$\hat{H}\psi_i = E_i\psi_i \tag{3-13}$$

where

$$\hat{H} = \hat{H}^\circ + \hat{H}' \tag{3-14}$$

and the term $\hat{H}'$ is a small perturbation on the system. It can be shown that the first-order correction to the energy is given by:

$$E_i^1 = \langle \psi_i^\circ | \hat{H}' | \psi_i^\circ \rangle \tag{3-15}$$

The correction to the old wave function, ψ_i^1, can be written as a linear combination of the k unperturbed wave functions of the original system:

$$\psi_i^1 = \sum_{k \neq i} a_{ik} \psi_k^\circ \tag{3-16}$$

By mixing the appropriate empty excited-state wave function into the ground state, it is possible to polarize or distort the ground state molecule in any way we desire. The contributions that the various ψ_k° functions make in our perturbed system are given by the coefficients a_{ik}, where

$$a_{ik} = -\frac{\langle \psi_k^\circ | \hat{H}' | \psi_i^\circ \rangle}{E_k^\circ - E_i^\circ} \quad (k \neq i) \tag{3-17}$$

Thus, our final expressions for the energy and wave function of the perturbed orbital, E_i and ψ_i, correct to first order, are

$$E_i = E_i^\circ + \langle \psi_i^\circ | \hat{H}' | \psi_i^\circ \rangle \qquad (3\text{--}18)$$

and

$$\psi_i = \psi_i^\circ + \sum_{k \neq i} \frac{\langle \psi_k^\circ | \hat{H}' | \psi_i^\circ \rangle}{E_i^\circ - E_k^\circ} \psi_k^\circ \qquad (3\text{--}19)$$

Note that the zero-order wave function gives us the first-order energy.

The formula for the second-order correction to the energy is given by

$$E_i^{(2)} = \sum_{k \neq i} \frac{\langle \psi_i^\circ | \hat{H}' | \psi_k^\circ \rangle \langle \psi_k^\circ | \hat{H}' | \psi_i^\circ \rangle}{E_i^\circ - E_k^\circ} \qquad (3\text{--}20)$$

It is informative to examine what we have done in the context of a rigorous solution of a 2 × 2 secular determinant. A 2 × 2 problem is selected because its secular determinant can be solved easily. We shall again use our original wave functions and the new Hamiltonian $\hat{H}$. We write $\langle \psi_1^\circ | \hat{H} | \psi_1^\circ \rangle$ as H_{11} and $\langle \psi_2^\circ | \hat{H} | \psi_2^\circ \rangle$ as H_{22}, leading to

$$\begin{vmatrix} H_{11} - E & H_{12} \\ H_{21} & H_{22} - E \end{vmatrix} = 0$$

which in turn leads to

$$E^2 - (H_{11} + H_{22})E + (H_{11}H_{22} - H_{12}^2) = 0$$

with roots

$$E = \frac{H_{11} + H_{22} \pm \sqrt{H_{11}^2 + H_{22}^2 - 2H_{11}H_{22} + 4H_{12}^2}}{2}$$

Ignoring H_{12}, we obtain the solutions

$$E_1 = H_{11} = E_1^\circ + \langle \psi_1^\circ | \hat{H}' | \psi_1^\circ \rangle \quad \text{and} \quad E_2 = H_{22} = E_2^\circ + \langle \psi_2^\circ | \hat{H}' | \psi_2^\circ \rangle$$

Using the perturbation equations given above, we write directly

$$E_1 = E_1^\circ + \langle \psi_1^\circ | \hat{H}' | \psi_1^\circ \rangle \quad \text{and} \quad E_2 = E_2^\circ + \langle \psi_2^\circ | \hat{H}' | \psi_2^\circ \rangle$$

We now see that first-order perturbation theory is permissible only when the off-diagonal elements can be ignored, compared with the diagonal ones.

The second-order correction to the energy gives us:

$$E_1 = H_{11} - \frac{H_{12}{}^2}{H_{22} - H_{11}}$$

$$E_2 = H_{22} - \frac{H_{12}{}^2}{H_{11} - H_{22}}$$

which is an even closer approximation to the exact solution.

3–4 WAVE FUNCTIONS AS A BASIS FOR IRREDUCIBLE REPRESENTATIONS

Symmetry in Quantum Mechanics

Since the energy of a molecule is time invariant and independent of the position of the molecule, the Hamiltonian must be unchanged by carrying out a symmetry operation on the system. (Recall that the symmetry operation produces an equivalent configuration that is physically indistinguishable from the original.) Thus, any symmetry operation must commute with the Hamiltonian; *i.e.*, $\hat{H}\hat{R} = \hat{R}\hat{H}$. Like the irreducible representations, the eigenfunctions are constructed to be orthonormal:

$$\int \psi_i{}^* \psi_j \, d\tau = \delta_{ij}$$

i.e., $\delta = 0$ when $i \neq j$ and $\delta = 1$ when $i = j$. It is a simple matter to show that *the wave functions form a basis for irreducible representations.*

If we take the wave equation:

$$\hat{H}\psi_i = E_i \psi_i$$

and multiply each side by the symmetry operation $\hat{R}$, we get (for non-degenerate eigenvalues):

$$\hat{H}\hat{R}\psi_i = E_i \hat{R}\psi_i$$

where $\hat{R}\hat{H}\psi = \hat{H}\hat{R}\psi$, and E commutes with $\hat{R}$ because E_i is a constant. Consequently, $\hat{R}\psi_i$ is an eigenfunction. Since ψ_i is normalized, $\hat{R}\psi_i$ must also be, for we do not change the length of a vector by a symmetry operation; therefore,

$$\hat{R}\psi_i = \pm 1 \psi_i$$

We have just shown that by applying each operation of the point group to ψ_i (non-degenerate), we generate a representation of the group with all characters equal to ± 1, that is, a one-dimensional and therefore irreducible representation. Thus, the eigenfunctions (non-degenerate) for a molecule are a basis for irreducible representations. It can also be shown that this is true for degenerate eigenfunctions.

Any integrals involving products of molecular orbitals that belong to two different irreducible representations must be zero. To show this, let ψ_a and ψ_b be such molecular orbitals that are eigenfunctions of $\hat{R}$ with eigenvalues a and b. Thus, $\langle \psi_a | \hat{H}\hat{R} | \psi_b \rangle = b \langle \psi_a | \hat{H} | \psi_b \rangle$. Since $\hat{H}\hat{R} = \hat{R}\hat{H}$, we have

$$\langle \psi_a | \hat{H}\hat{R} | \psi_b \rangle = \langle \psi_a | \hat{R}\hat{H} | \psi_b \rangle = \langle \hat{R}\psi_a | \hat{H} | \psi_b \rangle$$

$\hat{R}$ (like other operators with physical significance) is Hermitian, so

$$\langle \hat{R}\psi_a | \hat{H} | \psi_b \rangle = a \langle \psi_a | \hat{H} | \psi_b \rangle$$

We have just shown that

$a \langle \psi_a | \hat{H} | \psi_b \rangle = b \langle \psi_a | \hat{H} | \psi_b \rangle$; since $a \neq b$, the integral $\langle \psi_a | \hat{H} | \psi_b \rangle$ must be zero.

3–5 PROJECTING MOLECULAR ORBITALS

In quantum mechanical calculations, we begin with LCAO's that themselves form a basis for an irreducible representation. Although this is not necessary, it is clearly often convenient to do so. We shall show next how symmetry considerations alone provide much information about the atomic orbitals that contribute to the various molecular orbitals in a molecule. In the last chapter, it was shown that any mathematical function (*e.g.*, points, $3N$ Cartesian vectors, or bond vectors) could be used as a basis set for a reducible representation. It is a simple matter to extrapolate to the case in which the valence atomic orbitals for a given molecule are used as a basis set for a reducible representation. As a simple example, we shall work out the traces of the reducible representation for the three a.o.'s comprising the π-orbitals of the nitrite ion, shown in a coordinate system in Fig. 3–1.

Recall that earlier, when we carried out symmetry operations on the two vectors **a** and **b**, many symmetry operations transformed one vector into the other and *the trace of the representation matrix was thus zero, the only non-zero elements being off-diagonal*. Accordingly, referring to Fig. 3–1, O_1 is moved into O_3 with C_2 and σ_v, so contributions to the total character from the oxygen orbitals are zero. This is true whenever atoms are interchanged. The C_2 and σ_{xz} operations on the nitrogen p_z orbital lead to -1 and $+1$, respectively. Reflection in the yz-plane changes the sign of the p orbitals, making a contribution of -1 for each of the three p orbitals. The total representation is (C_{2v} point group; recall that x is perpendicular to the plane).

C_{2v}	E	C_2	$\sigma_v(xz)$	$\sigma_v'(yz)$
	3	-1	$+1$	-3

Factoring the total representation gives A_2 and $2B_1$; *i.e.*, the symmetries of the three m.o.'s (from three a.o.'s, we must get three m.o.'s) that result from these a.o.'s must be A_2, B_1, and B_1.

We can now use a technique involving *projection operators*, $\hat{P}$, to give us more information about the wave functions. The projection operators work on the basis set (in this example, the three p orbitals) and annihilate (convert to zero) any element in the basis set that does not contribute to a given irreducible representation. In using the projection operators, we generally work on one atomic orbital (or a linear combination of them) (*e.g.*, φ_1 the p orbital of atom 1 of Fig. 3–1) at a time with the operator

$$\hat{P} = \frac{l}{h} \sum_R \chi_R \hat{R} \qquad (3\text{–}21)$$

where $\hat{R}$ is the symmetry operation, χ_R is the character of that operation in the appropriate point group, l is the dimension of the irreducible representation, and h is the order of the group. The sum is taken over all symmetry operations in the point group. Accordingly, if we operate on φ_1 of the nitrite ion with the projection operator of the A_2 irreducible representation, we get:

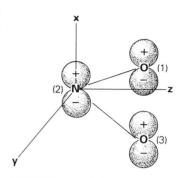

FIGURE 3–1 A.O. basis set for the π-orbitals of the nitrite ion. (x is perpendicular to the NO_2 plane.)

$$\psi \text{ (for } A_2) = \hat{P}(A_2)\varphi_1 = \frac{1}{4}[(1)E\varphi_1 + (1)C_2\varphi_1 + (-1)\sigma_v\varphi_1 + (-1)\sigma_v'\varphi_1]$$

The numbers in parentheses are characters of the A_2 irreducible representation. Next, we carry out the specified operations on φ_1, and indicate the result in terms of what the orbital φ_1 is changed into. The result is

$$\hat{P}(A_2)\varphi_1 = \frac{1}{4}[(1)(1)\varphi_1 + (1)(-1)\varphi_3 + (-1)(1)\varphi_3 + (-1)(-1)\varphi_1]$$

$$= N[\varphi_1 - \varphi_3]$$

The result is the molecular orbital corresponding to the A_2 irreducible representation. We shall ignore the factor $1/4$, for we will want to obtain the absolute values for the normalized wave function, and not the relative values resulting from the operators.

If we had selected the orbital φ_2 to operate on, we would have obtained

$$\hat{P}(A_2)\varphi_2 = \frac{1}{4}[(1)(1)\varphi_2 + (1)(-1)\varphi_2 + (-1)(1)\varphi_2 + (-1)(-1)\varphi_2] = 0$$

This result could mean that there is no A_2, but we know better, from having factored the trace of the reducible representation. In this case, a node exists at atom 2 in the m.o. with A_2 symmetry.

In order to normalize the wave function that we have just obtained, we realize that

$$\int (C_1\varphi_1 - C_1\varphi_3)^2 \, d\tau = C_1^2 \int \varphi_1^2 \, d\tau - 2C_1^2 \int \varphi_1\varphi_3 \, d\tau + C_1^2 \int \varphi_3^2 \, d\tau = 1$$

Recall that φ_1 and φ_3 have the same coefficient. Using an orthonormal basis set,

$$\int \varphi_1^2 \, d\tau = 1; \quad \int \varphi_1\varphi_3 \, d\tau = 0; \quad \text{and} \quad \int \varphi_3^2 \, d\tau = 1.$$

These substitutions convert the normalization equation to

$$2C_1^2 = 1 \quad \text{or} \quad C_1 = \frac{1}{\sqrt{2}}$$

The normalized function is

$$\frac{1}{\sqrt{2}}(\varphi_1 - \varphi_3)$$

From factoring the total representation, we know that there is no A_1 symmetry m.o. We shall use $\hat{P}(A_1)$ to show that a consistent result is obtained. First, consider:

$$\hat{P}(A_1)\varphi_1 = \frac{1}{4}[(1)(1)\varphi_1 + (1)(-1)\varphi_3 + (1)(1)\varphi_3 + (1)(-1)\varphi_1] = 0$$

Try the other p orbitals, and you will see that there is no A_1 symmetry m.o. Next try $\hat{P}(B_2)$ on φ_1:

$$\hat{P}(B_2)\varphi_1 = \frac{1}{4}[(1)\varphi_1 + (+1)\varphi_3 + (-1)\varphi_3 + (-1)\varphi_1] = 0$$

We know that there are two m.o.'s with B_1 symmetry. Accordingly,

$$\hat{P}(B_1)\varphi_1 = \frac{1}{4}[(1)(1)\varphi_1 + (-1)(-1)\varphi_3 + (1)(1)\varphi_3 + (-1)(-1)\varphi_1]$$

$$= N[\varphi_1 + \varphi_3]$$

When we use $\hat{P}(B_1)$ on φ_2, the following result is obtained:

$$\hat{P}(B_1)\varphi_2 = \frac{1}{4}[(1)(1)\varphi_2 + (-1)(-1)\varphi_2 + (1)(1)\varphi_2 + (-1)(-1)\varphi_2] = \varphi_2$$

Thus, the projection operators produce two m.o.'s of B_1 symmetry, $\psi = N[\varphi_1 + \varphi_3]$ and $\psi = \varphi_2$. The A_2 and two B_1 molecular orbital wave functions that we have generated are referred to as *symmetry adapted linear combinations* of atomic orbitals. Since the B_1 m.o.'s have similar symmetries, they can mix so that any linear combination of these two m.o.'s also has the appropriate symmetry to be one of the two B_1 m.o.'s; i.e., the two m.o.'s can be:

$$\psi_1 = a\varphi_1 + b\varphi_2 + a\varphi_3$$
$$\psi_2 = a'\varphi_1 - b'\varphi_2 + a'\varphi_3$$

The magnitudes of a and b will depend on the energies of the φ_1 and φ_2 atomic orbitals and their overlap integrals. Thus, we cannot go any further in determining the wave function without a molecular orbital calculation. We shall indicate approximate ways of doing this in the next sections. We can at this point make one further generalization. Since there is no symmetry operation that interchanges

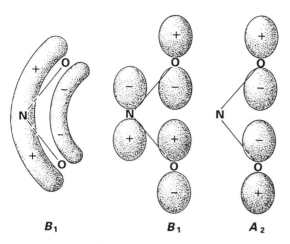

FIGURE 3–2 Shapes of the π-molecular orbitals of the nitrite ion.

the nitrogen orbital with oxygen, the nitrogen must form a separate irreducible representation.

We are now in a position to draw pictures of the m.o.'s we have constructed. They are shown in Fig. 3–2.

If one carries out the operations of the C_{2v} point group on the m.o.'s shown in Fig. 3–2, it will be found that these shapes transform according to the symmetry labels.

We shall work through the C_{3v} ammonia molecule as our next example, for it has more than one element in some classes and also has a doubly degenerate irreducible representation. The coordinate system is given in Fig. 3–3. Our procedure is as follows:

1. First work out the total representation and factor it:

Orbital Shapes	E	$2C_3$	$3\sigma_v$
	1	1	1
	3	0*	1
	3	0	1
Total	7	1	3

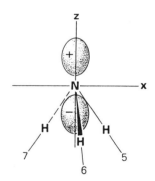

FIGURE 3–3 The location of NH$_3$ in a coordinate system. The xz-plane passes through H$_5$ and N.

$$* C_3 \begin{bmatrix} x \\ y \\ z \end{bmatrix} = \begin{bmatrix} -\frac{1}{2} & +\frac{\sqrt{3}}{2} & 0 \\ -\frac{\sqrt{3}}{2} & -\frac{1}{2} & 0 \\ 0 & 0 & 1 \end{bmatrix} \begin{bmatrix} x \\ y \\ z \end{bmatrix}$$

(trace is 0)

Factoring the total representation, we obtain:

$$a_{A_1} = \frac{1}{6}[1 \times 1 \times 7 + 2 \times 1 \times 1 + 3 \times 1 \times 3] = 3$$

$$a_{A_2} = \frac{1}{6}[1 \times 1 \times 7 + 2 \times 1 \times 1 + 3 \times (-1) \times 3] = 0$$

$$a_E = \frac{1}{6}[1 \times 2 \times 7 + 2 \times (-1) \times 1 + 0] = 2$$

No symmetry operation can take a hydrogen atom into a nitrogen atom, so we can factor the total reducible representation formed from the hydrogen orbitals (3, 0, 1) and find an A_1 and E. In this way we can work on the nitrogen atom and hydrogen atoms separately.

2. Next, work with projection operators $\hat{P}_{A_1}$ and $\hat{P}_E$. There is no need to work with $\hat{P}_{A_2}$, for factoring the total representation told us that there are no molecular orbitals with this symmetry. For this purpose, we label our atomic orbitals as follows:

$$\varphi_1 = N2s$$
$$\varphi_1 = N2p_x$$
$$\varphi_3 = N2p_y$$
$$\varphi_4 = N2p_z$$
$$\varphi_5 = H_5 1s \quad (\sigma_v \text{ contains } H_5)$$
$$\varphi_6 = H_6 1s \quad (\sigma_v' \text{ contains } H_6)$$
$$\varphi_7 = H_7 1s \quad (\sigma_v'' \text{ contains } H_7)$$

$\hat{P}_{A_1}$ on φ_5

$$\psi_1 = \frac{1}{6}[1E\varphi_5 + 1C_3\varphi_5 + 1C_3{}^2\varphi_5 + 1\sigma_v\varphi_5 + 1\sigma_v'\varphi_5 + 1\sigma_v''\varphi_5]$$

$$= \frac{1}{6}[\varphi_5 + \varphi_7 + \varphi_6 + \varphi_5 + \varphi_6 + \varphi_7] = \frac{1}{6}[2\varphi_5 + 2\varphi_6 + 2\varphi_7]$$

$\hat{P}_{A_1}$ operating on φ_6 and φ_7 will, of course, give the same wave function, so there is no need to carry these out. We need linear combinations that are not identical; i.e., they must not be related to one another by multiplication by -1 or by interchange of the labels of equivalent atoms.

$\hat{P}_{A_1}$ on φ_4. None of the operations take φ_4 into anything else, so it is not necessary to use the projection operator on φ_4. It is a basis set by itself and must belong to the irreducible representation A_1, for it is not changed by any of the symmetry operations. If we use the projection operator, we obtain:

$$\psi_2 = \frac{1}{6}[\varphi_4 + \varphi_4 + \varphi_4 + \varphi_4]$$

$$\psi_2 = \frac{4}{6}\varphi_4$$

$\hat{P}_{A_1}$ **on** φ_1. We recognize by inspection that φ_1 is also a basis set by itself and must belong to the irreducible representation A_1.

$$\psi_3 = \frac{4}{6}\varphi_1$$

$\hat{P}_{A_1}$ **on** φ_2. We know by inspection that φ_2 and φ_3 form a doubly degenerate basis set that must belong to an E irreducible representation. Thus, it is not necessary to carry out this operation. We shall do so here to show that we get the right answer with the operator and to demonstrate how the symmetry operations work on a doubly degenerate basis set.

$$\frac{1}{6}[1E\varphi_2 + 1C_3\varphi_2 + 1C_3^2\varphi_2 + 1\sigma_v\varphi_2 + 1\sigma_v'\varphi_2 + 1\sigma_v''\varphi_2]$$

$$= \frac{1}{6}\left[\varphi_2 - \frac{1}{2}\varphi_2 - \frac{\sqrt{3}}{2}\varphi_3 - \frac{1}{2}\varphi_2 + \frac{\sqrt{3}}{2}\varphi_3 + \varphi_2 - \frac{1}{2}\varphi_2 + \frac{\sqrt{3}}{2}\varphi_3 - \frac{1}{2}\varphi_2 - \frac{\sqrt{3}}{2}\varphi_3\right]$$

$$= 0$$

For C_3, we use

$$\begin{bmatrix} \cos\varphi & \sin\varphi \\ -\sin\varphi & \cos\varphi \end{bmatrix}\begin{bmatrix} \vec{x} \\ \vec{y} \end{bmatrix} = \begin{bmatrix} \vec{x}' \\ \vec{y}' \end{bmatrix}$$

For a clockwise C_3 rotation, $\varphi = -120°$; for C_3^2, we have $\varphi = -240°$.

$$\vec{x}' = \vec{x}\cos\varphi + \vec{y}\sin\varphi$$
$$\vec{y}' = -\vec{x}\sin\varphi + \vec{y}\cos\varphi$$

For the mirror planes, we use the matrices developed in Chapter 2 and the coordinate system shown in Fig. 3–4.

Likewise, the result of $\hat{P}_{A_1}$ on φ_3 is 0.

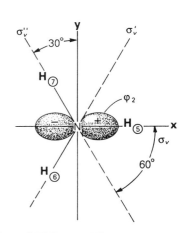

FIGURE 3–4 Effect of reflections through σ_v on φ_2 and the ammonia hydrogen.

Now we apply the projection operator for the degenerate irreducible representation E.

$$\hat{P}_E \text{ on } \varphi_2 = \psi_4 = \frac{1}{3}[2E\varphi_2 - 1C_3\varphi_2 - 1C_3{}^2\varphi_2]$$

$$= \frac{1}{3}\left\{2\varphi_2 - \left[-\frac{1}{2}\varphi_2 - \frac{\sqrt{3}}{2}\varphi_3\right] - \left[-\frac{1}{2}\varphi_2 + \frac{\sqrt{3}}{2}\varphi_3\right]\right\} = \varphi_2$$

$\hat{P}_E$ on $\varphi_3 = \psi_5 = \varphi_3$

$$\hat{P}_E \text{ on } \varphi_5 = \frac{1}{3}[2E\varphi_5 - 1C_3\varphi_5 - 1C_3{}^2\varphi_5]$$

$$= \frac{1}{3}[2\varphi_5 - \varphi_6 - \varphi_7]$$

$$\hat{P}_E \text{ on } \varphi_6 = \frac{1}{3}[2\varphi_6 - \varphi_7 - \varphi_5]$$

$$\hat{P}_E \text{ on } \varphi_7 = \frac{1}{3}[2\varphi_7 - \varphi_5 - \varphi_6]$$

Our resulting orbitals from φ_5, φ_6, and φ_7 are not orthogonal, and we know that there are only two wave functions in E. Appropriate linear combinations must be taken to get an orthonormal basis. There are various ways to do this. One that works for this system involves working with the matrix elements, instead of with the trace as we did above. The diagonal elements are:

$$
\begin{array}{cccccc}
E & C_3 & C_3{}^2 & \sigma_v & \sigma_v' & \sigma_v'' \\
\begin{bmatrix} 1 & 0 \\ 0 & 1 \end{bmatrix} &
\begin{bmatrix} -\frac{1}{2} & 0 \\ 0 & -\frac{1}{2} \end{bmatrix} &
\begin{bmatrix} -\frac{1}{2} & 0 \\ 0 & -\frac{1}{2} \end{bmatrix} &
\begin{bmatrix} 1 & 0 \\ 0 & -1 \end{bmatrix} &
\begin{bmatrix} -\frac{1}{2} & 0 \\ 0 & \frac{1}{2} \end{bmatrix} &
\begin{bmatrix} -\frac{1}{2} & 0 \\ 0 & \frac{1}{2} \end{bmatrix}
\end{array}
$$

where a_{11} corresponds to $\hat{P}_{E_a}$ and a_{22} to $\hat{P}_{E_b}$.

$\hat{P}_{E_a}$ on φ_5 gives:

$$\psi_a = N\left[E\varphi_5 - \frac{1}{2}C_3\varphi_5 - \frac{1}{2}C_3{}^2\varphi_5 + 1\sigma_v\varphi_5 - \frac{1}{2}\sigma_v'\varphi_5 - \frac{1}{2}\sigma_v''\varphi_5\right]$$

$$= N[2\varphi_5 - \varphi_6 - \varphi_7]$$

$\hat{P}_{E_b}$ on φ_5 gives:

$$\psi_b = N\left[E\varphi_5 - \frac{1}{2}C_3\varphi_5 - \frac{1}{2}C_3{}^2\varphi_5 - 1\sigma_v\varphi_5 + \frac{1}{2}\sigma_v'\varphi_5 + \frac{1}{2}\sigma_v''\varphi_5\right] = 0$$

There is a node in ψ_b at φ_{5}. The result of similar operations with $\hat{P}_{E_b}$ on φ_6 leads to the other wave function:

$$\psi_b = N(\varphi_6 - \varphi_7)$$

One additional way to get wave functions for degenerate irreducible representations involves working with the character table for the rotation group involving only the principal axis, i.e., C_n.

For ammonia, we would use C_3. In C_3, the E representations are given terms of the imaginaries; e.g., for C_3,

$$
\begin{array}{c c c c}
 & E & C_3 & C_3{}^2 \\
E & \begin{Bmatrix} 1 & \varepsilon & \varepsilon^* \\ 1 & \varepsilon^* & \varepsilon \end{Bmatrix}
\end{array}
$$

where $\varepsilon = \exp(2\pi i/3)$. Thus,

$$\hat{P}_a \varphi_6 = 1E\varphi_6 + \varepsilon C_3 \varphi_6 + \varepsilon^* C_3{}^2 \varphi_6 = \varphi_6 + \varepsilon\varphi_7 + \varepsilon^*\varphi_5 = \psi_a'$$
$$\hat{P}_b \varphi_6 = 1E\varphi_6 + \varepsilon^* C_3 \varphi_6 + \varepsilon C_3{}^2 \varphi_6 = \varphi_6 + \varepsilon^*\varphi_7 + \varepsilon\varphi_5 = \psi_b'$$

If we add the two together, we get one wave function; when we subtract the two and divide by i, we get the second wave function. Upon addition, we get

$$2\varphi_6 + (\varepsilon^* + \varepsilon)\varphi_7 + (\varepsilon + \varepsilon^*)\varphi_5$$

where $\varepsilon = \exp(2\pi i/h)$ and h is the order of the rotation group. Abbreviating these as $\varepsilon = \exp(i\Phi)$ and $\varepsilon^* = \exp(-i\Phi)$ and using the trigonometric identities

$$\exp(i\Phi) = \cos\Phi + i\sin\Phi$$
$$\exp(-i\Phi) = \cos\Phi - i\sin\Phi$$

we see that $\varepsilon + \varepsilon^* = 2\cos\Phi$.

In the C_3 group, the rotation axis is 120°, so $2\cos(-120°) = 2(-1/2) = -1$. Accordingly, we have

$$\psi_a = 2\varphi_6 - \varphi_7 - \varphi_5$$

(This is equivalent to our previous result, $\psi_a = 2\varphi_5 - \varphi_6 - \varphi_7$, obtained by using the C_{3v} point group.)

To get the other degenerate wave function, we subtract ψ_a' from ψ_b' and divide by i:

$$\frac{\psi_b' - \psi_a'}{i} = 0 + \frac{(\varepsilon^* - \varepsilon)}{i}\varphi_7 + \frac{(\varepsilon - \varepsilon^*)}{i}\varphi_5$$

Substituting the trigonometric functions for ε and ε^* produces

$$\frac{(\varepsilon - \varepsilon^*)}{i} = \frac{2i}{i}\sin\Phi \quad \text{and} \quad \frac{(\varepsilon^* - \varepsilon)}{i} = -\frac{2i}{i}\sin\Phi$$

Since $\sin(-120°) = -\sqrt{3}/2$, we obtain

$$\psi_b = \sqrt{3}\varphi_7 - \sqrt{3}\varphi_5$$

The final wave functions should be normalized to complete the problem. Upon normalization of ψ_b, we obtain:

$$\psi_b = \frac{1}{\sqrt{2}}(\varphi_7 - \varphi_5)$$

The two E functions from φ_5, φ_6, and φ_7 can be obtained in yet another way. First, one projects $\hat{P}_E$ on φ_5 to obtain $2\varphi_5 - \varphi_6 - \varphi_7$. Now, if one can judiciously select for projection another function involving φ_6 and φ_7 (or sometimes φ_5, φ_6, and φ_7) that is orthogonal to the one just obtained, the desired function will result. First, we try φ_6 and φ_7 and realize that $\varphi_6 - \varphi_7$ will be orthogonal to $2\varphi_5 - \varphi_6 - \varphi_7$.

$$\hat{P}_E(\varphi_6 - \varphi_7) = \hat{P}_E\varphi_6 - \hat{P}_E\varphi_7 = \varphi_6 - \varphi_7$$

With these two examples, we see that the symmetry of the molecule determines many aspects of our resulting wave functions. Next, we shall consider some approximate molecular orbital calculations that complete the problem (but not always necessarily correctly).

Molecular Orbital Calculations

3–6 HÜCKEL PROCEDURE

The various ways to guesstimate values for the integrals presented in the introduction correspond to various ways to do m.o. calculations. First, we shall consider some very crude approximations, the so-called Hückel m.o. calculations, which give a fair account of the properties of many hydrocarbons and which will make the solution of the quantum mechanical problem in the introduction more specific while the math remains simple. *Generally, only the π orbitals of hydrocarbon systems are treated.* The secular equations are written directly by generalizing our earlier result for a two-atomic-orbital system as:

$$C_{11}(H_{11} - ES_{11}) + \cdots C_{1n}(H_{1n} - ES_{1n}) = 0$$
$$\vdots \qquad \qquad \vdots$$
$$C_{n1}(H_{n1} - ES_{n1}) + \cdots C_{nn}(H_{nn} - ES_{nn}) = 0$$

The following Hückel approximations are then introduced:

1. When two atoms i and j are not directly bonded, we take $H_{ij} = 0$.
2. When all atoms are alike and if all bond distances are equal (e.g., the carbon $2p_z$ orbitals in benzene), all neighboring atom H_{ij} values are taken as equal and symbolized by β (i.e., $H_{12} = H_{23} = \beta$).
3. If all atoms are alike, the H_{ii} integrals are symbolized by α ($\int \varphi_1 \hat{H} \varphi_1 \, d\tau = \int \varphi_2 \hat{H} \varphi_2 \, d\tau = \alpha$).

4. Since we use normalized atomic orbitals in the basis set, $S_{ii} = 1$.
5. All S_{ij}'s are set equal to zero.

The secular equations become

$$C_{11}(\alpha - E) + C_{12}\beta_{12} \cdots\cdots \quad C_{1n}\beta_{1n} = 0$$
$$\vdots \qquad\qquad\qquad \vdots$$
$$C_{n1}\beta_{n1} + \cdots\cdots \qquad C_{nn}(\alpha - E) = 0$$

The secular determinant now becomes

$$\begin{vmatrix} \alpha - E & \beta_{12} \cdots\cdots \beta_{1n} \\ \vdots & \vdots \\ \beta_{n1} \cdots\cdots\cdots \alpha - E \end{vmatrix} = 0 \qquad (3\text{–}22)$$

where all but directly bonded β's are zero. As a specific example, we shall consider the π-system of the allyl radical in Fig. 3–5. (All sigma bonds are ignored.)

The 3 × 3 secular determinant can be directly written from equation (3–22) as:

$$\begin{vmatrix} \alpha - E & \beta & 0 \\ \beta & \alpha - E & \beta \\ 0 & \beta & \alpha - E \end{vmatrix} = 0$$

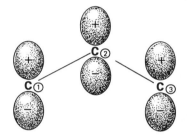

FIGURE 3–5 The π-system of the allyl radical.

Zeroes arise for the β_{13} matrix element, for the carbon atoms 1 and 3 are not bonded. Dividing by β and letting $(\alpha - E)/\beta = X$, we obtain the determinant:

$$\begin{vmatrix} X & 1 & 0 \\ 1 & X & 1 \\ 0 & 1 & X \end{vmatrix} = 0$$

This determinant can be solved by expanding by the method of cofactors. The procedure involves an expansion as shown for the general case:

$$\begin{vmatrix} a_1 & b_1 & c_1 \\ a_2 & b_2 & c_2 \\ a_3 & b_3 & c_3 \end{vmatrix} = a_1 \begin{vmatrix} b_2 & c_2 \\ b_3 & c_3 \end{vmatrix} - b_1 \begin{vmatrix} a_2 & c_2 \\ a_3 & c_3 \end{vmatrix} + c_1 \begin{vmatrix} a_2 & b_2 \\ a_3 & b_3 \end{vmatrix} = 0$$

The expansion can be done by row or column. In expanding by row, we take each element separately and multiply it by the determinant that results when the row and column containing the element are eliminated. Labeling the row i and the column j, the sign is positive whenever $i \times j$ is odd and negative when $i \times j$ is even.

For the allyl radical determinant, we expand to get:

$$X \begin{vmatrix} X & 1 \\ 1 & X \end{vmatrix} - 1 \begin{vmatrix} 1 & 1 \\ 0 & X \end{vmatrix} = 0$$

Solving these determinants gives $X^3 - 2X = 0$ with solutions $X = 0, \pm\sqrt{2}$, i.e.,

$$\frac{\alpha - E}{\beta} = 0, \sqrt{2}, -\sqrt{2}.$$

These three roots produce the three energies:

$$E_1 = \alpha + \sqrt{2}\beta$$
$$E_2 = \alpha$$
$$E_3 = \alpha - \sqrt{2}\beta$$

For the allyl radical, the secular equations are easily written by taking the general set of secular equations and writing explicit ones for the three-atom allyl π system or by multiplying the column matrix $\{C_1, C_2, C_3\}$ by the 3×3 secular determinant:

$$C_1 X + C_2 = 0 \quad (3\text{–}23)$$

$$C_1 + C_2 X + C_3 = 0 \quad (3\text{–}24)$$

$$C_2 + C_3 X = 0 \quad (3\text{–}25)$$

where $X = (\alpha - E)/\beta$.

For ψ_1, the bonding molecular orbital, $X = -\sqrt{2}$, so substituting this into (3–23) and (3–25) gives

$$-\sqrt{2}C_1 + C_2 = 0 \quad \text{and} \quad C_2 - \sqrt{2}C_3 = 0$$

Rearranging, we get:

$$C_1 = \frac{C_2}{\sqrt{2}} \quad \text{and} \quad C_3 = \frac{C_2}{\sqrt{2}}$$

Using the normalization criterion:

$$C_1^2 + C_2^2 + C_3^2 = \left(\frac{C_2}{\sqrt{2}}\right)^2 + C_2^2 + \left(\frac{C_2}{\sqrt{2}}\right)^2 = 1$$

or

$$2C_2^2 = 1 \quad \text{and} \quad C_2 = \frac{1}{\sqrt{2}}$$

Now

$$C_1 = C_3 = \frac{C_2}{\sqrt{2}}$$

and since

$$C_2 = \frac{1}{\sqrt{2}}$$

we obtain

$$C_1 = C_3 = \frac{1}{2}.$$

With the coefficients defined, we can write

$$\psi_1 = \frac{1}{2}\varphi_1 + \frac{1}{\sqrt{2}}\varphi_2 + \frac{1}{2}\varphi_3 \qquad \text{(Bonding)}$$

For ψ_2

$$X = 0 \qquad C_2 = 0 \quad \text{and} \quad C_1 = -C_3$$

$$C_1^2 + C_3^2 = 1 \quad \text{or} \quad C_1 = \frac{1}{\sqrt{2}}$$

Accordingly, we can write

$$\psi_2 = \frac{1}{\sqrt{2}}\varphi_1 - \frac{1}{\sqrt{2}}\varphi_3 \qquad \text{(Non-bonding)}$$

Similarly

$$\psi_3 = \frac{1}{2}\varphi_1 - \frac{1}{\sqrt{2}}\varphi_2 + \frac{1}{2}\varphi_3 \qquad \text{(Antibonding)}$$

The radical, cation, and anion of allyl are all described by adding the appropriate number of electrons to these molecular orbitals.

Pictures of these orbitals can be drawn by realizing that a node exists every time the wave function changes sign. In general, as the number of nodes in the wave function increases, its energy increases. Furthermore, we see that all of the symmetry information about the molecule is built into the molecular orbital calculation. Using projection operators on the allyl radical, one can show that:

$$\psi_2(A_2) = \frac{1}{\sqrt{2}}(\varphi_1 - \varphi_3) \quad \text{and}$$

$$\psi_1(B_1) = \frac{1}{\sqrt{2}}(\varphi_1 + \varphi_3) \quad \text{and} \quad \psi_3(B_1) = \varphi_2$$

The two B_1 m.o.'s have the same symmetry and can mix. Projection operators cannot tell us the extent of mixing. Using a and b to describe the mixing, we can write from symmetry the two wave functions:

$$a\varphi_1 + b\varphi_2 + a\varphi_3 \quad \text{and} \quad a'\varphi_1 - b'\varphi_2 + a'\varphi_3$$

The Hückel calculation gives the a's and b's, and also gives the same form for the wave functions as the projection operator because all of the symmetry is automatically included in the Hückel Hamiltonian.

The symmetry considerations for the nitrite ion are the same as for allyl, but the molecular orbital coefficients are different. Hückel calculations have been carried out on systems containing heteroatoms, X. These are treated by using

$$\alpha_x = \alpha_c + k\beta_{c-c}$$

and

$$\beta_{c-x} = k'\beta_{c-c}$$

Tables of values to use for the quantities k and k' are listed in books on Hückel m.o. calculations but will not be presented here because the general agreement between fact and prediction deteriorates rapidly when Hückel calculations are employed on molecules containing carbon bound to other kinds of atoms.

3–7 PROPERTIES DERIVED FROM WAVE FUNCTIONS

There are many quantities of chemical interest that can be calculated from the wave functions. A few are listed and defined:

1. ELECTRON DENSITY AT AN ATOM r, q_r

The electron density q_r at an atom r in a molecule is given by:

$$q_r = -\sum_j n_j C_{jr}^2 \tag{3–26}$$

where n_j is the number of electrons in the jth m.o. The summation is carried out over all of the j occupied m.o.'s and C_{jr} is the coefficient of the atomic orbital r in the jth m.o.

Using the Hückel wave functions for the filled molecular orbitals of the allyl anion (two electrons in each ψ_1 and ψ_2):

$$\psi_1 = \frac{1}{2}\varphi_1 + \frac{1}{\sqrt{2}}\varphi_2 + \frac{1}{2}\varphi_3$$

$$\psi_2 = \frac{1}{\sqrt{2}}\varphi_1 - \frac{1}{\sqrt{2}}\varphi_3$$

We sum (with $n = 2$) over the bonding and non-bonding m.o.'s and obtain for the electron density on carbon 1:

$$q_1 = -2\left(\frac{1}{2}\right)^2 - 2\left(\frac{1}{\sqrt{2}}\right)^2 = -1\frac{1}{2}$$

At carbon 2 we find:

$$q_2 = -2\left(\frac{1}{\sqrt{2}}\right)^2 - 2(0) = -1$$

The sum of all the q's equals the total number of electrons in the system.

2. Formal Charge, δ

The electron density at an atom in the molecule minus that on a free atom is referred to as the formal charge on the atom in the molecule. Considering carbon, we place one electron in each of the orbitals of the atom to form four bonds. In Hückel theory, where we worry only about the π system, we consider only the p_z orbital on a carbon. Since it contains one electron, the charge on the "free atom" is -1. If the charge on this atom in a molecule (*i.e.*, the electron density at an atom r) is -1.4, δ, the formal charge, is -0.4, that is, the difference between the charge on this atom in the molecule and that on the neutral atom. The sum of all the formal charges equals the total charge on the molecule.

For the allyl anion we obtain, using the charge density calculated above, a charge of -0.5 at carbon 1 and a charge of zero at carbon 2.

3. Bond Order, P_{AB}

The bond order between two adjacent atoms A and B is given by:

$$P_{AB} = \sum_j n_j C_{Aj} C_{Bj} \qquad (3\text{--}27)$$

i.e., just take the product of the coefficients of atoms A and B in each m.o. times the number of electrons in the m.o. and sum over all of the m.o.'s.

Using this formula, the pi-bond order between carbon atoms 1 and 2 of the allyl anion is given by

$$P_{12} = 2\left(\frac{1}{2}\right)\left(\frac{1}{\sqrt{2}}\right) + 2\left(\frac{1}{\sqrt{2}}\right)(0) = \frac{\sqrt{2}}{2} = 0.7$$

Hückel theory contains many crude approximations which can be justified[1] well for hydrocarbon molecules. With more polar molecules, they are very crude. There have been many attempts to improve on these calculations. Extended Hückel, CNDO, INDO, and others constitute various semi-empirical procedures that have been used to approximate various integrals. In all of these procedures, the overlap integral is evaluated and not set equal to zero. To go through the various procedures in detail would take more time than we can allot to this topic in this text. We will provide only a very rough overview of the calculations, and will concern ourselves more with what can be done with the output.

3–8 EXTENDED HÜCKEL PROCEDURE

In the extended Hückel method, the H_{ii}'s are approximated by valence state ionization potentials. The H_{ij}'s, given by the Wolfsberg-Helmholz formula, are:

$$H_{ij} = \frac{1}{2} k S_{ij}(H_{ii} + H_{jj}) \tag{3–28}$$

where k is a constant usually taken to be 1.8. All overlaps are explicitly evaluated. Slater or Clementi atomic orbitals are generally used.

We shall briefly describe the criteria used to describe the electron density in a molecule when $S \neq 0$ for $i \neq j$. With a LCAO wave function

$$\psi_1 = C_1 \varphi_1 + C_2 \varphi_2$$

we see that the total electron density, ψ_1^2, is given by

$$\int \psi_1^2 \, d\tau = C_1^2 \int \varphi_1^2 \, d\tau + 2C_1 C_2 \int \varphi_1 \varphi_2 \, d\tau + C_2^2 \int \varphi_2^2 \, d\tau$$

$$= C_1^2 + 2C_1 C_2 S_{12} + C_2^2$$

Using normalized a.o.'s $\varphi_1^2 = 1$, we also see that $\int \varphi_1 \varphi_2 \, d\tau$ is the overlap integral S_{12}. We shall now define some quantities that are important results of calculations in which $S \neq 0$.

1. NET ATOMIC POPULATION

The quantity $n_i C_i^2$ is referred to as the net atomic population (n_i equals the number of electrons in the m.o.). It is that part of the total electron density in the m.o. that can be assigned to atom 1.

2. OVERLAP POPULATION

The quantity $2n C_1 C_2 S_{12}$ is referred to as the overlap population between *atomic orbitals 1 and 2* for one molecular orbital. In a polyatomic molecule, the overlap population between atomic orbitals i and j is obtained by summing over all occupied molecular orbitals k:

$$\sum_k 2n C_i C_j S_{ij} \tag{3–29}$$

It is often convenient to have the *atom-atom* overlap population. This corresponds to the total overlap of all the atomic orbitals on atom x with all the atomic orbitals on atom y. Many m.o. programs present this as a reduced overlap population matrix. This is

$$\sum_{k(xy)} 2nC_iC_jS_{ij} \qquad (3\text{–}30)$$

where $k(xy)$ refers to the xy overlap population of every atomic orbital on x and y summed over all filled molecular orbitals. When one compares bonds with similar ionic character, the overlap population is related to the bond strength.

3. Gross Atomic Population

This is a procedure for assigning all the electron density in a molecule to each of the individual atoms. Half of the bonding electron density (*e.g.*, half of $2nC_1C_2S_{12}$) is given to each of the atoms involved in bonding.

$$\text{Gross atomic population} = \sum_{\substack{\text{all} \\ \text{m.o.'s}}} (nC_i^2 + nC_iC_jS_{ij}) \qquad (3\text{–}31)$$

4. Formal Charge

The formal charge on an atom is equal to the core charge (nucleus plus all non-valence electrons) minus the gross atomic population.

To give you some appreciation for the kind of information available from a semi-empirical molecular orbital calculation, the output from an extended Hückel m.o. calculation is presented in Table 3–1. The matrix summarizes the wave function; *e.g.*, from m.o. 12, we see that

$$\psi_{12} = -0.40\ C(2)2p_z - 0.56\ C(1)2p_z - 0.40\ C(3)2p_z$$

(Numbers in parentheses refer to the numbering in the figure of allyl on the second page of computer output.) This is the lowest-energy filled pi-orbital. The odd electron in m.o. 9 is delocalized over both of the terminal carbons, and these are the only two orbitals contributing to this molecular orbital.

One important concept to be gained from this calculation is the extent of delocalization of the bonding in the sigma system, as can be seen by looking at a sigma m.o. For example, none of the filled m.o.'s even vaguely resemble a localized carbon-hydrogen sigma bond. All of the σ m.o.'s are very extensively delocalized. The π-type atomic orbitals are orthogonal to the σ m.o.'s, and they contain only contributions from the carbon p_z orbitals. Net charges, orbital occupations, and an atom-by-atom overlap population are evaluated in this program from the wave functions calculated.

Extended Hückel calculations give reasonable wave functions for systems in which the formal charges on the atoms of the molecule are small, $\sim \pm 0.5$. As an example of the utility of such calculations, the results from a study of several metallocenes[2] and bis-arene complexes[3] will be discussed. In all systems studied, the correct ground state results; *i.e.*, the unpaired electrons end up being placed in molecular orbitals that are essentially the correct d orbitals. Certain ground state spectroscopic properties (nmr and epr) are also predicted. This agreement with experiment suggests that the wave functions are reasonably good. They are

Chapter 3 Molecular Orbital Theory and Its Symmetry Aspects

TABLE 3–1 Wave Function from Extended Hückel Calculation of the Allyl Radical

(The vertical numbers correspond to A.O.'s and the horizontal ones to M.O.'s)

A.O.'s ↓	1	2	3	4	5	6	7	8_π	9_π
1	−1.80	−0.00	0.24	−0.00	0.27	−0.00	0.25	−0.00	0.00
2	0.00	0.00	0.00	−0.00	0.00	−0.00	0.00	−0.97	−0.00
3	−0.51	−0.00	−0.35	0.00	−1.31	0.00	−0.24	0.00	−0.00
4	0.00	−1.28	−0.00	−0.68	−0.00	−0.29	−0.00	−0.00	−0.00
5	0.76	1.12	0.87	−0.38	−0.01	0.22	0.01	−0.00	0.00
6	−0.00	−0.00	−0.00	0.00	0.00	−0.00	0.00	0.65	−0.74
7	−0.43	−0.53	0.31	−0.58	0.30	0.62	−0.65	0.00	−0.00
8	−0.68	−0.13	0.58	−0.89	−0.53	−0.54	0.15	−0.00	0.00
9	0.76	−1.12	0.87	0.33	−0.01	−0.22	0.01	−0.00	−0.00
10	−0.00	−0.00	−0.00	0.00	−0.00	0.00	−0.00	0.65	0.74
11	−0.43	0.53	0.31	0.58	0.30	−0.62	−0.65	−0.00	−0.00
12	0.68	−0.13	−0.58	−0.89	0.53	−0.54	−0.15	0.00	0.00
13	0.28	0.00	−0.30	0.00	−1.17	0.00	−0.61	−0.00	0.00
14	−0.07	−0.48	−0.61	0.49	0.46	0.68	−0.59	0.00	0.00
15	0.15	−0.09	−0.75	0.82	−0.03	−0.55	0.61	−0.00	−0.00
16	0.15	0.09	−0.75	−0.82	−0.03	0.55	0.161	0.00	0.00
17	−0.07	0.48	−0.61	−0.49	0.46	−0.68	−0.59	0.00	−0.00

		10	11	12_π	13	14	15	16	17
C	1	−0.00	0.01	−0.00	0.00	−0.09	0.33	−0.00	0.48
C	2	−0.00	−0.00	−0.56	0.00	−0.00	−0.00	−0.00	0.00
	3	−0.00	0.33	−0.00	−0.00	−0.36	−0.17	−0.00	0.01
	4	−0.49	−0.00	−0.00	−0.30	0.00	−0.00	0.16	−0.00
	5	−0.14	0.05	0.00	−0.03	−0.03	−0.20	0.44	0.36
C	6	−0.00	0.00	−0.40	−0.00	0.00	−0.00	0.00	0.00
	7	−0.12	−0.35	−0.00	0.37	−0.12	−0.12	0.01	−0.00
	8	0.41	0.00	−0.00	0.07	0.29	−0.14	0.02	−0.01
	9	0.14	0.05	−0.00	0.03	−0.03	−0.20	−0.44	0.36
C	10	0.00	0.00	−0.40	0.00	−0.00	−0.00	0.00	0.00
	11	0.12	−0.35	−0.00	−0.37	−0.12	−0.12	−0.01	−0.00
	12	0.41	−0.00	−0.00	0.07	−0.29	0.14	0.02	0.01

TABLE 3–1 (*continued*)

	13	0.00	−0.40	−0.00	0.00	0.27	0.32	0.00	0.08
	14	0.37	0.26	0.00	−0.15	0.27	−0.15	0.17	0.04
H	15	−0.07	−0.33	0.00	0.36	0.02	−0.23	0.19	0.05
	16	0.07	−0.33	0.00	−0.36	0.02	−0.23	−0.19	0.05
	17	−0.37	0.26	−0.00	0.15	0.27	−0.15	−0.17	0.04

M.O.'s 10 to 17 have 2e⁻
" 9 has 1e⁻
" 1 to 8 empty
Atomic Orbitals 1–4, 5–8, and 9–12 are carbon orbitals in the order $2s, 2p_z, 2p_x, 2p_y$ for each atom.
A.O.'s 13 to 17 are hydrogen $1s$.

M.O.	Energy (eV)	Occ.
1	53.22	0
2	36.55	0
3	28.31	0
4	22.97	0
5	18.76	0
6	7.30	0
7	5.19	0
8	−4.63	0
9	−10.16	1
10	−12.67	2
11	−13.83	2
12	−14.03	2
13	−15.01	2
14	−15.96	2
15	−19.40	2
16	−22.83	2
17	−27.54	2

Coordinate System

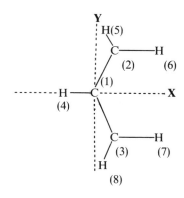

Gross Atomic Population

Atomic Orbital	Occupations
1	1.13
2	0.98
3	0.91
4	0.96
5	1.17
6	1.01
7	0.96
8	0.89
9	1.17
10	1.01
11	0.96
12	0.89
13	0.98
14	1.00
15	0.98
16	0.98
17	1.00

Atom Number	Net Charges
1	0.013
2	−0.020
3	−0.030
4	0.015
5	0.000
6	0.016
7	0.017
8	0.000

(*Continued*)

TABLE 3–1 (continued)

	Reduced Overlap Population Matrix Atom by Atom							
	1	2	3	4	5	6	7	8
1	5.236	1.411	1.141	0.836	−0.117	−0.075	−0.075	−0.117
2	1.141	5.604	−0.216	−0.097	0.821	0.812	−0.013	0.009
3	1.141	−0.216	5.604	−0.097	0.009	−0.013	0.812	0.821
4	0.836	−0.097	−0.097	1.360	−0.020	0.005	0.005	−0.020
5	−0.117	0.821	0.009	−0.020	1.402	−0.094	0.000	−0.000
6	−0.075	0.812	−0.013	0.005	−0.094	1.334	−0.000	0.000
7	−0.075	−0.013	0.812	0.005	0.000	−0.000	1.334	−0.094
8	−0.117	0.009	0.821	−0.020	−0.000	0.000	−0.094	1.402

then analyzed in detail to provide information about the bonding interactions responsible for bonding the hydrocarbon ring to the metal. An interesting result involves the extensive mixing of the ring sigma orbitals with the metal orbitals in these systems.

3–9 SCF-INDO (INTERMEDIATE NEGLECT OF DIFFERENTIAL OVERLAP)

Before discussing the INDO calculation, it is appropriate to discuss in more detail some of the approximations that have been employed without mention in the methods discussed for many-electron systems. The wave function will be considered first. The ground state of our molecule is described by a wave function Ψ, which, when substituted into equation (3–4), would give us the energy, if the equation could be rigorously solved. In the LCAO approach, we are attempting to represent Ψ by a combination of one-electron functions (i.e., molecular orbitals), each of which depends only on the coordinates of one electron. For an n-electron system, the ground state wave function is given by simply taking the product of the occupied one-electron orbitals, i.e.,

$$\Psi(1, 2 \cdots n) = \psi_1(1)\psi_2(2) \cdots \psi_n(n) \qquad (3\text{–}32)$$

The subscript on ψ labels the orbital used and the number in parentheses describes the electron. Physically, this amounts to making the untenable assumption that electron 1 is independent of the $n-1$ other electrons in the system. We shall have to correct or pay the consequences for this later.

If the $\hat{H}$'s were written as the sum of one-electron operators, the Schrödinger equation could be solved by a simple separation of variables. This would amount to ignoring completely the existence of the other electrons; e.g., the e^2/r_{ij} electron repulsion terms would not be included. What one does is to use one-electron operators, and correct for the potential experienced at an electron p from the field of the atomic nuclei and the average field of the other $n-1$ electrons. We label this field as $V_{(p)}$, and it can be approximated as will be discussed shortly. Essentially, then, our Hamiltonian is $\mathfrak{F}_{(p)}$, given by

$$\mathfrak{F}_{(p)} = \sum_p F_{(p)} = \sum_p \left[-\frac{1}{2}\frac{\hbar^2}{m}\nabla_{(p)}^2 + V_{(p)} \right] \tag{3-33}$$

When $\mathfrak{F}_{(p)}$ is used in a Schrödinger-type equation, we have:

$$\mathfrak{F}_{(p)}(1, 2 \cdots n)\Psi(1, 2 \cdots n) = \mathscr{E}\Psi(1, 2 \cdots)$$

Note that our operator $\mathfrak{F}_{(p)}$ can be separated (separation of variables) into a sum of one-electron operators $F_{(p)}$, which give rise to a set of equations of the form

$$F_{(1)}\psi_i(1) = E_i\psi_i(1) \tag{3-34}$$

The ψ_i functions are the one-electron molecular orbitals obtained by the variation procedure. These are the wave functions we have been talking about when we do Hückel or extended Hückel calculations. In these cases, all of the electron-electron repulsions are very indirectly lumped into the parameters employed, the charge correction and the Wolfsberg-Helmholtz type of approximation. The Hamiltonian is not explicitly defined.

In self-consistent field (SCF) methods, the Hamiltonian is explicitly written and used. The $V_{(p)}$ term of equation (3–33) is an e^2/r_{ij} operator. In an INDO calculation, the formalism is basically SCF (*vide infra*), but many of the resulting integrals are ignored. In SCF calculations, the Hamiltonian for an *n*-electron problem is written as:

$$\hat{H} = \sum_{i=1}^{n} \hat{H}_i + \sum_{i<j} \frac{e^2}{r_{ij}} \tag{3-35}$$

Here, r_{ij} is the interelectronic distance and the $i < j$ index ensures that electron-electron interactions are counted only once. $\hat{H}_i$, the operator for electron i in the field of the nuclei, consists of a kinetic energy term and an electron-nuclear attraction term:

$$\hat{H}_i = -\frac{\hbar^2}{2m}\nabla_i^2 - \sum_k \frac{Z_k e^2}{r_{ik}} \tag{3-36}$$

where ∇_i^2 is the familiar kinetic energy operator for electron i, r_{ik} is the electron-nucleus distance, and Z_k is the charge on the kth nucleus. The energy is found by a variational procedure, and successive sets of trial wave functions are used until this energy is minimized. These calculations are made more difficult because the operators depend on the wave functions. That is, in order to evaluate e^2/r_{ij} we must know the electron distribution, so we must know the wave function before we calculate the wave function. A starting wave function is guessed, the calculations are carried out, and a new wave function is obtained. The new wave function is then used to repeat the calculation. The procedure is iterated until there are no further changes in the wave functions, *i.e.*, we have a self-consistent field. This still is not a rigorous procedure because we are treating the electrons in a molecule as if they experience a charge cloud from all the other electrons. In actual fact, the electron motion is correlated; *i.e.*, the motion of one electron is governed by the instantaneous positions of the other electrons and not their average. This effect, called *electron correlation*, is often ignored, leading to erroneous wave functions.

In any molecule of more than four or five atoms, even a limited basis set (*i.e.*, one in which only the valence atomic orbitals on each atom are employed) gives rise to a huge number of integrals when all interactions (kinetic energy, nuclear-electron, nuclear-nuclear, and electron-electron) are taken into account. The various calculation procedures CNDO (complete neglect of differential overlap) CNDO-2, INDO (intermediate neglect of differential overlap), and so forth, differ in the types of integrals that are ignored. At present, INDO is the most popular of these approximate methods.

We can illustrate the types of integrals encountered by considering the following four orbitals taken from a large molecule:

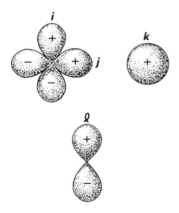

Electron repulsion integrals of the form

$$\iint \varphi_i^*(1)\varphi_j^*(1) \frac{1}{r_{12}} \varphi_k(2)\varphi_l(2) \, d\tau_1 \, d\tau_2$$

are the most numerous and require the most computer time to integrate numerically. This integral represents the $1/r$ repulsion between the electron density in the region where φ_i and φ_j have differential overlap and the electron density in the region where φ_k and φ_l have differential overlap. The *differential overlap* for electron 1, for example, is simply defined as

$$\psi_i(1)\psi_j(1) \, d\tau_1$$

It is important to emphasize that although an overlap integral may be zero,

$$\int \psi_k(1)\psi_l(1) \, d\tau_1 = 0$$

an integral such as

$$\int \hat{O}_p \psi_k(1)\psi_l(1) \, d\tau_1$$

3-9 SCF-INDO (Intermediate Neglect of Differential Overlap)

TABLE 3–2. INDO Calculation for Allyl
(*One atomic unit of energy is defined as twice the energy of the first Bohr orbit in the hydrogen atom; i.e., twice the ionization potential of a hydrogen atom.*)

Electronic Energy	−60.0089
Electronic Energy	−60.0340
Electronic Energy	−60.0352
Electronic Energy	−60.0353
Electronic Energy	−60.0353
Electronic Energy	−60.0353
Electronic Energy	−60.0353
Energy Satisfied	

Eigenvalues and Eigenvectors

				eV →	−52.19	−42.62	−36.32	−33.74	−30.21	−26.56	−23.36	−23.17
		Eigenvalues A.U. (orbital energies)		→	−1.915	−1.566	−1.335	−1.240	−1.110	−0.976	−0.859	−0.851
					1	2	3	4	5	6	7	8
1	1	C	S		−0.62	−0.00	−0.34	0.02	0.00	0.00	0.00	−0.10
2	1	C	Px		−0.13	−0.00	0.29	0.48	−0.00	−0.00	−0.00	0.40
3	1	C	Py		−0.00	0.42	−0.00	0.00	−0.45	−0.00	0.33	0.00
4	1	C	Pz		−0.00	0.00	0.00	−0.00	0.00	−0.72	0.00	−0.00
5	2	C	S		−0.45	0.51	0.27	−0.06	0.03	−0.00	0.16	0.06
6	2	C	Px		0.08	−0.08	0.30	0.20	0.44	0.00	0.24	−0.36
7	2	C	Py		0.19	0.11	0.21	−0.43	0.25	0.00	−0.42	0.02
8	2	C	Pz		0.00	0.00	−0.00	−0.00	−0.00	−0.49	−0.00	−0.00
9	3	C	S		−0.45	−0.51	0.27	−0.06	−0.03	−0.00	−0.16	0.06
10	3	C	Px		0.08	0.08	0.30	0.20	−0.44	0.00	−0.24	−0.36
11	3	C	Py		−0.19	0.11	−0.22	0.43	0.25	−0.00	−0.42	−0.02
12	3	C	Pz		0.00	−0.00	−0.00	−0.00	0.00	−0.49	0.00	−0.00
13	4	H	S		−0.20	−0.00	−0.33	−0.28	0.00	0.00	0.00	−0.47
14	5	H	S		−0.13	0.30	0.15	0.33	0.00	−0.00	−0.40	0.27
15	6	H	S		−0.13	0.23	0.34	−0.05	0.38	0.00	0.17	−0.31
16	7	H	S		−0.13	0.23	0.34	−0.05	−0.38	0.00	−0.17	−0.31
17	8	H	S		−0.13	0.30	0.15	−0.33	−0.00	0.00	0.40	0.27

TABLE 3-2 (continued)

Eigenvalues →				-8.53 -0.314	-7.05 -0.068	-7.07 -0.051	-0.08 -0.033	-0.20 -0.029	0.81 0.030	1.79 0.066	4.48 0.165	5.76 0.212
				9	10	11	12	13	14	15	16	17
1	1	C	S	0.00	0.43	0.00	-0.19	0.00	0.00	0.43	0.30	-0.00
2	1	C	Px	0.00	-0.09	-0.00	-0.20	0.00	0.00	-0.31	0.60	-0.00
3	1	C	Py	-0.00	0.00	-0.29	-0.00	0.00	-0.12	0.00	0.00	0.65
4	1	C	Pz	0.00	-0.00	0.00	0.00	0.69	0.00	-0.00	0.00	0.00
5	2	C	S	0.00	-0.26	0.35	0.37	-0.00	0.25	0.07	-0.10	-0.18
6	2	C	Px	0.00	-0.22	0.29	-0.24	0.00	-0.23	0.37	0.02	0.32
7	2	C	Py	-0.00	0.08	-0.08	0.14	-0.00	0.37	0.16	0.42	0.34
8	2	C	Pz	0.71	0.00	-0.00	-0.00	-0.51	0.00	0.00	-0.00	0.00
9	3	C	S	-0.00	-0.26	-0.35	0.37	-0.00	-0.25	0.07	-0.01	0.18
10	3	C	Px	-0.00	-0.22	-0.29	-0.24	0.00	0.23	0.36	0.02	-0.32
11	3	C	Py	-0.00	-0.08	-0.08	-0.14	0.00	0.37	-0.16	-0.42	0.34
12	3	C	Pz	-0.71	0.00	0.00	-0.00	-0.51	-0.00	0.00	-0.00	-0.00
13	4	H	S	-0.00	-0.50	-0.00	-0.16	0.00	0.00	-0.43	0.33	-0.00
14	5	H	S	0.00	-0.08	0.08	-0.48	0.00	-0.49	0.01	-0.16	-0.04
15	6	H	S	-0.00	0.39	-0.49	-0.04	0.00	-0.06	-0.33	-0.08	-0.19
16	7	H	S	0.00	0.39	0.49	-0.04	-0.00	0.06	-0.33	-0.08	0.19
17	8	H	S	-0.00	-0.08	-0.08	-0.48	0.00	0.49	0.01	-0.19	0.04

Binding Energy = -2.6884 A.U. = -73.15 eV   Dipole Moment = 3.09 Debyes

is not necessarily zero. For example, when the operator $\hat{O}_p$ is $1/r_{1N}$, we have the electron-nuclear attraction integral, which cannot be assumed to be zero.

Note that φ_k and φ_l overlap very little, so the differential overlap $\int \varphi_k(2)\varphi_l(2)\,d\tau_2$ is almost zero everywhere. The value of the integral is thus very small. If a systematic method were developed to set certain of these differential overlaps to zero, the number of such three- and four-center integrals would be drastically reduced. This is just what the various NDO schemes do.

For purposes of comparison with the results from the Hückel calculation and the extended Hückel output, the computer output from the INDO calculation on allyl is presented in Table 3-2. The eigenvalues of the energy are given above each molecular orbital in units of electron volts and atomic units.

It is of interest to examine ψ_9 in the INDO output:

$$\psi_9 = 0.71(\varphi_8 - \varphi_{12})$$

It is normalized, since

$$\int \psi_9^2 \, d\tau = (0.71)^2 \left[\int \varphi_8 \varphi_8 \, d\tau - 2 \int \varphi_8 \varphi_{12} \, d\tau + \int \varphi_{12} \varphi_{12} \, d\tau \right]$$

$$= (0.71)^2(1 - 0 + 1)$$
$$= 1$$

Now look at ψ_9 in the extended Hückel calculation:

$$\psi_9 = 0.74(\varphi_{10} - \varphi_6)$$

It is the same orbital with the same symmetry, but it has a different coefficient. Is it still normalized? Yes, because extended Hückel does not neglect differential overlaps, so

$$\int \varphi_{10} \varphi_6 \, d\tau \neq 0$$

and the normalization procedure as used above will yield a different coefficient.

Of the various INDO programs available, the one referred[4] to as INDO/1 or ZINDO is particularly good for reproducing geometries of inorganic systems.[5] The method has been used[6] to calculate points on the potential energy surface of alkane hydroxylations and alkene epoxidations by metal oxo species. The results lead to reaction mechanisms consistent with those proposed from experimental observations.

3–10 SOME PREDICTIONS FROM M.O. THEORY ON ALTERNATELY DOUBLE BONDED HYDROCARBONS

You probably have seen examples of the successes of m.o. theory for homonuclear diatomics, *e.g.*, the prediction of paramagnetic O_2. One further success involves the prediction of the structures of even numbered, alternately double bonded hydrocarbons. The systems C_4H_4, C_8H_8, and $C_{16}H_{16}$ are more stable as a set of alternating double bonds than they are as a delocalized pi-system. On the other hand, C_6H_6, $C_{10}H_{10}$, $C_{14}H_{14}$, and $C_{18}H_{18}$ (*i.e.*, $4n + 2$, when $n = 1, 2, 3, \ldots$) are aromatic.

The m.o. results for these even numbered, alternately double bonded hydrocarbons show a singly degenerate low energy orbital and a singly degenerate high energy level with doubly degenerate levels in between:

Note that the molecules with two unpaired electrons are stabilized much less than the other molecules because the highest energy electrons are in non-bonding m.o.'s. These molecules will in fact be more stable if they undergo a distortion in which the delocalized molecular orbital is converted into two localized double bonds, as shown below:

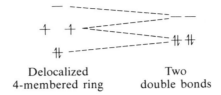

Delocalized　　　　　Two
4-membered ring　　double bonds

3–11 MORE ON PRODUCT GROUND STATE WAVE FUNCTIONS

The product wave function written in equation (3–32) is not complete. If we write a wave function, interchange of the labels on the electrons 1, 2, ... n cannot change any physical property associated with the electron density, Ψ^2, of the system. In order for Ψ^2 to be unaffected, the effect of the interchange on Ψ must produce $\pm \Psi$ so that Ψ^2 is unchanged. To demonstrate the problem, we shall define a permutation operator which switches the electron labels. For a two-electron system α and β in orbital ψ_1, we can write

$$\Psi = \psi_1(1)\alpha(1)\psi_1(2)\beta(2) \tag{3–37}$$

The permutation operator switches the electron labels in parentheses to

$$\hat{P}_{12}\Psi(1,2) = \psi_1(2)\alpha(2)\psi_1(1)\beta(1) \tag{3–38}$$

The result is not related to our initial wave function by ± 1.
However, consider the wave function

$$\Psi = \psi_1(1)\alpha(1)\psi_1(2)\beta(2) - \psi_1(2)\alpha(2)\psi_1(1)\beta(1) \tag{3–39}$$

Now

$$\hat{P}_{12}\Psi = \psi_1(2)\alpha(2)\psi_1(1)\beta(1) - \psi_1(1)\alpha(1)\psi_1(2)\beta(2) \tag{3–40}$$

This result (equation (3–40)) is related to our starting wave function by -1. Since we have the result $\hat{P}\Psi = -1\Psi$, we have an *antisymmetrized* wave function. Had we chosen

$$\Psi = \psi_1(1)\alpha(1)\psi_1(2)\beta(2) + \psi_1(2)\alpha(2)\psi_1(1)\beta(1)$$

the result would have been $+1$. This is a *symmetrized* wave function. Later we shall show that the antisymmetrized function must be chosen in order that the Pauli exclusion principle apply.

One can write the correct antisymmetrized wave function for a $2n$-electron system by constructing a Slater determinant and expanding it. The determinant is constructed as follows:

$$\Psi = N \begin{vmatrix} \psi_1(1)\alpha(1) & \psi_1(1)\beta(1) & \psi_2(1)\alpha(1) & \cdots & \psi_n(1)\beta(1) \\ \psi_1(2)\alpha(2) & \psi_1(2)\beta(2) & \cdots & & \\ \vdots & & & & \\ \psi_1(2n)\alpha(2n) & \cdots & & & \psi_n(2n)\beta(2n) \end{vmatrix}$$

Expanding this determinant by the method of cofactors produces the antisymmetrized wave function. Try it for the two-electron system just discussed.

The properties of determinants are consistent with several properties of electrons in molecules. For example, if rows of a determinant are changed, so is the sign. Exchanging rows corresponds to the use of the permutation operator, *e.g.*, equation (3–40). Thus, the antisymmetry property is built into the determinant. Further, there is a theorem stating that if two columns of a determinant are identical, the determinant vanishes. Thus, a non-zero wave function cannot be constructed if two electrons with the same spin are placed in the same orbital.

REFERENCES CITED

1. M. J. S. Dewar, "The Molecular Orbital Theory of Organic Chemistry," McGraw-Hill, New York, Chap. 5, 1969.
2. M. F. Rettig and R. S. Drago, J. Amer. Chem. Soc., *91*, 3432 (1969).
3. S. E. Anderson, Jr., and R. S. Drago, Inorg. Chem., *11*, 1564 (1972).
4. W. Anderson, W. D. Edwards, and M. C. Zerner, Inorg. Chem., *25*, 2728 (1986) and references therein. The program is available by contacting M. C. Zerner.
5. **a.** T. R. Cundari, M. C. Zerner, and R. S. Drago, Inorg. Chem., *27*, 4239 (1988).
   b. W. P. Anderson, T. R. Cundari, R. S. Drago, and M. C. Zerner, Inorg. Chem., *29*, 1 (1990).
6. **a.** T. R. Cundari and R. S. Drago, Intern. J. Quantum Chem., *24*, 665 (1990).
   b. T. R. Cundari and R. S. Drago, Inorg. Chem., *29*, 487 (1990).

ADDITIONAL READING

M. J. S. Dewar, "The Molecular Orbital Theory of Organic Chemistry," McGraw-Hill, New York, 1969.
J. M. Anderson, "Introduction to Quantum Chemistry," W. A. Benjamin, New York, 1969.
A. Streitwieser, "Molecular Orbital Theory for Organic Chemists," Wiley, New York, 1961.
A. Liberles, "Introduction to Molecular Orbital Theory," Holt, Rinehart and Winston, New York, 1966. (A simple approach to Hückel theory with many worked examples.)
W. Kauzman, "Quantum Chemistry," Academic Press, New York, 1957.
J. A. Pople and D. L. Beveridge, "Approximate Molecular Orbital Theory," McGraw-Hill, New York, 1970.
W. G. Richards and J. A. Horsley, "Ab Initio Molecular Orbital Calculations for Chemists," Oxford University Press, New York, 1970.

COMPILATIONS

W. G. Richards, T. E. H. Walker, and R. K. Hinkley, "A Bibliography of Ab Initio Molecular Wave Functions," Oxford University Press, New York, 1971.

W. G. Richards et al., "Bibliography of Ab Initio Molecular Wave Functions—Supplement for 1970–73," Oxford University Press, New York, 1974.

A. Hammett, Specialist Periodical Reports, The Chemical Society, London, England.

H. Hulbronner and C. Straub, "Hückel Molecular Orbitals, Springer-Verlag, New York, 1966.

A. Streitwieser and J. Brauman, "Supplemental Tables of Molecular Orbital Calculations," Volumes I and II, Pergamon Press, New York, 1965.

EXERCISES

1. Using the Hückel wave functions reported in this chapter, calculate the formal charges on the carbons and the C—C bond orders for

   a. the allyl cation

   b. the allyl radical

2. a. Using projection operators, work out the symmetry-adapted linear combinations of pi orbitals for the allyl radical.

   b. Use these linear combinations to obtain a set of orthonormal molecular orbitals.

   c. Sketch the shapes of the π and π^* molecular orbitals.

   d. Compare your results with those obtained from the Hückel calculation described in this chapter.

3. a. Using projection operators, work out the symmetry-adapted linear combinations of pi orbitals for cyclobutadiene.

   b. Use these linear combinations to obtain a set of orthonormal molecular orbitals.

   c. Sketch the shapes of the π and π^* molecular orbitals.

   d. Would any more information be obtained from a Hückel calculation? If so, what?

4. For the π *system* of borazole, $B_3N_3H_6$, obtain:

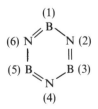

   a. the total representation

b. all of the wave functions

c. the normalization factor for each wave function; and indicate which wave functions can mix.

5. Propargylene is a linear diradical with the following valence bond structure:

$$H-C\equiv C-\ddot{C}-H \rightarrow Z\text{-axis}$$

Using the three $2p_y$ and three $2p_x$ orbitals of the carbons as a basis set,

a. How many m.o.'s will be produced?

b. Do a simple Hückel calculation to determine the energies of the m.o.'s from part a.

c. What are the a.o. coefficients of the two largest energy m.o.'s?

d. Calculate the bond order between two adjacent carbons.

6. The ion $Re_2Cl_8^{2-}$ is made up of two square planar $ReCl_4$ units joined by a Re—Re bond. The two planes are parallel and the chlorines are eclipsed (as viewed down the Re—Re bond, which we define as the z-axis). The symmetry is therefore D_{4h}.

a. Determine the symmetries of the m.o.'s obtained using the two Re d_{z^2} orbitals as the basis set. *Use projection operators* to determine these m.o.'s and sketch them.

b. Repeat part a using the two $d_{x^2-y^2}$ orbitals as the basis set. These orbitals point directly at the chlorines.

c. Applying this same procedure to the d_{xy} orbitals, one obtains

$$\psi(b_{2g}) = \frac{1}{\sqrt{2}}(d_{xy}^{(1)} + d_{xy}^{(2)})$$

$$\psi(b_{1u}) = \frac{1}{\sqrt{2}}(d_{xy}^{(1)} - d_{xy}^{(2)})$$

For d_{xz} and d_{yz}:

$$\psi_1(e_g) = \frac{1}{\sqrt{2}}(d_{xz}^{(1)} + d_{xz}^{(2)})$$

$$\psi_2(e_g) = \frac{1}{\sqrt{2}}(d_{yz}^{(1)} + d_{yz}^{(2)})$$

$$\psi_1(e_u) = \frac{1}{\sqrt{2}}(d_{xz}^{(1)} - d_{xz}^{(2)})$$

$$\psi_2(e_u) = \frac{1}{\sqrt{2}}(d_{yz}^{(1)} - d_{yz}^{(2)})$$

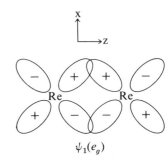

$\psi_1(e_g)$

Sketch a plausible energy level diagram. Fill in the electrons and determine the Re—Re bond order. (Hint: the bonding orbital obtained in part b participates in dsp^2 hybrids that are filled by chlorine electron pairs. The other orbitals must be filled by Re electrons.)

d. Ignore the chlorines and treat the ion as if it belonged to $D_{\infty h}$. What are the symmetries of the 10 m.o.'s obtained above? (Note: this illustrates the origin of the terms σ-bond, π-bond, and so forth.)

7. Consider the H_3^+ molecule in both the linear $(H\!-\!H\!-\!H)^+$ and triangular $(\overset{H}{\underset{H\ \ H}{\triangle}})^+$ states.

   a. Utilizing simple Hückel m.o. theory with the hydrogen 1s orbitals as a basis set, set up the secular determinant and compute the energies of the m.o.'s for each geometry.

   b. From your calculation, predict which geometry would be most stable for H_3^+ and for H_3^-.

   c. Consider only the linear geometry again, using the hydrogen 1s orbitals as a basis set. How many molecular orbitals will be formed? Substitute the energies found for the m.o.'s in part a into the secular determinant and find the *orthonormal* set of LCAO m.o.'s.

8. a. Two possible structures for the ion I_3^- are

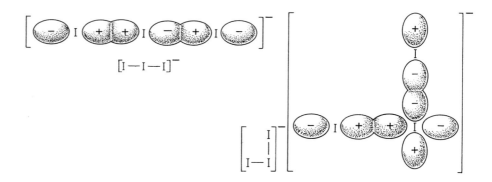

Using a basis set of p_z and p_y orbitals, simple Hückel molecular orbitals can be calculated for the two structures. Ignoring any π bonding, calculate the energies of the three sigma m.o.'s for the linear case, using the three atomic p_z orbitals shown above. (Do not calculate coefficients.)

b. Making the assumptions that $\beta = \int \varphi_i \hat{H}_{\text{bent}} \varphi_j = \int \varphi_i \hat{H}_{\text{linear}} \varphi_j$ and $\alpha = \int \varphi_j \hat{H}_{\text{bent}} \varphi_j = \int \varphi_j \hat{H}_{\text{linear}} \varphi_j$, which structure is more stable if the total energy of the p-orbital lone pairs and the m.o.'s in the bent configuration is $16\alpha + 2\beta$? (Hint: There will be six non-bonding p-orbital lone pairs, which will contribute 12α to the total energy of the linear structure.)

9. Consider the trigonal bipyramidal $CuCl_5^{3-}$ anion:

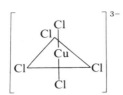

Choose axes at the Cl^- ligands such that z points upward, parallel to C_3. For the in-plane Cl^- nuclei, let x point toward the Cu^{2+}. Thus, from the bottom the xy-system for the ligands looks like the following:

Choose the x and y axes as you wish for the axial Cl^- ligands.

a. Find the total representation of the p orbitals of the five chlorines and decompose it into irreducible representations.

b. Use *a priori* reasoning to determine the effect on an equatorial p_x or p_y orbital of the projection operator $\hat{P}_{\chi''}$, where χ'' denotes any double primed irreducible representation. (Hint: What do these irreducible representations have in common?)

c. Operate on an equatorial p_z with $\hat{P}_{E'}$ to find a symmetry-adapted linear combination with E' symmetry.

4 General Introduction to Spectroscopy

4-1 NATURE OF RADIATION

There are many apparently different forms of radiation, *e.g.*, visible light, radio waves, infrared, x-rays, and gamma rays. According to the wave model, all of these kinds of radiation may be described as oscillating electric and magnetic fields. Radiation, traveling in the z-direction for example, consists of electric and magnetic fields perpendicular to each other and to the z-direction. These fields are represented in Fig. 4–1 for plane-polarized radiation. Polarized radiation was selected for simplicity of representation, since all other components of the electric field except those in the x-z-plane have been filtered out. The wave travels in the z-direction with the velocity of light, $c(3 \times 10^{10}$ cm sec$^{-1})$. The intensity of the radiation is proportional to the ampltiude of the wave given by the projection on the x- and y-axes. At any given time, the wave has different electric and magnetic field strengths at different points along the z-axis. The wavelength, λ, of the radiation is indicated in Fig. 4–1, and the variation in the magnitude of this quantity accounts for the apparently different forms of radiation listed above. If the radiation consists of only one wavelength, it is said to be monochromatic. Polychromatic radiation can be separated into essentially monochromatic beams. For visible, u.v., or IR radiation, prisms or gratings are employed for this purpose.

Radiation consists of energy packets called photons, which travel with the velocity of light. The different forms of radiation have different energies.

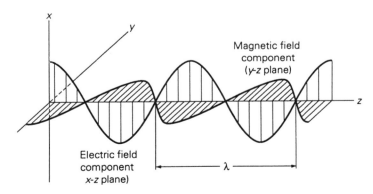

FIGURE 4–1 Electric and magnetic field components of plane-polarized electromagnetic radiation.

In our discussion of rotational, vibrational, and electronic spectroscopy, our concern will be with the interaction of the electric field component of radiation with the molecular system. This interaction results in the absorption of radiation by the molecule. In epr and nmr the concern is with the interaction with the magnetic component of radiation.

In order for absorption to occur, the energy of the radiation must match the energy difference between the quantized energy levels that correspond to different states of the molecule. If the energy difference between two of these states is represented by ΔE, the wavelength of the radiation, λ, necessary for matching is given by the equation:

$$\Delta E = hc/\lambda \quad \text{or} \quad \lambda = hc/\Delta E \tag{4-1}$$

where h is Planck's constant, 6.623×10^{-27} erg sec molecule^{-1}, and c is the speed of light in cm sec^{-1}, giving ΔE in units of erg molecule^{-1}. Equation (4-1) relates the wave and corpuscular models for radiation. Absorption of one quantum of energy, hc/λ, will raise one molecule to the higher energy state.

As indicated by equation (4-1), the different forms of electromagnetic radiation (*i.e.*, different λ) differ in energy. By considering the energies corresponding to various kinds of radiation and comparing these with the energies corresponding to the different changes in the state that a molecule can undergo, an appreciation can be obtained for the different kinds of spectroscopic methods.

4-2 ENERGIES CORRESPONDING TO VARIOUS KINDS OF RADIATION

Radiation can be characterized by its wavelength, λ, its wave number, $\bar{v}$, or its frequency, v. The relationship between these quantities is given by equations (4-2a) and (4-2b):

$$v(\text{sec}^{-1}) = \frac{c(\text{cm sec}^{-1})}{\lambda(\text{cm})} \tag{4-2a}$$

$$\bar{v}(\text{cm}^{-1}) = \frac{1}{\lambda(\text{cm})} \tag{4-2b}$$

The quantity $\bar{v}$ has units of reciprocal centimeters, for which the official IUPAC nomenclature is a kaiser; 1000 cm^{-1} are equal to a kilokaiser (kK). From equations (4-1) and (4-2), the relationship of energy to frequency, wave number, and wavelength is:

$$\Delta E(\text{ergs molecule}^{-1}) = hv = hc/\lambda = hc\bar{v} \tag{4-3}$$

In describing an absorption band, one commonly finds several different units being employed by different authors. Wave numbers, $\bar{v}$, which are most commonly employed, have units of cm^{-1} and are defined by equation (4-2b). Various units are employed for λ. These are related as follows: 1 cm = 10^8 Å (angstroms) = 10^7 nanometers = 10^4 μ (microns) = 10^7 mμ (millimicrons). The relationship to various common energy units is given by: 1 cm^{-1} = 2.858 cal/mole of par-

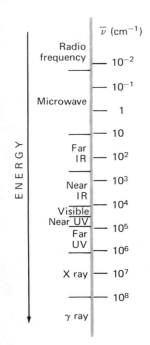

FIGURE 4–2 Wave numbers of various types of radiation.

ticles = 1.986×10^{-16} erg/molecule = 1.24×10^{-4} eV/mole. These conversion units can be employed to relate energy and wavelength; or the equation

$$\Delta E(\text{kcal mole}^{-1}) \times \lambda(\text{Å}) = 2.858 \times 10^5 \tag{4-4}$$

can be derived to simplify the calculation of energy from wavelength.

Wave numbers corresponding to various types of radiation are indicated in Fig. 4–2. The small region of the total spectrum occupied by the visible portion is demonstrated by this figure. The higher energy radiation has the smaller wavelength and the larger frequency and wave number (equation (4–3)). The following sequence represents decreasing energy:

ultraviolet > visible > infrared > microwave > radio-frequency

4–3 ATOMIC AND MOLECULAR TRANSITIONS

In an atom, the change in state induced by the quantized absorption of radiation can be regarded as the excitation of an electron from one energy state to another. The change in state is from the ground state to an excited state. In most cases, the energy required for such excitation is in the range from 60 to 150 kcal mole^{-1}. Calculation employing equation (4–3) readily shows that radiation in the ultraviolet and visible regions will be involved. Atomic spectra are often examined as emission spectra. Electrons are excited to higher states by thermal or electrical energy, and the energy emitted as the atoms return to the ground state is measured.

In molecular spectroscopy, absorption of energy is usually measured. Our concern will be with three types of molecular transitions induced by electromagnetic radiation: electronic, vibrational, and rotational. A change in *electronic state* of a molecule occurs when a bonding or nonbonding electron of the molecule in the ground state is excited into a higher-energy empty molecular orbital. For example, an electron in a π-bonding orbital or a carbonyl group can be excited into a π^* orbital, producing an excited state with configuration $\sigma^2 \pi^1 \pi^{*1}$. The electron distributions in the two states (ground and excited) involved in an electronic transition are different.

The *vibrational energy states* are characterized by the directions, frequencies, and amplitudes of the motions that the atoms in a molecule undergo. As an example, two different kinds of vibrations for the SO_2 molecule are illustrated in Fig. 4–3. The atoms in the molecule vibrate (relative to their center of mass) in the directions indicated by the arrows, and the two extremes in each vibrational mode are indicated. In the vibration indicated in (A), the sulfur-oxygen bond length is varying, and this is referred to as the *stretching vibration*. In (B), the motion is perpendicular to the bond axis and the bond length is essentially constant. This is referred to as a *bending vibration*. In these vibrations, the net effect of all atomic motion is to preserve the center of mass of the molecule so that there will be no net translational motion. The vibrations indicated in Fig. 4–3 are drawn to satisfy this requirement. Certain vibrations in a molecule are referred to as *normal vibrations* or *normal modes*. These are independent, self-repeating displacements of the atoms that preserve the center of mass. In a normal vibration, all the atoms vibrate in phase and with the same frequency. It is possible to resolve the most complex molecular vibration into a relatively

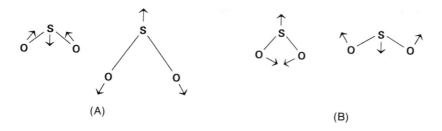

FIGURE 4-3 Two different vibrations for the SO_2 molecule. (The amplitudes and angle deformations are exaggerated to illustrate the mode.)

small number of such normal modes. For a non-linear molecule, there are $3N - 6$ such modes, where N is the number of atoms in the molecule. In Chapter 6, the procedure for calculating the total number of normal modes and their symmetries will be given. The normal modes can be considered as the $3N - 6$ internal degrees of freedom that (in the absence of anharmonicity) could take up energy independently of each other. The motion of the atoms of a molecule in the different normal modes can be described by a set of *normal coordinates*. These are a set of coordinates defined so as to describe the normal vibration most simply. They often are complicated functions of angles and distances.

The products of the normal modes of vibration are related to the total vibrational state, ψ_v, of a non-linear molecule as shown in equation (4-5).

$$\psi_v = \prod_{n=1}^{3N-6} \psi_n \qquad (4\text{-}5)$$

where Π indicates that the product of the n vibrational modes is to be taken and ψ_n is the wave function for a given normal mode. There exists for each normal vibration (ψ_n) a whole series of excited vibrational states, i, whose harmonic oscillator wave functions are given by:

$$\psi_i = N_i \exp[(-\tfrac{1}{2})a_i q^2] H_i(\sqrt{a_i} q) \qquad (4\text{-}6)$$

where $i = 0, 1, 2 \ldots$; H_i is the Hermite polynomial of degree i; $a_i = 2\pi v_i/h$; $N_i = [\sqrt{a_i}/(2^i i! \sqrt{\pi})]^{1/2}$; and q is the normal coordinate.

From this, the following wave functions are obtained for the ground, first, and second excited states:

$$\psi_0 = \left(\frac{a_0}{\pi}\right)^{1/4} \exp\left[\left(-\frac{1}{2}\right)a_0 q^2\right] \qquad (4\text{-}7)$$

$$\psi_1 = 2\left(\frac{a_1}{\pi}\right)^{1/4} q \exp\left[\left(-\frac{1}{2}\right)a_1 q^2\right] \qquad (4\text{-}8)$$

$$\psi_2 = 2\left(\frac{a_2}{\pi}\right)^{1/4} (2a_2 q^2 - 1)\exp\left[\left(-\frac{1}{2}\right)a_2 q^2\right] \qquad (4\text{-}9)$$

When ψ_0 is plotted as a function of displacements about the normal coordinate, q, with zero taken as the equilibrium internuclear distance, Fig. 4-4 is obtained. This symmetric function results because q appears only as q^2. The

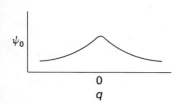

FIGURE 4-4 A plot of the ground vibrational wave function versus the normal coordinate.

same type of plot is obtained for the ground state of all normal modes. It is totally symmetric; accordingly, the total vibrational ground state of a molecule must belong to the totally symmetric irreducible representation, for it is the product of only totally symmetric vibrational wave functions.

The vibrational excited state function ψ_1 has a functional dependence on q that is not of an even power, and accordingly it does not necessarily have a_1 symmetry. When one normal mode is excited in a vibrational transition, the resulting total state is a product of all the other totally symmetric wave functions and the wave function for this first vibrational excited state for the excited normal mode. Thus, the total vibrational state has symmetry corresponding to the normal mode that is excited.

The *rotational states* correspond to quantized molecular rotation around an axis without any appreciable change in bond lengths or angles. Different rotational states correspond to different angular momenta of rotation or to rotations about different axes. Rotation about the C_2 axis in SO_2 is an example of rotational motion.

In the treatment of molecular spectra, the *Born-Oppenheimer approximation* is invoked. This approximation proposes that the total energy of a system may be regarded as the sum of three independent energies: electronic, vibrational, and rotational. For example, the electronic energy of the system does not change as vibration of the nuclei occurs. The wave function for a given molecular state can then be described by the product of three independent wave functions: ψ_{el}, ψ_{vib}, and ψ_{rot}. As we shall see later, this approximation is not completely valid.

The relative energies of these different molecular energy states in typical molecules are represented in Fig. 4–5. Rotational energy states are more closely spaced than are vibrational states, which, in turn, have smaller energy differences than electronic states. The letters v_0, v_1, etc., and v_0', v_1', etc., each represent the vibrational levels of one vibrational mode in the ground and first electronic excited states, respectively. *Ultraviolet or visible radiation is commonly required to excite the molecule into the excited electronic states. Lower-energy infrared suffices for vibrational transitions, while pure rotational transitions are observed in the still lower-energy microwave and radio-frequency regions.*

Electronic transitions are usually accompanied by changes in vibration and rotation. Two such transitions are indicated by arrows (a) and (b) of Fig. 4–5. In the vibrational spectrum, transitions to different rotational levels also occur. As a result, vibrational fine structure is often detected in electronic transitions.

FIGURE 4-5 Energy states of a diatomic molecule.

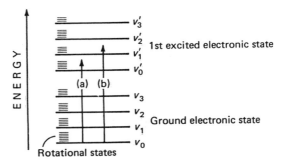

Rotational fine structure in electronic transitions can be detected in high resolution work in the gas phase. Rotational fine structure in vibrational transitions is sometimes observed in the liquid state and generally in the gaseous state.

The energy level diagram in Fig. 4–5 is that for a diatomic molecule. For a polyatomic molecule, the individual observed transitions can often be described by diagrams of this type, each transition in effect being described by a different diagram.

4–4 SELECTION RULES

In order for matter to absorb the electric field component of radiation, another requirement in addition to energy matching must be met. The energy transition in the molecule must be accompanied by a change in the electrical center of the molecule in order that electrical work can be done on the molecule by the electromagnetic radiation field. Only if this condition is satisfied can absorption occur. Requirements for the absorption of light by matter are summarized in the *selection rules*. Transitions that are possible according to these rules are referred to as *allowed* transitions, and those not possible as *forbidden* transitions. It should be noted, however, that the term "forbidden" refers to rules set up for a simple model and, while the model is a good one, "forbidden" transitions may occur by mechanisms not included in the simple model. The intensity of absorption or emission accompanying a transition is related to the probability of the transition, the more probable transitions giving rise to more intense absorption. Forbidden transitions have low probability and give absorptions of very low intensity. This topic and the symmetry aspects of this topic will be treated in detail when we discuss the various spectroscopic methods.

4–5 RELAXATION AND CHEMICAL EXCHANGE INFLUENCES ON SPECTRAL LINE WIDTH

Spectral transitions involve different atomic or molecular energy states. The Boltzmann expression indicates the probability, P_i, that any molecule is in the ith state as a function of the ground state energy E_0 and the excited state energy E_i.

$$P_i = \frac{e^{-(E_i - E_0)/kT}}{\sum_i e^{-(E_i - E_0)/kT}}$$

Here, k is the Boltzmann constant (1.3807×10^{-23} JK^{-1}) and T the absolute temperature.

When the energy differences of the states are large as they are in most electronic and vibrational transitions, only the ground state is populated at room temperature. When a photon is absorbed, the molecule changes to the excited state and then relaxes back to its ground state by a radiative process that emits radiation or increases the kinetic energy (temperature) of the sample. In time, the Boltzmann equilibrium is reestablished. When the energy difference of the states is small compared to kT at room temperature (as is the case for states involved in an nmr transition), the states are populated in accord with the Boltzmann expression. A dynamic system results in which molecules in the excited state exchange energy with other identical molecules. The fraction of molecules in the two states is constant, but a given molecule exchanges between the different states. The process of switching energy states is called relaxation and the efficiency of interchange is governed by the relaxation mechanism, which in turn governs the lifetime of the molecule in a given state. When a molecule is excited by the absorption of energy, these relaxation mechanisms determine the lifetime of the excited state as the system returns to the Boltzmann equilibrium. The natural line width of a spectral line depends upon the lifetime of the excited state via the uncertainty principle.

$$\Delta E \Delta t \sim \hbar$$

or

$$\Delta \nu \Delta t \sim \frac{1}{2\pi} \qquad (4\text{--}10)$$

As the lifetime, Δt, decreases, the range of frequencies possible, *i.e.*, the natural line width, increases.

Next, consider the consequences of chemical reactions on the widths of spectral lines. When two or more chemically distinct species coexist in rapid equilibrium (two conformations, or rapid chemical exchange, *e.g.*, proton exchange between NH_3 and NH_4^+, or other equilibria), one often sees absorption peaks corresponding to the individual species in some forms of spectroscopy; but with other methods, only a single average peak is detected. A process may cause a broadening of the spectral line in some spectral regions, while the same process in some other spectral region has no effect.

Consider the case in which two different sites give rise to two distinct peaks. As the rate of the chemical exchange comes close to the frequency $\Delta \nu$ of the spectroscopic method, the two peaks begin to broaden. As the rate becomes faster, they move together, merge, and then sharpen into a single peak. In subsequent chapters, we shall discuss procedures for extracting rate data over this entire range. Here we shall attempt to gain an appreciation for the time scales corresponding to the various methods by examining the rates that result in the line broadening of a sharp single resonance and the merging of two distinct resonances.

First, we shall be concerned with the broadening of a resonance line of an individual species. If Δt is the lifetime of the excited state, we have:

$$\Delta \nu (\sec^{-1}) \sim \frac{1}{2\pi \Delta t (\sec)}$$

If there is some chemical or physical process that is fast compared to the excited state lifetime, Δt can be shortened by this process and the line will be broadened.

In infrared, for example, it is possible to resolve* two bands corresponding to different sites that are separated by 0.1 cm^{-1}, so we can substitute this value into equation (4–10) and find that lifetimes, Δt, corresponding to this are given by:

$$\Delta v = 0.1 \text{ cm}^{-1} \times 3 \times 10^{10} \text{ cm/sec} = 3 \times 10^9 \text{ sec}^{-1}$$

$$\Delta t = \frac{1}{(2)(\pi)(3 \times 10^9 \text{ sec}^{-1})} = 5 \times 10^{-11} \text{ sec}$$

Therefore, we need a process that will give rise to lifetimes of $\sim 10^{-11}$ sec or less to cause broadening. Since lifetimes are the reciprocals of first order rate constants, this means we need a process whose rate constant is at least 10^{11} sec^{-1} in order to detect a broadening in the infrared band. Diffusion-controlled chemical reactions have rate constants of only 10^{10}, so for systems undergoing chemical exchange, the chemical process will have no influence on the infrared line shape. Rotational motion occurs on a time scale that causes broadening of an infrared or Raman line, and it is possible to obtain information about the motion from the line shape.[1] The inversion doubling of ammonia is also fast enough to affect the Raman and microwave spectra.

In nmr, typical resolution is ~ 0.1 Hz (cycle per second), which, when substituted into the above equation, shows that a process with a lifetime of about 2 sec (or less) or a rate constant of about 0.5 sec^{-1} (or more) is needed to broaden the spectral line. This is in the range of many chemical exchange reactions.

Next, we shall consider the rate at which processes must occur to result in a spectrum in which only an average line is detected for two species. In order for this to occur, the species must be interconverting so fast that the lifetime in either one of the states is less than Δt, so that only an average line results. This is calculated by substituting the difference between the frequencies of the two states for Δv in equation (4–10). In the infrared, a rate constant of 5×10^{13} sec^{-1} or greater would be required to merge two peaks that are separated by 300 cm^{-1}.

$$\Delta t = \frac{1}{(2)(\pi)(9.0 \times 10^{12} \text{ sec}^{-1})} = 2 \times 10^{-14} \text{ sec}$$

In nmr, for a system in which, for example, the two proton peaks are separated by 100 Hz, one obtains

$$\Delta t = \frac{1}{(2)(\pi)(100 \text{ sec}^{-1})} = 2 \times 10^{-3} \text{ sec}$$

Thus, an exchange process involving this proton that had a rate constant of 5×10^2 sec^{-1} would cause these two resonances to appear as one broad line. As the rate constant becomes much larger than 5×10^2 (e.g., by raising the temperature for a process with a positive activation enthalpy), the broadened

---

*By resolve, we mean that one can detect the existence of two distinct maxima. We are using this quantity to roughly estimate Δv, the time scale of the spectroscopic method (i.e., the natural line width).

TABLE 4–1 Kinetic Techniques and the Appropriate Lifetimes*

conventional kinetics	10 sec or longer (determined by ability to mix reagents)
stop flow	10^{-3} sec or longer (faster mixing is possible)
nmr	10^{-1} to 10^{-5} sec
esr	10^{-4} to 10^{-8} sec
temperature jump	0.1 to 10^{-6} sec
Mössbauer (iron)	10^{-7} sec
ultrasonics	10^{-4} to 10^{-8} sec
fluorescent polarization (the depolarization is measured)	10^{-8} to 10^{-9} sec
IR and Raman line shapes	10^{-11} sec
photoelectron spectroscopy	10^{-18} sec

*If some of these techniques are unfamiliar at present, do not be concerned. This will provide a ready reference after you encounter them in your study.

single resonance begins to sharpen. Eventually, as the process becomes very fast, the line width is no longer influenced by the chemical process. The radiation is too slow to detect any chemical changes that are occurring and a sharp average line results. This is analogous to the eye being too slow to detect the electronic sweep that produces a television picture.

In x-ray diffraction, the frequency of the radiation is 10^{18} sec^{-1}. Since this is too fast, compared with molecular rearrangements, all we would detect for a dynamic system would be disorder, *i.e.*, contributions from each of the dominant conformations.

In Mössbauer spectroscopy, we observe a very high energy process in which a nucleus in the sample absorbs a γ-ray from the source. The lifetime of the nuclear excited state, for iron is constant at 10^{-7} sec. Thus, if there is a chemical process occurring in an iron sample that equilibrates two iron atoms with a first order rate constant greater than 10^7 sec^{-1}, the Mössbauer spectrum will reveal only an average peak. In dichlorobisphenanthroline iron (III), the equilibrium mixture of high spin (five unpaired electrons) and low spin (one unpaired electron) iron complexes interconvert so rapidly (they have a lifetime $< 10^{-7}$ sec) that only a single iron species is observed in the spectrum. For a ruthenium nucleus, the lifetime of the nuclear excited state is 10^{-9} sec; this determines the time scale for experiments with this nucleus. A similar situation pertains in x-ray photoelectron spectroscopy. The time scale of an electron is not influenced by chemical processes and the lifetime for the resulting excited state is constant at about 10^{-18} sec.

Table 4–1 summarizes the methods used in kinetic studies and the ranges of lifetimes that can be investigated.

General Applications

The following general applications of spectroscopy are elementary and pertain to both vibrational and electronic spectroscopic methods: (1) determination of concentration, (2) "fingerprinting," and (3) determination of the number of species in solution by the use of isosbestic points.

4-6 DETERMINATION OF CONCENTRATION

Measurement of the concentrations of species has several important applications. If the system is measured at equilibrium, equilibrium constants can be determined. By evaluating the equilibrium constant, K, at several temperatures, the enthalpy ΔH^0 for the equilibrium reaction can be calculated from the van't Hoff equation:

$$\log K = \frac{-\Delta H^0}{2.3RT} + C \qquad (4\text{-}11)$$

Determination of the change in concentration of materials with time is the basis of kinetic studies that give information about reaction mechanisms. In view of the contribution of results from equilibrium and kinetic studies to our understanding of chemical reactivity, the determination of concentrations by spectroscopic methods will be discussed.

The relationship between the amount of light absorbed by certain systems and the concentration of the absorbing species is expressed by the Beer-Lambert law:

$$A = \log_{10} \frac{I_0}{I} = \varepsilon c b \qquad (4\text{-}12)$$

where A is the absorbance, I_0 is the intensity of the incident light, I is the intensity of the transmitted light, ε is the molar absorptivity (sometimes called extinction coefficient) at a given wavelength and temperature, c is the concentration (the molarity if ε is the molar absorptivity), and b is the length of the absorbing system. The molar absorptivity varies with both wavelength and temperature, so these must be held constant when using equation (4-12). When matched cells are employed to eliminate scattering of the incident beam, there are no exceptions to the relationship between absorbance and b (the Lambert law). For a given concentration of a certain substance, the absorbance is always directly proportional to the length of the cell. The part of equation (4-12) relating absorbance and concentration ($\log_{10} I_0/I = \varepsilon c$, for a constant cell length) is referred to as Beer's law. Many systems have been found that do not obey Beer's law. The anomalies can be attributed to changes in the composition of the system with concentration (*e.g.*, different degrees of ionization or dissociation of a solute at different concentrations). For all systems, the Beer's law relationship must be demonstrated, rather than assumed, over the entire concentration range to be considered. If Beer's law is obeyed, it becomes a simple matter to use equation (4-12) for the determination of the concentration of a known substance if it is the only material present that is absorbing in a particular region of the spectrum. The Beer's law relationship is tested and ε determined by measuring the absorbances of several solutions of different known concentrations covering the range to be considered. For each solution, ε can be calculated from

$$\varepsilon = \frac{A(1 \text{ cm cell})}{c, \text{ molarity}} \qquad (4\text{-}13)$$

where the units of ε are liters mole^{-1} cm^{-1}. For a solution of this material of unknown concentration, the absorbance is measured, ε is known (4–13), and c is calculated from equation (4–12).

There are some interesting variations on the application of Beer's law. If the absorption of two species should overlap, this overlap can be resolved mathematically and the concentrations determined. This is possible as long as the two ε values are not identical at all wavelengths. Consider the case in which the ε values for two compounds whose spectra overlap can be measured for the pure compounds. The concentration of each component in a mixture of the two compounds can be obtained by measuring the absorbance at two different wavelengths, one at which both compounds absorb strongly and a second at which there is a large difference in the absorptions. Both wavelengths should be selected at reasonably flat regions of the absorption curves of the pure compounds, if possible. Consider two such species B and C, and wavelengths λ_1 and λ_2. There are molar absorptivities of $\varepsilon_{B_{\lambda_1}}$ for B at λ_1, $\varepsilon_{B_{\lambda_2}}$ for B at λ_2, and similar quantities $\varepsilon_{C_{\lambda_1}}$ and $\varepsilon_{C_{\lambda_2}}$ for C. The total absorbance of the mixture at λ_1 is A_1, and that at λ_2 is A_2. It follows that:

$$A_1 = x\varepsilon_{B_{\lambda_1}} + y\varepsilon_{C_{\lambda_1}} \tag{4-14}$$

and

$$A_2 = x\varepsilon_{B_{\lambda_2}} + y\varepsilon_{C_{\lambda_2}} \tag{4-15}$$

where x = molarity of B and y = molarity of C. The two simultaneous equations (4–14) and (4–15) have only two unknowns, x and y, and can be solved.

A situation often encountered in practice involves an equilibrium of the type

$$D + E \rightleftarrows DE$$

Here the spectra of, say, D and DE overlap but E does not absorb; if the equilibrium constant is small, pure DE cannot be obtained, so its molar absorptivity cannot be determined directly. It is possible to solve this problem for the equilibrium concentrations of all species, *i.e.*, to determine the equilibrium constant:

$$K = \frac{[DE]}{[D][E]} \tag{4-16}$$

Let $[D]_0$ be the initial concentration of D. This can be measured, as can $[E]_0$, the initial concentration of E. The molar absorptivity of DE, ε_{DE}, cannot be determined directly, but it is assumed that Beer's law is obeyed. It can be seen from a material balance that:

$$[D] = [D]_0 - [DE] \tag{4-17}$$

$$[E] = [E]_0 - [DE] \tag{4-18}$$

so

$$K = \frac{[DE]}{[D]_0 - [DE])([E]_0 - [DE])} \qquad (4\text{--}19)$$

The total absorbance consists of contributions from [D] and [DE] according to

$$A = \varepsilon_D[D] + \varepsilon_{DE}[DE] \qquad (4\text{--}20)$$

Substituting equation (4–17) into (4–20) and solving for [DE], one obtains

$$[DE] = \frac{A - \varepsilon_D[D]_0}{\varepsilon_{DE} - \varepsilon_D}$$

but $\varepsilon_D[D]_0$ is the initial absorbance, A^0, of a solution of D with concentration $[D]_0$ without any E in it, so

$$[DE] = \frac{A - A^0}{\varepsilon_{DE} - \varepsilon_D} \qquad (4\text{--}21)$$

When equation (4–21) is substituted into the equilibrium constant expression, equation (4–19), and this is rearranged, we have[2]

$$K^{-1} = \frac{A - A^0}{\varepsilon_{DE} - \varepsilon_D} - [D]_0 - [E]_0 + \frac{[D]_0[E]_0(\varepsilon_{DE} - \varepsilon_D)}{A - A^0} \qquad (4\text{--}22)$$

The advantage of this equation is that it contains only two unknown quantities, ε_{DE} and K^{-1}. Furthermore, these unknowns are constant for any solution of different concentrations $[D]_0$ and $[E]_0$ that we care to make up. For two different sets of experimental conditions (different values of $[D]_0$ and $[E]_0$) two simultaneous equations can be solved; ε_{DE} is eliminated, and K is obtained. If several sets of experimental conditions are employed, all possible combinations of simultaneous equations can be considered by employing a least-squares computer analysis[3-7] that finds the best values of K^{-1} and ε_{DE} to reproduce the experimental absorbances. These least-squares procedures usually provide an error analysis that is most important to have (*vide infra*).

There are several advantages to a graphical display of the simultaneous equations that are being analyzed. The graph is constructed[2] by taking a solution of known $[D]_0$, $[E]_0$, and A, and calculating the values of K^{-1} that would result by selecting several different values of ε_{DE} near the expected value. These results are plotted, as in Fig. 4–6, by the line $[D]_0[E]_0^*$. The procedure is repeated for other initial concentrations, e.g., $[D]_0'[E]_0'$ and $[D]_0''[E]_0''$. The intersection of any two of these lines is a graphical representation of the solution of two simultaneous equations. The intersection of all the calculated curves should occur at a point whose values of K^{-1} and ε satisfy all the experimental data. This common intersection justifies the Beer's law assumption used in the derivation because it indicates a unique value for ε for all concentrations. As a result of experimental error, a triangle usually results instead of a point. The best K^{-1} and ε to fit the data are then found by a least-squares procedure. References 6 and

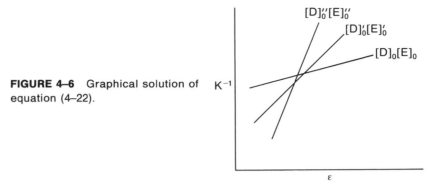

FIGURE 4–6 Graphical solution of equation (4–22).

7 describe the least-squares procedure and error analysis. When the experiment described above is not designed properly, a set of concentrations of D and E are employed that result in a set of parallel lines for the K^{-1} vs. ε plots. The slope of one of these lines is given by taking the partial derivative of the K^{-1} expression in (4–22) with respect to $\varepsilon_{DE} - \varepsilon_D$:

$$\frac{\partial K^{-1}}{\partial \varepsilon_{DE} - \varepsilon_D} = -\frac{A - A^0}{(\varepsilon_{DE} - \varepsilon_D)^2} + \frac{[D]_0[E]_0}{A - A^0} \tag{4–23}$$

Since the first term is generally small, we see from equation (4–23) that if experimental conditions are selected for a series of experiments in which [DE] or the value of A nearly doubles every time $[D]_0$ or $[E]_0$ is doubled, then the various resulting K^{-1} vs. ε plots will be nearly parallel. The values of K^{-1} and ε obtained from the computer analysis of this kind of data[6] are thus highly correlated and should be considered undefined, for the simultaneous equations are essentially dependent ones. The K^{-1} vs. ε plots have been described in terms of the results from the error analysis of the least-squares calculation, and the reader is referred to reference 7 for details.

The approach described above for the evaluation of equilibrium constants is a general one that applies to any form of spectroscopy or any kind of measurement in which the measured quantity is linearly related to concentration (i.e., in which a counterpart to equation (4–20) exists). Similar analyses have been described for calorimetric data[8] and for nmr spectral data.[7]

There have been many reports in the literature of attempts to solve simultaneous equations for equilibrium constants for two or more consecutive equilibria:

$$A + B \rightleftharpoons AB \quad \text{step 1}$$

$$AB + B \rightleftharpoons AB_2 \quad \text{step 2}$$

Usually ε_{AB}, ε_{AB_2}, and the stepwise equilibrium constants K_1 and K_2 are unknown. In most instances, even though the reported parameters fit the experimental data well, the system is undefined. Careful examination often shows that many other, very different combinations of parameters also fit the data. To solve this problem, one must find a region of the spectrum in which AB makes the main contribution to the absorbance and another in which AB_2 makes the predominant contribution.

Only by working at both wavelengths can one solve for all the unknowns in a rigorous fashion. This is impossible in many nmr, u.v.-visible, and calorimetric experiments because properties related predominantly to the individual AB and AB_2 species cannot be monitored separately. This problem has been discussed in detail in the literature[4] and several examples are given there.

4–7 ISOSBESTIC POINTS

If two substances at equal concentrations have absorption bands that overlap, there will be some wavelength at which the molar absorptivities of the two species are equal. If the sum of the concentrations of these two materials in solution is held constant, there will be no change in absorbance at this wavelength as the ratio of the two materials is varied. For example, assume a reaction $A + B \rightarrow AB$, in which only A and AB absorb; the sum of [A] plus [AB] will be constant as long as the initial concentration of A is held constant as B is varied. Since the absorbance of the solution is given by

$$\text{abs} = \varepsilon_A[A] + \varepsilon_{AB}[AB] \qquad (4\text{–}24)$$

the absorbance will be constant when $\varepsilon_A = \varepsilon_{AB}$ and [A] plus [AB] is held constant. The invariant point obtained for this system is referred to as the *isosbestic point*. The existence of one or more isosbestic points in a system provides information regarding the number of species present. For example, the spectra in Fig. 4–7 were obtained[9] by keeping the total iron concentration constant in a system consisting of:

Curve 1 2.1 moles of LiCl per $Fe(DMA)_6(ClO_4)_3$
Curve 2 2.6 moles of LiCl per $Fe(DMA)_6(ClO_4)_3$
Curve 3 3.1 moles of LiCl per $Fe(DMA)_6(ClO_4)_3$
Curve 4 4.1 moles of LiCl per $Fe(DMA)_6(ClO_4)_3$

where DMA is the abbreviation for *N,N*-dimethylacetamide. Points A and B are isosbestic points; their existence suggests that the absorption in this region is essentially accounted for by two species. Curve 4 is characteristic of $FeCl_4^-$. A study of solutions more dilute in chloride establishes the existence of a species $Fe(DMA)_4Cl_2^+$, which exists at a 2:1 ratio of Cl^- to Fe^{III} and absorbs in this region. The isosbestic points indicate that the system can be described by the species $Fe(DMA)_4Cl_2^+$ and $FeCl_4^-$ over the region from 2:1 to 4:1 ratios of Cl^- to Fe^{III}. The addition compound $FeCl_3 \cdot DMA$ probably does not exist in appreciable concentration in this system, *i.e.*, at total Fe^{III} concentrations of $2 \times 10^{-4} M$. Curve 4 in this spectrum does not pass through the isosbestic points. Small deviations (*e.g.*, curve 4 at point A) may be due to experimental inaccuracy, changes in solvent properties in different solutions, or a small concentration of a third species, probably $FeCl_3 \cdot DMA$, present in all systems except that represented by curve 4.

The conclusion that only two species are present in appreciable concentrations could be in error if $FeCl_3 \cdot DMA$ or other species were present that had molar absorptivities identical to that of the above two ions at the isosbestic points. However, if equilibrium constants for the equilibrium

$$Fe(DMA)_4Cl_2^+ + 2Cl^- \rightleftharpoons FeCl_4^- + 4DMA$$

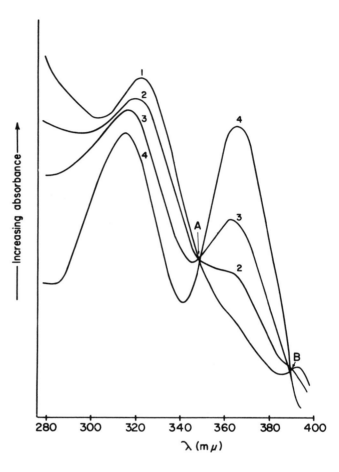

FIGURE 4–7 Spectra of the Fe(DMA)$_4$Cl$_2{}^+$-LiCl system in N,N-dimethylacetamide as solvent.

are calculated from these different curves at different wavelengths, the possibility of a third species is eliminated if the constants agree.

An interesting situation results in which an isosbestic point can be obtained in solution when more than two species with differing extinction coefficients exist if the base (or acid) employed has two donor (or acceptor) sites. For example, if DMA were to coordinate to an acid A to produce an oxygen-bound complex and a nitrogen-bound one, the mixture of complexes AN (nitrogen bound), AO (oxygen bound), and free A (an absorbing Lewis acid) will give rise to an isosbestic point.[10] The absorbance for such a system is given by equation (4–25):

$$\text{abs} = \varepsilon_A[A] + \varepsilon_{AO}[AO] + \varepsilon_{AN}[AN] \tag{4-25}$$

The equilibrium constant expressions are given by

$$K_O = \frac{[AO]}{[A][B]} \quad \text{and} \quad K_N = \frac{[AN]}{[A][B]}$$

$$K_O + K_N = \frac{[AO] + [AN]}{[A][B]} = \frac{[AB]}{[A][B]} \tag{4-26}$$

where we define [AO] + [AN] = [AB]. The fraction of complex that is oxygen-coordinated, X_{AO}, is given by

$$X_{AO} = \frac{K_O}{K_O + K_N} = \frac{\frac{[AO]}{[A][B]}}{\frac{[AB]}{[A][B]}} = \frac{[AO]}{[AB]} \qquad (4\text{-}27)$$

The fraction that is nitrogen-coordinated is similarly derived as:

$$X_{AN} = \frac{[AN]}{[AB]} \qquad (4\text{-}28)$$

Now the total absorbance, abs, becomes

$$\text{abs} = \varepsilon_A[A] + \varepsilon_{AO}X_{AO}[AB] + \varepsilon_{AN}X_{AN}[AB]$$

or

$$\text{abs} = \varepsilon_A[A] + (\varepsilon_{AO}X_{AO} + \varepsilon_{AN}X_{AN})[AB] = \varepsilon_A[A] + \varepsilon'[AB] \qquad (4\text{-}29)$$

Since the sum of [A] and [AB] is constant, and there must be a point in the overlapping spectra at which $\varepsilon = \varepsilon'$, an isosbestic point will be obtained even though three absorbing species exist. This is true because

$$\frac{X_{AN}}{X_{AO}} = \frac{K_N}{K_O} = \text{a constant}$$

We have taken two absorbing species whose ratio is a constant, independent of the parameter being varied, and we have translated them into what is effectively a single absorbing species *via* equation (4–29). The general criterion is thus that $2 + N$ absorbing species will give rise to an isosbestic point if there are N *independent* equations of the form

$$\frac{[Y]}{[Z]} = k \qquad (4\text{-}30)$$

where Y and Z are two of the absorbing species in the system and the value of k is independent of the parameter being varied.

As an example of the application of this criterion, consider an acid that can form two isomers with a base, AB and A*B (*e.g.*, Cu(hfac)$_2$ forming basal and apical adducts); let this acid form an adduct with a base that can form two isomers with an acid, AB and AB* (*e.g.*, N-methyl imidazole bound through the amine and imine nitrogens). Five absorbing species (A, AB, A*B, AB*, and A*B*) can exist,

$$A + B \rightarrow AB$$
$$A + B \rightarrow A^*B$$
$$A + B \rightarrow AB^*$$
$$A + B \rightarrow A^*B^*$$

so three constants are needed. The equilibrium constants are

$$K_1 = \frac{[AB]}{[A][B]} \qquad K_2 = \frac{[A^*B]}{[A][B]} \qquad K_3 = \frac{[AB^*]}{[A][B]} \qquad K_4 = \frac{[A^*B^*]}{[A][B]}$$

We now note that three independent ratios exist, which are independent of [B]:

$$\frac{K_1}{K_2} = \frac{[AB]}{[A^*B]} \qquad \frac{K_2}{K_3} = \frac{[A^*B]}{[AB^*]} \qquad \frac{K_3}{K_4} = \frac{[AB^*]}{[A^*B^*]}$$

Therefore, an isosbestic point is expected if $[A]_0$ is held constant and $[B]_0$ is varied. Any other ratio of equilibrium constants, e.g., K_1/K_3, is not independent of the three ratios written. The general rules presented here apply to a large number of systems. However, rote application of rules is no substitute for an understanding of the systems under consideration.

4-8 JOB'S METHOD OF ISOMOLAR SOLUTIONS

By examining the spectra of a series of solutions of widely varying mole ratios of A to B, but with the same total number of moles of A + B, the stoichiometry of complexes formed between A and B in solution can often be determined. Absorbance at a wavelength of maximum change is plotted against the mole ratio of A to B; the latter is usually used as the abscissa. The plots have at least one extremum, often a maximum. In simple cases, the extrema occur at mole fractions corresponding to the stoichiometry of the complexes that form in solution.[11-13]

4-9 "FINGERPRINTING"

This technique is useful for the identification of an unknown compound that is suspected to be the same as a known compound. The spectra are compared with respect to ε values, wavelengths of maximum absorption, and band shapes.

In addition to this direct comparison, certain functional groups have characteristic absorptions in various regions of the spectrum. For example, the carbonyl group will generally absorb at certain wavelengths in the ultraviolet and infrared spectra. Its presence in an unknown compound can be determined from these absorptions. Often one can even determine whether or not the carbonyl group is in a conjugated system. These details will be considered later when the spectroscopic methods are discussed individually.

Spectroscopic methods provide a convenient way of detecting certain impurities in a sample. For example, the presence of water in a system can easily be detected by its characteristic infrared absorption. Similarly, a product can be tested for absence of starting material if the starting material has a functional group with a characteristic absorption that disappears during the reaction. Spectral procedures are speedier and less costly than elemental analyses. The presence of contaminants in small amounts can be detected if their molar absorptivities are large enough.

REFERENCES CITED

1. E. F. Johnson and R. S. Drago, J. Amer. Chem. Soc., 95, 1391 (1973), and references therein.
2. N. J. Rose and R. S. Drago, J. Amer. Chem. Soc., 81, 6138 (1959).
3. T. D. Epley and R. S. Drago, J. Amer. Chem. Soc., 91, 2883 (1969), and references therein.
4. T. O. Maier and R. S. Drago, Inorg. Chem., 11, 1861 (1972).
5. P. J. Lingane and Z. Z. Hugus, Jr., Inorg. Chem., 9, 757 (1970).
6. R. M. Guidry and R. S. Drago, J. Amer. Chem. Soc., 95, 6645 (1973).
7. F. L. Slejko, R. S. Drago, and D. G. Brown, J. Amer. Chem. Soc., 94, 9210 (1972).
8. T. F. Bolles and R. S. Drago, J. Amer. Chem. Soc., 87, 5015 (1965)..
9. R. L. Carlson, K. F. Purcell, and R. S. Drago, Inorg. Chem., 4, 15 (1965).
10. R. G. Mayer and R. S. Drago, Inorg. Chem., 15, 2010 (1976).
11. W. Likussar and D. F. Boltz, Anal. Chem., 43, 1265 (1971).
12. M. M. Jones and K. K. Innes, J. Phys. Chem., 62, 1005 (1958).
13. E. Asmus, Z. Anal. Chem., 183, 321, 401 (1961).

EXERCISES

1. A series of different molecular transitions require the following energies in order to occur. Indicate the spectral region in which you would expect absorption of radiation to occur, and the wavelength of the radiation

   a. 0.001 kcal mole^{-1}.

   b. 100 kcal mole^{-1}.

   c. 30 kcal mole^{-1}.

2. Convert the wavelength units in part a of exercise 1 to wavenumbers (cm^{-1}).

3. Convert the following wavenumbers to μ and Å:

   a. 3600 cm^{-1}.

   b. 1200 cm^{-1}.

4. Convert 800 mμ to cm^{-1}.

5. The ε value for compound X is 9000 liters mole^{-1} cm^{-1}. A 0.1 molar solution of X in water, measured in a 1 cm cell, has an absorbance of 0.542. It is known that X reacts according to the equation:

$$X \rightleftharpoons Y + Z$$

   X obeys Beer's law, and Y and Z do not absorb in this region. What is the equilibrium constant for this reaction?

6. For the equilibrium

$$A + B \rightleftharpoons C$$

   assume that only C absorbs and that ε_C cannot be measured. Derive an equation similar to (4–22) for this system.

7. **a.** Use equations (4–14) and (4–15) and values of $A_1 = 0.3$, $A_2 = 0.7$, $\varepsilon_{B_{\lambda_1}} = 5$, $\varepsilon_{B_{\lambda_2}} = 12$, $\varepsilon_{C_{\lambda_1}} = 13$, and $\varepsilon_{C_{\lambda_2}} = 2$ to construct a plot similar to that in Fig. 4–6 to illustrate the graphical solution of the two simultaneous equations.

b. Construct the same plot as in part a, using $A_1 = 0.3$, $A_2 = 0.7$, $\varepsilon_{B_{\lambda_1}} = 5$, $\varepsilon_{B_{\lambda_2}} = 12$, $\varepsilon_{C_{\lambda_1}} = 3$, and $\varepsilon_{C_{\lambda_2}} = 5$. Compare your confidence in the two answers, realizing that there are 2 to 3% error in the A and ε values.

c. Equation (4–22) is somewhat analogous to that used in parts a and b, for the first term is generally small under the conditions used in a spectroscopic study. Often investigators work in a concentration range in which, holding $[D]_0$ constant and doubling $[E]_0$, they double the amount of complex formed or they double $A - A^0$. What would the two lines in the K^{-1} vs. ε plot look like?

8. A proton is being rapidly exchanged between B and B', i.e.,

$$BH^+ + B' \rightleftharpoons B'H^+ + B$$

The frequency difference between the H^+ peaks of BH^+ and $B'H^+$ in the nmr spectrum is 100 Hz (cycles per second).

a. What rate of exchange would be required for the two peaks to appear as a single peak?

b. Would a slower or faster rate be required to observe two separate resonances that are broadened by the exchange process?

c. What is the approximate maximum rate of chemical exchange that could occur and still have no effect on the nmr spectrum?

9. A system $A + B \rightleftharpoons C + D$ is investigated spectroscopically. Species A, C, and D all have absorbances that overlap. Since there are more than two absorbing species, would one expect to find an isosbestic point if this were the only equilibrium involved? Prove your answer.

Electronic Absorption Spectroscopy 5

Introduction

5–1 VIBRATIONAL AND ELECTRONIC ENERGY LEVELS IN A DIATOMIC MOLECULE

Prior to a discussion of electronic absorption spectroscopy, the information summarized by a potential energy curve for a diatomic molecule (indicated in Fig. 5–1) will be reviewed. Fig. 5–1 is a plot of E, the total energy of the system, *versus r*, the internuclear distance, and is one of many types of potential functions referred to as a Morse potential. The curve is expressed mathematically by

$$V = D\{1 - \exp[-v_0(2\pi^2\mu D)^{1/2}(r - r_e)]\}^2$$

All terms are defined in Fig. 5–1 except μ, which is the reduced mass, $(m_1 m_2)/(m_1 + m_2)$.

As the bond distance is varied in a given vibrational state, *e.g.*, along A–B in the v_2 level, the molecule is in a constant vibrational energy level. At A and B we have, respectively, the minimum and maximum values for the bond distance in this vibrational level. At these points the atoms are changing direction, so the vibrational kinetic energy is zero and the total vibrational energy of the system is potential. At r_e, the equilibrium internuclear distance, the vibrational kinetic

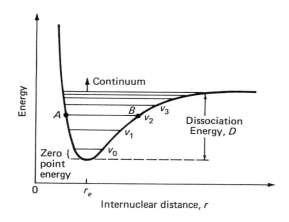

FIGURE 5–1 Morse energy curve for a diatomic molecule.

energy is a maximum and the vibrational potential energy is zero. Each horizontal line represents a different vibrational energy state. The ground state is v_0 and excited states are represented as v_1, v_2, etc. Eventually, if enough energy can be absorbed in vibrational modes, the molecule is excited into the continuum and it dissociates. For most compounds, nearly all the molecules are in the v_0 level at room temperature because the energy difference $v_1 - v_0$ is usually much larger than kT (thermal energy), which has a value of 200 cm^{-1} at 300 K. (Recall the Boltzmann expression.)

Each excited electronic state also contains a series of different vibrational energy levels and may be represented by a potential energy curve. The ground electronic state and one of the many excited electronic states for a typical diatomic molecule are illustrated in Fig. 5–2. Each vibrational level, v_n, is described by a vibrational wave function, ψ_{vib}. For simplicity only four levels are indicated. The square of a wave function gives the probability distribution, and in this case ψ_{vib}^2 indicates probable internuclear distances for a particular vibrational state. This function, ψ_{vib}^2, is indicated for the various levels by the dotted lines in Fig. 5–2. The dotted line is not related to the energy axis. The higher this line, the more probable the corresponding internuclear distance. The most probable distance for a molecule in the ground state is r_e, while there are two most probable distances corresponding to the two maxima in the next vibrational energy level of the ground electronic state, three in the third, etc. In the excited vibrational levels of the ground and excited electronic states, there is high probability of the molecule having an internuclear distance at the ends of the potential function.

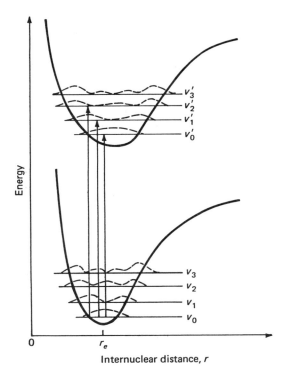

FIGURE 5–2 Morse curves for a ground and excited state of a diatomic molecule, showing vibrational probability function, ψ_{vib}^2, as dotted lines.

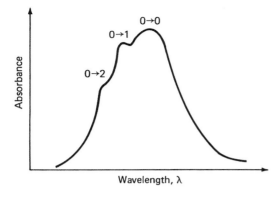

FIGURE 5–3 Spectrum corresponding to the potential energy curves indicated in Fig. 5-2.

5–2 RELATIONSHIP OF POTENTIAL ENERGY CURVES TO ELECTRONIC SPECTRA

An understanding of electronic absorption spectroscopy requires consideration of three additional principles:

1. In the very short time required for an electronic transition to take place (about 10^{-15} sec), the atoms in a molecule do not have time to change position appreciably. This statement is referred to as the *Franck-Condon principle*.* Since the electronic transition is rapid, the molecule will find itself with the same molecular configuration and vibrational kinetic energy in the excited state that it had in the ground state at the moment of absorption of the photon. As a result, all electronic transitions are indicated by a vertical line on the Morse potential energy diagram of the ground and excited states (see the arrows in Fig. 5–2); *i.e.*, there is no change in internuclear distance during the transition.

2. There is no general selection rule that restricts the change in vibrational state accompanying an electronic transition. Frequently transitions occur from the ground vibrational level of the ground electronic state to many different vibrational levels of a particular excited electronic state. Such transitions may give rise to vibrational fine structure in the broader peak of the electronic transition.

The three transitions indicated by arrows in Fig. 5–2 could give rise to three peaks. Since nearly all of the molecules are present in the ground vibrational level, nearly all transitions that give rise to a peak in the absorption spectrum will arise from v_0.† Transitions from this ground level (v_0) to v_0', v_1', or v_2' are referred to as $0 \rightarrow 0$, $0 \rightarrow 1$, or $0 \rightarrow 2$ transitions, respectively. It can be shown[1] that the relative intensity of the various vibrational sub-bands depends upon the vibrational wave function for the various levels. A transition is favored if the

---

*This principle was originally proposed by Franck. It was given a quantum mechanical interpretation and extended by Condon.

†In the gas phase, various rotational levels in the ground vibrational state will be populated and transitions to various rotational levels in the excited state will occur, giving rise to fine structure in the spectrum. This fine structure is absent in solution because collision of the solute with a solvent molecule occurs before a rotation is completed. Rotational fine structure will not be discussed.

probabilities of the ground and excited states of the molecule are both large for the same internuclear distance. Three such transitions are indicated by arrows in Fig. 5–2. The spectrum in Fig. 5–3 could result from a substance in solution undergoing the three transitions indicated in Fig. 5–2. The 0 → 0 transition is the lowest energy–longest wavelength transition. The differences in wavelength at which the peaks occur represent the energy differences of the vibrational levels in the excited state of the molecule. Much information about the structure and configuration of the excited state can be obtained from the fine structure.

Electronic transitions from bonding to antibonding molecular orbitals are often encountered. In this case the potential energy curve for the ground state will be quite different from that of the excited state because there is less bonding electron density in the excited state. As a result, the equilibrium internuclear distance will be greater and the potential energy curve will be broader for the excited state. Because of this displacement of the excited state potential energy curve, the 0 → 0 and transitions to other low vibrational levels may not be observed. A transition to a higher vibrational level becomes more probable. This can be visualized by broadening and displacing the excited state represented in Fig. 5–2.

3. There is an additional *symmetry requirement*, which was neglected for the sake of simplicity in the above discussion. It has been assumed that this symmetry requirement is satisfied for the transitions involved in this discussion. This will be discussed subsequently in the section on selection rules.

The above discussion pertains to a diatomic molecule, but the general principles also apply to a polyatomic molecule. Often the functional group in a polyatomic molecule can be treated as a diatomic molecule (for example, $>$C=O in a ketone or aldehyde). The electronic transition may occur in the functional group between orbitals that are approximated by a combination of atomic orbitals of the two atoms, as in a diatomic molecule. The actual energies of the resulting molecular orbitals of the functional group will, of course, be affected by electronic, conjugative, and steric effects arising from the other atoms. This situation can be understood qualitatively in terms of potential energy curves similar to those discussed for the diatomic molecules. For more complex cases in which several atoms in the molecule are involved (*i.e.*, a delocalized system), a polydimensional surface is required to represent the potential energy curves.

The energies required for electronic transitions generally occur in the far u.v., u.v., visible, and near infrared regions of the spectrum, depending upon the energies of the molecular orbitals in a molecule. For molecules that contain only strong sigma bonds (*e.g.*, CH_4 and H_2O), the energy required for electronic transitions occurs in the far u.v. region of the spectrum and requires specialized instrumentation for detection. In fact, in the selection of suitable transparent solvents for study of the u.v. spectrum of the dissolved solute, such transitions come into play in determining the cutoff point of the solvent. On the other hand, the colored dyes used in clothing have extensively conjugated pi-systems and undergo electronic transitions in the visible region of the spectrum. Standard instruments cover the region from 50,000 cm^{-1} to 5,000 cm^{-1}. This spectral region is subdivided as follows:

Ultraviolet	50,000 cm^{-1} to 26,300 cm^{-1}	(2000 to 3800 Å)
Visible	26,300 cm^{-1} to 12,800 cm^{-1}	(3800 to 7800 Å)
Near infrared	12,800 cm^{-1} to 5,000 cm^{-1}	(7800 to 20,000 Å)

5–3 NOMENCLATURE

In our previous discussion we were concerned only with transitions of an electron from a given ground state to a given excited state. In an actual molecule there are electrons in different kinds of orbitals (σ bonding, non-bonding, π bonding) with different energies in the ground state. Electrons from these different orbitals can be excited to higher-energy molecular orbitals, giving rise to many possible excited states. Thus, many transitions from the ground state to different excited states (each of which can be described by a different potential energy curve) are possible in one molecule.

There are several conventions[2,3] used to designate these different electronic transitions. A simple representation introduced by Kasha[2] will be illustrated for the carbonyl group in formaldehyde. A molecular orbital description of the valence electrons in this molecule is:

$$\sigma^2 \quad \pi^2 \quad n_a^2 \quad n_b^2 \quad (\pi^*)^0 \quad (\sigma^*)^0$$

The n_a and n_b orbitals are the two non-bonding molecular orbitals containing the lone pairs on oxygen. Symmetry considerations do not require the lone pairs to be degenerate, for there are no doubly degenerate irreducible representations in C_{2v}. They are not accidentally degenerate either, but differ in energy. The relative energies of these orbitals are indicated in Fig. 5–4. The ordering of these orbitals can often be arrived at by intelligent guesses and by looking at the spectra of analogous compounds. Also indicated with arrows are some transitions chosen to illustrate the nomenclature. The transitions (1), (2), (3), and (4) are referred to as $n \rightarrow \pi^*$, $n \rightarrow \sigma^*$, $\pi \rightarrow \pi^*$, and $\sigma \rightarrow \sigma^*$, respectively. The $n \rightarrow \pi^*$ transition is the lowest energy–highest wavelength transition that occurs in formaldehyde and most carbonyl compounds.

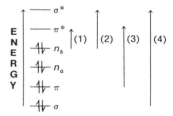

FIGURE 5–4 Relative energies of the carbonyl molecular orbitals in H_2CO.

Electron excitations can occur with or without a change in the spin of the electron. If the spin is not changed in a molecule containing no unpaired electrons, both the excited state and ground state have a multiplicity of one; these states are referred to as singlets. The *multiplicity* is given by two times the sum of the individual spins, m_s, plus one: $2S + 1 = 2\Sigma m_s + 1$. If the spin of the electron is changed in the transition, the excited state contains two unpaired electrons with identical magnetic spin quantum numbers, has a multiplicity of 3, and is referred to as a triplet state.

There are some shortcomings of this simple nomenclature for electronic transitions. It has been assumed that these transitions involve a simple transfer of an electron from the ground state level to an empty excited state level of our ground state wave function. For many applications, this description is precise enough. In actual fact, such transitions occur between states, and the excited state is not actually described by moving an electron into an empty molecular orbital of the ground state. The excited state has, among other things, different electron-electron repulsions than those in the system represented by simple excitation of an electron into an empty ground state orbital. In most molecules, various kinds of electron-electron interactions in the excited state complicate the problem; in addition to affecting the energy, they give rise to many more transitions than predicted by the simple picture of electron promotion because levels that would otherwise be degenerate are split by electronic interactions. This problem is particularly important in transition metal ion complexes. We shall return to discuss this problem more fully in the section on configuration interaction.

In a more accurate[4] system of nomenclature the symmetry, configuration, and multiplicity of the states involved in the transition are utilized in describing the transition. This system of nomenclature can be briefly demonstrated by again considering the molecular orbitals of formaldehyde. The diagrams in Fig. 5-5 qualitatively represent the boundary contours of the molecular orbitals. The solid line encloses the positive lobe and the dashed line the negative lobe. The larger π and π^* lobes indicate lobes above the plane of the paper, and the smaller ones represent those below the plane; the two lobes actually have identical sizes. To classify these orbitals it is necessary first to determine the overall symmetry of the molecule, which is C_{2v}. The next step is to consult the C_{2v} character table.

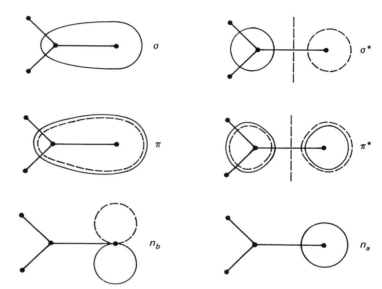

FIGURE 5–5 Shapes of the molecular orbitals of formaldehyde.

Character tables for several common point groups are listed in Appendix A. The C_{2v} character table is duplicated here in Table 5–1. By convention, the yz-plane is selected to contain the four atoms of formaldehyde. The symmetry operations E, C_2, $\sigma_{v(xz)}$, and $\sigma'_{v\,(yz)}$ performed on the π orbital produce the result $+1$, -1, $+1$, -1. This result is identical to that listed for the irreducible representation B_1 in the table. The orbital is said to belong to (or to transform as) the symmetry species b_1, the lowercase letter being employed for an orbital and the uppercase

TABLE 5–1. Character Table for the C_{2v} Point Group[a]

	E	C_2	$\sigma_{v(xz)}$	$\sigma'_{v\,(yz)}$	
A_1	1	1	1	1	z
A_2	1	1	-1	-1	R_z
B_1	1	-1	1	-1	x, R_y
B_2	1	-1	-1	1	y, R_x

[a] The x-axis is perpendicular to the plane of the molecule.

letters being reserved to describe the symmetry of the entire ground or excited state. Similarly, if the n_a, n_b, π^*, σ, and σ^* orbitals of formaldehyde are subjected to the above symmetry operations, it can be shown that these orbitals belong to the irreducible representations a_1, b_2, b_1, a_1, and a_1, respectively. The two n orbitals can be viewed as s and p_y orbitals (p_z is used in the σ bond and p_x in the π). As a result, they lie in the yz-plane and possess a_1 and b_2 symmetry. (The a_1 s-orbital can mix with p_z in the sigma bonding.) The difference in s-orbital character causes the energies of the a_1 and b_2 lone pairs to differ. Molecular orbital calculations are consistent with these ideas. As mentioned in Chapter 2, a and b indicate single degeneracy. The a representation does not change sign with rotation about the n-fold axis, but b does.

The symmetry species of a state is the product of the symmetry species of each of the odd electron orbitals. In the state that results from the $n \to \pi^*$ transition in formaldehyde, there is one unpaired electron in the n-orbital with b_2 symmetry and one in the π^*-orbital with b_1 symmetry. The direct product is given by:

$$\begin{array}{c|cccc} & E & C_2 & \sigma_{v(xz)} & \sigma'_{v\,(yz)} \\ \hline b_1 \times b_2 = & (1)(1) & (-1)(-1) & (1)(-1) & (-1)(1) \\ \text{result} = & +1 & +1 & -1 & -1 \end{array} = A_2$$

The resulting irreducible representation is A_2.* The excited state from this transition is thus described as A_2 and the transition as $A_1 \to A_2$. A common convention involves writing the high energy state first and labeling the transition $A_2 \leftarrow A_1$. The spin multiplicity is usually included, so the complete designation becomes $^1A_2 \leftarrow {}^1A_1$. The ground state is A_1 because there is a pair of electrons in each orbital. Commonly, the orbitals involved in the transitions are indicated, and the symbol becomes $^1A_2(n, \pi^*) \leftarrow {}^1A_1$. If a non-specific, general symbol is needed for a state symmetry species, Γ is employed.

Instead of representing the orbitals of formaldehyde symbolically, as in Fig. 5–5, we could simply have been given the wave functions. We can deduce the symmetry from ψ by converting the wave functions into a physical picture. The following equations describe the formaldehyde π and π^* orbitals:

$$\psi_\pi = a\varphi_{p^\circ} + b\varphi_{p^c}$$
$$\psi_{\pi^*} = b'\varphi_{p^\circ} - a'\varphi_{p^c}$$

where φ_{p° and φ_{p^c} are the wave functions for the atomic oxygen and carbon p-orbitals, respectively. The atomic orbitals are mathematically combined to produce the π and π^* orbitals. Since oxygen is more electronegative (*i.e.*, $a > b$), it becomes clear why it is often stated that an electron is transferred from oxygen to carbon in the $\pi \to \pi^*$ transition. We shall shortly return to a discussion of the experimental spectrum of formaldehyde.

---

*The actual procedure for determining the symmetry of this state involves multiplication of the orbital symmetries for all the electrons in the carbonyl group. However, all filled orbitals contain two electrons whose product must be A_1. Thus, paired electrons have no effect on the final result for the excited state symmetry.

Some of the molecular orbitals for benzene are represented mathematically and pictorially in Fig. 5–6. Notice that a difference in sign between adjacent atomic orbitals of the wave function represents a node (point of zero probability) in the molecular orbital. Using the D_{6h} character table, the symmetries of the orbitals can be shown to be a_{2u}, e_{1g}, e_{2u}, and b_{2g}, for ψ_1, $\psi_2 + \psi_3$, $\psi_4 + \psi_5$ and ψ_6, respectively.

In addition to the above conventions used to label ground and excited states, a convention used for diatomic molecules will be described. With this terminology the electronic arrangement is indicated by summing up the contributions of the separate atoms to obtain the net orbital angular momentum. If all electrons are paired, the sum is zero. Contributions are counted as follows: one unpaired electron in a σ orbital is zero, one in a π orbital one, and one in a δ orbital two. For more than one electron, the total is $|\Sigma m_l|$. If the total is zero, the state is described as Σ, one as Π, and two as Δ. The multiplicity is indicated by a superscript, e.g., the ground state of NO is $^2\Pi$. A plus or minus sign often follows the symbol to illustrate, respectively, whether the molecular orbitals are symmetric or antisymmetric to a plane through the molecular axis.

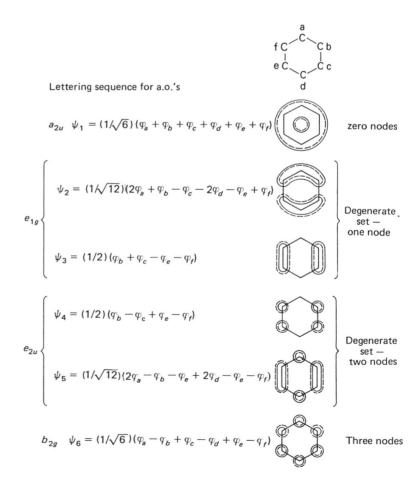

FIGURE 5–6 Shapes of the benzene molecular orbitals.

Assignment of Transitions

If the energies of electronic transitions were related to the ground state molecular orbital energies, the assignment of transitions to the observed bands would be simple. In formaldehyde (Fig. 5–4) the $n \rightarrow \pi^*$ would be lower in energy than the $\pi \rightarrow \pi^*$, which in turn would be lower in energy than $\sigma \rightarrow \pi^*$. In addition to different electron-electron repulsions in different states, two other effects complicate the picture by affecting the energies and the degeneracy of the various excited states. These effects are *spin-orbit coupling* and *higher-state mixing*.

5–4 SPIN-ORBIT COUPLING

There is a magnetic interaction between the electron spin magnetic moment (signified by quantum number $m_s = \pm^1/_2$) and the magnetic moment due to the orbital motion of an electron. To understand the nature of this effect, consider the nucleus as though it were moving about the electron (this is equivalent to being on earth and thinking of the sun moving across the sky). We consider the motion from this reference point because we are interested in effects at the electron. The charged nucleus circles the electron, and this is equivalent in effect to placing the electron in the middle of a coil of wire carrying current. As moving charge in a solenoid creates a magnetic field in the center, the orbital motion described above causes a magnetic field at the electron position. This magnetic field can interact with the spin magnetic moment of the electron, giving rise to spin-orbit interaction. The orbital moment may either complement or oppose the spin moment, giving rise to two different energy states. The doubly degenerate energy state of the electron (previously designated by the spin quantum numbers $\pm^1/_2$) is split, lowering the energy of one and raising the energy of the other. Whenever an electron can occupy a set of degenerate orbitals that permit circulation about the nucleus, this interaction is possible. For example, if an electron can occupy the d_{yz} and d_{xz} orbitals of a metal ion, it can circle the nucleus around the z-axis. (A more complete discussion will be given in Chapter 11.)

5–5 CONFIGURATION INTERACTION

As mentioned before, electronic transitions do not occur between empty molecular orbitals of the ground state configuration, but between states. The energies of these states are different from those of configurations derived by placing electrons in the empty orbitals of the ground state, because electron-electron repulsions in the excited state differ from those in the simplified, "ground state orbital description" of the excited state. A further complication arises from configuration interaction. One could attempt to account for the different electron-electron repulsions in the excited state by doing a molecular orbital calculation on a molecule having the ground state geometry but with the electron arrangement of the excited configuration. This would not give the correct energy of the state because this configuration can mix with all of the other configurations in the molecule of the same symmetry by configuration interaction.

This mixing is similar in a mathematical sense (though much smaller in magnitude) to the interaction of two hydrogen atoms in forming the H_2 molecule.

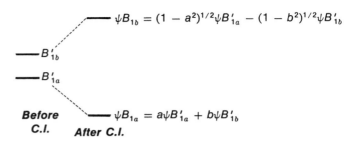

FIGURE 5–7 Energy levels before and after configuration interaction.

Thus, we could write a secular determinant to account for the mixing of two B_1 states, B'_{1a} and B'_{1b}, as:

$$\begin{vmatrix} E_1^0 - E & H_{12} \\ H_{12} & E_2^0 - E \end{vmatrix} = 0$$

where H_{12} is the difficult-to-solve integral $\int \psi(B_{1a}) \hat{H} \psi(B_{1b}) \, d\tau$, whose value is dependent upon the interelectron repulsions in the various states. The closer in energy the initial states E_1^0 and E_2^0, the more mixing occurs. Solution of the secular determinant gives us the two new energies after mixing:

$$E_1 = \tfrac{1}{2}[E_1^0 + E_2^0 + ((E_1^0)^2 + (E_2^0)^2 - 2E_1^0 E_2^0 + 4H_{12}^2)^{1/2}]$$
$$E_2 = \tfrac{1}{2}[E_1^0 + E_2^0 - ((E_1^0)^2 + (E_2^0)^2 - 2E_1^0 E_2^0 + 4H_{12}^2)^{1/2}]$$

The energies of the initial (B'_{1a}, B'_{1b}) and final (B_{1a}, B_{1b}) states are illustrated in Fig. 5–7 along with the new wave functions to describe the final state. The wave function is seen to be a linear combination of the two initial molecular orbitals. Interactions of this sort can occur with all the molecular orbitals of B_1 symmetry in the molecule, complicating the problem even further than is illustrated in Fig. 5–7.

5–6 CRITERIA TO AID IN BAND ASSIGNMENT

An appreciation for some of the difficulties encountered in assigning transitions can be obtained from reading the literature[5] and noting the changes in the assignments that have been made over the years. Accordingly, many independent criteria are used in making the assignment. These include the intensity of the transition and the behavior of the absorption band when polarized radiation is employed. Both of these topics will be considered in detail shortly. Next we shall describe some simple observations that aid in the assignment of the $n \to \pi^*$ and $\pi \to \pi^*$ transitions.

For $n \to \pi^*$ transitions, one observes the following characteristics:

1. The *molar absorptivity* of the transition is generally less than 2000. An explanation for this is offered in the section on intensity.
2. A *blue shift* (hypsochromic shift, or shift toward shorter wavelengths) is

observed for this transition in high dielectric or hydrogen-bonding solvents. This indicates that the energy difference between the ground and excited state is increased in a high dielectric or hydrogen-bonding solvent. In general, for solvent shifts it is often difficult to ascertain whether the excited state is raised in energy or the ground state lowered. A blue shift may result from a greater lowering of the ground state relative to the excited state or a greater elevation of the excited state relative to the ground state. It is thought that the solvent shift in the $n \to \pi^*$ transition results from both a lowering of the energy of the ground state and an elevation of the energy of the excited state. In a high dielectric solvent the molecules arrange themselves about the absorbing solute so that the dipoles are properly oriented for maximum interaction (*i.e.*, solvation that lowers the energy of the ground state). When the excited state is produced, its dipole will have an orientation different from that of the ground state. Since solvent molecules cannot rearrange to solvate the excited state during the time of a transition, the excited state energy is raised in a high dielectric solvent.[6]

Hydrogen bonding solvents cause pronounced blue shifts. This is reported to be due to hydrogen bonding of the solvent hydrogen with the lone pair of electrons in the *n* orbital undergoing the transition. In the excited state there is only one electron in the *n* orbital, the hydrogen bond is weaker and, as a result, the solvent does not lower the energy of this state nearly as much as that of the ground state. In these hydrogen bonding systems an adduct is formed, and this specific solute-solvent interaction is the main cause of the blue shift.[7] If there is more than one lone pair of electrons on the donor, the shift can be accounted for by the inductive effect of hydrogen bonding to one electron pair on the energy of the other pair.

3. The $n \to \pi^*$ band often disappears in acidic media owing to protonation or upon formation of an adduct that ties up the lone pair, *e.g.*, $BCH_3^+I^-$ (as in $C_5H_5NCH_3^+I^-$), where B is the base molecule containing the *n* electrons. This behavior is very characteristic if there is only one pair of *n* electrons on B.

4. Blue shifts occur upon the attachment of an electron-donating group to the chromophore. For example in the series

$$CH_3C(O)H < CH_3C(O)CH_3 < CH_3C(O)OCH_3 < CH_3C(O)N(CH_3)_2$$

an increasing blue shift is observed in the carbonyl absorption band. A molecular orbital treatment[8] indicates that this shift results from raising the excited π^* level relative to the *n* level.

5. The absorption band corresponding to the $n \to \pi^*$ transition is absent in the hydrocarbon analogue of the compound. This would involve, for example, comparison of the spectra of benzene and pyridine or of $H_2C=O$ and $H_2C=CH_2$.

6. Usually, but not always, the $n \to \pi^*$ transition gives rise to the lowest energy singlet-singlet transition.

In contrast to the above behavior, $\pi \to \pi^*$ transitions have a high intensity. A slight red (bathochromic) shift is observed in high dielectric solvents and upon introduction of an electron-donating group. It should be emphasized that in the above systems only the difference in energy between the ground and excited states can be measured from the frequency of the transition, so only the relative energies of the two levels can be measured. Other considerations must be invoked to determine the actual change in energy of an individual state.

The Intensity of Electronic Transitions

5–7 OSCILLATOR STRENGTHS

As mentioned in the previous chapter, the intensity of an absorption band can be indicated by the molar absorptivity, commonly called the extinction coefficient. A parameter of greater theoretical significance is f, the *oscillator strength* of integrated intensity, often simply called the *integrated intensity*:

$$f = 4.315 \times 10^{-9} \int \varepsilon \, d\bar{v} \qquad (5\text{–}1)$$

In equation (5–1), ε is the molar absorptivity and $\bar{v}$ is the frequency expressed in wave numbers. The concept of oscillator strength is based on a simple classical model for an electronic transition. The derivation* indicates that $f = 1$ for a fully allowed transition. The quantity f is evaluated graphically from equation (5–1) by plotting ε, on a linear scale, *versus* the wavenumber $\bar{v}$ in cm^{-1}, and calculating the area of the band. Values of f from 0.1 to 1 correspond to molar absorptivities in the range from 10,000 to 100,000, depending on the width of the peak.

For a single, symmetrical peak, f can be approximated by the expression:

$$f \approx (4.6 \times 10^{-9}) \varepsilon_{max} \, \Delta v_{1/2} \qquad (5\text{–}2)$$

where ε_{max} is the molar absorptivity of the peak maximum and $\Delta v_{1/2}$ is the half intensity bandwidth, *i.e.*, the width at $\varepsilon_{max}/2$.

5–8 TRANSITION MOMENT INTEGRAL

The integrated intensity, f, of an absorption band is related to the transition moment integral as follows:

$$f \propto \left| \int_{-\infty}^{+\infty} \psi_{el} \hat{M} \psi_{el}^{ex} \, dv \right|^2 = D \qquad (5\text{–}3)$$

where D is called the dipole strength, ψ_{el} and ψ_{el}^{ex} are electronic wave functions for the ground and excited states respectively, $\hat{M}$ is the electric dipole moment operator (*vide infra*), and the entire integral is referred to as the *transition moment integral*. To describe $\hat{M}$, one should recall that the electric dipole moment is defined as the distance between the centers of gravity of the positive and negative charges times the magnitude of these charges. The center of gravity of the positive charges in a molecule is fixed by the nuclei, but the center of gravity of the electrons is an average over the probability function. The vector for the average distance from the nuclei to the electron is represented as $\vec{r}$. The electric dipole moment vector, $\vec{M}$, is given by $\vec{M} = \Sigma e \vec{r}$, with the summation carried out over

---

*For this derivation, see Additional References, Barrow, pages 80 and 81.

all the electrons in the molecule. For the ground state, the electric dipole moment is given by the average over the probability function, or

$$\int \psi_g \sum e \vec{r} \psi_g \, d\tau$$

By comparison of this equation for the ground state dipole moment with the transition moment integral:

$$\int \psi_g \hat{M} \psi^{ex} \, d\tau$$

this integral can be seen roughly to represent charge migration or displacement during the transition.

When the integral in equation (5–3) is zero, the intensity will be zero and, to a first approximation, the transition will be forbidden. In general, we do not have good state wave functions to substitute into this equation to calculate the intensity, for reasons discussed in the previous section. However, symmetry can tell us if integrals are zero. It is important to examine the symmetry properties of the integrand of equation (5–3), for this enables us to make some important predictions. These symmetry considerations will also lead to selection rules for electronic transitions. The quantity $\hat{M}$ is a vector quantity and can be resolved into x, y, and z components. The integral in equation (5–3) then has the components:

$$\int \psi_{el} \hat{M}_x \psi_{el}^{ex} \, dv \qquad (5\text{–}4)$$

$$\int \psi_{el} \hat{M}_y \psi_{el}^{ex} \, dv \qquad (5\text{–}5)$$

$$\int \psi_{el} \hat{M}_z \psi_{el}^{ex} \, dv \qquad (5\text{–}6)$$

In order to have an allowed transition, at least one of the integrals in equations (5–4) to (5–6) must be non-zero. If all three of these integrals are zero, the transition is called forbidden and, according to approximate theory, should not occur at all. Forbidden transitions do occur; more refined theories (discussed in the section on Spin-Orbit and Vibronic Coupling) give small values to the intensity integrals.

Symmetry considerations can tell us if the integral is zero in the following way. An integral can be non-zero only if the direct product of the integrand belongs to symmetry species A_1. Another way of saying this is that such integrals are different from zero only when the integrand remains unchanged for any of the symmetry operations permitted by the symmetry of the molecule. The reason for the above statements can be seen by looking at some simple mathematical functions. First, consider a plot of the curve $y = x$, shown in Fig. 5–8(A). Symmetry operations of the C_{2v} point group on this figure tell us that this function does not have A_1 symmetry. The integral $\int y(x) \, dx$ represents the area under the curve. Since $y(x)$ is positive in one quadrant and negative in the other, the area from

$+a$ to $-a$ or from $+\infty$ to $-\infty$ is zero. Now, plot a function $y(x)$ such that y is symmetric in x. One of many such functions is $y = x^2$, shown in Fig. 5–8(B). This function is called "even" and does have A_1 symmetry, because it is unchanged by any of the operations of the symmetry group to which the function belongs. This can be shown by carrying out the symmetry operations of the C_{2v} point group on Fig. 5–8(B). The integral does not vanish. The shaded area gives the value for the integral $\int_{-a}^{a} y\,dx$ for $y = x^2$, and it is seen not to cancel to zero. Since y is a function of x^2, we can take the direct product of the irreducible representation of x with itself, $\Gamma_x \times \Gamma_x$, and reduce this to demonstrate that this is a totally symmetric irreducible representation. Next, the function $y = x^3$, in Fig. 5–8(C), will be examined. Symmetry considerations tell us that this is an odd function, and we see that $\int y\,dx$ (for $y = x^3$) $= 0$. Does $\Gamma_x \times \Gamma_x \times \Gamma_x = A_1$? The integral over all space of an odd function always vanishes, as shown for the examples in Figs. 5–8(A) and 5–8(C). The integral over all space of an even function *generally* does not vanish.

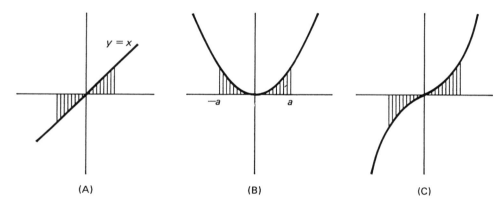

FIGURE 5–8 Plots of some simple functions, $y = f(x)$, and their symmetries.

Thus, to determine whether or not the integral is zero, we take the direct product of the irreducible representations of everything in the integrand. If the direct product is or contains A_1, the integral is non-zero and the transition is allowed.

The application of these ideas is best illustrated by treating an example. Consider the $\pi \rightarrow \pi^*$ transition in formaldehyde. The ground state, like all ground states containing no unpaired electrons, is A_1. The excited state is also A_1 ($b_1 \times b_1 = A_1$). The components $\hat{M}_x$, $\hat{M}_y$, and $\hat{M}_z$ transform as the x, y, and z-vectors of the point group. The table for the C_{2v} point group indicates that $\hat{M}_z$, the dipole moment vector lying along the z-axis, is A_1. Since $A_1 \times A_1 \times A_1 = A_1$, the integrand $\psi_{el}\hat{M}_z\psi_{el}^{ex}$ for the $\pi \rightarrow \pi^*$ is A_1, and the $\pi \rightarrow \pi^*$ transition is allowed.

For an $n \rightarrow \pi^*$ transition, the ground state is A_1 and the excited state is A_2. The character table indicates that no dipole moment component has symmetry A_2. Therefore, none of the three integrals [equations (5–4) to (5–6)] can be A_1, and the transition is forbidden ($A_2 \times A_2$ is the only product of A_2 that is A_1).

As was mentioned when we introduced this topic, the transition moment integral can be used to derive some important selection rules for electronic transitions.

5–9 DERIVATION OF SOME SELECTION RULES

1. For molecules with a center of symmetry, allowed transitions are $g \to u$ or $u \to g$. (The abbreviations g and u refer to *gerade* and *ungerade*, which are German for even and odd, respectively.) The d- and s-orbitals are g, and p-orbitals are u. All wave functions in a molecule with a center of symmetry are g or u. All components of the vector $\hat{M}$ in a point group containing an inversion center are necessarily *ungerade*.

$$\Gamma\psi_g \times \Gamma_{op} \times \Gamma\psi_{ex} = \Gamma$$

u	$\times \; u \; \times$	u	$= u$	forbidden
u	$\times \; u \; \times$	g	$= g$	allowed
g	$\times \; u \; \times$	g	$= u$	forbidden
g	$\times \; u \; \times$	u	$= g$	allowed

This leads to the selection rule that $g \to u$ and $u \to g$ are allowed, but $g \to g$ and $u \to u$ are forbidden. Therefore, $d \to d$ transitions in transition metal complexes with a center of symmetry are forbidden. Values of ε for the d-d transitions in $Ni(H_2O)_6^{2+}$ are ~ 20.

2. Transitions between states of different multiplicity are forbidden. Consider a singlet $\to$ triplet transition. Focusing on the electron being excited, we have in the singlet ground state $\psi\alpha\psi\beta$ and, in the excited state, $\psi\alpha\psi\alpha$ or $\psi\beta\psi\beta$, where α and β are the spin coordinates. The dipole strength is given by

$$D = \left| \int \psi_i \alpha \hat{M} \psi_f \beta \, d\tau \, d\sigma \right|^2$$

(where $d\sigma$ is the volume element in the spin coordinates and the i and f subscripts refer to initial and final states). We can rewrite the integral corresponding to D as

$$\left| \int \psi_i \hat{M} \psi_f \, d\tau \int \alpha\beta \, d\sigma \right|^2$$

Since the second term is the product of $+1/2$ and $-1/2$ spins, it is always odd and zero, *i.e.*, the spins are orthogonal. The ε for absorption bands involving transitions between states of different multiplicity is generally less than one. Since $\int \alpha\alpha \, d\sigma = 1$ and $\int \beta\beta \, d\sigma = 1$, in working out the intensity integral we only have to worry about the electron that is undergoing the transition, and we can ignore all the electrons in the molecule that do not change spin.

3. Transitions in molecules without a center of symmetry depend upon the symmetries of the initial and final states. If the direct product of these and any one of $\hat{M}_x$, $\hat{M}_y$, or $\hat{M}_z$ is A_1, the transition is allowed. If all integrals are odd, the transition is forbidden.

5–10 SPECTRUM OF FORMALDEHYDE

We can summarize the above ideas and illustrate their utility by returning again to the ultraviolet spectrum of formaldehyde. The various possible excited states arising from electron excitations from the highest-energy filled orbitals (n_a, n_b, and π) are given by:

$$a_1{}^2 b_1{}^2 b_2{}^1 b_1{}^{1*} = {}^1A_2 \quad (n_b \to \pi^*)$$
$$a_1{}^2 b_1{}^2 b_2{}^1 a_1{}^{1*} = {}^1B_2 \quad (n_b \to \sigma^*)$$
$$a_1{}^2 b_1{}^1 b_2{}^2 b_1{}^{1*} = {}^1A_1 \quad (\pi \to \pi^*)$$
$$a_1{}^2 b_1{}^1 b_2{}^2 a_1{}^{1*} = {}^1B_1 \quad (\pi \to \sigma^*)$$
$$a_1{}^1 b_1{}^2 b_2{}^2 b_1{}^{1*} = {}^1B_1 \quad (n_a \to \pi^*)$$
$$a_1{}^1 b_1{}^2 b_1{}^2 a_1{}^{1*} = {}^1A_1 \quad (n_a \to \sigma^*)$$

Two bands are observed, one with $\varepsilon = 100$ at 2700 Å and an extremely intense one at 1850 Å. We see from Fig. 5–4 that the lowest-energy transitions are $n_b \to \pi^*$ and $\pi \to \pi^*$, better expressed as ${}^1A_1 \to {}^1A_2$ and ${}^1A_1 \to {}^1A_1$. The ${}^1A_1 \to {}^1A_2 (n_b \to \pi^*)$ is forbidden, and accordingly is assigned to the band at 2700 Å. Both ${}^1A_1 \to {}^1B_1 (n_a \to \pi^*)$ and ${}^1A_1 \to {}^1A_1 (\pi \to \pi^*)$ are allowed. The former may contribute to the observed band at 1850 Å or may be in the far u.v.

The integrands for ${}^1A_1 \to {}^1B_1 (\pi \to \sigma^*)$, ${}^1A_1 \to {}^1B_2 (n_b \to \sigma^*)$, and ${}^1A_1 \to {}^1A_1 (n_a \to \sigma^*)$ are all A_1, leading to allowed transitions. These are expected to occur at very short wavelengths in the far ultraviolet region. This is the presently held view of the assignment of this spectrum, and it can be seen that the arguments are not rigorous. We shall subsequently show how polarization studies aid in making assignments more rigorous.

Next, it is informative to discuss the u.v. spectrum of acetaldehyde, which is quite similar to that of formaldehyde. The $n_b \to \pi^*$ transition has very low intensity. However, acetaldehyde has C_s symmetry; this point group has only two irreducible representations, A and B, with the x- and y-vectors transforming as A and the z-vector as B. Accordingly, *all* transitions will have an integrand with A_1 symmetry and will be allowed. Though the $n_b \to \pi^*$ transition is allowed by symmetry, the value of the transition moment integral is very small and the intensity is low. The intensity of this band in acetaldehyde is greater than that in formaldehyde. We can well appreciate the fact that although monodeuteroformaldehyde [DC(O)H] does not have C_{2v} symmetry, it will have an electronic spectrum practically identical to that of formaldehyde. These are examples of a rather general type of result, which leads to the idea of *local symmetry*. According to this concept, even though a molecule does not have the symmetry of a particular point group, if the groups attached to the chromophore have similar bonding interactions, (e.g. CH_3 and C_2H_5) the molecule for many purposes can be treated as though it had this higher symmetry.

5–11 SPIN-ORBIT AND VIBRONIC COUPLING CONTRIBUTIONS TO INTENSITY

The discrepancy between the theoretical prediction that a transition is forbidden and the experimental detection of a weak band assignable to this transition is attributable to the approximations of the theory. More refined calculations that include effects from *spin-orbit coupling* often predict low intensities for otherwise forbidden transitions. For example, a transition between a pure singlet state and a pure triplet state is forbidden. However, if spin-orbit coupling is present, the singlet could have the same total angular momentum as the triplet and the two states could interact. The interaction is indicated by equation (5–7):

$$\psi = a^1\psi + b^3\psi \tag{5-7}$$

where $^1\psi$ and $^3\psi$ correspond to pure singlet and triplet states, respectively, ψ represents the actual ground state, and a and b are coefficients indicating the relative contributions of the pure states. If $a \gg b$, the ground state is essentially singlet with a slight amount of triplet character and the excited state will be essentially triplet. This slight amount of singlet character in the predominantly triplet excited state leads to an intensity integral for the singlet-triplet transition that is not zero; this explains why a weak peak corresponding to the multiplicity-forbidden transition can occur.

Another phenomenon that gives intensity to some forbidden transitions is *vibronic coupling*. We have assumed until now that the wave function for a molecule can be factored into an electronic part and a vibrational part, and we have ignored the vibrational part. When we applied symmetry considerations to our molecule, we assumed some symmetrical, equilibrium internuclear configuration. This is not correct, for the molecules in our system are undergoing vibrations and during certain vibrations the molecular symmetry changes. For example, in an octahedral complex, the T_{1u} and T_{2u} vibrations shown in Fig. 5–9 remove the center of symmetry of the molecule. Since electronic transitions occur much more rapidly than molecular vibrations, we detect transitions occurring in our sample from many geometries that do not have high symmetry, *e.g.*, the vibrationally distorted molecules of the octahedral complex shown in Fig. 5–9. The local symmetry is still very close to octahedral, so the intensity gained this way is not very great; but it is large enough to allow a forbidden transition to occur with weak intensity.

The electronic transition can become allowed by certain vibrational modes but not by all. We can understand this by rewriting the transition moment integral to include both the electronic and the vibrational components of the wave function as in equation (5–8):

$$f \propto D = \left| \int \psi_{el} \psi_{vib} \hat{M} \psi_{el}^{ex} \psi_{vib}^{ex} \, d\tau \right|^2 \tag{5-8}$$

As we mentioned in Chapter 4, all ground vibrational wave functions are A_1, so the symmetry of $\psi_{el}\psi_{vib}$ becomes that of ψ_{el}, which is also A_1 for molecules with no unpaired electrons. (In general discussion, we shall use the symbol A_1 to represent the totally symmetric irreducible representation, even though this is not the appropriate label in some point groups.) To use this equation to see whether a forbidden transition can gain intensity by vibronic coupling, we must take a product $\hat{M}_{(x,y,\text{or }z)}\psi_{el}^{ex}$ that is not A_1 and see whether there is a vibrational mode with symmetry that makes the product $\hat{M}_{(x,y,\text{or }z)}\psi_{el}^{ex}\psi_{vib}^{ex}$ equal to A_1. When ψ_{vib}^{ex} has the same symmetry as the product $\hat{M}_{(x,y,\text{or }z)}\psi_{el}^{ex}$, the product will be A_1.

This discussion can be made clearer by considering some examples. We shall consider vibrational spectroscopy in more detail in the next chapter. A non-linear

FIGURE 5–9 T_{1u} and T_{2u} vibrations of an octahedral complex.

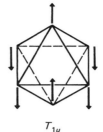
T_{1u}

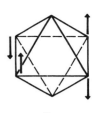
T_{2u}

molecule has $3N - 6$ internal vibrations; for formaldehyde these are $3a_1$, b_1, and $2b_2$. For the forbidden transition $^1A_1 \rightarrow {}^1A_2(n_b \rightarrow \pi^*)$, the vibrational wave function of a_1 symmetry does not change the direct product $\hat{M}_{(x,y,\text{or }z)}{}^1A_2$ so no intensity can be gained by this mode. Excitation of the b_1 vibrational mode leads to a direct product $\psi_{el}{}^{ex}\psi_{vib}{}^{ex}$ of $b_1 \times A_2 = B_2$. Since $\hat{M}_y$ has B_2 symmetry, the total integral ($\int \psi_{el}\psi_{vib}\hat{M}_y\psi_{el}{}^{ex}\psi_{vib}{}^{ex}\, d\tau$) has A_1 symmetry, and the electronic transition becomes allowed by vibronic coupling to the b_1 mode.

It is informative to consider $\text{Co(NH}_3)_6{}^{3+}$ as an example, for it contains triply degenerate irreducible representations. The ground state is $^1A_{1g}$ (a strong field O_h d^6 complex). The excited states from d-d transitions are $^1T_{1g}$ and $^1T_{2g}$. $\hat{M}_x$, $\hat{M}_y$, and $\hat{M}_z$ transform as T_{1u}. For the $^1A_{1g} \rightarrow {}^1T_{1g}$ transition one obtains:

$$A_{1g} \times T_{1u} \times T_{1g}$$

The resulting direct product representation has a dimensionality of nine (the identity is $1 \times 3 \times 3 = 9$) and the total representation is reduced into a linear combination of $A_{1u} + E_u + T_{1u} + T_{2u}$ irreducible representations. With no A_{1g} component, the $^1A_{1g} \rightarrow {}^1T_{1g}$ transition is forbidden. However, the vibrations for an octahedral complex have the symmetries of a_{1g}, e_g, $2t_{1u}$, t_{2g}, t_{2u}. Since the direct products $t_{1u} \times T_{1u}$ and $t_{2u} \times T_{2u}$ have A_{1g}-components, this transition becomes allowed by vibronic coupling. For practice, the reader should take the direct products and factor the reducible representations discussed here.

5–12 MIXING OF d AND p ORBITALS IN CERTAIN SYMMETRIES

There is one further aspect of the intensity of electronic transitions that can be understood *via* the symmetry aspects of electronic transitions. The electronic spectra of tetrahedral complexes of cobalt(II) contain two bands assigned to d-d transitions at $\sim 20{,}000$ cm^{-1} and ~ 6000 cm^{-1}, assigned as $A_2 \rightarrow T_1$ and $A_2 \rightarrow T_2$ transitions respectively, with molar absorptivities of 600 and 50. Since the $\hat{M}$-components transform as T_2, we obtain for the $A_2 \rightarrow T_1$ transition

$$A_2 \times T_2 \times T_1 = A_1 + E + T_1 + T_2$$

so the transition is allowed. However, if only the d-orbitals were involved in this transition, the intensity would be zero for the integrals

$$\int \psi_{d_{xy}} \hat{M} \psi_{d_{xz}} \, d\tau = 0$$

However, in the T_d point group, the d_{xy}, d_{xz} and d_{yz} orbitals and the p orbitals transform as T_2 and therefore can mix. If the two states involved in the transition, A_2 and T_1, have differing amounts of p-character, intensity is gained by having some of the highly allowed $p \to d$ or $d \to p$ character associated with the transition.

Consider the consequences of this mixing on the $A_2 \to T_2$ transition. The transition moment integrand for this transition is

$$A_2 \times T_2 \times T_2$$

which, as the reader should verify, can be reduced to $A_2 + E + T_1 + T_2$. Since there is no A_1 component, the transition is forbidden. Mixing p-character into the wave functions will not help, for this type of transition is still forbidden. Accordingly, the ε for the $A_2 \to T_1$ transition is ten times greater than that of $A_2 \to T_2$. The latter transition gains most of its intensity by vibronic coupling.

5-13 MAGNETIC DIPOLE AND ELECTRIC QUADRUPOLE CONTRIBUTIONS TO INTENSITY

So far, our discussion of the intensity of electronic transitions has centered on the electric dipole component of the radiation, with the transition moment integral involving the electric dipole operator, $e\vec{\mathbf{r}}$. There is also a magnetic dipole component. The magnetic dipole operator transforms as a rotation $R_x R_y R_z$, and the intensity from this effect may be regarded as arising from the rotation of electron density. Transition moment integrals similar to those for electric dipole transitions can be written for the contribution from both magnetic dipole and electric quadrupole effects. In a molecule with a center of symmetry, both of these operators are symmetric with respect to inversion, so $g \to g$ and $u \to u$ transitions are allowed. Approximate values of the transition moment integral for allowed transitions for these different operators are: 6×10^{-36} cgs units for an electric dipole transition, 9×10^{-41} cgs units for a magnetic dipole transition, and 7×10^{-43} cgs units for a quadrupole transition. Thus, we can see that these latter two effects will be important only when electric dipole transitions are forbidden. They do complicate the assignment of very weak bands in the spectrum.

5-14 CHARGE TRANSFER TRANSITIONS

A transition in which an electron is transferred from one atom or group in the molecule to another is called a *charge-transfer* transition. More accurately stated, the transition occurs between molecular orbitals that are essentially centered on different atoms. Very intense bands result, with molar absorptivities of 10^4 or greater. The frequency at maximum absorbancy, v_{max}, often, but not always, occurs in the ultraviolet region. The anions ClO_4^- and SO_4^{2-} show very intense bands. Since MnO_4^- and CrO_4^{2-} have no d electrons, the intense colors of these ions cannot be explained on the basis of d-d transitions; they are attributed to charge-transfer transitions.[9] The transitions in MnO_4^- and CrO_4^{2-} are most simply visualized as an electron transfer from a non-bonding orbital of an oxygen atom to the manganese or chromium ($n \rightarrow \pi^*$), in effect reducing these metals in the excited state.[10] An alternate description for this transition involves excitation of an electron from a π bonding molecular orbital, consisting essentially of oxygen atomic orbitals, to a molecular orbital that is essentially the metal atomic orbital.

In the case of a pyridine complex of iridium(III), a charge-transfer transition that involves oxidation of the metal has been reported.[11] A metal electron is transferred from an orbital that is essentially an iridium atomic orbital to an empty π^* antibonding orbital in pyridine.

In gaseous sodium chloride, a charge-transfer absorption occurs from the ion pair Na^+Cl^- to an excited state described as sodium and chlorine atoms having the same internuclear distance as the ion pair. A charge-transfer absorption also occurs in the ion pair, N-methylpyridinium iodide[36] (see Fig. 5–13) in which an electron is transferred from I^- to a ring antibonding orbital. The excited state is represented in Fig. 5–13. A very intense charge-transfer absorption is observed in addition compounds formed between iodine and several Lewis bases. This phenomenon will be discussed in more detail in a later section.

5–15 POLARIZED ABSORPTION SPECTRA

If the incident radiation employed in an absorption experiment is polarized, only those transitions with similarly oriented dipole moment vectors will occur. In a powder, the molecules or complex ions are randomly oriented. All allowed transitions will be observed, for there will be a statistical distribution of crystals with dipole moment vectors aligned with the polarized radiation. However, suppose, for example, that a formaldehyde crystal, with all molecules arranged so that their z-axes are parallel, is examined. As indicated in the previous section, the integrand $\psi^* \hat{M}_z \psi$ has appropriate symmetry for the $^1A_{(\pi, \pi^*)} \leftarrow {}^1A$ transition, but $\psi^* \hat{M}_x \psi$ and $\psi^* \hat{M}_y \psi$ do not. When the z-axes of the molecules in the crystal are aligned parallel to light that has its electric vector polarized in the z-direction, light will be absorbed for the $^1A_{(\pi, \pi^*)} \leftarrow {}^1A$ transition. Light of this wavelength polarized in other planes will not be absorbed. If this crystal is rotated so that the z-axis is perpendicular to the plane of polarization of the light, no light is absorbed. This behavior supports the assignment of this band to the transition $^1A_1 \leftarrow {}^1A_1$. To determine the expected polarization of any band, the symmetry species of the product $\psi_a \psi_b$ is compared with the components of $\hat{M}$, as was done before for formaldehyde. The polarization experiment is schematically illustrated in Fig. 5–10.

In Fig. 5–10(A), absorption of radiation will occur if $\hat{M}_z$ results in an A_1 transition moment integrand for equation (5–6). No absorption will occur if it is not A_1 regardless of the symmetries of the integrand for the $\hat{M}_x$ or $\hat{M}_y$ components [*i.e.*, equations (5–4) and (5–5)]. In Fig. 5–10B, absorption will occur if the $\hat{M}_y$ component gives an A_1 transition moment integral. Even if $\hat{M}_z$ has an integrand with A_1 symmetry, no absorption of the z-component will occur for this orientation and absorption will not occur unless the $\hat{M}_y$ integrand is A_1.

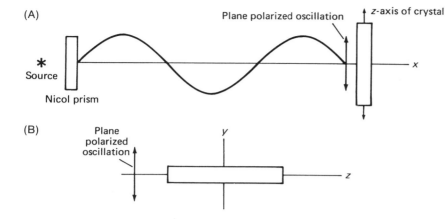

FIGURE 5–10 Schematic illustration of a polarized single crystal study. (A) The z-axis of the crystal is parallel to the oscillating electromagnetic plane polarized component. (B) The z-axis of the crystal is perpendicular to the oscillating electromagnetic plane polarized component, and the y-axis is parallel to it.

We can further illustrate these ideas by considering the electronic absorption spectrum of $PtCl_4^{2-}$. The transitions are charge transfers involving electron excitation from a mainly chlorine m.o. to the empty $d_{x^2-y^2}$ orbital on Pt(II). The symmetry is D_{4h}; using the same approach as that employed in Chapter 3 on the NO_2^- ion, on a basis set of four p_z orbitals on chlorine, we obtain symmetry orbitals for chlorine of b_{2u}, e_u, and a_{2u} symmetry. This leads to the following possible charge-transfer transitions:

$b_{2u}(\pi) \rightarrow b_{1g}(d_{x^2-y^2})$ with state labels $^1A_{1g} \rightarrow {}^1A_{2u}$ (here A_{2u} is the direct product of $b_{2u} \times b_{1g}$)

$e_u(\pi) \rightarrow b_{1g}(d_{x^2-y^2})$ with state labels $^1A_{1g} \rightarrow {}^1E_u$

$a_{2u}(\pi) \rightarrow b_{1g}(d_{x^2-y^2})$ with state labels $^1A_{1g} \rightarrow {}^1B_{2u}$

In the D_{4h} point group, $\hat{M}_x$ and $\hat{M}_y$ transform as E_u, and $\hat{M}_z$ as A_{2u}. If we first consider the $A_{1g} \rightarrow A_{2u}$ transition, we get for $\hat{M}_z$:

$$A_{1g} A_{2u} A_{2u} = A_{1g}$$

and for $\hat{M}_x$ and $\hat{M}_y$ we get:

$$A_{1g} E_u A_{2u} \neq A_{1g}$$

Accordingly, this transition is allowed and is polarized in the z-direction. If we use polarized light and a single crystal, light will be absorbed when the z-axis of the crystal is parallel to the z-direction of the light; but there will be no absorption when the z-axis is perpendicular to the light because the $\hat{M}_x$ and $\hat{M}_y$ integrands are not A_1.

For the band assigned to $^1A_{1g} \rightarrow {}^1E_u$, the $A_{1g} E_u E_u$ product has an A_{1g} component, so this transition is also allowed. Since M_z yields $A_{1g} A_{2u} E_u$, which does not have an A_{1g} component, there will be no absorption when the z-component is parallel to the plane of the polarized light but absorption will occur when the x and y-axes of the crystal are parallel to the light.

The $A_{1g} \rightarrow B_{2u}$ transition turns out to be forbidden. Thus, we see that by employing polarized single crystal spectroscopy, we can rigorously assign the two intense charge-transfer bands observed in the electronic spectrum of $PtCl_4^{2-}$. *If the single crystal employed in these experiments did not have all of the molecular z-axes aligned, the polarization experiments would not work.*

Applications

Most applications of electronic spectroscopy have been made in the wavelength range from 2100 to 7500 Å, for this is the range accessible with most recording spectrophotometers. Relatively inexpensive commercial instruments can now be obtained to cover the range from 1900 to 8000 Å. The near infrared region, from 8000 to 25,000 Å, has also provided much useful information. Spectra can be examined through the 1900 to 25,000 Å region on samples of vapors, pure liquids, or solutions. Solids can be examined as single crystals or as discs formed by mixing the material with KCl or NaCl and pressing with a hydraulic press until a clear disc is formed.[12] Spectra of powdered solids can also be examined over a more limited region (4000 to 25,000 Å) as reflectance spectra or on mulls of the solid compounds.[12]

5–16 FINGERPRINTING

Since many different substances have very similar ultraviolet and visible spectra, this is a poor region for product identification by the "fingerprinting" technique. Information obtained from this region should be used in conjunction with other evidence to confirm the identity of the compound. Evidence for the presence of functional groups can be obtained by comparison of the spectra with reported data. For this purpose, v_{max}, ε_{max}, and band shapes can be employed. It is also important that the spectra be examined in a variety of solvents to be sure that the band shifts are in accord with expectations (see discussion of blue shifts).

Spectral data have been compiled by Sadtler (see Additional References), Lang,[13] and Hershenson,[14] and in "Organic Electronic Spectral Data."[15,16] A review article by Mason[17] and the text by Jaffe and Orchin[1] are excellent for this type of application. If a functional group (chromophore) is involved in conjugation or steric interactions, or is attached to electron-releasing groups, its spectral properties are often different from those of an isolated functional group. These differences can often be predicted semiquantitatively for molecules in which such effects are expected to exist.[17]

The spectra of some representative compounds and examples of the effect of substituents on the wavelength of a transition will be described briefly.

SATURATED MOLECULES

Saturated molecules without lone pair electrons undergo high-energy $\sigma \rightarrow \sigma^*$ transitions in the far ultraviolet. For example, methane has a maximum at 1219 Å and ethane at 1350 Å corresponding to this transition. When lone pair

TABLE 5–2. Frequencies of Electronic Transitions in Some Saturated Molecules

Compound	λ_{max} Å	ε_{max}	Medium
H_2O	1667	1480	vapor
MeOH	1835	150	vapor
Me_2O	1838	2520	vapor
Me_2S	2290, 2100	140, 1020	ethanol
S_8	2750	8000	ethanol
F_2	2845	6	vapor
Cl_2	3300	66	vapor
Br_2	4200	200	vapor
I_2	5200	950	vapor
ICl	~4600	153	CCl_4
SCl_2	3040	1150	CCl_4
PI_3	3600	8800	Et_2O
AsI_3	3780	1600	pet. ether

electrons are available, a lower-energy $n \to \sigma^*$ transition is often detected in addition to the $\sigma \to \sigma^*$. For example, in triethylamine two transitions are observed, at 2273 and 1990 Å.

Table 5–2 contains a listing of absorption maxima for some saturated compounds and gives some indication of the variation in the range and intensity of transitions in saturated molecules.

CARBONYL COMPOUNDS

The carbonyl chromophore has been very extensively studied. Upon conjugation of the carbonyl group with a vinyl group, four π energy levels are formed. The highest occupied π level has a higher energy, and one of the lowest empty π^* levels has a lower energy, than the corresponding levels in a nonconjugated carbonyl group. The lone pair and σ electrons are relatively unaffected by conjugation. As a result, the $\pi \to \pi^*$ and $n \to \pi^*$ transition energies are lowered and the absorption maxima are shifted to longer wavelengths when the carbonyl is conjugated. The difference is greater for the $\pi \to \pi^*$ than for the $n \to \pi^*$ transition. The $n \to \sigma^*$ band is not affected appreciably and often lies beneath the shifted $\pi \to \pi^*$ absorption band. As stated earlier, electron-donating groups attached to the carbonyl cause a blue shift in the $n \to \pi^*$ transition and a red shift in $\pi \to \pi^*$.

It is of interest to compare the spectra of thiocarbonyl compounds with those of carbonyl compounds. In the sulfur compounds, the carbon-sulfur π interaction is weaker and, as a result, the energy difference between the π and π^*-orbitals is smaller than in the oxygen compounds. In addition, the ionization potential of the sulfur electrons in the thiocarbonyl group is less than the ionization potential of oxygen electrons in a carbonyl. The n electrons are of higher energy in the thiocarbonyl and the $n \to \pi^*$ transition requires less energy in these compounds than in carbonyls. The absorption maximum in thiocarbonyls occurs at longer wavelengths and in some compounds is shifted into the visible region.

INORGANIC SYSTEMS

The SO$_2$ molecule has two absorption bands in the near ultraviolet at 3600 Å ($\varepsilon = 0.05$) and 2900 Å ($\varepsilon = 340$) corresponding to a triplet and singlet $n \rightarrow \pi^*$ transition. The gaseous spectrum shows considerable vibrational fine structure, and analysis has produced information concerning the structure of the excited state.[18]

In nitroso compounds, an $n \rightarrow \pi^*$ transition involving the lone pair electrons on the nitrogen occurs in the visible region. An $n \rightarrow \pi^*$ transition involving an oxygen lone pair occurs in the ultraviolet.

The nitrite ion in water has two main absorption bands at 3546 Å ($\varepsilon = 23$) and 2100 Å ($\varepsilon = 5380$) and a weak band at 2870 Å ($\varepsilon = 9$). The assignment of these bands has been reported,[19] and this article is an excellent reference for gaining an appreciation of how the concepts discussed in this chapter are used in band assignments. The band at 3546 Å is an $n \rightarrow \pi^*$ transition ($^1B_1 \leftarrow {}^1A_1$) involving the oxygen lone pair. The band at 2100 Å is assigned as $\pi \rightarrow \pi^*(^1B_2 \leftarrow {}^1A_1)$, and the band at 2870 Å is assigned to an $n \rightarrow \pi^*$ transition ($^1A_2 \leftarrow {}^1A_1$) involving the oxygen lone pair.

The absorption peaks obtained for various inorganic anions in water or alcohol solution are listed in Table 5–3. For the simple ions (Br$^-$, Cl$^-$, OH$^-$) the absorption is attributed to charge transfer in which the electron is transferred to the solvent.

There are many more examples of applications of electronic spectroscopy to inorganic and organometallic systems. These are reviewed on a regular basis in the Specialist Periodical Reports of the Chemical Society (London). When well armed with the fundamentals, this source[20] provides more examples.

TABLE 5–3. Characteristic Absorption Maxima for Some Inorganic Anions

	λ (Å)	ε
Cl$^-$	1810	10^4
Br$^-$	1995	11,000
	1900	12,000
I$^-$	2260	12,600
	1940	12,600
OH$^-$	1870	5000
SH$^-$	2300	8000
S$_2$O$_3^{2-}$	2200	4000
S$_2$O$_8^{2-}$	2540	22
NO$_2^-$	3546	23
	2100	5380
	2870	9
NO$_3^-$	3025	7
	1936	8800
N$_2$O$_2^{2-}$	2480	4000

5–17 MOLECULAR ADDITION COMPOUNDS OF IODINE

The absorption band maximum for iodine (core plus $\sigma^2\pi^4n^4\pi^{*4}$) occurs at about 5200 Å in the solvent CCl_4 and is assigned to a $\pi^* \to \sigma^*$ transition. When a donor molecule is added to the above solution, two pronounced changes in the spectrum occur (see Fig. 5–11).

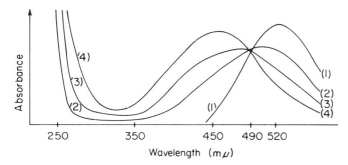

FIGURE 5–11 Spectra of iodine and base-iodine solutions. (1) I_2 in CCl_4; (2, 3, 4) same I_2 concentration but increasing base concentration in the order 2 < 3 < 4.

A blue shift is detected in the iodine peak, and a new peak arises in the ultraviolet region that is due to a charge-transfer transition.[21] The existence of an isosbestic point at 490 mμ indicates that there are only two absorbing species in the system; namely, free iodine and the complex B:Î—Ī|. As indicated in Chapter 4, a 1:1 equilibrium constant can be calculated from absorbance measurements for this system. The constant value for K obtained over a wide range of donor concentrations is evidence for the existence of a 1:1 adduct.

The bonding in iodine adducts can be described by the equation:

$$\psi^0 = a\psi_{cov} + b\psi_{el}$$

where ψ_{el} includes contributions from purely electrostatic forces while ψ_{cov} includes contributions from covalent interactions (these are described as charge-transfer interactions). In most adducts $b > a$ in the ground state and the band around 2500 Å arises from a charge-transfer transition in which an electron from this ground state is promoted to an excited state in which $a > b$. In view of these coefficients, the charge-transfer band assignment can be approximated by a transfer of a base electron, n_b, to the iodine σ^* orbital. These facts and the blue shift that occurs in the normal $\pi^* \to \sigma^*$ iodine transition upon complexation can be explained by consideration of the relative energies of the molecular orbitals of iodine and the complex (Fig. 5–12). In Fig. 5–12, n_b refers to the donor orbital on the base, and $\sigma_{I_2}^*$ and $\pi_{I_2}^*$ refer to the free iodine antibonding orbitals involved in the transition leading to iodine absorption. The σ_c, π_c^*, and σ_c^* are labels for molecular orbitals in the complex that are very much like the original base and iodine orbitals because of the weak Lewis acid-base interaction (2 to 10 kcal). The orbitals n_b and $\sigma_{I_2}^*$ combine to form molecular orbitals in the complex, σ_c and σ_c^*, in which σ_c, the bonding orbital, is essentially n_b and σ_c^* is essentially

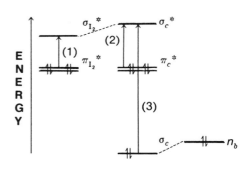

FIGURE 5–12 Some of the molecular orbitals in a base-iodine addition compound.

$\sigma_{I_2}^*$. Since σ_c^* is slightly higher in energy than the corresponding $\sigma_{I_2}^*$, the transition in complexed iodine [arrow (2) in Fig. 5–12] requires slightly more energy than the corresponding transition in free I_2 [arrow (1)] and a blue shift is observed. The charge-transfer transition occurs at higher energy in the ultraviolet region and is designated in Fig. 5–12 by arrow (3).

Some interesting correlations have been reported, which claim that the blue shift is related to the magnitude of the base-iodine interaction, *i.e.*, the enthalpy of adduct formation.[21] This would be expected qualitatively from the treatment in Fig. 5–12 as long as the energy of π_c^* differs very little from that of $\pi_{I_2}^*$ or else its energy changes in a linear manner with the enthalpy, ΔH. A rigorous evaluation of this correlation with accurate data on a wide range of different types of Lewis bases indicates that a rough general trend exists, but that a quantitative relation (as good as the accuracy of the data) does not exist. A relationship involving the charge transfer band, the ionization potential of the base, I_b, and the electron affinity of the acid, E_a, is also reported[22,23]:

$$v = I_b - E_a - \Delta \qquad (5\text{–}9)$$

where Δ is an empirically determined constant for a related series of bases.

The enthalpies for the formation of these charge-transfer complexes are of interest and significance to both inorganic and organic chemists. For many inorganic systems, especially in the areas of coordination chemistry and nonaqueous solvents, information about donor and acceptor interactions is

TABLE 5–4. Equilibrium Constants and Enthalpies of Formation for Some Donor-I_2 Adducts

Donor	K(liter mole^{-1})	$-\Delta H$(kcal mole^{-1})
C_6H_6	0.15 (25°)	1.4
Toluene	0.16 (25°)	1.8
CH_3OH	0.47 (20°)	1.9
Dioxane	1.14 (17°)	3.5
$(C_2H_5)_2O$	0.97 (20°)	4.3
$(C_2H_5)_2S$	180 (25°)	8.3
$CH_3C(O)N(CH_3)_2$	6.1 (25°)	4.7
Pyridine	270 (20°)	7.8
$(C_2H_5)_3N$	5130 (25°)	12.0

essential to an understanding of many phenomena. Since the above adducts are soluble in CCl$_4$ or hexane, the thermodynamic data can be interpreted more readily than results obtained in polar solvents, where large solvation enthalpies and entropies are encountered. Some typical results from donor-I$_2$ systems in which such solvation effects are minimal illustrate the wide range of systems that can be studied, and are contained in Table 5–4.

The following few examples illustrate the information that can be obtained by studying enthalpies of association in non-polar, weakly basic solvents.

    1. The donor properties of the π electron systems of alkyl-substituted benzenes have been reported.[24]

    2. A correlation of the heat of formation of iodine adducts of a series of *para*-substituted benzamides with the Hammett substituent constants[25] of the benzamides is reported.

    3. The donor properties of a series of carbonyl compounds [(CH$_3$)$_2$CO, CH$_3$C(O)N(CH$_3$)$_2$, (CH$_3$)$_2$NC(O)N(CH$_3$)$_2$, CH$_3$C(O)OCH$_3$, CH$_3$C(O)SCH$_3$] have been evaluated and interpreted[26] in terms of conjugative and inductive effects of the group attached to the carbonyl functional group.

    4. The donor properties of sulfoxides, sulfones, and sulfites have been investigated.[27] The results are interpreted to indicate that sulfur-oxygen π bonding is less effective in these systems than carbon-oxygen π bonding is in ketones and acetates.

    5. The effect of ring size on the donor properties of cyclic ethers and sulfides has been investigated.[28] It was found that for saturated cyclic sulfides, of general formula (CH$_2$)$_n$S, the donor properties of sulfur are in the order $n = 5 > 6 > 4 > 3$. The order for the analogous ether compounds is $4 > 5 > 6 > 3$. Explanations of these effects are offered.

    6. The donor properties of a series of primary, secondary, and tertiary amines have been evaluated.[29,30] The order of donor strength of amines varies with the acid studied. Explanations have been offered, which are based upon the relative importance of covalent and electrostatic contributions to the bonding in various adducts.

In addition to iodine, several other Lewis acids form charge-transfer complexes that absorb in the ultraviolet or visible regions. For example, the relative acidities of I$_2$, ICl, Br$_2$, SO$_2$, and phenol toward the donor N,N-dimethylacetamide have been evaluated. Factors affecting the magnitude of the interaction[31] and information regarding the bonding in the adducts are reported. Good general reviews of charge-transfer complexes are available.[9,32,33,34,35]

5–18 EFFECT OF SOLVENT POLARITY ON CHARGE-TRANSFER SPECTRA

The ion pair N-methylpyridinium iodide undergoes a charge-transfer transition that can be represented[36] as in Fig. 5–13. It has been found that the position of the charge-transfer band is a function of the solvating ability of the solvent. A shift to lower wavelengths is detected in the better solvating solvents. The positions of the bands are reported as transition energies, E_T. Transition energies (kcal mole^{-1}) are calculated from the frequency as described in Chapter 4. The transition energy is referred to as the Z-value. Some typical data are reported

FIGURE 5–13 Ion pair and charge transfer excited state of N-methylpyridinium iodide.

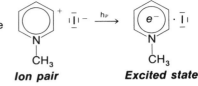

in Table 5–5. An explanation for the observed shift has been proposed.[36a] The dipole moment of the ion pair, $C_5H_5NCH_3{}^+I^-$, is reported to be perpendicular to the dipole moment of the excited state (Fig. 5–13). Polar solvent molecules will align their dipole moments for maximum interaction with the ground state, lowering the energy of the ground state by solvation. The dipole moment of the solvent molecules will be perpendicular to the dipole moment of the excited state, producing a higher energy for the excited state than would be found in the gas phase. Since solvent molecules cannot rearrange in the time required for a transition, the relative lowering of the ground state and raising of the excited state increases the energy of the transition, E_T, over that in the gas phase (Fig. 5–14), shifting the wavelength of absorption to higher frequencies. Hydrogen-bonding solvents are often found to increase E_T more than would be expected by comparing their dielectric constants with those of other solvents. This is due to the formation of hydrogen bonds with the solute. The use of the dielectric constant to infer solvating ability can lead to difficulty because the local dielectric constant in the vicinity of the ion may be very different from the bulk dielectric constant.

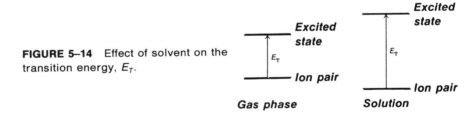

FIGURE 5–14 Effect of solvent on the transition energy, E_T.

TABLE 5–5.
Z-Values for Some Common Solvents

Solvent	E_T or Z value[a]
H_2O	94.6
CH_3OH	83.6
C_2H_5OH	79.6
CH_3COCH_3	65.7
$(CH_3)_2NCHO$	68.5
CH_3CN	71.3
Pyridine	64.0
CH_3SOCH_3	71.1
H_2NCHO	83.3
CH_2Cl_2	64.2
Isoctane	60.1

[a] The E_T or Z value is the transition energy in kcal mole^{-1} at 25°C, 1 atm pressure, for the compound 1-ethyl-4-carbomethoxypyridinium iodide.

The data obtained from these spectral shifts are employed as an empirical measure of the ionizing power of the solvent. The results can be correlated with a scale of "solvent polarities" determined from the effect of solvent on the rate of solvolysis of t-butyl chloride.[36a] Other applications of these data to kinetic and spectral studies are reported.[36b] Solvent effects are quite complicated, and these correlations at best provide a semiquantitative indication of the trends expected.

Significant differences exist between "solvating power" inferred from the dielectric constant and the results from spectral and kinetic parameters. Although methanol and formamide are found to have similar Z-values, the dielectric constants are 32.6 and 109.5, respectively. Solvent effects cannot be understood solely on the basis of the dielectric constant. Specific Lewis acid-base interactions make the problem more complex than the simple dielectric model.

The band positions for the $n \to \pi^*$ transitions in certain ketones in various solvents are found to be linearly related to the Z-values for the solvents. A constant slope is obtained for a plot of E_T versus Z for many ketones. Deviations from linearity by certain ketones in this plot can be employed to provide interesting structural information about the molecular conformation. Cycloheptanone, for example, does not give a linear plot of E_T versus Z. The deviation is attributed to solvent effects on the relative proportion of the conformers present in solution.

5–19 STRUCTURES OF EXCITED STATES

Considerable information is available about the structure of excited states of molecules from analysis of the rotational band contours in the electronic spectra. Both geometrical information and vibrational information about the excited states of large molecules can be obtained.[37] By studying and analyzing the perturbation made on the vibrational fine structure of an electronic transition by an electric field, the dipole moment of the excited state can be obtained.[38]

Optical Rotary Dispersion, Circular Dichroism, and Magnetocircular Dichroism

5–20 INTRODUCTION

Plane-polarized light consists of two circularly polarized components of equal intensity. The two types of circularly polarized light correspond to right-handed and left-handed springs. Circularly polarized light is defined as right-handed when its electric or magnetic vector rotates clockwise as viewed by an observer facing the direction of the light propagation (*i.e.*, the source). The frequency of the rotation is related to the frequency of the light. Plane polarized light can be resolved into its two circular components, and the two components when added together produce plane-polarized light in an optically isotropic medium. If plane-polarized light is passed through a sample for which the refractive indices of the left- and right-polarized components differ, the components will, upon recombination, give plane-polarized radiation in which the plane of the polarization has been rotated through an angle α, given by

$$\alpha = \frac{n_l - n_r}{\lambda} \qquad (5\text{–}10)$$

where the subscripts refer to left and right, n is the appropriate refractive index, and λ is the wavelength of light employed. The units are radians per unit length, with the length units given by those used for λ.

If the concentration of an optically active substance, c', is expressed in units of g cm^{-3} (corresponding to the density for a pure substance), the specific rotation $[\alpha]$ is defined as:

$$[\alpha] = \frac{\alpha}{c'd'} \qquad (5\text{–}11)$$

where d' is the thickness of the sample in decimeters. The molar rotation $[M]$ is defined as:

$$[M] = M[\alpha] \times 10^{-2} = M\alpha \times 10^{-2}/c'd' \tag{5-12}$$

where M is the molecular weight of the optically active component. (The quantity 10^{-2} is subject to convention and not always included in $[M]$.)

The *optical rotatory dispersion* curve, ORD, is a plot of the molar rotation, $[\alpha]$ or $[M]$, against λ. When the plane of polarization rotates clockwise as viewed by an observer facing the direction of propagation of the radiation, $[\alpha]$ or $[M]$ is defined as positive; a counterclockwise rotation is defined as negative.

The technique whereby one determines that an optically active substance *absorbs* right and left circularly polarized light differently is called *circular dichroism, CD*. All optically active substances exhibit CD in the region of appropriate electronic absorption bands. The molar circular dichroism $\varepsilon_l - \varepsilon_r$ is defined as

$$\varepsilon_l - \varepsilon_r = \frac{k_l - k_r}{c} \tag{5-13}$$

where k, the absorption coefficient, is defined by $I = I_0 \, 10^{-kd}$ with I_0 and I being the intensity of the incident and resultant light and d being the cell thickness. The CD curve results when we plot $\varepsilon_l - \varepsilon_r$ versus λ.

Wherever circular dichroism is observed in a sample, the resulting radiation is not plane polarized, but is elliptically polarized. The quantity α in the above equations is then the angle between the initial plane of polarization and the major axis of the ellipse of the resultant light. One can define a quantity φ' (in radians), the tangent of which is the ratio of the major to minor axes of the ellipse. The quantity φ' (in radians), the tangent of which is the ratio of the major to minor axes of the ellipse. The quantity φ' is used to approximate the ellipticity; when it is expressed in degrees, it can be converted to a specific ellipticity $[\varphi]$ or molar ellipticity $[\theta]$ by

$$[\varphi] = \frac{\varphi'}{c'd'} \tag{5-14}$$

and

$$[\theta] = M[\varphi]10^{-2} \tag{5-15}$$

where the symbols are as defined in equations (5-11) and (5-12). The quantity $[\theta]$ is related to $\varepsilon_l - \varepsilon_r$ by the following equation:

$$\varepsilon_l - \varepsilon_r = 0.3032 \times 10^{-3}[\theta] \tag{5-16}$$

Thus, one often sees the CD curve plotted as $[\theta]$ versus λ.

With CD one can measure only the optical activity if there is an accompanying electronic absorption band. On the other hand, ORD is measurable

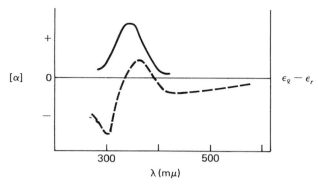

FIGURE 5–15 ORD (---) and CD (—) curves for D-(—)-[Rh(en)₃]³⁺ (en is an abbreviation for ethylenediamine).

both inside and outside the absorption band.* The ORD and CD curves of D-(−)-[Rh(en)₃]³⁺ are illustrated[39] in Fig. 5–15. Throughout most of the visible region, the ORD curve is negative. However, the CD curve associated with the visible d-d transitions at ∼300 mμ is clearly positive. All the chromophores in a molecule contribute to the rotatory power at a given wavelength, but only the chromophore that absorbs at the given wavelength contributes to the CD. Thus, a transition in the far u.v. can make a significant contribution to the rotation in the region of d-d transitions in the ORD. The negative effect in the u.v. dominates the d-d contribution through most of the visible region, and the negative ORD curve results. For most of the applications to be discussed here, CD is the method of choice.

5–21 SELECTION RULES

We have previously given [equation (5–3)] the transition moment integral for an electric dipole transition, and we mentioned that a magnetic dipole transition integral has a similar form. In order for an electronic transition to give rise to optical activity, the transition must be both electric and magnetic dipole allowed, i.e.,

$$R \propto \left[\int \psi_{el} \hat{M} \psi_{el}^{ex} \, dv \right] \left[\int \psi_{el} \widehat{MD} \psi_{el}^{ex} \, dv \right]$$

where R is the rotational strength, $\widehat{MD}$ is the magnetic dipole operator, and $\hat{M}$ is the electric dipole operator. If an electronic absorption band is observed, there must be some mechanism for making this allowed; it then becomes important

---

*CD is the absorption difference between left and right circularly polarized light for an electronic transition. ORD is related to the difference between the indices of refraction for left and right circularly polarized light. The two effects are interrelated via the Kronig-Kramers relation.[46]

to be concerned with the magnetic dipole selection rules. The sign and magnitude of the activity can be calculated[40] by evaluating both the electric and magnetic dipole integrals.

5-22 APPLICATIONS

The use of CD in band assignments is an obvious application of our previous discussion of the selection rules. For example, octahedral nickel(II) complexes have three bands assigned as $^3A_{2g}$ to $^3T_{2g}$, $^3T_{1g}(F)$, and $^3T_{1g}(P)$ in order of increasing energy. Since the magnetic dipole operator (the magnetic dipole transforms like the rotations $R_x R_y R_z$) in O_h is T_{1g}, only the $^2A_{2g} \to {}^3T_{2g}$ transition is magnetic dipole allowed. Accordingly, it is found in the CD spectrum of $Ni(pn)_3^{2+}$ (pn = propylene diamine) that the low-energy band has a maximum $(\varepsilon_l - \varepsilon_r)$ value of 0.8, while those for the other bands are less than 0.04. This confirms the original band assignments.

A second type of application involves using CD to show that certain absorption bands have contributions from more than one electronic transition. The CD bands are usually narrower and can be positive or negative. This idea is illustrated in Fig. 5–16, where the absorption curve and CD curve for $\Delta(+)\text{-Co(en)}_3^{3+}$ in aqueous solution are shown.[41]

Co(III) complexes with O_h symmetry commonly have two absorptions assigned to $^1T_{1g}$ and $^1T_{2g}$. The symmetry of Co(en)_3^{3+} is D_3, so the transitions to the T states are expected to show some splittings. This is not detected in the absorption spectrum, as seen in Fig. 5–16. However, the effects of lower symmetry are observed in the CD spectrum. The magnetic dipole selection rules for D_3 predict that the low-energy band will have two magnetic dipole allowed components, $^1A_1 \to {}^1A_2$ and $^1A_1 \to {}^1E_a$. The $^1A_1 \to {}^1E_b$ transition of the high-energy band is magnetic dipole allowed, but the $^1A_1 \to {}^1A_1$ transition is not. In the CD, two components (+ and −) are seen in the low-energy band and one in the high-energy band, as predicted for a D_3 distortion. Applications of these ideas can be used to indicate the symmetry of molecules in solution and in uniaxial single crystals.

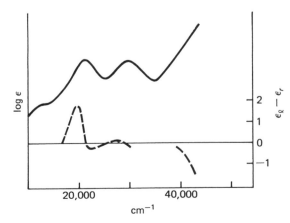

FIGURE 5–16 The absorption spectrum (solid line) and CD curve (dashed line) of (+)-Co(en)$_3^{3+}$ in aqueous solution.

Another application involves using the sign of the CD to obtain the absolute configuration of a molecule.[42] This application has been particularly successful for organic compounds.[43] In organic systems, absolute configurations are often assigned by analogy to known systems. Particular care must be employed in determining what constitutes an analogous compound.[44,45] By using complete operator-matrices for the electric and magnetic dipole components, the signs of the trigonal components in some Co(III) and Cr(III) complexes have been related to the absolute configurations of the complexes.[40].

Finally, optically active transitions are polarized, and the polarization information can be used to support the assignment of the electronic spectrum.

5–23 MAGNETOCIRCULAR DICHROISM

When plane-polarized light is passed through any substance in a magnetic field, H_0, whose component in the direction of the light propagation is non-zero, the substance appears to be optically active. Left and right circularly polarized light do not interact in equivalent ways. For atoms, for example, left circularly polarized (lcp) light induces a transition in which Δm_J is -1, while for right circularly polarized (rcp) light Δm_J is $+1$. Note that m_J has the same relationship to J as m_l has to l. If one observed a transition from an S state where $J = 0$ to a ^1P state where $J = 1$, the two transitions in Fig. 5–17 would occur as a consequence of this selection rule.[46]

If the absorptions of left and right circularly polarized light corresponding to these transitions are measured separately, the curves in Fig. 5–17(B) are obtained. Here v_0 is the band maximum for the absorption band. When the mcd curve is plotted, the result in Fig. 5–17(C) is obtained, provided that the band width is much greater than the Zeeman splitting of the excited state. A curve of this sort is referred to as an *A*-term and can arise only for a transition in which $J > 0$ for one of the states involved. The sign of the *A*-term in molecules depends upon the sign of the Zeeman splitting and the molecular selection rules for circularly polarized light.

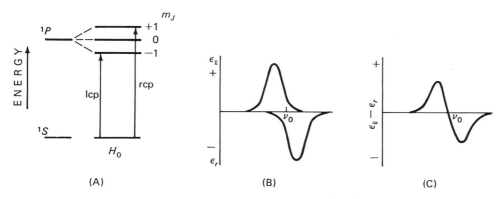

FIGURE 5–17 The transitions and expected spectrum for $^1S \rightarrow {}^1P$ in the mcd experiment. (A) the transitions; (B) spectra for left (ε_l) and right (ε_r) circularly polarized radiation; (C) the mcd spectrum ($\varepsilon_l - \varepsilon_r$) circularly polarized radiation; (C) the mcd spectrum ($\varepsilon_l - \varepsilon_r$); *A*-term behavior.

Next consider a transition from a ^1P to a ^1S state. Figure 5–18 summarizes this situation. The $\Delta m_J = +1$ transition for rcp light is now that of m_J from -1 to 0. Because the m_J states are not equally populated, but Boltzmann populated, the two transitions will not have equal intensity, as shown in Fig. 5–18B. The relative intensities will be very much temperature dependent. The resultant mcd curve, shown in Fig. 5–18C, is referred to as a *C*-term. The band shape and intensity are very temperature dependent. An *A*-term curve usually occurs superimposed upon a *C*-term curve.

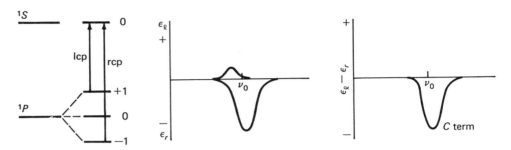

FIGURE 5–18 The transitions and expected spectrum for $^1P \to {}^1S$ in the mcd experiment. (A) the transitions; (B) spectra for left and right circularly polarized radiation; (C) the mcd spectrum; *C*-term behavior.

A third type of curve (*B*-term) results when there is a field-induced mixing of the states involved (this phenomenon also creates temperature independent paramagnetism, TIP, and will be discussed in more detail in the chapter on magnetism). This is manifested in a curve that looks like a *C*-curve but that is temperature independent. Since this mixing is present to some extent in all molecules, *all substances have mcd activity*. The magnitude of the external magnetic field intensity will determine whether or not the signal is observed.

The following characteristics summarize the basis for detecting and qualitatively interpreting mcd curves:

1. An *A*-term curve changes sign at the absorption maximum, while *B*- and *C*- curves maximize or minimize at the maximum of the electronic absorption band.
2. A *C*-term curve's intensity is inversely proportional to the absolute temperature, while a *B*-term is independent of temperature.
3. An *A*-term spectrum is possible only if the ground or excited state involved in the electronic transition is degenerate and has angular momentum.
4. A *C*-term spectrum is possible only if the ground state is degenerate and has angular momentum.

As can be anticipated, mcd measurements are of considerable utility[46,47] in assigning the electronic spectrum of a compound. Furthermore, the magnitude of the parameters provides information about many subtle electronic effects.[47] The molecular orbital origin of an electron involved in a transition can be determined. The lowest energy band in RuO_4[48] is clearly an oxygen to ruthenium charge-transfer band. One cannot determine from the electronic spectrum whether the oxygen electron involved in the transition came from a $t_1\pi$ or $t_2\pi$ type of oxygen molecular orbital. The sign of the mcd A-term established[48] the transition as $t_1\pi$(oxygen) → $e_{x^2-y^2,z^2}$ (ruthenium). Another significant advantage of mcd is in the assignment of spin-forbidden electronic transitions that have very low intensity in the electronic absorption spectrum. The assignments of the components in a six-coordinate chromium(III) complex have been made with this technique.[49] Other applications have been summarized in review articles.[46,47] Mcd has been extensively applied[46b] to provide information regarding the symmetries, angular momenta, electronic splittings, and vibrational-electronic interactions in excited electronic states.

REFERENCES CITED

1. H. H. Jaffe and M. Orchin, "Theory and Applications of Ultraviolet Spectroscopy," Wiley, New York, 1962.
2. M. Kasha, Discussions Faraday Soc., *9*, 14 (1950).
3. M. Kasha, Chem. Revs., *41*, 401 (1947); J. R. Platt, J. Opt. Soc. Amer., *43*, 252 (1953): A. Terenin, Acta Physica Chim. USSR, *18*, 210 (1943) (in English).
4. M. Kasha, "The Nature and Significance of $n \to \pi^*$ Transitions," in Light and Life, ed. by W. D. McElroy and B. Glass, Johns Hopkins, Baltimore, 1961.
5. J. Sidman, Chem. Revs., *58*, 689 (1958),.
6. H. McConnell, J. Chem. Phys., *20*, 700 (1952).
7. G. J. Brealey and M. Kasha, J. Amer. Chem. Soc., *77*, 4462 (1955).
8. S. Nagakura, Bull. Chem. Soc. Japan, *25*, 164 (1952) (in English).
9. L. E. Orgel, Quart. Revs., *8*, 422 (1954). For a good discussion of charge-transfer transitions, see A. B. Lever, J. Chem. Ed., *51*, 612 (1974).
10. S. P. McGlyn and M. Kasha, J. Chem. Phys., *24*, 481 (1956).
11. C. K. Jorgenson, Acta Chem. Scand., *11*, 166 (1957); R. J. P. Williams, J. Chem. Soc., *1955*, 137.
12. R. P. Bauman, "Absorption Spectroscopy," Wiley, New York, 1962.
13. L. Lang, "Absorption Spectra in the Ultraviolet and Visible Region," Vols. 1–20, Academic, New York, 1961–76.
14. H. M. Hershenson, "Ultraviolet and Visible Absorption Spectra—Index for 1930–54," Academic, New York, 1956.
15. "Organic Electronic Spectral Data," Vols. 1–10, Interscience, New York. Compilation of spectral data from 1946 to 1968.
16. "ASTM (American Society for Testing Materials) Coded IBM Cards for Ultraviolet and Visible Spectra" ASTM, Philadelphia, 1961.
17. S. F. Mason, Quart. Revs., *15*, 287 (1961).
18. N. Metropolis, Phys. Rev., *60*, 283, 295 (1941).
19. S. J. Strickler and M. Kasha, J. Amer. Chem. Soc., *85*, 2899 (1963).
20. "Spectroscopic Properties of Inorganic and Organometallic Compounds." Volumes 1– (1967–), Specialist Periodical Report, Chemical Society, London.
21. R. Foster, "Organic Charge Transfer Complexes," Academic Press, London, 1969.
22. H. McConnell, J. S. Ham and J. R. Platt, J. Chem. Phys., *21*, 66 (1953).
23. G. Briegleb and J. Czekalla, Z. Phys. Chem. (Frankfurt), *24*, 37 (1960).
24. R. M. Keefer and L. J. Andrews, J. Amer. Chem. Soc., *77*, 2164 (1955).
25. R. L. Carlson and R. S. Drago, J. Amer. Chem. Soc., *85*, 505 (1963).
26. R. L. Middaugh, R. S. Drago and R. J. Niedzielski, J. Amer. Chem. Soc., *86*, 388 (1964).
27. R. S. Drago, B. Wayland and R. L. Carlson, J. Amer. Chem. Soc., *85*, 3125 (1963).
28. Sister M. Brandon, O. P., M. Tamres and S. Searles, Jr., J. Amer. Chem. Soc., *82*, 2129 (1960); M. Tamres and S. Searles, Jr., J. Phys. Chem., *66*, 1099 (1962).

29. H. Yada, J. Tanaka and S. Nagakura, Bull. Chem. Soc. Japan, *33*, 1660 (1960).
30. R. S. Drago, D. W. Meek, R. Longhi and M. Joesten, Inorg. Chem., *2*, 1056 (1963).
31. R. S. Drago and D. A. Wenz, J. Amer. Chem. Soc., *84*, 526 (1962). Other donors studied toward these acids are summarized here.
32. L. J. Andrews and R. M. Keefer, "Advances in Inorganic Chemistry and Radiochemistry," Vol. 3, eds. H. J. Emeleus and A. G. Sharpe, pp. 91–128, Academic, New York, 1961.
33. O. Hassel and Chr. Rømming, Quart. Revs., *16*, 1 (1962).
34. R. Foster, "Molecular Complexes," Crane, Russak and Co., New York, 1974.
35. G. Briegleb, "Electronen Donator-Acceptor-Komplexes," Springer-Verlag, Berlin, 1963.
36. **a.** E. M. Kosower, et al., J. Amer. Chem. Soc., *83*, 3142, 3147 (1961); E. M. Kosower, J. Amer. Chem. Soc., *80*, 3253, 3261, 3267 (1958); E. M. Kosower, "Change Transfer Complexes," in "The Enzymes," Vol. 3, eds. Boyer, Lardy, and Myrbach, p. 171, Academic Press, New York, 1960. The Spectrum of *N*-methylpyridinium iodide in many solvents is treated exhaustively in these references.
    b. C. Reichardt, "Solvents and Solvent Effects in Organic Chemistry," VCH Publishers, New York, 1988.
37. J. M. Hollas, "Molecular Spectroscopy," Vol. 1, pp. 62–112, Specialist Periodical Reports, Chemical Society, London (1972).
38. D. E. Freeman and W. Klemperer, J. Chem. Phys., *45*, 52 (1966).
39. J. P. Mathieu, J. Chim. Phys., *33*, 78 (1936).
40. R. S. Evans, A. F. Schreiner, and J. Hauser, Inorg. Chem., *13*, 2185 (1974) and references therein.
41. A. J. McCaffery and S. F. Mason, Mol. Phys., *6*, 359 (1963).
42. E. J. Corey and J. C. Bailar, Jr., J. Amer. Chem. Soc., *81*, 2620 (1959).
43. L. Velluz, M. Legrand and M. Grosjean, "Optical Circular Dichroism," Academic Press, New York, 1965.
44. F. Woldbye, "Technique of Inorganic Chemistry," Vol. 4, eds. H. B. Jonassen and A. Weissberger, p. 249, Interscience, New York, 1965.
45. A. M. Sargeson, Transition Metal Chemistry, *3*, 303 (1966).
46. (a) A. D. Buckingham and P. J. Stephens, Ann. Rev. Phys. Chem., *17*, 399 (1966); (b) P. J. Stephens, Ann. Rev. Phys. Chem., *25*, 201 (1974).
47. P. N. Schatz and A. J. McCaffery, Quart. Revs., *23*, 552 (1969), and references therein.
48. A. H. Bowman, R. S. Evans and A. F. Schreiner, Chem. Phys. Letters, *29*, 140 (1974), and references therein.
49. P. J. Hauser, A. F. Schreiner and R. S. Evans, Inorg. Chem., *13*, 1925 (1974), and references therein.

ADDITIONAL REFERENCES

A. The following references may be consulted for further study:

H. H. Jaffe and M. Orchin, "Theory and Applications of Ultraviolet Spectroscopy," Wiley, New York, 1962.

M. Orchin and H. H. Jaffe, "Symmetry, Orbitals and Spectra," Wiley Interscience, New York, 1971.

C. Sandorfy, "Electronic Spectra and Quantum Chemistry," Prentice-Hall, Englewood Cliffs, N. J., 1964.

R. P. Bauman, "Absorption Spectroscopy," Wiley, New York, 1962.

E. A. Braude and F. C. Nachod, "Determination of Organic Structures by Physical Methods," pp. 131–195, Academic, New York, 1955.

G. Herzberg, "Molecular Spectra and Molecular Structure," Vols. 1 and 3. D. Van Nostrand Co., Princeton, N.J., 1966.

J. N. Murrell, "The Theory of Electronic Spectra of Organic Compounds," Wiley, New York, 1963.

A. E. Gillam and E. S. Stern, "Introduction to Electronic Absorption Spectroscopy," 2nd ed., Arnold, London, 1957.

G. W. King, "Spectroscopy and Molecular Structure," Holt, Rinehart, and Winston, New York, 1964.

A. B. P. Lever, "Inorganic Electronic Spectroscopy," 2nd ed. Elsevier, New York, 1984.

A. P. Demchenko, "Ultraviolet Spectroscopy of Proteins," Springer-Verlag, Berlin, 1986.

"Spectroscopy in Biochemistry" ed. J. E. Bell, CRC Press, Boca Raton, Fla., 1980.

EXERCISES

1. The compound $(C_6H_5)_3As$ is reported [J. Mol. Spectr., 5, 118–132 (1960)] to have two absorption bands, one near 2700 Å and a second around 2300 Å. One band is due to the $\pi \to \pi^*$ transition of the phenyl ring and the other is a charge-transfer transition from the lone pair electrons on arsenic to the ring. The 2300 band is solvent dependent, and the 2700 band is not.

   a. Which band do you suspect is $\pi \to \pi^*$?

   b. What effect would substitution of one of the phenyl groups by CF_3 have on the frequency of the charge-transfer transition?

   c. In which of the following compounds would the charge-transfer band occur at highest frequency, $(C_6H_5)_3P$, $(C_6H_5)_3Sb$, or $(C_6H_5)_3Bi$?

2. What effect would changing the solvent from a nonpolar to a polar one have on the frequency for the following:

   a. Both ground and excited states are neutral (*i.e.*, there is no charge separation)?

   b. The ground state is neutral and the excited state is polar?

   c. The ground state is polar, the excited state has greater charge separation, and the dipole moment vector in the excited state is perpendicular to the ground state moment?

   d. The ground state is polar and the excited state is neutral?

3. Would you have a better chance of detecting vibrational fine structure in an electronic transition of a solute in liquid CCl_4 or CH_3CN solution? Why?

4. Do the excited states that result from $n \to \pi^*$ and $\pi \to \pi^*$ transitions in pyridine belong to the same irreducible representation? To which species do they belong?

5. Under what conditions can electronic transitions occur in the infrared spectrum? Which compounds would you examine to find an example of this?

6. Explain why the transition that occurs in the ion pair *N*-methylpyridinium iodide does not occur in the solvent-separated ion pair.

7. Recall the center of gravity rule and explain why the blue shift in the iodine transition should be related to the heat of interaction of iodine with a donor (see Fig. 5–12).

8. What is the polarization expected in the $^1A_1 \to {}^1A_2(n \to \pi^*)$ transition of formaldehyde from vibronic coupling? The vibrational modes have a_1, b_1, and b_2 symmetry.

9. Two observed transitions in octahedral cobalt(II) complexes occur at 20,000 cm^{-1} and 8000 cm^{-1} with extinction coefficients of $\varepsilon \sim 50$ and 8, respectively. These have been assigned to $^4T_{1g} \to {}^4T_{1g}$ and $^4T_{1g} \to {}^4T_{2g}$. Indicate whether or not vibronic coupling can account for the intensity difference. (The vibrations are a_{1g}, e_g, t_{1u}, t_{2g}, and t_{2u}).

10. The vibrations of $CoCl_4^{2-}$ have symmetries corresponding to a_1, e, and $2t_2$. Explain whether or not the $A_2 \to T_2$ and $A_2 \to T_1$ d-d transitions can gain intensity by vibronic coupling.

11. Consider the cyclopropene cation shown below:

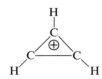

a. Using the three out-of-plane p orbitals (labeled φ_1, φ_2, and φ_3) as a basis set, do a simple Hückel calculation to obtain the energies of the resultant m.o.'s in terms of α and β.

b. The symmetries of these m.o.'s in the D_{3h} point group are A_2'' and E''. Briefly describe two methods you could use to find $\psi(A_2'')$ in terms of φ_1, φ_2, and φ_3.

c. Show whether the electronic transition to the first excited state is allowed. If so, what is its polarization?

d. What is the energy of this transition (in terms of α and β)?

e. What complications are likely in the approach used in part d?

12. Pyrazine has D_{2h} symmetry. The six pi-levels are shown below with a rough order of energies. The two unlabeled levels are the non-bonding m.o.'s from the nitrogen lone pairs.

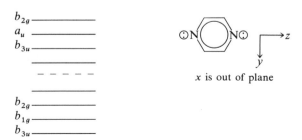

a. What are the symmetries of the two non-bonding m.o.'s?

b. What is the symmetry of the state of lowest energy that arises from a $\pi \to \pi^*$ transition? Is it allowed? If so, what is its polarization?

c. What is the symmetry of the states that arise from $n \to \pi^*$ transitions from the two non-bonding levels to the lowest π^* level? Determine whether either is allowed, and, if so, give the polarization. If either is forbidden, give the symmetry that a vibrational mode would need in order to lend intensity via the vibronic mechanism.

13. In the $Re_2Cl_8^{2-}$ ion, the transition of an electron from the b_{2g} orbital to the b_{1u} orbital is a d-d transition in a molecule with a center of inversion. Is it allowed? Explain.

$$\psi(b_{2g}) = \frac{1}{\sqrt{2}} (d_{xy}^{(1)} + d_{xy}^{(2)})$$

$$\psi(b_{1u}) = \frac{1}{\sqrt{2}} (d_{xy}^{(1)} - d_{xy}^{(2)})$$

14. In several coordination complexes of cis-butadiene, the 2,3 carbon-carbon bond distance was observed to be appreciably shorter than the normal single bond distance. The possibility of back donation of electron density from the metal into the π^* orbital of the butadiene was proposed to account for this shortening. This exercise is designed to provide some insights into the chemistry of this ligand.

   a. Do a simple Hückel calculation on the π orbitals of cis-butadiene. Obtain the energies of the π-symmetry orbitals. (Hint: set $y = x^2$.)

   b. Work out the symmetries of the π and π^* orbitals.

   c. The coefficients of the π molecular orbitals are given as follows.

	C_1	C_2	C_3	C_4
ψ_a	.60	−.37	−.37	.60
ψ_b	.37	.60	.60	.37
ψ_c	.37	−.60	.60	−.37
ψ_d	.60	.37	−.37	−.60

   (1) Assign the symmetry species of each molecular orbital.
   (2) Arrange them in order of increasing energy.

   d. Determine the symmetries of all four singly excited states of cis-butadiene. Which transitions are allowed? Give their polarizations.

   e. The lowest-energy $\pi \to \pi^*$ transition is observed at $\lambda = 217$ mμ in cis-butadiene. Calculate β.

   f. Calculate the 2,3 bond order for cis-butadiene and then calculate the 2,3 bond order with two additional electrons in the lowest π^* orbital. Do your results agree with the conclusions regarding π back bonding?

15. For trans-butadiene,

   a. determine the symmetries of the π molecular orbitals.

   b. determine the symmetries of all four singly excited states. Which transitions are allowed? Give their polarizations.

   c. compare the results in parts a and b with those obtained for cis-butadiene. Would an HMO calculation differentiate between these two rotamers?

16. Consider the π molecular orbitals obtained from the p_z atomic orbitals in NO_3^-. The ground state orbital energies and symmetries are:

   a. Obtain the state symmetries of the singly excited states.

   b. Which electronic transitions are allowed? Explain.

   c. NO_3^- has four vibrational modes, A_1', $2E'$, and A_2''. Does vibronic coupling provide a mechanism by which any forbidden transitions become allowed? Explain.

17. Show that the following two statements are equivalent for the point groups D_4, C_{3h}, and T_d:

   a. In order to observe circular dichroism, the transition from ground state to excited state must be simultaneously electric dipole-allowed and magnetic dipole-allowed. (The magnetic dipole moment along the α axis transforms as a rotation about the α axis, R_α.)

   b. In order to observe circular dichroism, the molecule must be optically active.

Vibration and Rotation Spectroscopy: Infrared, Raman, and Microwave

6

Introduction

6–1 HARMONIC AND ANHARMONIC VIBRATIONS

As discussed earlier (Chapter 4), quanta of radiation in the infrared region have energies comparable to those required for vibrational transitions in molecules. Let us begin this discussion by considering the classical description of the vibrational motion of a diatomic molecule. For this purpose it is convenient to consider the diatomic molecule as two masses, A and B, connected by a spring. In Fig. 6–1A the equilibrium position is indicated. If a displacement of A and B is carried out, moving them to A' and B', [as in Fig. 6–1B], there will be a force acting to return the system to the equilibrium position. If the restoring force exerted by the spring, f, is proportional to the displacement Δr, i.e.,

$$f = -k\,\Delta r \tag{6-1}$$

the resultant motion that develops when A' and B' are released and allowed to oscillate is described as *simple harmonic motion*. In equation (6–1), the Hooke's law constant for the spring, k, is called the *force constant* for a molecular system held together by a chemical bond.

For harmonic oscillation of two atoms connected by a bond, the potential energy, V, is given by

$$V = \tfrac{1}{2} k X^2$$

where X is the displacement of the two masses from their equilibrium position. A plot of the potential energy of the system as a function of the distance x between the masses is thus a parabola that is symmetrical about the equilibrium internuclear distance, r_e, as the minimum (see Fig. 6–2). The force constant. k, is a measure of the curvature of the potential well near r_e.

This classical springlike model does not hold for a molecule because a molecular system cannot occupy a continuum of energy states, but can occupy only discrete, quantized energy levels. A quantum mechanical treatment of the molecular system yields the following equation for the permitted energy states of a molecule that is a simple harmonic oscillator:

$$E_v = h\nu(v + \tfrac{1}{2}) \tag{6-2}$$

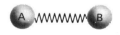

(A)

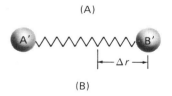

(B)

FIGURE 6–1 Displacement of the equilibrium position of two masses connected by a spring.

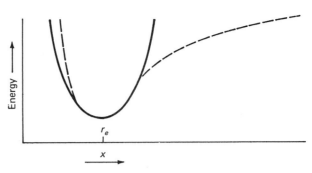

FIGURE 6–2 Potential energy versus distance, x, for (A) a harmonic oscillator (solid line) and (B) an anharmonic oscillator (dotted line).

where v is an integer 0, 1, 2, ..., representing the vibrational quantum number of the various states, E_v is the energy of the vth state, h is Planck's constant, and v is the *fundamental vibration frequency* (sec^{-1}) (*i.e.*, the frequency for the transition from state $v = 0$ to $v = 1$). These states are indicated for a harmonic oscillator in Fig. 6–3.

The potential energy curve of a real molecule (see Fig. 5–1), reproduced as a dotted line in Fig. 6–2, is not a perfect parabola. The vibrational energy levels are indicated in Fig. 5–1; they are not equally spaced, as equation (6–2) required, but converge. The levels converge because the molecule undergoes anharmonic rather than harmonic oscillation, *i.e.*, at large displacements the restoring force is less than predicted by equation (6–1). Note that as the molecule approaches dissociation, the bond becomes easier to stretch than the harmonic oscillator function would predict. This deviation from harmonic oscillation occurs in all molecules and becomes greater as the vibrational quantum number increases. As will be seen later, the assumption of harmonic oscillation will be sufficiently accurate for certain purposes (*e.g.*, the description of fundamental vibrations) and is introduced here for this reason.

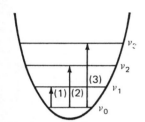

FIGURE 6–3 Vibrational state corresponding to a normal vibrational mode in a harmonic oscillator.

6–2 ABSORPTION OF RADIATION BY MOLECULAR VIBRATIONS—SELECTION RULES

The interaction of electromagnetic, infrared radiation with a molecule involves interaction of the oscillating electric field component of the radiation with an oscillating electric dipole moment in the molecule. Thus, *in order for molecules to absorb infrared radiation as vibrational excitation energy, there must be a change in the dipole moment of the molecule as it vibrates.* Consequently, the stretching of homonuclear diatomic molecules will not give rise to infrared absorptions. According to this selection rule, any change in direction or magnitude of the dipole during a vibration gives rise to an oscillating dipole that can interact with the oscillating electric field component of infrared radiation, giving rise to absorption of radiation. A vibration that results in a change in direction of the dipole is illustrated by the N—C—H bending mode of HCN. There is little change in the magnitude of the dipole, but an appreciable change in direction occurs when the molecule bends.

The second selection rule can be derived from the harmonic oscillator approximation. This selection rule, which is rigorous for a harmonic oscillator, states that in the absorption of radiation only transitions for which $\Delta v = +1$ can occur. Since most molecules are in the v_0 vibrational level at room temperature, most transitions will occur from the state v_0 to v_1. The transition is indicated by arrow (1) of Fig. 6–3. The frequency corresponding to this energy is called the *fundamental frequency*. According to this selection rule, radiation with energy corresponding to transitions indicated by arrows (2) and (3) in Fig. 6–3 will not induce transitions in the molecule. Since most molecules are not perfect harmonic oscillators, this selection rule breaks down and transitions corresponding to (2) and (3) do occur. The transition designated as (2) occurs at a frequency about twice that of the fundamental (1), whereas (3) occurs at a frequency about three times that of the fundamental. Transitions (2) and (3) are referred to as the *first* and *second overtones*, respectively. The intensity of the first overtone is often an order of magnitude less than that of the fundamental, and that of the second overtone is an order of magnitude less than the first overtone.

6–3 FORCE CONSTANT

The difference in energy, ΔE, between two adjacent levels, E_v and E_{v+1}, is given by equation (6–3) for a harmonic oscillator:

$$\Delta E = \left(\frac{h}{2\pi}\right)\left(\frac{k}{\mu}\right)^{1/2} \tag{6-3}$$

where k is the stretching force constant and μ is the reduced mass [$\mu = m_A m_B/(m_A + m_B)$ for the diatomic molecule A—B]. The relationship between energy and frequency, $\Delta E = hv = hc\bar{v}$, was presented in Chapter 4. The symbol v will be used interchangeably for frequency (sec^{-1}) or wavenumber (cm^{-1}), but the units will be indicated when necessary. In the HCl molecule, the absorption of infrared ratiation with $v = 2890$ cm^{-1} corresponds to a transition from the ground state to the first excited vibrational state. This excited state corresponds to a greater amplitude and frequency for the stretching of the H—Cl bond. Converting v to energy produces ΔE of equation (6–3). Since all other quantities are known, this equation can be solved to produce: $k = 4.84 \times 10^5$ dynes cm^{-1} or, in other commonly used units, 4.84 md Å^{-1}. In this calculation the reduced mass is expressed in grams. Stretching force constants for various diatomic molecules are summarized in Table 6–1.

The force constants in Table 6–1 are calculated by using equation (6–3), which was derived from the harmonic oscillator approximation. When an anharmonic oscillator model is employed, somewhat different values are obtained. For example, a force constant of 5.157×10^5 dynes cm^{-1} results for HCl. The latter value is obtained by measuring the first, second, and third overtones and evaluating the anharmonicity from the deviation of these frequencies from 2, 3, and 4 times the fundamental, respectively. Since these overtones are often not detected in larger molecules, we shall not be concerned with the details of the anharmonicity calculation.

The force constants for some other stretching vibrations of interest are listed in Table 6–2. For larger molecules, the nature of the vibration that gives rise to a particular peak in the spectrum is quite complex. Accordingly, one cannot calculate a force constant for a bond by substituting the "carbonyl frequency,"

TABLE 6–1 Stretching Force Constants for Various Diatomic Molecules (Calculated by the Harmonic Oscillator Approximation)

Molecule	v (cm^{-1})	k (dynes cm^{-1})
HF	3958	8.8×10^5
HCl	2885	4.8×10^5
HBr	2559	3.8×10^5
HI	2230	2.9×10^5
F$_2$[a]	892	4.5×10^5
Cl$_2$[a]	557	3.2×10^5
Br$_2$[a]	321	2.4×10^5
I$_2$[a]	213	1.7×10^5
CO	2143	18.7×10^5
NO	1876	15.5×10^5

[a] Observed by Raman spectroscopy.

TABLE 6–2. Stretching Force Constants for Various Stretching Vibrations (Harmonic Oscillator Approximation)

Bond	k (dynes cm^{-1})
⩾C—C⩽	4.5×10^5
⩾C—C≡	5.2×10^5
⩾C=C⩽	9.6×10^5
—C≡C—	15.6×10^5
⩾C=O	12.1×10^5
—C≡N	17.7×10^5
≡C—H	5.9×10^5
⩾C—H	4.8×10^5

for example, of a complex molecule into equation (6–3). This will become clearer as we proceed and is mentioned here as a note of caution. The force constants in Table 6–2 result from a normal coordinate analysis, which will also be discussed in more detail shortly. A larger force constant is often interpreted as being indicative of a stronger bond, but *there is no simple relation between bond dissociation energy and force constant*. We defined the force constant earlier as a measure of the curvature of the potential well near the equilibrium internuclear configuration. The curvature is the rate of change of the slope, so the force constant is the second derivative of the potential energy as a function of distance:

$$k = \left(\frac{\partial^2 V}{\partial r^2}\right)_{r \to 0} \quad (6\text{–}4)$$

Here V is the potential energy and r is the deviation of the internuclear distance, at which $r = 0$. In a more complicated molecule, r is replaced by q, which is a composite coordinate that describes the vibration.

Triple bonds have stretching force constants of 13 to 18×10^5, double bonds about 8 to 12×10^5, and single bonds below 8×10^5 dynes cm^{-1}. In general, force constants for bending modes are often about a tenth as large as those for stretching modes.

The bands in the 4000 cm^{-1} to 600 cm^{-1} region of the spectrum mostly involve stretching and bending vibrations. Most of the intense bands above 2900 cm^{-1} involve hydrogen stretching vibrations for hydrogen bound to a low-mass atom. This frequency range decreases as the X—H bond becomes weaker and the atomic weight of X increases. Triple bond stretches occur in the 2000 to 2700 cm^{-1} region. Absorption bands assigned to double bond stretches occur in the 1500 to 1700 cm^{-1} region. Bands in this region of the spectrum of an unknown molecule are a considerable aid in structure determination. Bands in the 1500 to 400 cm^{-1} region find utility in the fingerprint type of application.

Metal-ligand vibrations usually occur below 400 cm^{-1} and into the far infrared region. They are very hard to assign, since many ligand ring deformation and rocking vibrations as well as lattice modes (vibrations involving the whole crystal) occur in this region.

As indicated by equation (6–3), the reduced mass is important in determining the frequency of a vibration. If, for example, a hydrogen bonded to carbon is replaced by deuterium, there will be a *negligible change in the force constant but an appreciable change in the reduced mass.* As indicated by equation (6–3), the frequency should be lower by a factor of about $1/\sqrt{2}$. The frequencies for C-H and C-D vibrations are proportional to $[12 \times 1/(12 + 1)]^{-1/2}$ and $[12 \times 2/(12 + 2)]^{-1/2}$, respectively. Normally, a vibration involving hydrogen will occur at 1.3 to 1.4 times the frequency of the corresponding vibration in the deuterated molecule. This is of considerable utility in confirming assignments that involve a hydrogen atom. The presence of the natural abundance of ^{13}C in a metal carbonyl gives separate bands due to $^{13}C\equiv O$ and $^{12}C\equiv O$ stretching vibrations in metal carbonyls (*vide infra*) because of the difference in reduced mass. Use of metal isotopes also has utility in confirming the assignment of vibrations involving the metal-ligand bond.[1]

6–4 THE 3N – 6(5) RULE

Vibrations in a Polyatomic Molecule

The positions of the N atoms in a molecule can be described by a set of Cartesian coordinates, and the general motion of each atom can be described by utilizing three displacement coordinates. The molecule is said, therefore, to have $3N$ degrees of freedom. Certain combinations of these individual degrees of freedom correspond to translational motion of the molecule as a whole without any change in interatomic dimensions. There are three such combinations which represent the x, y, and z components of translational motion, respectively. For a nonlinear molecule there are three combinations that correspond to rotation about the three principal axes of the molecule without change in interatomic dimensions. Therefore, *for a nonlinear molecule there are $3N - 6$ normal modes of vibration* that result in a change in bond lengths or angles in the molecule. *Normal modes represent independent self-repeating motions in a molecule.* They correspond to $3N - 6$ degrees of freedom that, in the absence of anharmonicity, could take up energy independently of each other. *These modes form the bases for irreducible representations.* Since a molecule is fundamentally not changed by applying a symmetry operation R, the normal mode $R\hat{Q}$ must have the same frequency as the normal mode $\hat{Q}$. Thus, if $\hat{Q}$ is non-degenerate, $R\hat{Q} = \pm 1\hat{Q}$ for all R's. Consequently, $\hat{Q}$ forms the basis for a one-dimensional representation in the molecular symmetry group. It can be shown that degenerate modes transform according to irreducible representations of dimensionality greater than one. The center of mass of the molecule does not change in the vibrations associated with the normal mode, nor is angular momentum involved in these vibrations. *All general vibrational motion that a molecule may undergo can be resolved into either one or a combination of these normal modes.*

For a linear molecule all the vibrations can be resolved into $3N - 5$ normal modes. The additional mode obtained for a linear molecule is indicated in Fig. 6–4(B), where plus signs indicate motion of the atoms into the paper and minus

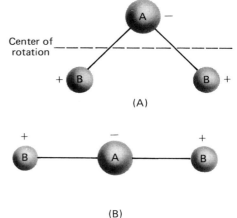

FIGURE 6–4 Rotational and bending modes for (A) non-linear and (B) linear molecules.

signs represent motion out of the paper. For the nonlinear molecule [Fig. 6–4(A)] the motion indicated corresponds to a rotation. For a linear molecule a similar motion corresponds to a bending of the bonds, and hence this molecule has an additional normal vibrational mode ($3N - 5$ for a linear molecule vs. $3N - 6$ for a nonlinear one).

As will be seen later, there are many applications for which we need to know which bands correspond to the fundamental vibrations.

6–5 EFFECTS GIVING RISE TO ABSORPTION BANDS

Sulfur dioxide is predicted to have three normal modes from the $3N - 6$ rule. The spectral data (Table 6–3) show the presence of more than three bands. The three bands at 1361, 1151, and 519 cm^{-1} are the fundamentals and are referred to as the v_3, v_1, and v_2 bands, respectively (see Fig. 6–5). The v_n symbolism is used to label the various frequencies of fundamental vibrations and should not be confused with the symbols v_0, v_1, v_2, etc., used to designate various vibrational levels of one mode in a molecule. By convention the highest-frequency totally symmetric vibration is called v_1, the second highest totally symmetric vibration v_2, etc. When the symmetric vibrations have all been assigned, the highest-frequency asymmetric vibration is counted next, followed by the remaining asymmetric vibrations in order of decreasing frequency. An exception is made to this rule for the bending vibration of a linear molecule, which is labeled v_2. Another common convention involves labeling stretching vibrations v, bending vibrations δ, and out-of-plane bending vibrations π. Subscripts, as, for asymmetric; s, for symmetric; and d, for degenerate, are employed with these symbols.

The v_1 mode in SO$_2$ is described as the *symmetric stretch*, v_3 as the *asymmetric stretch*, and v_2 as *the O—S—O bending mode*. In general, the asymmetric stretch will occur at higher frequency than the symmetric stretch, and stretching modes occur at much higher frequencies than bending modes. There is a slight angle change in the stretching vibrations in order for the molecule to retain its center of mass. The other absorption frequencies in Table 6–3 are assigned as indicated. The overtone of v_1 occurs at about $2v_1$ or 2305 cm^{-1}. The bands at 1871 and 2499 cm^{-1} are referred to as *combination bands*. Absorption of radiation of these

TABLE 6–3. Infrared Spectrum of SO$_2$

v (cm^{-1})	Assignment
519	v_2
606	$v_1 - v_2$
1151	v_1
1361	v_3
1871	$v_2 + v_3$
2305	$2v_1$
2499	$v_1 + v_3$

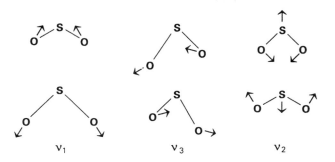

FIGURE 6-5 The three fundamental vibrations for sulfur dioxide (The amplitudes are exaggerated to illustrate the motion.)

energies occurs with the simultaneous excitation of both vibrational modes of the combination. The 606 cm^{-1} band is a *difference band*, which involves a transition originating from the state in which the v_2 mode is excited and changing to that in which the v_1 mode is excited. Note that all the bands in the spectrum are accounted for by these assignments. Making the assignments is seldom this simple; as will be shown later, much other information, including a normal coordinate analysis, is required to substantiate these assignments.

A more complicated case is the CO_2 molecule, for which four fundamentals are predicted by the $3N - 5$ rule. A single band results from the two degenerate vibrations v_2 of Fig. 6-6, which correspond to bending modes at right angles to each other. Later we shall see how symmetry considerations aid in predicting the number of degenerate bands to be expected. In more complex molecules some of the fundamentals may be accidentally degenerate because two vibration frequencies just happen to be equal. This is not easily predicted, and the occurrence of this phenomenon introduces a serious complication. The assignment of the fundamentals for CO_2 is more difficult than for SO_2 because many more bands appear in the infrared and Raman spectra. Bands at 2349, 1340, and 667 cm^{-1} have been assigned to v_3, v_1, and v_2, respectively. The tests of these assignments have been described in detail by Herzberg (see Additional References) and will not be repeated here. In this example the fundamentals are the three most intense bands in the spectrum. In some cases, there is only a small dipole moment change in a fundamental vibration, and the corresponding absorption band is weak (see the first selection rule).

The above discussion of the band at 1340 cm^{-1} has been simplified. Actually, it is an intense doublet with band maxima at 1286 and 1388 cm^{-1}. This splitting is due to a phenomenon known as *Fermi resonance*. The overtone $2v_2$ ($2 \times 667 = 1334$ cm^{-1}) and the fundamental v_1 should occur at almost the same

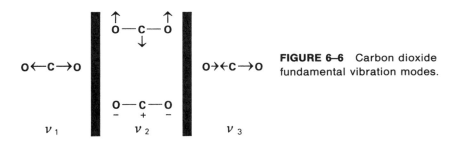

FIGURE 6-6 Carbon dioxide fundamental vibration modes.

frequency. The two vibrations interact by a typical quantum mechanical resonance, and the frequency of one is raised while the frequency of the other is lowered. The wave function describing these states corresponds to a mixing of the wave functions of the two vibrational excited states (v_1 and $2v_2$) that arise from the *harmonic oscillator approximation*. We cannot say that one line corresponds to v_1 and the other to $2v_2$, for both are mixtures of v_1 and $2v_2$. This interaction also accounts for the high intensity of what, in the absence of interaction, would have been a weak overtone ($2v_2$). The intensity of the fundamental is distributed between the two bands, for both bands consist partly of the fundamental vibration.

The presence of Fermi resonance can sometimes be detected in more complex molecules by examining deuterated molecules or by determining the spectrum in various solvents. Since the Fermi resonance interaction requires that the vibrations involved have nearly the same frequency, the interaction will be affected if one mode undergoes a frequency shift from deuteration or a solvent effect while the other mode does not. The two frequencies will no longer be equivalent, and the weak overtone will revert to a weak band or not be observed in the spectrum. Other requirements for the Fermi resonance interaction will be discussed in the section on symmetry considerations.

6-6 NORMAL COORDINATE ANALYSES AND BAND ASSIGNMENTS

Degeneracy, vibration frequencies outside the range of the instruments, low intensity fundamentals, overtones, combination bands, difference bands, and Fermi resonance all complicate the assignment of fundamentals. The problem can sometimes be resolved for simple molecules by a technique known as *normal coordinate analysis*. Normal coordinate analysis involves solving the classical mechanical problem of the vibrating molecule, assuming a particular form of the potential energy (usually the valence force field). The details of this calculation are beyond the scope of this text,[2,3] but it is informative to outline the problem briefly so the reader can assess the value and the limitations of the approach. Furthermore, several important qualitative ideas will be developed that we shall use in subsequent discussion. When we have finished, you will not know how to do a normal coordinate analysis, but hopefully you will have a rough idea of what is involved.

Just as the electronic energy and electronic wave functions of a molecule are related by a secular determinant and secular equations (Chapter 3), the vibrational energies, vibrational wave functions, and force constants are related by a secular determinant and a series of secular equations. The vibrational secular determinant will be given here as (for the derivation see references 2 and 3):

$$\begin{vmatrix} F_{11} - (G^{-1})_{11}\lambda & F_{12} - (G^{-1})_{12}\lambda & \cdots & F_{1n} - (G^{-1})_{1n}\lambda \\ \vdots & & & \vdots \\ F_{n1} - (G^{-1})_{n1}\lambda & F_{n2} - (G^{-1})_{n2}\lambda & \cdots & F_{nn} - (G^{-1})_{nn}\lambda \end{vmatrix} = 0$$

Here $\lambda = 4\pi^2 v^2$, where v is the vibrational frequency (note the resemblance of this $G\lambda$ term to the energy term in a molecular orbital calculation); this quantity is known, for one begins by making a tentative assignment of all the normal modes to the bands in the spectrum. The basis set for this calculation (*i.e.*, the counterpart of atomic orbitals in an m.o. calculation) is the set of internal coordinates of the molecule expressed in terms of $3N - 6$ atom displacements, L. Three such coordinates, needed to describe the normal modes for the water molecule, are illustrated in Fig. 6–7. The selection of these internal ccordinates is complicated in larger highly symmetric molecules because redundant coordinates can result. One proceeds by selecting an internal displacement vector for a change in bond length for each bond in the molecule and then selecting *independent* bond angles to give the $3N - 6$ basis set. The normal mode is going to be some combination of this basis set of internal displacement coordinates, and the vibrational wave function will tell us what this combination will be (again note the resemblance of this to an electronic wave function for a molecule consisting of the atomic orbital bases set). In the secular determinant given above, F_{11} is the force constant for stretching the O—H bond along L_{11}, F_{22} is the force constant for stretching along L_{22}, and F_θ is related to the bend along L_θ. The off-diagonal element F_{21} in the vibrational secular determinant is called an *interaction force constant*, and it indicates how the two isolated stretches interact with one another. When, for example, L_{22} is subjected to a unit displacement, the bond along L_{11} will distort to minimize the potential energy of the strained molecule. F_{21} is roughly proportional to the displacement of the bond along L_{22} resulting from minimization of the energy of the molecule after displacement along L_{11}. Part of the interaction relates to how the bond strength along L_{22} changes as the oxygen rehybridizes when the bond along L_{11} is stretched. F_{ij} is not necessarily identical to F_{ji}.

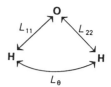

FIGURE 6–7 Internal coordinates for the H$_2$O molecule.

The F-matrix elements account for the potential energies of the vibration. The G-matrix elements contain information about the kinetic energy of the molecule. The latter can be written exactly for a molecule from formulas given by Wilson, Decius, and Cross[2] if we know the atom masses, the molecular bond distances, and the bond angles.* This is a symmetric matrix. In practice, the force constants are usually the only unknowns in the secular determinant, and they can be determined. (Now the problem differs from the format of the molecular orbital calculation.)

The secular determinant given above can be written in matrix notation as:

$$|\mathbf{F} - \mathbf{G}^{-1}\lambda| = 0$$

If we multiply by $\mathbf{G}$ we get

$$|\mathbf{GF} - \mathbf{E}\lambda| = 0 \qquad (6\text{–}5)$$

---

*The evaluation of the G-matrix for a bent XY_2-molecule has been worked out in detail in reference 18, Section I–11. The reader is referred to this treatment for further details.

where **E** is the unit matrix. Furthermore, the **G** matrix has the property that the λ's in the off-diagonal elements of the previous secular determinant have been eliminated. The problem now is to solve

$$\begin{vmatrix} G_{11} & \cdots & G_{1n} \\ \vdots & & \vdots \\ G_{n1} & & G_{nn} \end{vmatrix} \begin{vmatrix} F_{11} & \cdots & F_{1n} \\ \vdots & & \vdots \\ F_{n1} & & F_{nn} \end{vmatrix} - \begin{vmatrix} 1 & \cdots & 0 & \cdots & 0 \\ \vdots & & 1 & & \\ 0 & & & & 1 \end{vmatrix} \lambda = 0$$

where the G elements are all known, as are the λ's. There are in general n values of λ that satisfy this equation, and they are known. The difficulty with normal coordinate analyses on most molecules is that there are more unknown force constants than there are frequencies. For the water molecule, the force constant matrix is 3×3, and there are four unknowns* for the three frequencies. If one deuterates the water molecule, new frequencies are obtained, but no new force constants are introduced. The ^{16}O, ^{17}O, and ^{18}O molecules could also be studied. The problem here is whether or not the new equations we obtain in trying to get more simultaneous equations than force constants are different enough to allow for a unique, meaningful solution. (This is similar to the K^{-1} vs. ε problem discussed in Chapter 4. Two parallel or nearly parallel equations do not solve the problem, even though in the latter case a computer can pick a minimum.) In the simpler molecules, this is often not a problem; but in more complex molecules, the analysis is not to be believed unless errors are reported and some statistical criterion of the significance of the fit is presented. This is readily accomplished with a correlation matrix that indicates how extensively the individual force constants are correlated. This problem is potentially so severe that one should not accept the results from a force constant analysis unless the correlation matrix is reported.

Suppose we were to do a normal coordinate analysis for Mn(CO)$_5$X. We would have 30 internal displacement coordinates ($3N - 6$) and a 30×30 force constant matrix, or many force constants. Symmetry can reduce this matrix to smaller blocks by requiring certain interaction constants to be zero (*vide infra*), but there remain several approximations that must be introduced to solve the problem. In one analysis of the problem, one could assume that the carbonyl stretches are so far removed in frequency from any of the other vibrations that they can be treated separately.[4] This is equivalent to claiming that the metal-carbon stretch does not influence the C—O vibration and setting all interaction constants of the C≡O stretches with anything else equal to zero. This leads to a 5×5 block for the total secular determinant of force constants. Isotopic substitution is then employed to solve the problem. In a more thorough analysis,[5] the interaction constants between M—C and C—O stretch coordinates that are *trans* to each other have been determined and found to be significant.

Cotton and Kraihanzel[4] have proposed a crude approximation for these systems. They arbitrarily set any interaction force constant F_{trans} for carbonyl moieties that are trans to one another equal to $2F_{cis}$, where F_{cis} corresponds to interaction constants involving groups that are *cis*. The numbers of unknowns and frequencies become comparable in many carbonyl compounds when this is done. Jones[5] has carried out a more rigorous evaluation of this problem and

---

* The symmetry of the water molecule leads to this.

suggests that severe limitations be placed on systems for which the Cotton-Kraihanzel approximations are invoked. It has been shown in the more complete study that F_{cis} is considerably larger than F_{trans}. Another approach that is being investigated to facilitate normal coordinate analyses involves transferring interaction force constants from a simple molecule, in which they are well known, to larger, similar systems.

Let us next consider a simple system, which we assume to have been solved for meaningful force constants. The force constant values can be substituted into the **G** and **F** matrices previously given, and the secular equations are now written by multiplying by a matrix of the basis set **L**, the internal displacements; *i.e.*,

$$|\mathbf{GF} - \mathbf{E}\lambda_n| |\mathbf{L}| = 0 \qquad (6\text{-}6)$$

In the case of water, **F** is 3×3, **G** is 3×3, **E** is 3×3 unit matrix, **L** is 3×1 (L_{11}, L_{22}, and L_{33}), and λ_n is the eigenvalue for which we wish to determine the normal mode. The energies are now substituted one at a time. Matrix multiplications then yield the contribution of the basis set (*i.e.*, the contribution of the individual internal displacements) to the vibrational wave function for the normal mode corresponding to that frequency. This is done three times for the water molecule with the three frequencies, leading to the wave functions for the three normal modes.

For the water molecule, the wave functions for the $3N - 6$ vibrations are:

$$\psi_1 = NL_\theta + N'(L_{11} + L_{22}) \qquad N \gg N'$$
$$\psi_2 = N(L_{11} + L_{22}) + N'L_\theta$$
$$\psi_3 = \frac{1}{\sqrt{2}}(L_{11} - L_{22})$$

The first two have A_1 symmetry, and the last has B_1 symmetry.

If we had a diatomic molecule (*e.g.*, HCl), all of the above matrices would be 1×1 and we would have from equation (6-6)

$$F_{11}G_{11} - \lambda = 0$$

where the G matrix element is the reduced mass and F_{11} is the only unknown. Rearranging, we see that

$$\frac{F_{11}}{\mu_{HCl}} = 4\pi^2 v^2$$

or

$$v = \sqrt{\frac{F_{11}}{4\pi^2 \mu}} \qquad (6\text{-}7)$$

The resemblance of this result to equation (6-3) is clear for $\Delta E = hv$. Now we see why equation (6-3) can be used to give the force constant for a bond in a diatomic molecule but not in a more complex molecule. In larger, *more complex* molecules, the observed frequency corresponds to a more complex vibration that depends upon several bond force constants. The results of a normal coordinate analysis produce the force constants and indicate exactly the form of each normal mode in terms of the internal coordinates. For water, for example, the internal coordinates in Fig. 6–7 are combined to produce normal modes that are similar in form to those shown for SO_2 in Fig. 6–5. This is expected because the internal displacement coordinates (which are of the same form for SO_2 and H_2O) can form the basis for producing a total representation by employing the symmetry operations of the point group on them. Factoring the total representation produces the symmetry of the irreducible representations, which in this case are the normal modes. For the water molecule, the total representation obtained by operating with the C_{2v} point group on the internal displacements shown in Fig. 6–7 is 3 1 3 1 (the angle θ is not changed by σ_v or C_2). Factoring the total representation produces $2A_1 + B_1$. Next, projection operators can be used on this basis set just as on an a.o. basis set to produce

$$\hat{P}(A_1)L_{11} \cong L_{11} + L_{22}$$
$$\hat{P}(A_1)L_\theta \cong L_\theta \qquad\qquad (6\text{–}8)$$
$$\hat{P}(B_1)L_{11} \cong L_{11} - L_{22}$$

This problem nicely illustrates the analogy between the symmetry aspects of the molecular orbital problem and those of the vibrational problem.

The inverse of the force constant matrix produces a matrix of *compliance constants*.[6] A diagonal compliance constant is a measure of the displacement that will take place in a coordinate as a result of a force imposed upon this coordinate, if the other coordinates are allowed to adjust to minimize the energy. There are reported advantages to employing compliance constants instead of force constants, and the reader is referred to the literature for details.[6]

6–7 GROUP VIBRATIONS AND THE LIMITATIONS OF THIS IDEA

The idea that we can look at the spectrum of a complex molecule and assign bands in the spectrum to various functional groups in the molecule is called the *group vibration* concept. This approach arose from the experimental observation that many functional groups absorb in a narrow region of the spectrum regardless of the molecule that contains the group. In the acetone molecule,* for example, one of the normal modes of vibration consists of a C—O stretching motion with negligible motion of the other atoms in the molecule. Similarly, the methyl groups can be considered to undergo vibrations that are independent of the motions of the carbonyl group. In various molecules it is found that carbonyl absorptions due to the stretching vibration occur in roughly the same spectral region (~ 1700 cm^{-1}). As will be seen in the section on applications, the position of the band does vary slightly (± 150 cm^{-1}) because of the mass, inductive, and conjugative effects of the groups attached. The methyl group has five characteristic

---

* This molecule has been analyzed by a normal coordinate treatment.

absorption bands: two bands in the region from 3000 to 2860 (from the asymmetric and symmetric stretch), one around 1470 to 1400 (the asymmetric bend), one around 1380 to 1200 (the symmetric bend), and one in the region from 1200 to 800 cm^{-1} (the rocking mode, in which the whole methyl group twists about the C—C bond, undergoing a rocking-chair type of motion). The concept of group vibrations involves dissecting the molecule into groups and assigning one or more bands in the spectrum to the vibrations of each group. Many functional groups in unknown compounds have been identified by using this assumption. Unfortunately, in many complicated molecules there are many group vibrations that overlap, and assignment of the bands in a spectrum becomes difficult. It is possible to perform experiments that help resolve this problem: deuteration will cause a shift in vibrations involving hydrogen (*e.g.*, C—H, O—H, or N—H stretches and bends) by a factor of 1.3 to 1.4; characteristic shifts, which aid in assignments, occur with certain donor functional groups (*e.g.*, C=O) when the spectrum is examined in hydrogen-bonding solvents or in the presence of acidic solutes.

The limitations of the group vibration concept should be emphasized so that incorrect interpretations of infrared spectral data can be recognized. The group vibration concept implies that the vibrations of a particular functional group are relatively independent of the rest of the molecule. If the center of mass is not to move, this is impossible. All of the nuclei in a molecule must undergo their harmonic oscillations in a synchronous manner in normal vibrations. In view of the discussion in the previous section, we can see that the group vibration concept will be good when the normal mode wave function consists largely (80 to 90%) of the internal displacement in which only the group is involved, *i.e.*, the functional group vibration is the main internal displacement coordinate. This will occur when the symmetry properties of the molecule are such as to restrict the combination of the internal displacements of the group with other internal displacement coordinates; *e.g.*, L_θ in the water molecule cannot contribute to the B_1 asymmetric stretch. When the vibrational motions involving the two internal displacement coordinates are very different in energy, then the off-diagonal interaction constants are small, the two internal displacement coordinates are not mixed extensively in the vibrational wave functions, and the vibrations can be treated separately as group vibrations. This is equivalent to the assumption made when we treated the carbonyl vibrations in M(CO)$_5$X separately to obtain the 5 × 5 matrix.

When the atoms in a molecule are of similar mass and are connected by bonds of comparable strength (*e.g.*, all single bonds as in BF$_3$·NH$_3$), all the normal modes will be mixtures of several internal displacement vectors. For example, in BF$_3$·NH$_3$ it would be impossible to assign a band to the B—N stretching vibration because none of the normal modes correspond predominantly to this sort of motion. When this occurs, the various group vibrations are said to be coupled. This term simply describes a situation in which the very crude group vibration concept is not applicable to the description of the normal mode corresponding to a particular absorption band. This discussion should be reread if the reader cannot distinguish coupling from the phenomena of Fermi resonance or combination bands.

The HCN molecule is interesting to consider in the context of the above discussion. The frequencies of a pure C—H and pure C—N stretch are similar in energy. The C—H and C—N group stretching vibrations in the H—C—N molecule are coupled. The absorption band attributed to the C—H stretch actually involves C—N motion to some extent and vice versa; *i.e.*, the observed frequencies cannot be described as being a pure C—H and a pure C—N stretch. Evidence to support this comes from deuteration studies, in which it is found that deuteration affects the frequency of the band that might otherwise be assigned to the C—N stretch. This absorption occurs at 2089 and 1906 cm^{-1} in the molecules H—C—N and D—C—N, respectively. Equation (6–3) would predict that deuteration should have very little effect on the C—N frequency, for it has little effect on the C—N reduced mass. The only way the frequency can be affected is to have a coupling of the C—D stretch with the C—N stretch. Upon deuteration, the C—H stretch at 3312 cm^{-1} is replaced by a C—D stretch at 2629 cm^{-1}. Since the C—D frequency is closer to the C—N frequency than is that of C—H, there is more extensive coupling in the deuterated compound. Because the symmetries of the normal modes differ, the C—D bending vibration does not couple with the "C—N stretch."

It would be improper to draw conclusions concerning the strength of the C—H bond from comparison of the frequency for the C—H stretch in H—CN with data for other C—H vibrations. The force constants for the C—H bonds can be calculated for the various compounds by a normal coordinate analysis. These values should be employed for such comparisons.

As another example of the difficulties encountered in the interpretation of frequencies, consider the stretching vibration for some carbonyl groups. Absorption bands occur at 1928, 1827, and 1828 cm^{-1} in F_2CO, Cl_2CO, and Br_2CO, respectively.[7] One might be tempted to conclude from the frequencies that the high electronegativity of fluorine causes a pronounced increase in the C—O force constant. However, a normal coordinate analysis[7] shows that the C—O and C—F vibrations are coupled and that the C—O stretch also involves considerable C—F motion. The frequency of this normal mode is higher than would be expected for an isolated C—O stretching vibration with an equivalent force constant. There is a corresponding lowering of the C—F stretching frequency. Since chlorine and bromine are heavier, they make little contribution to the carbonyl stretch. The normal coordinate analysis indicates that the C—O stretching force constants are 12.85, 12.61, and 12.83 millidynes/Å in F_2CO, Cl_2CO, and Br_2CO, respectively.

A normal coordinate analysis and infrared studies of ^{12}C-, ^{13}C-, ^{16}O-, ^{18}O-, and H- and D-substituted ketones containing an ethyl group attached to the carbonyl group indicate[8] that the band near 1750 cm^{-1} assigned to the C—O stretch in these ketones corresponds to a normal mode that consists of about 75% C—O and 25% C—C stretching motions. These examples should be kept in mind whenever one is tempted to interpret the frequencies of infrared bands in terms of electronic effects. Force constants obtained from normal coordinate analyses should be compared; frequencies read directly from spectra should not be. Unfortunately, reliable force constants are in limited supply.

Raman Spectroscopy

6–8 INTRODUCTION

Raman spectroscopy is concerned with vibrational and rotational transitions, and in this respect it is similar to infrared spectroscopy. Since the selection rules are different, the information obtained from the Raman spectrum often comple-

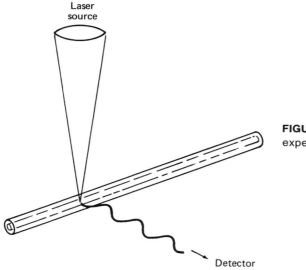

FIGURE 6–8 The Raman experiment (schematic).

ments that obtained from an infrared study and provides valuable structural information.

In a Raman experiment, a monochromatic beam of light illuminates the sample, and observations are made on the scattered light at right angles to the incident beam (Fig. 6–8). Monochromatic light sources of different frequencies can be employed; gas lasers are commonly used for this purpose. Various sample holder configurations are employed[9]; the only requirement is that the detector must be at right angles to the source. This configuration is illustrated in Fig. 6–8 for a capillary cell. Absorption of the monochromatic light beam, leading to decomposition, can be a problem, as can fluorescence. These problems are minimized by choice of an appropriate gas laser[10] line: the He—Ne laser gives a line at 6328 Å (red); the Ar laser gives lines at 4579, 4658, 4765, 4880, 4915, and 5145 Å (blue-green) (the Ar^+ 4880 Å and 5145 Å lines are usually used); and the Kr laser gives 5682 and 6471 Å lines. The introduction of tunable dye lasers has extended the range of usable wavelengths even further. Use of a narrow bandpass filter for each laser line reduces effects from the large number of laser ghosts. A valuable discussion of various aspects of this experiment is contained in reference 11.

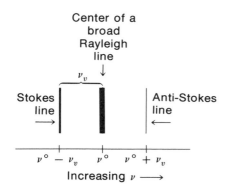

FIGURE 6–9 Lines observed in a Raman spectrum (schematic).

We should wonder how we obtain vibrational transitions when we employ a visible or u.v. source. A quantum of the incident light of frequency $v°$ and energy $hv°$ can collide with a molecule and be scattered with unchanged frequency. This is referred to as Rayleigh scattering. The mechanism involves inducing a dipole moment, D, in the molecule when it is in the field of the electric vector of the radiation. The electrons in the molecule are forced into oscillations of the same frequency as the radiation. This oscillating dipole radiates energy in all directions and accounts for the Rayleigh scattering. If the photon is actually absorbed in the process and re-emitted, this is fluorescence. The difference between scattering and fluorescence is thus a subtle one, having to do with the lifetime of the species formed in the photon-molecule collision.

The scattering process just described corresponds to an elastic collision of the molecule with the photon. Raman scattering involves an inelastic collision. The molecule in its ground vibrational state accepts energy from the photon being scattered, exciting the molecule into a higher vibrational state, while the incident radiation now becomes scattered with energy $h(v° - v_v)$. In the Raman spectrum of the scattered light measured at right angles, we now detect a frequency $v° - v_v$, called the Stokes line, as shown in Fig. 6–9. The measured value of v_v is identical to the infrared frequency that would excite this vibrational mode if it were infrared-active.

A molecule in the vibrationally excited state $v = 1$ can collide with an incident light quantum of frequency $v°$. The molecule can return to the ground state by giving its additional energy hv_v to the photon. This photon, when scattered, will have a frequency $v° + v_v$. The spectral line with this frequency in the Raman spectrum is referred to as an anti-Stokes line (see Fig. 6–9). Because of the Boltzmann distribution, there are fewer molecules in the $v = 1$ state than in $v = 0$, and the intensity of the anti-Stokes line is much lower than that of the Stokes line.

Both Rayleigh and Raman scattering are relatively inefficient processes. Only about 10^{-3} of the intensity of the incident exciting frequency will appear as Rayleigh scattering, and only about 10^{-6} as Raman scattering. As a result, very intense sources are required in this experiment. Laser beams provide the required intensity (~ 100 milliwatts to ~ 1 watt of power) and produce good spectra even with very small samples.

6–9 RAMAN SELECTION RULES

The molecule interacts with electromagnetic radiation in the Raman experiment via the oscillating induced dipole moment or, more accurately, the oscillating molecular polarizability. The intensity, I, of the scattered radiation, v_s, is a function of the polarizability, α_{ij}, of the molecule which, for a randomly oriented solid, is given by:

$$I = \frac{2^7 \pi^5}{3^2 c^4} I_0 v_s^4 \sum_{ij} \alpha_{ij} \tag{6–9}$$

where I_0 is the intensity of the incident radiation and α_{ij} is an element of the molecular polarizability tensor. Molecular polarizability can be represented by an ellipsoid; the change in polarizability (as the CO_2 molecule, for example, undergoes the symmetric stretch) can be schematically illustrated as in Fig. 6–10.

symmetric stretching displacements

(extended)   (equilibrium)   (compressed)

corresponding polarizability ellipsoid

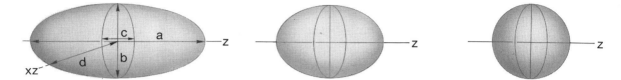

FIGURE 6–10 The change in polarizability for the CO_2 molecule for various displacements of the atoms during the symmetric stretching vibration. The double-headed arrows in the ellipsoid indicate (a) the z-axes of the ellipsoid, (b) the x, (c) the y, and (d) the xz.

When a molecule is placed in a static electric field, the nuclei are attracted toward the negative pole and the electrons toward the positive pole, inducing a dipole moment in the molecule. If D represents the induced dipole moment and E the electric field, the polarizability α is defined by the following equation:

$$D = \alpha E \tag{6-10}$$

The magnitude of this polarizability depends upon the orientation of the bonds in the molecule with respect to the electric field direction; *i.e.*, it is anisotropic. In most molecules it is a tensor quantity. The induced dipole moment in the molecule will depend upon the orientation of the molecule with respect to the field and, since the dipole moment is a vector, it will have three components D_x, D_y, and D_z. The polarizability tensor can be represented physically as an ellipsoid (see Fig. 6–10), and it has nine components that can be described by a 3 × 3 matrix. The induced dipole moments are related to the polarizability tensor by the following equations:

$$\begin{aligned} D_x &= \alpha_{xx} E_x + \alpha_{xy} E_y + \alpha_{xz} E_z \\ D_y &= \alpha_{yx} E_x + \alpha_{yy} E_y + \alpha_{yz} E_z \\ D_z &= \alpha_{zx} E_x + \alpha_{zy} E_y + \alpha_{zz} E_z \end{aligned} \tag{6-11}$$

These equations state that if we locate a molecule in an electric field so that the ellipsoid axes point between the Cartesian axes, then the x-component of the induced dipole moment will depend upon the magnitude of the electric field along the y-axis because there is a polarizability component along xy that can interact with E_y and make a contribution to the change observed along x. The same is true for the interaction of E_z with α_{xz}. Equations (6–11) can be written in matrix form as

$$\begin{bmatrix} D_x \\ D_y \\ D_z \end{bmatrix} = \begin{bmatrix} \alpha_{xx} & \alpha_{xy} & \alpha_{xz} \\ \alpha_{yx} & \alpha_{yy} & \alpha_{yz} \\ \alpha_{zx} & \alpha_{zy} & \alpha_{zz} \end{bmatrix} \begin{bmatrix} E_x \\ E_y \\ E_z \end{bmatrix}$$

For an optically inactive molecule, the α tensor is symmetric and

$$\alpha_{xy} = \alpha_{yx}$$

If we choose a proper set of axes, we can carry out a transformation that diagonalizes the tensor previously given. The new axes are called the principal axes of the polarizability tensor with components $\alpha_{x'x'}$, $\alpha_{y'y'}$, and $\alpha_{z'z'}$ whose trace is equal to that in the previous coordinate system.

Quantum mechanically, the polarizability of a molecule by electromagnetic radiation along the direction ij is given by

$$(\alpha_{ij})_{mn} = \frac{1}{h} \sum_e \left[\frac{(M_j)_{me}(M_i)_{en}}{\nu_e - \nu_0} + \frac{(M_i)_{me}(M_j)_{en}}{\nu_e + \nu_s} \right] \qquad (6\text{–}12)$$

where m and n are the initial and final states of the molecule and e represents an excited state. M_i and M_j are electric dipole transition moments along i and j, whereas ν_e is the energy of the transition to e and ν_0 and ν_s are the frequencies of the incident and scattered radiation.

As previously mentioned, the selection rules for Raman spectroscopy are different from those for infrared. *In order for a vibration to be Raman-active, the change in polarizability of the molecule with respect to vibrational motion must not be zero at the equilibrium position of the normal vibration;* i.e.,

$$\left(\frac{\partial \alpha}{\partial r} \right)_{r_e} \neq 0 \qquad (6\text{–}13)$$

where α is the polarizability and r represents the distance along the normal coordinate. If the plot of polarizability, α, versus distance from the equilibrium distance, r_e, along the normal coordinate is that represented by Fig. 6–11(A), the vibration will be Raman active. If the plot is represented by the curves 1 or 2 of Fig. 6–11(B), $\partial \alpha / \partial r$ will be zero at or near the equilibrium distance, r_e, and the vibration will be Raman-inactive.

Small amplitudes of vibration (as are normally encountered in a vibration mode) are indicated by the region on the distance axis on each side of zero and between the dotted lines. As can be seen from Fig. 6–11, the vibration in (A)

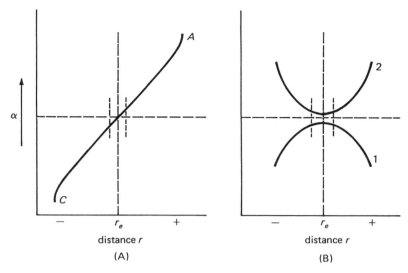

FIGURE 6–11 Polarizability as a function of distance for some hypothetical molecules.

corresponds to an appreciable change in polarizability occurring in this region, while that in (B) corresponds to practically no change. Hence, the Raman selection rule (6–13) is often stated: *In order for a vibration to be Raman-active, there must be a change in polarizability during the vibration.*

The completely symmetric stretch for CO_2 is represented in Fig. 6–10. The points labeled A and C in Fig. 6–11(A) correspond to extreme stretching of the bonds for the symmetric stretch. A and C represent, respectively, more and less polarizable structures than the equilibrium structure. A plot of α versus r for the symmetric stretch is represented by Fig. 6–11(A). The change in polarizability for the asymmetric stretch is represented by one of the curves in Fig. 6–11(B). As a result, this vibration is Raman-inactive. Infrared activity is just the opposite of Raman activity in this case. *In general, for any molecule that possesses a center of symmetry, there will be no fundamental lines in common in the infrared and Raman spectra.* This is a very valuable generalization for structure determination. If the same absorption band is found in both the infrared and Raman spectra, it is reasonably certain that the molecule lacks a center of symmetry. It is possible that in a molecule lacking a center of symmetry no identical lines appear because of the low intensity of one of the corresponding lines in one of the spectra.

This discussion serves to illustrate qualitatively the selection rules for Raman spectroscopy. It is often difficult to tell by inspection the form of the polarizability curve for a given vibration. As a result, this concept is more difficult to utilize qualitatively than is the dipole moment selection rule in infrared. In a later section it will be shown how character tables and symmetry arguments can provide information concerning the infrared and Raman activity of vibrations. For now, we summarize the above discussion by stating that in order for a vibration to be Raman active, the shape, size, or orientation of the ellipsoid must change during the vibration, as shown in Fig. 6–11(A).

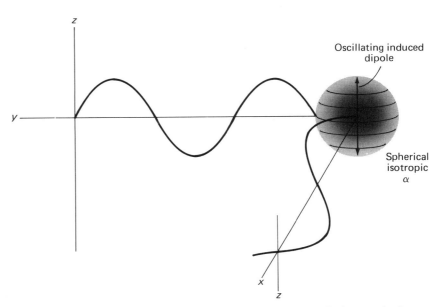

FIGURE 6–12 Raman scattering of *yz* plane-polarized radiation producing an *xz*-polarized scattered wave when the polarizability is isotropic.

6–10 POLARIZED AND DEPOLARIZED RAMAN LINES

Valuable information can be obtained by studying the polarized components of a Raman line. In order to understand how the Stokes lines can have a polarization, we will first consider some scattering experiments with polarized light. In Fig. 6–12 the interaction of *yz*-polarized radiation with a molecule that has an isotropic polarizability tensor is indicated. The oscillation induced in the molecule will be in the same plane as the electric field and will have a *z*-component as shown by the double-headed arrow. If we observe emitted radiation along the *x*-axis, it will have to be *xz* polarized. Since α is isotropic, the direction of the induced dipole is unchanged as the molecule tumbles, and only *xz* scattered radiation results. If we had used *xy*-polarized radiation as our source, the oscillating induced dipoles in a molecule with an isotropic polarizability would be oscillating parallel to the *x*-axis. In Fig. 6–13 the *x*-axis is perpendicular to the page, so the *xy*-plane polarized radiation is also perpendicular to the page. No component is detected

FIGURE 6–13 Raman scattering of *xy*-plane polarized radiation produces no *xz* or *yz* components for the scattered wave when the polarizability is isotropic.

along x, for the molecular induced dipole moment oscillates parallel to x for all orientations of the molecule.

As a result of the above discussion, it should now be clear that—even if our source is non-polarized—the scattered radiation will be *polarized* if our molecule is isotropic in α, for there will be no xy-component.

The polarizability tensor is usually anisotropic, as shown in Fig. 6–14. Accordingly, the induced moment will not be coincident with the plane of the electric field but will tend to be oriented in the direction of greatest polarizability. The scattered light vibrates in the same plane as the induced dipole. As the molecule tumbles, the orientation of the induced dipole moment relative to the x-axis changes. Therefore, there are both xz and xy-components in the scattered beam giving rise to radiation that is *depolarized*. Even if the incident radiation is polarized, an anisotropic polarizability tensor will give rise to scattered radiation that is depolarized. In the actual experiment, polarized radiation from a laser source is employed and the polarization of the radiation scattered along the x-axis is determined.

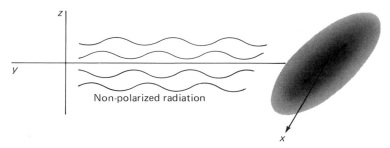

FIGURE 6–14 Raman scattering of non-polarized radiation from an anisotropic polarizability tensor.

The above considerations apply to the Stokes lines, for which it is found that a *totally symmetric vibration mode gives rise to a polarized scattered line and that a vibration with lower symmetry is depolarized. This property can be used to confirm whether or not a vibration has A_1 symmetry.* Thus, by using an analyzer, the scattered Stokes radiation traveling along the x-axis is resolved into two polarized components: radiation polarized in the y-direction and that in the z-direction. Radiation traveling along x and polarized in the y-direction is referred to as perpendicular and that component polarized in the z-direction is called parallel. The *depolarization ratio*, ρ, is defined as the ratio of the intensity, I, of the perpendicular ($y(\perp)$) to the parallel, $z(\|)$, components of the Stokes line:

$$\rho = \frac{I_{y(\perp)}}{I_{z(\|)}} \qquad (6\text{–}14)$$

Those Raman lines for which $\rho = 3/4$ are referred to as depolarized lines and correspond to vibrations of the molecule that are not totally symmetric. Those Raman lines for which $0 < \rho < 3/4$ are referred to as polarized lines. The vibrations of the molecule must transform as A_1 in order to be polarized. The use of this information in making band assignments will be made clearer after a discussion of symmetry considerations.

6-11 RESONANCE RAMAN SPECTROSCOPY

As the laser exciting frequency in the Raman experiment approaches an allowed electronic transition in the molecule being investigated, those normal modes that are vibronically active in the electronic transition exhibit a pronounced enhancement in their Raman intensities.[12] There are two applications of this fact. One involves utilizing the intensity enhancement to study materials that can be obtained only at low concentrations. The second application involves the fact that only those few vibrational modes near the site of an electronic transition in a complex molecule will be enhanced. Most examples of resonance Raman spectroscopy involve the enhancement of totally symmetric modes and arise from intense electronic transitions. There are recent examples in which vibronically allowed electronic transitions enhance the Raman bands of modes that are not totally symmetric, making them vibronically allowed.

It is possible for more than one electronic transition to be responsible for the resonance Raman effect, complicating the symmetry-based application just discussed. By studying the intensity of the resonance Raman bands as a function of the frequency of the exciting laser line, one can determine whether one electronic transition (the so-called A-mechanism) or two (B-mechanism) are involved. The intensity of a particular band is proportional to the square of the frequency factor, F. For the two mechanisms above, equations (6–15) and (6–16) have been proposed, respectively:

$$F_A = \frac{v^2(v_e^2 + v_0^2)}{(v_e^2 - v_0^2)^2} \quad (6-15)$$

where v_e is the frequency for the electronic transition, v_0 is the source frequency and $v = v_0 - \Delta v_{m,n}$ with $\Delta v_{m,n}$ equal to the frequency for the vibrational transition; and

$$F_B = \frac{2v^2(v_e v_s + v_0^2)}{(v_e^2 - v_0^2)(v_s^2 - v_0^2)} \quad (6-16)$$

where v_e and v_s are the frequencies for the two electronic transitions involved and the other terms are as defined for equation (6–15). A large number of other mechanisms and equations have recently been proposed.

An example of the enhanced intensity[13] from the resonance Raman effect is shown in Fig. 6–15. In (A), a typical electronic spectrum for a heme chromophore is shown. In (B), parallel and perpendicular components of the resonance Raman spectrum of a 5×10^{-4} M solution of oxyhemoglobin are illustrated. Without the resonance enhancement associated with the 568 nm exciting wavelength, this concentration of material would not have yielded a Raman spectrum. The label dp refers to a depolarized band (depolarization ratio of $3/4$), p to a polarized band, and ip to a band in which the depolarization ratio is greater than one (i.e., the scattered light is polarized perpendicular to the polarization of the incident light). An ip band requires that the scattering tensor be antisymmetric (i.e., $\alpha_{ij} = -\alpha_{ji}$), which is possible when v_0 approaches v_e. When the exciting line is far from an electronic transition, the ratio cannot be larger than $3/4$. The scattering tensor elements discussed here and symbolized by α are not to be confused with the polarizability tensor elements discussed earlier.

An assignment of the heme Raman spectrum has been made that is consistent with the assigned electronic transitions. The spectral features common to many

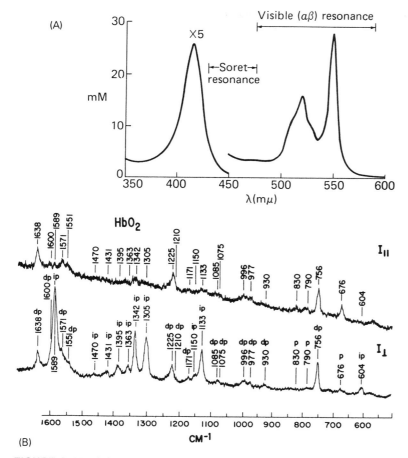

FIGURE 6–15 (A) Ultraviolet (Soret) and visible ($\alpha - \beta$) chromophores in hemes. Spectrum of ferrocytochrome c. (B) Resonance Raman spectra of oxyhemoglobin. Both the direction and the polarization vector of the incident laser radiation are perpendicular to the scattering direction. The scattered radiation is analyzed into components perpendicular ($I_\perp$) and parallel ($I_\parallel$) to the incident polarization vector. The exciting wavelength was 568.2 nm (5682 Å) for HbO_2. Concentrations are about 0.5 mM for HbO_2. [From T. G. Spiro and T. C. Strekas, *Proc. Natl. Acad. Sci.*, **69**, 2622 (1972).]

heme proteins are discussed, as is the use of the depolarization ratios to infer local symmetry of the heme unit.[13]

In another application of this technique,[14] a resonance Raman analysis of oxyhemerythrin was carried out using the oxygen → iron charge-transfer band at 5000 Å to intensify the O—O stretching vibrations. Two Raman frequencies were observed at 844 cm^{-1} ($\rho = 0.33$) and at 500 cm^{-1} ($\rho = 0.4$). When $^{18}O_2$ is employed, the bands shift to 798 cm^{-1} and 478 cm^{-1}, respectively. These bands do not appear in the deoxygenated protein. The band at 844 cm^{-1} has been assigned to the O—O stretch and is not coupled to the other vibrational modes in the molecule, since the predicted frequency for isotopic substitution (796 cm^{-1}) is observed. The 500 cm^{-1} band is assigned to an Fe—O—O stretching mode. The spectra indicate only one type of O_2 complex.

Symmetry Aspects of Molecular Vibrations

6–12 SIGNIFICANCE OF THE NOMENCLATURE USED TO DESCRIBE VARIOUS VIBRATIONS

In this section the conventions used to label the vibrations A_1, A_2, etc., will be reviewed to refresh the reader's memory. If the vibration is symmetric with respect to the highest-fold rotation axis, the vibration is designated by the letter A. If it is antisymmetric, the letter B is employed. E stands for a doubly degenerate vibration and T for a triply degenerate mode (F is commonly used instead of T in vibrational spectroscopy). The subscripts g and u refer to symmetry with respect to an inversion through a center of symmetry and are used only for molecules with a center of symmetry. If a vibration is symmetric with respect to the horizontal plane of symmetry, this is designated by $'$, while $''$ indicates antisymmetry with respect to this plane. Subscripts 1 and 2 (as in A_1 and A_2) indicate symmetry and antisymmetry, respectively, toward a twofold axis that is perpendicular to the principal axis.

6–13 USE OF SYMMETRY CONSIDERATIONS TO DETERMINE THE NUMBER OF ACTIVE INFRARED AND RAMAN LINES

In this section we shall be concerned with classifying the $3N - 6$ (or $3N - 5$ for a linear molecule) vibrations in a molecule to the various irreducible representations in the point group of the molecule. This information will then be used to indicate the degeneracy and number of infrared and Raman-active vibrations. The procedure is best illustrated by considering the possible structures for SF$_4$ (C_{2v}, C_{3v}, and T_d) represented in Fig. 6–16.

We first consider the C_{2v} structure. An x, y, z-coordinate system is constructed for each atom, and we proceed to determine the total representation for this basis set using the C_{2v} character table reproduced in Table 6–4.

We know that when an atom is moved by a symmetry operation, the contribution to the total representation will be zero; so we need only worry about the x, y, and z coordinates on atoms not moved. For E, none of the atoms or coordinates are moved, giving $\chi_T(E) = 15$. Only the sulfur atom is not moved by the C_2 operation, but the x and y coordinates of the sulfur each contribute -1 to the total representation while the z contributes $+1$. This leads to $\chi_T(C_2) = -1$. Sulfur and two of the fluorines are not moved by reflection in the σ_{xz} plane, while two fluorines are moved. On the sulfur atom this reflection changes the sign of y but leaves x and z unchanged for a contribution of $+1$. The same is true for each

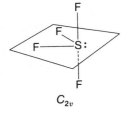

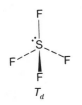

FIGURE 6–16 Some possible structures for SF$_4$.

TABLE 6–4. C_{2v} Character Table[a]

	E	C_2	$\sigma_{v(xz)}$	$\sigma'_{v(yz)}$	
A_1	1	1	1	1	$z, \alpha_x^2, \alpha_y^2, \alpha_z^2$
A_2	1	1	-1	-1	R_z, α_{xy}
B_1	1	-1	1	-1	x, R_y, α_{zx}
B_2	1	-1	-1	1	y, R_x, α_{yz}

[a] For planar molecules the x-axis is perpendicular to the plane.

of the unshifted fluorines, leading to $\chi_T(\sigma_{v(xz)}) = 3$. Reflection through yz changes the x-coordinate to -1 for all three atoms not moved but does not change y or z coordinates, leading to $\chi_T(\sigma'_{v(yz)}) = 3$. Summing up these results, we obtain:

$$\chi_T = 15 \quad -1 \quad 3 \quad 3$$

We can factor the total representation as follows:

$$\begin{array}{cccc} \text{for } E & \text{for } C_2 & \text{for } \sigma_{v(xz)} & \text{for } \sigma'_{v(yz)} \end{array}$$

$$n^{A_1} = \frac{1}{4}[1 \cdot 1 \cdot 15 + 1 \cdot 1 \cdot (-1) \quad + 1 \cdot 1 \cdot 3 \quad + 1 \cdot 1 \cdot 3] = 5$$

$$n^{A_2} = \frac{1}{4}[1 \cdot 1 \cdot 15 + 1 \cdot 1 \cdot (-1) \quad + 1 \cdot (-1) \cdot 3 + 1 \cdot (-1) \cdot 3] = 2$$

$$n^{B_1} = \frac{1}{4}[1 \cdot 1 \cdot 15 + 1 \cdot (-1) \cdot (-1) + 1 \cdot (-1) \cdot 3 + 1 \cdot 1 \cdot 3] = 4$$

$$n^{B_2} = \frac{1}{4}[1 \cdot 1 \cdot 15 + 1 \cdot (-1) \cdot (-1) + 1 \cdot 1 \cdot 3 \quad + 1 \cdot (-1) \cdot 3] = 4$$

The results $n^{A_1} = 5$, $n^{A_2} = 2$, $n^{B_1} = 4$, and $n^{B_2} = 4$ repesent the $3N$ degrees of freedom for a pentatomic C_{2v} structure.

To get the total number of vibrations, the three degrees of translational freedom and three of rotational freedom must be subtracted (giving $3N - 6$ for a nonlinear molecule). Translation along the x-axis can be represented by an arrow lying along this axis, which, as seen from the C_{2v} character table, belongs to the species B_1. Simiarly, translation along the z and y axes transforms according to species A_1 and B_2. Rotations along the x, y, and z axes are represented by the symbols R_x, R_y, and R_z and belong to species A_2, B_1, and B_2, respectively. These six degrees of freedom ($1A_1$, $1A_2$, $2B_1$, and $2B_2$) are subtracted from the representation of the total degrees of freedom, producing $n^{A_1} = 4$, $n^{A_2} = 1$, $n^{B_1} = 2$, and $n^{B_2} = 2$, corresponding to the nine vibrations predicted from the $3N - 6$ rule.

We have now calculated the species to which the $3N - 6$ vibrations of SF_4 belong if it has a C_{2v} structure. Four are of species A_1, one of A_2, two of B_1, and two of B_2. The next job is to determine which vibrations are infrared active and which are Raman active. For a fundamental transition to occur by absorption of infrared electromagnetic radiation, one of the three integrals

$$\int \psi_v^\circ \hat{x} \psi_v^{ex} \, d\tau \quad \int \psi_v^\circ \hat{y} \psi_v^{ex} \, d\tau \quad \int \psi_v^\circ \hat{z} \psi_v^{ex} \, d\tau \quad (6\text{--}17)$$

must be non-zero or A_1. The $\hat{x}$, $\hat{y}$, and $\hat{z}$ operators correspond to the orientation of the electric field vector relative to a molecular Cartesian coordinate system. The operator is identical to that discussed in electronic absorption spectroscopy. Since all ground vibrational wave functions are A_1 (if necessary, reread Chapter 4, the section on atomic and molecular transitions), this amounts to having a component of the transition dipole operator $\hat{x}$, $\hat{y}$, or $\hat{z}$ that has the same symmetry as ψ_{vib}^{ex}, where the total vibrational excited state for a fundamental transition has the same symmetry as the normal mode excited. Accordingly, *a fundamental will be infrared* active *if the excited normal mode belongs to one of the irreducible representations corresponding to the x, y, and z vectors, and will be inactive if it does not.*

In order for a vibration to be Raman active, it is necessary that one of the integrals of the type

$$\int \psi_v^\circ \hat{P} \psi_v^{ex} \, d\tau \qquad (6\text{--}18)$$

be non-zero, *i.e.*, the integrand A_1. Here the operator $\hat{P}$ is one of the quadratic functions of the x, y, and z-vectors (*i.e.*, x^2, y^2, z^2, xy, yz, xz), simply or in combination (*i.e.*, $x^2 - y^2$). The symmetries of these functions are listed in the character tables opposite their irreducible representations. Since these quantities are components of the polarizability tensor, we find the following rule stated: *A fundamental transition will be Raman active if the corresponding normal mode belongs to the same irreducible representations as one or more of the components of the polarizability tensor.*

For the C_{2v} structure of SF_4, the A_1, B_2, and B_1 modes are infrared active, while A_2 is infrared inactive (neither x, y, nor z has A_2 symmetry). Eight infrared bands are expected: $4A_1$, $2B_1$, and $2B_2$. All nine of the fundamental vibrations are Raman active, and the four A_1 modes will be polarized.

When we carry out the above procedure on the T_d structure for SF_4, we obtain a total representation $\chi_T = 15(E)\ 0(8C_3)\ 3(6\sigma_d) - 1(6S_4) - 1(3S_4^2 = 3C_2)$. Factoring, we obtain:

$$n^{A_1} = \frac{1}{24} [1 \cdot 1 \cdot 15 + 8 \cdot 1 \cdot 0 \quad + 6 \cdot 1 \cdot 3 \quad + 6 \cdot 1 \cdot (-1) \quad + 3 \cdot 1 \cdot (-1)] \quad = 1$$

$$n^{A_2} = \frac{1}{24} [1 \cdot 1 \cdot 15 + 8 \cdot 1 \cdot 0 \quad + 6 \cdot (-1) \cdot 3 + 6 \cdot (-1) \cdot (-1) + 3 \cdot 1 \cdot (-1)] \quad = 0$$

$$n^{E} = \frac{1}{24} [1 \cdot 2 \cdot 15 + 8 \cdot (-1) \cdot 0 + 6 \cdot 0 \cdot 3 \quad + 6 \cdot 0 \cdot (-1) \quad + 3 \cdot 2 \cdot (-1)] \quad = 1$$

$$n^{T_1} = \frac{1}{24} [1 \cdot 3 \cdot 15 + 8 \cdot 0 \cdot 0 \quad + 6 \cdot (-1) \cdot 3 + 6 \cdot 1 \cdot (-1) \quad + 3 \cdot (-1) \cdot (-1)] = 1$$

$$N^{T_2} = \frac{1}{24} [1 \cdot 3 \cdot 15 + 8 \cdot 0 \cdot 0 \quad + 6 \cdot 1 \cdot 3 \quad + 6 \cdot (-1) \cdot (-1) + 3 \cdot (-1) \cdot (-1)] = 3$$

The T_d character table indicates that the three translations are of species T_2 and the three rotations of species T_1. Since T_1 and T_2 are triply degenerate, we need subtract only one of each from the total degrees of freedom to remove three degrees of translation and three of rotation. The total number of vibrations belong to species A_1, E, and $2T_2$ giving rise to the nine modes predicted from the $3N - 6$ rule. The six T_2 vibrations (two triply degenerate sets) are infrared active and give rise to two fundamental bands in the infrared spectrum. All modes are Raman active, giving rise to four spectral bands corresponding to fundamentals. Of these, A_1 is polarized. It can be shown similarly that the C_{3v} structure of SF_4 leads to six infrared lines (three A_1 and three $\dot{E}$) and six Raman lines (three A_1 and three E), three of which are polarized (A_1).

TABLE 6–5. Summary of Active Modes Expected for Various Configurations of SF_4

	C_{2v}	T_d	C_{3v}	Found[a]
Infrared-active modes	8($4A_1$, $2B_1$, $2B_2$)	2($2T_2$)	6($3A_1$, $3E$)	5(or 7)
Raman modes	9($4A_1$, A_2, $2B_1$, $2B_2$)	4(A_1, E, $2T_2$)	6($3A_1$, $3E$)	5(or 8)
Polarized modes	4($4A_1$)	1(A_1)	3($3A_1$)	1

[a] R. E. Dodd, L. A. Woodward, and H. L. Roberts, *Trans. Faraday Soc.*, **52**, 1052 (1956).

The actual spectrum is found to contain five infrared fundamentals, and five Raman lines, one of which is polarized. These results, summarized in Table 6–5, eliminate the T_d structure, leaving either C_{2v} or C_{3v} of the possible structures considered. This example demonstrates the point that although there cannot be more fundamental vibrations than allowed by a symmetry type, often not all vibrations are detected. The problems in application of these concepts involve separating overtones and combinations from the fundamentals and, as is often the case, not finding some "active" fundamental vibrations because they have low intensities.

Table 6–5 should be verified by the reader by using the character tables contained in Appendix A and the procedure described above. By a detailed analysis of the band contours and an assignment of the frequencies, it was concluded that the C_{2v} structure best fits the observed spectrum. This structure was subsequently confirmed by other physical methods.

The data available on $CHCl_3$ and $CDCl_3$ are summarized in Table 6–6 and will be discussed briefly because this example serves to review many of the principles discussed. The $3N - 6$ rule predicts nine normal vibrations. Just as with the C_{3v} structure of SF_4, the total representation shows that these nine vibrations consist of three A_1 and three E species or a total of six fundamental frequencies. All six fundamentals will be IR and Raman active, and the three A_1 fundamentals will give polarized Raman lines. The observed bands that have been assigned to fundamental frequencies are summarized in Table 6–6. These data indicate that the 3033, 667, and 364 cm^{-1} bands belong to the totally symmetric vibrations (species A_1), for these bands are Raman polarized. They are labeled v_1, v_2, and v_3, respectively. The other three bands are of species E, and the 1205, 760, and 260 cm^{-1} bands are v_4, v_5, and v_6, respectively. Note that since the molecule does not have a center of symmetry, the IR and Raman

TABLE 6–6. Infrared and Raman Fundamentals for $CHCl_3$

IR-Active Vibrations for $CHCl_3$ (cm^{-1})	Raman-Active Vibrations for $CHCl_3$ (cm^{-1})	Raman Spectra of $CDCl_3$ (cm^{-1})	Designation
260	262	262	v_6
364	366 (polarized)	367	v_3
667	668 (polarized)	651	v_2
760	761	738	v_5
1205	1216	908	v_4
3033	3019 (polarized)	2256	v_1

spectra have lines in common. The exact form of these fundamental vibrations is illustrated in Appendix C. The bands v_1 and v_4 are the C—H stretching and bending vibrations, respectively. Note how deuteration has a pronounced effect on these frequencies but almost no effect on the others. The vibrations at 3033 and 1205 cm^{-1} are almost pure C—H modes. Since the very light hydrogen or deuterium atom is moving, very little C—Cl motion is necessary to retain the center of mass (to prevent net translation of the molecule).

The above manipulations are quite simple and yet yield valuable information about the structures of simple molecules. There are many applications of these concepts. It is strongly recommended that the reader carry out the exercises at the end of this chapter.

Spectroscopists often use the v_n symbolism in assigning vibrations. This can be translated into the language of stretches, bends, and twists by referring to the diagrams in Appendix C or in Herzberg (see Additional References) for an example of a molecule with similar symmetry and the same number of atoms. The irreducible representation of a given normal vibration can be determined from the diagrams by the procedure outlined in Chapter 2.

6–14 SYMMETRY REQUIREMENTS FOR COUPLING COMBINATION BANDS, AND FERMI RESONANCE

There are some important symmetry requirements regarding selection rules for overtone and combination bands. These can be demonstrated by considering BF$_3$, a D_{3h} molecule, as an example. The D_{3h} character table indicates that the symmetric stretch, v_1, of species A_1' is infrared inactive (there is no dipole moment change). The species of the combination band $v_1 + v_3$ (where v_3 is of species E') is given by the product of $A_1' \times E' = E'$. The combination band is infrared active. The v_2 vibration is of symmetry A_2'' and is infrared active. The overtone $2v_2$ is of species $A_2'' \times A_2'' = A_1'$, which is infrared inactive; $3v_2$ is of species A_2'' and is observed in the infrared spectrum. This behavior is strong support for the initial structural assignment, that of a planar molecule, and serves here as a nice example to demonstrate the symmetry requirements for overtones and combination bands (recall the discussion in Section 6–5).

In the case of Fermi resonance it is necessary that the states which are participating have the same symmetry. For example, if $2v_2$ is to undergo Fermi resonance with v_1, one of the irreducible representations for the direct product $2v_2$ must be the same as for v_1.

Coupling of group vibrations was mentioned earlier in this chapter. In order for coupling to occur, the vibrations must be of the same symmetry type. For example, in acetylene the symmetric C—H stretching vibration and the C—C stretching vibrations are of the same symmetry type and are highly coupled. The observed decrease in the frequency of the symmetric C—H stretch upon deuteration is smaller than expected, because of this coupling. Since the asymmetric C—H stretch and the "C—C" stretch are of different symmetry, they do not couple, and deuteration has the normal effect on the asymmetric C—D stretch (a decrease from 3287 to 2427 cm^{-1} is observed).

6-15 MICROWAVE SPECTROSCOPY

Pure rotational transitions in a molecule can be induced by radiation in the far infrared and microwave regions of the spectrum. Extremely good precision for frequency determination is possible in the microwave region. Compared to the infrared region of the spectrum, where measurements are routinely made to about 1 cm^{-1}, one can get resolution of about 10^{-8} cm^{-1} in the mirowave region. A wide spectral range plus resolution and accuracy to 10^{-8} cm^{-1} make this a very valuable region for fingerprint applications. A list of frequencies has been tabulated[15,16] consisting of 1800 lines for about 90 different substances covering a span of 200,000 MHz.* Only in 10 of the cases reported were two of the 1800 lines closer than 0.25 MHz.

There are two requirements that impose limitations on microwave studies. (1) The spectrum must be obtained on the material in the gaseous state. For conventional instrumentation, a vapor pressure of at least 10^{-3} torr is required. (2) The molecule must have a permanent dipole moment in the ground state in order to absorb microwave radiation, since rotation alone cannot create a dipole moment in a molecule.

In addition to the fingerprint application, other useful data can be obtained from the microwave spectrum of a compound. Some of the most accurate bond distance and bond angle data available have been obtained from these studies. Let us first consider a diatomic molecule. The rotational energy, E, for a diatomic molecule is given by the equation:

$$E = hBJ(J + 1) \qquad (6-19)$$

where J is an integer, the rotational quantum number; h is Planck's constant (6.62×10^{-27} erg sec); and $B = h/8\pi^2 I$ where I is the moment of inertia. Since $E = hv$, we obtain the relationship for the frequency, v, corresponding to this energy:

$$v = BJ(J + 1) \qquad (6-20)$$

The energy of the transition from state J to $J + 1$ can be determined by substitution in equation (6–19), leading to

$$\Delta E = 2Bh(J + 1)$$

or

$$\Delta v = 2B(J + 1)$$

There is another selection rule (in addition to the dipole moment requirement) for microwave absorption, which states that $\Delta J = \pm 1$. Therefore, the longest wavelength (lowest energy) absorption band in the spectrum will correspond to

---

*A wave number of 1 mm^{-1} is equivalent to a frequency of 299,800 MHz.

the transition from $J = 0$ to $J = 1$, for which the frequency of the absorbed energy is:

$$\Delta v = 2B = \frac{2h}{8\pi^2 I} \tag{6-21}$$

Δv is measured and all other quantities in equation (6–21) are known except I, which can then be calculated.

All other lines in the spectrum will occur at shorter wavelengths and will be separated from each other by $2B$ [i.e., $\Delta\Delta v = 2B(J + 2) - 2B(J + 1) = 2B$, where $(J + 1)$ is the quantum number for the higher rotational state and J that for the lower; see Fig. 6–17]. Once the moment of inertia is determined, the equilibrium internuclear separation, r_0, in the diatomic molecule can be calculated with equation (6–22):

$$I = \mu r_0^2 \tag{6-22}$$

where μ is the reduced mass.

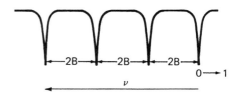

FIGURE 6–17 Illustration of the $0 \rightarrow 1$ and other rotational transitions (idealized).

In the above discussions we have assumed a rigid rotor as a model. This means that the atom positions in the molecule are not influenced by the rotation. Because of the influence of centrifugal force on the molecule during the rotation, there will be a small amount of distortion. As a result, the bond distance (and I) will be larger for high values of J, and the spacings between the peaks will decrease slightly as J increases.

In a more complex molecule the moment of inertia is related to the bond lengths and angles by a more complex relationship than (6–22). A whole series of simultaneous equations has to be solved to determine all the structural parameters. In order to get enough experimental observations to solve all these equations, isotopic substitution is employed. For pyridine, the microwave spectra of six isotopically substituted compounds (pyridine, 2-deuteropyridine, 3-deuteropyridine, 4-deuteropyridine, pyridine-2-^{13}C, and pyridine-3-^{13}C) were employed. For a more complete discussion of this problem, see Gordy, Smith, and Trambarulo.[15]

By a detailed analysis of the microwave spectrum, one can obtain information about internal motions in molecules. The torsional barrier frequency and energy can be obtained.[17a]

Double resonance experiments have also proved valuable for assigning peaks and for determining relaxation times of states. A given transition is saturated with microwave power, producing a non-equilibrium distribution in the populations of the two levels involved. One then looks for an increase or decrease in other transitions that are thought to involve one of these levels.[17b]

Accurate dipole moment measurements can be made from microwave experiments. When an electric field is applied to the sample being studied, the rotational line is split (the Stark effect). The magnitude of the splitting depends upon the product of the dipole moment (to be evaluated) and the electric field strength (which is known).

Much information is available from the Zeeman splitting of rotational lines.[17c,d] The components of the magnetic susceptibility anisotropy can be evaluated, as can the expectation value $\langle r^2 \rangle$. As we shall see in the chapter on nuclear quadrupole resonance, nqr, much the same information obtained on solids by nqr can be obtained on gases by microwave spectroscopy.

6–16 ROTATIONAL RAMAN SPECTRA

Information equivalent to that obtained in the microwave region can be obtained from the rotational Raman spectrum, for which the permanent dipole selection rule does not hold. As a result, very accurate data on homonuclear diatomic molecules can be obtained from the rotational Raman spectrum. Experimentally, the bands are detected as Stokes lines with frequencies corresponding to rotational transitions.

In order for a molecule to exhibit a rotational Raman spectrum, the polarizability perpendicular to the axis of rotation must be anisotropic; *i.e.*, the polarizability must be different in different directions in the plane perpendicular to the axis. If the molecule has a threefold or higher axis of symmetry, the polarizability will be the same in all directions and rotational modes about this axis will be Raman inactive. Other rotations in the molecule may be Raman active.

For diatomic molecules the selection rule $\Delta J = \pm 2$ now applies:

$$\Delta E = Bh(4J' - 2) \qquad (6\text{--}23)$$

and the frequency separation between the lines is $4B$. If this separation is not at least ~ 0.1 cm^{-1} the lines cannot be resolved in the Raman spectrum. The rest of the calculation is identical to that described previously. There has been only limited application of this technique to more complex molecules.

6–17 PROCEDURES

a. In Infrared

The most common infrared equipment covers the wavelength region from 4000 to 400 cm^{-1}. The grating resolves polychromatic radiation into monochromatic radiation so that variations in absorption of a sample with change in wavelength can be studied. The resulting spectrum is a plot of sample absorbance or percent transmission versus wavelength (see Fig. 6–26).

Applications of Infrared and Raman Spectroscopy*

---

* The reader is referred to the text by Nakamoto[18] and to reference 11 for extensive compilations of examples of infrared and Raman spectroscopy, respectively.

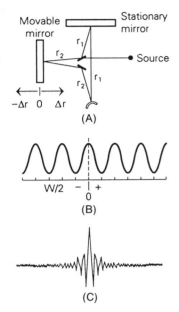

FIGURE 6–18 (A) Schematic of the FTIR technique. (B) Radiation sensed at the detector for a continuously moving mirror. (C) Interferogram from a polychromatic source.

The Fourier transform (FTIR) technique,[19] which will be discussed in more detail in Chapter 7 on nmr, is now very common in infrared instrumentation.[19] The method greatly decreases the amount of time needed to accumulate a spectrum. Thus, multiple scanning of the spectrum and computer storage of the scans become feasible, greatly enhancing the signal-to-noise ratio. Furthermore, filtering is not required, eliminating the need for energy-wasting slits. Enhanced resolution in a shorter time period is possible. A schematic illustration of the method is shown in Fig. 6–18(A).

The monochromatic source is split into two components. One is sent to a stationary mirror traveling a distance r. The second component is sent to a mirror that moves, traveling a distance of $r_2^{ave} \pm \Delta r$. The reflected components are refocused at the detector. When $\Delta r = 0$, $r_1 = r_2^{ave}$, the beams reinforce. When the mirror is moved so that $r_2 = r_2^{ave} + \lambda/4$, the radiation travels $r_2^{ave} + \lambda/2$ ($\lambda/4$ going and coming back) and is 180° out of phase with the component that traveled r_1. The beams cancel. If the mirror is moved continuously, the intensity, I_b, of the radiation at the detector varies as shown in Fig. 6–18(B). If the mirror is moved with a constant velocity, V_M, over a distance $\lambda/2$, one cycle of a sine wave is produced.

The frequency at the detector, v_D, is related to the wavelength by

$$v_D = V_M/\lambda_{cm} \qquad (6\text{-}24)$$

With a wavelength λ of 10 μ (i.e., a frequency of 1000 cm^{-1}) and a mirror velocity of 0.5 mm/sec^{-1}, a frequency of 50 Hz (cps) results. The low frequency signal that results from the high frequency source is termed an interferogram. Polychromatic radiation is employed in FTIR. Each component produces a wave with a unique frequency. The summation of all such waves produces an interferogram as shown in Fig. 6–18(C). All frequencies add only at $r_1 = r_2$. The Fourier transform of the interferogram converts the time axis to a frequency axis and gives all the frequency components of the source along with their intensities. If the beam passes through a sample before hitting the detector, some frequencies will be absorbed and missing in the Fourier transform. The spectrum is obtained by difference. A laser is used as a reference source to constantly calibrate the frequency scale of the instrument.

Infrared spectra can be measured on gases, liquids, or solids. Liquids and gases are generally studied by absorbance measurements. Certain wavelengths are absorbed by the sample and the others are transmitted. The materials used to construct the cells must not absorb in the regions of interest. Special cells with a long path length are needed for most gaseous samples.

Solutions are commonly studied. The solvent absorption must be subtracted out. Different techniques are used depending on the instrumentation. In regions where the solvent absorption is very large almost all of the radiation is absorbed by the solvent and the sample and no sample absorptions can be detected in these blank regions. The open regions for several common liquids used for solution work are illustrated in Fig. 6–19.

Absorbance spectra of solid samples are often examined as mulls. The mull is prepared by first grinding the sample to a fine particle size and then adding enough oil or mulling agent to make a paste. The paste is examined as a thin layer between sodium chloride (or other optical material) plates. The quality of the spectrum obtained is very much dependent on the mulling technique. When

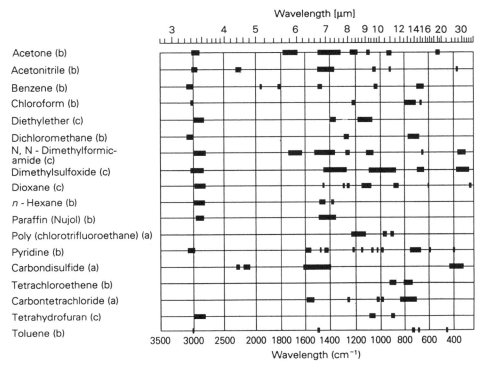

FIGURE 6–19 Regions in which various solvents transmit less than 20% of the incident radiation (indicated by solid bars). (A) 200 μm cell thickness; (B) 100 μm; (C) 20 μm. See references 18 and 19 for other solvents.

the spectrum is examined, peaks from the mulling material will appear in the spectrum and possibly mask sample peaks. If two spectra are obtained, one in Nujol and one in hexachlorobutadiene, all wavelengths in the 5000 to 650 cm^{-1} region can be examined. Solid samples are sometimes examined as KBr discs. The sample and KBr are intimately mixed, ground, and pressed into a clear disc that is mounted and examined directly. Care should be exercised in this procedure, for anion exchange and other reactions may occur with the bromide ion during grinding or pressing.

Transmission IR can be used[20] to study surfaces of solids that have transparent windows in the infrared, *e.g.*, oxides of silica and alumina. High surface area materials should be used so that the spectra have an appreciable contribution from the surface as opposed to the bulk. These samples can be pressed into discs and inserted into the beam path.

Reflectance techniques[21] can be applied to samples that cannot be studied by transmission. These methods can be classified into general categories—specular and diffuse reflectance. Specular reflectance[22] is the front surface, mirrorlike reflection off a smooth surface such as a metal foil or single crystal surface. Diffuse reflectance is observed with rough surfaces and powders. Since the incident radiation is reflected diffusely, it must be collected and refocused on a detector. The reader is referred to reference 22 for the factors that must be considered in comparing absorption and reflection spectra.

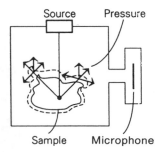

FIGURE 6–20 Schematic illustration of the photoacoustic technique.

An additional technique, referred to as IR-External Reflection spectroscopy,[22] involves studying films coated on highly reflecting surfaces, *e.g.*, gold. Large angles of incidence enhance the sensitivity of the technique. A polarization dependence of the vibrational modes of the adsorbed molecule provides orientational information about the adsorbed molecule when polarized light is employed.

Photoacoustic Fourier transform spectroscopy (PAS)[23] provides an additional technique for obtaining the infrared spectrum of surface molecules. Absorption of light energy by a confined gas heats it and increases its pressure. If the light source is modulated there will be a modulation of the gas pressure. Sound is generated at the frequency of the light modulation and can be detected with a microphone. Modulated light absorbed by a solid heats its surface, which in turn conductively heats a boundary layer of air next to the surface as shown in the Fig. 6–20. Sound is again generated, but the signal detected is reduced in amplitude and delayed in phase because the heat must be conductively transmitted to the surface. For solids, only the energy absorbed near the sample surface is effective in heating the surface quickly enough to contribute to the photoacoustic signal. Thus, PAS is a true surface technique. In the FTIR/PAS experiment, each wave number of light is modulated at a unique frequency and this frequency information is preserved in the microphone signal. Thus, PA detection is a sort of thermal detection in which the light absorbed heats up the sample. PA spectra are qualitatively and quantitatively similar to transmission spectra and are commonly ratioed to a saturated reference (carbon black). Sample surface morphology has a minor qualitative effect on the spectra. Depending on the source employed, one can do PAS in the IR, near IR, u.v., or visible. Diffuse reflectance infrared Fourier transform spectroscopy (DRIFTS), is compared[22,23] with PAS to enable one to choose the best method for the problem at hand.

The spectrum obtained for a given sample depends upon the physical state of the sample. Gaseous samples usually exhibit rotational fine structure. This fine structure is damped in solution spectra because collisions of molecules in the condensed phase occur before a rotation is completed. In addition to the difference in resolved fine structure, the number of absorption bands and the frequencies of the vibrations vary in the different states. As an illustration, the fundamentals of SO_2 are reported in Table 6–7 for the different physical states.

Other effects not illustrated by the data in Table 6–7 are often encountered. Often there are more bands in the liquid state than in the gaseous state of a substance. The stronger intermolecular forces that exist in the solid and liquid states compared with those in the gaseous state are the cause of slight shifts

TABLE 6–7. Frequencies (in cm^{-1}) for SO_2 Fundamentals in Different Physical States

	Gas	Liquid[a]	Solid
$v_2(a_1)$	518	525	528
$v_1(a_1)$	1151	1144	1144
$v_3(b_1)$	1362	1336	1322
			1310

[a] Raman spectrum.

(assuming that there are no pronounced structural changes with change of phase) in frequencies. Frequently, new bands below 300 cm^{-1} appear in the solid state spectrum. The causes of these frequency shifts, band splittings, and new bands in the condensed states are well understood[25] and will be discussed briefly here. Bands below 300 cm^{-1} are often caused by *lattice vibrations* in solids, *i.e.*, translational and torsional motions of the molecules in the lattice. These are called *phonon modes* and are responsible for the infrared cutoff of NaCl cells at low wave numbers. Heavier atom alkali halides have phonon modes at lower energies. These vibrations can form combination bands with the intramolecular vibrations and cause pronounced frequency shifts in higher frequency regions of the spectrum. An additional complication arises if the unit cell of the crystal contains more than one chemically equivalent molecule. When this is the case, the vibrations in the individual molecules can couple with each other. This intermolecular coupling can give rise to frequency shifts and band splitting.

As mentioned earlier, molecular symmetry is very important in determining the infrared activity and degeneracy of a molecular vibration. When a molecule is present in a crystal, the symmetry of the surroundings of the molecule in the unit cell, the so-called *site symmetry*, determines the selection rules. Often bands forbidden in the gaseous state or in solution appear in the solid, and degenerate vibrations in the gaseous state are split in the solid. The general problem of the effect of site symmetry on the selection rules has been treated theoretically.[26] As a simple illustration of this effect, the infrared spectra of materials containing carbonate ion will be considered. The infrared and Raman spectra of $CaCO_3$ in calcite, where the carbonate ion is in a site of D_3 symmetry, contain the following bands (in cm^{-1}): v_1, 1087 (R); v_2, 879 (IR); v_3, 1432 (IR, R); v_4, 710 (IR, R). The (IR) indicates infrared activity and the (R) Raman activity. The infrared spectra of $CaCO_3$ in aragonite, where the site symmetry for carbonate is C_s, differs in that v_1 becomes infrared active and v_3 and v_4 each split into two bands. By using the symmetry considerations previously discussed on the CO_3^{2-} ion, the following results can be obtained:

	v_1	v_2	v_3	v_4
D_{3h}	$A_1'(R)$	$A_2''(IR)$	$E'(IR, R)$	$E'(IR, R)$
D_3	$A_1(R)$	$A_2(IR)$	$E(IR, R)$	$E(IR, R)$
C_s	$A'(IR, R)$	$A''(IR, R)$	$A'(IR, R) + A'(IR, R)$	$A'(IR, R) + A'(IR, R)$

As a result of all these possible complications, the interpretation of spectra obtained on solids is difficult.

The spectrum of a given solute often varies in different solvents. In hydrogen-bonding solvents the shifts are, in part, due to specific solute-solvent interactions that cause changes in the electron distribution in the solute. The frequency of the band is also dependent upon the refractive index of the solvent[27] and other effects.[28]

Matrix isolation experiments combined with infrared and Raman studies have led to interesting developments. Unstable compounds, radicals, and intermediates are trapped in an inert or reactive solid matrix by co-condensing the matrix (*e.g.*, argon) and the species to be studied at low temperatures (often 4.2 to 20 K). Uranium, platinum, and palladium carbonyls have been prepared[29a,b] by allowing controlled diffusion of CO into an argon matrix of the metal. The compounds LiO_2, NaO_2, KO_2, RbO_2, and LiN_2 have also been made[29c] and investigated by infrared. The findings are consistent with C_{2v} symmetry. Several interesting species containing bound O_2, CO, and N_2 have been made.[29d]

b. In Raman

Raman spectra are routinely run on gases, liquids, solutions, or pressed pellets. Liquids can be purified for spectral studies by distillation into a sample cell. Removal of dust is essential, since such contamination increases the background near the exciting line. Turbid solutions are to be avoided. Water and D_2O are excellent Raman solvents because, in contrast to their behavior in the infrared, they have good transparency in the Raman vibrational region. Figure 6–21 illustrates the cutoff region for various solvents in the Raman region. A complete Raman spectrum can be obtained by using CS_2, $CHCl_3$, and C_2Cl_4. For colored solutions, an exciting line should be used that is not absorbed. Fluorescence and photolysis are problems when the sample absorbs the exciting line.

Solids are best investigated as pellets or mulls. The techniques for preparing these are the same as those for infrared. If powders are examined, the larger the crystals the better. For colored materials, rotating the sample during the experiment prevents sample decomposition and permits one to obtain a good spectrum. Surface scanning techniques are also helpful.[9]. Experimental aspects are described in more detail in references 10 and 11.

6–18 FINGERPRINTING

If an unknown is suspected to be a known compound, its spectrum can be compared directly with that of the known. The more bands the sample contains, the more reliable the comparison.

The presence of water in a sample can be detected by its two characteristic absorption bands in the 3600 to 3200 cm^{-1} region and in the 1650 cm^{-1} region.

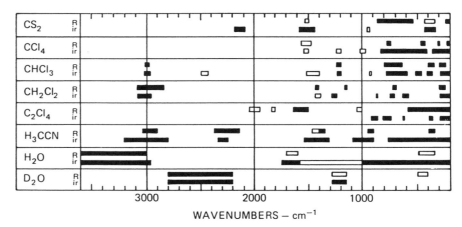

FIGURE 6–21 Solvent interference regions in the Raman region. Solid regions are completely obscured, open rectangles partially obscured.

If the water is present as lattice water, these two bands and one in the 600 to 300 cm^{-1} region are observed. If the water is coordinated to a metal ion, an additional band in the 880 to 650 cm^{-1} region is often observed.[18] In the clathrate compound Ni(CN)$_2$NH$_3 \cdot$ C$_6$H$_6 \cdot x$H$_2$O, the infrared spectrum[23b] clearly showed the presence of water, although it was not detected in a single crystal x-ray study. In connection with the earlier discussion on the effect of physical state on the spectra of various substances, it is of interest to mention in passing that the benzene is so located in the crystal lattice of this clathrate that the frequencies of the out-of-plane vibrations are increased over those in the free molecule, while the in-plane vibrations are not affected.

If the product of a reaction is suspected of being contaminated by starting material, this can be confirmed by the presence of a band in the product due to the starting material that is known to be absent in the pure product. Even if spectra are not known, procedures aimed at purification, e.g., recrystallization, may be attempted. Changes in relative band intensities may then indicate that partial purification has been achieved. Consistent spectra on repeated purification steps may be used as a check on completeness of purification.

A more common application employs the group vibration concept to ascertain the presence or absence of various functional groups in the molecule. The following generalizations aid in this application:

1. Above 2500 cm^{-1} nearly all fundamental vibrations involve a hydrogen stretching mode. The O—H stretching vibration occurs around 3600 cm^{-1}. Hydrogen bonding lowers the frequency and broadens the band. The N—H stretch occurs in the 3300 to 3400 cm^{-1} region. These bands often overlap the O—H bands, but the N—H peaks are usually sharper. The N—H stretch in ammonium and alkylammonium ions occurs at lower frequencies (2900 to 3200 cm^{-1}). The C—H stretch occurs in the 2850 to 3000 cm^{-1} region for an aliphatic compound and in the 3000 to 3100 cm^{-1} region for an aromatic compound. Absorptions corresponding to S—H, P—H, and Si—H occur around 2500, 2400, and 2300 cm^{-1}, respectively.

2. The 2500 to 2000 cm^{-1} region involves the stretching vibration for triply bonded molecules. The C≡N group gives rise to a strong, sharp absorption that occurs in the 2200 to 2300 cm^{-1} region.

3. The 2000 to 1600 cm^{-1} region contains stretching vibrations for doubly bonded molecules and bending vibrations for the O—H, C—H, and N—H groups. The carbonyl group in a ketone absorbs around 1700 cm^{-1}. Conjugation in an amide $\left(\begin{matrix} \text{O} \\ \| \\ \text{RC—N(CH}_3)_2 \end{matrix} \leftrightarrow \begin{matrix} \text{O}^- \\ | \\ \text{R}\overset{+}{\text{C}}\text{=N(CH}_3)_2 \end{matrix} \right)$ decreases the C—O force constant and lowers the highly coupled (with C—N) carbonyl absorption to the 1650 cm^{-1} region. Hydrogen bonding lowers the carbonyl vibration frequency. Stretching vibrations from C=C and C=N occur in this region.

4. The region below 1600 cm^{-1} is referred to as the fingerprint region for many organic compounds. In this region significant differences occur in the spectra of substances that are very much alike. This is the single bond region and, as mentioned in the section on coupling, it is very common to get coupling of individual single bonds that have similar force constants and connect similar masses (*e.g.*, C—O, C—C, and C—N stretches often couple). The absorption

FIGURE 6–22(A), (B), and (C) Range of infrared group frequencies for some inorganic and organic materials. The symbol v_{M-X}, where M and X are general symbols for the atoms involved, corresponds to a stretching vibration. The symbols v_1, v_2, etc, where previously defined. The symbol δ corresponds to an in-plane bending vibration (δ_s is a symmetric bend, δ_d is an asymmetric bend), π to an out-of-plane bend, ρ to rocking and wagging vibrations.

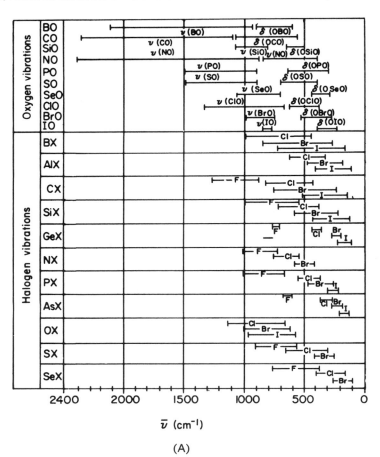

bands in this region for a given functional group occur at different frequencies depending upon the skeleton of the molecule, because each vibration often involves oscillation of a considerable number of atoms of the molecular skeleton.

Many characteristic infrared group frequencies have been compiled to aid in this application. Figure 6–22 shows several typical plots.

In the far region, it becomes very difficult to assign the observed spectrum to group frequencies. Figure 6–23 is a compilation that illustrates typical regions in which absorption bands are found for molecules containing particular functional groups.

6–19 SPECTRA OF GASES

The spectra of gases are often very much different from the spectra of materials in the condensed phase or in solution. As can be seen in Fig. 6–20, a significant difference is the presence of considerable fine structure in the gaseous spectrum. The fine structure is due to a combination of vibrational and rotational transitions. For example, in a diatomic molecule, there are not only transitions corresponding to the pure vibrational mode, v_0, but also absorption corresponding to $v_0 \pm v_r$

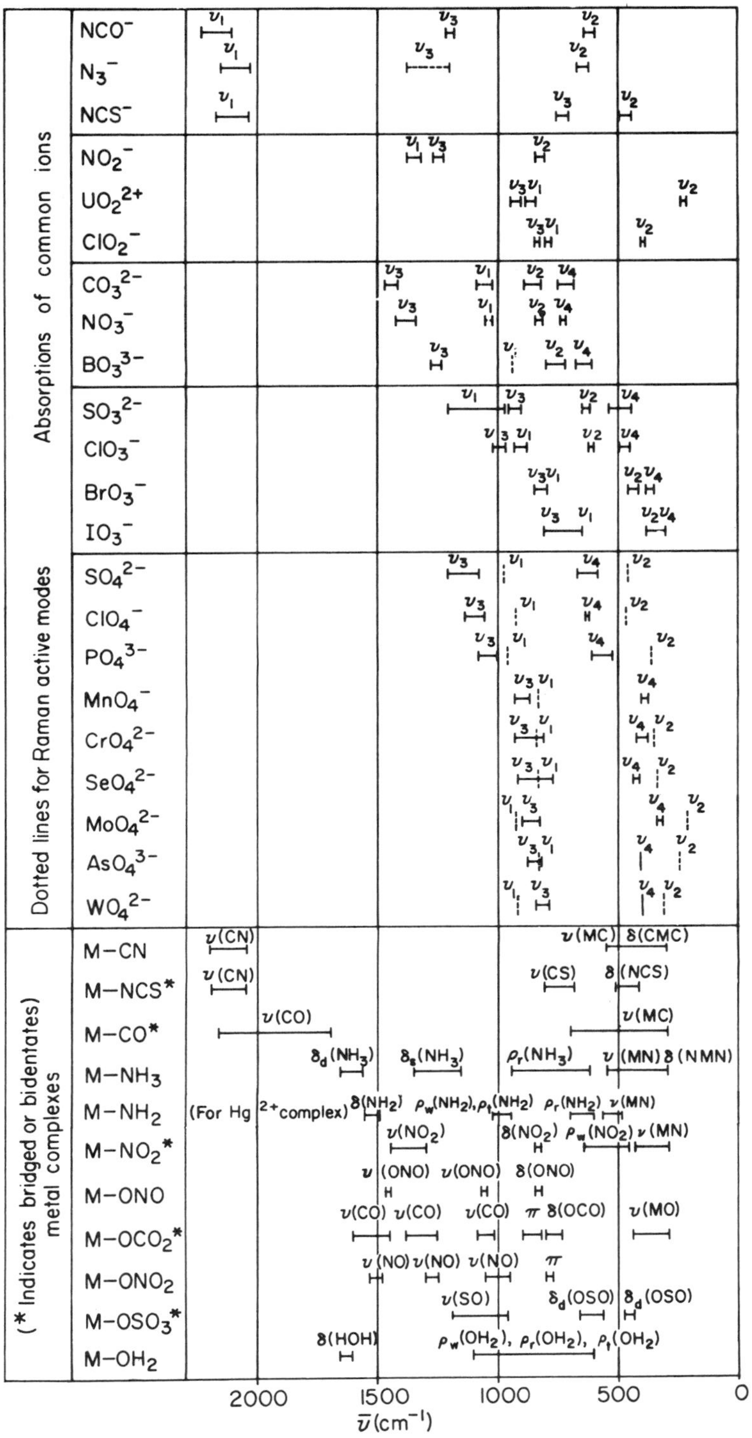

(B)

FIGURE 6–22 *(continued)*

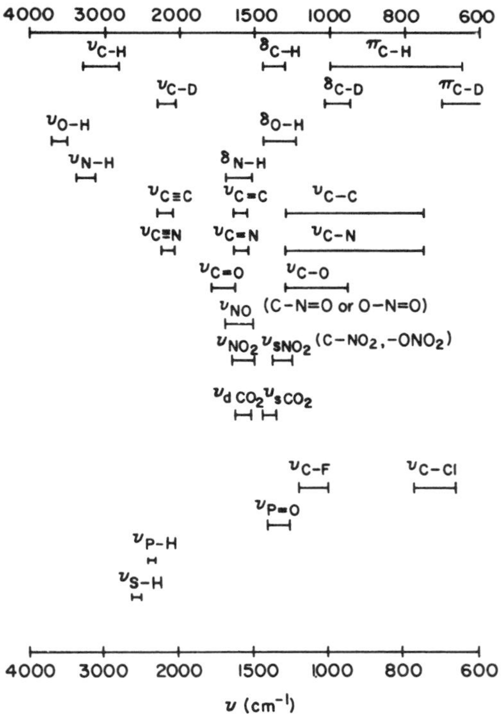

(C)

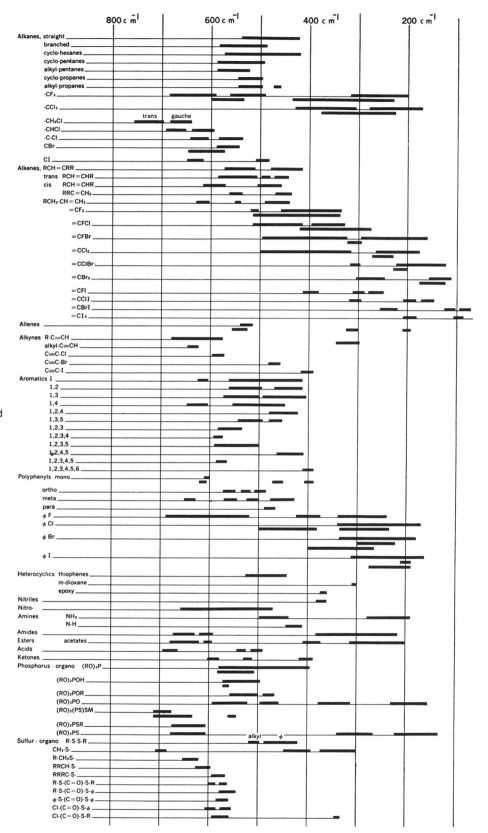

FIGURE 6–23 Far infrared vibrational frequency correlation chart.

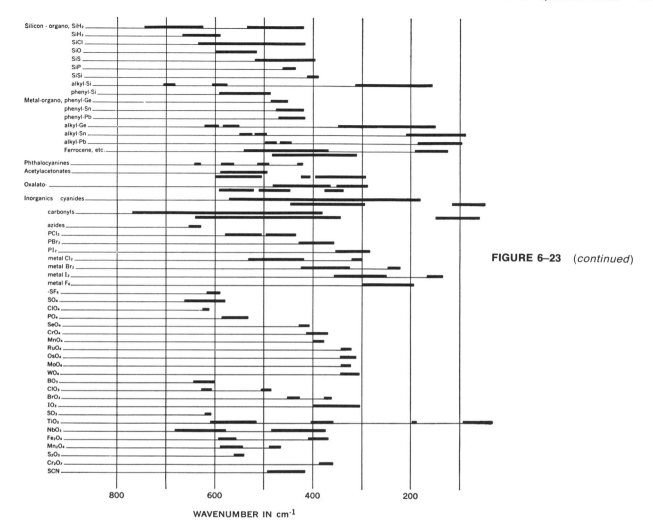

FIGURE 6–23 (continued)

where v_r represents the rotational frequency detected. Since any finite sample contains many molecules, many different rotational states will be populated and there will be a whole series of lines corresponding to different v_r values (*i.e.*, transitions between many different rotational states). This phenomenon is illustrated in Fig. 6–24. The Q-branch corresponds to the transition in which v_r is zero (*i.e.*, a transition with no change in rotational quantum number), the R-branch to $v_0 + v_r$, and the P-branch to $v_0 - v_r$.

The frequencies of all the bands in Fig. 6–24 can be expressed by the following equation:

$$v = v_0 + 2hm_J/8\pi^2 I$$

where m_J has all integral values, including zero, from $+J$ to $-J$ (where J is the rotational quantum number) depending on the selection rules. When $m_J = 0$, the

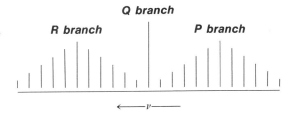

FIGURE 6–24 Schematic of the transitions giving rise to *P*, *Q*, and *R* branches in a spectrum of a gas.

vibrational transition is one that occurs with no change in rotational quantum number. A *Q*-branch results. When m_J is less than zero ($J_{n+1} \to J_n$ transitions), lines in the *P*-branch result, whereas lines in the *R*-branch correspond to m_J greater than zero.

A series of selection rules for the combination of vibration and rotation transitions in various molecules is helpful in deducing structure:

   1. **Diatomic Molecules.** Most diatomic molecules do not possess a *Q*-branch. Nitric oxide (NO) is the only known example of a stable diatomic molecule that has a *Q*-branch. The diatomic molecule must possess angular momentum about the molecular axis in order to have a *Q*-branch. (Σ states, $L = 0$, have no *Q*-branch.)

   2. **Linear Polyatomic Molecules.** If the changing dipole moment for a given vibrational mode is parallel to the principal rotation axis in the molecule, a so-called *parallel band* results, which *has no Q-branch*. The selection rule for this case is $\Delta J = \pm 1$; ΔJ cannot be zero. If the dipole moment change for the vibration has any vector component perpendicular to the principal axis, a *perpendicular band* will result *with a Q-branch*, and ΔJ can be 0, ± 1. The asymmetric C—O stretch in CO_2 is a parallel band, while the C—O bending vibration is a perpendicular band. The utilization of these criteria in making and substantiating assignments of vibrations is apparent. If, in the spectrum of a triatomic molecule, any of the infrared bands (v_1, v_2, or v_3) consists of a single *P*- and *R*-branch without any *Q*-branch (*i.e.*, a zero-gap separation), the molecule must be linear. This type of evidence can be used to support linear structures for N_2O, HCN, and CO_2. The rotational spacings can be employed, as discussed under the section on microwave spectroscopy, to evaluate the moment of inertia. Note that even though CO_2 does not have a permanent dipole (and hence is microwave inactive), the rotational spacings can be obtained from the fine structure in the infrared spectrum.

   3. **Non-linear Polyatomic Molecules.** For the discussion of vibration-rotation coupling in non-linear polyatomic molecules it is necessary to define the terms *spherical*, *symmetric*, and *asymmetric top*. Every non-linear molecule has three finite moments of inertia. In a spherical top, *e.g.*, CCl_4, all three moments are equal. In a symmetric top, two of the three moments are equal. For example, of one selects the C_3-axis of CH_3Br as the *z*-axis, then the moment of inertia along the *x*-axis is equal to that along the *y*-axis. Any molecule with a threefold or higher rotation axis is a symmetric top molecule (unless it is a spherical top). In an asymmetric top, no two of the three moments are equal. Molecules with no symmetry or those with only a twofold rotation axis (H_2O, for example) are asymmetric top molecules.

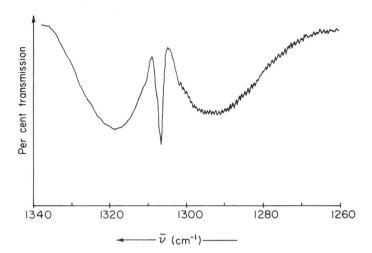

FIGURE 6-25 A parallel band for CH$_3$Br (symmetric top). [From G. M. Barrow, "Introduction to Molecular Spectroscopy". Copyright © 1962 by McGraw-Hill, Inc. Used by permission of McGraw-Hill Book Company.]

For symmetric top molecules, those vibrations having oscillating dipole moments that are parallel to the principal axis produce a *parallel absorption band with P-, Q-, and R-branches*. The symmetric C—H stretching and bending frequencies in CH$_3$Br are examples of parallel bands. The type of spectrum obtained for a parallel band is illustrated in Fig. 6–25. In this example the rotational fine structure in the *R*-branch is not resolved. The parallel band for a symmetric top molecule is similar to the perpendicular band for a linear molecule. For a perpendicular absorption band in a symmetric top molecule, several *Q*-bands are detected, often with overlapping unresolved *P*- and *R*-branches. The C—Cl bending vibration in CH$_3$Cl is an example of a perpendicular band in a symmetric top molecule. A typical spectrum for this case is illustrated in Fig. 6–26. In a spherical top, the selection rule for a *perpendicular band is* $\Delta J = 0, \pm 1$. This information can be employed to support band assignments.

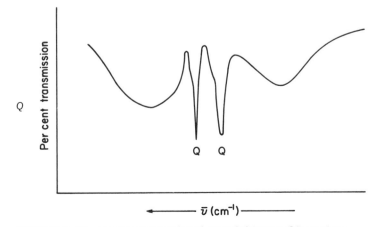

FIGURE 6-26 Perpendicular band containing two *Q* branches.

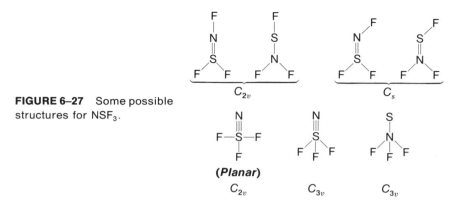

FIGURE 6–27 Some possible structures for NSF$_3$.

It becomes very difficult in some cases to distinguish between symmetric and asymmetric top molecules. As the symmetry of the asymmetric top becomes lower, the task becomes easier. Considerable information can be obtained from the band shape to support vibrational assignments.*

6–20 APPLICATION OF RAMAN AND INFRARED SELECTION RULES TO THE DETERMINATION OF INORGANIC STRUCTURES

1. The structures in Fig. 6–27 represent some of the possibilities that should be considered for a material with an empirical formula NSF$_3$. Table 6–8 summarizes the calculated number and symmetry of bands for these possible structures. These results and the infrared activity of the bands should be determined for practice, employing the procedure in the section on symmetry considerations and the character tables in Appendix A.

The infrared spectrum has been reported.[30a] Six intense bands are found, in agreement with the C_{3v} structure but certainly not in conflict with other possible structures. Some of the fundamentals may be of such low intensity that they are not seen. However, it was found that four of the bands (v_1, v_2, v_3, v_5) have P-, Q-, and R-branches. This is good evidence that the molecule being investigated is a symmetric top molecule and supports a C_{3v} structure. The spectral data are contained in Table 6–9, where they are compared with data reported for POF$_3$.[30b] The forms of the vibrations are diagrammed in Appendix C (see the discussion of the ZXY$_3$-molecule). These data, combined with an nmr study, have been employed to support the C_{3v} structure F$_3$S—N. Polarized Raman studies were not carried out. These would not add a great deal of structural information but would help confirm the assignments. The A_1 bands should be polarized and the E bands depolarized.

The F$_3$N—S structure has not been eliminated by these data. The frequencies of the observed bands provide information to favor the F$_3$SN structure. The

---

*See Herzberg (in Additional References), pp. 380–390, for an explanation of the phenomena just discussed. Many applications of these principles are discussed in this reference.

6-20 Application of Raman and Infrared Selection Rules to the Determination of Inorganic Structures

TABLE 6–8. Calculated Number of Bands for Various Structures of NSF$_3$

C_{3v}	C_{2v} (C_2-axis is the z-axis)	C_s
$3A_1$ (IR active)	$4A_1$ (IR active)	$7A'$ (IR active)
	$3B_1$ (IR active)	$2A''$ (IR active)
$3E$ (IR active)	$2B_2$ (IR active)	

S—N stretch is the only fundamental vibration that could be expected at a frequency of 1515 cm^{-1}. [A very rough approximation for this frequency can be obtained by employing the C≡C force constant and the S and N masses in equation (6–3).] The NF stretching vibrations in NF$_3$ occur at 1031 cm^{-1} and would not be expected to be as high as 1515 cm^{-1} in F$_3$N—S. The F$_3$SN structure receives further support from microwave and mass spectroscopic studies. The low moment of inertia calculated from the microwave studies cannot be explained unless the sulfur atom is at the center of mass.

2. The same approach as that outlined above was employed to prove that perchlorylfluoride, ClO$_3$F, has a C_{3v} structure instead of O$_2$ClOF.[30c]

3. An appreciation for some of the difficulties encountered in making spectral assignments can be obtained from an article by Wilson and Hunt[31a] on the reassignment of the spectrum of SO$_2$F$_2$. Often a set of band assignments may appear self-consistent, but these assignments may not be a unique set. The effect of isotopic substitution on spectral shifts can be employed to provide additional support for the assignments made. A Raman study has also been reported.[31b]

4. The infrared and Raman spectra have been interpreted to indicate a planar structure for N$_2$O$_4$.[31c] Structures corresponding to the planar D_{2h} model and the staggered D_{2d} configuration were considered. Since the infrared and Raman spectra have no lines in common, it is proposed that the molecule has a center of symmetry. A rigorous assignment of the bands could not be made with the available data.

5. The spectral data for B$_2$H$_6$ support the bridged hydrogen structure.[32a] The entire spectrum is analyzed, assignments are made, and conclusions regarding the structure are drawn from the rotational fine structure of some of the perpendicular bands.

6. An infrared and Raman study[32b] supports a C_{2v} planar T-shaped structure for ClF$_3$. Too many fundamentals were observed for either a D_{3h} or C_{3v} structure. Some of the bands had P-, Q-, and R-branches, while others had only P- and R-branches. Since all fundamentals for a tetraatomic C_{3v} molecule (*e.g.*, PF$_3$) have P-, Q-, and R-branches, this structure is eliminated.

7. The infrared and Raman spectra of B$_2$(OCH$_3$)$_4$ have been interpreted to indicate[32c] a planar arrangement for the boron and oxygen atoms. Many bands were found in the spectrum. Assignments were made that are consistent with a planar C_{2h} molecule. More fundamentals were found than would be predicted from a D_{2d} model. The infrared and Raman spectra do not have a coincidence of any of the vibrations asigned to B—B or B—O modes; but, as would be expected for the proposed structure, the CH$_3$ and O—C vibrations are found in both the infrared and Raman. Good agreement is found for the bands

TABLE 6–9. Fundamental Vibration Frequencies for NSF$_3$ and OPF$_3$

	NSF$_3$	OPF$_3$
v_1	1515	1415
v_2	775	873
v_3	521	473
v_4	811	990
v_5	429	485
v_6	342	345

assigned to the O—C and —CH$_3$ modes in this molecule and in the B(OCH$_3$)$_3$ molecule, illustrating the applicability of the group frequency concept. The above assignments indicate a planar structure, but since the interpretation is very much dependent on the correctness of the assignments, they should be confirmed by isotopic substitution. The band assignments made are consistent with those calculated from a normal coordinate analysis assuming a C_{2h} model.

One final advantage of combining Raman studies with infrared will be mentioned. Often, in the infrared spectra, it is difficult to distinguish combination bands or overtones from fundamentals. This is usually not a problem in the Raman, for the fundamentals are much more intense. An extensive review of the inorganic applications of Raman spectroscopy[11] should be consulted for other applications of this technique.

Bond Strength Frequency Shift Relations

There are many examples in the literature of attempts to infer the bond strength for coordination of a Lewis acid to Lewis bases from the magnitude of the infrared shift of some acid (or base) functional group upon coordination. Usually a relationship between these two quantities is tacitly assumed, and the infrared shift interpreted as though it were a bond strength in terms of the electronic structures of the acid and base. This approach is to be discouraged, for the problem is too complex to permit such an assumption.

The systems that have been most thoroughly studied are the hydrogen bonding ones. In general, hydrogen bonding[33] to an X—H molecule results in a decrease in the frequency and a broadening of the absorption band that is assigned to the X—H stretching vibration. The spectra of free phenol, A (where X = C$_6$H$_5$O$^-$), and a hydrogen-bonded phenol, B, are indicated in Fig. 6–28. The magnitude of Δv_{OH}, the frequency shift upon formation of a 1:1 complex with a base, is related to $-\Delta H$ for the formation of the 1:1 adduct measured in a poorly solvating solvent, for a large series of different bases as shown[34] in Fig. 6–29. Different alcohols require different lines, as illustrated by the separate

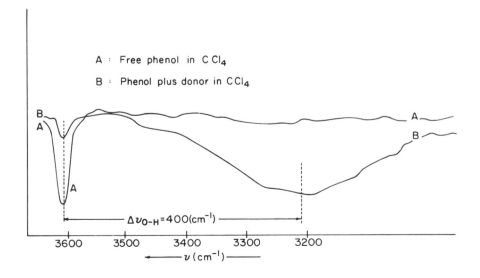

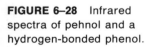

FIGURE 6–28 Infrared spectra of pehnol and a hydrogen-bonded phenol.

6-20 Application of Raman and Infrared Selection Rules to the Determination of Inorganic Structures   195

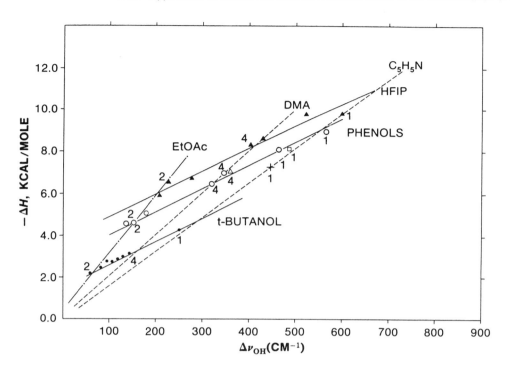

FIGURE 6–29 Constant-acid.constant-base frequency shift-enthalpy relations. Solid lines are constant-acid lines. The phenol line also contains values for p-t-butylphenol (+), phenol (○), p-chlorophenol (□), and m-trifluoromethylphenol (△). The acid butanol is represented by ● and $(CF_3)_2CHOH$ by ▲. The number 1 represents all points on the pyridine constant-base line; 2, ethyl acetate; and 4, N,N-dimethylacetamide. (The pyridine-$(CF_3)_2CHOH$ Δv is estimated from the Δv vs. Δv plot of this acid and phenol.)

solid lines in this figure. The solid lines are referred to as constant-acid lines. The dotted lines represent data in which the base is held constant and the acid is varied. Over this range of enthalpies, a linear constant-acid equation exists for various alcohols:

$$-\Delta H(\text{kcal mole}^{-1}) = (0.0105 \pm 0.0007) \Delta v_{OH} + 3.0(\pm 0.2)(\text{phenols}) \quad (6\text{-}25a)$$

$$-\Delta H(\text{kcal mole}^{-1}) = (0.0115 \pm 0.0008) \Delta v_{OH} + 3.6(\pm 0.03)[(CF_3)_2CHOH] \quad (6\text{-}25b)$$

$$-\Delta H(\text{kcal mole}^{-1}) = (0.0106 \pm 0.0005) \Delta v_{OH} + 1.65(\pm 0.09)(t\text{-butanol}) \quad (6\text{-}25c)$$

$$-\Delta H(\text{kcal mole}^{-1}) = (0.0123 \pm 0.0006) \Delta v_{OH} + 1.8(\pm 0.1)(\text{pyrrole}^{[34]}) \quad (6\text{-}25d)$$

The authors[34–38] caution against the use of this relationship for a new class of donors (i.e., a functional group not yet studied) without first evaluating $-\Delta H$ and Δv_{OH} for at least one system involving this functional group. The enthalpy for a given acid has been shown to be related to two different properties[35] of the base via the following relation:

$$-\Delta H = E_A E_B + C_A C_B \quad (6\text{-}26)$$

where E_A and E_B are roughly related to the tendency of the acid and base to undergo electrostatic interaction and C_A and C_B to their tendency to undergo covalent interaction. Parameters for many different acids and bases have been reported. When the range of C_B/E_B is between 1.5 and 6.0, the enthalpy-frequency shift correlations hold to within 0.3 kcal mole^{-1}. For sulfur donors, where this ratio is ~20, the plots of Fig. 6-29 do not hold.[36] Unless Δv_{OH} should have the same functional dependence on E_B and C_B as $-\Delta H$ does, there will have to be a limitation on the C/E ratio of the base that can be used. Furthermore, there is an experimental point at which ΔH and Δv_{OH} are both equal to zero. Accordingly, the line will have to curve; this curvature is predicted by theory.[37] The theoretical treatment also predicts curvature on the high frequency end of the curve. As a result, extrapolation of the curve in either direction is dangerous. The analysis[37] indicates that frequency shifts can be employed instead of force constants because the high-energy O—H stretching vibration is far enough removed in energy from other vibrations that the band assigned to it is a nearly pure O—H stretching vibration.

In view of the considerable effort involved in an enthalpy measurement (often one or two days of work), this relationship is a welcome one for ascertaining the strength of the hydrogen bond. However, much confusion has been generated by the inappropriate use of this relation. Inappropriate solvent selection, and enthalpies that are a composite of things other than the simple 1:1 adduct formation reaction (*e.g.*, two different donor sites in the same molecule[38]), should be avoided.

Additional correlations of this sort involving different types of acids, and others involving some spectroscopic change in the base upon coordination, would be most useful. However, they cannot be naively assumed but must be tested against measured enthalpies and the limitations sought. In a study of the Lewis acid chloroform,[39] it was found that the enthalpy of adduct formation did not correlate with the observed frequency shift.

6-21 CHANGES IN THE SPECTRA OF DONOR MOLECULES UPON COORDINATION

The infrared spectrum of *N,N*-dimethylacetamide in the solvent CCl_4 has an absorption band at 1662 cm^{-1} that is due to a highly coupled carbonyl absorption. The low frequency compared to acetone (1715 cm^{-1}) is attributed to a resonance interaction with the lone pair on the nitrogen (see Fig. 6-30). Upon complexation with several Lewis acids, a decrease in the frequency of this band is observed.[40] This decrease has been attributed to the effect of oxygen coordination to the acid. Oxygen coordination could have several effects upon the vibration:

FIGURE 6-30 Resonance structures for *N,N*-dimethylacetamide.

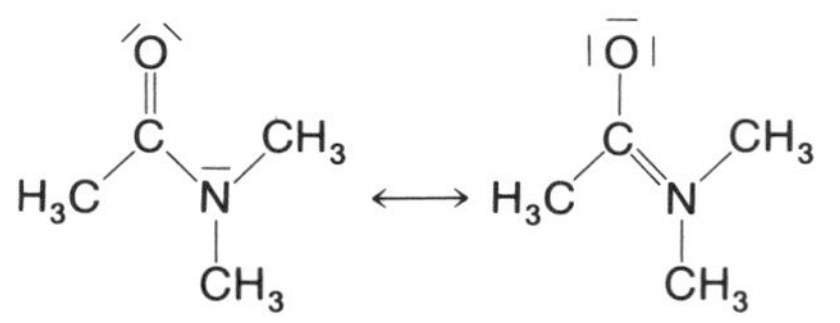

1. Since the oxygen atom has to move against the atom to which it is coordinated, an increase in frequency should result. In effect this is to say that for the system X—O=C<, the C—O and X—O vibrations couple, producing a higher-energy carbonyl absorption.

2. A change in oxygen hybridization could increase (or decrease) the C—O σ bond strength and increase (or decrease) the force constant of the C—O bond.

3. The most important effect in this case involved decreasing the carbonyl force constant by draining π electron density out of the carbonyl group. This causes the observed decrease in the carbonyl frequency and indicates oxygen coordination. The absence of any absorption in the carbonyl region on the high-frequency side of the uncomplexed carbonyl band is further support for only oxygen coordination. If there were nitrogen coordination in the complexes, the nitrogen lone pair would be involved, resulting in a decreased C—N vibration frequency and a higher-energy carbonyl absorption.

The decrease in the carbonyl stretching frequency of urea upon complexation to Fe^{3+}, Cr^{3+}, Zn^{2+}, or Cu^{2+} is interpreted as indicating oxygen coordination in these complexes.[41] The explanation is similar to that described for the amides. This conclusion is supported by x-ray studies on the structure of the iron and chromium complexes.[42] Nitrogen coordination is observed in the compounds $Pd(NH_2CONH_2)_2Cl_2$ and $Pt(NH_2CONH_2)_2Cl_2$, and the spectra show the expected increase in the C—O stretching frequency as well as a decrease in the C—N frequency.

Decreases in the P—O stretching frequencies indicative of oxygen coordination are observed when triphenylphosphine oxide[43] and hexamethylphosphoramide, $OP[N(CH_3)_2]_3$,[44] are coordinated to metal ions, phenol, or iodine. A decrease in the S—O stretching frequency, indicative of oxygen coordination, is observed when dimethyl sulfoxide or tetramethylene sulfoxide is complexed to many metal ions, iodine, and phenol.[45] The S—O stretching frequency increases in the palladium complex of dimethyl sulfoxide, compared to free sulfoxide. This is an indication of sulfur coordination in this complex. The N—O stretching frequency of pyridine N-oxide is decreased upon complexation.[46]

The infrared spectra of some ethylenediaminetetraacetic acid complexes indicate that this ligand behaves as a tetradentate, pentadentate, or hexadentate ligand in various complexes. The interpretation is based upon absorption bands in the carbonyl region corresponding to free and complexed carbonyl groups.[47]

The change in C≡N stretching frequency of nitriles[48] and metal cyanides[49] resulting from their interaction with Lewis acids has attracted considerable interest. When such compounds are coordinated to Lewis acids that are not able to π back-bond, the C≡N stretching frequency increases. This was originally thought to be due to the coupling effect described above in the discussion of the coordination of amides. A combined molecular orbital and normal coordinate analysis of acetonitrile and some of its adducts[48] indicates that a slight increase is expected from this effect but that the principal contribution to the observed shift arises from an increase in the C≡N force constant. This increase is mainly due to an increase in the C≡N sigma bond strength from nitrogen rehybridization.[48] In those systems where there is extensive π back-bonding from the acid into the π* orbitals of the nitrile group, the decreased pi bond energy accounts for the decreased frequency.

There have been many papers[50-52] on the assignment of the absorption bands in simple ammine complexes. The spectra have been analyzed in terms of a single M—NH$_3$ group, *i.e.*, a local C_{3v} structure. In the sodium chloride range, absorption is usually found in four regions near 3300, 1600, 1300, and 825 cm^{-1}. These have been assigned to N—H stretching, NH$_3$ degenerate deformation, NH$_3$ symmetric deformation, and —NH$_3$ rocking, respectively. Deuteration studies[51] support these assignments. It was concluded that no metal-nitrogen vibrations in these transition metal ion complexes occur in the sodium chloride region. The frequencies of the bands are dependent upon the metal ion, the type of crystal lattice, and the anion.[52] The anion effect is attributed to hydrogen bonding. References by Chatt *et al.*[52] are highly recommended readings for more details on this subject.

Infrared spectroscopy has been very valuable[53–55] in determining whether the thiocyanate ion, NCS$^-$, is bonded in a metal complex via the nitrogen atom or the sulfur atom. Since both modes of bonding have been detected, the anion is said to be *ambidentate*. The CN stretch generally occurs at a lower frequency in the N-bonded isomer than in the S-bonded one. The vibration assigned to "C—S" is a better guide, occurring in the 780 to 760 cm^{-1} region for nitrogen-coordinated complexes and in the 690 to 720 cm^{-1} region for sulfur-bonded ones. The "NCS bend" is also diagnostic, occurring in the 450 to 440 cm^{-1} region for the N-bonded isomer and in the 400 to 440 cm^{-1} region for the S-bonded isomer. The observed intensity differences for the different isomers are also diagnostic. The reader is referred to literature reviews[56,57] for many examples of ambidentate coordination by this anion.

It should be pointed out that solid materials can be obtained that contain noncoordinated donor molecules trapped in the crystal lattice. By virtue of the lattice environment, changes in the spectrum of the donor can occur in these instances without coordination. It is thus not possible to prove coordination by a frequency shift in the infrared spectrum of the donor taken on a mull of a solid. This problem can be resolved if the solid can be put in solution and the spectrum compared with that of the donor in this solvent.

Caution should also be exercised in drawing conclusions regarding the strength of the interaction between a donor and a Lewis acid (including metal ions) from the magnitude of the frequency shift of the donor. Usually, this relationship has not been adequately demonstrated with reliable, easily interpreted enthalpy data. There will very probably be cases in which good correlations are obtained, but no general statements can be made at present.

Excellent discussions of the important role infrared spectroscopy has played in elucidating the structure of many transition metal complexes are available.[18] The spectra of metal carbonyls, metal acetylacetonates, cyanide, thiocyanate, and several other types of complexes are discussed.

6–22 CHANGE IN SPECTRA ACCOMPANYING CHANGE IN SYMMETRY UPON COORDINATION

Some of the most useful applications of infrared spectroscopy in the area of coordination and organometallic chemistry occur when the symmetry of a ligand is changed upon complexation. For example, the symmetry change accompanying the binding of small molecules (*e.g.*, N$_2$, O$_2$, and H$_2$) to transition metal ions has a dramatic influence on the infrared spectra of these materials. When azide ion is added to Ru(NH$_3$)$_5$(H$_2$O)$^{3+}$, a product containing coordinated dinitrogen, N$_2$, is obtained.[58] The stretching vibration of free N$_2$ is infrared inactive but Raman active, with a band at 2331 cm^{-1}. In Ru(NH$_3$)$_5$N$_2$$^{2+}$, a sharp intense line appears[59] at 2130 cm^{-1} in the infrared spectrum that is attributed to the N—N stretching vibration. In IrN$_2$Cl[P(C$_6$H$_5$)$_3$]$_2$ and HCoN$_2$[P(C$_6$H$_5$)$_3$]$_3$,[60] the same vibration gives rise to bands at 2095 cm^{-1} and 2088 cm^{-1}, respectively.

The complex IrCOCl[P(C$_6$H$_5$)$_3$]$_2$ picks up O$_2$ and binds it in such a way that the O—O bond axis is perpendicular to a line drawn from iridium to the center of the O—O bond. An infrared band assigned[61] to the O—O stretching vibration in the complex occurs at 857 cm^{-1}. This corresponds to an infrared inactive stretch in free O$_2$, which does have a Raman band at 1555 cm^{-1}. The lower frequency in the complex is attributed to electron transfer making the O—O bond peroxide-like.

When H$_2$ is added to IrCOCl[P(C$_6$H$_5$)$_3$]$_2$, two new sharp absorption bands arise in the infrared at 2190 and 2100 cm^{-1}. The infrared bands at 2160 and 2101 cm^{-1} in IrH$_2$CO[P(C$_6$H$_5$)$_3$]$_2$ are shifted to 1620 and 1548 cm^{-1} in the deuterated analogue.[62] The agreement of these shifts with those predicted from the change in reduced mass indicates that these vibrations are mainly metal-hydrogen modes.

Evidence to support coordination of the nitrate ion can also be obtained from infrared studies[63,64] because of the symmetry change accompanying coordination. Free nitrate ion has D_{3h} symmetry, but upon coordination of one of the oxygens, the symmetry is lowered to C_{2v} or to C_s as shown in Fig. 6–31.

By determining the total representation and the symmetries of the irreducible representations comprising the $3N - 6$ vibrations in D_{3h} NaNO$_3$, one finds A_1'(R), A_2''(IR), $2E'$(R)(IR). The Raman and infrared activity is indicated in parentheses. The bands in D_{3h} NaNO$_3$ are assigned as v_1 1068 cm^{-1} (R)A_1', v_2 831 cm^{-1} (IR) A_2'', v_3 1400 cm^{-1} (R)(IR)E', and v_4 710 cm^{-1} (R)(IR)E'. The forms of these vibrations are shown in Appendix B. In C_{2v} and C_s symmetry, A_1 becomes infrared active and each of the E' modes splits into two bands, as shown in Table 6–10.

The E' asymmetric stretch in the nitrate ion is split into a high-frequency N—O asymmetric stretch and a lower-frequency symmetric stretch (see Appendix B for planar XY$_3$ and ZXY$_2$ molecules) upon coordination. These occur at 1460 and 1280 cm^{-1} (± 20 cm^{-1}), respectively. An additional NO stretching frequency, corresponding roughly to the inactive symmetric stretch in nitrate ion, also appears. Thus, in the absence of complications from site symmetry problems, one can readily tell from the infrared spectrum whether or not the nitrate ion is coordinated. The magnitude of the splitting of the E' mode is generally greater in C_{2v} symmetry than in C_s, and this criterion has been used to support this mode of coordination.

FIGURE 6–31 C_s and C_{2v} structures of coordinated nitrate.

TABLE 6–10. Correlation Chart Connecting the Vibrational Mode Symmetries for Various Kinds of NO$_3$$^-$ Coordination

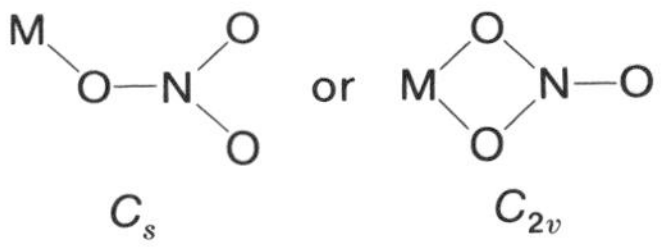

	v_1	v_2	v_3	v_4
D_{3h}	A_1(R)	A_2''(IR)	E'(IR, R)	E'(IR, R)
C_{2v}	A_1(IR, R)	B_1(IR, R)	A_1(IR, R) + B_2(IR, R)	A_1(IR, R) + B_2(IR, R)
C_s	A_1'(IR, R)	A''(IR, R)	A'(IR, R) + A'(IR, R)	A'(IR, R) + A'(IR, R)

The sulfate ion is another good example to demonstrate the effect of change in symmetry on spectra. The sulfate ion (T_d symmetry) has two infrared bands in the sodium chloride region: one assigned to v_3 at 1104 cm^{-1} and one to v_4 at 613 cm^{-1}. (See Appendix C for the forms of these vibrations.) In the complex [Co(NH$_3$)$_5$OSO$_3$]Br the coordinated sulfate group has lower symmetry, C_{3v}. Six bands now appear at 970 (v_1), 438 (v_2), 1032 to 1044 and 1117 to 1143 (from v_3), and 645 and 604 cm^{-1} (from v_4). In a bridged sulfate group the symmetry is lowered to C_{2v} and even more bands appear. For a bridged group the v_3 band of SO$_4^{2-}$ is split into three peaks and the v_4 band into three peaks.[65] Infrared spectroscopy is thus a very effective tool for determining the nature of the bonding of sulfate ion in complexes.

The infrared spectra of various materials containing perchlorate ion have been interpreted[66] to indicate the existence of coordinated perchlorate. As above, the change in symmetry brought about by coordination increases the number of bands in the spectrum. References to similar studies on other anions are contained in the text by Nakamoto.[18]

In another application of this general idea, it was proposed that the five-coordinate addition compound, (CH$_3$)$_3$SnCl·(CH$_3$)$_2$SO, had a structure in which the three methyl groups were in the equatorial positions because the symmetric Sn—C stretch present in (CH$_3$)$_3$SnCl disappeared in the addition compound.[67] In (CH$_3$)$_3$SnCl the asymmetric and symmetric stretches occur at 545 cm^{-1} and 514 cm^{-1}, respectively.[68] In the adduct a single Sn—C vibration due to the asymmetric Sn—C stretch is detected at 551 cm^{-1}. This is expected for the isomer with three methyl groups in the equatorial position because a small dipole moment change is associated with the symmetric stretch in this isomer. For all other possible structures that can be written, at least two Sn—C vibrations should be observed. There is also a pronounced decrease in the frequency of the Sn—Cl stretching mode in the addition compound.

Infrared spectroscopy has been used to excellent advantage in a study of the influence of pressure on molecular structural transformations. Several systems were found to change reversibly with pressure. Changes in the infrared spectrum accompanying high spin–low spin interconversions of transition metal ion systems with pressure are also reported.[69]

A very elegant illustration of the consequences of symmetry change on the infrared spectrum involves the metal carbonyl compounds of general formula M(CO)$_5$X and the ^{13}C-substituted derivatives.[70] Our main interest is the high wavenumber region associated mainly with the carbonyl stretching vibrations. This, coupled with the fact that the C≡O stretch and the M—C stretch are widely separated in energy, enables us to use as a crude basis set for the vibrational problem the five C—O bond displacement vectors shown in Fig. 6–32A. The operations of the C_{4v} point group for the all-^{12}C—O compound produce the result $2A_1$ (one radial and one axial), B_1 (radial), and E (radial). Three are infrared active and four are Raman active. The forms of these vibrations are illustrated in Fig. 6–32B.

The infrared spectrum of Mo(CO)$_5$Br is shown in Fig. 6–33. The band labeled a is assigned to A_1 (radial), e to E_1, and g to A_1 (axial). The assignment of e to E_1 is confirmed by studying[71] derivatives in which Br is replaced by an asymmetric group that lowers the symmetry from C_{4v}. This results in splitting of this band into two peaks separated by 3 to 12 cm^{-1} in various derivatives. The other bands and shoulders in the spectrum have been assigned to ^{13}C

6-22 Change in Spectra Accompanying Change in Symmetry upon Coordination

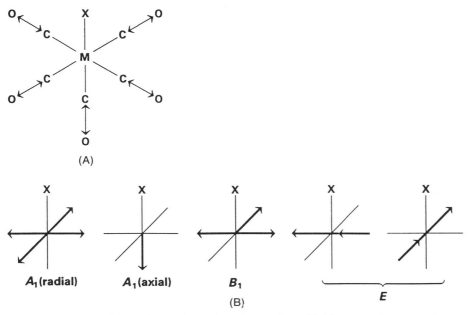

FIGURE 6–32 (A) CO basis set for carbonyl stretches. (B) The resulting normal modes.

derivatives, for there is 1.1% ^{13}C in naturally occurring carbon compounds. Consequently, 1.1% of the molecules have an axial ^{13}CO and 4.4% have a radial ^{13}CO. There will be very few molecules with two ^{13}CO groups in them.

First, we shall consider the consequences of having an axial ^{13}CO in the molecule. The symmetry is still C_{4v}, and the B_1 and E modes (which have no contribution from the axial group) will be unaffected by the mass change. The two A_1 modes having the same symmetry will mix, so isotopic substitution could affect both; but it will have a major effect on A_1 (axial). The band h (at 1958 cm^{-1}) in the spectrum is assigned to A_1 (axial), for it is roughly what can be expected for the change in g from the mass change.

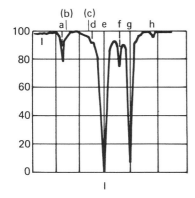
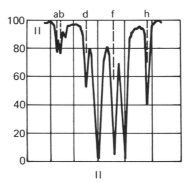

FIGURE 6–33 (A) High resolution infrared spectrum of Mo(CO)$_5$Br in the 2200 to 1900 cm^{-1} region. (B) Spectrum after three hours of exchange with ^{13}CO.

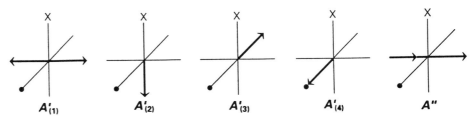

FIGURE 6–34 Infrared allowed vibrations in M(CO)$_4$(^{13}CO)X with ^{13}CO in the radial position.

For molecules in which a ^{13}CO is located in the radial position, the symmetry is lowered to C_s, leading to the possible vibrations (the solid circle indicates ^{13}CO) illustrated in Fig. 6–34. Only the A'' will rigorously be unaffected by the mass change, and it resembles the E mode in the all-^{12}CO molecule. All of the other four modes can mix and could be shifted by distributing the mass effect over all four modes. This would shift the vibrations to lower frequencies than those in the all ^{12}CO molecule. Band f is assigned as an A' mode in the radial ^{13}CO monosubstituted molecules that is largely A'. Band d is either a radial A_1 mode of the axially substituted molecule or the highest A' of the C_s molecule. The former assignment would involve an intensity that is about 1% of the A_1 in the all ^{12}CO molecule; since the intensity is found to be about 10%, it is assigned to an A' mode in the C_s molecule. A normal coordinate analysis[4] has been carried out to substantiate these assignments. An important part of this work is the assignment of a particular band to radial ligands and another to axial ligands. When ^{13}CO is added, the change in intensity of these bands can be followed and the relative rates of axial versus radial substitution determined.

Jones *et al.*[5] have carried out a normal coordinate analysis on systems of this type, employing a large number of isotopes. They conclude that the π-bond orders on these systems obtained by using the Cotton-Kraihanzel[4] approximations must be viewed with suspicion. The more complete analysis indicates that as the formal positive charge on the central atom increases in forming M(CO)$_5$Br from M(CO)$_6$, the σ-bonding increases and the extent of π back-bonding decreases.[5] This study[5] provides a good reference to illustrate the information that can be reliably obtained from a normal coordinate analysis.

REFERENCES CITED

1. K. Nakamoto et. al., Chem. Comm., 1451 (1969).
2. E. B. Wilson, J. C. Decius, and P. C. Cross, "Molecular Vibrations," McGraw-Hill, New York, 1955.
3. G. Barrow, "Introduction to Molecular Spectroscopy," pp. 146–130, McGraw-Hill, New York, 1962.
4. F. A. Cotton and C. Kraihanzel, J. Amer. Chem. Soc., *84*, 4432 (1962); Inorg. Chem., *2*, 533 (1963).
5. L. H. Jones *et al.*, Inorg. Chem., *6*, 1269 (1967); ibid., *7*, 1681 (1968); ibid., *8*, 2349 (1969); ibid., *12*, 1051 (1973).
6. a. L. H. Jones and R. R. Ryan, J. Chem. Phys., *52*, 2003 (1970).
   b. L. H. Jones, "Inorganic Vibrational Spectroscopy," p. 40, M. Dekker, New York, 1971.
   c. L. H. Jones and B. I. Swanson, Accts. Chem. Res., *9*, 128 (1976).

7. J. Overend and J. R. Scherer, J. Chem. Phys., *32*, 1296 (1960).
8. J. O. Halford, J. Chem. Phys., *24*, 830 (1956); S. A. Francis, J. Chem. Phys., *19*, 942 (1951); G. Karabatsos, J. Org. Chem., *25*, 315 (1960).
9. J. A. Gachter and B. F. Koningstein, J. Opt. Soc. Amer., *63*, 892 (1973).
10. J. A. Loader, "Base Laser Raman Spectroscopy," Heyden-Sadtler, London, 1970.
11. **a.** S. Tobias, "The Raman Effect," Volume 2, Chapter 7, ed. A. Anderson, M. Dekker, New York, 1973.
    b. T. G. Spiro, "Chemical and Biochemical Applications of Lasers," Chapter 2, C. B. Moore, Ed., Academic Press, New York, 1974.
12. J. Tang and A. C. Albrecht, "Raman Spectroscopy," Chap. 2, ed. H. A. Szymanski, Plenum Press, New York, 1970.
13. T. G. Spiro and T. C. Strekas, Proc. Nat. Acad. Sci., *69*, 2622 (1972); Accts. Chem. Res., *7*, 339 (1974).
14. J. B. Dunn, D. F. Shriver, and I. M. Klotz, Proc. Nat. Acad. Sci., *70*, 2582 (1973)
15. National Bureau of Standards, Monograph 70, I to IV.
16. C. H. Townes and A. L. Schawlow, "Microwave Spectroscopy," pp. 56–59, 1110–114, McGraw-Hill, New York, 1955; Ann. Rev. Phys. Chem.—see the indices for all volumes.
17. **a.** W. Gordy and R. L. Cook, "Techniques of Organic Chemistry," Volume IX, Part II, Microwave Molecular Spectra, ed. A. Weissberger, Interscience, New York, 1970.
    b. J. E. Parkin, Ann. Reports Chem. Soc., *64*, 181 (1967); ibid., *65*, 111 (1968)
    c. W. H. Flygare, Ann. Rev. Phys. Chem., *18*, 325 (1967).
    d. H. D. Rudolph, Ann. Rev. Phys. Chem., *21*, 73 (1970).
18. K. Nakamoto, "Infrared Spectra of Inorganic and Coordination Compounds," 2nd ed., Wiley, New York, 1970. An excellent reference for inorganic compounds.
19. **a.** M. D. Low, J. Chem. Educ., *47*, A 349, A 415 (1970).
    b. R. J. Bell, "Introductory Fourier Transform Spectroscopy," Academic Press, New York, 1972.
    c. J. R. Ferraro and L. J. Basils, "FTIR-Applications of Chemical Systems," Volumes 1 and 2, Academic Press, New York, 1978, 1979.
20. M. L. Hair, "Infrared Spectroscopy in Surface Chemistry," M. Dekker, New York, 1967.
21. **a.** A. T. Bell in "Vibrational Spectroscopies for Adsorbed Species," ed. A. T. Bell and M. L. Hair, American Chemical Society Symposium Series, *137*, 1980.
    b. G. Kortüm, "Reflectance Spectroscopy: Principles, Methods and Applications," s Springer-Verlag, New York, 1969.
    c. W. W. Wendlandt and H. G. Hecht, "Reflectance Spectroscopy," Wiley, New York, 1961.
    d. W. W. Wendlandt, "Modern Aspects of Reflectance Spectroscopy," Plenum Press, New York, 1961.
22. **a.** M. D. Porter, Anal. Chem. *60*, 1143A (1988).
    b. N. J. Harrick, "Internal Reflection Spectroscopy," Wiley, New York, 1967.
23. **a.** A. Rosencwaig, "Optoacoustic Spectroscopy and Detection," ed. Y. H. Pao, Academic Press, New York; Science *181*, 657 (1973).
    b. D. W. Vidrine, in "Fourier Transform Infrared Spectroscopy," ed. J. R. Ferraro and I. S. Basile, vol. 3, Chapter 4, Academic Press, New York, 1982.
    c. J. W. Chiders and R. W. Palmer, American Laboratory, 22 (1986).
24. **a.** W. J. Potts, Jr., "Chemical Infrared Spectroscopy," Volume I, Wiley, New York, 1963.
    b. R. G. Miller, "Laboratory Methods in Infrared Spectroscopy," Heyden, London, 1965.
25. S. Krimm, "Infrared Spectra of Solids," in "Infrared Spectroscopy and Molecular Structure," Chapter 8, ed. M. Davies, Elsevier, New York, 1963.
26. **a.** H. Winston and R. S. Halford, J. Chem. Phys., *17*, 607 (1949).
    b. D. F. Hornig, J. Chem. Phys., *16*, 1063 (1948).
27. M.-L. Josien and N. Fuson, J. Chem. Phys., *22*, 1264 (1954).

28. a. A. Allerhand and P. R. Schleyer, J. Amer. Chem. Soc., *85*, 371 (1963).
    b. R. S. Drago, J. T. Kwon, and R. D. Archer, J. Amer. Chem. Soc., *80*, 2667 (1958).
29. a. J. L. Slater et al., J. Chem. Phys., *55*, 5129 (1971).
    b. H. Huber et al., Nature Phys. Sci., *235*, 98 (1972).
    c. R. C. Spiker et al., J. Amer. Chem. Soc., *94*, 2401 (1972).
    d. See G. A. Ozin and A. V. Voet, Accts. Chem. Res., *6*, 313 (1973) for a review.
30. a. H. Richert and O. Glemser, Z. Anorg. Allgem. Chem., *307*, 328–344 (1961).
    b. N. J. Hawkins, V. W. Coher and W. S. Koski, J. Chem. Phys., *20*, 258 (1952).
    c. F. X. Powell and E. R. Lippincott, J. Chem. Phys., *32*, 1883 (1960).
31. a. M. K. Wilson and G. R. Hunt, Spectrochim. Acta, *16*, 570 (1960).
    b. E. L. Pace and H. V. Samuelson, J. Chem. Phys., *44*, 3682 (1966).
    c. R. G. Snyder and J. C. Hisatsune, J. Mol. Spectr., *1*, 139 (1957).
32. a. W. C. Price, J. Chem. Phys., *16*, 894 (1948).
    b. H. H. Claassen, B. Weinstock, and J. G. Malm, J. Chem. Phys., *28*, 285 (1958).
    c. H. Becher, W. Sawodny, H. Nöth, and W. Meister, Z. Anorg. Allgem. Chem., *314*, 226 (1962).
33. M. D. Joesten and L. J. Schaad, "Hydrogen Bonding," M. Dekker, New York, 1974 an excellent treatment of the work on hydrogen bonding.
34. M. S. Nozari and R. S. Drago, J. Amer. Chem. Soc., *92*, 7086 (1970), and references therein.
35. R. S. Drago, G. C. Vogel, and T. E. Needham, J. Amer. Chem. Soc., *93*, 6014 (1971); A. P. Marks and R. S. Drago, J. Amer. Chem. Soc., *97*, 3324 (1975).
36. G. C. Vogel and R. S. Drago, J. Amer. Chem. Soc., *92*, 5347 (1970).
37. K. F. Purcell and R. S. Drago, J. Amer. Chem. Soc., *89*, 2874 (1967).
38. R. S. Drago, B. Wayland, and R. L. Carlson, J. Amer. Chem. Soc., *85*, 3125 (1963).
39. F. L. Slejko, R. S. Drago, and D. G. Brown, J. Amer. Chem. Soc., *94*, 9210 (1972).
40. C. D. Schmulbach and R. S. Drago, J. Amer. Chem. Soc., *82*, 4484 (1960); R. S. Drago and D. A. Wenz, J. Amer. Chem. Soc., *84*, 526 (1962).
41. R. B. Penland, S. Mizushima, C. Curran, and J. V. Quagliano, J. Amer. Chem. Soc., *79*, 1575 (1957).
42. Y. Okaya et al., "Abstracts of Papers of 4th International Congress of the International Union of Crystallography," p. 69, Montreal, 1957.
43. F. A. Cotton *et al.*, J. Chem. Soc., *1961*, 2298, 3735, and references contained therein.
44. J. T. Donoghue and R. S. Drago, Inorg. Chem., *1*, 866 (1962).
45. R. S. Drago and D. W. Meeks, J. Phys. Chem., *65*, 1446 (1961), and papers cited therein.
46. J. V. Quagliano *et al.*, J. Amer. Chem. Soc., *83*, 3770 (1961).
47. M. L. Morris and D. H. Busch, J. Amer. Chem. Soc., *78*, 5178 (1956).
48. K. F. Purcell and R. S. Drago, J. Amer. Chem. Soc., *88*, 919 (1966).
49. D. F. Shriver and J. Posner, J. Amer. Chem. Soc., *88*, 1672 (1966).
50. M. G. Miles *et al.*, Inorg. Chem., *7*, 1721 (1968), and references therein.
51. S. Mizushima, I. Nakagawa, and J. V. Quagliano, J. Chem. Phys., *23*, 1367 (1955); G. Barrow, R. Krueger, and F. Basolo, J. Inorg. Nucl. Chem., *2*, 340 (1956).
52. J. Chatt, L. A. Duncanson, and L. M. Venanzi, J. Chem. Soc., *1955*, 4461; *1956*; 2712.
53. J. Lewis, R. S. Nyholm, and P. W. Smith, J. Chem. Soc., *1961*, 4590.
54. A. Sabatini and I. Bertini, Inorg. Chem., *4*, 959 (1965).
55. N. J. DeStefano and J. L. Burmeister, Inorg. Chem., *10*, 998 (1971).
56. J. L. Burmeister, Coord. Chem. Rev., *3*, 225 (1968).
57. A. H. Norbury and A. I. P. Sinha, Quart. Rev. Chem. Soc., *24*, 69 (1970).
58. A. D. Allen and C. V. Senoff, Chem. Comm., 621 (1965); A. D. Allen, et al., J. Amer. Chem. Soc., *89*, 5595 (1967); A. D. Allen and F. Bottomley, Accts. Chem. Res., *1*, 360 (1968).
59. J. P. Collman and Y. Kang, J. Amer. Chem. Soc., *88*, 3459 (1966).
60. I. Yamamoto *et al.*, Chem. Comm., 79 (1967); Bull. Chem. Soc. Japan, *40*, 700 (1967).
61. P. B. Chock and J. Halpern. J. Amer. Chem. Soc., *88*, 3511 (1966).
62. L. Vaska, Chem. Comm., 614 (1960).

63. B. M. Gatehouse, S. E. Livingston, and R. S. Nyholm, J. Chem. Soc., *1957*, 4222; C. C. Addison and B. M. Gatehouse, J. Chem. Soc., *1960*, 613.
64. J. R. Ferraro, J. Inorg. Nucl. Chem., *10*, 319 (1959).
65. K. Nakamoto *et al.*, J. Amer. Chem. Soc., *79*, 4904 (1957); C. G. Barraclough and M. L. Tobe, J. Chem. Soc., *1961*, 1993.
66. B. J. Hathaway and A. E. Underhill, J. Chem. Soc., *1961*, 3091.
67. N. A. Matwiyoff and R. S. Drago, Inorg. Chem., *3*, 337 (1964).
68. H. Kriegsman and S. Pischtschan, Z. Anorg. Allgem. Chem., *308*, 212 (1961); W. F. Edgell and C. H. Ward, J. Mol. Spectr., *8*, 343 (1962).
69. J. R. Ferraro and G. J. Long, Accts. Chem. Res., *8*, 171 (1975).
70. H. D. Kaesz *et al.*, J. Amer. Chem. Soc., *89*, 2844 (1967).
71. J. B. Wilford and F. G. A. Stone, Inorg. Chem., *4*, 389 (1965).

ADDITIONAL REFERENCES

A. In the area of infrared, highly recommended references dealing with theory are:

Introductory

N. B. Colthup, L. H. Daly, and S. E. Wiberley, "Introduction to Infrared and Raman Spectroscopy," 3rd ed., Academic Press, New York, 1990.

L. H. Jones, "Inorganic Vibrational Spectroscopy," Vol. 1, M. Dekker, New York, 1971.

W. G. Fately *et al.*, "Infrared and Raman Selection Rules for Molecular and Lattice Vibrations: The Correlation Method," Wiley-Interscience, New York, 1972.

L. A. Woodward, "Introduction to the Theory of Molecular Vibrations and Vibrational Spectroscopy," Oxford University Press, New York, 1972.

G. Barrow, "Introduction to Molecular Spectroscopy," McGraw-Hill, New York, 1962.

Advanced

G. Herzberg, "Spectra of Diatomic Molecules," 2nd ed., Van Nostrand, Princeton, 1950.

G. Herzberg, "Infrared and Raman Spectra of Polyatomic Molecules," Van Nostrand, Princeton, 1945.

E. B. Wilson, J. C. Decius, and P. C. Cross, "Molecular Vibrations," McGraw-Hill, New York, 1955.

D. Steele, "Theory of Vibrational Spectroscopy," W. B. Saunders, Philadelphia, 1971.

B. References dealing with applications are:

L. H. Jones, "Inorganic Vibrational Spectroscopy," Vol. 1, M. Dekker, New York, 1971.

M. Davies, ed., "Infrared Spectroscopy and Molecular Structure," Elsevier, Amsterdam, 1963.

K. Nakamoto, "Infrared and Raman Spectra of Inorganic and Coordination Compounds," 4th ed., Wiley, New York, 1986.

K. Nakanishi and P. H. Solomon, "Infrared Absorption Spectroscopy" 2nd ed., Holden-Day, San Francisco, 1977. This reference contains many examples of involving structure determination of organic compounds from infrared spectra.

A. Finch *et al.*, "Chemical Applications of Far Infrared Spectroscopy," Academic Press, New York, 1970.

L. J. Bellamy, "The Infrared Spectra of Complex Molecules," 3rd ed., Wiley, New York, 1975.

L. J Bellamy, "Advances in Infrared Group Frequencies," Halsted Press, New York, 1968.

J. R. Ferraro, "Low Frequency Vibrations of Inorganic Coordination Compounds," Plenum Press, New York, 1971.

"Practical Fourier Transform Infrared Spectroscopy," eds. J. R. Ferraro and K. Krishman, Academic Press, San Diego, 1990.

Compilations

"Sadtler Standard Spectra," Sadtler Research Laboratories, Philadelphia, Pa. (A collection of spectra for use in identification, determination of the purity of compounds, etc.)

Raman Spectroscopy

H. A. Szymanski, ed., "Raman Spectroscopy," Plenum Press, New York, 1967; Vol. 2, 1970.

A. Anderson, ed., "The Raman Effect," Vols. 1 and 2, M. Dekker, New York, 1971.

M. C. Tobin, "Laser Raman Spectroscopy," Wiley, New York, 1971; Vol. 35 in "Chemical Analysis," P. J. Elving and I. M. Kolthoff, ed.

R. S. Tobias, J. Chem. Educ., *44*, 2, 70 (1967).

J. A. Koningstein, "Introduction to the Theory of the Raman Effect," D. Reidel, Dordrecht, 1972.

D. A. Long, "Raman Spectroscopy," McGraw-Hill, New York, 1977.

J. M. Stencel, "Raman Spectroscopy for Catalysis," Van Nostrand Reinhold, New York, 1990.

A. T. Tu, "Raman Spectroscopy in Biology" Wiley, New York, 1982.

C. B. Moore, ed., "Chemical and Biochemical Applications of Lasers," Academic Press, New York, 1974.

EXERCISES

1. Consider the C_{3v} structure of N—SF$_3$. (a) Show that the total representation is $E = 15$; $2C_3 = 0$; $3\sigma_v = 3$. (b) Indicate the procedure that shows that the irreducible representations $4A_1$, $1A_2$ and $5E$ result from this total representation. (c) Show that $4A_1 + 1A_2 + 5E$ equals $E = 15$; $2C_3 = 0$; $3\sigma_v = 3$. (d) Indicate the species for the allowed: (e) infrared, (f) Raman, and (g) Raman-polarized bands. (h) How many bands would be seen for (e), (f), and (g), assuming all allowed fundamentals are observed?

2. Consider XeF$_4$ as having D_{4h} symmetry. Using the procedure in exercise 1, show that the total number of vibrations is indicated by $A_{1g} + B_{1g} + B_{2g} + A_{2u} + B_{2u} + 2E_u$. Also indicate the species for the allowed (a) infrared, (b) Raman, and (c) Raman-polarized bands. (d) How many bands would be seen for (a), (b), and (c), assuming all allowed fundamentals are observed?

3. Consider the molecule *trans*-N$_2$F$_2$.

    a. To which point group does it belong?

    b. How many fundamental vibrations are expected?

    c. To what irreducible representations do these belong?

    d. What is the difference between the A_u and B_u vibrations?

    e. Which vibrations are infrared active and which are Raman active?

    f. How many polarized lines are expected?

g. How many lines are coincident in the infrared and Raman? Does this agree with the center of symmetry rule?

h. To which species do the following vibrations belong? N—N stretch, symmetric N—F stretch, and asymmetric N—F stretch.

i. Can the N—N stretch and N—F stretch couple?

4. Indicate to which irreducible representations the vibrations belong and indicate how many active IR, active Raman, and polarized Raman lines are expected for:

   a. cis-N_2F_2 (C_{2v})

   b. A linear N_2F_2 molecule ($D_{\infty h}$)

5. a. Why is cis-N_2F_2 in the point group C_{2v} instead of D_{2h}?

   b. What symmetry element is missing from the pyramidal structure of SF_4 that is required for D_{4h} (i.e., which one element would give rise to all missing operations)?

6. The infrared spectrum of gaseous HCl consists of a series of lines spaced 20.68 cm^{-1} apart. (Recall that wavenumbers must be converted to frequency to employ the equations given in this text.)

   a. Calculate the moment of inertia of HCl.

   b. Calculate the equilibrium internuclear separation.

7. Suppose that valence considerations enabled one to conclude that the following structures were possible for the hypothetical molecule X_2Y_2:

$$Y-X-X-Y \qquad X-X\genfrac{}{}{0pt}{}{\diagup Y}{\diagdown Y} \qquad \genfrac{}{}{0pt}{}{Y\diagdown}{Y\diagup}X-X\genfrac{}{}{0pt}{}{\diagup Y}{} \qquad \genfrac{}{}{0pt}{}{\diagup Y}{Y\diagup}X-X\genfrac{}{}{0pt}{}{\diagup Y}{}$$

The infrared spectrum of the gas has several bands with P-, Q-, and R-branches. Which structures are eliminated?

8. An X—H fundamental vibration in the linear molecule A—X—H is found to occur at 3025 cm^{-1}. At what frequency is the X—D vibration expected? It is found that the X—D vibration is lowered by only about half the expected amount, and the A—X stretching frequency is affected by deuteration. Explain. Would you expect the A—X—H bending frequency to be affected by deuteration? Why?

9. How many normal modes of vibration does a planar BCl_3 molecule have? Refer to Appendix C and illustrate these modes. Indicate which are infrared inactive. Confirm these conclusions by employing symmetry considerations to determine the number of active infrared and raman lines. Indicate the Raman polarized and depolarized lines and the parallel and perpendicular vibrations.

10. The following assignments are reported for the spectrum of GeH_4: 2114, T_2; 2106, A_1; 931, E; 819, T_2. Label these using the v_n symbolism. (E modes are numbered after singly degenerate symmetric and asymmetric vibrations, and T modes are numbered after E modes.) Refer to Appendix C to illustrate the form for each of these vibrations and label them as bends or stretches.

11. Refer to the normal modes for a planar ZXY_2-molecule (Appendix C). Verify the assignment of v_5 as B_2 and v_6 as B_1 by use of the character tables. Explain the procedure.

12. a. Indicate the infrared and Raman activities for the modes of the trigonal bipyramidal XY_5-molecule (Appendix C).

    b Does this molecule possess a center of symmetry?

13. The spectrum of $Co(NH_3)_6(ClO_4)_3$ has absorption bands at 3320, 3240, 1630, 1352, and 803 cm^{-1}. For purposes of assignment of the ammonia vibrations, the molecule can be treated as a C_{3v}-molecule, $\left[Co-N\begin{smallmatrix}H\\H\\H\end{smallmatrix}\right]^{3+}$. Refer to Appendix C for a diagram of these modes; utilizing the material in this chapter and Chapter 4, assign these modes. Use the v_n symbolism to label the bands and also describe them as bends, stretches, etc.

14. What changes would you expect to see in the infrared spectrum of CH_3COSCH_3 if

    a. coordination occurred on oxygen?

    b. coordination occurred on sulfur?

15. In which case would the spectrum of coordinated $[(CH_3)_2N]_3PO$ be most likely to resemble the spectrum of the free ligand:

    a. Oxygen coordination?

    b. Nitrogen coordination?

    Why?

16. Using a value of $k = 4.5 \times 10^5$ dynes cm^{-1} for the C—C bond stretching force constant, calculate the wavenumber (cm^{-1}) for the C—C stretching vibration.

17. The compound $HgCl_2 \cdot O\begin{smallmatrix}CH_2-CH_2\\CH_2-CH_2\end{smallmatrix}O$ could be a linear polymer with dioxane in the chair form or a monomer with bidentate dioxane in the boat form. How could infrared and Raman be used to distinguish between these possibilities?

18. Which complex would have more N—O vibrations: $(NH_3)_5CoNO_2^{2+}$ or $(NH_3)_5CoONO^{2+}$?

19. What differences (number of bands and frequencies) are expected between the C—O absorptions of the following:

$$CH_3COO^-, Zn\left(\begin{smallmatrix}O\\O\end{smallmatrix}C-CH_3\right)_2 \cdot 2H_2O, CH_3COOH, \text{ and } CH_3C\begin{smallmatrix}O\\O\end{smallmatrix}Ag?$$

20. Given the NH_3 molecule below, determine the total representation for the vibrations using r's and θ's (θ_2 not shown). Use projection operators to determine the vibrational wave functions.

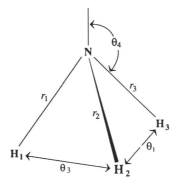

21. Consider the tetrahedral anion BCl_4^-.

   a. How many bands do you expect in the infrared spectrum?

   b. How many bands do you expect in the Raman spectrum?

   c. How many Raman-polarized bands do you expect?

   d. How many B—Cl stretches do you expect in the IR?

22. The hydrogen bonding Lewis acid $CHCl_3$ has a C—H stretching frequency that is lowered upon deuteration.

   a. What is responsible for the theoretical factor of 1.4?

   b. What could be responsible if the observed factor were less than 1.4?

   c. When the compound is dissolved in the Lewis base NEt_3, the C—H stretching frequency is lowered. Rationalize the lowering.

   d. If one examines the C—D shift upon adduct formation with Et_3N, the percentage lowering is smaller than that in (b). Rationalize this finding.

23. The compound $(CH_3)_3SnCl$ exhibits two infrared active Sn—C stretches at 545 cm^{-1} and 514 cm^{-1}. The IR spectrum on the five-coordinate adduct $B \cdot (CH_3)_3SnCl$ (where B is a base) shows a single Sn—C stretching band around 550 cm^{-1}.

   a. Consider a tetrahedral structure for $(CH_3)_3SnCl$ and predict the number of Sn—C stretches expected.

   b. Assume a trigonal bipyramidal structure for the adduct with the methyl groups lying in a plane. How many Sn—C stretching bands are expected?

   c. How do you explain the actual spectrum?

24. a. How many infrared-active normal modes containing Sn—Cl stretching would be expected for the six-coordinate species *cis*-$SnCl_4X_2$?

   b. Describe briefly the selection rules for Raman spectrosocopy and for infrared spectroscopy.

25. Iron pentacarbonyl, Fe(CO)$_5$, possesses D_{3h} symmetry.

   a. Determine the number and irreducible representations of the IR-active and Raman-active fundamentals to be expected for this compound.

   b. Determine the number of IR-active carbonyl stretching bands to be expected.

26. Consider the following IR data, taken from Nakamoto's book[18]:

   $K_3[Mn(CN)_6]$ 2125 cm^{-1}
   $K_4[Mn(CN)_6]$ 2060 cm^{-1}
   $K_5[Mn(CN)_6]$ 2048 cm^{-1}

   Discuss this trend in CN stretching frequencies in terms of bonding in the complexes, and the relationship this has on force constants and frequencies. What do you predict for the Mn—C force constants in this series?

27. Use of IR and Raman spectroscopy is valuable in determining the stereochemistry of metal carbonyl complexes and various substituted metal carbonyls. C=O stretches are often analyzed separately from the rest of the molecule (explain why this assumption is or is not valid).

   a. For Cr(CO)$_6$ and Cr(CO)$_5$L (treat L as a point ligand), work out the symmetry of the IR- and Raman-allowed CO stretches.

   b. Do the same as in (a) for Mo(CO)$_4$DTH (DTH is the bidentate ligand CH$_3$SCH$_2$CH$_2$SCH$_3$), and for *trans*-Mo(CO)$_4$[P(OC$_6$H$_5$)$_3$]$_2$. Do your findings explain the IR spectra below? (Hint: remember asumptions implicit in the analysis.) [Spectra reported by M. Y. Darensbourg and D. J. Darensbourg, J. Chem. Ed., *47*, 33 (1970).]

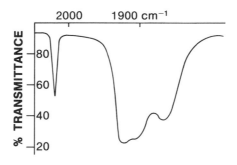
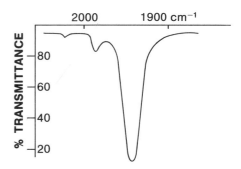

Nuclear Magnetic Resonance Spectroscopy— Elementary Aspects

7

Introduction[1,2,3]

In this chapter, the principles necessary for an elementary appreciation of nuclear magnetic resonance[1-3] (nmr) will be presented. The reader mastering this chapter will have a minimum knowledge of those principles required for spectral interpretation of the results from slow passage as well as Fourier transform experiments. Chapter 8 expands upon several of the important concepts introduced here and illustrates several other types of applications of nmr. Varying appreciations of this subject can be obtained by complete reading of Chapter 7 and selected readings from Chapter 8.

Protons and neutrons both have a spin quantum number of $1/2$ and, depending on how these particles pair up in the nucleus, the resultant nucleus may or may not have a net non-zero nuclear spin quantum number, I. If the spins of all the particles are paired, there will be no net spin and the nuclear spin quantum number I will be zero. This type of nucleus is said to have zero spin and is represented in Fig. 7–1(A). When I is $1/2$, there is one net unpaired spin and *this unpaired spin imparts a nuclear magnetic moment, μ, to the nucleus.* The distribution of positive charge in a nucleus of this type is spherical. The properties for $I = 1/2$ are represented symbolically in Fig. 7–1(B) as a spinning sphere (*vide infra*). When $I \geq 1$, the nucleus has spin associated with it and the nuclear charge distribution is non-spherical; see Fig. 7–1(C). The nucleus is said to possess an electric quadrupole moment eQ, where e is the unit of electrostatic charge and

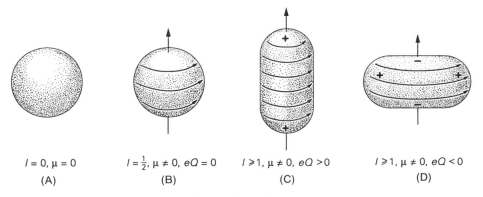

$I = 0, \mu = 0$ $I = \frac{1}{2}, \mu \neq 0, eQ = 0$ $I \geq 1, \mu \neq 0, eQ > 0$ $I \geq 1, \mu \neq 0, eQ < 0$
(A)   (B)   (C)   (D)

FIGURE 7–1 Various representations of nuclei.

Q is a measure of the deviation of the nuclear charge distribution from spherical symmetry. For a spherical nucleus, eQ is zero. A positive value of Q indicates that charge is oriented along the direction of the principal axis (Fig. 7–1(C)), while a negative value of Q indicates charge accumulation perpendicular to the principal axis (Fig. 7–1(D)).

Nuclei with even numbers of both protons and neutrons all belong to the type represented in Fig. 7–1(A). Typical examples include ^{16}O, ^{12}C, and ^{32}S. The values for I, μ, and eQ for many other nuclei have been tabulated,[1,2,3] and are more easily looked up than figured out. A table summarizing these properties is included inside the front cover.

Nuclear magnetic resonance spectroscopy is often concerned with nuclei with $I = {}^1\!/_2$, examples of which include ^1H, ^{13}C, ^{31}P, and ^{19}F. Spectra also result from nuclei for which $I \geq 1$, but cannot be obtained on nuclei with $I = 0$.

Unpaired nuclear spin leads to a nuclear magnetic moment. The allowed orientations of the nuclear magnetic moment vector in a magnetic field are indicated by the *nuclear spin angular momentum quantum number*, m_I. This quantum number takes on values $I, I - 1, \ldots, (-I + 1), -I$. When $I = {}^1\!/_2$, $m_I = \pm {}^1\!/_2$ corresponding to alignments of the magnetic moment with and opposed to the field. When $I = 1$, m_I has values of 1, 0, and -1, corresponding, respectively, to alignments with, perpendicular to, and opposed to the field. In the absence of a magnetic field, all orientations of the nuclear moment are degenerate. In the presence of an external field, however, this degeneracy will be removed. For a nucleus with $I = {}^1\!/_2$, the $m_I = + {}^1\!/_2$ state will be lower in energy and the $- {}^1\!/_2$ state higher, as indicated in Fig. 7–2. The energy difference between the two states, at magnetic field strengths commonly employed, corresponds to radio frequency radiation; it is this transition that occurs in the nmr experiment.

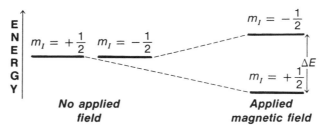

FIGURE 7–2 Splitting of the $m = \pm {}^1\!/_2$ states in a magnetic field.

Classical Description of the NMR Experiment—The Bloch Equations

7–1 SOME DEFINITIONS

We begin our classical description of nmr by reviewing some of the background physics of magnetism and defining a few terms necessary to understand nmr. *It is very important to appreciate what is meant by angular momentum.* Circulating charges have angular momentum associated with them. Planar angular momentum, $\vec{p}(\varphi)$ is given by $\vec{p}(\varphi) = \vec{r} \times m\vec{v}$ where $\vec{r}$, shown in Fig. 7–3, is the position vector of the particle e; $\vec{v}$ its linear momentum vector; m is the mass; and φ is the angular change that signifies an angular momentum. The angular momentum ρ is perpendicular to the plane of the circulating change, and the direction of the angular momentum vector is given by the right-hand rule. For the situation shown in Fig. 7–3, the vector points into the page.

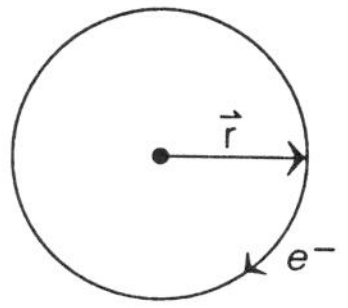

The nucleus is a more complicated three-dimensional problem. The total angular momentum of a nucleus is given by $\vec{J}$, but it is convenient to define a dimensionless angular momentum operator $\hat{I}$ by

$$\hat{J} = \hbar \hat{I} \quad \text{or} \quad \hat{I} = \hat{J}/\hbar \qquad (7\text{--}1)$$

FIGURE 7–3 Circulating charge giving rise to planar angular momentum.

Associated with the angular momentum is a classical magnetic moment, $\vec{\mu}_N$, which can be taken as parallel to $\vec{J}$, so:

$$\vec{\mu}_N = \gamma \vec{J} = \gamma \hbar \vec{I} \qquad (7\text{--}2)$$

where γ, the magnetogyric (sometimes called gyromagnetic) ratio is a constant characteristic of the nucleus. From equation (7–2), we see that the magnetogyric ratio represents the ratio of the nuclear magnetic moment to the nuclear angular momentum.

7–2 BEHAVIOR OF A BAR MAGNET IN A MAGNETIC FIELD

A nucleus with a magnetic moment can be treated as though it were a bar magnet. If a bar magnet were placed in a magnetic field, the magnet would precess about the applied field, $\vec{H}_0$, as shown for a spinning nuclear moment in Fig. 7–4. Here θ is the angle that the magnetic moment vector makes with the applied field, and ω, the Larmor frequency, is the frequency of the nuclear moment precession. The instantaneous motion of the nuclear moment (indicated by the arrowhead on the dashed circle) is tangential to the circle and perpendicular to $\vec{\mu}$ and $\vec{H}_0$. The magnetic field is exerting a force or torque on the nuclear moment, causing it to precess about the applied field. For use later, we would like to write an equation to describe the precession of a magnet in a magnetic field. The applied field $\vec{H}_0$ exerts a torque $\vec{\tau}$ on $\vec{\mu}$ which is given by the *cross product*:

$$\vec{\tau} = \vec{\mu} \times \vec{H}_0 \qquad (7\text{--}3)$$

The right-hand rule tells you that *the moment is precessing clockwise in Fig. 7–4, with the torque and instantaneous motion perpendicular to $\vec{\mu}$ and $\vec{H}_0$*. According to Newton's law, force is equal to the time derivative of the momentum, so the torque is the time derivative of the angular momentum.

$$\vec{\tau} = \frac{d}{dt}(\hbar \vec{I})$$

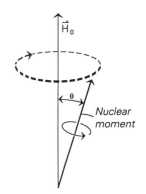

FIGURE 7–4 Precession of a nuclear moment in an applied field of strength H_0.

With the equation $\vec{\mu} = \gamma \hbar \vec{I}$, *the equation for precession of the moment is given by:*

$$\frac{d\vec{\mu}}{dt} = \frac{\gamma d(\hbar \vec{I})}{dt} = \gamma \vec{\tau}$$

or (7–4)

$$\frac{d\vec{\mu}}{dt} = \dot{\vec{\mu}} = -\gamma \vec{H}_0 \times \vec{\mu}$$

(The dot is an abbreviation for the time derivative of some property, in this case $\vec{\mu}$; the minus sign arises because we have changed the order for taking the cross product.) Thus, ω, the precession frequency or the *Larmor frequency*, is given by:

$$\omega = |\gamma| H_0 \tag{7-5}$$

where the sign of γ determines the sense of the precession. According to equation (7–5), *the frequency of the precession depends upon the applied field strength and the magnetogyric ratio of the nucleus.* The energy of this system is given by the dot product of $\vec{\mu}$ and $\vec{H}_0$.

$$E = -\vec{\mu} \cdot \vec{H}_0 = -|\mu||H_0|\cos\theta \tag{7-6}$$

7–3 ROTATING AXIS SYSTEMS

There is one more mathematical construct that greatly aids the analysis of the nmr experiment, and this is the idea of a rotating coordinate system or *rotating frame*. In Fig. 7–5, a set of x-, y-, and z-coordinates is illustrated. The rotating frame is described by the rotating axes u and v, which rotate at some frequency ω_1 in the xy-plane. The z-axis is common to both coordinate systems. If the rotating frame rotates at some frequency less than the Larmor frequency, it would appear to an observer on the u,v frame that the precessional frequency has slowed, which would correspond to a weakening of the applied field. Labeling this apparently weaker field as $\vec{H}_{\text{eff}}$, our precessing moment is described in the rotating frame by

$$\dot{\vec{\mu}} = -\gamma \vec{H}_{\text{eff}} \times \vec{\mu} \tag{7-7}$$

where $|H_{\text{eff}}| < |H_0|$. $\vec{H}_{\text{eff}}$ is seen to be a function of ω_1, and if ω_1 is faster than the precessional frequency of the moment, ω, it will appear as though the z-field has changed direction. The effective field, $\vec{H}_{\text{eff}}$, can be written as:

$$\vec{H}_{\text{eff}} = \left[H_0 - \frac{\omega_1}{\gamma} \right] \vec{e}_z \tag{7-8}$$

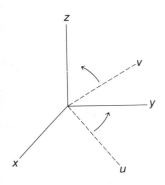

FIGURE 7–5 The rotating frame u, v, z in a Cartesian coordinate system x, y, z.

where $\vec{e}_z$ is a unit vector along the z-axis. The time dependence of the moment, $\vec{\mu}$, can be rewritten as

$$\dot{\vec{\mu}} = -\gamma\left[\left(H_0 - \frac{\omega_1}{\gamma}\right)\vec{e}_z\right] \times \vec{\mu} \tag{7-9}$$

When $\omega_1 = \omega = \gamma H_0$, then $|H_{\text{eff}}| = 0$ and $\vec{\mu}$ appears to be stationary. Thus, in a frame rotating with the Larmor frequency, we have eliminated time dependency and, in so doing, greatly simplified the problem.

7-4 MAGNETIZATION VECTORS AND RELAXATION

The concepts in this section are vital to an understanding of nmr. Of these, the concept of *relaxation* will be developed in stages as the necessary background information is covered. For practical purposes, it is necessary to consider an ensemble or large number of moments because the nmr experiment is done with bulk samples. The individual moments in the sample add vectorially to give a *net magnetization*, $\vec{M}$.

$$\vec{M} = \sum_i \vec{\mu}_i \tag{7-10}$$

In an ensemble of spins in a field, those orientations aligned with the field will be lower in energy and preferred. However, thermal energies oppose total alignment; experimentally, only a small net magnetization is observed. The equation for the motion of precession of $\vec{M}$ is similar to that for $\vec{\mu}$, *i.e.*,

$$\dot{\vec{M}} = -\gamma \vec{H}_0 \times \vec{M} \tag{7-11}$$

If we place a sample in a magnetic field at constant temperature and allow the system to come to equilibrium, the resulting system is said to be at thermal equilibrium. At thermal equilibrium, the magnetization, M_0, is given by

$$|M_0| = \frac{N_0 \mu^2}{3kT} I(I+1) H_0 \tag{7-12}$$

where k is the Boltzmann constant and N_0 is the number of magnetic moments per unit volume (*i.e.*, the number of spins per unit volume). This equation results from equation (7-6) and the assumption of a Boltzmann distribution. For H_2O at 300 K and $H_0 = 10,000$ gauss (1 tesla) the value of M_0 is about 3×10^{-6} gauss (3×10^{-10} T).

The description of the magnetization is not yet complete for, at equilibrium, we have a dynamic situation in which any one given nuclear moment is rapidly changing its orientation with respect to the field. The mechanism for reorientation involves time-dependent fields that arise from the molecular motion of magnetic nuclei in the sample. Suppose, as in Fig. 7-6, that nucleus B, a nucleus that has a magnetic moment, passed by nucleus A, whose nuclear moment is opposed to the field. The spin of A could be oriented with the field via the fluctuating field from moving B. In the process, the translational or rotational energy of the

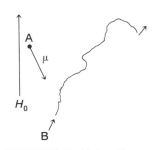

FIGURE 7-6 Relaxation mechanisms.

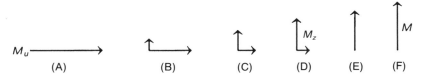

FIGURE 7–7 The decay of the M_u-component and growth of the M_z-component when a field in the u-direction is turned off and one in the z-direction is turned on.

molecule containing B would be increased, since A gives up energy on going to a more stable orientation aligned with the field. This process is referred to as *spin-lattice relaxation*. Nucleus B is the lattice; it can be a magnetic nucleus in the same or another sample molecule or in the solvent. These nuclei do not have to be hydrogens and may even be unpaired electrons in the same or other molecules. The fluctuating field from motion of the magnetic nucleus must have the same frequency as that of the nmr transition in order for this relaxation process to occur. However, moving magnetic nuclei have a wide distribution of frequency components, and the required one is always present.

Another process can also occur when the two nuclei interact, whereby nucleus B, with $m_I = {}^1/_2$ goes to the higher-energy $m_I = -{}^1/_2$ state while nucleus A changes from $-{}^1/_2$ to $+{}^1/_2$ There is no net change in spin from this process, and it is referred to as a *spin-spin relaxation* mechanism.

To gain more insight into the nature of relaxation processes, let us examine the decay in magnetization as a function of time. Consider an experiment in which we have our magnetization aligned along the u axis as in Fig. 7–7(A). We then switch on a field aligned with the z-axis; M_u diminishes and M_z grows, generally at different rates. In Fig. 7–7(E), for example, the u-component has completely decayed, but the z-component has not yet reached equilibrium, for it gets larger in Fig. 7–7(F). The growth and decay are first-order processes, and Bloch proposed the following equation for the three-dimensional case (i.e., u-, v-, and z-magnetizations are involved and the field is located along the z-axis):

$$\dot{M}_u = -\frac{M_u}{T_2} \qquad \dot{M}_v = -\frac{M_v}{T_2} \qquad (7\text{–}13)$$

and

$$\dot{M}_z = -\frac{1}{T_1}(M_z - M_0) \qquad (7\text{–}14)$$

where $1/T_2$ and $1/T_1$ are first-order rate constants, M_0 is the equilibrium value of the z-magnetization, and u- and v-components vanish at equilibrium. In general, it is found that $1/T_2 \geqslant 1/T_1$. Instead of rate constants, $1/T$, it is more usual to refer to their reciprocals (i.e., the T's), which are lifetimes or relaxation times. Since T_1 refers to the z-component, it is called the *longitudinal relaxation time*, while T_2 is called the *transverse relaxation time*. The spin-lattice mechanism contributes to T_1, and the spin-spin mechanism is one of several contributions to T_2.

7–5 THE NMR TRANSITION

Before proceeding with our classical description of the nmr experiment, it is advantageous to introduce a few concepts from the quantum mechanical description of the experiment. When the bare nucleus (no electrons around it) is placed in a magnetic field, H_0, the field and the nuclear moment interact [see equation (7–6)] as described by the Hamiltonian for the system

$$\hat{H} = -\vec{\mu}_N \cdot \vec{H}_0 \qquad (7\text{--}15)$$

where, as shown in equation (7–2), $\vec{\mu}_N = \gamma \hbar \vec{I}$ with the N subscript denoting a nuclear moment. When it is obvious that we are referring to a nuclear moment, the N subscript will be dropped. Combining equations (7–2)* and (7–15) yields

$$\hat{H} = -\gamma_N \hbar H_0 \hat{I}_z = -g_N \beta_N H_0 \hat{I}_z \qquad (7\text{--}16)$$

where γ_N is constant for a given nucleus, g_N is the nuclear g factor and:

$$\beta_N = e\hbar/2mc \qquad (7\text{--}17)$$

In equation (7–17), m is the mass of the proton, e is the charge of the proton, and c is the speed of light.

The expectation values of the operator $\hat{I}_z$, where z is selected as the applied field direction, are given by m_I where $m_I = I, I-1, \ldots, -I$. The degeneracy of the m_I states that existed in the absence of the field is removed by the interaction between the field H_0 and the nuclear magnetic moment μ_N. The quantized orientations of these nuclear moments relative to an applied field $\vec{H}_0$ are shown in Fig. 7–8 for $I = \frac{1}{2}$ and $I = 1$. Since the eigenvalues of the operator $\hat{I}_z$ are m_I, the eigenvalues of $\hat{H}$ (i.e., the energy levels) are given by equation (7–18).

$$E = -\gamma \hbar m_I H_0 \qquad (7\text{--}18)$$

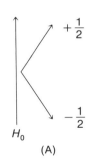

The energy as a function of the field is illustrated for $I = \frac{1}{2}$ in Fig. 7–9. The quantity g_N is positive for a proton, as is β_N, so the positive m_I value is lowest in energy. The nuclear wave functions are abbreviated as $|\alpha\rangle$ and $|\beta\rangle$ for the $+\frac{1}{2}$ and $-\frac{1}{2}$ states, respectively. By inserting values of $m_I = +\frac{1}{2}$ and $m_I = -\frac{1}{2}$ into equation (7–18), we find that the energy difference, ΔE, between these states or the energy of the transition $h\nu$ for a nucleus of spin $\frac{1}{2}$ is given by $g_N \beta_N H_0$ or $\gamma \hbar H_0$.

$$\Delta E = g_N \beta_N H_0 \qquad (7\text{--}18a)$$

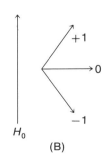

FIGURE 7–8 Quantized orientation of m_I for (A) $I = \frac{1}{2}$ and (B) $I = 1$.

---

* We locate the field H_0 along the z-axis and, since H_y and H_x are zero, only $H_0 \hat{I}_z$ is non-zero; i.e., $H_x \hat{I}_x$ and $H_y \hat{I}_y$ are zero because H_x and H_y are zero.

FIGURE 7–9 The field dependence of the energies of $m_I = \pm 1/2$.

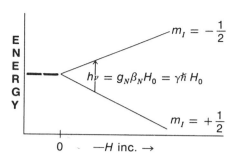

Since $m_I = +1/2$ is lower in energy than $m_I = -1/2$, there will be a slight excess population of the low energy state at room temperature, as described by the Boltzmann distribution expression in equation (7–19):

$$\frac{N(-1/2)}{N(+1/2)} = \exp(-\Delta E/kt) \cong 1 - \frac{\Delta E}{kT} \text{ when } \Delta E \ll kT \qquad (7\text{–}19)$$

For a proton, ΔE is $\sim 10^{-3}$ cm^{-1} in a 1 T field, and $kT \sim 200$ cm^{-1}. At room temperature, there is a ratio of 1.0000066 $(+1/2)$ spins to one $(-1/2)$. The probability of a nuclear moment being in the $+1/2$ state is $(1/2)\left(1 + \dfrac{\mu H_0}{kT}\right)$ and that of it being in the $-1/2$ state is $(1/2)\left(1 + \dfrac{\mu H_0}{kT}\right)$ (recall our discussion of equation (7–12)).

The energy separation corresponding to $h\nu$ occurs in the radio-frequency region of the spectrum at the magnetic field strengths usually employed in the experiment. One applies a circularly polarized radio frequency (r.f.) field at right angles to $\vec{H}_0$ (*vide infra*), and the magnetic component of this electromagnetic field, $\vec{H}_1$, provides a torque to flip the moments from $m_I = +1/2$ to $-1/2$, causing transitions to occur.

7–6 THE BLOCH EQUATIONS

In order to understand many of the applications of nmr, it is necessary to appreciate the change in magnetization of the system with time as the H_1 field is applied. This result is provided by the Bloch equation. Incorporating equations (7–13) and (7–14), describing relaxation processes, into equation (7–11), which describes the precession of the magnetization, and converting to the rotating frame gives the Bloch equation:

$$\dot{\vec{M}} = \underbrace{-\gamma \vec{H}_{\text{eff}} \times \vec{M}}_{\substack{\text{torque from} \\ \text{the magnetic} \\ \text{field}}} \underbrace{- \frac{1}{T_2}(M_u \vec{e}_u + M_v \vec{e}_v) - \frac{1}{T_1}(M_z - M_0)\vec{e}_z}_{\text{relaxation effects}} \qquad (7\text{–}20)$$

In the presence of H_1 and in the rotating frame, equation (7–8)—which describes $\vec{H}_{eff}$ in the rotating frame—becomes:

$$\vec{H}_{eff} = \left(H_0 - \frac{\omega_1}{\gamma}\right)\vec{e}_z + H_1\vec{e}_u \qquad (7\text{–}21)$$

where the frame is rotating with the frequency ω_1 corresponding to the frequency of the H_1 field, the oscillating field at right angles to $\vec{H}_0$. Equation (7–20) is a vector equation in the rotating frame that can best be written in terms of the components of $\vec{M}$, which are M_u, M_v, and M_z.

The three components of the Bloch equation are:

$$\dot{M}_u = \frac{dM_u}{dt} = -(\omega_0 - \omega_1)M_v - \frac{M_u}{T_2} \qquad (7\text{–}22)$$

$$\dot{M}_v = \frac{dM_v}{dt} = (\omega_0 - \omega_1)M_u - \frac{M_v}{T_2} + \gamma H_1 M_z \qquad (7\text{–}23)$$

$$\dot{M}_z = \frac{dM_z}{dt} = \gamma H_1 M_v + (M_0 - M_z)/T_1 \qquad (7\text{–}24)$$

In these equations, ω_0 is the Larmor frequency, which equals γH_0, and the u, v reference frame is rotating at angular velocity ω_1.

Experimentally, we monitor the magnetization in the xy-plane, referred to as the transverse component. Using a phase-sensitive detector, we monitor the component of magnetization induced along the u-axis. In a slow-passage or steady-state experiment, only a u-component of magnetization exists; but because of a 90° phase lag associated with the electronic detection system, a component 90° out of phase with u is measured. Slow-passage or steady-state conditions require H_1 to be weak (of the order of milligauss) compared to H_0 (which is of the order of kilogauss). Then, according to equation (7–21), the z-component dominates unless ω_1 is close to ω_0 so that the first term ($H_0 - \omega_0/\gamma$) becomes small. (Note that ω_1 is the frequency of the r.f. field and *not* the Larmor frequency of precession about H_1; i.e., $\omega_1 \neq \gamma H_1$.) When ω_1 is close to the Larmor frequency, ω_0, then $\vec{H}_{eff}$ is tipped toward the u-axis. Since H_0 is being changed slowly, the net effect is to change H_{eff} slowly. The individual moments continue to precess about $\vec{H}_{eff}$ as a consequence of the torque, which is perpendicular to $\vec{H}_{eff}$. As a result, $\vec{M}$ remains parallel to $\vec{H}_{eff}$ and a u-component results. Figure 7–10 represents the tipping of $\vec{M}$ to remain aligned with $\vec{H}_{eff}$ as we pass through resonance. The $\vec{H}_1$ field is along the u-axis. It is also an alternating field with the frequency of the rotating frame. A static field will not tip the magnetization vector significantly because H_1 is so small.

Relaxation processes, T_1, try to preserve the orientation along the strong z-field, as shown by the arrow labeled T_1 in Fig. 7–10. Under steady-state

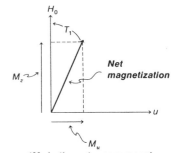

(M_z is the net z component of the magnetization and not a precessing magnetic moment)

FIGURE 7–10 Effect of H_1 on the net magnetization of an H_0 field.

conditions, all time derivatives are zero, so equations (7–22), (7–23), and (7–24) are all equal to zero and can be solved to produce:

$$M_u = M_0 \frac{\gamma H_1 T_2^2 (\omega_0 - \omega_1)}{1 + T_2^2(\omega_0 - \omega_1)^2 + \gamma^2 H_1^2 T_1 T_2} \qquad (7\text{–}25)$$

$$M_v = M_0 \frac{\gamma H_1 T_2}{1 + T_2^2(\omega_0 - \omega_1)^2 + \gamma^2 H_1^2 T_1 T_2} \qquad (7\text{–}26)$$

$$M_z = M_0 \frac{1 + T_2^2(\omega_0 - \omega_1)^2}{1 + T_2^2(\omega_0 - \omega_1)^2 + \gamma^2 H_1^2 T_1 T_2} \qquad (7\text{–}27)$$

7–7 THE NMR EXPERIMENT

Equations (7–25) to (7–27) describe the magnetization of our sample in the so-called slow passage experiment, which is schematically illustrated in Fig. 7–11. In this method, one applies a strong homogeneous magnetic field, causing the nuclei to precess. Radiation of energy comparable to ΔE is then imposed with a radio frequency transmitter, producing H_1. When the applied frequency from the radio transmitter is equal to the Larmor frequency, the two are said to be in resonance, and a u,v-component is induced which can be detected. This is the condition in (7–21) when $H_0 \approx \omega_1/\gamma$. Quantum mechanically, this is equivalent to some nuclei being excited from the low energy state ($m_I = +^1/_2$) to the high energy state ($m_I = -^1/_2$; see Fig. 7–9) by absorption of energy from the r.f. source at a frequency equal to the Larmor frequency. Since $\Delta E = h\nu$ and $\omega = 2\pi\nu$, ΔE is proportional to the Larmor frequency, ω. Energy will be extracted from the r.f. source only when this resonance condition ($\omega = 2\pi\nu$) is fulfilled. With an electronic detector (see Fig. 7–11), one can observe the frequency at which a u,v-component is induced or at which the loss of energy from the transmitter occurs, allowing the resonance frequency to be measured.

It is possible to match the Larmor frequency and the applied radio frequency either by holding the field strength constant (and hence ω constant) and scanning

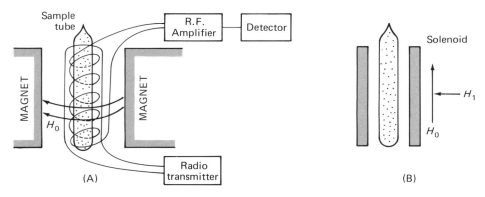

FIGURE 7–11 Schematic diagram of a simple nmr spectrometer.

a variable applied radio frequency until matching occurs or, by varying the field strength until ω becomes equal to a fixed applied frequency. In the latter method, fixed frequency probes (source and detector coils) are employed and the field strength at which resonance occurs is measured. Two experimental configurations are used in this experiment. In Fig. 7–11(A) the H_0 field direction (z-direction) is perpendicular to the sample tube, while in Fig. 7–11(B) the H_0 field and the sample are coaxial. There are some applications in which this difference is important (*vide infra*).

We can now see why the relaxation processes discussed in equations (7–13) and (7–14) had to be added to our complete equation ((7–20), which led to (7–25) to (7–27) for the slow passage experiment) for the behavior of the magnetization. If the populations of nuclei in the ground and excited states were equal, then the probability that the nucleus would emit energy under the resonance condition would equal the probability that the nucleus would absorb energy (*i.e.*, transitions $m_I(+1/2) \to m_I(-1/2)$ would be as probable as $m_I(-1/2) \to m_I(+1/2)$. No net change would then be detected by the radio-frequency probe. As mentioned earlier, in a strong magnetic field there will be a slight excess of nuclei aligned with the field (lower energy state) and consequently a net absorption of energy results. In the absence of relaxation, energy is absorbed from the r.f. signal until the population in the lower state equals that in the higher state. This equilibrium position is attained asymptotically. Initially, absorption might be detected but this absorption would gradually disappear as the populations of ground and excited states became equal. When this occurs, the sample is said to be saturated. Without saturation, the relaxation mechanisms allow nuclei to return to the lower energy state without emitting radiation. As a result there is always an excess of nuclei in the lower energy state, and a continuous absorption of energy from the r.f. source by the sample can occur.

The nmr experiment has significance to the chemist because nuclei exist in atoms linked by bonds into molecules. The energy of the nuclear resonance (*i.e.*, the field strength required to attain a Larmor frequency equal to the fixed frequency) is dependent upon the electronic environment about the nucleus. The electrons shield the nucleus, so that the magnitude of the field seen at the nucleus, H_N, is different from the applied field, H_0:

$$H_N = H_0(1 - \sigma) \qquad (7\text{–}28)$$

where σ, the shielding constant, is a dimensionless quantity that represents the shielding of the nucleus by the electrons. The value of the shielding constant depends on several factors, which will be discussed in detail later.

In a slow sweep of H_0, the various magnetizations of the different nuclei are sampled individually, because the nucleus is shielded or deshielded causing H_N to differ. For a molecule with two different kinds of hydrogen atoms this will lead to a spectrum like that shown in Fig. 7–12.

One other point is worth making here. The differences in the magnetogyric ratios of different kinds of nuclei are much larger than the effects from σ, so there is no trouble distinguishing signals from the different kinds of nuclei in a sample; *e.g.*, ^{19}F and ^1H are never confused. The frequency ranges are so vastly separated that different instrumentation is required to study different kinds of nuclei. In Table 7–1, the resonance conditions ($h\nu = g_N \beta_N H_0$) for various nuclei are given for an applied field of 10,000 gauss.

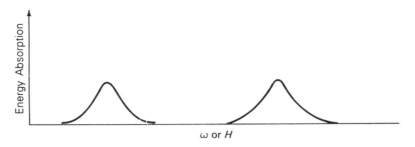

FIGURE 7–12 A low resolution nmr spectrum of a sample containing two different kinds of protons.

The frequencies in Table 7–1 are in MHz (10^6 Hertz or 10^6 c/s), and the variation in the proton resonances of typical organic compounds from different shielding constants is only 600 Hz at 10 kilogauss. In some paramagnetic complexes, shifts of the order of magnitude of 840,000 Hz have been observed; but even these could not be confused with a fluorine resonance. The relative sensitivities of some nuclei in the nmr experiment are also listed in Table 7–1, with a more complete compilation given in Table 8–2.

The most common nmr experiment involves the pulse method. We shall discuss this experiment in more detail later, but now briefly consider it in the context of equation (7–21). Suppose that a single strong H_1 pulse is imposed (e.g., $H_1 \approx 100$ gauss $\gg H_0 - \omega_1/\gamma$ for all of the protons). Referring to the equation for $\vec{H}_{eff}$.

$$\vec{H}_{eff} = H_1 \vec{e}_u + \left(H_0 - \frac{\omega_1}{\gamma}\right)\vec{e}_z$$

TABLE 7–1. Important Nuclei in NMR

Isotope	Abundance (%)	NMR Frequency in 10 Kilogauss Field[d]	Relative[a] Sensitivity (constant H_0)	Magnetic[b] Moment (μ)	Spin[c] (I)
^{1}H	99.9844%	42.577	1.0000	2.7927	$^1/_2$
^{2}H (D)	0.0156	6.536	0.0096	0.8574	1
^{10}B	18.83	4.575	0.0199	1.8006	3
^{11}B	81.17	13.660	0.165	2.6880	$^3/_2$
^{13}C	1.108	10.705	0.0159	0.7022	$^1/_2$
^{14}N	99.635	3.076	0.0010	0.4036	1
^{15}N	0.365	4.315	0.0010	−0.2830	$^1/_2$
^{19}F	100.	40.055	0.834	2.6273	$^1/_2$
^{29}Si	4.70	8.460	0.0785	−0.5548	$^1/_2$
^{31}P	100.	17.235	0.0664	1.1305	$^1/_2$
^{117}Sn	7.67	15.77	0.0453	−0.9949	$^1/_2$
^{119}Sn	8.68	15.87	0.0518	−1.0409	$^1/_2$

[a] For equal numbers of nuclei, where ^{1}H equals one.
[b] In multiples of the nuclear magneton, $eh/4\pi Mc$.
[c] In multiples of $h/2\pi$.
[d] MHz.

the local H_0 will vary slightly from nucleus to nucleus, but $\vec{H}_1$ is so large that the term $H_1\vec{e}_u$ is completely dominant, and $\vec{H}_{eff}$ is almost the same for all nuclei present. All nuclei are sampled simultaneously. Since $\vec{H}_1$ is directed along **u** and the net magnetization along **z**, the cross product requires that $\vec{M}$ begin to precess about **u**. The pulse time is short, compared to the time for one full precession, so the magnetization does not have time to precess around **u**, but merely tips toward **v**. Accordingly, the v-component is measured in this experiment as $\vec{M}$ is tipped toward **v**. Most experiments are now being done by Fourier transform procedures. If a strong pulse were employed for a very long time, the nuclei would precess around $\vec{H}_{eff}$ as in the slow passage experiment; a u-component would arise and v would disappear. Pulse experiments are not usually done this way. When a strong pulse is employed for a short enough duration* that there is no time for any relaxation to occur during the pulse ($t_p \approx 10$ microseconds), the Bloch equations predict the following expression for the v-component, as a function of time, t:

$$M_v = \sin(\gamma H_1 t_p) M_0 \exp(-t/T_2) \tag{7-29}$$

This expression results from the Fourier transform of equation (7–23), which allows conversion from the frequency domain (7–23) to the time domain (7–29). We shall discuss this topic in more detail later. The quantity dM_v/dt cannot be set equal to zero in this experiment. The v-component of the magnetization decays with time according to equation (7–29), as shown in Fig. 7–13. This is called a free induction decay curve and it can be analyzed by computer (via Fourier transformation, *vide infra*) to give a frequency spectrum indentical to the Lorentzian slow passage result. A comparison of the slow passage and Fourier transform experiments is provided in Table 7–2. We shall cover fundamental properties of nmr in this chapter that are independent of the way in which the nmr experiment is carried out, *i.e.*, slow passage or Fourier transform.

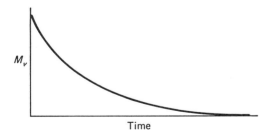

FIGURE 7–13 Free induction decay curve; a plot of the change in magnetization of the sample with time.

---

*In Chapter 8, we shall provide an *equivalent* description of this experiment in terms of the distribution of frequencies that make up the square wave produced from this pulsing.

TABLE 7–2. A Comparison of the Fourier Transform and the Slow-Passage Experiment (100 MHz) in Terms of the Equation $\gamma \vec{H}_{eff}{}^a = (\gamma H_0 - \omega_1)\vec{e}_z + \gamma H_1 \vec{e}_u$

Experiment	Order of Magnitude of Frequencies in sec^{-1} (Hz)			What Happens?	What Is Measured?
	γH_0 and ω_1	$\gamma H_0 - \omega_1{}^b$	γH_1		
slow passage	10^8	10^0 (at resonance)	10^0	$\vec{\mu}_i$'s precess about $\vec{H}_{eff}$ many times as we sweep through resonance, so $\vec{M}$ tips toward **u**.	M_u
Fourier transform (FT)	10^8	10^4	10^6	$\vec{\mu}_i$'s (and hence $\vec{M}$) process about $\vec{H}_{eff}$ $\frac{1}{4}$ of a revolution or less during the pulse time, so $\vec{M}$ tips toward **v**.	M_v

[a] Multiplying by γ puts $\vec{H}_{eff}$ in units of frequency (Hz).
[b] If you are accustomed to thinking in terms of parts per million, note that $10^4/10^8 = 100 \times 10^{-6} = 100$ ppm.

The Quantum Mechanical Description of the NMR Experiment

7–8 PROPERTIES OF $\hat{I}$

Now, to gain valuable insight, let us reexamine this whole problem using a quantum mechanical instead of a classical mechanical approach. Quantum mechanics shows us that for $I = {}^1/_2$, there are two allowed orientations of the spin angular momentum vector in a magnetic field (Fig. 7–8) and will also indicate the necessary requirements to induce transitions between these energy states by an appropriate perturbation, *i.e.*, the application of an oscillating magnetic field with energies corresponding to r.f. radiation. The necessary direction for this field can be determined from a consideration of the spin angular momentum operators.

The $\hat{I}^2$ operator has eigenvalues $I(I + 1)$ (as in an atom, where the orbital angular momentum operator $\hat{L}^2\psi = l(l + 1)\hbar^2\psi$). Any one of the components of $\hat{I}$ (e.g., $\hat{I}_z$) commutes with $\hat{I}^2$, so we can specify eigenvalues of both $\hat{I}^2$ and $\hat{I}_z$. The eigenvalues of $\hat{I}_z$ are given by $I, (I - 1), \ldots - I$ (like m_l values in an atom, where the z-component of angular momentum, $\hat{L}_z$, is given by $\hat{L}_z\psi = m_l\hbar\psi$ with $m_l = \pm l, \ldots, 0$). *In general, if two operators commute, then there exist simultaneous eigenfunctions of both operators for which eigenvalues can be specified.*

It is a simple matter to determine whether two operators commute. If they do, then by our earlier definition of commutation (Chapter 2)

$$\hat{I}^2\hat{I}_z - \hat{I}_z\hat{I}^2 = 0$$

The above difference is abbreviated by the symbol $[\hat{I}^2, \hat{I}_z]$. Similar equations can be written for $\hat{I}_x$ and $\hat{I}_y$, *i.e.*,

$$[\hat{I}^2, \hat{I}_x] = [\hat{I}^2, \hat{I}_y] = 0$$

However, $\hat{I}_z$ does not commute with $\hat{I}_x$ or $\hat{I}_y$, e.g.,

$$\hat{I}_z\hat{I}_y - \hat{I}_y\hat{I}_z \neq 0$$

Eigenvalues for $\hat{I}^2$ exist, and if we decide to specify eigenvalues for $\hat{I}_z$, then eigenvalues for $\hat{I}_x$ and $\hat{I}_y$ do not exist.

Average values (Chapter 3) could be calculated for the $\hat{I}_x$ and $\hat{I}_y$ operators. We shall make these concepts more specific by applying these operators to the spin wave functions α and β for $I = {}^1/_2$ nuclei and showing the results:

$$\hat{I}_z\alpha = (+{}^1/_2)\alpha$$
$$\hat{I}_z\beta = (-{}^1/_2)\beta \qquad (7\text{–}30)$$

$$\hat{I}_x\alpha = ({}^1/_2)\beta \qquad \hat{I}_x\beta = (-{}^1/_2)\alpha$$
$$\hat{I}_y\alpha = ({}^1/_2)i\beta \qquad \hat{I}_y\beta = (-{}^1/_2)i\alpha \qquad (7\text{–}31)$$

Thus, the $\hat{I}_z$ operator yields eigenvalues, since operation on α gives back α and operation on β gives β. The $\hat{I}_x$ and $\hat{I}_y$ operators do not yield eigenvalues, since operation on α produces β and operation on β yields α. The average value for the property $\hat{I}_x$ or $\hat{I}_y$ is given by an equation of the sort $\int \psi^* \text{Op}\psi \, d\tau / \int \psi^2 \, d\tau$. The following relations hold for α and β (as they do for orthonormal electronic wave functions):

$$\int \alpha^2 \, d\tau = \int \beta^2 \, d\tau = 1 \quad \text{and} \quad \int \alpha\beta \, d\tau = 0$$

As mentioned in Chapter 3, the integrals encountered in quantum mechanical descriptions of systems are written employing the *bra* and *ket notation* for convenience. Recall that the symbol $\langle \, |$ is referred to as a bra and $| \, \rangle$ as a ket. An integral of the form $\int(\psi^* \text{ Op } \psi)d\tau$ is written as $\langle\psi|\text{Op}|\psi\rangle$, whereas an integral of the form $\int \psi_1^*\psi_2 \, d\tau$ is written as $\langle\psi_1|\psi_2\rangle$.

7–9 TRANSITION PROBABILITIES

Consider the effect of the H_1 field in the quantum mechanical description. If the alternating field is written in terms of an amplitude H_x°, we get a perturbing term in the Hamiltonian of the form of equation (7–32):

$$\hat{H}_{\text{pert}} = -\gamma\hbar H_x^\circ \hat{I}_x \cos \omega_1 t \qquad (7\text{–}32)$$

Recall from equation (7–16) that $\hat{H}$ for a nucleus in a z-field was $\hat{H} = -\gamma\hbar H_0 \hat{I}_z = -g_N\beta_N H_0 \hat{I}_z$, so now the perturbation is of a similar form but for an x-field that is alternating.

The equation describing the probability of a transition in the nmr, P, is similar to that in u.v. and IR, and is given by

$$P = 2\pi\gamma_N^2 H_1^2 |\langle\varphi^{\text{ex}}|\hat{I}_x|\varphi\rangle|^2 g(\omega) \qquad (7\text{–}33)$$

where $g(\omega)$ is a general line shape function, which is an empirical function that describes how the absorption varies near resonance. To apply equation (7–33), we need to evaluate matrix elements of the form $\langle \varphi^{\text{ex}} | \hat{I}_x | \varphi \rangle$ and determine whether they are zero or non-zero.

The solution is best accomplished by constructing a matrix that summarizes all of the integrals that must be evaluated to describe the system in a magnetic field with and without H_1. Rows and columns are constructed, which are headed by the basis set. In this case, we have one nucleus with α and β nuclear spin wave functions leading to:

	α	β
α	$\langle \alpha | \hat{H} + \hat{H}_{\text{pert}} | \alpha \rangle$	$\langle \alpha | \hat{H} + \hat{H}_{\text{pert}} | \beta \rangle$
β	$\langle \beta | \hat{H} + \hat{H}_{\text{pert}} | \alpha \rangle$	$\langle \beta | \hat{H} + \hat{H}_{\text{pert}} | \beta \rangle$

By this procedure, we have considered all possible matrix elements and also have them in such a form that if E were subtracted from the diagonal elements we would have the secular determinant, which can be solved to give the energies of these states in an applied field $(H_0 + H_1)$.* The resulting energies could then be used in the secular equations (produced by matrix multiplication of the secular determinant with a matrix of the basis set) to give the wave functions in the field. (Notice the analogy of this to our handling of the secular determinant and equations in the section on Hückel calculations in Chapter 3.)

We begin by evaluating the elements in the secular determinant when the applied field is H_0 (with a z-component only), i.e., the Zeeman experiment. Since there is no x- or y-field component (only z), there are no $\hat{I}_x$ or $\hat{I}_y$ operators and all matrix elements of the form $\langle |\hat{I}_x| \rangle$ or $\langle |\hat{I}_y| \rangle$ are zero. The off-diagonal elements, $\langle \alpha | \hat{I}_z | \beta \rangle$ and $\langle \beta | \hat{I}_z | \alpha \rangle$ are also zero because $\hat{I}_z | \beta \rangle = -{}^1/_2 \beta$, and $\langle \alpha | \beta \rangle$ and $\langle \beta | \alpha \rangle$ are zero. The only non-zero matrix elements are $\langle \alpha | \hat{I}_z | \alpha \rangle$ and $\langle \beta | \hat{I}_z | \beta \rangle$, with the former corresponding to stabilization, $+{}^1/_2$ (i.e., ${}^1/_2 \langle \alpha | \alpha \rangle$), and the latter to destabilization, $-{}^1/_2$. With no off-diagonal elements, the eigenvalues are obtained directly and the basis set is not mixed, so the two wave functions are α and β. When these are substituted into equation (7–33) for φ and φ^{ex}, the matrix element is zero with $\hat{I}_z$ and the transition is not allowed, $\langle \alpha | \beta \rangle = 0$.

Next, we shall consider what happens when an H_1 field along the x-axis is added to the Zeeman experiment described above. We must now worry about matrix elements involving $\hat{I}_x$. The diagonal elements $\langle \alpha | \hat{I}_x | \alpha \rangle$ are zero $[\langle \alpha | \hat{I}_x | \alpha \rangle = {}^1/_2 \langle \alpha | \beta \rangle = 0]$, but the off-diagonal elements $\langle \alpha | \hat{I}_x | \beta \rangle$ are non-zero. Since the H_1 field is small compared to H_0 (z-component), these off-diagonal matrix elements are so small as to have a negligible effect on the energies (the effect of $\hat{I}_z$ on the diagonal is the same as in the Zeeman experiment). However, the small off-diagonal matrix elements are important because they provide a mechanism for inducing transitions from α to β *because the new wave functions for the system with H_1 present (i.e., obtained after diagonalizing our matrix) mix a little β character into the α-Zeeman state and a little α into the β-Zeeman state.* The new wave function for the α-Zeeman state now is $\varphi = \sqrt{1 - a^2} | \alpha \rangle + a | \beta \rangle$,

---

*Recall that $\langle \psi | \hat{H} | \psi \rangle = E \langle \psi | \psi \rangle$. For an orthonormal basis set, $\langle \psi_n | \psi_m \rangle$ equals 1 when $n = m$ but equals zero when $n \neq m$. Thus, E appears only on the diagonal of the energy determinant.

where $a \ll 1$, and a similar change occurs in the β state. When these new wave functions φ are substituted into equation (7–33), P is non-zero, making the transition allowed. This corresponds to the classical picture of H_1 exerting a torque to give a transverse component to the magnetization. Thus, we see that the probability, P, of a transition occurring depends upon the off-diagonal matrix elements in the α,β basis being non-zero, so that equation (7–33) is non-zero.

Next, consider what would happen if the r.f. perturbing field H_1 was placed along the z-axis. The off-diagonal matrix elements would again be zero, so equation (7–33) becomes

$$P = \langle \alpha | \hat{I}_z | \beta \rangle = 0$$

This would correspond to a slight reinforcement of H_0, but would not allow transitions. This exercise shows that H_1 cannot be colinear with H_0 to get an nmr transition. (In the classical model, there has to be a perpendicular component for H_1 to exert a torque.)

Finally, consider the case in which $I = 1$. Matrix elements $\langle m' | \hat{I}_x | m \rangle$ vanish unless $m' = m \pm 1$. Consequently, for $I = 1$, the allowed transitions are between adjacent levels with $\Delta m = \pm 1$, giving $\hbar \omega = \Delta E = \gamma \hbar H_0$.

In summary, simply placing a sample in a magnetic field, H_0, removes the degeneracy of the m_I states. Now, a radio-frequency source is needed to provide $h\nu$ to induce the transition. Absorption of energy occurs provided that the magnetic vector of the oscillating electromagnetic field, H_1, has a component perpendicular to the steady field, H_0, of the magnet. Otherwise (i.e., if H_1 is parallel to H_0), the oscillating field simply modulates the applied field, slightly changing the energy levels of the spin system, but no energy absorption occurs.

Relaxation Effects and Mechanisms

The influence of relaxation effects on nmr line shapes leads to some very important applications of nmr spectroscopy. Accordingly, it is worthwhile to summarize and extend our understanding of these phenomena. We begin with a discussion of relaxation processes and their effect on the shapes of the resonance line. The lifetime of a given spin state influences the spectral line width via the Uncertainty Principle, which is given in equation (7–34):

$$\Delta E \, \Delta t \approx \hbar \tag{7–34}$$

Since $\Delta E = h \, \Delta \nu$ and $\Delta t = T_2$, the lifetime of the excited state, the range of frequencies is given by $\Delta \nu \approx 1/T_2$. The quantity $1/T_2$ as employed here lumps together all of the factors influencing the line width (i.e., all the relaxation processes) and is simply one-half the width of the spectral line at half-height. When the only contribution to T_2 is from spin-lattice effects, then $T_2 = T_1$. Most molecules contain magnetic nuclei; in the spin-lattice mechanism a local fluctuating field arising from the motion of magnetic nuclei in the lattice couples the energy of the nuclear spin to other degrees of freedom in the sample, e.g., translational or rotational energy. The lattice refers to other atoms in the molecule or other molecules, including the solvent. For liquids, T_1 values are usually between 10^{-2} and 10^2 seconds but approach values of 10^{-4} seconds if certain paramagnetic ions are present. The extent of spin-lattice relaxation depends upon (1) the magnitude of the local field and (2) the rate of fluctuation. Paramagnetic ions have much more intense magnetic fields associated with them and are very efficient at causing relaxation. The water proton signal in a 0.1-M solution of $Mn(H_2O)_6^{2+}$ is so extensively broadened by efficient relaxation from Mn^{2+} that a proton signal cannot be detected in the nmr spectrum. As mentioned before, the spin-lattice process can be described by a first order rate constant $(1/T_1)$ for the decay of the z-component of the magnetization, say, after the field is turned off, and is referred to as *longitudinal relaxation*.

Next, we shall discuss some processes that affect the xy-components of the magnetization and are referred to as *transverse relaxation processes*. The spin-lattice effect discussed above always contributes to randomization of the xy-component; therefore,

$$\frac{1}{T_2} = \frac{1}{T_1} + \frac{1}{T_2'}$$

where $1/T_2'$ includes all effects other than the spin-lattice mechanism. When the dominant relaxation mechanism is spin-lattice for both longitudinal and transverse processes, i.e., $1/T_1 = 1/T_2$, then $1/T_2'$ is ignored. The quantity we used in the Bloch equations is $1/T_2$ and not $1/T_2'$. In solution, these other effects, $1/T_2'$, are small for a proton compared to $1/T_1$, so $(1/T_2) = (1/T_1)$. The other effects include field inhomogeneity, spin-spin exchange, and the interaction between nuclear moments. We shall discuss these in detail, beginning with field inhomogeneity.

When the field is not homogeneous, protons of the same type in different parts of the sample experience different fields and give rise to a distribution of frequencies, as shown in Fig. 7–14. This causes a broad band. The effect of inhomogeneity can be minimized by spinning the sample.

Interaction between nuclear moments is also included in T_2'. When a neighboring magnetic nucleus stays in a given relative position for a long time, as in solids or viscous liquids, the local field felt by the proton has a *zero frequency* contribution; i.e., it is not a fluctuating field as in the T_1 process from the field of the neighboring magnetic dipoles. A given type of proton could have neighbors with, for instance, a $+ + - + -$ combination of nuclear moments in one molecule, $+ - + - -$ in another, and so forth. Variability in the static field experienced by different protons of the same type causes broadening just like field inhomogeneity did. To give you some appreciation for the magnitude of this effect, a proton, for example, creates a field of about 10 gauss when it is 1 Å away. This could cause a broadening of 10^5 Hz. As a result of this effect, extremely broad lines are observed when the nmr spectra of solids are taken. This effect is averaged to zero in non-viscous solutions by molecular rotation.

The process of spin-spin exchange contributes to T_2 but not to T_1, for it does not influence the z-component of the magnetization. In this process, a nucleus in an excited state transfers its energy to a nucleus in the ground state. The excited nucleus returns to the ground state in the process and simultaneously converts the ground state nucleus to the excited state. No net change in the z-component results, but the u- and v-components are randomized.

In common practice, $1/T_2'$ is never really used. Either $1/T_2 = 1/T_1$ (with $1/T_2'$

FIGURE 7–14 The same nmr transition occurring in a bulk sample in an inhomogeneous field.

negligible) or we just discuss $1/T_2$. In a typical nmr spectrum, $1/T_2$ is obtained from the line shape (as will be shown) and it either equals $1/T_1$ or it does not.

As mentioned earlier, the true form of a broadened line is described empirically by a shape function $g(\omega)$, which describes how the absorption of energy varies near resonance according to equation (7–33). Since magnetic resonance lines in solution have a Lorentzian line shape

$$g(\omega) = \frac{T_2}{\pi} \frac{1}{1 + T_2^2(\omega - \omega_0)^2} \tag{7-35}$$

The width of the band between the points where absorption is half its maximum height is $2/T_2$ in units of radians sec^{-1}. In units of hertz, the full bandwidth at half height is given by $1/\pi T_2$.

7–10 MEASURING THE CHEMICAL SHIFT

The shielding constant, equation (7–28), makes nmr of interest to a chemist. In the typical nmr spectrum, the magnetic field is plotted versus absorption as illustrated by the low resolution spectrum of ethanol shown in Fig. 7–15. The

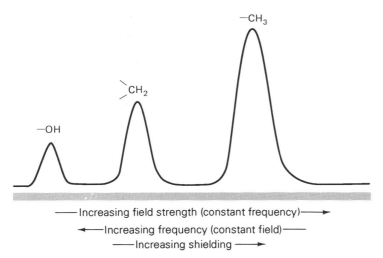

FIGURE 7–15 Low resolution proton nmr spectrum of C_2H_5OH.

least shielded proton (smallest σ) on the electronegative oxygen atom interacts with the field at lowest applied field strength. The areas under the peaks are in direct proportion to the numbers of equivalent hydrogens, 1:2:3, on the hydroxyl, methylene, and methyl groups. Note that the separation of absorption peaks from —CH_3 and —CH_2— hydrogens in this spectrum is much greater than that in the infrared spectrum.

We wish to calibrate the horizontal axis of Fig. 7–15 so that the field strength (or some function of it) at which the protons absorb energy from the radio frequency probe can be recorded. Equation (7–28) could be employed, but accurate measurement of H_N and H_0 is difficult. Instead, a reference material is employed, and the difference between the field strength at which the sample nucleus absorbs and that at which the nucleus in the reference compound absorbs is measured.

The reference compound is added to the sample (*vide infra*), so it must be unreactive. Furthermore, it is convenient if its resonance is in a region that does not overlap other resonances in the molecules typically studied. Tetramethylsilane, TMS, has both properties and is a very common reference material for non-aqueous solvents. In view of the limited solubility of TMS in water, salts of the anion $(CH_3)_3SiCD_2CD_2CO_2^-$ are commonly used. The position of its resonance is set at zero on the chart paper.

The magnetic field differences indicated by the peaks in Fig. 7–15 are very small, and it is difficult to construct a magnet that does not drift on this scale. Accordingly, most instruments pick a resonance and electronically adjust the field circuit to maintain or lock this peak at a constant position. This can be accomplished by having some material in a sealed capillary in the sample tube or in the instrument to lock on, or else by picking a resonance in the spectrum of the solution being studied for the purpose. The former procedure is described as an *external lock* and the latter as an *internal lock*. The internal lock produces more accurate results. Temperature and bulk magnetic susceptibility effects are minimized. The field is being locked on a resonance that has all of the advantages of an internal standard (*vide infra*). In a typical experiment employing an internal lock, the ^2H signal of a deuterated solvent is utilized via a separate transmitter-receiver system.

With TMS at zero, it is possible to measure the differences in peak maxima, Δ. Although the field is being varied in the slow passage experiment, the abscissa and hence the differences in peak positions are calibrated in a frequency unit of cycles per second, referred to as hertz (Hz). This should cause no confusion, for frequency and field strength are related by equation (7–5):

$$\omega = \gamma H$$

According to equation (7–28), the shielding of the various nuclei, σH_0, depends upon the field strength, H_0. If a fixed frequency probe of 60 MHz is employed, the field utilized will be different than if a 100-MHz probe is employed. As shown in Fig. 7–16, the peak separations are 10/6 as large at 100 MHz as at

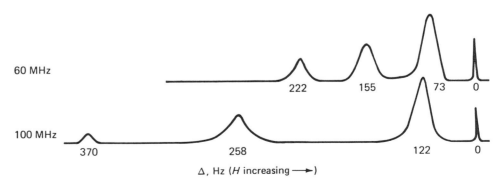

FIGURE 7–16 The nmr spectrum of CH_3CH_2OH at 60 Mhz and at 100 MHz.

60 MHz. To overcome this problem and to obtain values for the peak positions, which are independent of field strength, the chemical shift, δ, is defined as

$$\delta = \frac{\Delta \times 10^6}{\text{fixed frequency of the probe, } \nu_0} \quad (7\text{--}36)$$

Since the probe frequency of ν_0 is in units of megahertz (MHz), and Δ has units of hertz, the fraction is multiplied by the factor 10^6 to give convenient numbers for δ in units of parts per million, ppm. The chemical shift, δ, is independent of the probe frequency employed.

If the sample nuclei are deshielded relative to the reference peak, Δ is positive* by convention. When $Si(CH_3)_4$ is employed as a standard, most proton δ values are positive and the larger positive numbers refer to lesser shielding.

For several reasons, it is advantageous to add the TMS directly to the solution being studied as opposed to having it in a separate sealed capillary tube, *i.e.*, an *internal* vs. an *external standard*. Magnetic field-induced circulations of the paired electron density in a molecule give rise to a magnetic moment that is opposed to the applied field. In diamagnetic substances, this effect accounts for the repulsion, or diamagnetism, experienced by these materials when placed in a magnetic field. The magnitude of this effect varies in different substances, giving rise to varying diamagnetic susceptibilities. This diamagnetic susceptibility in turn gives rise to magnetic shielding of a molecule in a solvent, and is variable for different solvents. The effect is referred to as the volume diamagnetic susceptibility of the solvent. Thus, the chemical shift of a solute molecule in a solvent will be influenced not only by shielding of electrons but also by the volume diamagnetic susceptibility of the solvent. The δ values obtained for a *solution* and a *pure* liquid solute (this is often referred to as the neat liquid) relative to an external standard (Fig. 7–17) would be different to the extent that the volume diamagnetic susceptibilities of the solvent and neat liquid were different. In solution, the diamagnetic contributions to the shielding of the solute depend upon the average number of solvent and solute neighbors. Consequently, the chemical shift will be concentration dependent. To get a meaningful value for δ with an external standard, it is necessary to measure δ at different concentrations and extrapolate to infinite dilution to produce a value for δ under volume susceptibility conditions of the pure solvent.

Contributions to the measured δ from the volume diamagnetic susceptibility of the solvent can be more easily minimized by using an *internal standard*. The internal standard must, of course, be unreactive with the solvent and sample. Furthermore, the sample and standard must be uniformly dispersed. Under these conditions the standard is subjected to the same volume susceptibility (from solvent and solute molecules) as is the solute, and the effects will tend to cancel when the difference, Δ, is calculated. (Due to variation in the arrangement of solvent around different solutes, an exact cancellation may not result.) For accurate work, spectra at two different concentrations should be run as checks and the results relative to two internal standards compared to insure that the

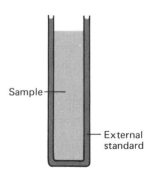

FIGURE 7–17 Concentric nmr sample tube containing an external standard. For illustrative purposes the external standard compartment size is exaggerated. This is really just large enough for two to three drops of standard.

---

* In the early literature and even as late as 1970 in the area of ^{19}F and ^{31}P nmr, the convention used is the opposite of that described here. The reader must exercise considerable care in ascertaining the convention used.

internal standard approximations are working. Molecular interactions of the solvent with the standard or solute cause complications.

7–11 INTERPRETATION OF THE CHEMICAL SHIFT

a. Local Effects

The range of proton chemical shifts measured on pure liquids[4] for a series of organic compounds is illustrated in Fig. 7–18. Proton shifts outside this total range have been reported, and sometimes shifts outside the range indicated for a given functional group occur. In general, the data in Fig. 7–18 serve to give a fairly reliable means of distinguishing protons of functional groups. Note the difference in $CH_3-C\lessgtr$, $CH_3C=$, CH_3-O, $HC=$, HCO, etc. Very extensive compilations of proton chemical shifts have been reported,[5] which can be employed for the fingerprint type of application. The reference section at the end of this chapter lists compilations of shifts for various nuclei. The shift of OH protons is very concentration and temperature dependent. For example, the spectrum of ethanol changes as a function of concentration in an inert solvent (e.g., CCl_4), by 5 ppm in going from concentrated to dilute solution. At infinite dilution, the hydroxyl proton appears at a higher field than the methyl protons, in contrast to the spectrum of pure ethanol represented in Fig. 7–16. There is more hydrogen bonding in concentrated than in dilute solutions and hydrogen bonding of the proton causes a shift to a lower field. This behavior on dilution can be employed to verify the assignment of a peak to a hydroxyl group, to investigate the existence of steric effects in hydrogen bonding[6] and to distinguish between intermolecular and intramolecular hydrogen bonding. Solvent effects are quite large whenever specific interactions occur and, as will be shown later, nmr has been very valuable in establishing the existence or absence of these interactions. The differential shift of the C—H and O—H protons of methanol or ethylene glycol is the basis of a standard technique for estimating probe temperature.[7]

The interpretation of proton chemical shifts is complicated by the existence of contributions from local (i.e., the atom undergoing the transition) shielding and neighbor atom anisotropic effects. Local shielding arises from magnetic field-induced electron circulation on the atom undergoing the transition. This shielding is a tensor quantity with σ_{zz} giving the contribution parallel to the field when the z-axis of the molecule is aligned with the field. Using perturbation theory and treating the magnetic field as a perturbation on the ground-state molecular wave function, $|0\rangle$, leads to the Ramsey equation[8]:

$$\sigma_{zz} = \frac{e^2}{2mc^2}\left\langle 0 \left| \frac{x^2+y^2}{r^3} \right| 0 \right\rangle$$
(diamagnetic term)

$$-\left(\frac{e\hbar}{2mc}\right)^2 \sum_n \left\{ \frac{\langle 0|\hat{L}_z|n\rangle\left\langle n\left|\frac{2\hat{L}_z}{r^3}\right|0\right\rangle}{E_n - E_0} + \frac{\left\langle 0\left|\frac{2\hat{L}_z}{r^3}\right|n\right\rangle\langle n|\hat{L}_z|0\rangle}{E_n - E_0} \right\} \quad (7\text{–}37)$$
(paramagnetic term)

7–11 Interpretation of the Chemical Shift   233

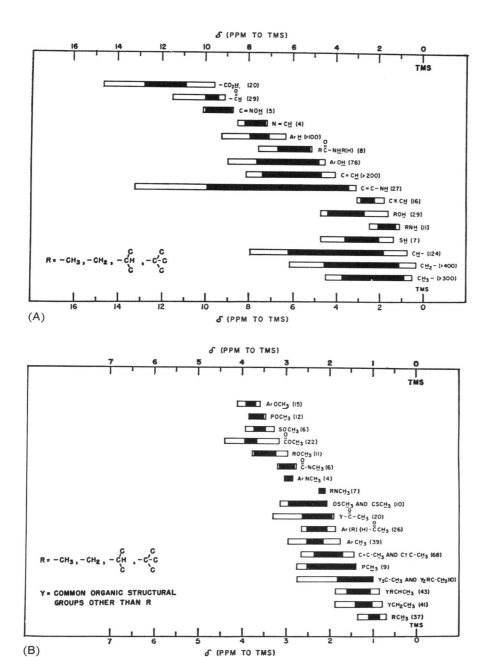

FIGURE 7–18 Hydrogen nuclear magnetic resonance chemical shift correlation charts, (A) for main proton groups, (B) for CH, OH, and NH subgroups. Open bar denotes extreme 10% of data. [from M. W. Dietrich and R. E. Keller, Anal. Chem., *36*, 258 (1964).] *Illustration continued on following page.*

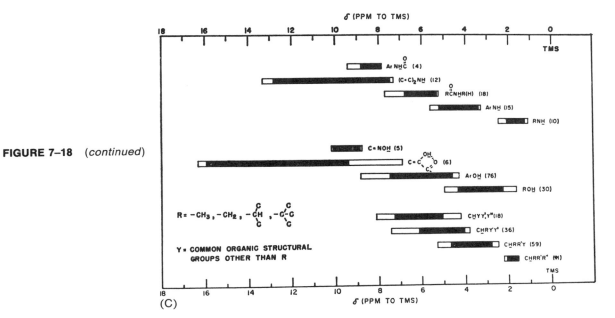

FIGURE 7-18 (continued)

This equation is similar to the Lamb equation for an atom, and it tells us what the applied field does to our molecule. A positive sign indicates shielding, i.e., an upfield or a diamagnetic contribution. The negative sign indicates deshielding, i.e., a downfield or a paramagnetic contribution.

The first term on the right-hand side of the equality is the so-called diamagnetic term. The symbol $\langle 0|$ corresponds to the ground state wave function, and r is the distance from the electron to the nucleus undergoing the transition. Since only $\langle 0|$, the ground state wave function, is involved in this first term, no excited states are mixed in by this term. The field does not distort the electron distribution in the molecule, but just induces a spherical electron circulation. If this were the only effect, the molecular wave function would be independent of the magnetic field.

The matrix element corresponding to this first term reduces to:

$$\left(\frac{2}{3}\right)(e^2/2mc^2)\lambda \int (\psi_{1s}^2/r)d\tau \qquad (7\text{--}38)$$

The term λ refers to the effective number of electrons in the s-orbital. This effect gives rise to the normal diamagnetism observed for $S = 0$ molecules and can be pictured as the shielding of the nucleus resulting from field-induced circulation of electron density as illustrated in Fig. 7–19.

The field generated at the nucleus by electron circulations is directly proportional to the strength of the applied field and will also depend on the electron density surrounding the nucleus. If this were the only effect of the magnetic field on a molecule, the chemical shift of a hydrogen nucleus would parallel the electronegativity of the groups attached. This correlation is not observed because of the neighboring anisotropic contributions to be discussed shortly.

For simplicity, we have treated only the σ_{zz} component to illustrate the factors that contribute to the shielding. There are similar expressions for σ_{xx} and

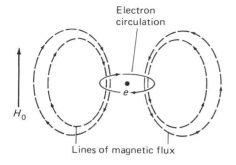

FIGURE 7–19 Local diamagnetic contribution to the shielding. Electron circulations occur in a plane perpendicular to the plane of the page. The induced field given by the arrow in the middle is opposed to H_0.

σ_{yy}, which, in general, are not necessarily equal to σ_{zz} (for σ_{xx}, the matrix element is $\langle 0|(y^2 + z^2)/r^3|0\rangle$) and the three quantities must be averaged to produce the shielding observed when the molecule is rapidly tumbling.

$$\sigma = \frac{1}{3}(\sigma_{xx} + \sigma_{yy} + \sigma_{zz}) \qquad (7\text{–}39)$$

For a hydrogen atom, the three contributions are equal for this local effect.

The range of chemical shifts for nuclei other than hydrogen is considerably larger. For example, the difference between the fluorine resonances in F_2 and in F^-_{aq} is 542 ppm, compared to the range of about 12 ppm for proton shifts (see Fig. 7–18). A brief compilation of fluorine shifts is contained in Table 7–3 to illustrate this point. A wide range (~ 500 ppm) of phosphorus chemical shifts and ^{13}C shifts are also observed. References are given at the end of this chapter for compilations. The fingerprint application is immediately obvious for compounds containing these elements.

The large range in chemical shifts for these nuclei arises from a local effect corresponding to the second term of the Ramsey equation. This is called the *paramagnetic* term.* We are *not* using the word paramagnetic here to connote the same thing as when it is used in connection with molecules containing

TABLE 7–3. Fluorine Chemical Shifts of Selected Compounds (in ppm Relative to $CFCl_3$—Larger Negative Numbers Indicate Higher Field)

CH_3F	-272	CF_4	-70	SOF_2	$+70$
HF	-203	TeF_6	-64	TiF_6^{2-}	$+75$
$(CF_3)_3CF$	-188	$(CF_3)_4C$	-61.5	ClF_3	$+80$
C_6F_6	-163	AsF_3	-48	BrF_5	$+132, +269$
SiF_4	-177	BrF_3	-38	NF_3	$+140$
BeF_2	-155	CF_3Cl	-32	CF_3OF	$+142$
BF_3OEt_2	-153	CF_2Cl_2	-9	SF_5OF	$+178$
BF_4^-	-149	CF_3I	-4	$FOClO_3$	$+225$
BF_3	-131	$CFCl_3$	0	IF_7	$+238, +274$
SiF_6^{2-}	-127	IF_5	$+4$	OF_2	$+250$
$(CF_3CF_2)_2$	-127	SO_2F_2	$+36$	$FClO_2$	$+288$
$F^-(aq)$	-120	SF_6	$+42$	FNO_2	$+394$
SbF_6^-	-109	SF_5OF	$+48$	F_2	$+422$
SbF_5	-108	SeF_6	$+55$	XeF_4	$+438$
CF_3H	-88	IF_5	$+53$	FNO	$+478$
SbF_3	-86	N_2F_4	$+60$	XeF_6	$+550$
$(CF_3)_2CO$	-82	NSF_3	$+66$	UF_6	$+746$
CF_3COOH	-77				

unpaired electrons. It is used here to describe the contributions from the field-induced nonspherical circulation of the electron density, *i.e.*, the second term on the right-hand side of the Ramsey equation. The term involves field-induced mixing in of excited states with the ground state, and this gives rise to a mechanism for non-spherical electron circulation and the accompanying paramagnetic contribution to the shielding. The equation for the paramagnetic effect corresponds very closely to the paramagnetic term in the Van Vleck equation for magnetic susceptibility (*vide infra*). The following quantities appear: $\langle 0|$ ground state and $\langle n|$ excited state wave functions; $\hat{L}_z$, the orbital angular momentum operator; and $E_n - E_0$, the energy difference between the ground state and the excited state being mixed in. The summation is carried out over all excited states; thus, to use the equation one needs wave functions and energies of all excited states, including those in the continuum. The energies of virtual orbitals (empty ones) are difficult to calculate, and thus it becomes impossible to employ the Ramsey equation in the rigorous calculation of the chemical shift of most molecules.

Evaluation of the entire second term of equation (7–37) and replacement of the $E_n - E_0$ term in the denominator by an average energy for the excited states in the molecule (*i.e.*, use of some electronic transition in the molecule that is felt to be an average of all possible transitions to excited states) yields the paramagnetic contribution, σ_p, by averaging σ_{xx}, σ_{yy}, and σ_{zz}:

$$\sigma_p = -\frac{2}{3}\left(\frac{e\hbar}{mc}\right)^2 \left\langle \frac{1}{r^3} \right\rangle \frac{1}{\Delta E} \qquad (7\text{–}40)$$

The average energy approximation is often referred to as the *closure approximation*. Substituting appropriate values for F_2 into equation (7–40) (*i.e.*, the energy of 4.3 eV for the $\pi \rightarrow \sigma^*$ transition and $\langle 1/r^3 \rangle$ for a fluorine atom 2p orbital) one obtains $\sigma_p \approx 2000$ ppm. Since F^- has spherical symmetry, there can be no angular momentum associated with the electron density in this species, and all the paramagnetic terms in the Ramsey equation must be zero. A chemical shift difference between F^- and F_2 of about 2000 ppm is expected compared to only 542 ppm observed. However, it is impossible to obtain a fluoride ion that is not solvated or ion paired. The F_2 chemical shift is found to be 625 ppm downfield from HF. This is due in part to the poor nature of the average energy approximation and in part to the use of an atomic $\langle 1/r^3 \rangle$ value. however, in spite of the poor quantitative calculation of σ_p, there are several important generalizations that can be drawn from this discussion. Since $\hat{L}_z$ is zero for an s orbital in hydrogen compounds, one must invoke transitions to the 2p orbital in order to obtain a paramagnetic contribution. Now the $E_n - E_0$ value is so large that paramagnetic contributions are not very important in the proton nmr spectrum. These paramagnetic contributions become very large when both an asymmetric distribution of p and d electrons in the molecule and low-lying excited

---

* The distinction between diamagnetic and paramagnetic contributions in equation (8–31) is somewhat artificial, for it depends upon the gauge selected for the vector potential (see Slichter text, p. 76, under Additional References). The traditional gauge corresponds to the measurement of angular momentum about the nucleus whose chemical shift is being calculated. The total shielding is gauge invariant.

states exist. The paramagnetic term gives rise to the principal contribution to the shift in ^{19}F and ^{13}C nmr, and a large chemical shift range is observed. The chemical shifts in ^{11}B nmr have significant contributions from both local paramagnetic and local diamagnetic effects.

There is one additional bit of insight provided by the second term of the Ramsey equation that will concern us. Since this term is important in atoms having accessible p orbitals, our concern will be with the behavior of the $\hat{L}$ operator on these orbitals. The behavior of $\hat{L}_z$ is summarized as follows:

$$\hat{L}_z|p_z\rangle = m_l|p_z\rangle = 0|p_z\rangle \qquad (7\text{–}41a)$$

$$\hat{L}_z|p_x\rangle = i|p_y\rangle \qquad (7\text{–}41b)$$

$$\hat{L}_z|p_y\rangle = -i|p_x\rangle \qquad (7\text{–}41c)$$

The results for the $\hat{L}_x$ and $\hat{L}_y$ operators will be presented as needed. We see from equation (7–37) that we need to evaluate matrix elements of the type

$$\left\langle 0 \left| \frac{\hat{L}_\alpha}{r^3} \right| n \right\rangle \langle n|\hat{L}_\alpha|0\rangle \qquad (7\text{–}42)$$

where $\alpha = z$, x, and y for the σ_{zz}, σ_{xx}, and σ_{yy} components of the shielding tensor for all possible excitations of an electron from the ground state m.o.'s to $\sigma_{2p_z}*$. (To simplify the problem, higher excited states will be omitted because the $E_n - E_0$ term becomes very large for molecular orbitals derived from 3s, 3p, and higher atomic orbitals.) We shall illustrate the evaluation of these matrix elements for the F$_2$ molecule (molecular orbital description, valence orbitals

$$\sigma_{2s}{}^2\sigma_{2s}*{}^2\sigma_{2p_z}{}^2(\pi_x = \pi_y)^4(\pi_x* = \pi_y*)^4\sigma_{2p_z}{}^{0*})$$

To be systematic, we shall consider the three different orientations in which the molecular axes are aligned with the z-axis parallel to the field, with the x-axis parallel to the field, and with the y-axis parallel to the field to evaluate σ_{zz}, σ_{xx}, and σ_{yy}, respectively. This requires the use of the $\hat{L}_z$, $\hat{L}_x$, and $\hat{L}_y$ operators, respectively, in the Ramsey equation.

Consider the cases where:

1. **The z-axis of the molecule is aligned with the field.** When the z-axis of the molecule is parallel to the applied field, the molecule senses no H_x or H_y field, so there are no field-induced $\hat{L}_x$ or $\hat{L}_y$ components. All one-electron excitations place an electron into the antibonding molecular orbital, leading to matrix elements such as:

$$\langle \sigma_{2s}|\hat{L}_z|\sigma*\rangle, \quad \langle \sigma_{2p}|\hat{L}_z|\sigma*\rangle, \quad \langle \pi_x|\hat{L}_z|\sigma*\rangle \quad \text{or} \quad \langle \pi_y|\hat{L}_z|\sigma*\rangle$$

where $\langle \sigma_{2p}|\hat{L}_z|\sigma*\rangle = (^1/_2)\langle(p_{za} + p_{zb})|\hat{L}_z|(p_{za} - p_{zb})\rangle$ for a bond between fluorine atoms A and B evaluating the contribution at A. We know that $\hat{L}_z|p_z\rangle = 0$. Thus, all the matrix elements for this orientation of the molecule are zero, and σ_{zz} is zero. This leads to a generalization that will become important in subsequent discussion of the chemical shift. Namely, *the contribution to the chemical shift from paramagnetic terms, σ_p, is zero when the highest-fold symmetry axis (z-axis) is parallel to the field.*

2. **The x-axis of the molecule is aligned with the field.** That is, the z-axis is perpendicular to the field. For the x-axis aligned this corresponds to evaluating σ_{xx}, and the $\hat{L}_x$ operator. Considering one of the possible transitions, we have

$$\langle \sigma | \hat{L}_x | \sigma^* \rangle = (1/2)[\langle p_{za} | \hat{L}_x | p_{za} \rangle + \langle p_{za} | \hat{L}_x | p_{zb} \rangle - \langle p_{zb} | \hat{L}_x | p_{za} \rangle - \langle p_{zb} | \hat{L}_x | p_{zb} \rangle]$$

Since the last term involves the $\hat{L}_x$ operator centered on a operating on b, integrals of this sort will generally be small; they are dropped here and in subsequent discussion. The $\langle p_{zb} | \hat{L}_x | p_{za} \rangle$ and $\langle p_{za} | \hat{L}_x | p_{zb} \rangle$ terms are two-center integrals; they, too, generally are small and will not be considered in future discussion. In the case under consideration, all of these matrix elements are zero because $\hat{L}_x | p_z \rangle$ equals $i | p_x \rangle$, so:

$$\langle p_z | \hat{L}_x | p_z \rangle = \langle p_z | i p_x \rangle = 0$$

Only those matrix elements corresponding to $\langle p_x | \hat{L}_x | p_z \rangle$ can be non-zero and these arise from

$$\langle \pi | \hat{L}_x | \sigma^* \rangle \quad \text{and} \quad \langle \pi^* | \hat{L}_x | \sigma^* \rangle$$

There is only one such matrix element to evaluate, namely, $(1/2)\langle p_{xa} | \hat{L}_x | p_{za} \rangle$ (recall that $\langle p_{xa} | \hat{L}_x | p_{zb} \rangle$ is a two-center integral and is small in comparison), which is equal to $i/2$. This value corresponds to the matrix element $\langle \pi | \hat{L}_x | \sigma^* \rangle$. The element $\langle \pi^* | \hat{L}_x | \sigma^* \rangle$ gives the same result. Thus, *there is a contribution to the paramagnetic term for the field along the x-axis. We see that the field has coupled the ground state with the excited state; i.e., we have field-induced mixing.*

3. **The y-axis of the molecule is aligned with the field.** This gives the same result. We must now concern ourselves with $\hat{L}_y$ in the molecular coordinate system. The result is

$$\hat{L}_y | p_z \rangle = i | p_y \rangle$$

By analogy with our discussion of $\hat{L}_x$, only the matrix element $\langle p_{ya} | \hat{L}_y | p_{za} \rangle$ is non-zero, and we also expect a paramagnetic contribution to the shielding at fluorine when the field is along the y-axis. As before, the other matrix elements are zero.

b. Remote Effects

Protons attached to metal ions are in general very highly shielded,[10] the resonance often occurring 5 to 15 ppm to the high field side of TMS and, in some cases, occurring over 60 ppm upfield from TMS. A local contribution cannot explain these large shifts. The existence of a field-induced moment on a neighboring atom can be felt at the atom whose nmr is being studied. These are referred to as *remote effects* or as *neighbor anisotropic contributions*. The field at this remote atom may be dominated by either a paramagnetic or a diamagnetic effect but may make a different contribution at the atom being studied. We shall work through the case of the HX molecule first, assuming that the diamagnetism of the remote atom X is dominant. As illustrated in Fig. 7–20, the field at the

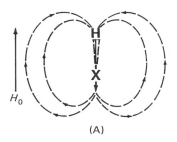 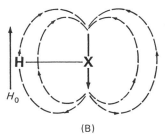

FIGURE 7–20 Neighbor anisotropy in HX for different orientations.

proton in HX from the diamagnetic effect at X will be strongly dependent upon the orientation of the HX molecule with respect to the direction of the applied field, H_0. When the applied field is parallel to the internuclear axis, the magnetic field generated from diamagnetic electron circulations on X (indicated by the dotted lines) will shield the proton (Fig. 7–20(A)), while in a perpendicular orientation (Fig. 7–20(B)) this same effect at X will result in deshielding at the proton. The magnitude of the induced moment on atom X (and hence the field at the proton from this neighbor effect) for the parallel and perpendicular orientations will depend upon the susceptibility of X for parallel and perpendicular orientations, $\chi_\parallel$ and $\chi_\perp$ respectively. The susceptibility, χ, is related to the intensity of the magnetization, M, by

$$\chi = \frac{M}{H_0} \tag{7-43}$$

When HX is parallel to the field, the induced moment on X is given by $\chi_{\parallel(X)} H_0$. The contribution this remote effect makes to the shielding *at the proton*, $\sigma_\parallel$, is given by $-\sigma_\parallel H_0$; expressed in terms of the susceptibility of X, $\sigma_\parallel$ is:

$$\sigma_\parallel = -2R^{-3}\chi_{\parallel(X)} \tag{7-44}$$

where R is the distance from X to the proton, χ is negative (and consequently the proton is shielded), and $\chi_{\parallel(X)}$ is the parallel component of the susceptibility at X.

For the perpendicular ($\perp$) orientation of HX, the contribution from X at the proton is given by

$$\sigma_\perp = R^{-3}\chi_{\perp(X)} \tag{7-45}$$

Note that equations (7–44) and (7–45) give the correct signs for the shielding at the proton, as illustrated in Fig. 7–20 ($\chi_\parallel$ and $\chi_\perp$ are both negative). We have two perpendicular components to consider, χ_x and χ_y, which are equal here and in any molecule with a threefold or higher symmetry axis. In solution, the molecule is rapidly tumbling, so the concern is with the average value of σ given by equation (7–46):

$$\sigma = -\tfrac{1}{3}R^{-3}(2\chi_\parallel - 2\chi_\perp) \tag{7-46}$$

According to equation (7–46), if the susceptibility is isotropic (*i.e.*, $\chi_\parallel = \chi_\perp$), there will be no contribution to the shielding at the nucleus of interest from diamagnetic effects on the neighbor. In CH_4, for example, $\chi_\parallel = \chi_\perp$ and neither the carbon nor the C—H bonds can make a neighbor contribution to the proton shift. In HX molecules, the remote contribution from X arises because $\chi_\parallel \neq \chi_\perp$. The value to use for R in equations (7–45) and (7–46) is a problem, for one must decide whether the susceptibility arises on the remote atom or in the bond. This question is not a real one but results from our arbitrary factoring of the molecular susceptibility into parts due to different atoms in the molecule. In spite of this simplification, the approach is valuable and one often selects the atom or center of the bond as the point from which to measure R.

If one had a molecule in which paramagnetic contributions were dominant, the same equations would be employed, except that the signs of χ would be positive; *i.e.*, the boldface arrow on X of Fig. 7–20 would point in the opposite direction. Paramagnetic contributions from X (or the H—X bond) will be zero for the orientation in which the bond axis is parallel to the field and will be a maximum for the perpendicular orientation. Thus, the paramagnetic term will be very anisotropic and often dominates any neighbor anisotropic contributions to the proton shifts. Refer to Fig. 7–20(B) and change the direction of the arrows for this paramagnetic effect at X. The paramagnetic effect at X is shielding (diamagnetic) at the proton. When discussing a remote effect, we label it according to the direction of the effect on the remote atom. If $\chi_\parallel$ and $\chi_\perp$ were known, we would know its sign and which effect is dominant. This information is seldom available and both paramagnetic and diamagnetic neighbor effects must be qualitatively considered when interpreting proton shifts.

Equations (7–44) and (7–45) apply only for linear molecules. For the general case

$$\sigma = (1/3)R^{-3}[(1 - 3\cos^2\theta_x)\chi_{xx} + (1 - 3\cos^2\theta_y)\chi_{yy} + (1 - 3\cos^2\theta_z)\chi_{zz}] \quad (7\text{–}47)$$

where χ_{xx}, χ_{yy}, and χ_{zz} are the values along the three principal axes of the susceptibility tensor and θ_x, θ_y, and θ_z correspond to the angles between these axes and a line drawn from the center of the anisotropic contributor (the neighbor atom or bond) to the atom being investigated. The angle is illustrated for a molecule of C_{3v} symmetry in Fig. 7–21, where the source of the neighbor effect has been taken at the iodine atom, χ_{zz} is taken along the threefold axis, and $\chi_{xx} = \chi_{yy}$. When $\chi_{xx} \neq \chi_{yy} \neq \chi_{zz}$, we can select our coordinate system such that the radius vector R lies in the yz- (or xz-)plane. Equation (7–47) then becomes

$$\sigma = \tfrac{1}{3}R^{-3}(2\Delta\chi_1 - \Delta\chi_2 - \Delta\chi_1 3\cos^2\theta_z) \quad (7\text{–}48)$$

where $\Delta\chi_1 = \chi_{zz} - \chi_{yy}$ and $\Delta\chi_2 = \chi_{zz} - \chi_{xx}$. When axial symmetry pertains, $\chi_{zz} \neq \chi_{xx} = \chi_{yy}$ and equation (7–48) becomes (7–49):

$$\sigma = (1/3)R^{-3}(\chi_\parallel - \chi_\perp)(1 - 3\cos^2\theta_z) = (1/3)R^{-3}\Delta\chi(1 - 3\cos^2\theta_z) \quad (7\text{–}49)$$

Values for $\Delta\chi$ for various bonds can be found in the literature.

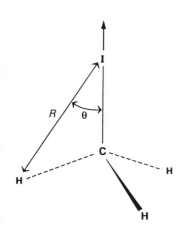

FIGURE 7–21 Illustration of the parameters θ and R of equation (8-37).

7-12 INTERATOMIC RING CURRENTS

Interatomic ring currents develop in cyclic, conjugated systems. Field-induced electron circulations occur in a loop around the ring and extend over a number of atoms. Analogous to the circulation of electrons in a wire, a magnetic moment is induced by the effect. The moment induced at the center of the ring is opposed to the field, but in benzene, for example, the magnetic lines of flux at the protons are parallel to the applied field, and these protons are deshielded as shown in Fig. 7-22. For another orientation of benzene relative to the field, the ring current shifts the protons to a smaller extent. The average overall orientation is deshielding. This is the explanation for the large deshielding observed for the protons in benzene.[11] Johnson and Bovey[11] have calculated the ring current contribution to the chemical shift for a proton located at any position relative to the benzene ring. The acidic hydrogen, upon hydrogen bonding to benzene, is shielded, whereas deshielding is observed with most donors other than benzene. This indicates that the proton of the Lewis acid is located on the C_6 axis of benzene.

7-13 EXAMPLES OF CHEMICAL SHIFT INTERPRETATION

One important generalization can be drawn from the preceding and subsequent discussion, *viz.*, chemical shift data are not reliable indications of the electron density around the nucleus being measured. Many effects contribute to δ. Contributions from neighbor anisotropy are qualitatively invoked when needed to account for differences between measured δ values and those expected on the basis of chemical intuition. The molecule is examined to see what property it possesses that could account for the observed discrepancies. For example, the δ value for HCl (in the vapor phase) indicates that this proton is more shielded than those in methane. This is in contrast to the greater formal positive charge on the proton of HCl than on that of CH_4. Thus, the local diamagnetic effect does not explain this behavior. The *local* paramagnetic effect is not expected to be significant for a proton. Since HCl is linear and cylindrically symmetric about the hydrogen-chlorine bond, a neighbor paramagnetic contribution is expected when the molecule is perpendicular to the applied field but not when it is parallel to the field. The net effect will be a shielding of the proton in HCl from a

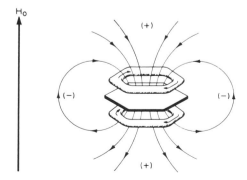

FIGURE 7-22 Ring currents in benzene.

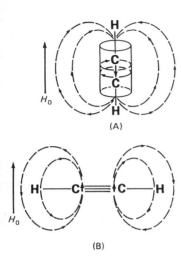

FIGURE 7–23 Contributions to the shielding in acetylene. (A) The effect here is a diamagnetic effect at carbon, which is a maximum for this orientation. (B) The effect here is a paramagnetic effect at carbon, which is a maximum for this orientation.

paramagnetic, neighbor anisotropic effect. The effect is paramagnetic (deshielding) at X but shielding at the proton. The neighboring anisotropic effect dominates the difference in CH_4 and HCl shifts.

In the series HF, HCl, HBr, and HI, the magnitude of the neighboring paramagnetic contribution, $\Delta\sigma_p$, giving rise to anisotropy increases with increasing atomic number of the halogen. This parallels the decreased difference in the energy of the ground and excited states, ΔE, making a field-induced mixing of ground and excited states in the compounds of the higher atomic number atoms more favorable. The Ramsey equation states that the paramagnetic contribution to $\Delta\sigma$ will be proportional to $-1/\Delta E$.

The formal positive charge on the proton for the compounds in the series $C_2H_6 < C_2H_4 < C_2H_2$ increases in the order listed. The δ values increase in the order $C_2H_6 < C_2H_2 < C_2H_4$, indicating decreased shielding. Acetylene is more highly shielded than is expected on the basis of its acidity, and the shielding is attributed to two effects: (1) Remote diamagnetic shielding from electron circulations in the triple bond (Fig. 7–23(A)). (This effect is a maximum when the molecular axis is aligned with the field, and it is classified here as a diamagnetic effect because the moment arising from the electron circulations opposes the applied field. Recall that the paramagnetic contribution is zero for this orientation and is a maximum when the axis is perpendicular to the field.) (2) Remote paramagnetic shielding from higher state mixing (Fig. 7–23(B)). Both the remote diamagnetic and paramagnetic effects influence $\Delta\chi$ so as to provide shielding at the proton. The measured proton chemical shift of acetylene relative to ethane is thus explained by the remote diamagnetic and paramagnetic contributions arising from the triple bond.

In addition to the trends in the population of the hydrogen 1s orbital as a consequence of the electronegativity of the attached group, Buckingham has proposed[12] *electric field* and reaction field models to account for changes in the electron density of a bound hydrogen atom.

Suffice it to say that in view of the many contributions described above, even the qualitative interpretation of proton chemical shifts is tenuous. When ^{13}C and ^{19}F chemical shifts are considered, the dominance of the local paramagnetic term gives rise to such large shift differences in series of molecules that neighbor anisotropy effects become small in comparison.[13] The qualitative interpretation of the shifts in these compounds has met with more success.[14] The problem is still not simple by any means. Changes in the average excitation energy, one-center terms, and two-center terms in a molecular orbital description are all important,[15,16] leading to the situation where there are many more parameters influencing the measurement than there are experimental observables. The reader is referred to references 13, 15, and 16 in the event one is ever tempted to propose a rationalization of chemical shift data.

The chemical shifts of analogous atoms in two different enantiomers are identical. However, as with many physical properties, differences are detected in diastereoisomers. Advantage has been taken of this fact[17] to utilize nmr for the determination of optical purity. For example, when a *d,l* mixture of the phosphine oxide, $C_6H_5CH_2-P(=O)(C_6H_5)(CH_3)$, is added to the enantiomeric solvent *d*-2,2,2-trifluoro-1-phenyl ethanol $(C_6H_5-C(H)(CF_3)(OH))$ the hydrogen bonded adducts are diastereoisomers in which the chemical shifts of the methyl groups differ by 1.4 Hz. The difference in shift arises from differences in the neighbor anisotropic contributions in the diastereoisomers. These peaks can be integrated and the relative amounts of the *d,l* isomers in the phosphine oxide determined. The optical purity of a resolved product can be ascertained by the absence of one of the peaks.

7–14 EFFECT OF SPIN-SPIN SPLITTING ON THE SPECTRUM

Spin-Spin Splitting

When an nmr spectrum is examined under high resolution, considerable fine structure is often observed. The difference in the high and low resolution spectra of ethanol can be seen by comparing Fig. 7–24 with Fig. 7–15.

The chemical shift of the CH_2 group relative to the CH_3 group in Fig. 7–24 is indicated by Δ measured from the band centers. The fine structure in the CH_3 and CH_2 peaks arises from the phenomenon known as *spin-spin splitting*, and the separation, J, between the peaks comprising the fine structure is referred to as the *spin-spin coupling constant*. J is usually expressed in hertz, and typically J_{HCCH} is about 7 Hz and J_{HOCH} about 5.5 Hz. As mentioned earlier, the magnitude of Δ depends upon the applied field strength. However, the magnitude of the spin-spin coupling constant in Hz is field independent.

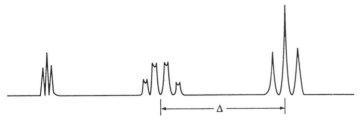

FIGURE 7–24 High resolution nmr spectrum of ethanol (facsimile). Compare with Fig. 7-15, p. 229.

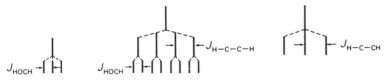

FIGURE 7–25 The interpretation in terms of the appropriate coupling constants of the spectrum in Fig. 7–24. The meaning of this stick-type spectrum will be made clear shortly in the text.

The cause of the fine structure and the reason for the field-independent character of J can be understood by considering an H—D molecule. If by some mechanism the magnetic moment of the proton can be transmitted to the deuteron, the field strength at which the deuteron precesses at the probe frequency will depend upon the magnetic quantum number of the neighboring hydrogen nucleus. If the proton nucleus has a spin of $+1/2$, its magnetic moment is aligned with the field so that the field experienced by the deuterium is the sum of the proton and applied fields. A lower applied field strength, H_0, will be required to attain the precession frequency of the deuterium nucleus in this molecule than in the one in which the hydrogen has a magnetic quantum number of $-1/2$. In the latter case the field from the proton opposes the applied field and must be overcome by the applied field to attain a precessional frequency equal to the probe frequency. The simulated spectrum that would result for the deuterium resonance is indicated in Fig. 7–26(A). The m_I values for the hydrogen nuclei in the different molecules that give rise to different peaks are indicated above the respective peaks. The two peaks are of equal intensity because there is practically equal probability that the hydrogen will have $+1/2$ or $-1/2$ magnetic quantum numbers. The proton resonance spectrum is indicated in Fig. 7–26(B). These proton resonance peaks correspond to magnetic quantum numbers of $+1$, 0, and -1 for the attached deuterium nuclei in different molecules. The spin-spin coupling constants J_{DH} and J_{HD} in Fig. 7–26(A) and 7–26(B), respectively, have the same value. Subsequently, we shall discuss mechanisms for transmitting the magnetic moment of a neighboring atom to the nucleus undergoing resonance.

Returning to ethyl alcohol, we shall examine the splitting of the methyl protons by the methylene protons. The two equivalent protons on the CH_2 group can have the various possible combinations of nuclear orientations indicated by the arrows in Fig. 7–27(A). In case 1, both nuclei have m_I values of $+1/2$, giving a sum of $+1$ and accounting for the low field peak of the CH_3 resonance. Case 2 is the combination of CH_2 nuclear spins that gives rise to the middle peak, and case 3 causes the high field peak. The probability that the spins of both nuclei will cancel (case 2) is twice as great as that of either of the combinations represented by case 1 and case 3. (There are equal numbers of $+1/2$ and $-1/2$ spins.) As a result, the area of the central peak will be twice that of the others (see Fig. 7–24).

The nuclear configurations of the CH_3 group that cause splitting of the CH_2 group are indicated in Fig. 7–27(B). The four different total net spins give rise to four peaks corresponding to the larger separation in this multiplet, shown in Fig. 7–24. The relative areas are in the ratio 1:3:3:1. The separation between these peaks in units of Hz is referred to as $J_{H-C-C-H}$ or as $^3J_{H-H}$. The superscript in the latter symbol indicates H—H coupling between three bonds. The peak separation in the methylene resonance from this coupling is equal to the peak separation in the methyl group. The spectrum of the CH_2 resonance is further complicated by the fact that each of the peaks in the quartet from the methyl splitting is further split into a smaller doublet by the hydroxyl proton. In the actual spectrum, some of the eight peaks expected from this splitting overlap, so they are not all clearly seen.

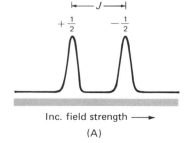
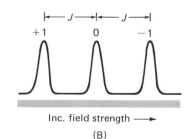

FIGURE 7–26 NMR spectra for the hypothetical HD experiment. (A) Deuterium resonance for HD; (B) proton resonance for HD.

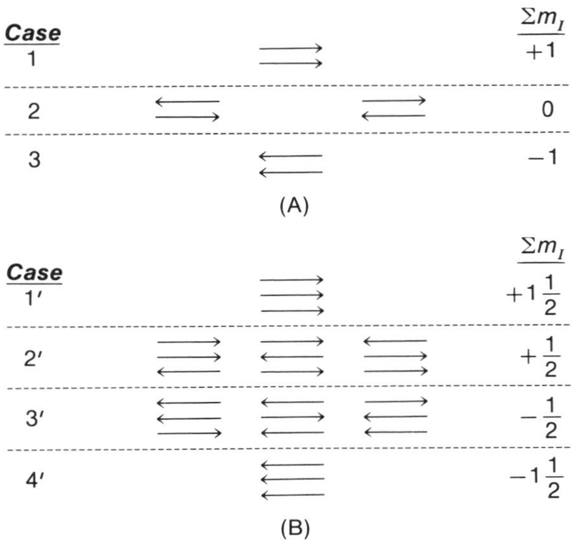

7–14 Effect of Spin-Spin Splitting on the Spectrum   245

Case				Σm_I
1	$\Rightarrow$			+1
2	$\rightleftarrows$	$\leftrightarrows$		0
3	$\Leftarrow$			−1

(A)

Case					Σm_I
1'	$\Rrightarrow$				$+1\frac{1}{2}$
2'	$\rightleftarrows\rightarrow$	$\leftrightarrows\rightarrow$	$\Rightarrow\leftarrow$		$+\frac{1}{2}$
3'	$\Leftarrow\rightarrow$	$\rightleftarrows\leftarrow$	$\leftrightarrows\leftarrow$		$-\frac{1}{2}$
4'	$\Lleftarrow$				$-1\frac{1}{2}$

(B)

FIGURE 7–27 Possible orientations of proton nuclear magnetic moments for (A) —CH$_2$— and (B) —CH$_3$ groups.

The OH peak is split into a triplet by the CH$_2$ protons with the same separation as H_{H-C-OH} in the methylene resonance. Usually, though not always, the effects of spin-spin coupling are not seen over more than three sigma bonds. Accordingly, the interaction of the OH proton with the methyl group is not seen. The spectral interpretation is illustrated in Fig. 7–25 by showing which lines arise from which couplings. This "stick-type spectrum" is constructed by drawing a line for each chemically shifted different nucleus. On the next row, the effect of the largest J is shown. Additional lines are added for each J until the final spectrum is obtained.

Because of the selection rules for this process (*vide infra*), equivalent nuclei do not split each other; *e.g.*, one of the protons in the CH$_3$ group cannot be split by the other two protons. A general rule for determining the splitting pattern can be formulated that eliminates the necessity for going through a procedure such as that in Fig. 7–27. For the general case of the peak from nucleus A being split by a non-equivalent nucleus B, the number of peaks, n, in the spectrum of A is given by the formula

$$n_A = 2\Sigma S_B + 1 \tag{7–50}$$

TABLE 7–4. Spin-Spin Splitting in Various Molecules[a]

Molecule	Groups Being Split (A)	Groups Doing the Splitting (B)	ΣS_B	n
CH$_3$CH$_2$OH	CH$_3$	CH$_2$	1	3
CH$_3$CH$_2$OH	CH$_2$	CH$_3$	$3/2$	4
PF$_3$	P	F	$3/2$	4
PF$_3$	F	P	$1/2$	2
(CH$_3$)$_4$N$^+$	CH$_3$	N	1	3

[a] Spin quantum numbers for the most abundant isotopes of nuclei contained in the above compounds are H = $1/2$, P = $1/2$, F = $1/2$, N = 1.

where ΣS_B equals the sum of the spins of equivalent B nuclei. The application of this formula is illustrated by the examples in Table 7–4. The relative intensities of the peaks can be obtained from the coefficients of the terms that result from the binomial expansion of $(r + 1)^m$, where $m = n - 1$ and r is an undefined variable; e.g., when there are four peaks, $n = 4$ and $m = 3$, leading to $r^3 + 3r^2 + 3r + 1$ from the expansion of $(r + 1)^3$. The coefficients $1:3:3:1$ produce the relative intensities. The Pascal triangle is a convenient device for remembering the coefficients of the binomial expansion for nuclei with $I = {}^1/_2$.

```
                1
              1   1
            1   2   1
          1   3   3   1
        1   4   6   4   1
      1   5  10  10   5   1
    1   6  15  20  15   6   1
```

The triangle is readily constructed, for the sum of any two elements in a row equals the element between them in the row below.

When two groups of non-equivalent nuclei B and C split a third nucleus A, the number of peaks in the A resonance is given by

$$n_A = (2\Sigma S_B + 1)(2\Sigma S_C + 1) \tag{7-51}$$

This is equivalent to what was done in the discussion of the methylene resonance of ethanol. Each of the four peaks from spin coupling by CH_3 was further split into a doublet from coupling to OH, leading to a total of eight peaks. When a nucleus with $I \geqslant 1$ is coupled to the observed nucleus, the number of lines is still given by equation (7–50); however, the intensities are no longer given by the binomial expansion. For instance, in Fig. 7–26(B), the hydrogen signal is split into three lines by the deuterium, for which $I = 1$. The intensities of the three lines are all equal and not in the ratio $1:2:1$, since there is near equal probability that the deuterium will have $+1$, 0, and -1 magnetic quantum numbers. The same situation occurs with splitting from ^{14}N.

For both chemical shift and spin-spin coupling applications, one must recognize when nuclei in a molecule are non-equivalent. This will be the subject of the next section.

7–15 DISCOVERING NON-EQUIVALENT PROTONS

In some molecules, a certain proton appears to be non-equivalent to others in a particular rotamer but becomes equivalent when rapid rotation occurs; e.g., consider the non-equivalence of the three methyl protons in a staggered configuration of CH_3CHCl_2. An interesting problem in which the nonequivalence is not removed even with rapid rotation is observed in the chemical shift of substituted ethanes of the type:[18]

```
       A     X
        \   /
     B—C—C—H
        /   \
       D     H
```

The two protons are not equivalent from symmetry arguments, and *no conformation can be found in which the two protons can be interchanged by a symmetry operation.* This intrinsic asymmetry will result in different chemical shifts which, if large enough, will be observable. This asymmetry can be seen if you label the two hydrogens and make a table indicating which atoms are staggered on each side of the two hydrogens for all possible staggered rotamers of (ABD)C—C(H₂X).

Early interpretations of the nonequivalence in the nmr assigned the different shifts to different rotamer populations. Even in the absence of population differences the protons are intrinsically non-equivalent in all rotamers.

The criterion to be applied in recognizing this phenomenon is the absence, in any conformer that can be drawn, of a symmetry operation that interchanges the two protons. In CH_3CH_2Cl, all the methyl protons are interconvertible in the various rotamers that can be drawn, as are the methylene protons. Another interesting example[19] of non-equivalent protons involves the methylene protons of $(CH_3CH_2O)_2SO$. The two protons of a given methylene group are not stereochemically equivalent because of the lack of symmetry of the sulfur atom with respect to rotation about the S—O—C bonds. One of the possible rotamers is depicted in Fig. 7–28. The small dot in the center represents the sulfur with lines connecting the oxygen, the lone pair, and the ethoxyl groups. The large circle represents the methylene carbon with two hydrogens, an oxygen, and a methyl group attached. The molecule is so oriented that we are looking along a line from sulfur to carbon. The non-equivalence of the two methylene hydrogens of a given CH_2 group is seen in this rotamer, and they cannot be interchanged by a symmetry operation in any other rotamer that can be drawn. Two non-equivalent nuclei, which cannot be interchanged by any symmetry element the molecule possesses, are called *diastereotopic* nuclei.

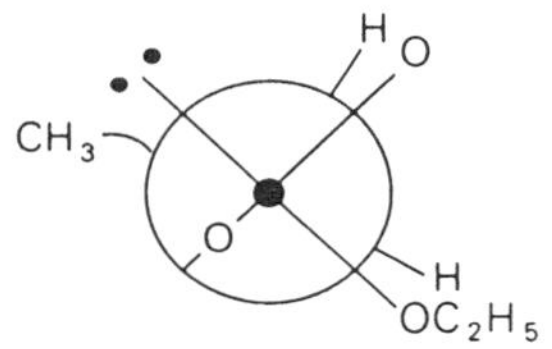

FIGURE 7–28
Non-equivalence of the two protons of a given methylene in one of the rotamers of $(CH_3CH_2O)_2SO$.

7–16 EFFECT OF THE NUMBER AND NATURE OF THE BONDS ON SPIN-SPIN COUPLING

The nature of the spectra of complex molecules will depend on the number of bonds through which spin-spin coupling can be transmitted. For proton-proton coupling in saturated molecules of the light elements, the magnitude of J falls off rapidly as the number of bonds between the two nuclei increases and, as mentioned earlier, usually is negligible for coupling of nuclei separated by more than three bonds. Long-range coupling (coupling over more than three bonds) is often observed in unsaturated molecules. A number of examples are contained and discussed in the literature.[20] In unsaturated molecules, the effects of nuclear spin are transmitted from the C—H σ bond by coupling of the resulting electron spin on carbon from the sigma bond with the π electrons. This π electron spin polarization is easily spread over the whole molecule because of the extensive delocalization of the π electrons. Spin polarization at an atom in the π-system can couple back into the σ-system, in the same way that it coupled into the π-system initially. In this way, the effect of a nuclear spin is transmitted several atoms away from the splitting nucleus. This problem has been treated quantitatively.[21]

When spin-spin coupling involves an atom other than hydrogen, long-range coupling can occur by through-space coupling. This interaction is again one of

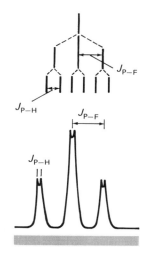

FIGURE 7–29 ^{31}P nmr spectrum expected for HPf$_2$ if $^1J_{P-F} > {}^1J_{P-H}$. (A) Stick interpretation. (B) Spectrum.

spin polarization, but it occurs through non-bonding pairs of electrons, and the coupling is through space instead of through σ or π bonds. In the compound $CF_3CF_2SF_5$ the coupling constant for the fluorines of the CF_2 group and the *trans* fluorine of the SF_5 group is about 5 Hz, much smaller than that for the CF_2 group coupling to the *cis* fluorines (~ 16 Hz), because of the reported contribution from through-space coupling in the latter case.$^{(22a)}$ This effect has been invoked rather loosely, and it is suggested that an abnormal temperature dependence be demonstrated before such an explanation is accepted.$^{(22b)}$

In order to demonstrate one further point, it is interesting to consider the phosphorus nuclear magnetic resonance spectrum of the compound HPF_2. The "stick spectrum" in Fig. 7–29 represents the splitting of the phosphorus signal by the two fluorine atoms followed by the smaller $^1J_{P-H}$ splitting. The resulting spectrum is indicated.

If instead of $^1J_{P-F} > {}^1J_{P-H}$ we have $^1J_{P-H} > {}^1J_{P-F}$, the spectrum in Fig. 7–30 arises. The fact that $^1J_{P-H}$ is larger than $^1J_{P-F}$ is illustrated in the "stick spectrum." The spectrum is expected to be one of those shown in Figs. 7–29 or 7–30, depending upon the magnitude of $^1J_{P-F}$ vs. $^1J_{P-H}$. If the (P—F coupling constant is greater than the P—H coupling constant, Fig. 7–29 will result, whereas Fig. 7–30 results if $^1J_{P-H}$ is much greater than $^1J_{P-F}$. If the two coupling constants were similar, the spectrum would be a complex pattern intermediate between those shown. In most compounds studied, $^1J_{P-F} > {}^1J_{P-H}$ (~ 1500 Hz vs. ~ 200 Hz), and as expected, the spectrum in Fig. 7–29 is found experimentally.

If two nuclei splitting another group in a molecule are magnetically non-equivalent, the spectrum will be very much different from that in which there is splitting by two like nuclei. Four lines of equal intensity will result from splitting by two non-equivalent nuclei with $I = {}^1/_2$, and three lines with an intensity ratio $1:2:1$ will be observed for splitting by two equivalent nuclei. Consequently, we must determine what constitutes magnetically non-equivalent nuclei. Non-equivalence may arise because of differences in the chemical shifts of the two splitting nuclei or because of differences in the J values of the two splitting nuclei with the nucleus being split. Equivalent nuclei have identical chemical shifts and coupling constants to all other nuclei. Isochronous nuclei have the same chemical shift and different couplings. In the molecule $H_2C=CF_2$, where the two hydrogens and the two fluorines are isochronous and non-equivalent. Each proton sees two non-equivalent fluorines (one *cis* and one *trans*) which have identical chemical shifts but different J_{H-F} coupling constants. The simple $1:2:1$ triplet expected when both protons and both fluorines are equivalent is not observed in either the fluorine or the proton nmr because of the non-equivalence that

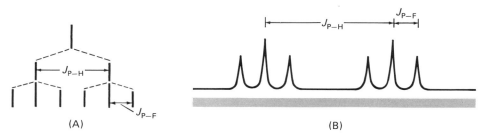

FIGURE 7–30 ^{31}P nmr spectrum expected for HPF_2 if $^1J_{P-H} > {}^1J_{P-F}$. (A) Stick interpretation. (B) Spectrum.

exists in the J values. We shall consider this effect in more detail in the section on second order effects in the next chapter.

Another effect that gives rise to spectra other than those predicted by equation (7–50) is nuclear quadrupole relaxation. Often splittings do not occur because the quadrupolar nucleus to which the element being investigated is attached undergoes rapid relaxation, which causes a rapid change in the spin state of the quadrupolar nucleus. Only the average spin state is detected; in some instances this relaxing quadrupolar nucleus gives rise to very broad lines in the nmr spectrum of the $I = {}^1/_2$ nucleus bonded to it, and sometimes a proton resonance absorption is broadened by this effect to such an extent that the signal cannot be distinguished from the background. We shall discuss this in more detail shortly in Section 7–25.

7–17 SCALAR SPIN-SPIN COUPLING MECHANISMS

In the preceding section the splitting in the spectrum of the HD molecule by the magnetic moment of the attached nucleus was discussed. We shall be concerned here with the mechanism whereby information regarding the spin of the nucleus causing the splitting is transmitted to the nucleus whose resonance is split. Consider a nucleus A split by the nucleus B (which has a spin of $^1/_2$) in the molecule AB. The nuclear spin of B is transmitted to A by polarization of the bonding electrons. Various processes cause polarization, and they constitute the mechanisms for spin-spin splitting of hydrogen in solution. They also contribute to the magnitude of the scalar spin-spin coupling constant, J, in various molecules. Contributions to J are transmitted via the electron density in the molecule and consequently are not averaged to zero as the molecule tumbles. Three contributions will be considered:

1. Spin-orbital effects
2. Dipolar coupling, indirect, or through-space coupling
3. Fermi contact coupling

It is worth emphasizing that all three effects are transmitted via the electron density in the molecule.

Spin-orbital effects involve the perturbation that the nuclear spin moment makes on the orbital magnetic moments of the electrons around the nucleus. For an $I = {}^1/_2$ nucleus which we label B, for example, the magnetic field of the nuclear dipole interacts differently for $m_I = +{}^1/_2$ than for $m_I = -{}^1/_2$, with the orbital magnetic moment of an appropriate electron around B causing a change in the magnetic field from the orbital contribution of these electrons. The new field from the electrons at B produces a field at the nucleus being split, A, that depends upon whether the nuclear moment at B is $+{}^1/_2$ or $-{}^1/_2$. Consequently, a splitting of the nmr resonance of A results. The Hamiltonian for the interaction on B that is felt at A is

$$g_N \beta_N \beta_e \frac{2\hat{L}\cdot\hat{I}}{r^3} \qquad (7\text{–}52)$$

where $\hat{L}$ is the electron orbital angular momentum operator, $\hat{I}$ is the nuclear spin moment operator, and r is the distance from A to B.

Dipolar coupling, often referred to as indirect coupling, is analogous to the classical dipolar interaction of two bar magnets. Since the classical situation is simpler, we shall treat it first. It is essential that you obtain a complete understanding of this interaction, for we shall encounter many phenomena in which this type of interaction is important in the remainder of this text. The classical interaction energy, E, between two magnetic moments (which we shall label $\vec{\mu}_e$ and $\vec{\mu}_N$) is given by:

$$E = \frac{\vec{\mu}_N \cdot \vec{\mu}_e}{r^3} - \frac{3(\vec{\mu}_N \cdot \vec{r})(\vec{\mu}_e \cdot \vec{r})}{r^5} \tag{7-53}$$

where $\vec{r}$ is a radius vector from $\vec{\mu}_e$ to $\vec{\mu}_N$ and r is the distance between the two moments.

The indirect dipolar coupling mechanism corresponds to a polarization of the paired electron density in a molecule by the nuclear moment. The polarization of this electron density depends on whether $m_I = +{}^1/_2$ or $-{}^1/_2$, and the modified electron moment is felt through space by the second nucleus. Replacing $\vec{\mu}_e$ with $-g\beta\hat{S}$ for the electron magnetic moment and $\vec{\mu}_N$ with $g_N\beta_N\hat{I}$ for the nuclear magnetic moment gives the dipolar interaction Hamiltonian:

$$\hat{H} = -g\beta g_N \beta_N \left\{ \frac{\hat{I}\cdot\hat{S}}{r^3} - \frac{3(\hat{I}\cdot\vec{r})(\hat{S}\cdot\vec{r})}{r^5} \right\} \tag{7-54}$$

The interaction between the electron spin moment and the nuclear moment polarizes the spin in the parts of the molecule near the splitting nucleus B. When this effect is averaged over the entire wave function, the field at B is modified and this modified field of the electron density acts directly through space on the nucleus A, which is being split. The direction of the effect depends upon the m_I value of the B nucleus, and thus this effect makes a contribution to J at the A nucleus.

The important point is that this indirect dipolar coupling of the nucleus B and the paired electron density, which modifies the electron moment felt at A, is not averaged to zero by rotation of the molecule. This is illustrated in Fig. 7–31, where an $m_I = +{}^1/_2$ value is illustrated at B. Three different orientations of the molecule are shown. The lines of flux from the moment on B are shown affecting the moment at the electron. The lines of flux associated with the change in the moment at the electron from B are shown at nucleus A. For the three orientations shown, the direction at A is the same.

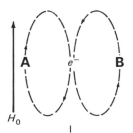 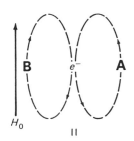 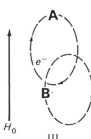

FIGURE 7–31 The indirect dipolar coupling of the nuclear moment B to A for different orientations of the molecule, AB.

The *Fermi contact term* is the final coupling mechanism we shall consider for molecules rapidly rotating in solution. This mechanism involves a direct interaction of the nuclear spin moment with the electron spin moment such that there is increased probability that the electron near B will have spin that is antiparallel to the nuclear spin. Since the electron pair in the bond have their spins paired, a slight increase in the probability of finding an electron of one spin near B will result in there being an increased probability of finding an electron of opposite spin near A, as shown in Fig. 7–32. Thus, A receives information about the spin of nucleus B; since the effect at A is in opposite directions when m_I equals $+1/2$ or $-1/2$, this interaction contributes to the magnitude of J. The effect is a direct interaction of the nuclear spin and the electron spin moment, $a\hat{I} \cdot \hat{S}$, where a is the coupling constant.

FIGURE 7–32 The antiparallel alignment of the nuclear moment (solid arrow) and the electron spin moment (hollow arrow).

If we consider the spin of nucleus B to be quantized along the z-axis for simplicity of presentation, we can describe this effect in more detail. The contact Hamiltonian becomes:

$$\hat{H}_A = \frac{8\pi}{3} g\beta g_N \beta_N \hat{I}_{zB} \hat{S}_{zB} \tag{7-55}$$

where $\hat{S}_{zB}$ has contributions from both electrons 1 and 2 at B given by

$$\hat{S}_{zB} = \hat{S}_{z1}\delta(r_1 - r_B) + \hat{S}_{z2}\delta(r_2 - r_B) \tag{7-56}$$

Here r_B is the radius of the nucleus and r_1 is the distance from nucleus B to the electron. The symbol $\delta(r_1 - r_B)$ is the Dirac delta function and has a value of zero unless electron 1 is at nucleus B. Thus, if *p*-orbitals were used to bond two atoms together, there would be nodes at the nuclei and this term would be zero. Accordingly, in molecules where the Fermi contact term dominates, *e.g.*, $J_{^{13}C-H}$, correlations of J with the amount of *s* character in the bond have been reported.

The mechanism for uncoupling the spins in the σ-bond involves field-induced mixing of ground and excited states. Perturbation theory produces:

$$J = -\left(\frac{8\pi}{3} g\beta g_N \beta_N\right)^2 \sum_n \frac{\langle 0|\hat{S}_{zA}|n\rangle\langle n|\hat{S}_{zB}|0\rangle + \langle 0|\hat{S}_{zB}|n\rangle\langle n|\hat{S}_{zA}|0\rangle}{E_n - E_0} \tag{7-57}$$

In the H_2 molecule, J_{H-H} (estimated from J_{H-D}) has an experimental value of 280 Hz. About 200 Hz comes from the Fermi contact term, 20 from the dipolar contribution, and 3 from the nuclear spin-orbital effect. In general, the Fermi contact contribution dominates most coupling constants involving hydrogen, *e.g.*, $^1J_{^{13}C-H}$, $^2J_{Sn-H}$, etc.

It can also be appreciated that the value of J will simply depend upon the energy difference of the two different kinds of molecules containing B in different spin states. This energy difference will be independent of the field strength. The field strength independence of J provides a criterion for determining whether two peaks in a spectrum are the result of two non-equivalent protons or spin-spin coupling. The peak separation in the spectrum will be different when 60-MHz and 100-MHz probes are employed if the two peaks are due to non-equivalent protons, but the separation will be the same at different frequencies if the two peaks arise from spin-spin splitting.

7-18 APPLICATIONS OF SPIN-SPIN COUPLING TO STRUCTURE DETERMINATION

There have been many applications of spin-spin coupling to the determination of structures. For example, if the spectrum of a sample contains the very characteristic fine structure of the CH_2 and CH_3 resonances of an ethyl group, this is a good indication of the presence of this group in the molecule. Other applications involve variations in the magnitude of J in different types of compounds. For two non-equivalent protons on the same sp^2 carbon, e.g., $ClBrC=CH_2$, proton-proton spin coupling constants, J_{H-H}, of 1 to 3 Hz are observed. Coupling constants for non-equivalent *trans* ethylenic protons have values in the range of 17 to 18 Hz, whereas *cis* protons give rise to coupling constants of 8 to 11 Hz. These differences aid in determining the structures of isomers.

The characteristic chemical shift of hydrogens attached to phosphorus occurs in a limited range, and the peaks have fine structure corresponding to $^1J_{P-H}$ (phosphorus $I = {}^1/_2$). The phosphorus resonance[23] of $HPO(OH)_2$ and $H_2PO(OH)$ is a doublet in the former and a triplet in the latter compound, supporting the structures in Fig. 7-33. The coupling of the hydroxyl protons with phosphorus is either too small to be resolved, or it is not observed because of a fast proton exchange reaction, a phenomenon to be discussed shortly. Similar results obtained from phosphorus nmr establish the structures of $FPO(OH)_2$ and $F_2PO(OH)$ as containing, respectively, one and two fluorines attached to phosphorus. The ^{31}P resonance in P_4S_3 consists of two peaks with intensity ratios of three to one.[24] The more instense peak is a doublet and the less intense is a quadruplet. Since $I = 0$ for ^{32}S, both the spin-spin splitting and the relative intensities of the peaks indicate three equivalent phosphorus atoms and a unique one. It is concluded that P_4S_3 has the structure in Fig. 7-34.

The fluorine resonance in BrF_5 consists of two peaks with an intensity ratio of four to one. The intense line is a doublet and the weak line is a quintet (1:4:6:4:1). Quadrupole relaxation eliminates splitting by bromine (I for $^{79}Br = {}^{81}Br = {}^3/_2$). This indicates that the molecule is a symmetrical tetragonal pyramid.

Solutions of equimolar quantities of TiF_6^{2-} and TiF_4 in ethanol give fluorine nmr spectra[25] consisting of two peaks with the intensity ratio 4 : 1 ($I = 0$ for ^{48}Ti). The low intensity peak is a quintuplet and the more intense peak a doublet. The structure $[TiF_5(HOC_2H_5)]^-$ containing octahedrally coordinated titanium was proposed.

The factors that determine the magnitude of the coupling constant are not well understood for most systems. It has been shown that in the Hamiltonian describing the interaction between the ^{13}C nucleus and a directly bonded proton, the *Fermi contact term* is the dominant one. Qualitatively, this term is a measure of the probability of the bonding pair of electrons existing at both nuclei. The need for this can be appreciated from the earlier discussion of the Fermi contact coupling mechanism. The greater the electron density at both nuclei, the greater the interaction of the nuclear moments with the bonding electrons and hence with each other through electron spin polarization. Since an electron in an *s* orbital has a finite probability at the nucleus and *p*, *d*, and higher orbitals have nodes (zero probability) at the nucleus, the Fermi contact term will be a measure of the *s* character of the bond at the two nuclei.

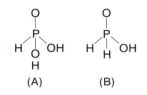

FIGURE 7-33 Structures of (A) phosphorous and (B) hypophosphorous acids.

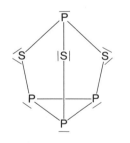

FIGURE 7-34 The structure of P_4S_3.

Since the s orbital of hydrogen accommodates all of the proton electron density, the magnitude of $^1J_{^{13}C-H}$ for directly bonded carbon and hydrogen will depend upon the fraction of s character, ρ, in the carbon hybrid orbital bonding the hydrogen. The following equation permits calculation of ρ from $^1J_{^{13}C-H}$ data.

$$^1J_{^{13}C-H}(\text{Hz}) = 500\rho_{C-H} \qquad (7\text{-}58)$$

Since $\rho = 0.33$ for an sp^2 hybrid and 0.25 for an sp^3 hybrid, it can be seen that $^1J_{^{13}C-H}$ should be a sensitive measure of carbon hybridization. The use of $^1J_{^{13}C-H}$ to measure the hybridization of carbon in a C—H bond was studied in detail[26] and is supported by valence bond calculations.[27] As expected from equation (7-58) the ^{13}C—H coupling constant for a hydrogen in a saturated hydrocarbon is about 125 Hz, that for the ethylenic hydrogen of a hydrocarbon is around 160 Hz, and that for an acetylenic hydrogen is around 250 Hz, corresponding to ρ's of 0.25, 0.33, and 0.50, respectively.

It has been shown that a linear relationship exists between $^1J_{^{13}C-H}$ and the proton shift, δ, for a series of methyl derivatives in which the contribution to τ from magnetic anisotropy is approximately constant,[28, 29] or varies with the inductive properties of the group attached to methyl. Those compounds that deviate from this line have appreciable contributions to δ from anisotropy. The following explanation is offered for the existence of this relationship. Isovalent hybridization arguments[30] indicate that as X becomes more electron withdrawing in the compounds CH$_3$X, more p character is employed in the C—X orbital, and there is a corresponding increase in the s character in the C—H orbitals. Carbon becomes more electron withdrawing toward hydrogen as the hybridization changes from sp^3 to sp^2 to sp.[31] If there are no anisotropic contribution, a correlation is expected between *carbon* hybridization and δ. Since carbon hybridization, is in turn related to $^1J_{^{13}C-H}$, a correlation is expected between $^1J_{^{13}C-H}$ and δ for compounds in which the contributions to δ from magnetic anisotropy are constant (or vary linearly with the electron-withdrawing properties of group X). In view of the difficulty of assessing the existence of even large anisotropic contributions to the chemical shift (see section on chemical shifts which are greatly influenced by anisotropy), this relationship is of considerable utility. Certainly the chemical shifts of compounds that do not fall on the line cannot be interpreted in terms of inductive arguments, *i.e.*, a local diamagnetic effect.

The constancy of the $^1J_{^{13}C-H}$ coupling constants of the alkyl halides (149, 150, 152, and 152 Hz respectively for F, Cl, Br, and I) is interesting to note. Even though there is a pronounced change in the electron withdrawal by F, Cl, Br, and I, $^1J_{^{13}C-H}$ does not change[26] as would be predicted from the isovalent hybridization argument. In this series, there is an effect on the carbon hybridization that is opposite in direction to electron withdrawal by the halogen. These two effects cause the hybridization and electron-withdrawing properties of carbon toward hydrogen to remain constant. This other effect is based on orbital overlap, and one should think in terms of the s orbital dependence of the bond strength for the three hydrogens and X. As X becomes larger, the orbital used in sigma bonding becomes more diffuse and the overlap integral becomes smaller. Consequently, the energy of the C—X bond becomes less sensitive to the amount of s-orbital used to bond X, and the total energy of the system is lowered by using more of the s orbital in the bonds to hydrogen. Thus, as the principal quantum number of the atom bonded to carbon increases, we expect to see, from this latter effect, an increase in $^1J_{^{13}C-H}$.

It has been shown that the carbon hybridization in $Si(CH_3)_4$, $Ge(CH_3)_4$, $Sn(CH_3)_4$, and $Pb(CH_3)_4$ changes in such a way as to introduce more s character into the C—H bond as the atomic number of the central element increases. An explanation for this change in terms of changing bond energies is proposed, and it is conclusively demonstrated that chemical shift data for these compounds cannot be correlated with electronegativities of the central element,[29] because large anisotropic contributions to the proton chemical shift exist in $Ge(CH_3)_4$, $Sn(CH_3)_4$, and $Pb(CH_3)_4$.

The ^{13}C—H coupling constant in the carbonium ions $(CH_3)_2CH^+$ and $(C_6H_5)_2CH^+$ are 168 and 164 Hz, respectively, compared to values of 123 and 126 Hz for propane and $(C_6H_5)_2CH_2$. This is consistent with a planar sp^2 hybridized carbon.

The magnitude of the coupling constant between two hydrogens on adjacent carbons, $^3J_{H-H}$, is a function of the dihedral angle between them[32-34] as shown in Fig. 7–35. The dihedral angle θ is shown at the top of Fig. 7–35, where the C—C bond is being viewed end-on with a dot representing the front carbon and the circle representing the back one. The function is of the form:

$$^3J_{H-H} = A + B \cos \theta + C \cos 2\theta \qquad (7\text{–}59)$$

For hydrocarbons, A is 7 Hz, B is 1 Hz, and C is 5 Hz. The end values change in different systems, but the shape of the curve is the same. Difficulties are encountered with electron-withdrawing substituents and with systems in which there are changes in the C—C bond order.

The $^1J_{^{13}C-H}$ values can be of use in assigning peaks in a spectrum.[35] For example, the proton chemical shifts of the two methyl resonances in $CH_3C(O)SCH_3$ are very similar, as shown in Fig. 7–36, and impossible to assign directly to methyl groups in the compound. However, by virtue of the fact that sulfur is a larger atom than carbon and has a comparable electronegativity, the

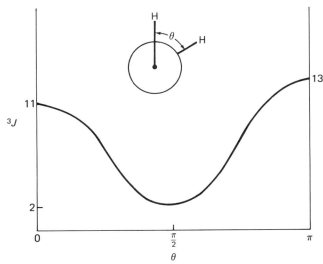

FIGURE 7–35 variation in $^3J_{H-H}$ with the dihedral angle θ shown at the top.

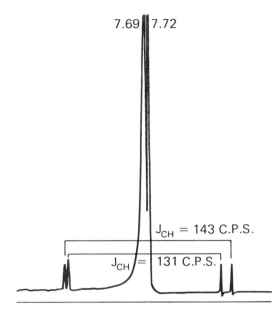

FIGURE 7–36 Proton nmr spectrum of S-methyl thioacetate at high spectrum amplitude showing satellites arising from ^{13}C. [From R. L. Middaugh and R. S. Drago, J. Amer. Chem. Soc., *85*, 2575 (1963).]

$^1J_{^{13}C-H}$ values for methyl groups attached to sulfur [$J = 138$ to 140 Hz in $CH_3SH, (CH_3)_2SO, (CH_3)_2SO_2$] are larger than those for methyl groups attached to carbonyls ($J = 125$ to 130 Hz). Since the proton resonance on a ^{12}C will be in the center of the ^{13}C satellites, the less shielded peak is unequivocally assigned[35] to the acetyl methyl.

It has been proposed[36,37] that a relationship exists between J_{Sn-H} and the hybridization of tin in the tin-carbon bonds of compounds of the type $CH_{34-n}SnX_n$. This relationship and results from other physical methods were employed to establish the existence and structure of five-coordinate tin addition compounds $(CH_3)_3SnCl \cdot B$,[37] where B is a Lewis base [*e.g.*, $(CH_3)_2SO$ or $CH_3C(O)N(CH_3)_2$]. The $^2J_{Sn-H}$ values suggest a trigonal bipyramidal geometry in which the tin employs essentially sp^2-orbitals in bonding to carbon, and, consequently, a three-center molecular orbital using a tin p or p–d orbital in bonding to the Lewis base and chlorine. Similar results were obtained for $^2J_{Pb-H}$ in the analogous lead compounds.[38]

Investigation[39] of a whole series of addition compounds between $(CH_3)_3SnCl$ and various bases indicated that the $^2J_{Sn-H}$ coupling constant changed in direct proportion to $-\Delta H$ of adduct formation with the base. This was interpreted to indicate that as the tin-base bond became stronger, the hybrids used to bond the methyl groups approached sp^2 from the $\sim sp^3$ hybrids used in $(CH_3)_3SnCl$. The relation found was:

$$^2J_{^{119}Sn-H} = 216\, \rho_{Sn} \qquad (7-60)$$

The $^2J_{Sn-H}$ values for $Sn(CH_3)_4$, $Sn(CH_3)_3Cl$, and $Sn(CH_3)_3Cl_2^-$ were 54, 57.6 (in keeping with the isovalent hybridization prediction), and 72 [which when substituted into equation (7–60) gives $\rho_{Sn} = 0.33$]. The coupling constants for the adducts ranged from 64.2 to 71.6; the weakest base studied was CH_3CN, and the strongest was $[(CH_3)_2N]_3PO$. The plot of $-\Delta H$ vs. $^2J_{Sn-H}$ was extrapolated to the value for free trimethyltin chloride at $-\Delta H = 0$.

Coupling constants of equivalent hydrogens cannot be determined directly. Deuteration of a molecule has only a slight effect on the molecular wave function, so this technique can be used to gain information about the coupling of equivalent protons. For example, the H—H coupling in H_2 cannot be directly measured, but that in HD is shown to be 45.3 Hz. Since the effect of isotopic substitution on the magnitude of J is proportional to the magnetogyric ratio, we have

$$\frac{J_{HH}}{J_{HD}} = \frac{\gamma_H}{\gamma_D} = 6.515 \tag{7-61}$$

From this, we calculate $^1J_{H-H} = 295.1$ Hz.

There have been several very interesting applications of phosphorus nmr in the determination of structures of complexes of phosphorus ligands.[40] The nmr spectrum and its interpretation are illustrated in Fig. 7–37 for the complex $Rh(\varphi_3P)_3Cl_3$ [I for ^{103}(Rh) = $^1/_2$]. Two isomers are possible, facial and meridional. All of the phosphorus ligands are equivalent in the facial isomer, so the spectrum in Fig. 7–37 substantiates that the complex studied was meridional. In this isomer two phosphorus atoms are *trans* to one another (labeled P_b), and one is *trans* to a chlorine (labeled P_a). The splittings are interpreted in the stick diagram at the top of Fig. 7–37.

The magnitude of $^2J_{P-P}$ often provides interesting information about the stereochemistry of complexes. The magnitude of this coupling is usually much larger when two phosphorus atoms are *trans* to one another than when they are

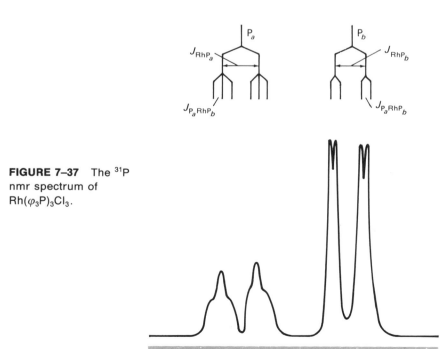

FIGURE 7–37 The ^{31}P nmr spectrum of $Rh(\varphi_3P)_3Cl_3$.

cis. In *cis*-PtCl$_2$(bu$_3$P)[(C$_6$H$_5$O)$_3$P], the value of $^2J_{P-P}$ is 20 Hz, while a value of 758 Hz is found in *trans*-PdI$_2$(bu$_3$P)[(C$_6$H$_5$O)$_3$P] and 565 Hz is found in *cis*-PdI$_2$[(CH$_3$)$_3$P][(C$_2$H$_5$)$_3$P]. Some exceptions to this rule are found in Cr^0, Mo^0, W^0, and Mn(I) compounds. For example, in *trans*-W(CO)$_4$[(C$_6$H$_5$)$_3$P](bu$_3$P), $^2J_{P-P}$ is found to be 65 Hz.

A very interesting result is obtained in the proton or carbon-13 nmr spectra of phosphorus complexes. In the proton nmr of *trans*-PdI$_2$[P(CH$_3$)$_3$]$_2$, we might expect to find a doublet methyl resonance with perhaps a small splitting of each peak from the second phosphorus. Although the two phosphorus atoms are *chemically equivalent* (*isochronous*) they are not *magnetically equivalent*. Any given methyl group would experience two different phosphorus couplings, $^2J_{P-H}$ and $^4J_{P-H}$. When the $^2J_{P-P}$ is much larger than $^2J_{P-H}$, as it is in the above complex, the proton nmr spectrum observed is a 1:2:1 triplet; *i.e.*, the two phosphorus nuclei behave as though they were two equivalent nuclei splitting the proton resonance. The two phosphorus nuclei are said to be *virtually coupled*.[41] In a similar *cis* complex where $^2J_{P-P} \ll {}^2J_{P-H}$, a doublet is obtained. In view of the very common, large $^2J_{P-P}$ for *trans* phosphorus ligands and the small $^2J_{P-P}$ values for *cis* phosphines, this phenomenon can be used to distinguish these two kinds of isomers. The use of ^{13}C nmr in this type of application[42] has been criticized.[42] Triplets occur in the ^{13}C nmr with much smaller values of J_{P-P}, so it is more probable that *cis* complexes will show virtual coupling. The reasons for virtual coupling have been discussed in the literature and will be considered in more detail in the section on second order effects. Virtual coupling is an attempt to treat second-order spectra in a first-order way. The second-order analysis should be done.

7–19 EFFECT OF FAST CHEMICAL REACTIONS ON THE SPECTRUM

Factors Influencing the Appearance of the NMR Spectrum

If one examines the high resolution spectrum of ethanol in an acidified solution, the result illustrated in Fig. 7–38 is obtained, in contrast to the spectrum shown in Fig. 7–24. The difference is that the spin-spin splitting from the hydroxyl proton has disappeared. Acid catalyzes a very rapid exchange of the hydroxyl proton. In the time it takes for a methylene proton to undergo resonance, many different hydrogen nuclei have been attached to the oxygen. As a result, the

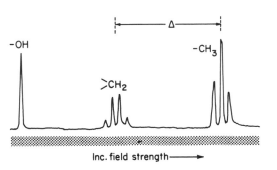

FIGURE 7–38 High resolution nmr spectrum of a sample of ethanol (acidified).

methylene proton experiences a field averaged to zero from the O—H nuclear moment, and the $^3J_{\text{HCOH}}$ coupling disappears. In a similar fashion, the hydroxyl proton is attached to many different ethanol molecules, averaging to zero the field it experiences from CH_2 protons, and only a single resonance is observed.

A very dramatic illustration of this effect is the spectrum of a solution of aqueous ammonia in which one does not see separate N—H and O—H protons, but only a single exchange-averaged line. When exchange is rapid, the chemical shift of this exchange-averaged line is found to be a mole-fraction-weighted average of the shifts of the different types of protons being exchanged:

$$\delta_{\text{AVG}} = N_{\text{NH}_3}\delta_{\text{NH}_3} + N_{\text{H}_2\text{O}}\delta_{\text{H}_2\text{O}} \qquad (7\text{–}62)$$

It is important to emphasize that N_{NH_3} is not the mole fraction of ammonia, but the mole fraction of N—H protons, *i.e.*, $N_{\text{NH}_3} = 3[\text{NH}_3]/(3[\text{NH}_3] + 2[\text{H}_2\text{O}])$. The ^{19}F spectra of solutions of TiF_4 in donor solvents taken at $-30°$ C consist of two triplets of equal intensity.[43] Six-coordinate complexes form by coordinating two solvent molecules, and the spectrum obtained is that expected for the *cis* structure. This structure contains two sets of non-equivalent fluorine atoms, with two equivalent fluorines in each set. At $0°$ C only a single fluorine peak is obtained. It is proposed that a rapid dissociation reaction occurs at $0°$ C, making all fluorines equivalent;

$$TiF_4 \cdot 2B \rightleftharpoons TiF_4B + B$$

At $-30°$ C this reaction is slowed down so that the non-equivalence can be detected by nmr. Internal rearrangements and ionic exchange mechanisms are also possible.

This example illustrates one of the possible pitfalls in structure determination using nmr spectroscopy. If only the high temperature ($0°$ C) spectrum had been investigated or if rapid exchange occurred at $-30°$ C, it could have been incorrectly assumed that the adduct had the *trans* structure on the basis of the single nmr peak. If the actual structure were *trans*, only a singlet fluorine resonance would be detected at all temperatures and it would have been difficult to draw any structural conclusion because of the possibility that rapid exchange might be occurring at both temperatures. Even in the present case, the possibility exists, on the basis of these data alone, that the *cis* isomer is the structure at $-30°$ C and that the *trans* isomer predominates at $0°$ C.

The fluorine nmr spectra of a large number of compounds of general formula $R_{5-n}PF_n$ (where R is a hydrocarbon, fluorocarbon, or halide other than fluorine) have been reported.[44] The number of peaks in the spectrum and the magnitude of the coupling constants are employed to deduce structures. For a series of compounds of the type R_2PF_3, a trigonal bipyramidal structure is proposed, and it is found that the $J_{\text{P–F}}$ values for axial fluorines are ~ 170 Hz less than those for equatorial ones. The most electronegative groups are found in the axial positions. The spectra and coupling constants, obtained on these compounds at low temperature, indicate that the two methyl groups in $(CH_3)_2PF_3$ are equatorial and the two trifluoromethyl groups in $(CF_3)_2PF_3$ are axial. At room temperature, rapid intramolecular exchange occurs and the effect of this exchange is to average the coupling constants.

7-20 QUANTUM MECHANICAL DESCRIPTION OF COUPLING

When the magnitude of the separation between two peaks in the nmr, expressed in Hz, is of the same order of magnitude as the coupling constant, so-called *second-order spectra* result. When this occurs, the peaks in the resulting spectrum cannot be assigned by inspection as we have done before. This is illustrated in Fig. 7–41, where the spectra of ClF_3, obtained by using two different probes, demonstrate these complications.

With a higher-frequency probe (40 MHz), $J \ll \Delta$ and the spectrum in Fig. 7–39(A) is obtained. The molecule ClF_3 has two long Cl—F bonds and one short one, giving rise to non-equivalent fluorines. The spectrum obtained at 40 MHz is that expected for non-equivalent fluorines splitting each other. As expected, the triplet is half the intensity of the doublet. Using a lower-frequency probe (10 MHz and the corresponding lower magnetic field), the difference in Δ for the non-equivalent fluorines is of the order of magnitude of J_{F-F} (recall that Δ is field dependent) and the complex spectrum in Fig. 7–39(B) is obtained. Complex patterns of this sort result whenever the coupling constants between non-equivalent nuclei are of the order of magnitude of the chemical shift.

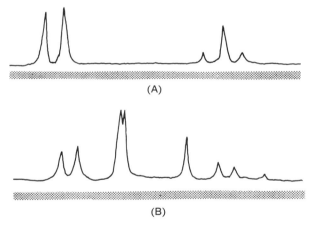

FIGURE 7–39 Fluorine nmr spectra of CiF_3. (A) at 40 MHz and (B) at 10 MHz.

When second-order spectra result, other strategies than going to a higher field can be used to obtain the chemical shifts and coupling constants. Our discussion will involve protons, but the treatment is perfectly general for any spin $1/2$ nucleus.

The energy for a proton, A, surrounded by paired electron density in a magnetic field is given by:

$$E = -g_N \beta_N H_0 (1 - \sigma_A) m_I$$

or the frequency of the transition, v, is given by:

$$v = v_0 (1 - \sigma_A) |\Delta m_I| \qquad (7\text{-}63)$$

where v_0 is the resonance frequency for a bare proton and σ_A the shielding constant.

Next consider a molecule containing two different hydrogens that are involved in spin-spin coupling. We shall focus attention only on the two hydrogens and ignore the rest of the molecule. When the chemical shift difference of the two hydrogens is very large compared to their J, we shall label such a system AX. The energy of a system of X_j protons whose shift differences are larger than J is given by equation (7–64).

$$E = -h \sum_j v_0(1 - \sigma_j) m_j + h \sum_{j<k} J_{jk} m_j m_k \tag{7-64}$$

The first summation on the right gives a different chemical shift term for each of the j different types of nuclei in the molecule. The second summation is taken only for pairs of nuclei in which $j < k$, to insure that each pair is considered only once. It is assumed in writing this equation that J is isotropic. The significance of equation (7–64) is illustrated in Fig. 7–40, where the energies of the various nuclear configurations of an AX system in a magnetic field are illustrated for the case in which J_{AX} is zero and for the case in which it is finite.

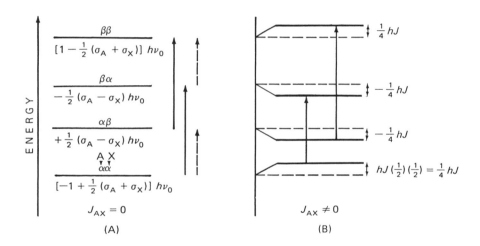

FIGURE 7–40 Energies for the AX system in a magnetic field [$I(A) = {}^1/_2$, $I(X) = {}^1/_2$]. (A) $J_{AX} = 0$. the wave function, written in the order AX, is written above the energy level; the energy is written below the level. (B) $J_{AX} \neq 0$. The J labeling indicates the change in energy that occurs from the dotted line, which represents the energy when $J_{AX} = 0$.

In Fig. 7–40A, where $J_{AX} = 0$, the energy levels for the $\alpha\alpha$, $\alpha\beta$, $\beta\alpha$, and $\beta\beta$ nuclear spin configurations are shown. The A nuclear spin state is listed first and that of the X nucleus second. The $\alpha\alpha$ energy is the sum of

$$-h[v_0(1 - \sigma_A)(^1/_2) + v_0(1 - \sigma_X)(^1/_2)] = [-1 + (\sigma_A + \sigma_X)(^1/_2)]hv_0$$

The energies of the other levels are calculated similarly by substituting the appropriate m_I values. The solid arrows indicate the transitions for the A type of nucleus. One arrow corresponds to A nuclei bonded to X nuclei with α spin and the other to those on X nuclei with β spin. The section rule is $\Delta m_I = 1$, so the A and X nuclear spins cannot change simultaneously. The dashed arrows indicate the transitions of the X nucleus. By referring to equation (7–64) and Fig. 7–40, we see that the frequency difference between $\alpha\alpha$ and $\beta\alpha$ ($v_0 - \sigma_A v_0$) is equal to the difference between $\alpha\beta$ and $\beta\beta$; i.e., the two transitions of nucleus A are degenerate. Thus without any coupling, a single peak would be observed for A. By similar reasoning, we expect a single peak for X.

Where J_{AX} is finite, and considerably smaller than the chemical shift difference of A and X, the $J_{jk}m_jm_k$ terms modify the energies shown in Fig. 7–40A, producing those shown in (B). Since the $\alpha\alpha$ energy is raised by $(^1/_4)Jh$ and $\beta\alpha$ is lowered by $(^1/_4)Jh$, the $\alpha\alpha \to \beta\alpha$ transition occurs at a frequency $J/2$ lower than the corresponding transition in the $J_{AX} = 0$ case. The $\alpha\beta \to \beta\beta$ transition is at a frequency $J/2$ higher than that of the $J_{AX} = 0$ transition. Thus, the frequency separation between the two peaks is equal to J in the $J_{AX} \neq 0$ case.

Spin-spin coupling constants can be positive or negative. In the above problem, a positive J was assumed. If J was taken as negative, the $\alpha\alpha$ energy would be lowered and the $\beta\alpha$ energy would be raised. This would shift the peak for the $\alpha\alpha \to \beta\alpha$ transition to a frequency $J/2$ higher than in the $J_{AX} = 0$ case, whereas the other transition would occur at a frequency $J/2$ lower. (A greater frequency corresponds to a lower field.) There would be no change in the appearance of the spectrum, so there would be no way to determine the sign of J as illustrated by referring to Fig. 7–41. In Fig. 7–41(A), line I represents the chemical shift difference between A and X and line II shows the spin-spin splitting of A by X when J_{AX} is positive. If J_{AX} were negative the result in Fig. 7–42(B) would be obtained. One cannot distinguish the two possibilities by examining the spectrum obtained in the normal nmr experiment.

Equation (7–64) is the so-called first-order solution of the chemical shift-coupling constant problem. It applies only when the chemical shift difference, Δ, of the two nuclei is large compared to their coupling constant, J (usually $\Delta \sim 5J$). We now treat the more general problem using a system where Δ and J have comparable magnitudes. The label AB will be used to describe this case, where $I(A) = {}^1/_2$ and $I(B) = {}^1/_2$. We shall solve for the energies using:

$$\int \psi^* \hat{H} \psi \, d\tau = \frac{E}{h} \int \psi^* \psi \, d\tau$$

With E/h being a frequency, the Hamiltonian is written in frequency units as:

$$\hat{H} = -v_0(1 - \sigma_A)\hat{I}_{ZA} - v_0(1 - \sigma_B)\hat{I}_{ZB} + J_{AB}\hat{I}_A \cdot \hat{I}_B \quad (7\text{–}65)$$

Our basis set will be the nuclear spin functions $\varphi_1 = |\alpha\alpha\rangle$, $\varphi_2 = |\alpha\beta\rangle$, $\varphi_3 = |\beta\alpha\rangle$, and $\varphi_4 = |\beta\beta\rangle$. To obtain the four energies, we need to solve for all the matrix elements in the secular determinant:

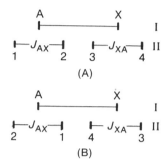

FIGURE 7–41 Effect of the sign of J_{AX} on the splitting of an A—X system. Here 1 refers to the $\alpha\alpha$ to $\beta\alpha$ transition, and 2 refers to the $\alpha\beta$ to $\beta\beta$ transition.

$$\begin{vmatrix} & |\alpha\alpha\rangle & |\alpha\beta\rangle & |\beta\alpha\rangle & |\beta\beta\rangle \\ |\alpha\alpha\rangle & \langle\alpha\alpha|\hat{H}|\alpha\alpha\rangle - \frac{E}{h} & \langle\alpha\alpha|\hat{H}|\alpha\beta\rangle & \langle\alpha\alpha|\hat{H}|\beta\alpha\rangle & \langle\alpha\alpha|\hat{H}|\beta\beta\rangle \\ |\alpha\beta\rangle & \langle\alpha\beta|\hat{H}|\alpha\alpha\rangle & \langle\alpha\beta|\hat{H}|\alpha\beta\rangle - \frac{E}{h} & \langle\alpha\beta|\hat{H}|\beta\alpha\rangle & \langle\alpha\beta|\hat{H}|\beta\beta\rangle \\ |\beta\alpha\rangle & \langle\beta\alpha|\hat{H}|\alpha\alpha\rangle & \langle\beta\alpha|\hat{H}|\alpha\beta\rangle & \langle\beta\alpha|\hat{H}|\beta\alpha\rangle - \frac{E}{h} & \langle\beta\alpha|\hat{H}|\beta\beta\rangle \\ |\beta\beta\rangle & \langle\beta\beta|\hat{H}|\alpha\alpha\rangle & \langle\beta\beta|\hat{H}|\alpha\beta\rangle & \langle\beta\beta|\hat{H}|\beta\alpha\rangle & \langle\beta\beta|\hat{H}|\beta\beta\rangle - \frac{E}{h} \end{vmatrix} = 0$$

The term $-E/h$ appears on the diagonal because only the diagonal $[-(E/h)\langle\varphi_n|\varphi_n\rangle]$ terms are non-zero; for example,

$$-(E/h)\langle\alpha\alpha|\alpha\alpha\rangle = -E/h$$

while

$$-(E/h)\langle\alpha\beta|\alpha\alpha\rangle = 0$$

The solution of this secular determinant for the four energies is shown in Appendix G. The four energies are given by:

$$\frac{E_1}{h} = v_0[-1 + (^1/_2)\sigma_A + (^1/_2)\sigma_B] + \frac{J}{4} \qquad (7\text{–}66a)$$

$$\frac{E_2}{h} = -(^1/_4)J - C \qquad (7\text{–}66b)$$

where $C = (^1/_2)(J^2 + \Delta^2)^{1/2}$

$$\frac{E_3}{h} = -(^1/_4)J + C \qquad (7\text{–}66c)$$

$$\frac{E_4}{h} = v_0[1 - (^1/_2)\sigma_A - (^1/_2)\sigma_B] + \frac{J}{4} \qquad (7\text{–}66d)$$

The corresponding wave functions are listed below.

$$\psi_1 = |\alpha\alpha\rangle \qquad (7\text{–}67a)$$

$$\psi_2 = \cos\theta|\alpha\beta\rangle - \sin\theta|\beta\alpha\rangle \qquad (7\text{–}67b)$$

$$\psi_3 = \sin\theta|\alpha\beta\rangle + \cos\theta|\beta\alpha\rangle \qquad (7\text{–}67c)$$

$$\psi_4 = |\beta\beta\rangle \qquad (7\text{–}67d)$$

We can summarize the results by constructing an energy diagram for this AB system in Fig. 7–42, as was previously done in Fig. 7–40 for the AX system. In Fig. 7–42(A), the situation with $J_{AB} = 0$ (which is identical to Fig. 7–40(A)) is presented as a starting place. The contribution to the total energy of the various levels from J_{AB} is shown in Fig. 7–42(B).

The arrows for the four transitions are indicated in Fig. 7–42. When A and B have similar chemical shifts, all four transitions will appear in a narrow region of the spectrum. The center of the spectrum is given by the average of the $E_1 \to E_2$ and $E_2 \to E_4$ transition energies or:

$$(^1/_2)(E_2 - E_1 + E_4 - E_2) = \frac{E_4 - E_1}{2} = v_0[1 - (^1/_2)\sigma_A - (^1/_2)\sigma_B]$$

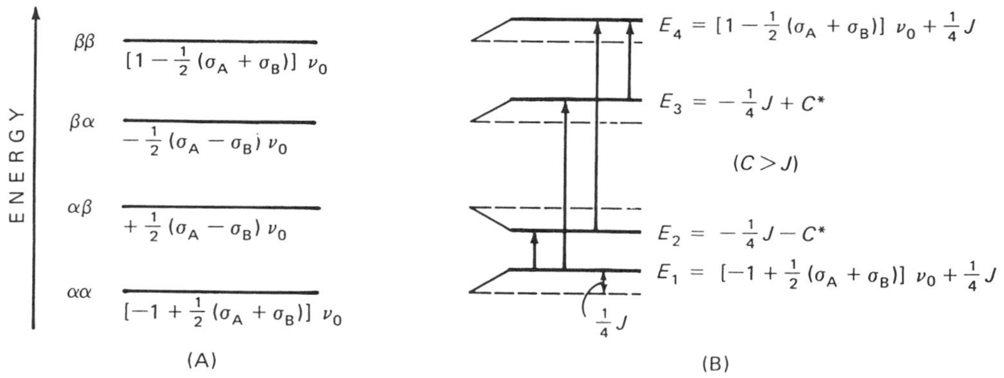

7-21 Effects of the Relative Magnitudes of J and Δ on the Spectrum of an AB Molecule

FIGURE 7–42 The second-order energies (written in frequency units) for an AB spin system. (A) $J_{AB} = 0$. (B) The second-order result. (*These are the total energies, where the quantity C contains the shielding constant terms.)

Energy levels shown:

(A) $J_{AB} = 0$:
- $\beta\beta$: $[1 - \frac{1}{2}(\sigma_A + \sigma_B)]\nu_0$
- $\beta\alpha$: $-\frac{1}{2}(\sigma_A - \sigma_B)\nu_0$
- $\alpha\beta$: $+\frac{1}{2}(\sigma_A - \sigma_B)\nu_0$
- $\alpha\alpha$: $[-1 + \frac{1}{2}(\sigma_A + \sigma_B)]\nu_0$

(B) Second-order result ($C > J$):
- $E_4 = [1 - \frac{1}{2}(\sigma_A + \sigma_B)]\nu_0 + \frac{1}{4}J$
- $E_3 = -\frac{1}{4}J + C^*$
- $E_2 = -\frac{1}{4}J - C^*$
- $E_1 = [-1 + \frac{1}{2}(\sigma_A + \sigma_B)]\nu_0 + \frac{1}{4}J$

The transition energies relative to this center can then be found using the energies given above and illustrated in Fig. 7–42(B). The results are presented in Table 7–5 along with the intensities of the various transitions. The intensities are dependent on $\hat{I}_X$ (recall that we discussed the fact that H_1 made the $\hat{I}_X$ matrix elements non-zero) and are determined from the evaluation of integrals of the form

$$|\langle \psi_j | \hat{I}_{XA} + \hat{I}_{XB} | \psi_i \rangle|^2 \qquad (7\text{–}68)$$

For example, the $1 \to 3$ (*i.e.*, E_1 to E_3) transition yields the result:

$$|\langle \alpha\alpha | \hat{I}_{XA} + \hat{I}_{XB} | \beta\alpha \cos\theta + \alpha\beta \sin\theta \rangle|^2 = (1/4)(\cos\theta + \sin\theta)^2$$
$$= (1/4)(1 + \sin 2\theta)$$

7-21 EFFECTS OF THE RELATIVE MAGNITUDES OF J AND Δ ON THE SPECTRUM OF AN AB MOLECULE

We are now in a position to use the results summarized in Table 7–5 to see what influence the relative magnitudes of Δ and J have on the appearance of the spectrum.

TABLE 7–5. Energies and Intensities for the Transitions in an AB Molecule

Transition	Separation from Center	Relative Intensity
$1 \to 2$	$-(J/2) - C$	$1 - \sin 2\theta$
$1 \to 3$	$-(J/2) + C$	$1 + \sin 2\theta$
$2 \to 4$	$(J/2) + C$	$1 - \sin 2\theta$
$3 \to 4$	$(J/2) - C$	$1 + \sin 2\theta$

where $\sin 2\theta = J/2C$ and $C = (1/2)(J^2 + \Delta^2)^{1/2}$

A. $J = 0$ and $\Delta \neq 0$

The $1 \to 2$ and $3 \to 4$ transitions are degenerate, as are $1 \to 3$ and $2 \to 4$. The spectrum consists of two peaks, one from A and one from B, as in the earlier discussion of Fig. 7–40(A).

B. $J \neq 0$, $\Delta \neq 0$, but $J \ll \Delta$—an AX System

When Δ dominates C, we have $\sin 2\theta = J/\Delta \approx 0$. With the sin 2θ equal to zero, all four transitions have equal intensity. The energies become those shown in Fig. 7–40(B). Relative to the center, we have $1 \to 2$ at $(-1/2)J - (1/2)\Delta$, $1 \to 3$ at $(-1/2)J + (1/2)\Delta$, $2 \to 4$ at $(1/2)J + (1/2)\Delta$, and $3 \to 4$ at $(1/2)J - (1/2)\Delta)$ for a spectrum that looks like that shown in Fig. 7–43.

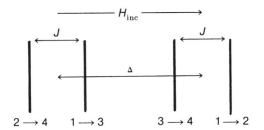

FIGURE 7–43 Spectrum of an AX system. (Greater frequency corresponds to lower field.)

C. $\Delta \neq 0$ and $J \neq 0$, but $J \approx \Delta$

When $J \approx \Delta$, sin 2θ is appreciable and positive. Therefore, the $1 \to 3$ and $3 \to 4$ transitions will have equal intensities which are greater than those of $1 \to 2$ and $2 \to 4$. The $1 \to 2$ and $3 \to 4$ transitions will differ by J, as will the $2 \to 4$ and $1 \to 3$; also, the $3 \to 4$ and $2 \to 4$ will differ by $2C$, as will the $1 \to 3$ and $1 \to 2$. The spectrum will look like that in Fig. 7–44.

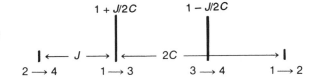

FIGURE 7–44 Spectrum of an AB system.

D. $J \neq 0$, Δ very small ($\Delta < J$)

When σ_A and σ_B are almost equal, the value of sin 2θ approaches unity and the $1 \to 2$ and $2 \to 4$ transitions can be easily lost in the baseline noise. The spectrum resembles that shown in Fig. 7–45.

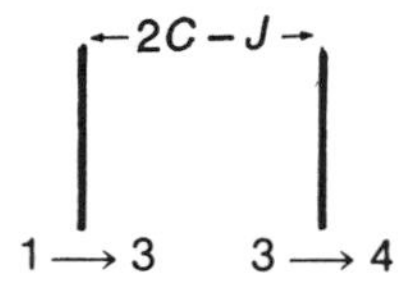

FIGURE 7–45 Spectrum of an AB molecule when σ_A and σ_B are close.

E. $J \neq 0$, but $\sigma_A = \sigma_B$, so $\Delta = 0$

This is the case of equivalent protons, and we are now in a position to see why equivalent protons do not split each other. When $\Delta = 0$, sin 2θ equals unity and, as can be seen from Table 7–5, the $1 \to 2$ and $2 \to 4$ transitions have zero intensity. The $1 \to 3$ and $3 \to 4$ transitions have intensity, but they have the same energy and occur at the center of the spectrum; i.e., $-(1/2)J + C = (1/2)J - C = 0$. This is a manifestation of a general rule: magnetically equivalent nuclei do not split each other.

7-22 MORE COMPLICATED SECOND-ORDER SYSTEMS

It is convenient to classify molecules according to the general type of Δ and J systems to which they belong, for this suffices to define the Hamiltonian for the coupling. The conventions that will be given are normally applied to spin $^1/_2$ systems.

1. Each type of magnetic nucleus in the molecule is assigned a capital letter of the Roman alphabet. All of one equivalent set of nuclei are given a single letter, and a subscript is used to indicate the number of such nuclei in the set; *e.g.*, benzene is A_6.

2. Roman letters are assigned to different nuclei in order of decreased shielding; *i.e.*, three different nuclei with similar shifts listed in order of decreased shielding would be labeled A, B, and C. If the second set of nuclei has a very different shift than the first (*i.e.*, $\Delta \gg J$), a letter far removed in the alphabet is used (X, Y, Z) to label the low field peaks. Nuclei in between would be labeled L, M, N, etc. For example, acetaldehyde, CH_3CHO, is an A_3X system. This implies a first-order system. HF would be an AX system. The molecule $HPFCl_2$ would be an AMX system. When one is not sure of the chemical shift difference, nearby letters are used to be safe, for this implies that the system could be second order.

3. The system of labeling must provide information about the magnetic, as well as the chemical, non-equivalence of nuclei in the molecule. If two atoms are equivalent by symmetry, they have the same chemical shift, but they could be involved in spin-spin coupling to other nuclei such that the J to one nucleus is different from the J to another; *e.g.*, in $C_6H_2F_2Cl_2$ (Fig. 7–46) the two hydrogens are isochronous, but two different HF couplings are needed for the full analysis. The H(1)-F(1) coupling constant is different from the H(1)-F(2) coupling constant. The protons and the fluorines are said to be chemically equivalent, but *magnetically non-equivalent*. All that is meant by this is that, in the Hamiltonian for this system, two different HF couplings must be considered even though the two hydrogens and two fluorines are related by symmetry. The dichloro-difluoro-benzene isomer discussed above is thus referred to as an AA'XX' system. The reader should now realize that in the non-planar (D_{2d}) molecule

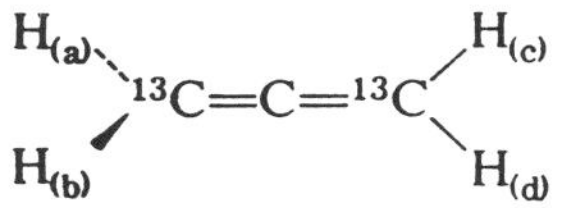

$$H_{(a)} \diagdown \diagup H_{(c)}$$
$$^{13}C=C=^{13}C$$
$$H_{(b)} \diagup \diagdown H_{(d)}$$

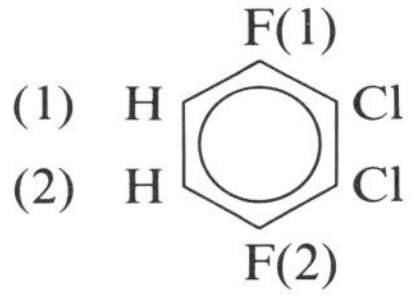

FIGURE 7–46 The molecule 1,2-dichloro, 3,6-fluoro benzene; an AA'XX' system.

the ab pair is magnetically non-equivalent to the cd pair because two different coupling constants to the ^{13}C atoms are involved. The system will be labeled $A_2A_2'XX'$. The molecule ClF_3 discussed in Chapter 6 (consider only fluorines) would be classified as an AB_2 molecule (at 10 MHz). The phosphorus nmr of the tetrapolyphosphate anion, $P_4O_{13}^{6-}$, would yield an A_2B_2 spectrum, and P_4S_3 is an A_3X system.

The energy levels of various kinds of coupled systems have been worked out in detail[8,45,46] in terms of the shift differences and coupling constants, and are reported in the literature. To use these results, one classifies the molecule of interest according to the scheme described above and looks up the analysis for this type of system; *e.g.*, one could find the energy levels for any AA'XX' system.[8] Computer programs are also available that find the best values of the J's and

Δ's that reproduce the chemical shifts and intensities of all the peaks in the experimental spectrum. The interpretation of a relatively simple second-order spectrum is illustrated for the ^{31}P nmr spectrum of the anion
$$H-\overset{\overset{O}{\|}}{\underset{\underset{O}{|}}{P}}-\overset{\overset{O^{(3-)}}{\|}}{\underset{\underset{O}{|}}{P}}-O$$
indicated in Fig. 7–47. The actual spectrum, IV, is interpreted by generating it in three stages. Consideration of the first stage, I, yields two lines from the two non-equivalent phosphorus atoms, $P_{(a)}$ and $P_{(b)}$, and their separation is the chemical shift difference. The second consideration, II, includes splitting by hydrogen. Since the hydrogen is on $P_{(a)}$, $J_{P_{(a)}-H} > J_{P_{(b)}-H}$. The four lines which result are included in II. The third consideration, III, included $P_{(a)}$-$P_{(b)}$ splitting and accounts for the final spectrum for this ABX case. Two of the expected lines are not detected in the final spectrum because they are too weak to be detected and another pair fall so close together that they appear as a single peak (the most intense peak). The analysis of the $HP_2O_5^{3-}$ spectrum to yield the interpretation contained in Fig. 7–47 was carried out[47] with a computer analysis that fitted the intensities and chemical shifts of the experimental spectrum.

The phenomenon of virtual coupling, which we discussed earlier, is a magnetically non-equivalent system of, for example, the type XAA'X' where $J_{AA'}$ is large.

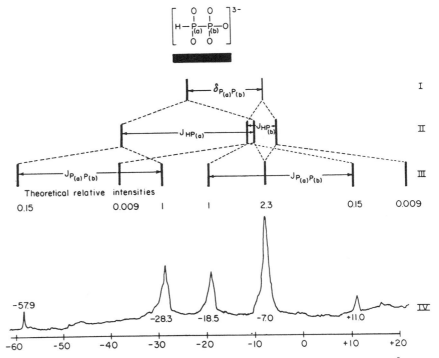

FIGURE 7–47 The phosphorus nmr spectrum of the diphosphate anion $HP_2O_5^{3-}$ and its interpretation. I, Chemical shift differences in $P_{(a)}$ and $P_{(b)}$. II, H—P splitting. III, P—P splitting. IV, Observed spectrum. [Reprinted with permission from C. F. Callis, *et al.*, J Amer. Chem. Soc., *79*, 2722 (1957). copyright by the American Chemical Society.]

7–23 DOUBLE RESONANCE AND SPIN-TICKLING EXPERIMENTS

In a double resonance experiment, the sample is subjected to a second r.f. source whose frequency corresponds to the Larmor frequency of one of the nuclei in the sample. This field causes the contribution to the spectrum from this nucleus to disappear, and this nucleus is said to be *decoupled*. The second r.f. field is applied with a large amplitude, at right angles to H_1 and orthogonal to the pickup coils. The net effect of this field is to cause nuclear transitions in, say, B of the AB system. Decoupling of the B nucleus is achieved, and B makes no contribution to the A spectrum. In practice, decoupling can be accomplished only if $v_A - v_B > 5|J_{AB}|$. When the spectrum of A is examined with B being irradiated, the symbol $A \to \{B\}$ is employed to indicate this fact.

For systems where $J \approx \Delta$, one can perform a *spin-tickling experiment*. In this experiment, a weak r.f. field is employed and all transitions having an energy level in common with the peak being irradiated will undergo a change. Referring to Figs. 7–42 and 7–44, the $1 \to 2$ transition is seen to have an energy level in common with the $1 \to 3$ and $2 \to 4$ transitions, but not the $3 \to 4$. Thus, the latter peak is not split by spin tickling the $1 \to 2$ peak, but the other spectral lines will be split. Experimentally, it is found that the tickling splits the lines that have an energy level in common with the line being saturated. Furthermore, if the transitions for the two peaks connected by a common energy level correspond to a consecutive change in spin of both nuclei, each by 1, as in $1 \to 2$ and $2 \to 4$ (*i.e.*, $\alpha\alpha$ to $\alpha\beta$ and $\alpha\beta$ to $\beta\beta$) or as in $1 \to 3$ and $3 \to 4$, a sharp doublet results. When this is not the case, as in $1 \to 2$ and $1 \to 3$ (*i.e.*, $\alpha\alpha$ to $\alpha\beta$ and $\alpha\alpha$ to $\beta\alpha$) or $3 \to 4$ and $2 \to 4$, a broad doublet results. We shall not go into the reasons for this, but simply point out that this is a valuable technique for spectral assignment and energy level ordering in second-order systems and for determining the relative signs of J in first-order systems.

The double resonance technique can be employed to evaluate chemical shifts for nuclei other than protons by using a proton probe.[48] If nucleus Y is splitting a proton, the frequency of the r.f. field that is most effective for decoupling Y from the protons is measured, and thus the chemical shift of Y is determined using a proton probe. This is the basis for the INDOR technique.

The proton nmr spectrum of diborane is illustrated in Fig. 7–48(A). This spectrum results from two sets of non-equivalent protons (bridge and terminal protons) being split by the ^{11}B nuclei. The asterisks indicate fine structure arising from the smaller abundance of protons on ^{10}B nuclei. (^{10}B has a natural abundance of 18.83% and $I = 3$ compared to 81.17% for ^{11}B with $I = {}^3/_2$.) In Fig. 7–48(B), the splitting caused by ^{11}B has been removed by saturation of the boron nuclei by the double resonance technique. Two peaks of intensity ratio 2:1 are obtained, corresponding to the four terminal and two bridge protons.[49]

Two isomers have been obtained in the preparation of N_2F_2. One definitely has a *trans* structure with one fluorine on each nitrogen. In conflicting reports, the structure of the second isomer has been reported to be the *cis* isomer and also $F_2N{=}N$. An excellent discussion of the results obtained by employing several different physical methods in an attempt to resolve this problem has been reported along with the fluorine nmr spectrum and results from a double resonance experiment.[50] Saturation of the ^{14}N nucleus in this second isomer with a strong r.f. field causes collapse of all nitrogen splitting. It is concluded that

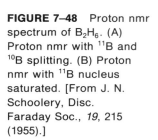

FIGURE 7–48 Proton nmr spectrum of B_2H_6. (A) Proton nmr with ^{11}B and ^{10}B splitting. (B) Proton nmr with ^{11}B nucleus saturated. [From J. N. Schoolery, Disc. Faraday Soc., *19*, 215 (1955).]

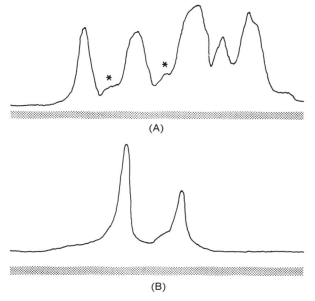

the chemical shift of the two nitrogens must be equivalent, and this eliminates the $F_2N{=}N$ structure. Additional evidence is obtained for the *cis* structure from a complete spectral interpretation. The value for J_{N-F} calculated in this study for a *cis* structure is reasonable when compared to J_{N-F} for NF_3.

A more complete discussion of the theory of the double resonance technique and many more examples of its application are contained in a review article[51] by Baldeschwieler and Randall.

7–24 DETERMINING SIGNS OF COUPLING CONSTANTS

The double resonance technique has been successfully employed to determine the relative sign of coupling constants. This can be illustrated by considering the proton nmr spectrum[52] of $(C_2H_5)_2Tl^+$ in Fig. 7–49 ($I = {}^1/_2$ for ^{205}Tl). If J_{Tl-CH_3} and J_{Tl-CH_2} are both positive, both low field peaks correspond to interaction with positive nuclear magnetic quantum numbers of Tl. If the signs of J are different, one low field peak corresponds to interaction with the moment from thallium nuclei where $m_I = +{}^1/_2$ and the other to $-{}^1/_2$. By irradiation at the center of each of the multiplets, it was shown that each CH_3 triplet was coupled to the

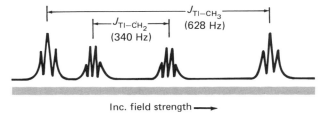

FIGURE 7–49 NMR spectrum of $(C_2H_5)_2Tl^+$ (facsimile).

distant methylene quartet and vice versa. For example, irradiation with a frequency corresponding to the low field triplet resulted in the disappearance of the fine structure of the high field methylene signal. This result indicates that J_{Tl-CH_3} and J_{Tl-CH_2} have opposite signs. If the sign were the same, the two low field multiplets would be coupled together as would the two high field multiplets, and saturation of the low field triplet would cause collapse of the fine structure in the low field methylene signal.

7–25 EFFECTS ON THE SPECTRUM OF NUCLEI WITH QUADRUPOLE MOMENTS

Quadrupolar nuclei are often very efficiently relaxed by the fluctuating electric fields that arise from the dipolar solvent and solute molecules. The mechanism of quadrupole relaxation depends upon the interaction of the quadrupolar nucleus with the electric field gradient at the nucleus. This gradient arises when the quadrupolar nucleus is in a molecule in which it is surrounded by a non-spherical distribution of electron density.

The *field gradient*, q, is used to describe the deviation of the electronic charge cloud about the nucleus from spherical symmetry. If the groups about the nucleus in question have cubic symmetry (*e.g.*, T_d or O_h point groups), the charge cloud is spherical and the value of q is zero. If the molecule has cylindrical symmetry (a threefold or higher symmetry axis), the deviation from spherical symmetry is expressed by the magnitude of q. If the molecule has less than cylindrical symmetry, two parameters are usually needed, q and η. The quantity η is referred to as the *asymmetry parameter*. The word "usually" is inserted because certain combinations of angles and charges can cause fortuitous cancellations of effects leading to $\eta = 0$. The axis of largest q is labeled z and is described by q_{zz}. The other axes, described by field gradients q_{xx} and q_{yy}, are described by the asymmetry parameter, which is defined as:

$$\eta = (q_{xx} - q_{yy})/q_{zz} \tag{7-69}$$

The effectiveness of the relaxation depends upon the magnitude of the field gradient. *Rapid nuclear quadrupole relaxation* has a pronounced effect on the linewidth obtained in the nmr spectrum of the quadrupolar nucleus, and it also influences the nmr spectra of protons or other nuclei attached to this quadrupolar nucleus. In the latter case, splittings of a proton from the quadrupolar nucleus may not be observed or the proton signal may be so extensively broadened that the signal itself is not observed. This can be understood by analogy to the effect of chemical exchange on the proton nmr spectra. Either rapid chemical exchange or rapid nuclear quadrupole relaxation in effect places the proton on a nucleus (or nuclei, for chemical exchange) whose spin state is rapidly changing. Nuclear quadrupole relaxation rates often correspond to an intermediate rate of chemical exchange, so extensive broadening is usually observed. As a result of quadrupole relaxation, the proton nmr spectrum of $^{14}NH_3$ (^{14}N, $I = 1$) consists of three very broad signals; while in the absence of this effect, the spectrum of $^{15}NH_3$ (^{15}N, $I = 1/2$) consists of a sharp doublet. On the other hand, in $^{14}NH_4^+$, where a spherical distribution of electron density gives rise to a zero field gradient, a sharp three-line spectrum results. In a molecule with a very large field gradient, a broad signal with no fine structure is commonly obtained.

When one attempts to obtain an nmr spectrum of a nucleus with a quadrupole moment (e.g., ^{35}Cl and ^{14}N) that undergoes relaxation readily, the signals are sometimes broadened so extensively that no spectrum is obtained. This is the case for most halogen (except fluorine) compounds. Sharp signals have been obtained for the halide ions and symmetrical compounds of the halogens (e.g., ClO_4^-), where the spherical charge distribution gives rise to only small field gradients at the nucleus, leading to larger values for T_1.

Solutions of I^- (^{127}I, $I = ^5/_2$) give rise to an nmr signal. When iodine is added, the triiodide ion, I_3^-, is formed, destroying the cubic symmetry of the iodide ion so that quadrupole broadening becomes effective and the signal disappears. Small amounts of iodine result in a broadening of the iodide resonance, and the rate constant for the reaction $I^- + I_2 \rightarrow I_3^-$ can be calculated broadening.[53] It is interesting to note that chlorine chemical shifts have been observed[54] for the compounds: $SiCl_4$, CrO_2Cl_2, $VOCl_3$, and $TiCl_4$, where the chlorine is in an environment of lower than cubic symmetry.

An interesting effect has been reported for the fluorine nmr spectrum of NF_3. The changes in a series of spectra obtained as a function of temperature are opposite to those normally obtained for exchange processes. At $-205°$ C a sharp single peak is obtained for NF_3; as the temperature is raised the line broadens and a spectrum consisting of a sharp triplet ($I = 1$ for ^{14}N) results at 20° C. It is proposed that at low temperature the slow molecular motions are most effective for quadrupole relaxation of ^{14}N; as a result, a single line is obtained. At higher temperatures, relaxation is not as effective and the lifetime of a given state for the ^{14}N nucleus is sufficient to cause spin-spin splitting. A similar effect is observed for pyrrole.[55] The ^{14}N spectrum of azoxybenzene exhibits only a singlet. The nitrogens are not equivalent, and it is

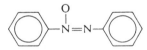

proposed that the field gradient at the N—O nitrogen is so large as to make this resonance unobservable.

The double resonance technique has been successfully used on the proton nmr spectrum of $Al(BH_4)_3$. This molecule contains six Al—H—B bridge bonds. Both B and ^{27}Al ($I = ^5/_2$) have quadrupole moments. The proton nmr at 30 MHz consists[56] of a single broad line (Fig. 7–50(A)). When the ^{11}B nucleus is saturated ($^1H \rightarrow \{^{11}B\}$), the proton resonance spectrum in Fig. 7–50(B) results. Fig. 7–50(C) represents the proton nmr spectrum when the sample is irradiated with frequency corresponding to that of ^{27}Al ($^1H \rightarrow \{^{27}Al\}$). The four large peaks in (C) arise from ^{11}B splitting of the proton and the smaller peaks from ^{10}B splitting. The bridging and terminal hydrogens are not distinguished because of a rapid proton exchange reaction that makes all hydrogens magnetically equivalent.

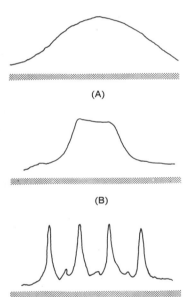

FIGURE 7–50 Proton nmr of $Al(BH_4)_3$. (A) Proton resonance. (B) Proton resonance, ^{11}B saturated. (C) Proton resonance, ^{27}Al saturated. [From R. A. Ogg, Jr., and J. D. Ray, Disc. Faraday Soc., 19, 239 (1955).]

REFERENCES CITED

1. H. S. Gutowsky, "Physical Methods of Organic Chemistry," 3rd Ed., part 4, A. Weissberger, ed. (Vol. 1 of "Techniques of Organic Chemistry"), Interscience, New York, 1960.
2. J. A. Pople, W. G. Schneider, and A. J. Bernstein, "High Resolution Nuclear Magnetic Resonance," McGraw-Hill, New York, 1959.
3. (a) J. W. Emsley, J. Feeney, and L. H. Sutcliffe, "High Resolution Nuclear Resonance Spectroscopy," Vol. 1, Pergamon Press, New York, 1966.
   (b) Same authors and title; Vol. 2.
4. M. W. Dietrich and R. E. Keller, Anal. Chem., *36*, 258 (1964) and references therein.
5. N. S. Bhacca. L. F. Johnson, and J. N. Shoolery, "NMR Spectra Catalog," Varian Associates, Palo Alto, Calif., 1962; "The Sadtler Standard Spectra; NMR," Sadtler Research Laboratories, Philadelphia, 1972; L. F. Johnson and W. C. Jankowski, "C-13 NMR Spectra" (a collection), Interscience, New York, 1972.
6. L. Eberson and S. Farsen, J. Phys. Chem., *64*, 767 (1960).
7. A. L. Van Gelb, Anal. Chem., *42*, 679 (1970).
8. (a) J. W. Emsley et al., "High Resolution Nuclear Magnetic Resonance Spectroscopy," Vol. 1, Pergamon Press, New York, 1966;
   (b) P. Diehl, E. Fluck, and R. Kosfeld, eds., "NMR Basic Principles and Progress," Vol. 5, Springer-Verlag, New York, 1971.
9. E. Fluck, "Anorganische und Allgemeine Chemie in Einzeldarstellungen. Band V. Die Kernmagnetische Resonanz und Ihre Anwendung in der Anorganische Chemie," Springer-Verlag, Berlin, 1963.
10. A. D. Buckingham and P. J. Stephens, J. Chem. Soc., *1964*, 4583.
11. C. E. Johnson and F. A. Bovey, J. Chem. Phys., *29*, 1012 (1958) and references therein.
12. (a) A. D. Buckingham, Can. J. Chem., *38*, 300 (1960);
    (b) J. I. Musher, J. Chem. Phys., *37*, 34 (1962).
13. W. N. Lipscomb et al., J. Amer. Chem. Soc., *88*, 5340 (1966); Adv. Mag. Res., *2*, 137 (1966).
14. D. Gurka and R. W. Taft, J. Amer. Chem. Soc., *91*, 4794 (1969).
15. F. Prosser and L. Goodman, J. Chem. Phys., *38*, 374 (1963).
16. (a) M. Karplus and T. P. Das, J. Chem. Phys., *34*, 1683 (1961);
    (b) R. J. Pugmire and D. M. Grant, J. Amer. Chem. Soc., *90*, 697 (1968).
17. W. H. Pirkle, S. D. Beare, and R. L. Muntz, J. Amer. Chem. Soc., *91*, 4575 (1969); *ibid.*, *93*, 2817 (1971) and references therein.
18. H. S. Gutowsky, J. Chem. Phys., *37*, 2136 (1962).
19. J. S. Waugh and F. A. Cotton, J. Phys. Chem., *65*, 562 (1961).
20. R. A. Hoffman and S. Gronowitz, Arkiv. Kemi., *16*, 471, 563, (1961); *17*, 1 (1961); Acta Chem. Scand., *13*, 1477 (1959).
21. M. Karplus, J. Chem. Phys., *33*, 1842 (1960).
22. (a) M. T. Rogers and J. D. Graham, J. Amer. Chem. Soc., *84*, 3666 (1962);
    (b) J. D. Kennedy et al., Inorg. Chem., *12*, 2742 (1973).
23. H. S. Gutowsky, D. W. McCall, and C. P. Slichter, J. Chem. Phys., *21*, 279 (1953).
24. C. F. Callis, J. R. Van Wazer, J. N. Shoolery, and W. A. Anderson, J. Amer. Chem. Soc., *79*, 2719 (1957).
25. R. O. Ragsdale, and B. B. Stewart, Inorg. Chem., *2*, 1002 (1963).
26. N. Muller, J. Chem. Phys., *36*, 359 (1962).
27. H. S. Gutowsky, and C. Juan, J. Amer. Chem. Soc., *84*, 307 (1962); J. Chem. Phys., *37*, 2198 (1962).
28. J. H. Goldstein and G. S. Reddy, J. Chem. Phys., *36*, 2644 (1962).
29. R. S. Drago and N. A. Matwiyoff, J. Chem. Phys., *38*, 2583 (1963); J. Organomet. Chem., *3*, 62 (1965).
30. H. A. Bent, Chem. Rev., *61*, 275 (1961).
31. J. Hinze and H. H. Jaffe, J. Amer. Chem. Soc., *84*, 540 (1962).

32. M. Karplus, J. Amer. Chem. Soc., *85*, 2870 (1963).
33. S. Sternhall, Quart. Rev. (London), *23*, 236 (1969) and references therein.
34. M. A. Cooper and S. L. Manatt, J. Amer. Chem. Soc., *92*, 1605 (1970).
35. R. L. Middaugh and R. S. Drago, J. Amer. Chem. Soc., *85*, 2575 (1963).
36. J. R. Holmes and H. D. Kaesz, J. Amer. Chem. Soc., *83*, 3903 (1961).
37. N. A. Matwiyoff and R. S. Drago, Inorg. Chem., *3*, 337 (1964).
38. G. D. Shier and R. S. Drago, J. Organomet. Chem., *6*, 359 (1966).
39. T. F. Bolles and R. S. Drago, J. Amer. Chem. Soc., *88*, 5730 (1966).
40. J. F. Nixon and A. Pidcock, Ann. Rev. NMR Spectroscopy, *2*, 345 (1969).
41. R. K. Harris, Canad. J. Chem., *42*, 2275, 2575 (1964).
42. **(a)** B. E. Mann, B. L. Shaw, and K. E. Stainbank, Chem. Comm., 151 (1972); **(b)** D. E. Axelson and C. E. Holloway, Chem. Comm., 455 (1973).
43. E. L. Muetterties, J. Amer. Chem. Soc., *82*, 1082 (1960).
44. E. L. Muetterties, W. Mahler, and R. S. Schmutzler, Inorg. Chem., *2*, 613 (1963).
45. J. D. Roberts, "An Introduction to the Spin-Spin Splitting in High Resolution NMR Spectra," W. A. Benjamin, New York, 1961.
46. P. L. Corio, Chem. Rev., *60*, 363 (1960).
47. C. F. Callis *et al.*, J. Amer. Chem. Soc., *79*, 2719 (1957).
48. G. C. Levy and G. L. Nelson, "Carbon-13 Nuclear Magnetic Resonance for Organic Chemists," Wiley-Interscience, New York, 1972.
49. J. N. Shoolery, Disc. Faraday Soc., *19*, 215 (1955).
50. J. H. Noggle, J. D. Baldeschwieler, and C. B. Colburn, J. Chem. Phys., *37*, 182 (1962).
51. J. Baldeschwieler and E. W. Randall, Chem. Rev. *63*, 82 (1963).
52. J. P. Maher and D. F. Evans, Proc. Chem. Soc., *1961*, 208; D. W. Turner, J. Chem. Soc., *1962*, 847.
53. O. E. Myers, J. Chem. Phys., *28*, 1027 (1958).
54. Y. Masuda, J. Phys. Soc. Japan, *11*, 670 (1956).
55. J. D. Roberts, J. Amer. Chem. Soc., *78*, 4495 (1956).
56. R. A. Ogg, Jr., and J. D. Ray, Disc. Faraday Soc., *19*, 239 (1955).

COMPILATIONS OF CHEMICAL SHIFTS

Proton Shifts

N. S. Bhacca, L. F. Johnson, and J. N. Shoolery, "NMR Spectra Catalog," Varian Associates, Palo Alto, Calif., 1962.

"The Sadtler Standard Spectra; NMR," Sadtler Research Laboratories, Philadelphia, 1972.

N. F. Chamberlain, "The Practice of NMR Spectroscopy with Spectra-Structure Correlations for Hydrogen-1," Plenum, London and New York, 1974.

^{13}C Shifts

L. F. Johnson and W. C. Jankowski, "C-13 NMR Spectra," Wiley-Interscience, New York, 1972.

G. C. Levy, R. L. Lichter, and G. L. Nelson, "Carbon-13 Nuclear Magnetic Resonance Spectroscopy," 2nd ed., Wiley-Interscience, 1989.

G. C. Levy, "Topics in ^{13}C NMR Spectroscopy," Wiley-Interscience, New York, volumes from 1974 to present.

B. E. Mann and B. F. Taylor, "^{13}C NMR Data for Organometallic Compounds," Academic Press, New York, 1981.

Q. T. Pham, R. Petiaud, and H. Waton, "Proton and Carbon NMR of Polymers," Wiley, New York, 1983.

^{31}P Shifts

E. Fluck, "Anorganische und Allgemeine Chemie in Einzeldrstellungen. Band V. Die Kernmagnetische Resonanz und Ihre Anwendung in der Anorganische Chemie," Springer-Verlag, Berlin, 1963.

"Methods in Stereochemical Analysis," Volume 8, by J. G. Verkade and L. D. Quin, eds., VCH Publishers, Deerfield Beach, Fla.

"^{31}P NMR, Principles and Applications," D. G. Gorestein, ed., Academic Press, New York, 1984.

^{19}F Shifts

C. H. Dungan and J. R. VanWazer, "Compilation of Reported ^{19}F Chemical Shifts," Wiley-Interscience, New York, 1970.

A. J. Gordon and R. A. Ford, "The Chemists Companion," Wiley, New York, 1972.

^{11}B Shifts

H. Noth, "NMR Spectroscopy of Boron Compounds," Springer-Verlag, New York, 1978.

Miscellaneous Nuclei Shifts

"NMR of Newly Accessible Nuclei," P. Laszlo, ed., Academic Press, New York, 1983.

R. K. Harris and B. E. Mann, "NMR and the Periodic Table," Academic Press, New York, 1978.

General Text

F. A. Bovey (with L. Jelinski and P. A. Mirav) "Nuclear Magnetic Resonance Spectroscopy," 2nd ed., Academic Press, New York, 1988.

For more practice with Spectral Interpretation

R. M. Silverstein, G. C. Bassler, and T. C. Morrill, "Spectrometric Identification of Organic Compounds," 4th ed., Wiley, New York, 1981.

EXERCISES

1. The δ value of a substance relative to the external standard, methylene chloride, is -2.5. Calculate δ relative to external standards (a) benzene and (b) water using the data in Appendix F.

2. Assuming the relationship discussed between J_{Sn-H} and hybridization, what would be the ratio of the coupling constants in a five-coordinate and six-coordinate complex of $(CH_3)_3SnCl$ [i.e., $(CH_3)_3SnCl \cdot B$ and $(CH_3)_3SnCl \cdot 2B$]?

3. The compound $B[N(CH_3)_2]_3$ is prepared and dissolved in a wide number of different solvents. Propose a method of determining in which ones the solvent is coordinated to the compound.

4. Consider the diamagnetic complex $(Me_3P)_4Pt^{2+}$. Sketch the phosphorus resonance signal if

   a. $J_{P-H} > J_{P-Pt}$.

   b. $J_{P-Pt} > J_{P-H}$.

5. Indicate the number of isomers for cyclic compounds of formulas $P_3N_3(CH_3)_2Cl_4$, and sketch the phosphorus resonance spectrum of each (assume $\Delta > J$, J_{P-H} is small, and J_{P-H} can be ignored for phosphorus atoms which do not contain methyl groups).

6. Would you expect the ^{14}N nmr spectrum to be sharper in NH_3 or NH_4^+? (For ^{14}N, $I = 1$.) Explain.

7. Consider all possible isomers that could be obtained for the eight-membered ring compound $P_4N_4Cl_6(NHR)_2$ and indicate the ideal phosphorus resonance spectrum expected for each. Which of the above are definitely eliminated if the phosphorus resonance consists of two triplets of equal intensity?

8. The proton nmr spectrum of

$$\begin{matrix} CH_2-O \\ | \quad\quad\quad >S-O \\ CH_2-O \end{matrix}$$

is not a singlet. Is the SO_3 group planar? What would the spectrum look like if the sulfur underwent rapid inversion?

9. It is found that the methylene groups in $(CH_3CH_2)_2SBF_3$ give rise to a single methylene resonance. Explain.

10. a. Are the methyl groups in $(CH_3)_2C\overset{H}{\underset{}{-}}P\overset{O}{\underset{Cl}{-}}C_6H_5$ equivalent?

   b. Ignore the splitting of the methyl groups by the phenyl protons in the above compound and assume $J < \Delta$. What would the spectrum of the methyl protons look like?

11. The proton nmr spectra of a series of compounds are given below. Assign their geometries and interpret the spectra.

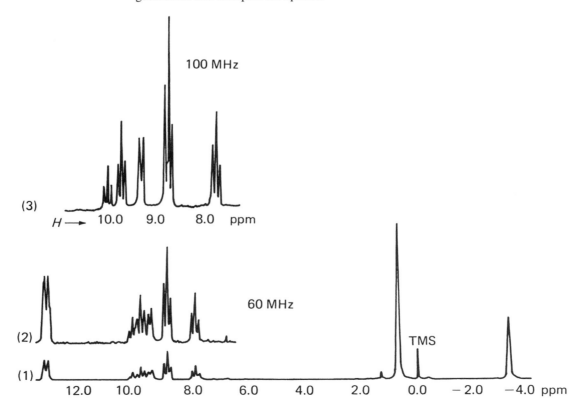

a. The compound is ReCl$_3$[P(CH$_3$)$_2$C$_6$H$_5$]$_3$. Curve (1) is the full spectrum, (2) is a more intense sweep of the 6.0 to 13.0 ppm region, and (3) is the 100 MHz spectrum of the 7.0 to 10 ppm region.

b. Why is the spectrum in part (a) at 100 MHz different from that at 60 MHz? Why are the chemical shifts in ppm relative to TMS the same?

c. A compound with empirical formula C$_4$H$_{11}$N.

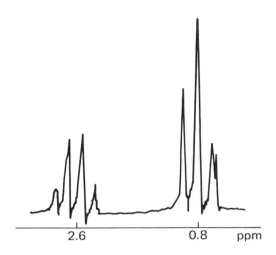

d. A compound with empirical formula C$_2$H$_3$F$_3$O.

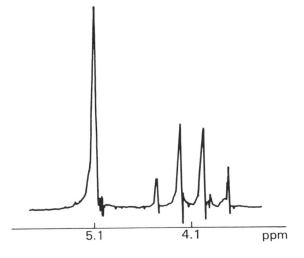

e. A compound with empirical formula C_2H_4O.

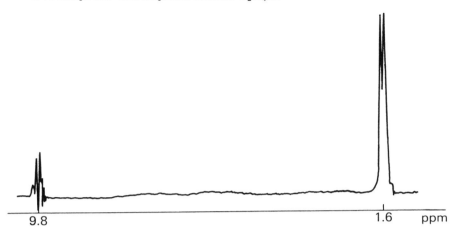

f. A compound with empirical formula C_4H_6.

g. A compound with empirical formula $C_4H_8O_2$.

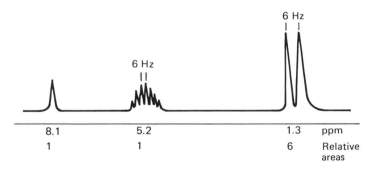

h. A compound with empirical formula $C_3H_5ClF_2$.

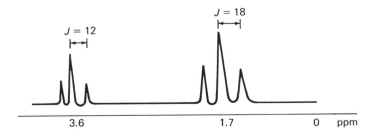

i. A compound with empirical formula $C_8H_{10}O$. The peak at 2.4 ppm vanishes in D_2O.

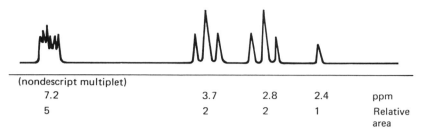

(nondescript multiplet)				
7.2	3.7	2.8	2.4	ppm
5	2	2	1	Relative area

j. A compound with empirical formula $C_6H_{15}O_4P$.

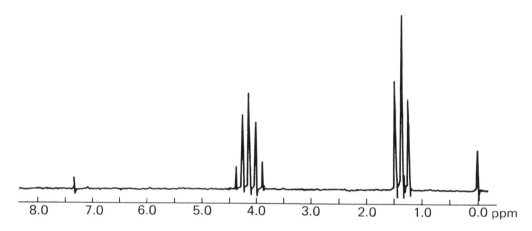

k. A compound with empirical formula $C_3H_5O_2Cl$.

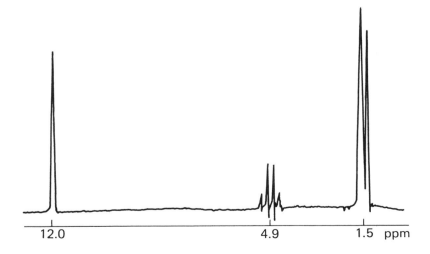

l. A compound with empirical formula C_2H_8NCl dissolved in water.

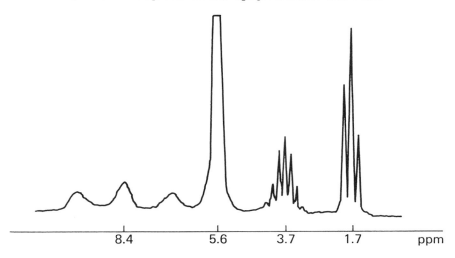

m. A compound with empirical formula $C_{14}H_{22}O_4$.

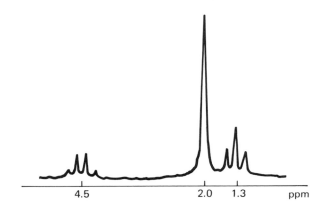

n. A compound with empirical formula $(SiH_3)_2PSiH_2CH_3$.

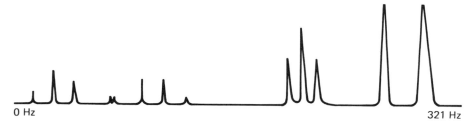

o. A compound with empirical formula Pt[P(C$_2$H$_5$)$_3$]$_2$HCl. (The peaks at 8 ppm are nondescript multiplets.)

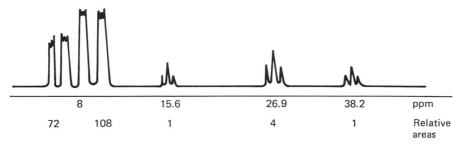

p. A compound with empirical formula C$_{11}$H$_{16}$. The curves above the peaks represent the integrated intensities of the peaks. The relative areas can be obtained by comparing the heights (number of squares) of the respective integration curves.

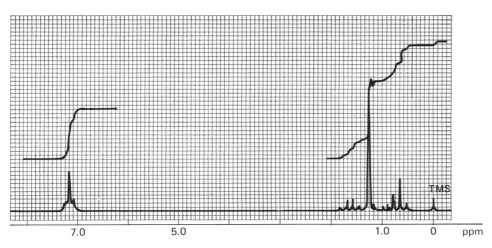

q. A compound with empirical formula C$_8$H$_{10}$O. The curves above the peaks represent the integrated intensities of the peaks. The relative areas can be obtained by comparing the heights, in squares, of the respective integration curves.

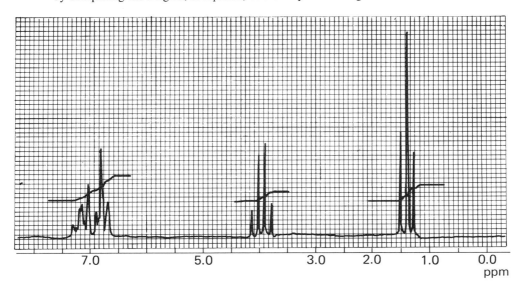

r. A compound with empirical formula $C_5H_{10}O$. The curves above the peaks represent the integrated intensities of the peaks. The relative areas can be obtained by comparing the heights, in squares, of the respective integration curves.

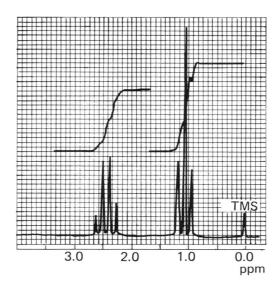

s. A compound with empirical formula C_8H_7OF.

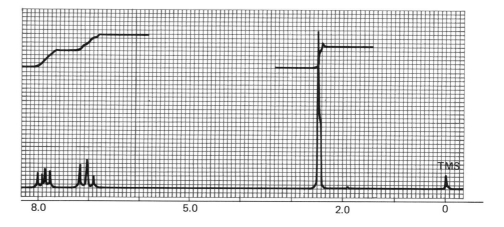

t. The ^{13}C nmr of OSC_2D_6 at 75.1 MHz (D = deuterium with a spin of 1).

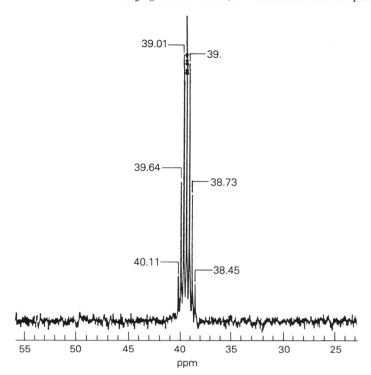

12. Spectrum (a) results for the methyl region of one isomer of $Pd[P(CH_3)_2C_6H_5]_2I_2$. Spectrum (b) is obtained in the methyl region for the other isomer of $Pt[P(CH_3)_2C_6H_5]_2I_2$.

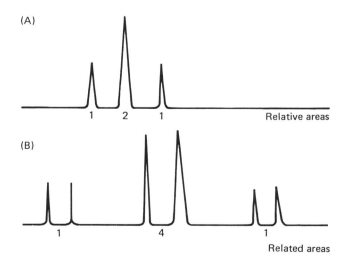

a. Which is *cis* and which is *trans*?

b. Explain why the areas in (b) are in the ratio of 1 to 4 to 1.

13. The ^{11}B nmr spectrum below is obtained for 5—Cl—2,4—C$_2$B$_5$H$_6$. Interpret the spectrum. (Hint: B$_7$H$_7^{2-}$ is a pentagonal bipyramid of B—H subunits in which the axial borons are numbered 1 and 7.)

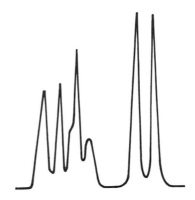

14. The ^{19}F nmr spectrum of PF$_3$(NH$_2$)$_2$ is reported in (a), and expanded spectra of the individual peaks are presented in (b) through (e). Propose a structure and interpret the spectrum.

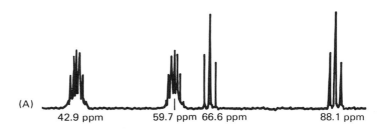

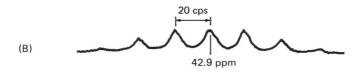

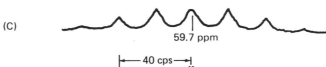

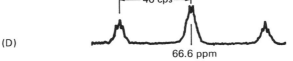

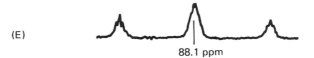

15. The proton resonance of the hydridic hydrogen of $HNi[OP(C_2H_5)_3]_4^+$ is given as follows. Propose a structure for this complex.

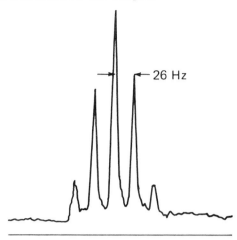

16. Given the following proton nmr spectra and molecular structures, assign all peaks:

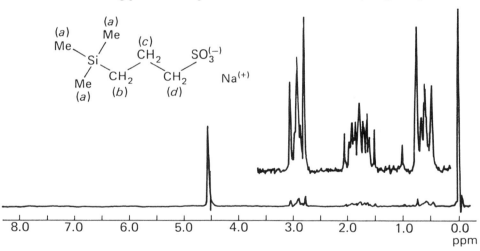

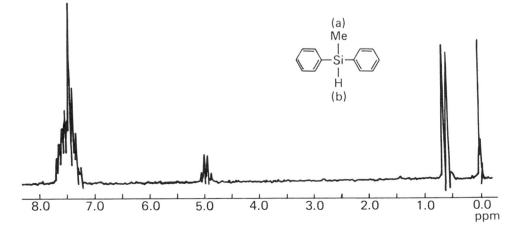

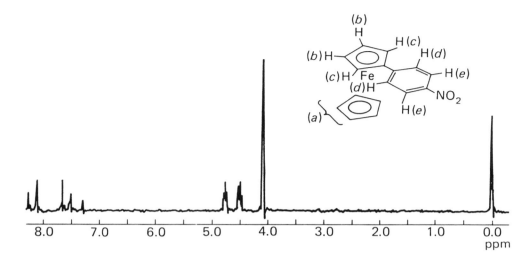

17. What properties in a functional group X (which can consist of more than one atom) attached to an isopropyl group can cause non-equivalent methyls to occur in the isopropyl group?

18. The following proton nmr spectrum is that of a compound with the molecular formula $C_8H_{11}NO$. Propose a structure consistent with the formula and spectrum. Insofar as possible, assign the peaks. Numbers in parentheses refer to relative peak areas.

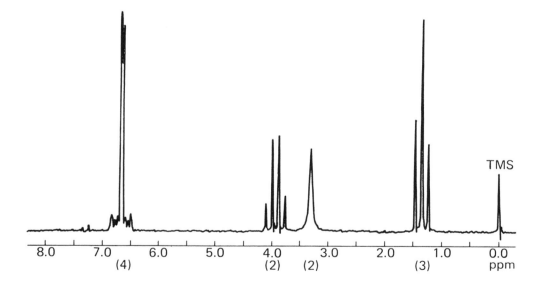

19. Interpret the spectrum below for the compound $CH_3Pt(CH_2\!=\!CH_2)[PC_6H_5(CH_3)_2]_2$. Justify all splittings. Is this the *cis* or *trans* isomer? (Hint: For ^{31}P, $I = 1/2$, abundance = 100%. For ^{195}Pt, $I = 1/2$, abundance = 34%.) [See Inorg. Chem., *12*, 994 (1973).]

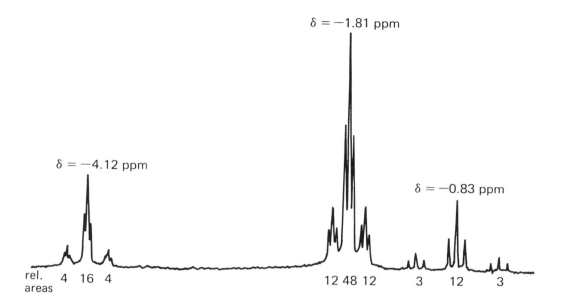

$\delta = -1.81$ ppm

$\delta = -4.12$ ppm

$\delta = -0.83$ ppm

rel. areas   4 16 4   12 48 12   3   12   3

20. Sketch the spectra for the following molecules and conditions (assume 100% abundances):

a. H_2PF_3, structure

$$F-P\begin{array}{c} H \\ | \\ \\ | \\ H \end{array}\begin{array}{c} F \\ \\ \\ F \end{array}$$

$J_{P-F} > J_{P-H} > J_{H-F}$ $I_H = {}^1/_2$
$I_F = {}^1/_2$
$I_P = {}^1/_2$

^1H, ^{19}F, and ^{31}P resonances

b. $N(CH_2F)_4{}^+$ $J_{FCH} > J_{FCN}$; $J_{FCNCH} = 0$
^{19}F only

21. The following proton nmr spectrum is reported [Inorg. Chem., 2, 939 (1963)] for $((NH_2)_3P)_2Fe(CO)_3$ (the two phosphorus ligands are *trans*).

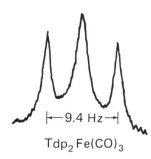

|—9.4 Hz—|

$Tdp_2 Fe(CO)_3$

a. Explain the splittings, assuming large P—P coupling, and that there is no proton-nitrogen coupling observed.

b. Why isn't the N—H coupling observed?

c. Using the A, B, X designation, label this system.

22. The spectrum of the following compound is a complex type. Explain how the double resonance technique could be employed to aid in interpreting the spectrum.

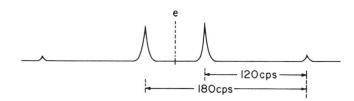

23. The spectrum in the following figure is that of an AB-type of molecule.

    a. Calculate J and Δ.

    b. A 40-MHz probe was employed. Calculate the difference in δ for the two peaks.

    c. If e occurs at a δ value of 2.8, what are the δ values for A and B?

24. Consider the series i-propyl X (where X = Cl, Br, I).

    a. In which compound would the remote paramagnetic effect from the halogen be largest? Why?

    b. In which would the remote diamagnetic effect be largest?

25. What would the nmr spectrum of PF_5 look like under the following conditions $(\Delta_{F(a)-F(b)} > J_{F(a)-F(b)})$?

    a. Very slow fluorine exchange.

    b. Rapid intermolecular fluorine exchange.

    c. Rapid intramolecular fluorine exchange.

26. Consider the molecule shown in problem 8. Using the A, B, X, ... terminology, classify this molecule and indicate the non-equivalent protons.

27. a. List and briefly describe in your own words the factors that influence the magnitudes of proton chemical shifts.

    b. Why do ^{19}F and ^{13}C chemical shifts cover a much larger range than do those of protons?

28. Ramsey's formula is used to calculate the local contributions to the chemical shielding of a nucleus; it is given as equation (7–37). The nuclear magnetic resonance spectrum of ^{59}Co has been observed in a variety of environments. A correlation has been proposed relating the chemical shift to the wavelength of an electronic absorption observed in these complexes.

Complex	λ (Ångstroms)
$K_3[Co(CN)_6]$	3110
$[Co(en)_3]Cl_3$	4700
$[Co(NH_3)_6]Cl_3$	4750
(en = ethylenediamine)	

a. Do you expect the diamagnetic term in Ramsey's formula to account for an 8150 ppm variation in the chemical shift observed in this series of complexes? Explain briefly.

b. Do you expect the paramagnetic contribution to the chemical shielding to vary in this series of complexes? Why?

c. List these cobalt complexes in order of increasing magnetic field for resonance. Give your reasoning.

29. Ramsey's equation [equation (7–37),] is used to calculate local contributions to chemical shifts. Consider the following boron compounds and their ^{11}B chemical shifts (in ppm relative to BF_3 etherate).

Three-coordinate	Four-coordinate
$BI_{3(melt)} = +5.5$	$NaBF_4 = +2.3$
$BF_{3(gas)} = -9.4$	$B(OH)_4^- = -1.8$
$BCl_3 = -47.7$	$BF_3 \cdot piperidine = +2.3$
$BBr_3 = 40.1$	$NaB(C_6H_5)_4 = +8.2$
$B(OCH_3)_3 = -18.1$	$LiB(OCH_3)_4 = -2.9$
$B(C_2H_5)_3 = -85.0$	$BF_3 \cdot P(C_6H_5)_3 = -0.4$

The generalized m.o. description for three-coordinate boron compounds consists of bonding, non-bonding (empty boron a.o.), and antibonding molecular orbitals (I). Four-coordinate boron compounds are described by bonding and antibonding m.o.'s (no non-bonding m.o.) (II).

a. Can the first term of the Ramsey equation explain the ^{11}B chemical shifts of the three-coordinate compounds? Why or why not?

b. Rationalize from the m.o. description the fact that the ^{11}B chemical shift of the three-coordinate complexes varies 125 ppm, while that of the four-coordinate species varies 11.1 ppm.

30. It was said that there is no contribution to the local paramagnetic shielding in a linear molecule aligned parallel to the applied magnetic field. Construct molecular orbitals for the molecule HF and show that matrix elements resulting from the Ramsey equation lead to the foregoing result.

31. Consider benzene and cyclohexane.

   a. In which do you think the proton resonances would be further upfield? Why?

   b. In which do you think the ^{13}C resonances would be further upfield? Why?

   c. In which would $J_{^{13}C-H}$ be larger? Why?

32. In compounds of the type CH_3HgX, the ^{199}Hg-^2H coupling constant is observed to vary by more than a factor of two, depending on the substituent X. Some examples are given below:

X	J(Hz)
CH_3	104
I	200
Br	212
Cl	215
ClO_4	233

   Propose an explanation for the observed splittings.

33. A number of ^{15}N-labeled aminophosphines have been synthesized and the ^{15}N-H nmr coupling constants have been measured. A sample of the results is as follows:

Compound	J(Hz)
$F_4P^{15}NH_2$	90.3
$(CF_3)_2P^{15}NH_2$	85.6
$F_2P^{15}NH_2$	82.7

   There are two possible explanations in terms of the bonding in this series.

   a. Give the explanation that is based on the extent of N → P π-bonding.

   b. Offer another explanation that was presented, which also accounts for these results.

34. a. Would you expect the difference $\delta_{CH_3} - \delta_{CH_2}$ in C_2H_5I to be independent of the remote anisotropy in the C—I bond? Why?

   b. Will this difference have a greater or smaller contribution from anisotropy than δ_{CH_2}?

   c. If in a series of compounds the δ_{CH_2} values showed a different trend than the $\delta_{CH_3} - \delta_{CH_2}$ values, what could you conclude about anisotropic contributions?

35. Combine in your thinking (reread if necessary) the discussion on interatomic ring currents with the relationship between $J_{^{13}C-H}$ and τ (discussed in section 7–18 on Applications of Spin-Spin Coupling to Structure Determination). Using these concepts, propose a method for determining if there is anisotropy in the methyl chemical shift of *B*-trimethyl borazine.

8 Dynamic and Fourier Transform NMR

Introduction

In this chapter we consider the applications of nmr to dynamic systems and then proceed to a discussion of Fourier transform nmr. The chapter concludes with a discussion of liquid crystal and solid nmr. The nmr spectra of paramagnetic ions (Chapter 12) is discussed after some fundamental background in magnetism is developed (Chapter 11).

Evaluation of Thermodynamic Data with NMR

As mentioned in Chapter 7, when two species undergo rapid exchange on the nmr time scale, the chemical shift observed is a mole-fraction weighted average of the two resonances. With rapid exchange:

$$A + X \rightleftharpoons AX \qquad (8-1)$$

the chemical shift of the A (or X) resonance will be a mole-fraction weighted average of the resonance of A (or X) and that of the analogous atom in the AX adduct, as shown in equation (8–2).

$$\delta_{obs(A)} = N_A \delta_A + N_{AX} \delta_{AX} \qquad (8-2)$$

N refers to mole fraction. An analogous equation could be written for the resonance, $\delta_{obs(X)}$.

This exchange averaging can be used[1] to evaluate the equilibrium constant for a reaction. The approach will be illustrated by deriving an expression for the equilibrium constant of the reaction illustrated in equation (8–1). Assume that we are observing the chemical shift of a proton of molecule A that is shifted considerably upon forming the complex AX and that there is rapid exchange between A and AX giving a mole-fraction weighted shift as in equation (8–2).

Expressing equation (8–2) in molarity units, we obtain for the reaction in equation (8–1):

$$\delta_{obs} = \frac{[A]}{[A] + [AX]} \delta_A + \frac{[AX]}{[A] + [AX]} \delta_{AX}$$

Rearranging, collecting terms, and subtracting $[AX]\delta_A$ from both sides of the equation produces:

$$[A](\delta_{obs} - \delta_A) + [AX](\delta_{obs} - \delta_A) = [AX](\delta_{AX} - \delta_A)$$

Defining $\Delta\delta_{obs}$ as $(\delta_{obs} - \delta_A)$ and $\Delta\delta_{CA}$ as $(\delta_{AX} - \delta_A)$, we can write:

$$[AX] = \frac{[A°]\,\Delta\delta_{obs}}{\Delta\delta_{CA}} \tag{8-3}$$

where $[A°]$, the initial concentration of A, equals $[A] + [AX]$.

Substituting equation (8–3) into the equilibrium constant expression

$$K = \frac{[AX]}{([A°] - [AX])([X°] - [AX])}$$

one obtains equation (8–4)

$$K = \frac{\Delta\delta_{obs}}{(\Delta\delta_{CA} - \Delta\delta_{obs})\left([X°] - \dfrac{\Delta\delta_{obs}[A°]}{\Delta\delta_{CA}}\right)} \tag{8-4}$$

In equation (8–4), all quantities are known except K and $\Delta\delta_{CA}$. The two unknowns are constant at a given temperature and can be obtained[1] by solving a series of simultaneous equations that result from measuring $\Delta\delta_{obs}$ in a series of experiments in which $[X°]$ and $[A°]$ are varied. This aspect of the problem is similar to that described in Chapter 4 for systems that obey Beer's law.

8–1 RATE CONSTANTS AND ACTIVATION ENTHALPIES FROM NMR

NMR Kinetics

The proton nmr spectrum of a deuterated cyclohexane molecule, (shown in Fig. 8–1(A)), as a function of temperature is shown in Fig. 8–1(B) to (G). Two isomers are possible where the hydrogen is in either the axial or equatorial position of the cyclohexane ring. At room temperature, the two forms are rapidly interconverting and there is no contribution to the observed line width from exchange effects. As the temperature is lowered, the band begins to broaden from chemical exchange contributions to the nuclear excited state lifetime. This range, down to the temperature at which the two separate peaks are just beginning to be resolved, is referred to as the *near fast exchange region*. As the temperature is lowered further, two peaks are seen and their chemical shifts change as a function of temperature until curve (F) is obtained. This range is referred to as the *intermediate exchange region*. As the temperature is lowered further, the chemical shifts of the peaks no longer change, but the resonances sharpen throughout the so-called *slow exchange region*. Finally, at $-79°$ and lower, there are no kinetic contribu-

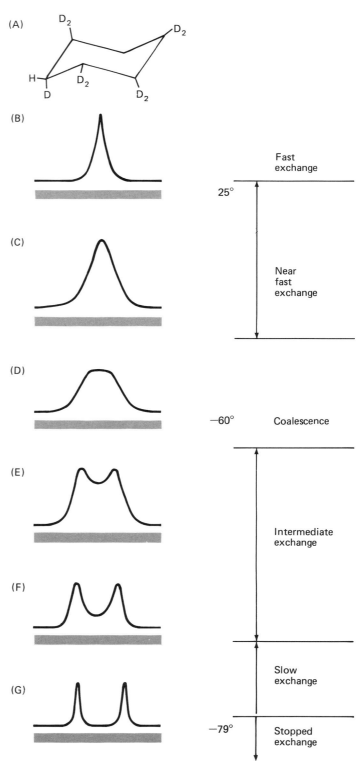

FIGURE 8–1 Deuterated cyclohexane, $C_6D_{11}H$, (A) and the temperature dependence of its nmr spectrum (B to G).

tions to the shape of the spectrum, and this is referred to as the *stopped exchange region*.

By a full analysis of the influence of chemical exchange on the magnetization via the Bloch equations, it is possible to derive equations for the evaluation of rate constants from the nmr spectra. The derivations are beyond the scope of this treatment and the reader is referred to Emsley, Feeney, and Sutcliffe, Volume 1, Chapter 9, or Pople, Schneider, and Bernstein, p. 218. for more details. We shall present the results of these derivations in a form useful for a kinetic analysis and comment on shortcomings of the various approaches that have been employed.

One of the simplest systems is one in which a given proton can be at either one of two molecular sites; the probability that it will be at one site is equal to the probability that it will be at the other, and it has the same lifetime at each. The cyclohexane interconversion and many other isomer or rotamer exchange problems satisfy these criteria. If one works with the chemical shift changes, crude rate data can be extracted from any two-site (A and B) equal-lifetime process in the intermediate exchange region by using equations (8–5).

$$\frac{(v_A - v_B)_{obs}}{v_A{}^0 - v_B{}^0} = \left[1 - \frac{1}{2\pi^2 \tau'^2 (v_A{}^0 - v_B{}^0)^2}\right]^{1/2} \tag{8–5}$$

In this equation $(v_A{}^0 - v_B{}^0)$ is the separation of peaks in the stopped exchange region in Hz; $(v_A - v_B)_{obs}$ is the separation of peaks at a particular temperature in the intermediate exchange region. Recall that we stipulated that the two lifetimes, τ_A' and τ_B', are equal. The lifetime τ' then is the sum of the two, or the lifetime at site A is simply $\tau'/2$.

At the coalescence temperature, the two peaks have just merged, so that $(v_A - v_B)_{obs}$ equals zero. The approximations used to derive equation (8–5) no longer apply, and equation (8–6) is used to obtain the lifetime:

$$\tau' = \frac{\sqrt{2}}{2\pi(v_A{}^0 - v_B{}^0)} \tag{8–6}$$

Equation (8–6) shows that the necessary condition for detecting two exchanging nuclei as separate resonances is given by:

$$\tau' > \frac{\sqrt{2}}{2\pi(v_A{}^0 - v_B{}^0)} \tag{8–7}$$

Thus, the farther apart the chemical shifts at sites A and B, the shorter the lifetime or the faster the kinetic process will have to be to average them. Recall the discussion of this effect in Chapter 4.

Utilization of the foregoing equations [especially equation (8–5)] leads to rate constants with large error limits. Activation enthalpies are obtained from the temperature dependence of the rate constant via the Arrhenius equation. Since the temperature range corresponding to the intermediate exchange region usually is very narrow, huge errors in the activation enthalpies can result. Considerably more accurate information is available about the kinetic process by using line width changes over the entire temperature region that the spectrum is influenced by the kinetic process. These regions are illustrated in Fig. 8–2, where a typical plot is shown for the change in the full width of the resonance line at half height as a function of $1/T$.

In the slow exchange region, the observed line width, $1/T_{2A}'$, of a given peak for site A has contributions to it from the natural line width, $1/T_{2A}$, and from chemical exchange, τ_A, according to:

$$\pi \Delta v_{1/2}(A) = \frac{1}{T_{2A}'} = \frac{1}{T_{2A}} + \frac{1}{\tau_A} \tag{8-8a}$$

Since this equation employs units for Δv of Hz, $\Delta v_{1/2}$ represents the full line width at half height.

This equation and all those in the equation (8–8) series are for a two-site problem with equal lifetimes at both sites, but not necessarily equal populations. This equation does not apply if there is spin-spin coupling of the protons, for contributions of this process to $1/T_2$ are not included. The $1/T_{2A}$ contribution to the width in the slow and intermediate exchange regions can be determined by extrapolating the line for no exchange effects (dotted line a). By difference, $(1/T_{2A}' - 1/T_{2A})$, one obtains $1/\tau_A$. The same calculation could be carried out with the B resonance.

In the near fast exchange region, the contributions to the line width are given by

$$\frac{1}{T_{2A}'} = \frac{N_A}{T_{2A}} + \frac{N_B}{T_{2B}} + N_A^2 N_B^2 (\omega_A^0 - \omega_B^0)^2 (\tau_A + \tau_B) \tag{8-8b}$$

where $\omega_A = 2\pi v_A$, and ω_A^0 and ω_B^0 are the chemical shifts in the absence of exchange. The first two terms on the right-hand side of the equality are

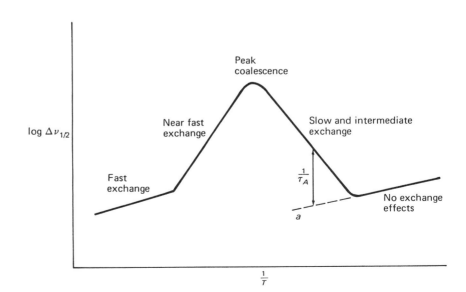

FIGURE 8–2 The influence of chemical exchange on the line width, $\Delta v_{1/2}$, of a spectral band as a function of absolute temperature, T. (The log is plotted to give linear activation energy plots for various regions.)

mole-fraction weighted averages of $1/T_{2A}$ and $1/T_{2B}$ for the A and B sites in the absence of exchange. The last term is the exchange-broadened contribution. The quantities N_A/T_{2A} and N_B/T_{2B} can also be obtained by virtue of the fact that in the fast exchange region

$$\pi \Delta v_{1/2} = \frac{1}{T_{2A}'} = \frac{N_A}{T_{2A}} + \frac{N_B}{T_{2B}} \qquad (8\text{–}8c)$$

Because one can work over a larger temperature range with the line width equations, one can obtain more accurate values for the activation enthalpies than with equations 8–5 and 8–6.

Neither of the above approaches (chemical shifts or line widths) are used any more to evaluate rate constants. Instead, the entire spectral line shape would be calculated using reported equations.[2] One can then obtain the exchange rate by varying τ until the calculated line shape fits the experimental curve. This analysis is generally carried out by a computer that calculates the line shape using an estimated value of τ, i.e., one from a rough $\Delta v_{1/2}$ calculation. The difference between the calculated and experimental intensities is determined. The computer then varies τ until the difference is minimized. Thus, from the line shape at a given temperature, the rate constant can be determined and the activation parameters obtained in the usual manner from the temperature dependence of the rate constant. A typical comparison between a calculated and an experimental spectrum is illustrated in Fig. 8–3. The spectrum[2] is that of the methyl proton of 2-picoline. The picoline is undergoing exchange with $Co(2\text{-pic})_2Cl_2$ in $(CD_3)_2CO$ solvent and the region is the near fast exchange.

When one wishes to study a multisite problem or when there is spin–spin coupling to one of the protons involved in the exchange, a better method involves use of the *density matrix approach*.[3–5] This approach is beyond the scope of our treatment and the reader is referred to references 3 to 5 for details.

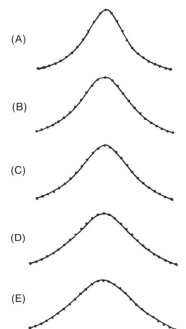

FIGURE 8–3 Comparison of experimental (solid line) and theoretical (dots) nmr methyl resonance spectra at $T = -44°$. All solutions are 0.09 M in $Co(2\text{-pic})_2Cl_2$ and contain the following concentrations of excess 2-picoline: (A) 3.54 M, (B) 2.64 M, (C) 2.40 M, (D) 2.11 M, (E) 1.85 M. [From S. S. Zumdahl and R. S. Drago, J. Amer. Chem. Soc., *89*, 4319 (1967).]

8–2 DETERMINATION OF REACTION ORDERS BY NMR

We return now to a more complete discussion of the τ' values in order to understand how to obtain information about reaction orders. In the nmr experiment, we are studying a reaction occurring at chemical equilibrium. A decay in the net magnetization of the sample results as the forward and reverse reactions occur. Suppose, for example, that the decay at site A is observed. The magnetization decays by a first-order decay process just like the decay of radioactive material. The rate constant, $1/\tau_A$, is a first-order rate constant for the decay of the initial number of protons in our sample at site A, i.e.,

$$-\frac{d[A]}{dt} = \frac{1}{\tau_A}[A]$$

or

$$-\frac{1}{[A]}\frac{d[A]}{dt} = \frac{1}{\tau_A} \qquad (8\text{–}9)$$

In these expressions [A] is the initial concentration of protons at this site and $d[A]/dt$ is the rate at which this initial concentration "disappears." Thus, we are concerned with the lifetime of the initial A and not with the bulk concentration of A. What we measure is $1/\tau_A$ and not the rate, so in a particular experiment a first order decay constant is always observed.

Now consider the mechanism for the chemical reaction that is occurring and causing A to leave site A. It could be a first-order process for which

$$-\frac{d[A]}{dt} = k[A] \quad \text{or} \quad k = -\frac{1}{[A]}\frac{d[A]}{dt} = \frac{1}{\tau_A} \tag{8–10}$$

Therefore, for a first-order reaction, the rate constant k equals the $1/\tau_A$ that is measured. If several experiments are carried out at different concentrations of A for a process that is first order in A, there will be a change in the rate; but there will be no change in the observable from the nmr experiment, which is $1/\tau_A$. Thus, an observation of *no change in lifetime with a change in concentration* corresponds to a first-order process.

Next consider a process that is second order in A:

$$-\frac{d[A]}{dt} = k[A]^2$$

or

$$-\frac{1}{[A]}\frac{d[A]}{dt} = k[A] = 1/\tau_A \tag{8–11}$$

Now in a series of experiments in which [A] is changed, $1/\tau_A$ will also change linearly with the concentration. Observation of this behavior suggests a second-order reaction.

Next consider a reaction that is first order in A and first order in B:

$$-\frac{d[A]}{dt} = k[A][B]$$

Now we must specify whether the A resonance or the B resonance is being studied. If we are studying the A resonance, then we must use

$$-\frac{1}{[A]}\frac{d[A]}{dt} = k[B] = 1/\tau_A \tag{8–12}$$

Now as the concentration of A is changed in a series of experiments, the value of $1/\tau_A$ remains constant; but as the concentration of B is changed in a set of experiments, the value of $1/\tau_A$ changes in direct proportion to the concentration of B. The opposite result is obtained if the B resonance is examined.

In the system illustrated in Fig. 8–3, the exchange of free 2-picoline with the four-coordinate complex, $Co(2\text{-pic})_2Cl_2$, was studied[2] in deuterated acetone as a function of the concentration of 2-picoline. The lifetime of the picoline on the metal, τ_M, was independent of the concentration of $Co(2\text{-pic})_2Cl_2$, but varied linearly with the free ligand concentration. A plot of $1/\tau_M$ vs. [2-picoline] passed through the origin. This indicates a reaction mechanism that is second order overall: first order in $Co(2\text{-pic})_2Cl_2$ and first order in 2-picoline. The activation parameters calculated in this article from a full line shape analysis were compared[2] with those from the line width equations. Errors in τ ranging from 5 to 15% were introduced by the line width approach.

8–3 SOME APPLICATIONS OF NMR KINETIC STUDIES

Ligand exchange reactions have been studied for many transition metal ion complexes. In many cases, the same compound serves as both the ligand and the solvent. One cannot determine the order of the reaction in ligand when the ligand and the solvent are the same, because its concentration cannot be varied. Furthermore, if one wishes to study a series of ligands in a series of experiments, one has varying contributions to the thermodynamic parameters obtained from changing the solvent when the ligand is changed, since the same compound serves both purposes. These problems are avoided, as in the system reported in Fig. 8–3, when the complex has solubility and can be studied in a non-coordinating solvent.

In a study[6] of the exchange of free L (where L is $[(CH_3)_2N]_3PO$) with CoL_2Cl_2 in $CDCl_3$, the resulting kinetic data indicated that the reaction proceeded by both first (*i.e.*, dependent only on CoL_2Cl_2) and second (*i.e.*, first order in CoL_2Cl_2 and first order in L) order reaction paths. Since $CDCl_3$ is a non-coordinating solvent, the first order path provides evidence for the existence of a three coordinate cobalt(II) complex as an intermediate.

The mechanisms for ligand substitution reactions in octahedral transition metal ion complexes are difficult to ascertain. A recent study[7a] of the exchange of CH_3OH with $[Co(CH_3OH)_6^{2+}](BF_4^-)_2$ illustrates the problem. The value of $1/\tau_M$ is independent of the metal complex or methanol concentration, supporting a rate law:

$$\text{Rate} = nk \, [\text{complex}]$$

where n represents the coordination number of the exchanging ligand. The following mechanisms[7a] are consistent with this rate law and should be considered:

Mechanism 1: $S_N1(\text{lim})$ or D-type

$$Co(MeOH)_6^{2+} \underset{k_{-1}}{\overset{k_1}{\rightleftharpoons}} Co(MeOH)_5^{2+} + MeOH \qquad \text{slow}$$

$$Co(MeOH)_5^{2+} + MeOH^* \xrightarrow{k_2} Co(MeOH)_5(MeOH^*)^{2+} \qquad \text{fast}$$

for which the rate law is

$$\text{Rate} = k_1[Co(MeOH)_6^{2+}]$$

Mechanism 2: $S_N(IP)$ or I_d

$$Co(MeOH)_6^{2+} + MeOH^* \underset{}{\overset{K_{os}}{\rightleftharpoons}} \{Co(MeOH)_6^{2+}\}\{MeOH^*\} \quad \text{fast}$$

$$\{Co(MeOH)_6^{2+}\}\{MeOH^*\} \xrightarrow{k_1} Co(MeOH)_5(MeOH^*) + MeOH \quad \text{slow}$$

which has the rate law

$$\text{Rate} = \frac{k_1 K_{OS} [Co]_T [MeOH]}{K_{OS}[MeOH] + 1}$$

where

$$[Co]_T = [Co(MeOH)_6^{2+}] + [\{Co(MeOH)_6^{2+}\}\{MeOH^*\}]$$

In the second mechanism, K_{OS} is the outer sphere complex formation constant. When $K_{OS}[CH_3OH] \gg 1$, the rate law for mechanism 2 reduces to a form identical to that for mechanism 1, and the two cannot be distinguished. If the $[CH_3OH]$ could be reduced to a small value, then $1 \gg K_{OS}[CH_3OH]$ and second-order kinetics would be observed, differentiating the two mechanisms. For a K_{OS} of about 1, changing the $[CH_3OH]$ from 3 M to 8 M would increase the rate constant by only 10% and experimental error would make this change hard to detect. A larger value of K_{OS} would make the difference even smaller. At lower $[CH_3OH]$, dissociation of some methanol from the complex, accompanied by anion coordination, becomes a problem. These problems are described here because they are common to many nmr studies on systems of this type. In order to truly distinguish these types of mechanisms, it will be necessary to study systems in which the free ligand concentration, $[L]$, can be made sufficiently low so that $1 \gg K_{OS}[L]$.

In an nmr kinetic study[7b] using a density matrix analysis the kinetic parameters and the thermodynamics of adduct formation, show that the reaction

$$\text{Ni(SDPT)} \cdot B + B' \longrightarrow \text{Ni(SDPT)} \cdot B' + B$$

(where SDPT is a pentadentate ligand with two negative charges, B and B' is 4-methyl pyridine) proceeds by a pure dissociative mechanism involving a five-coordinate, NiSDPT, intermediate. By comparing data obtained in toluene and in CH_2Cl_2 as solvent, the role played by the latter solvent in the reaction was clearly established.

The spectrum of N,N-dimethylacetamide, $CH_3C(O)N(CH_3)_2$, has three peaks at room temperature, two of which correspond to the different environments of the two methyl groups on the nitrogen (one *cis* and one *trans* to oxygen). The C—N bond has multiple bond character and gives rise to an appreciable barrier to rotation about this bond. As a result of this barrier, the two non-equivalent N—CH$_3$ groups are detected. As the temperature is increased, the rate of rotation about the C—N bond increases and the N-methyl resonances merge, giving rise to a series of spectra similar to those in Fig. 8–1. The lifetime of a particular configuration can be determined as a function of temperature, and the activation energy for the barrier to rotation can be evaluated.[8] Similar studies have been carried out on other amides[9] and on nitrosamines.[10] The quadrupolar ^{14}N nucleus makes a temperature dependent contribution to the line widths in some of these systems, introducing error into the resulting parameters. The rates of inversion of substituted 1,2-dithianes and 1,2-dioxanes[11] are among the many other rates that have been studied.

The mechanism of proton exchange for solutions of methyl ammonium chloride[12] in water as a function of pH was evaluated by an nmr procedure. In acidic solution (pH = 1.0) the nmr spectra consist of a quadruplet methyl peak (split by the three ammonium protons), a sharp water peak, and three broad peaks from the ammonium protons. The triplet for the ammonium protons results from nitrogen splitting. No fine structure is observed in the ammonium proton peak from the expected coupling to the methyl protons. Partial quadrupole relaxation by the nitrogen causes these peaks to be broader than $^3J_{HNCH}$. As the pH is increased, rapid proton exchange reactions begin to occur and the CH_3, H_2O, and NH_3^+ bands begin to broaden. Eventually, at about pH 5, two peaks with no fine structure remain, one from the protons on the CH_3 group and the other a broad peak from an average of all other proton shifts. As the pH is raised to 8, the broad proton peak sharpens again. The CH_3 broadening yields the exchange rate of protons on the amine nitrogen; the broadening of the water line measures the lifetime of the proton on water, and the broadening of the NH_3^+ triplet measures the lifetime of the proton on the ammonium nitrogen. An analysis of the kinetic data yields: (1) the rate law and a consistent mechanism for the exchange

$$CH_3NH_3^+ + B \xrightarrow{k_1} CH_3NH_2 + BH^+$$

where the base B = H_2O, CH_3NH_2, or OH^-; (2) the fraction of the above protolysis that involves water; and (3) the contribution to the exchange reaction from:

$$CH_3NH_2 + BH^+ \longrightarrow CH_3NH_3^+ + B$$

and

$$CH_3NH_2 + B \longrightarrow CH_3NH_2 + B'$$

(B' is B with one hydrogen replaced by a different hydrogen). The details of this analysis can be obtained from the reference by Grunwald et al.,[12] which is highly recommended reading. In subsequent studies the rates of proton exchange in aqueous solutions containing NH_4^+, $(CH_3)_2NH_2^+$, and $(CH_3)_3NH^+$ were measured and compared.[13]

Another example of information obtained from nmr rate studies is illustrated[14] by the ^{19}F spectrum of SiF_6^{2-} in aqueous solution, Fig. 8-4. The main peak arises from fluorines on ^{28}Si ($I = 0$), and the two small peaks result from spin-spin splitting of the fluorine by ^{29}Si ($I = ^1/_2$). The appearance of the satellites corresponding to J_{Si-F} indicates that the rate of exchange of fluorine atoms must be less than 10^3 sec^{-1} [$\tau' > 1/(v_A - v_B)$]. It is also found that the spectrum of solutions of SiF_6^{2-} containing added F^- contains two separate fluorine resonances. These are assigned to F^- and SiF_6^{2-}. There are satellites (J_{Si-F}) on the

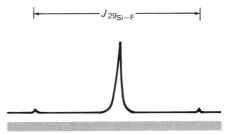

FIGURE 8–4 The fluorine nmr spectrum of SiF_6^{2-}. See table inside back cover (Properties of Selected Nuclei) for ^{29}Si natural abundance.

SiF_6^{2-} peak. When a solution of SiF_6^{2-} is acidified, rapid exchange occurs, the satellites disappear, and the central peak broadens. The following reactions are proposed:

$$SiF_6^{2-} + H_3O^+ \xrightarrow[-H_2O]{} HFSiF_5^- \xrightarrow[-HF]{H_2O} H_2OSiF_5^- \xrightarrow{HF} SiF_6^{2-} + H_3O^+$$

The ^{13}C spectrum of CO_2 in water gives rise to two peaks,[15] one from dissolved CO_2 and a second from H_2CO_3, HCO_3^-, and CO_3^{2-}. Rapid proton exchange gives rise to a single ^{13}C peak for these latter three species. The reaction $CO_2 + H_2O \rightarrow H_2CO_3$ has a half-life of about 20 sec, so a separate peak for dissolved CO_2 is detected.

8–4 INTRAMOLECULAR REARRANGEMENTS STUDIED BY NMR—FLUXIONAL BEHAVIOR

The nmr spectrum of $B_3H_8^-$ is of interest because it demonstrates the effect of intramolecular exchange on the nmr spectrum. The structure of $B_3H_8^-$ is illustrated in Fig. 8–5, along with a possible mechanism for the intramolecular hydrogen exchange.[16–18] The ^{11}B spectrum is a nonet that results from a splitting of three equivalent borons by eight equivalent protons. The process in Fig. 8–5 is very rapid, making the three borons equivalent and the eight protons equivalent in the nmr spectrum. The eight hydrogen atoms remain attached to the boron atoms of the same molecule during the kinetic process and the splitting does not disappear. J_{BH} is a time average of all the different B–H couplings in the molecule. Contrast this to rapid intermolecular exchange, in which a single boron resonance signal would result if exchange made all protons equivalent.

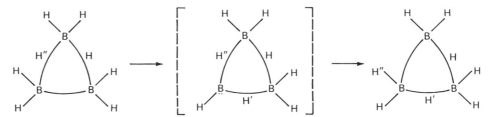

FIGURE 8–5 The structure of $B_3H_8^-$ and a proposed intermediate for the intramolecular exchange.

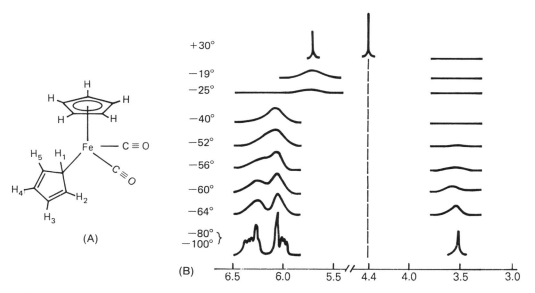

FIGURE 8–6 The proton magnetic resonance spectra of $(\pi\text{-}C_5H_5)Fe(CO)_2C_5H_5$ in CS_2 at various temperatures. The dotted line represents the resonance position of the $\pi\text{-}C_5H_5$ protons at each temperature. the amplitude of the $+30°$ spectrum is shown $\times 0.1$ relative to the others. [Part (B) reprinted with permission from M. J. Bennett, Jr., et al., J. Amer. Chem. Soc., **88**, 4371 (1966). Copyright by the American Chemical Society.]

Another possible explanation of the observed spectrum, that does not involve exchange, is based on virtual coupling. Virtual coupling of the eight protons would give rise to the observed nonet in the ^{11}B resonance. Strong boron coupling has not been observed in the spectra of other boron hydrides.

Considerable information is available regarding the mechanism of exchange processes from the non-symmetrical collapse of an nmr spectrum. An illustration of this basic idea is provided[19] by the temperature dependence of the nmr spectrum of $(\eta^5\text{-}C_5H_5)Fe(CO)_2C_5H_5$, whose structure is shown in Fig. 8–6(A). The spectrum obtained as a function of temperature is shown in Fig. 8–6(B). The pi-bonded cyclopentadiene resonance gives rise to a single sharp peak at $\delta = 4.4$ at all temperatures because rapid rotation of the ring causes the five protons to be equivalent. The remaining peaks in the spectrum arise from the sigma-bonded cyclopentadiene ring. At $-100°C$, three distinct groups of resonances are observed at ~ 6.3, ~ 6.0, and 3.5 ppm with relative intensities $2:2:1$. The most shielded resonance ($\delta = 3.5$) is assigned to H_1. The H_2 and H_5 protons are isochronous, as are H_3 and H_4. The two sets differ only slightly in chemical shift relative to the magnitude of their coupling constant and, as shown in a subsequent section, this situation gives rise to complicated *second-order spectra* that often contain more peaks than a simple analysis of the chemical shift differences and spin-spin coupling would produce. The peaks at ~ 6.3 and ~ 6.0 are due to the 2, 3, 4, and 5 protons. The splitting from the H_1 proton is small; in the limit where $^3J_{H_1-H_{2,5}}$ equals $^4J_{H_1-H_{3,4}}$ the two resonances at 6.3 and 6.0 would be mirror images. However, since $^3J_{H_1-H_{2,5}}$ would be expected to be larger than $^4J_{H_1-H_{3,4}}$, additional fine structure would be expected in the peak assigned to H_2 and H_5. With $^3J_{H_1-H_{2,5}}$ small, the additional fine structure would not be

FIGURE 8–7 Diagrams showing possible intramolecular paths leading to the nmr equivalence of the σ-C_5H_5 protons at room temperature. [Reprinted with permission from M. J. Bennett, Jr., et al., J. Amer. Chem. Soc., *88*, 4371 (1966). Copyright by the American Chemical Society.]

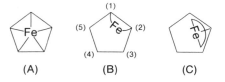

resolved but could be manifested in a broadening of the resonances. Using this criterion, the deshielded multiplet with the broader resonance ($\delta \approx 6.3$) is assigned to the 2 and 5 protons. Since these protons are isochronous, we shall label them as A and the 3,4 pair as B. As the temperature is raised, the A resonance collapses faster than the B resonance. At $-25\,°C$, they are broadened and have completely collapsed. At $+30\,°C$, a sharp peak corresponding to a mole-fraction weighted average resonance of the *three* types of protons is obtained.

Several mechanisms for this *fluxional behavior* are possible. The non-symmetrical collapse of the spectrum rules out any bimolecular process (dissociative or exchange in nature). A first-order dissociative process is ruled out because the experimentally found activation enthalpy is too low. This low activation enthalpy suggests some interaction between the pi orbitals of the ring and the metal in the transition state. The three structures shown in Fig. 8–7 are considered as possible transition states or intermediates and are referred to as the η^5-cp (A), η^2-olefin (B) and η^3-allyl (C) mechanisms. The terms 1,2 shift and 1,3 shift have also been applied to the latter two mechanisms respectively.

The mechanism involving (A) can be eliminated for it would result in a symmetrical collapse of the spectrum. The mechanisms involving (B) and (C) would result in a non-symmetrical collapse, as shown by considering how the labels in Fig. 8–7(B) are changed by a 1,2 shift. This is illustrated symbolically by

$$\begin{bmatrix} H_a \\ A \\ B \\ B \\ A \end{bmatrix} \longrightarrow \begin{bmatrix} A \\ H_a \\ A \\ B \\ B \end{bmatrix}$$

This analysis shows that all A's are changed into different types of protons, but only half the B's are, so the A resonance should collapse more rapidly.

On the other hand, the 1,3 shift is described by

$$\begin{bmatrix} H_a \\ A \\ B \\ B \\ A \end{bmatrix} \longrightarrow \begin{bmatrix} B \\ A \\ H_a \\ A \\ B \end{bmatrix}$$

For this mechanism, the B resonance should collapse faster. Thus, to the extent that the original assignment of the A and B resonances is correct, the 1,2 shift is established. The intermediate or transition state in the 1,2 shift could be thought of as one in which the π-electron density of the ring is arranged in a two-center π bond coordinated to the iron, while the other three carbon atoms have an allyl anion distribution of electron density.

Many fluxional systems have been discovered and are summarized in reference 20. A procedure for a complete line shape analysis on systems of this sort has been reported.[21]

There has been a very considerable effort devoted to nmr studies of the intramolecular rearrangements of trigonal bipyramidal and octahedral complexes. In 1951, Gutowsky and Hoffman[22] reported that the ^{19}F nmr spectrum of PF_5 was a doublet, even though electron diffraction had established the trigonal bipyramidal structure of this molecule. Since P-F coupling is maintained, an intramolecular process is required to equilibrate the fluorine atoms. Many other systems show similar behavior e.g., $Fe(CO)_5$ (^{13}C nmr), $CF_3Co(CO)_3PF_3$, several $HM(PF_3)_4$ species, and $HIr(CO)_2[P(C_6H_5)_3]_2$. There have been several attempts[23,24] to systematically enumerate all of the physically distinguishable intramolecular modes for interconverting groups on a trigonal bipyramid. The problem is a complex one, for it is necessary to insure that apparently different pathways are physically distinguishable. The reader is referred to references 23 and 24 for details.

We shall briefly consider the fluxional behavior of $CH_3Ir(COD)$-$[C_6H_5P(CH_3)_2]_2$ to provide an illustration of the detailed mechanistic information obtainable from work in this area of fluxionality.[25,26] The static structure of the molecule at low temperature is shown in Fig. 8–8(A). At $-3°$, the resonances are assigned as in Fig. 8–8(C). The two methyls on each dimethylphenylphosphine group are not equivalent (i.e., they are diastereotopic). Should the phosphorus ligands change sites, they would become equivalent. The vinyl hydrogens, H_1 and H_2, are also non-equivalent. The phenyl resonances are not shown. As the temperature is raised, the vinyl proton signal collapses into a singlet, but the P-CH_3 proton resonance is not affected up to $67°$.

The reasonable mechanistic paths that could exchange axial and equatorial positions in the trigonal bipyramid are illustrated in Fig. 8–9. Scheme A involves a twist of the diene about a pseudo-twofold axis through the metal and through a point midway between the two double bonds. The intervening intermediate or transition state is the distorted tetragonal pyramid labeled I.

Schemes B and C involve permutations of three sites. In B, one axial and two equatorial ligands are interchanged by a rotation about a pseudo-threefold axis constructed by drawing a line from the metal to the center of the face of the trigonal bipyramid defined by double bond 1, double bond 2, and P_1. This rotation leads to the transition state labeled II. The process in C is best described by considering the group R as being located in the center of the face of a tetrahedron formed by P_1, P_2, double bond 1, and double bond 2. The R group then moves through an edge and into the center of the face formed by P_1, P_2, and double bond 1. The resulting structure has interchanged the two double bonds. Scheme D proceeds through two trigonal bipyramid intermediate structures, Va and Vb, each of which involves a change of four ligands. Each change occurs by a so-called Berry pseudorotation mechanism: two equatorial groups open up their angle and the two axial groups move together in the direction in which the equatorial angle is increasing, to form a distorted tetragonal pyramid. The motion continues in this direction until the two axial bonds are equatorial and the two equatorial bonds become axial. Structures Va and Vb are enantiomers, and ready interconversion is expected. Paths in which the diene spans equatorial

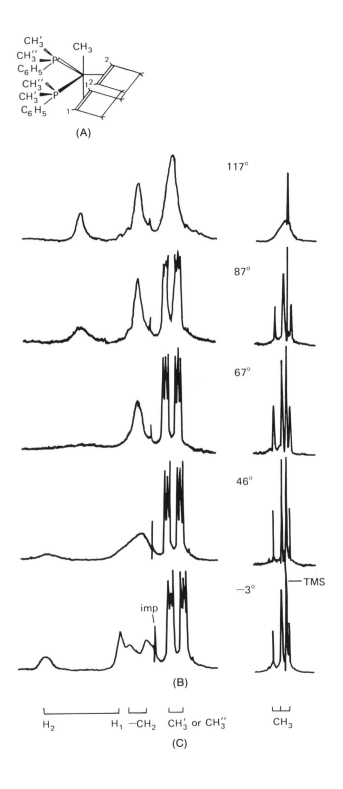

FIGURE 8–8 (A) Structure, (B) temperature dependent proton nmr spectra, and (C) assignment of low temperature spectrum of $CH_3Ir(COD)[(P(C_6H_5)(CH_3)_2]_2$. Solvent is chlorobenzene and IMP refers to acetone impurity present from recrystallization. [Reprinted with permission from J. R. Shapley and J. A. Osborn, Accts. Chem. Res., *6*, 305 (1973). Copyright by the American Chemical Society.]

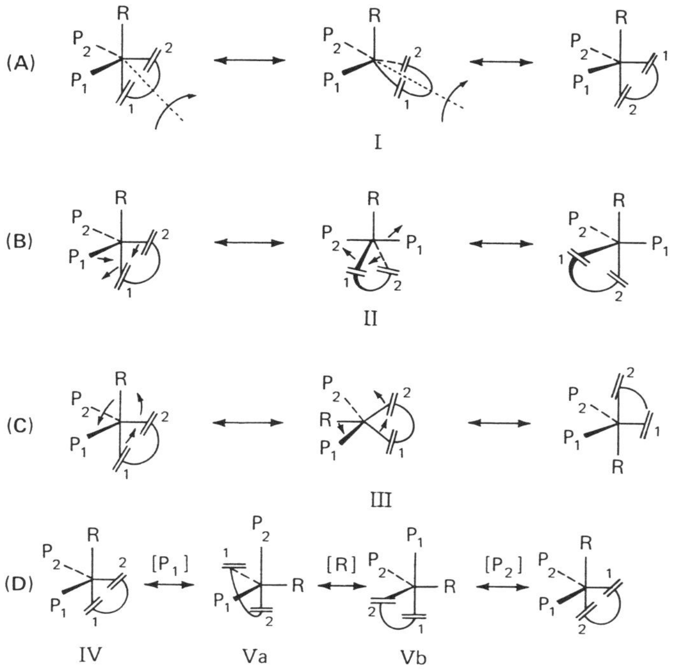

FIGURE 8–9 Mechanistic schemes to account for axial-equatorial equilibration of COD vinyl protons in the complexes RIr(COD)P$_2$, where R is CH$_3$. [Reprinted with permission from J. R. Shapley and J. A. Osborn, Accts. Chem. Res., *6*, 305 (1973). Copyright by the American Chemical Society.]

sites have been eliminated as energetically unfavorable because of the geometrical preference of this ligand for a 90° chelate angle.

Examination of these schemes indicate that B and C interchange the two phosphorus ligands, but A and D do not. Thus, only the latter modes are consistent with the observed spectral behavior. One cannot distinguish between the A and D modes with nmr. There is a very extensive literature on this subject and, with the example discussed here, we have indicated the kind of information that can be obtained. For more details, the reader should consult references 25 and 26. Studies involving fluxional behavior in six-coordinate complexes are described in references 27–29.

8–5 SPIN SATURATION LABELING[30]

The double-resonance experiment can be used to label a proton and obtain kinetic information. Consider molecules AD and A'D,

$$AD + A' \rightarrow A'D + A$$

where A and A' are the same molecules and the exchange rate is slow enough to give two peaks in the nmr for A and AD. If we saturate a proton resonance of A', this peak will disappear, but the exchange process will also cause a partial saturation, *i.e.*, a decrease in the intensity of the corresponding proton resonance in AD, if the exchange rate of A' with AD is comparable to the relaxation rate at the two sites. The lifetime for A, leaving a *particular spin state in AD*, τ_{1AD}, then has contributions from T_{1AD} and τ_{AD} (the lifetime of A at AD)

$$\tau_{1AD} = (T_{1AD}^{-1} + \tau_{AD}^{-1})^{-1} \qquad (8\text{--}13)$$

If a saturating r.f. field is turned on at resonance A′, saturation of this resonance occurs immediately and one can observe (by sitting on the resonance of AD) an asymptotic approach to a new equilibrium value of the magnetization at this point. The plot of intensity versus time and the equilibrium value for the magnetization can be analyzed for τ_{1AD}. When T_{1AD} is known, τ_{AD} can be calculated simply from the equilibrium value of the magnetization of AD.

When the field is turned off, the resonance at AD returns to its initial intensity asymptotically, and this curve can also be analyzed to yield T_{1AD}. This procedure is suited for the determination of reaction rates in the range between 10^{-3} and $1\ \text{sec}^{-1}$. It thus can provide data on the slow side of the nmr line shape experiment and is complementary to the line shape technique.

Spin saturation can also be thought of as equivalent to a deuterium labeling experiment, and some of the same mechanistic information is available from this technique as from the labeling experiment. For example, the heptamethylbenzenonium ion is fluxional:

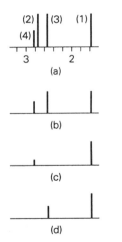

FIGURE 8–10 The spectrum of the heptamethylbenzenonium ion; (B) spectrum (A) with the (2) proton decoupled; (C) spectrum (A) with the (2) and (3) protons decoupled; (D) spectrum (A) with the (2) and (4) protons decoupled.

Does the interchange proceed by a 1,2 shift or by a random migration process? At 28°, four fairly sharp resonances result, as shown in Fig. 8–10A. [There are two types (1) CH_3 groups, two orthos, two metas, and one para.] Upon raising the temperature, the resonances broaden, etc., and a complete line shape analysis suggests a 1,2 shift; but the differences between the expected spectrum for a 1,2 shift and that for the random process are subtle, and a simultaneous operation of both processes was not ruled out. Saturation of the methyl resonance at site 2 produced spectrum (B) in Fig. 8–10, in which the intensity at site 1 is decreased. No additional decrease in intensity is observed at site 1 when sites 2 and 3 are saturated or when sites 2 and 4 are saturated, as shown in Fig. 8–10(C) and (D). If the random mechanism were operative, a further reduction in intensity at site 1 of 22% would be observed upon saturation at sites 3 and 4. The intensity doesn't change, within the 1% accuracy of its determination, so the random process is minor or absent.

8–6 THE NUCLEAR OVERHAUSER EFFECT

An interesting phenomenon associated with the double resonance experiment is the Overhauser effect, discovered in the course of studying free radicals.[31a] When there is a coupling of the nuclear and electron spins, an enhancement in the intensity of the nmr transitions is observed when the esr transitions are saturated. This same effect occurs in a nuclear-nuclear double-resonance experiment and is called the nuclear Overhauser effect, nOe.[31] Changes in the intensity of a given transition are observed when another transition is saturated or perturbed by the second r.f. field. Consider two uncoupled ($J = 0$) spin $^1/_2$ nuclei with the same γ but different chemical shifts. The energy level diagram in Fig. 8–11(A) results with populations indicated by N and δ. For a total of $4N$ nuclei, the populations of each level are indicated as N with $+\delta$ corresponding to the Boltzman excess and $-\delta$ corresponding to the deficiency. The two transitions of A are degenerate ($J = 0$) and give one line. The B transitions give another peak. The $\alpha\alpha \to \beta\beta$ and $\beta\alpha \to \alpha\beta$ transitions have $\Delta m = 2$ and 0 and are forbidden. The population differences for the transitions are δ for the allowed A and B transitions, 0 for $\Delta m = 0$ and 2δ for $\Delta m = 2$. Though transitions involving the latter processes ($\Delta m = 0$ and 2) do not occur, they can provide relaxation mechanisms if the Boltzmann distribution is disturbed. In Fig. 8–11(B), the assumed first-order rate constants for the various paths are indicated by W. We will label these transitions with the Δm value as a subscript and the nucleus involved as a superscript.

If we were concerned with the T_1 relaxation of the A transition, it would depend on the rates for W_1^A, W_2, and W_0. When W_2 and W_0 are zero we get our simple definition of T_1:

$$T_1 = \tfrac{1}{2} W_1^A$$

If W_2 and W_0 are non-zero in a multispin system, our T_1 experiment becomes more complex. Now consider the case of an nOe experiment where we saturate the A transition and observe B after saturation. After saturation the $\alpha_A\beta_B \to \beta_A\beta_B$ levels of Fig. 8–11 have populations $N - (^1/_2)\delta$ while $\alpha_A\alpha_B \to \beta_A\alpha_B$ populations are $N + (^1/_2)\delta$. These levels are given in parentheses in Fig. 8–11. Thus the B transitions have population differences of δ, whereas the A is saturated and zero. Instead of 0 for $\Delta m = 0$ and 2δ for the $\Delta m = 2$ transitions, their population differences are now δ. We shall consider the adjustment back to the equilibrium system shown in Fig. 8–11. W_1^A is of no concern for this is saturated. The population difference across B is not affected for it is still δ. Thus, without W_0 and W_2 there is no Overhauser effect. W_0 transfers population from the $\beta\alpha$ to the $\alpha\beta$ state to get back to the zero population difference. This increases the population of the excited state of one B transition and decreases the population of the ground state of the other one decreasing the intensity of the B transition. This is counterbalanced by W_1^B but if W_0 is dominant, this will lead to a negative nOe at B due to saturation at A.

The W_2 process acts to transfer population to the $\alpha\alpha$ state to restore a population difference of 2δ. This decreases an excited state and increases a ground state population for the B transition. If it dominates W_1^B, a positive nOe arises at B from saturation at A. Solving the differential equations for the system leads to

$$\eta_B^{(A)} = \frac{W_2 - W_0}{2W_1^B + W_2 + W_0} \qquad (8\text{–}14)$$

for the Overhauser enhancement with

$$\eta = \frac{I - I_0}{I} \qquad (8\text{–}15)$$

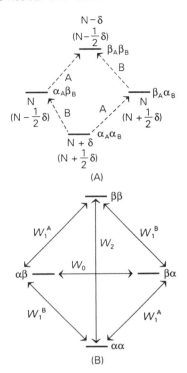

FIGURE 8–11 (A) Transition in an A-B spin system ($I_A = I_B = {}^1/_2$ and $J = 0$). The equilibrium populations are given for each level and those after saturation of the A transition are given in parentheses. (B) Rate processes that can contribute to relaxation.

where I_0 is the normal intensity and I is the intensity that is observed during the perturbation. If nOe is observed, the W_2 or W_0 processes, referred to as cross relaxation, must occur with the sign of the nOe indicating which one is dominant.

The cross-relaxation processes are dominated by dipolar coupling and W_2 is the dominant term for small molecules in non-viscous solvents, with W_0 becoming dominant for large molecules or in viscous solution. These are the most common cases, but there are other possibilities. Depending on the relative signs and magnitudes of the g_N values, there may be signal reduction instead of enhancement, and even negative Overhauser enhancements (emission instead of absorption). The enhancement has been treated quantitatively,[31b] and it can be shown that for the direct coupling mechanism the theoretical enhancement, η, is given by:

$$\eta = \left(1 + \frac{g_1 \beta_1}{g_2 \beta_2}\right) = \left(1 + \frac{\gamma_1}{\gamma_2}\right) \tag{8-16}$$

where the 1 subscript refers to the spin being saturated and the 2 subscript refers to the spin being observed. The ideal enhancement is seldom observed because of incomplete saturation and relaxation by other processes, *e.g.*, T_1 relaxation. If γ_1 or γ_2 is negative, a negative enhancement results for the dipolar mechanism leading in some cases to no peak ($\gamma_1/\gamma_2 = -1$) or to an inverted peak.

This effect can be used to indicate whether the coupling mechanism is direct or indirect. Furthermore, by systematically observing the intensity changes in a proton nmr spectrum as various protons are saturated, one can determine which nuclei are in close proximity. The magnitude of the indirect coupling decreases with the sixth power of the distance. Thus, one can distinguish *cis-trans* isomers this way.

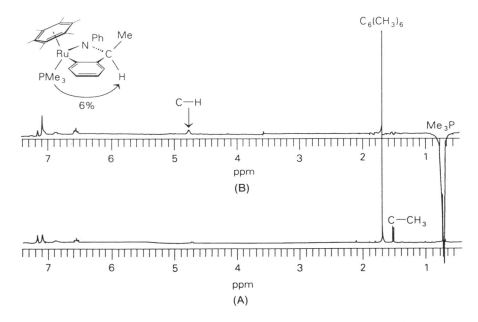

FIGURE 8–12 (A) The nmr spectrum of the complex indicated. (B) The nOe difference spectrum with PMe$_3$ saturation.

An example of the nOe enhancement is shown[32] in Fig. 8–12. The proton nmr [(Figure 8–12(A)], with the C—CH$_3$ doublet at ~1.5 ppm and the C—H resonance at 4.75 does not distinguish between the structure indicated and one in which the hydrogen and methyl of the chiral carbon are interchanged. The spectrum in Fig. 8–12B is the difference spectrum between that in Fig. 8–12 and one obtained when the methyl protons of the coordinated PMe$_3$ are irradiated. With no nOe, a straight line would be observed and with nOe a peak corresponding to enhancement results. The significant (6%) enhancement of the C—H peak in the difference spectrum (indicated with an arrow) as well as the disappearance of the C—CH$_3$ doublet at 1.5 ppm indicates the compound has the structure shown. Enhancements are also observed in the C$_6$(CH$_3$)$_6$ and phenyl resonances as would be expected for both isomers.

When pairs of free radicals are produced in solution, one may observe an effect called chemically induced dynamic nuclear polarization (CIDNP) without the need of saturating the electron spin transition to attain equal electron spin state populations. The mechanism for the CIDNP process is involved, and the reader is referred to reference 33 for more details.

8–7 PRINCIPLES

Fourier Transform NMR

Many magnetic nuclei are present in nature in low abundance and also have low sensitivity (*e.g.*, ^{13}C). In examining the spectra of these materials, the tendency is to increase the r.f. power, but this often saturates the signal. Alternatively, one can sweep the spectrum many times at low power and store the spectra in a computer. This is called a CAT (computer averaged transients) experiment. The noise is random and partially cancels (the signal is proportional to N and the noise to $N^{1/2}$). The signal is reinforced by each sweep and, eventually, on adding many sweeps, the spectrum emerges from the background. This process is time consuming. Pulse techniques, referred to as Fourier transform (FT) nmr, are very advantageous for this situation. In view of the many advantages of FTNMR to be discussed in this chapter, this instrumentation now dominates the market.

In a magnetic field H_0 (z-direction), the equilibrium static magnetization is shown in Fig. 8–13 as a bold arrow. The individual nuclei precess around the z-axis with a Larmor frequency, $\omega = \gamma H_N$, where H_N is the effective field felt at the nucleus ($H_0(1-\sigma)$ of equation 7–28). This leads to different frequencies of rotation for each chemically shifted nucleus. The rotating xy-components of two such nuclei, M_1 and M_2, are shown in Fig. 8–13 where they are rotating with Larmor frequencies ω_1 and ω_2. Now define a rotating coordinate system that rotates in the xy-plane at the same angular frequency, ω_0, as the Larmor frequency of a reference compound *e.g.*, TMS. If you were an observer on the rotating y-axis, for example, the TMS component and all nuclei with the same chemical shift (*i.e.*, same ω) would appear to be stationary. Nuclei in the sample that are deshielded compared to that from TMS, precess at higher frequencies, ω_1, and are moving slowly relative to the rotating frame at the rate $\omega_1 - \omega_0$. Since the nuclei are either standing still or moving slowly, this corresponds to an effective field along the z-axis that is either zero or some small value.

Since the effect of H_0 has vanished, we can examine the effect of a weaker applied magnetic field, H_1, on the nuclei. *In order to exert a constant force, this*

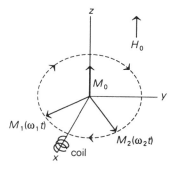

FIGURE 8–13 Rotating x,y-components, M_1 and M_2 for two nuclei with different chemical shifts. M_0 is the static magnetization aligned along z field direction H_0.

H_1 field has to be rotating at the frequency of the nuclei. This rotating magnetic field is applied by a coil of wire around the x-axis energized by an oscillating current at reference frequency ω_0. The rotating H_1 field is a static vector in the rotating frame and appears as such to the nuclei. When the H_1 field is on, it is the only field seen by M_0 and, as a result, M_0 begins to precess at the rate $\omega = \gamma H$, in the yz plane. For a ^{13}C nucleus, a 10-gauss rotating H_1 field would produce one full rotation of M_0 (from $+z$ to y to $-z$ to $-y$ back to z) in 100 microseconds. If the pulse is applied for 25 microseconds, the M_0 vector will rotate 90° and point along the y axis. The angle θ, in radians, through which the magnetization is tipped is given by:

$$\theta = \gamma H_1 t_p \qquad (8\text{--}17)$$

where t_p is the time of the pulse. The angle of tipping is called the flip angle and the pulse causing a 90° flip is called a $\pi/2$ or 90° pulse. Doubling the t_p of a $\pi/2$ pulse leads to a 180° or a π pulse.

In the nmr experiment, we monitor the y-component of M_0. After the H_1 field is shut off, the individual nuclei decay back to their H_0 precession. For a nucleus rotating at ω_0, our rotating frame and H_1 frequency, the y-component decays as shown by the dashed line in Fig. 8–14. This is referred to as a free induction decay curve and the y-component axis is labeled FID in Fig. 8–14. These nuclei are rotating in phase with the rotating frame and the y-magnetization slowly decays to zero as the z magnetization approaches its equilibrium value. Now consider a nucleus that is rotating in the xy plane at a frequency ω_1, different than ω_0 (M_1 of Fig. 8–13). These nuclei are moving faster or slower than the rotating x,y frame and move in and out of phase with the y axis. The component detected along the y axis during the time required for M_y to decay to zero is shown by the solid line in Fig. 8–14.

The FID curve is the superposition of four decay curves like that shown by the solid line in Fig. 8–15. These four decay curves correspond to the four ^{13}C resonances in the frequency spectrum that arise from the proton spin-spin coupling. The FID ^{13}C spectra of molecules containing more nuclei are even more complicated.[43,44] That of progesterone, with the protons decoupled, is given in Fig. 8–16.

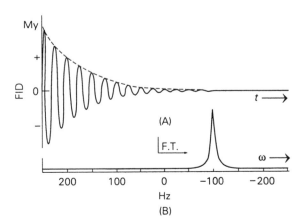

FIGURE 8–14 A free indication decay curve (A) and its Fourier transform for a single absorbing nucleus (B). (The dashed line is obtained when ω_0 equals the chemical shift, and the solid line when ω_0 is off resonance.)

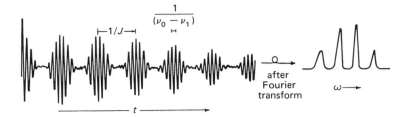

FIGURE 8–15 FID spectrum and its Fourier transform for $^{13}CH_3I$. (From T. C. Farrar and E. D. Becker, "Introduction to Pulse and Fourier Transform NMR Methods," Academic Press, New York, 1971.)

In view of the complexity of this pattern, the FID interferogram is seldom reported. When we are doing repeated scans, the FID curves are stored in the computer, added together and Fourier transformed. Adding the individual frequency spectra would be inefficient because of the computer time required to do a Fourier transform. For a typical compound, we pulse for 10^{-5} sec, measure the free induction decay curve, store it in the computer, pulse again, measure the FID curve, add it to the other one in the computer, and continue this for many pulses. If the nucleus were sensitive enough, just one pulse would give an FID curve that could be Fourier transformed to give the frequency spectrum. When it is not sensitive enough, the entire spectrum is run many times and since one FID curve is obtained in a few seconds, many spectra can be obtained in the time required for a slow passage experiment. Basically, then, we measure the decay of the magnetization in the u,v coordinates of the rotating frame in which the net magnetizations of the different kinds of nuclei are precessing at their Larmor frequencies.

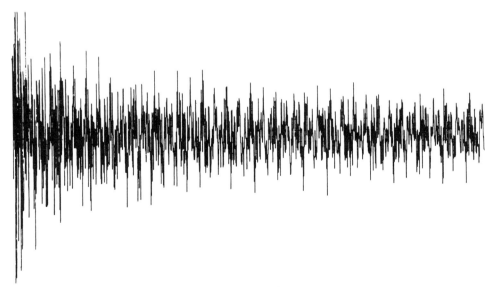

FIGURE 8–16 ^{13}C resonance of progresterone (proton noise decoupled). (From T. C. Farrar and E. D. Becker. "Introduction to Pulse and Fourier Transform NMR Methods," Academic Press, New York, 1971.)

FIGURE 8–17 A sequence of r.f. pulses.

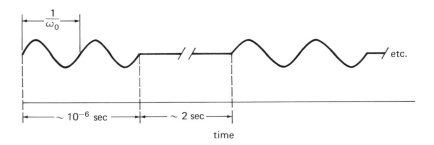

The r.f. pulse is applied for a time t_p (typical times are 10^{-5} to 10^{-6} sec) with a fixed frequency ω_0 and with very high power (100 watts). After waiting a couple of seconds, the pulse is applied again, as shown in Fig. 8–17.

The pulse, if properly selected, causes all of the magnetic nuclei of a certain element in the sample to absorb and eventually (after a transformation) leads to a typical nmr spectrum. The problem now is to show how all these nuclei with different Larmor frequencies can be made to undergo transitions by a pulse of a single frequency ω_0. The mathematical function of time, $f(t)$ in Fig. 8–17, corresponding to a single pulse can be reproduced by summing together a series of sine and cosine functions with different ω's. This can be shown with a Fourier transform. A Fourier transform corresponds to a generalized transformation in function space. In this case, we wish to convert the time plot in Fig. 8–17 to the corresponding frequency plot, using equation (8–18):

$$F(\omega) = 2\pi \int_{-\infty}^{\infty} e^{i\omega t} f(t)\, dt \tag{8–18}$$

where $f(t)$ is in terms of ω_0, t_p, and t. The result in Fig. 8–18 is obtained, where Δ is the range of the principal frequency components. Consequently, pulsing our frequency ω_0 is comparable to having a whole distribution of frequencies available to us from the r.f. source.

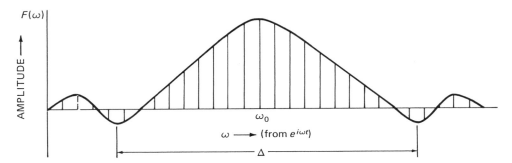

FIGURE 8–18 The frequency plot corresponding to the time plot of Fig. 8–17.

Probably the easiest way to see this is by considering how a square wave, $f(t)$, can be produced with a series of sine waves, i.e., a Fourier series. A square wave is illustrated in Fig. 8–19.

In the limit as $n \to \infty$, the following Fourier series will reproduce the square wave function:

$$f(t) = a_0 + \sum_{1}^{n} (a_n \cos 2\pi nt + b_n \sin 2\pi nt) \qquad (8\text{-}19)$$

Figure 8–20 shows how the superposition of the first three partial sums S_1, S_2, and S_3 corresponding to $n = 1$, 2, and 3, approaches a square wave. In Fig. 8–20(A), we illustrate S_1; in (B) the second term has been added to give S_2. Adding the third term in (C) produces S_3, which is beginning to resemble a square wave. As n becomes very large, the resemblance becomes better.

In a similar way to that just described for a square wave, the distribution of frequencies in Fig. 8–18 can be converted to A vs. t plots and added to give the curve in Fig. 8–19. If there were no pulse, but just one continuous wave approaching infinite time (the slow passage limit), only one frequency would be required to describe this continuous wave, ω_0. As the time, t_p, of a single pulse decreases, the span of frequencies needed to describe this pulse increases. The range of frequencies, Δ, in Fig. 8–18 is obtained from the Fourier transform of the wave in Fig. 8–17 and is given by:

$$\Delta = 4\pi/t_p \qquad (8\text{-}20)$$

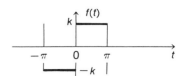

FIGURE 8–19 Graphical representation of a square wave.

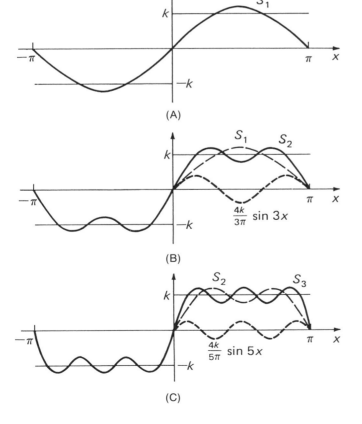

FIGURE 8–20 Addition of the first three waves in the Fourier series leading to a square wave.

where t_p is the duration of the pulse. In the description of the single pulse, the wave stops after time t_p. A given frequency continues for infinite time. When the pulse is passed through the sample, the appropriate frequencies in Fig. 8–18 are absorbed by the sample, causing transitions. Therefore, the pulse must be short enough to cover the distribution of expected spectral frequencies with similar intensity frequency components. As we can see in Fig. 8–18, the intensity falls off as one moves from ω_0. Accordingly, t_p should be much less than $4\pi/\Delta$ to get a reasonable distribution of similar intensity frequency components. For a typical pulse of 10 μsec, the flat central portion of Fig. 8–18, where the amplitude is about 1% of the peak value is about 16,000 Hz wide.

8–8 OPTIMIZING THE FTNMR EXPERIMENT

One of the major advantages to FTNMR is the ability to obtain spectra of dilute solutions of sensitive nuclei or concentrated solutions of insensitive nuclei (*i.e.*, nuclei with low magnetogyric ratios *e.g.*, ^{14}N.) Several variables need to be considered in order to optimize the experiment for difficult samples. These will be considered briefly here to give an appreciation of the problem. The reader is referred to references 35 and 36 for greater detail.

A single 90° pulse rotates the magnetization into the xy plane and produces the maximum signal in the FID. If one then collects the FID for a period τ equal to $5T_1$, 99.3% of the recovery back to the equilibrium z-component occurs. In the usual case where $T_1 \gg T_2$, most of the instrument time is spent waiting instead of collecting data. If the second pulse is followed before a τ equal to $5T_1$, a smaller magnetization is observed. When τ is $1.27T_1$, only 72% of M_0 is collected, but since more pulses can be employed in the same time, a 43% gain in sensitivity results. If a small flip angle, *e.g.*, 30°, is used, M_y is now 1/2 of M_0 (sin 30°), but since M_z is 85% (cos 30°) of its original value, it doesn't take as long to decay back to the equilibrium M_z and the second pulse can be applied faster. A gain of 50% is achieved for $\tau = T_1$ and even higher when $\tau < T_1$. The flip angle approach is used to enhance signal to noise with the optimum flip angle given by:

$$\cos \theta_{\text{opt}} = e^{-\tau/T} \qquad (8\text{–}21)$$

It is clear from this discussion that in a sample containing nuclei with varying T_1's, the intensity distribution of the peaks can be distorted because of variations in the amount of the FID curve obtained. It can also be appreciated that nuclei with large T_1's can be completely missed, *e.g.*, the carbons of metal carbonyls are often not seen using the standard acquisition times because only a small fraction of the FID curve is obtained.

Selection of the pulse frequency, ω_0, is an additional consideration. Changing ω_0 lets you vary the regions over which one would carry out a 500-Hz sweep, for example. If ω_0 is selected at a lower frequency than a peak in the spectrum, this peak can appear as an echo on the other side of ω_0. It often shows up as an inverted peak with an irregular pulse.

Our final consideration is variation of the spectral width. As shown in equation (8–20), shorter pulses give a wider range of frequencies enabling one to increase the spectral width. However, going to a larger spectral width leads to a decrease in digital resolution because of the limited number of storage channels

8-9 THE MEASUREMENT OF T_1 BY FTNMR

In a static field, the nuclear moments precess about the field direction as shown in Fig. 8–21. As described earlier, when a secondary field H_1 is applied, which is in phase with the Larmor frequency, a torque is exerted that tends to make the moment precess about H_1. If we define a rotating coordinate system that rotates as the Larmor frequency, we only have to worry about the torque from H_1. If we consider the H_1 direction to be perpendicular to the page, the cone is so tipped that projection of the magnetic moment vectors in the xy plane gives a net xy-component (see Fig. 8–22). Relaxation tends to restore the system to the situation in Fig. 8–21. The torque is the cross product of the magnetic moment vector and $\vec{H}_1$, so at resonance in the slow passage experiment, H_1 tends to tip the net magnetic moment vector (which has no xy-component in the absence of H_1), inducing an xy-component. (When H_1 is applied, α and β are not eigenfunctions, but some linear combination of them is.) This net xy-magnetization is detected when one passes through resonance in the slow passage experiment.

In a pulse experiment, it is possible to tip the magnetization vector 90°, 180°, or $n°$, depending on the duration of the pulse. In all but the 180° pulse experiments, an xy-component is induced. In a 180° pulse experiment, we invert the magnetic moment vector (180° inversion) from the position where $H_1 = 0$ (i.e., from positive m_z to negative m_z) and do not generate any xy-component as can be seen for the net moment in Fig. 8–23.

After the pulse is turned off and decay occurs, the magnitude of M_z just decreases at a rate governed by the longitudinal relaxation time T_1. The series of arrows in Fig. 8–24 represent the decay of the $\vec{M}_z$ vector with time. *Since the individual moments relax in a purely random manner, no xy-component results.*

It is impossible to detect this decay of M_z for it has no xy-component. Now consider an experiment in which we hit the sample with a 180° pulse, followed by a 90° monitoring pulse. We can then detect the magnetization. After waiting for thermal equilibrium to be reestablished (usually a time corresponding to $5T_1$ is employed), we can again subject the sample to the same 180° pulse, but now wait a while, and then follow with a 90° monitoring pulse. Such an experiment is referred to by the symbolism 180-τ-90. The process can be repeated by waiting for a longer time τ before applying the 90° monitoring pulse. The series of spectra that result from a sample is shown in Fig. 8–25 (page 316), where the delay time τ, before the 90° pulse, increases from left to right.

FIGURE 8–21 Precession of nuclear moments in a Zeeman experiment. (From T. C. Farra and E. D. Becker, "Introduction to Pulse and Fourier Transform NMR Methods," Academic Press, New York, 1971.)

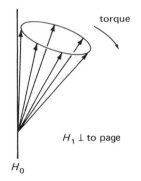

FIGURE 8–22 Effect of a secondary field H_1.

FIGURE 8–23 The effect on the magnetization of a 180° pulse.

FIGURE 8–24 Decay of the M_z component with time after a 180° pulse.

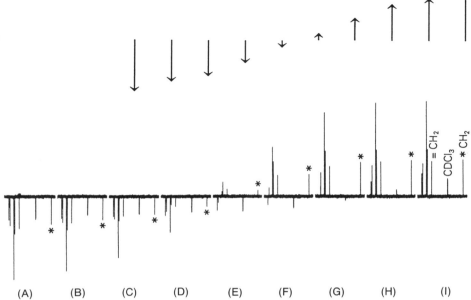

FIGURE 8–25 T_1 determination of the ^{13}C Resonances of allyl benzene ($C_6H_5CH_2CH=CH_2$). (A) 0.469 sec; (B) 0.938 sec, (C) 1.875 sec, (D) 3.75 sec, (E) 7.5 sec, (F) 15 sec, (G) 30 sec, (H) 60 sec, (I) 120 sec after the 180° pulse. The T_1 value for the CH_2 and $=CH_2$ resonances are 9.7 and 8.4 sec.

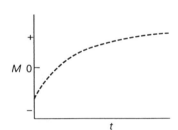

FIGURE 8–26 Change in magnetization (peak height) of the z-component with time following a 180° pulse.

The peak assigned to the CH_2 carbon bound to the phenyl ring is marked with an asterisk. The magnetization of this carbon is seen to follow the pattern shown in Fig. 8–24. If one plots the magnetization as a function of time, τ, the curve in Fig. 8–26 is obtained. This curve is of the form:

$$M = M_0(1 - 2e^{-t/T_1})$$

where M_0 is the initial magnetization. The equation can be solved for T_1. A value of 9.7 sec results for the carbon marked with the asterisk in Fig. 8–25. T_1 can also be estimated at zero signal intensity from the relation $t_{1/2} = T_1 \ln 2$. NMR instruments have the software to provide values of T_1 for all of the peaks in the spectrum. Proceeding from right to left in Fig. 8–25, values of 9.7 sec for CH_2; 68, 62, and 66 sec for the $CDCl_3$ triplet; 8.4, 7.4, 10.6, 10.6, 15.3, and 41.4 sec result. The last value listed corresponds to the ring carbon to which the allyl group is attached and is barely discernible in the spectra. The $CDCl_3$ solvent T_1 is not accurate because the delay time between experiments is not long enough.

An inverted resonance is obtained whenever the net magnetization of the sample is opposed to the field, for the detection system senses this as an emission (transition from a state opposed to the field to one aligned), and vice versa for net magnetization with the field. Another way of looking at this is that a 90° clockwise rotation about $\vec{H}_1$ of a magnetic moment vector opposed to the field gives rise to a different phase (180° different) than rotation by 90° of a vector aligned with the field. This is illustrated in Fig. 8–27, where $\vec{H}_1$ is to be considered perpendicular to the page.

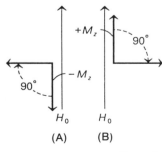

FIGURE 8–27 The effect of a 90° pulse on (A) $-M_z$ and (B) $+M_z$ (in the rotating frame).

8-10 USE OF T_1 FOR PEAK ASSIGNMENTS

We will digress for a moment from our discussion of nmr to make clear what is meant by a *correlation function*. Correlation functions are employed in the description of processes in which the value of x does not depend on t in a completely definite way, *i.e.*, a random time process. However, the average dependence of $x(t)$ on t can be written in terms of probability distributions. When discussing the self-correlation of a variable, the function is referred to as an *autocorrelation function*.

Correlation functions are useful for describing random processes *e.g.*, Brownian motion of the fluctuating fields arising from the random motion of molecules that cause relaxation. We can define a correlation function, $R(\tau)$, for a system in which the spin exchanges energy with some property (*e.g.*, a magnetic moment) that is in equilibrium with the lattice. When these motions correspond to a large number, n, of independent and uncorrelated events, the time-dependent correlation function is given by the relation:

$$T(\tau) = \left\langle \sum_i a_i(t)^2 \right\rangle e^{-\tau/\tau_C} \tag{8-22}$$

where $a_i(t)$ is the magnitude of the property (*e.g.*, the nuclear magnetic moment, describing the n particles, with the summation over all particles $i = 0 \ldots n$. The symbol t refers to time, τ to a time increment and τ_C a time constant whose reciprocal is the molecular reorientation rate constant. If we consider a case in which an assemblage of nuclear moments were all aligned, equation (8-22) describes how the vectoral sum is randomized with time.

A plot of a correlation function for some molecular process that gives a Lorentzian line in the frequency domain is given in Fig. 8-28. The correlation time τ_c for a process in which $R(\tau)$ is an exponential in the time domain is defined as the time required to get through $1/e$ of the curve (curve is $R(\tau) \approx e^{-\tau/\tau_c}$). As we shall see, the correlation time will be needed in order to interpret the values obtained from T_1 measurements.

The following principles underlie the use of T_1 in making peak assignments in the nmr spectrum of a complex molecule.

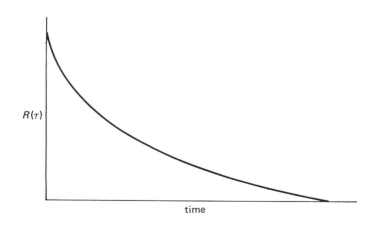

FIGURE 8-28 Correlation function for a process yielding a Lorentzian line in the frequency domain.

1. ^{13}C relaxation times of protonated carbons in large or asymmetric molecules are dominated by dipolar interactions with the directly bonded protons. The value of $1/T_1$ for ^{13}C relaxation by a hydrogen comes from use of equation (8–23) when molecular rotation is isotropic, the hydrogens are decoupled, and $(\omega_C + \omega_H)\tau_{eff} \ll 1$ (the so-called extreme narrowing limit).

$$1/T_1 = N\hbar^2 \gamma_C^2 \gamma_H^2 r_{CH}^{-6} \tau_{eff} \qquad (8\text{–}23)$$

Here γ_C and γ_H are the gyromagnetic ratios of ^{13}C and ^{1}H, N is the number of directly attached hydrogens, r_{CH} is the CH distance, and τ_{eff} is the effective correlation time for rotational reorientation. Assume that only directly bound hydrogens contribute. Otherwise, we have to sum the $r_{CH}^{-6}\tau_{eff}$ terms for any other proton that could contribute. In order for the limit $(\omega_C + \omega_H)\tau_{eff} \ll 1$ to hold, the atom must be rapidly rotating. The equation applies only under conditions of proton decoupling where the scalar coupling is removed. However, under these conditions, it should be emphasized that the magnetic moments of the protons are not undergoing transitions *fast enough to average the moment to zero* in the time required for a rotation, so dipolar relaxation still occurs. Under these decoupled conditions, the Fermi contact coupling makes no contribution to the relaxation mechanism and the dipolar effect dominates. It should be emphasized that because of the inverse sixth power in r_{CH}, dipolar contributions to $1/T_1$ from atoms a long distance away will be negligible. In proteins, long is usually considered to be ≥ 5 Å but is longer in rigid molecules.

2. If two carbon atoms in a molecule have the same τ_{eff}, but one of them is not protonated and the other is, then the non-protonated one will have a much longer T_1 than the protonated one.

3. Differences in τ_{eff} for different carbons in the same molecule may arise from anisotropic rotation of the molecule in solution and from the effects of internal reorientation. For a more complete discussion of the effects of internal rotation and of proton decoupling, the reader is referred to references 37 and 38.

Applications of these principles are illustrated in the T_1 measurements of some of the ^{13}C resonances of cholesteryl chloride.[39] The results are summarized in Fig. 8–29. The protonated carbons on the ring backbone (not shown) and other protonated carbons (also not shown) all have the same T_1 and hence the same τ_{eff}. The overall reorientation of the molecule is isotropic, and T_1 values can be used to distinguish CH and CH$_2$ protons. The carbon with no protons attached, C(13), is seen to have a much larger T_1 (*i.e.*, a sharper line when $T_1 = T_2$) than others in the molecule. The methyl carbons have long T_1's, considering that

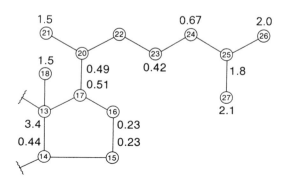

FIGURE 8–29 T_1 values in (sec for various ^{13}C atoms in cholesteryl chloride in CCl$_4$ at 42°. [Reprinted with permission from A. Allerhand and D. Doddrell, *J. Amer. Chem. Soc.*, **93**, 2777 (1971). Copyright by the American Chemical Society.]

there are three protons on such a carbon. This is due to the rapid rotation about the C—C bond that decreases τ_{eff}; as seen in equation (8–22) this decreases $1/T_1$ or increases T_1. By comparing T_1's down the long side chain at the top of the figure, one sees that the effect of internal motion increases toward the free end of the chain as expected.

The proton T_1 values of vinyl acetate are shown in Fig. 8–30. The r^{-6} dipolar effects of the protons on each other are seen in all three values.

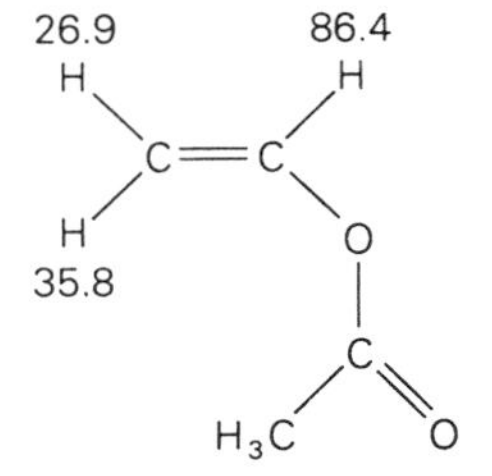

Advantage can be taken of the large differences in the T_1 values of protons to simplify the nmr spectrum. For a particular delay time following the initial 180° pulse, the magnetization vector can decay to zero intensity. No intensity will then be detected when this is followed by a monitoring 90° pulse (see Fig. 8–25 and the discussion of it). Thus, if there are two overlapping peaks with different T_1's, a 180° pulse followed by a properly timed wait before the 90° pulse will remove the peak with the shorter T_1. A threefold or greater differential in the relaxation times of the overlapping protons is ideal for this application. With T_1 for HOD about 7 sec and that of solutes usually less, this technique can be used for solvent suppression.

FIGURE 8–30 T_1 values for the protons of vinyl acetate.

8–11 NMR OF QUADRUPOLAR NUCLEI

The nmr lines of quadrupolar nuclei are very broad. For spherical rotation, the width is a function of the nuclear quadrupole moment and the correlation time τ_c as given in equation (8–24):

$$\frac{1}{T_1} = \frac{3}{40} \cdot \frac{2I+3}{I^2(2I-1)} \left(1 + \frac{1}{3}\eta^2\right)\left(\frac{e^2qQ}{\hbar}\right)^2 \tau_c \qquad (8\text{–}24)$$

Since the nuclear quadrupole effect on the relaxation process dominates the line width and since this effect is intramolecular, only the rotational contribution to T_1 is important. However, it is possible to have both isotropic and anisotropic rotation, and this complicates the line width interpretation. For example, in $CHCl_3$, the rotational diffusion constant at room temperature perpendicular to the C_3 axis, $D_\perp$, is 0.96×10^{11} sec^{-1}, whereas that parallel to this axis, $D_\parallel$, is 1.8×10^{11} sec^{-1}.

8–12 ^{13}C

A brief discussion of ^{13}C magnetic resonance, abbreviated cmr, will give us an opportunity to review and apply several of the principles and phenomena we have developed: the Fourier transform technique, the use of T_1 data to assign resonances, and the nuclear Overhauser enhancement that results in the ^{13}C spectrum by decoupling the proton nmr.

The total signal-to-noise gain when all of the techniques we have been discussing are employed is interesting to note. In Fourier transform spectroscopy, the multiplet collapse from double resonance, the nuclear Overhauser effect, the larger size of sample tube, and the accumulation that can be done in the same

Applications and Strategies in FTNMR

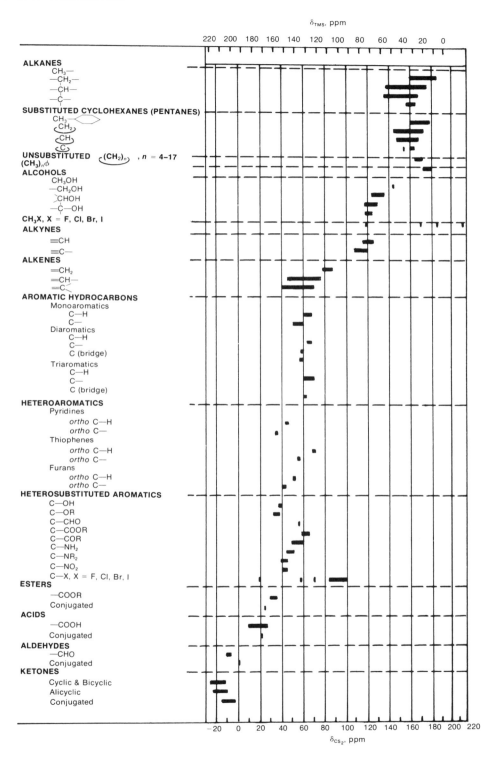

FIGURE 8–31 A ^{13}C correlation chart. [From R. K. Jensen and L. Petrakis, J. Mag. Res., 7, 106 (1972).]

time all lead to a total gain of $\sim 10^4$. Routine high resolution ^{13}C spectra of organic molecules can be obtained on 0.5-M solutions in 20 minutes and on as low a concentration as 0.05 M in 20 hours.

The cmr spectra are dominated by the local paramagnetic term leading to a wider range of shifts than protons. The shifts observed provide a very good indication of the functional groups present in a molecule, as can be seen from the correlation chart in Fig. 8–30. The fact that a given carbon-containing functional group (e.g., a substituted benzene or heterocycle containing ring carbons) shows a very characteristic shift has made the fingerprint type of application even more successful than in proton magnetic resonance.

Because of the nuclear Overhauser effect, the integrated intensity of a ^{13}C resonance is not necessarily proportional to the number of carbons under the peak. In some cases, this problem can be overcome by adding a free radical or a paramagnetic ion to the solution. Spin-spin coupling from ^{13}C-^{13}C is not seen in the spectrum of a molecule containing several C—C bonds if it is not enriched in ^{13}C because with a 1.1% natural abundance the probability that two ^{13}C nuclei would be next to each other in the molecule is very low, $\sim 10^{-4}$.

In a large organic molecule, there are many ^{13}C resonances to be assigned. The correlation chart in Fig. 8–31 will obviously not permit an assignment of all the resonances. As mentioned earlier, T_1 measurements are a considerable aid. There is another technique, referred to as *off-center double resonance*. As mentioned earlier, if all the protons are decoupled in an nmr experiment, all the information potentially available regarding the proton splitting is lost. If the proton decoupling frequency employed is off center with respect to the proton frequencies involved, only a partial collapse of the multiplet results and some Overhauser enhancement is obtained. The resulting coupling constants and peak intensities are distorted[40] compared to those in the spectrum obtained without any decoupling. Long-range coupling is eliminated. The partial multiplet structure helps assign the resonances. This technique is more difficult at fields of 4.7 T or more and has been replaced by polarization transfer and two-dimensional techniques.

Since a major relaxation path involves the dipolar coupling of bound protons, it is often found (particularly for non-protonated carbons, e.g., a metal carbonyl) that a carbon signal can be saturated and is missing from the spectrum. To observe nuclei with very long T_1's, small flip angles [θ of equation (8–15)] are used in running the spectra and long pauses between pulses must be employed to obtain the free induction decay curve. Even with these techniques, poor spectra are often obtained. It has been shown that T_1 can be drastically reduced by adding paramagnetic complexes, as mentioned earlier, for large fluctuating magnetic fields arise from the paramagnetic complex moving through the solvent and these lead to the very effective T_1 relaxation. Trisacetyl-acetonato chromium(III) is a paramagnetic complex that is quite unreactive and has been successfully employed this way.[41] A series of spectra in which the concentration of the relaxation agent is varied should be studied to determine whether the chemical shift of the ^{13}C resonance is being influenced by the paramagnetic species. Doddrell and Allerhand[42] have utilized a combination of these techniques in the assignment of the resonances of vitamin B_{12}, coenzyme B_{12}, and other corrinoids. The reader is referred to the discussion in this reference for a practical illustration of ^{13}C peak assignments.

Another interesting spectral enhancement technique is referred to as *gated decoupling*.[43] When proton decoupling is terminated immediately before pulsing, the ^{13}C—H coupling returns immediately, but the populations of the nuclear energy levels are not equilibrated as rapidly. Thus, some Overhauser enhancement remains. The technique thus involves decoupling, termination of decoupling, pulsing, storage of the free induction decay, and repetition of this process for all pulse cycles. Correct values of $J_{^{13}C-H}$ are obtained.

Alternately, if the decoupling pulse is turned on just after the pulse, decoupling occurs instantly but nOe develops slowly producing fully decoupled spectra with little nOe distortion. One must wait for the nOe to decay fully between pulses. This procedure has value in systems with negative nOe's

A very exciting application of cmr involves its use as a probe, permitting one to employ stable isotopes as tracers.[44] Because the ^{13}C resonances are extremely sensitive to the location of the atom in a molecule, the site of incorporation of the ^{13}C can often be easily identified without the time-consuming degradation required with radioactive isotopes. The incorporation of ^{13}C-labeled glycine into coproporphyrin-III by the bacterium *Rhodopseudomonas spheroides* is an excellent illustration of the method.[44] The bacteria were grown on a medium containing glycine that was 93% labeled in the α-position with ^{13}C. The resulting prophyrins had the ^{13}C atoms incorporated solely as pyrrole α-carbon atoms and methine bridge carbon atoms. This result is consistent with the sequence of reactions proposed for the biosynthesis of porphyrins. If this type of tracer application had been carried out using ^{14}C and subsequent radiochemical analysis, the study would have been less definitive and more time consuming. The product would have had to be degraded into small fragments to find the location of the radioactive ^{14}C. There are many potential applications of this type that involve using ^{13}C as a tracer atom. One of the main advantages to using ^{13}C labels in biological systems is that this label does not disturb the conformation of the biomolecule as a spin label or paramagnetic probe might.

In a ^{13}C study[45] of labeled CO binding to human hemoglobin, two separate resonances could be observed for the coordinated CO. These occurred at 207.04 ppm and 206.60 ppm. The former resonance was assigned to CO bound to the α-chain and the latter to that bound to the β-chain by studying an abnormal hemoglobin that contains normal β-chains but α-chains that do not bind CO. In rabbit hemoglobin, three distinct iron(II)-binding sites were found.[46] A functionally different hemoglobin subunit was established.

In another interesting application[47] of ^{13}C, T_1 measurements were carried out on selectively ^{13}C-labeled histidine bound to intracellular and extracellular mouse hemoglobin. The intracellular and extracellular T_1 values differed by only 25% suggesting that the viscosity of the intracellular fluid is not unusually large (at least in this system). There had been considerable controversy regarding this problem.

When molecules containing two directly bonded ^{13}C atoms are investigated, spin-spin splitting results. The deviation of the intensity of the multiplet pattern from statistical considerations in a biosynthetic experiment can provide a measure of the correlation in the enrichment.[48] Such information has important mechanistic implications. For example, mixtures of doubly labeled and unlabeled material can be studied. Dilution in the ^{13}C—^{13}C interaction would indicate cleavage of this bond in the biosynthesis.

Cmr has been of considerable utility[49] in identifying the existence and structure of carbonium ions and carbocations in solution. Typically, a positively charged carbon will be deshielded compared to the analogous carbon in a neutral reference compound, and it will cause an inductive deshielding of 5 to 15 ppm in a neighbor atom.

The cmr spectra of many organometallic compounds have been reported.[50-54] Complexation of olefins to $AgNO_3$ results[50] in 1 to 4 ppm shifts, whereas shifts of 30 to 110 ppm have been observed upon coordination to rhodium.[51] The benzene resonances in $C_6H_6Cr(CO)_3$ are shielded[52] some 30 ppm relative to free benzene, whereas those of bound CO are usually deshielded from free CO. The ^{13}CO resonances in diamagnetic metal carbonyls are relatively insensitive to substituent and metal. Some typical results[50] are listed in Table 8–1. The interpretation of the ^{13}C chemical shifts in organometallic compounds has recently been criticized by Evans and Norton.[55] A thorough analysis of the shift interpretation and some results on the calculation of the shielding constant have recently been summarized.[56] The interpretation of small shift differences is very difficult. Consistent interpretations based on increased deshielding with increased metal-to-ligand π back-bonding have been offered.

TABLE 8–1 Some Typical ^{13}C Resonances in Metal Carbonyls

Complex	δ ^{13}C of Carbonyl
$Ni(CO)_4$	+191.6
$Fe(CO)_5$	+212
$Cr(CO)_6$	+211.3
$Mo(CO)_6$	+200.8
$W(CO)_6$	+191.4
$V(CO)_6^-$	+225.7
$CpCr(CO)_3^-$	+246.8
$CpMn(CO)_3$	+225.1
$(C_6H_5)_3PW(CO)_5$	+221.3 cis; +216.5 trans
$(C_6H_5O)_3PW(CO)_5$	+217.6 cis; +213.9 trans
(Cp = cyclopentadienide)	

8–13 OTHER NUCLEI

The advent of FTNMR enables the study of nuclei with low sensitivity[57] and has led to a substantial increase in the number of elements readily studied. Various nuclear properties of the elements are listed inside the front cover. The relative receptivity of some of the more common nuclei are listed in Table 8–2. This quantity indicates the relative signal strengths of solutions of equal concentration and is proportional to $\gamma^3 N(I + 1)$ where N is the natural abundance.

Quadrupolar nuclei in environments with cubic symmetry give reasonably sharp lines that are broadened extensively in lower-symmetry environments. As seen in Fig. 8–32, the sharp line for $^{35}Cl^-$ in aqueous solution is broadened extensively by an effect as subtle as ion pairing in $R_3NR'^+Cl^-$ where $R = n$-octyl and $R' = CH_3$. The slower tumbling of the large ion pair leads to faster relaxation and this coupled with the lower symmetry of chloride in the ion pair leads to a broad line.

TABLE 8–2 NMR Properties of Selected Nuclei

Isotope	Spin	Relative nmr Frequency/MHz	Relative Receptivity	Isotope	Spin	Relative nmr Frequency/MHz	Relative Receptivity
^{107}Ag	1/2	4.0	3.5×10^{-5}	^{14}N	1	7.2	1.0×10^{-3}
^{109}Ag	1/2	4.7	4.9×10^{-5}	^{15}N	1/2	10.1	3.9×10^{-6}
^{27}Al	5/2	26.1	2.1×10^{-1}	^{23}Na	3/2	26.5	9.3×10^{-2}
^{75}As	3/2	17.2	2.5×10^{-2}	^{93}Nb	9/2	24.5	4.9×10^{-1}
^{197}Au	3/2	1.7	2.6×10^{-5}	^{61}Ni	3/2	8.9	4.1×10^{-5}
^{10}B	3	10.7	3.9×10^{-3}	^{17}O	5/2	13.6	1.1×10^{-5}
^{11}B	3/2	32.1	1.3×10^{-1}	^{187}Os	1/2	2.3	2.0×10^{-7}
^{137}Ba	3/2	11.1	7.9×10^{-4}	^{189}Os	3/2	7.8	3.9×10^{-4}
^{9}Be	3/2	14.1	1.4×10^{-2}	^{31}P	1/2	40.5	6.6×10^{-2}
^{209}Bi	9/2	16.2	1.4×10^{-1}	^{207}Pb	1/2	20.9	2.0×10^{-3}
^{79}Br	3/2	25.1	4.0×10^{-2}	^{105}Pd	5/2	4.6	2.5×10^{-4}
^{81}Br	3/2	27.1	4.9×10^{-2}	^{195}Pt	1/2	21.4	3.4×10^{-3}
^{13}C	1/2	25.1	1.8×10^{-4}	^{87}Rb	3/2	32.8	4.9×10^{-2}
^{43}Ca	7/2	6.7	8.7×10^{-6}	^{185}Re	5/2	22.7	5.1×10^{-2}
^{111}Cd	1/2	21.2	1.2×10^{-3}	^{187}Re	5/2	22.9	8.8×10^{-2}
^{113}Cd	1/2	22.2	1.3×10^{-3}	^{103}Rh	1/2	3.2	3.2×10^{-5}
^{35}Cl	3/2	9.8	3.6×10^{-3}	^{99}Ru	5/2	4.6	1.5×10^{-4}
^{37}Cl	3/2	8.2	6.7×10^{-4}	^{101}Ru	5/2	5.2	2.8×10^{-4}
^{59}Co	7/2	23.6	2.8×10^{-1}	^{33}S	3/2	7.7	1.7×10^{-5}
^{53}Cr	3/2	5.7	8.6×10^{-5}	^{121}Sb	5/2	24.0	9.3×10^{-2}
^{133}Cs	7/2	13.2	4.8×10^{-2}	^{123}Sb	7/2	13.0	2.0×10^{-2}
^{63}Cu	3/2	26.5	6.5×10^{-2}	^{45}Sc	7/2	24.3	3.0×10^{-1}
^{65}Cu	3/2	28.4	3.6×10^{-2}	^{77}Se	1/2	19.1	5.3×10^{-4}
^{19}F	1/2	94.1	8.3×10^{-1}	^{29}Si	1/2	19.9	3.7×10^{-4}
^{57}Fe	1/2	3.2	7.4×10^{-7}	^{117}Sn	1/2	35.6	3.5×10^{-3}
^{69}Ga	3/2	24.0	4.2×10^{-2}	^{119}Sn	1/2	37.3	4.5×10^{-3}
^{71}Ga	3/2	30.6	5.7×10^{-2}	^{87}Sr	9/2	4.3	1.9×10^{-4}
^{73}Ge	9/2	3.5	1.1×10^{-4}	^{181}Ta	7/2	12.0	3.7×10^{-2}
^{1}H	1/2	100.0	1.00	^{125}Te	1/2	31.5	2.2×10^{-3}
^{2}H	1	15.4	1.5×10^{-6}	^{47}Ti	5/2	5.6	1.5×10^{-4}
^{177}Hf	7/2	4.0	2.6×10^{-4}	^{49}Ti	7/2	5.6	2.1×10^{-4}
^{179}Hf	9/2	2.5	7.4×10^{-5}	^{203}Tl	1/2	57.1	5.7×10^{-2}
^{199}Hg	1/2	17.9	9.8×10^{-4}	^{205}Tl	1/2	57.6	1.4×10^{-1}
^{201}Hg	3/2	6.6	1.9×10^{-4}	^{169}Tm	1/2	8.3	5.7×10^{-4}
^{127}I	5/2	20.1	9.5×10^{-2}	^{51}V	7/2	26.3	3.8×10^{-1}
^{115}In	9/2	22.0	3.4×10^{-1}	^{183}W	1/2	4.2	1.1×10^{-5}
^{191}Ir	3/2	1.7	9.8×10^{-6}	^{129}Xe	1/2	27.8	5.7×10^{-3}
^{193}Ir	3/2	1.9	2.1×10^{-5}	^{131}Xe	3/2	8.2	5.9×10^{-4}
^{39}K	3/2	4.7	4.8×10^{-4}	^{89}Y	1/2	4.9	1.2×10^{-4}
^{139}La	7/2	14.2	6.0×10^{-2}	^{171}Yb	1/2	17.6	7.8×10^{-4}
^{6}Li	1	14.7	6.3×10^{-4}	^{67}Zn	5/2	6.3	1.2×10^{-4}
^{7}Li	3/2	38.9	2.7×10^{-1}	^{91}Zr	5/2	9.3	1.1×10^{-3}
^{25}Mg	5/2	6.1	2.7×10^{-4}				
^{55}Mn	5/2	24.7	1.8×10^{-1}				
^{95}Mo	5/2	6.5	5.1×10^{-4}				
^{97}Mo	5/2	6.7	3.3×10^{-4}				

Nitrogen nmr is an interesting example to consider[58] for it illustrates the advantages and difficulties associated with multinuclear magnetic resonance. Compounds containing nitrogen exist in nine stable oxidation states and five coordination numbers with lone pairs and π-systems leading to interesting perturbations. Two isotopes can be investigated ^{15}N($I = {}^1/_2$) and ^{14}N($I = 1$). Both have low sensitivity with receptivities of 3.9×10^{-6} and 1×10^{-3}, respectively, compared to ^{13}C at 1.8×10^{-4}. Higher fields help because sensitivity increases with $\gamma^3 H_0^{3/2}$. Relaxation is slow with ^{15}N slowing down spectral acquisition and nOe is negative because γ is negative leading to diminished signals with proton decoupling. Enriched samples or large samples at high fields are required. ^{14}N has a receptivity five times that of ^{13}C but is a quadrupolar nucleus. In addition to environments with cubic symmetry, reasonably sharp lines are found in nitrates, nitrites, nitro compounds, nitriles, and isonitriles. Correlation charts indicating ranges of chemical shifts for a variety of nitrogen compounds can be found in the literature.[58] The range in chemical shifts found in nitrogen compounds is about three times that found in ^{13}C. Deshielding of nitrogen is observed when it is involved in π-bonding to neighbor atoms and when it is bent as opposed to linear in its coordination to metal complexes, *e.g.*, bent *vs.* linear nitrosyl coordination.

Deuterium nmr is of interest because its relaxation behavior is dominated by a quadrupolar mechanism and consequently is indicative of molecular dynamics at the molecular position of substitution. In a simple case of rapid isotropic motion, the longer T_1 the more mobile the C—D bond. The relationship becomes more complicated for more complex slower motion. Significant findings involving the degree of molecular organization and dynamic processes occurring in membranes have been reported.[59]

The advantages of multinuclear nmr in structural elucidation are nicely illustrated by the nmr study[60] of:

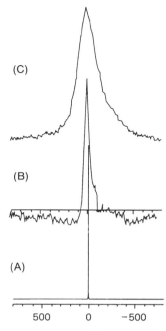

FIGURE 8-32 ^{35}Cl NMR of (A) a 5-*M* aqueous sodium cloride; (B) ion-paired, 0.1 *M* R$_4$N$^+$Cl$^-$ in butanol, and (C) 1.2 *M* R$_4$N$^+$Cl$^-$ in butanol.

The ^{1}H, ^{13}C, ^{31}P, and ^{199}Hg magnetic resonances can all be investigated. The proton nmr shows the three signals expected for the CH$_3$, CH$_2$, and CH protons with relative intensities of 12:4:1. Observation of one CH$_3$ and CH$_2$ resonance proves that the ligands are symmetrically bound to mercury. The CH$_3$ and CH$_2$ groups show couplings to ^{31}P and satellites from the 16.8% abundant ^{199}Hg, whereas the C—H peak is a broad resonance with no resolution of couplings. The proton decoupled ^{31}P spectrum has a single main peak with satellites from ^{199}Hg and the proton-decoupled ^{199}Hg resonance is a quintet arising from coupling to four equivalent ^{31}P nuclei. No other structure is consistent with all these observations.

More on Relaxation Processes

8–14 SPECTRAL DENSITY

In equation (8–22), the autocorrelation function for a particle undergoing random motion was discussed. As mentioned earlier, the Fourier transform converts a function from the time domain to the frequency domain. The Fourier transform of the correlation function produces a quantity called the spectral density, $J(\omega)$, which is defined by

$$J(\omega) = \int_{-\infty}^{\infty} R(\tau_c) \, e^{i\omega\tau_c} \, d\tau \tag{8-25}$$

The spectral density for any given frequency, $J(\omega)$, gives the intensity of the fluctuation at that particular frequency. The fluctuations of importance in nmr include the relative diffusion of one molecule past another, described by a correlation time, τ_D; the rotation of a molecule about its rotation axes, τ_R; a rapidly relaxing nucleus, τ_S; and chemical exchange, τ_E. Classical texts on nmr give equations for various contributions to relaxation, e.g., dipolar interaction, quadrupolar interaction, chemical shift anisotropy, scalar coupling, spin-rotation, and so on, in terms of the correlation function formalism. All of these phenomena cause a given nucleus to experience a fluctuating field from another nucleus in solution. The efficiency of nuclear relaxation will depend upon the intensity of the frequency arising from the fluctuation, which in turn depends upon the correlation times τ_c for the various processes. Accordingly, it is appropriate to examine in more detail the relationship between spectral density and τ_c.

If we plot the spectral density vs. ω, curves comparable to those shown in Fig. 8–33 are obtained. These different curves correspond to different correlation times, τ_c, for our systems. (Recall that R depends upon the value of τ_c.) The half-intensity height indicated by the dots corresponds to the frequency $\omega = \tau_c^{-1}$.

Therefore, only frequencies smaller than $\sim \tau_c^{-1}$ are available in the sample. The flat portion of the curve corresponds to $\omega\tau_c \ll 1$, whereas the region to the

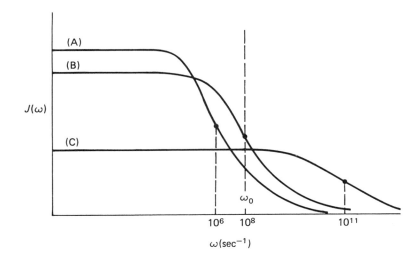

FIGURE 8–33 A plot of the spectral density of various frequencies for systems (A), (B), and (C) corresponding to different correlation times.

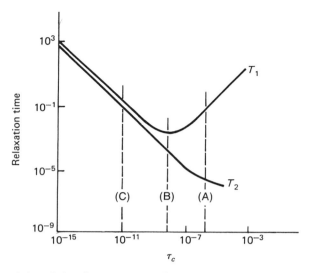

FIGURE 8-34 Dependences of the T_1 and T_2 relaxations.

right of the dot corresponds to $\omega\tau_c \gg 1$. The area under each of these curves is the same. This is equivalent to saying that the kinetic energy in any two samples at the same temperature is the same. This plot gives an intensity distribution for any particular frequency; e.g., in the case of a magnetic nucleus undergoing molecular motion, we have the intensity distribution for each frequency of the oscillating fields arising from the complex motion that an assemblage of molecules undergoes.

For the middle curve (B), a correlation time has been selected to maximize $J(\omega)$ at ω_0 (i.e., τ_c^{-1} is selected to be ω_0). Any curve with a larger or smaller τ_c will have a smaller value of $J(\omega)$ at ω_0. If we choose ω_0 at the Larmor frequency, the magnitude of $J(\omega)$ at ω_0 will be proportional to the relaxation efficiency; i.e., the greater the intensity of the fluctuating moment, the more effective the relaxation. Correlation times corresponding to curve (B) are most efficient, (C) next most of those shown, and (A) the least efficient.

The T_1 relaxation is caused by frequencies that correspond to ω_0, whereas T_2 is caused by $\omega = 0$ and ω_0 frequency components. This is consistent with our earlier description of the fluctuating field causing T_1 relaxation and the static ($\omega = 0$) field in solids causing T_2 relaxation. If we were to plot the behavior of T_1 as a function of τ_c, we would obtain the curve labeled T_1 in Fig. 8-34. The minimum corresponds to the solid dot of curve (B) in Fig. 8-33. Larger or smaller correlation times give a smaller spectral density at this frequency. Curve (C) of Fig. 8-33, corresponding to the dashed line labeled (C) in Fig. 8-34, gives the behavior at shorter correlation times; a longer relaxation time results. It is interesting to point out that T_1 is a double-valued function; i.e., a given T_1 can correspond to two possible values of τ_c. To obtain τ_c, one must know on what part of the curve the system is located.

The behavior of T_2 is described by the curve so labeled in Fig. 8-34. Note that for short correlation times, T_2 parallels T_1; i.e., the spin-lattice relaxation mechanism is randomizing the z-component and the xy-component equally. In curve (A) of Fig. 8-33, we have a situation that would correspond to correlation times in a viscous liquid or solid. There is a greatly reduced intensity of the frequency component ω_0 needed for longitudinal relaxation. However, the $\omega = 0$ component is very large and the transverse relation is enhanced. The zero frequency dipolar broadening thus decreases T_2, causing line broadening in viscous liquids and solids. The increased T_1 associated with the increased correlation time in this region of the curve explains why spectra of solids and viscous liquid are easily saturated.

Often in the following pages we will note that processes that increase or decrease the correlation time can sharpen nmr peaks, *i.e.*, increase T_1 for the protons. One should be able to deduce whether T_1 will increase or decrease by knowing whether the spectral density change caused by the perturbation moves the curve away from (B) toward (A) or (C). If the curve is broadened, the perturbation moves the system from the (A) or (C) direction toward (B).

Multipulse Methods

8–15 INTRODUCTION

The multipulse methods all depend on variation in the start time, duration, amplitudes, frequencies, and phases of a sequence of pulses. The measurement of T_1 discussed earlier in Section 8–9 is a simple example of a multipulse sequence to measure relaxation times. We can take advantage of pulse sequences to simplify spectra, gain intensity, separate overlapping peaks to assign resonances, study certain parts of the spectrum, suppress peaks, determine to which nuclei certain nuclei are coupled even in heavily overlapped spectra, and determine which nuclei are close by in space. The advances in this area have been phenomenal and promise to continue. We shall briefly cover a few of the more common applications here to illustrate the principles and the reader is referred to the literature[61] for more details.

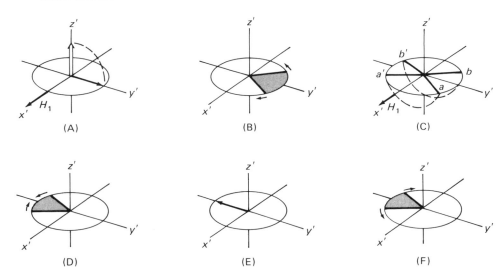

FIGURE 8–35 The Hahn spin-echo experiment. (A) A 90° pulse applied along x' at time 0 causes M to tip to the positive y' axis. (B) During the time τ, the microscopic moments in the xy-plane dephase because of field, H_0, inhomogeneity. (C) A 180° pulse is applied along x'. (D) The moments regroup. (E) After a time τ following the 180° pulse, the moments rephase. (F) At subsequent times, the moments dephase again. (From T. C. Farrar and E. D. Becker, "Introduction to Pulse and Fourier Transform NMR Methods," Academic Press, New York, 1971.)

8–16 SPIN ECHOES

We can begin our introduction to this subject by considering a pulse sequence that leads to a spin echo. A 90° pulse is applied along x' (Fig. 8–35) causing M to tip to y'. After a time τ, T_2 effects and field inhomogeneity at different places in the sample tube lead to different rates of precession in the xy plane and cause a fanning out of the magnetic moments in the xy plane. We shall refer to both field homogeneity and other T_2 effects as T_2^*, i.e., $1/T_2^* = 1/T_2 +$ other effects.

Those nuclei precessing faster than average (the rotation rate of the rotating frame) appear in the rotating frame to move toward the observer (i.e., looking down the z' axis they appear to move clockwise), while those slower than average move away [counterclockwise; Fig. 8–35(B)]. The shaded region in the figure can be thought of as a whole series of vectors corresponding to nuclei precessing at different frequencies. These are called isochromats. A 180° pulse is then applied along x', moving all the moments 180° about the x' axis [i.e., they are now fanned out about $-y'$ as shown in Fig. 8–35(C)]. It should be emphasized that a 180° rotation about x' is different from a 180° rotation about z. The former causes a to become a' and b to become b'. While we are waiting another time period corresponding to τ, the faster nuclei move clockwise and the slower nuclei move counterclockwise in the rotating frame, as shown by the arrows in Fig. 8–35(D). However, now the slower nuclei move toward the observer and the faster ones away from the observer, causing them to get back in phase. After the same time τ that was used after the 90° pulse, the isochromats have regrouped along y'. The signal detected is shown in Fig. 8–36 and the peak at 2τ is referred to as a spin echo. If there had been no T_2 relaxation by mechanisms other than field inhomogeneity, we would have the same signal intensity in the echo as at the start. We have, in effect, eliminated the field inhomogeneity contribution to T_2 and can use the decrease in intensity to calculate T_2. The process τ-π-τ can be repeated from the peak of the echo to produce another echo. In practice, there are two effects that reduce this amplitude: (1) T_2 of the nucleus, which fans out

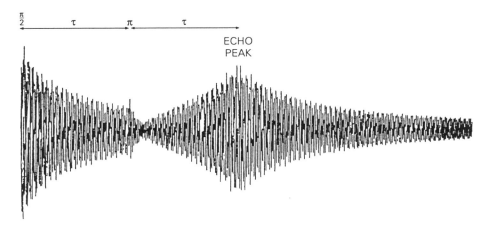

FIGURE 8–36 The FID Signal for an Echo Sequence of Pulses. The signal is recorded immediately after the $\pi/2$ pulse and after a time, τ, a π pulse is applied. The signal rebuilds to a new peak at time 2τ (i.e., τ after the π pulse) and then decays. The process can be repeated by applying a π pulse at time 3τ.

the magnetization randomly as opposed to the systematic nature of the applied field, and (2) diffusion of molecules to different parts of the sample tube in the time 2τ. The latter introduces error in T_2. We shall refer to the pulse sequence $(\pi/2)_x\text{-}\tau\text{-}\pi_x\text{-}\tau\text{-}acquire$ as an *echo pulse sequence*.

Next, we consider how nuclei with different chemical shifts respond to the echo pulse sequence. The 90° pulse rotates the magnetization of both nuclei to the y-axis. In a rotating frame chosen to rotate at the frequency of the most shielded a-nucleus, the other nucleus precesses faster leading after a time, τ, to the situation shown in Fig. 8–37(C). The b-nucleus is precessing relative to a at

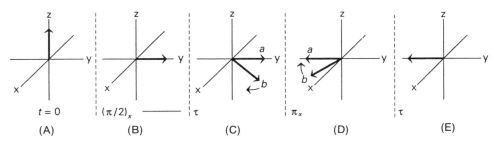

FIGURE 8–37 The effect of an echo pulse sequence on nuclei with two different chemical shifts. This is a simplified representation of Figure 8–35.

a frequency in Hz corresponding to the chemical shift difference. After a 180° pulse around x, b is behind a and catches up after an additional time τ as shown in Fig. 8–37(E). Thus, the chemical shifts are refocused at the peak of the echo. Thus, we can carry out experiments on nuclei that are independent of field inhomogeneity or chemical shift differences. We can now concern ourselves with coupled systems without worrying about chemical shift differences.

First, let us consider the case of homonuclear coupled systems *e.g.*, an A-X proton-coupled system whose nmr spectrum consists of a doublet of doublets. Focus on one of the doublets arising from coupling and let our frame rotate at a frequency corresponding to the center of the doublet. Both components are now precessing, one at $+J/2$ and the second at $-J/2$ Hz leading after a short time to the situation in Fig. 8–38(C). Now when a 180° pulse is applied it converts every α nucleus on A and X into a β and vice versa. All the nuclei precessing at $+J/2$ with α neighbors now have β neighbors and precess at $-J/2$. The same is true for the nuclei precessing at $-J/2$. Thus, the π-pulse has changed the labels as shown in Fig. 8–38(D). Now after a time τ the isochromats are not refocused. If we selected τ-values corresponding to $J/4$, the isochromats would be aligned along x in Fig. 8–38(E), and generate antiphase components at the echo. It is

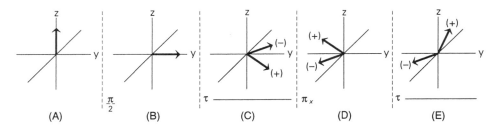

FIGURE 8–38 The behavior of the A doublet in an A-X system to an echo pulse sequence.

also important to point out that in a heteronuclear coupled system, the neighbor nucleus is not inverted, the labels are not interchanged upon application of the π-pulse (*i.e.*, in Fig. 8–38(D) above), and the components refocus in (E). With appropriate equipment, pulses could be applied to both nuclei to exchange the labels and lead to the case in Fig. 8–38. In all these experiments we should remember that the magnetization is reappearing along the z axis with a time constant T_1.

8–17 SENSITIVITY-ENHANCEMENT METHODS

As can be seen by its receptivity (Table 8–2), the proton is the most sensitive nucleus. When only small amounts of materials containing nuclei orders of magnitude less sensitive are available, it becomes impossible to obtain their nmr spectra by standard means. The nOe helps a little, but its advantages are limited. Methods that transfer the favorable properties of the proton to other nuclei are advantageous. Selective population inversion (SPI), insensitive nuclei enhanced by polarization transfer (INEPT), and distortionless enhanced polarization transfer (DEPT) are methods of accomplishing polarization transfer. DEPT, usually the method of choice, is more involved and requires an understanding of the principles of SPI and INEPT. We begin with SPI.

Consider two nuclei ^{13}C and ^{1}H whose energy diagram is shown in Fig. 8–39. Coupling energies are so small on the energy scale of the diagram that they are not seen. The magnetogyric ratio and Larmor frequency of hydrogen are about four times those of carbon, so the transitions are four times as energetic. With a total of N nuclei, each level would have a population of $N/4$ if they were equally populated. We shall concentrate on the deviation from $N/4$, which for the ground and excited states is represented by $2\Delta H$ for the hydrogen transition and $2\Delta C$ for carbon; where ΔH equals $4\Delta C$ from the magnetogyric ratios. If we were working with ^{13}C—H, the fourfold excess ^{1}H population of the lower state over that of carbon and the fourfold larger magnetic moment of the proton leads to a sixteenfold increase in the transverse magnetization when a $\pi/2$ pulse is applied to the proton compared to carbon. The signal depends on the rate of precession of the magnetization, which is four times greater for the proton than ^{13}C thus a 64-fold difference results. The ^{103}Rh nucleus has a γ that is 3% that of the proton leading to a sensitivity 32,000 times less. In general, the relative signals are proportional to γ^3 and acquisition times to γ^5.

Now consider the effect of inverting one line of the proton doublet H_1. This inverts the H_1 populations leading to the situation in Fig. 8–40. The population differences across the proton transitions are the same except one is negative. However, the carbon transitions that previously were $2\Delta C$ are now $(\Delta H + \Delta C) - (-\Delta H - \Delta C)$ or $2\Delta H + 2\Delta C$ for C_1 and $(-\Delta H + \Delta C) - (\Delta H - \Delta C)$ or $-2\Delta H + 2\Delta C$ for C_2. We have transferred the proton population differences to the carbon and added them to the existing differences. With $\Delta H = 4\Delta C$, we get $10\Delta C$ from $2\Delta H + 2\Delta C$ or five times the original intensity (which was $2\Delta C$) and we get $6\Delta C$ from $-2\Delta H + 2\Delta C$ or minus three times the original intensity. This leads to the ^{13}C spectrum shown in Fig. 8–40(B).

It would be tedious to analyze a complex spectrum this way for we would have to find and selectively invert one line of each proton doublet. Furthermore, we cannot obtain a ^{13}C spectrum that is proton decoupled. The INEPT

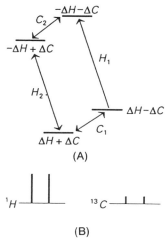

FIGURE 8–39 (A) Energy level diagram including population differences for an AX system. (B) Corresponding NMR spectra.

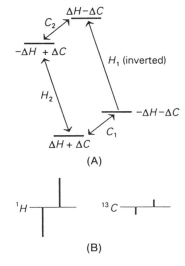

FIGURE 8–40 (A) Population differences for an AX system after saturation of H_1 (one component of the doublet). (B) The corresponding nmr spectra.

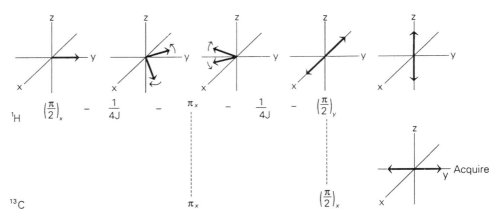

FIGURE 8–41 INEPT pulse sequence.

experiment puts pairs of proton transitions in antiphase independent of their chemical shift by using a non-selective pulse. This is accomplished by using a spin echo from the pulse sequence shown in Fig. 8–41.

After a $(\pi/2)_x$ pulse to the protons, we wait for a time of $J/4$ so doublet components from spin-spin splitting move through 1/8 of the cycle before the π-pulse is applied (the components precess at $\pm J/2$ Hz). The sensitive nucleus, S, (in this case the proton) is rotated into the other half of the xy-plane. A simultaneous application of the π-pulse to the insensitive nucleus (in this case the carbon) causes a reversal in the sense of the precession of the J components causing them to move as shown in Fig. 8–41. The chemical shifts and field inhomogeneity will be refocused in the next $J/4$, but the magnetization due to spin coupling is not, and ends up on the x-axis. A 90° pulse to the protons gives us the antiphase components and the population enhancements we had in SPI for all the protons. The simultaneous 90° pulse to ^{13}C enables us to acquire the ^{13}C signal. Each ^{13}C spin-spin doublet gives rise to a positive and negative peak. We have transferred the sensitive nucleus population difference to the insensitive nucleus, but the positive and negative peaks have unequal intensities, e.g., +5 and −3 for a ^{13}C doublet. Various pulse sequences enable us to eliminate this and to decouple proton splittings. The DEPT sequence has replaced INEPT and gives rise to a spectrum with the appearance of a normal nmr spectrum.

A further advantage of these polarization enhancement experiments arises from the fact that only the population difference of the sensitive nucleus, e.g., the protons, leads to the signal intensity, so it is the T_1 of this nucleus that determines the pulse repetition rate. This is a big advantage for the study of insensitive nuclei with long T_1's (e.g., ^{15}N and ^{29}Si). A further advantage arises in the application of polarization enhancement techniques to nuclei that have negative nOe enhancements. For nuclei that do not have large proton coupling constants (e.g., some rhodium-phosphine complexes), other sensitive nuclei (e.g., phosphorus) can be used if a probe tuned to the two frequencies and two broadband transmitters are available.

8–18 SELECTIVE EXCITATION AND SUPPRESSION

The DEPT pulse sequence can be used to select certain multiplets for study. Pulse sequences can be used to selectively study all the CH or CH$_2$ or CH$_3$

resonances; this is called spectrum editing. As an illustration of this application, consider the ^{13}C nmr of

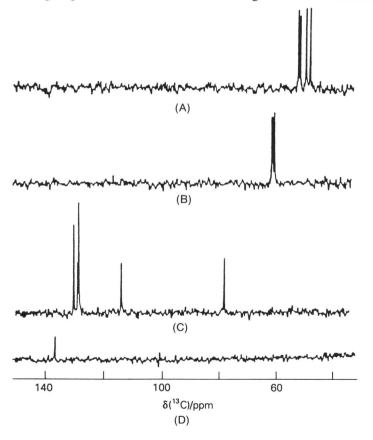

which contains thirteen ^{13}C resonances. Assignment is difficult. The chemical shifts of CH$_3$, CH$_2$, and CH protons can be obtained[62] by selectively studying each group of resonances as shown in Figure 8–42. All the coupling constants

FIGURE 8–42 The edited ^{13}C spectra of Pd(ϕCHCHCH$_2$)(CH$_3$)$_2$NCH$_2$-CH$_2$N(CH$_3$)$_2$ using the DEPT pulse sequence. (A) The CH$_3$ groups; (B) the CH$_2$ resonances; and (C) the CH resonances. This is a different pulse sequence than INEPT. The spectra come from mathematical operations on several spectra with different pulse lengths.

in each group should be within about 30% of their average value. The advantages illustrated are even more important in analyzing even more complicated spectra.

The digitizer ($\sim$12 bit) that converts the spectrum to digital form has a dynamic range of about 2000 to 1. If the largest signal in the spectrum is just equal to the capacity of the analogue-to-digital converter, signals 2000 times smaller than this will not be detected at all. No amount of time averaging will help. Because of the digitization process, a peak outside the spectral width will fold back into the spectrum. Peak suppression or tailored excitation can alleviate the problem. Peak suppression is best accomplished by pre-saturation if the other needed components of the spectrum are not undergoing exchange with the saturated peak. The irradiation of the resonance that induces saturation is turned off immediately before the application of the pulses and acquisition of the spectrum. This procedure can be used in conjunction with some of the pulse sequences discussed so far and with the $1\bar{3}3\bar{1}$ experiment to be discussed shortly [1 refers to $(^1/_4)\pi$, 3 to $(^3/_4)\pi$ and the line over the number indicates a negative pulse, e.g., along the $-x$-axis].

Tailored excitations attempt to tailor the frequency distribution of the r.f. excitation to avoid frequencies corresponding to certain regions of the spectrum (recall the distribution shown in Fig. 8–18). The sequence, $(\pi/2)_x$-τ-$(\pi/2)_{-x}$ acquire, known as jump and return (JR), illustrates the principles for several experiments of this type. If the transmitter frequency is set at the peak to be suppressed, it will remain static in the τ-interval while others will precess. The second $(\pi/2)_{-x}$ pulse returns the peak to be suppressed to the z-axis with no x-component generated, i.e., no peak. The peak of interest moves relative to the rotating frame during τ, thus an x-component remains after the second pulse. This is illustrated in Fig. 8–43 where A is the resonance to be suppressed and B the peak of interest.

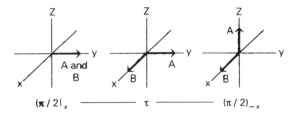

FIGURE 8–43 The effect of the JR pulse sequence on a suppressed resonance A and one of interest B.

Ideally, τ should be selected so the B magnetization lies along X. Other pulse sequences have been utilized that are more effective than JR. The $(\pi/4)_x$-τ-$(3\pi/4)_{-x}$-τ-$(3\pi/4)_x$-τ-$(\pi/4)_{-x}$ sequence, referred to as a $1\bar{3}3\bar{1}$ sequence, is particularly effective with τ selected to optimize the spectral region of interest. This sequence can provide a 1000-fold better reduction in a solvent resonance than pre-saturation. For nuclei other than protons, the DANTE sequence can be utilized for selective excitation. This consists of a train of hard pulses of very small flip angles with constant phase and constant separation.

8–19 TWO-DIMENSIONAL NMR

The two-dimensional, (2-D) nmr experiment involves the following sequence of events:

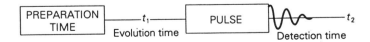

The detection period corresponds to the usual collection of the FID, whose transform leads to the usual frequency spectrum. To examine the rest of the

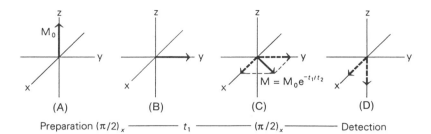

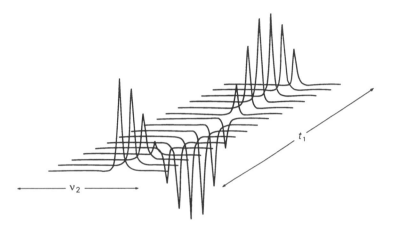

FIGURE 8–44 Behavior of the magnetization during a $(\pi/2)_x - t_1 - (\pi/2)_x$ sequence. The x- and y-components are shown in (C) and only the rotation of the components by the $\pi/2$ pulse is illustrated in (D).

FIGURE 8–45 Nuclear magnetic resonance spectra detected at increasing t_1 by increments of Δt_1 for the pulse scheme in Fig. 8-44.

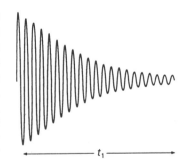

FIGURE 8–46 The interferrogram obtained by plotting the amplitude of a slice parallel to the t_1-axis of Fig. 8-45.

sequence and some important consequences, we shall consider a very simple experiment involving the proton nmr of chloroform, $CHCl_3$. Our preparation time will consist of a $(\pi/2)_x$ pulse. We will wait for a time t_1, after which time another $(\pi/2)_x$ pulse is carried out and the signal detected. The behavior of the magnetization during this pulse sequence is shown in Fig. 8-44. We shall ignore T_1 relaxation, for simplicity, during the time t_1 (which is not related to T_1) and concern ourselves only with the x,y-magnetization in the rotating frame. Fig. 8–44C shows the precession of the magnetization at v Hz forming an angle of $2\pi v t_1$ during the t_1 interval. The M_y and M_x components are shown and their rotation by a $(\pi/2)_x$ pulse is shown in (d). The M_x component, given by $M \sin 2\pi v t_1$, is detected, Fourier transformed, and a frequency spectrum is obtained. Now consider a series of n experiments in which t_1 is increased by an increment Δt_1 each time. The M_x magnetization and resulting spectral intensity will vary as t_1 is varied leading to the series of 17 spectra shown in Fig. 8–45 for 17 of our n experiments with regular Δt_1 increments.

Now consider a slice through Fig. 8–45 at fixed v_2 corresponding to the top of the $CHCl_3$ peak. Plotting the variation in amplitude vs. t_1, for not just the 17 slices but, for all n of the slices of our original experiment, gives Fig. 8–46. The amplitude is oscillating with a frequency v corresponding to the chemical shift of $CHCl_3$ because M_x is given by $M \sin 2\pi v t_1$. It decays exponentially with a time constant T_2. If we Fourier transform the curve in Fig. 8–46 we obtain the frequency spectrum in Fig. 8–47(A), with its peak at a frequency corresponding to v (the chemical shift of $CHCl_3$) because the transverse magnetization is oscillating sinusoidally at $M \sin 2\pi v t_1$. If we now take a whole series of slices through Fig. 8–45 at different values of v_2 and stack these spectra together, we get the plot shown in Fig. 8–47(B). The stacked plot is often presented as a contour plot as shown in Fig. 8–47(C). Generally two-dimensional (2-D) plots contain v_2 along the abscissa and v_1 along the ordinate and this change is made in Fig. 8–47. The symbols F_1 and F_2 are often employed to indicate the result after Fourier transform of t_1 and t_2 to the frequency domain.

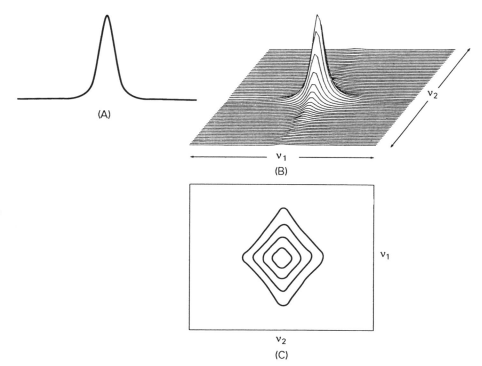

FIGURE 8–47 (A) The Fourier transfer of Fig. 8–46. (B) The stacked plot of Fourier transforms of various slices of Fig. 8–46. (C) Contour plot of Fig. 8–47B.

We have created a two-dimensional frequency spectrum $f(v_1, v_2)$ by Fourier transforming the two time dependencies t_1, t_2. In this example, we get a square spectrum with both axes representing chemical shifts and a peak on the *diagonal* in the frequency domain at v, v. We have not learned any new information from the simple example chosen but have selected it to illustrate the 2-D method. Now we move to more interesting 2-D systems and will begin by considering *homonuclear 2-D nmr*. In principle, we can perform a 2-D experiment on any system in which a modulation of the frequency spectrum occurs during the evolution time, t_1. In the above example, the same modulation occurred during both t_1 and t_2. Significant results are obtained when different processes occur during the two time intervals. Generally, normal chemical shifts and couplings correspond to t_2, and when a different modulation occurs on t_1, off-diagonal or cross peaks occur in the plot of v_1 vs. v_2. To interpret the resulting contour plot we need to know what the two axes represent and how the two magnetizations are related. In all 2-D experiments, the F_2 dimension contains information about the precession frequency of the observed nucleus. The F_1 dimension contains information about the interaction of the observed nucleus and other magnetic nuclei that develops during the evolution period.

Consider what happens during the second pulse for a case in which there is a homonuclear coupling. To simplify the discussion, consider an experiment in which one automatically increments the decoupling frequency across the region of interest. Each decoupling frequency would manifest itself by a change at coupled protons when their partner is irradiated. It would be time consuming to march across all the frequencies. Instead, we will do this experiment by using FT in the coupling dimension. This pulse sequence $((\pi/2)_x$-t_1-$(\pi/2)_x)$ is referred to as Jeener's or the COSY experiment. In a homonuclear, AX, system the second $(\pi/2)_x$ pulse causes magnetization to be redistributed among all the transitions with which a given transition is associated. In an A-X system (four peaks), a given A transition is associated with the other A peak and the two X peaks and likewise for the other A and two X transitions. Thus a given line in t_2 may have components of its amplitude modulated as functions of the frequencies of all the other lines in t_1. When the two Fourier transforms are carried out, this redistribution gives rise to cross peaks in the 2-D spectrum. The resulting contour plot for an AX system is shown in Fig. 8–48. The peaks along the diagonal arise from magnetization components that have the same frequency during t_1 and t_2 (recall CHCl$_3$). This is the portion of the magnetization that was not transferred elsewhere during the second pulse. Thus, looking along the diagonal, we see the normal spectrum—a doublet of doublets. The cross peaks corresponding to the small cluster arise from redistribution within a given multiplet (A_1A_2 or X_1X_2). The off-diagonal peaks farther off the diagonal correspond to redistribution between different nuclei. In a more complex molecule we can determine which protons are coupled to each other by the existence of off-diagonal peaks. We can think of the coupling as providing a pathway through which the magnetization can travel during the second pulse. Couplings that are less than the natural line width and hence unresolved in the one-dimensional experiment, are easily detected in COSY.

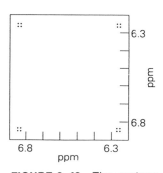

FIGURE 8–48 The contour plot of a homonuclear two-dimensional A-X system ($I = \frac{1}{2}$ for A and X).

COSY has had important applications in cluster chemistry. Typical examples include ^{11}B clusters[63] and ^{183}W clusters.[64] For quadrupolar nuclei, the relaxation times must be long enough that the coupling is not completely eliminated.

The INADEQUATE pulse sequence enables one to detect natural abundance ^{13}C—^{13}C couplings. In this experiment, the peaks from lone ^{13}C nuclei are eliminated and only those involved in coupling remain. At each existing frequency in the v_1 spectrum, a row exists parallel to v_2 corresponding to the coupling of adjacent ^{13}C atoms.

In the COSY experiment we have ignored the transverse magnetization returned to the z-axis by the second pulse. This magnetization is also modulated during t_1. Since the component is along the z-axis, it is not detected. If processes of interest modulated this z-component, a 90° pulse could be applied and its behavior studied. Two such effects are chemical exchange and nOe. If we wait a while τ_m after the second pulse, i.e., the sequence $\pi/2$-t_1-$\pi/2$-τ_m-$\pi/2$-t_2, a nucleus whose magnetization was modulated by one chemical shift during t_1 migrates to another site during τ_m. Thus, it has a different chemical shift during t_2 and the resulting 2-D spectrum will have cross peaks between the peaks of exchanging sites. Rate data is not obtained, but the peaks undergoing exchange are identified.

In a similar fashion pairs of nuclei, that show nOe in a one-dimensional experiment, show cross peaks in this two-dimensional experiment. Spatial

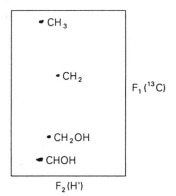

FIGURE 8–49 The correlated heteronuclear (^{1}H, ^{13}C) two-dimensional spectrum of butane-1,3-diol.

relationships between protons can be determined by knowing which ones undergo nOe. This technique is particularly useful in large molecules where nOe's are large.

The heteronuclear correlated 2-D experiment, is similar in concept to INEPT. A $\pi/2$-t_1 period on the sensitive, S, nucleus is followed by a $\pi/2$ pulse on both the sensitive and insensitive, I, nucleus followed by accumulation of the signal of I during t_2. This is referred to as the heteronuclear shift correlation (HSC) sequence. The 2-D plot gives $v_2(S)$ vs. $v_1(I)$ i.e., a peak appears in the contour plot for correlated nuclei at the chemical shift of I on the v_2 axis and S on the v_2 axis. In order to remove coupling, a π-pulse is applied to the I nucleus halfway through t_1. The advantage of HSC is in assigning coupling relationships in heteronuclear systems. If the proton spectra is assigned, assignment of the ^{13}C spectra in the 2-D plot is trivial. In weakly coupled systems, contour peaks appear only between protons and their directly bonded heteronuclear partners. In the presence of strong coupling between H_A and H_B in H_A—C—C—H_B, cross peaks to the non-bonded carbon can appear. Often the ^{13}C spectra or that of some nucleus other than H is easier to assign than hydrogen because of less homonuclear coupling in the former. These ideas are illustrated by the correlated, heteronuclear (^{1}H, ^{13}C) 2-D contour plot of butane-1,3-diol in Fig. 8–49. The ^{13}C peaks are well separated and their assignment leads to assignment of the directly bonded proton peaks. The 2-D spectra can then be used to assign the hydrogen spectra. Taking a slice through the HSC spectra parallel to v_1 at the chemical shift of a nucleus yields the proton spectra of its attached protons.[65] This method is particularly useful for locating proton peaks that are obscured by an unrelated multiplet whose attached ^{13}C resonances are well separated. Combination of the COSY experiment (homonuclear correlated) on the protons with the HSC spectra often allows the complete framework of a molecule to be traced.

Our final topic in 2-D nmr will involve J-spectroscopy. This involves spin echo pulse sequences on the nucleus, observed during t_2, to provide information about coupling. Recall our discussion in Section 8–16 where we showed the echo could be used to cancel out chemical shift differences to let us study couplings. The proton decoupled spectrum will be along one axis and the carbon-proton or proton-proton coupling constants spread out along the other. The decoupling is modulated during the evolution period of the echo sequence with gated decoupling or other techniques. The technique aids in the assignment of the resonances and peaks due to spin-spin coupling in badly overlapping spectra. The proton decoupled ^{13}C spectrum of a mixture of deuterated derivatives of toluene is shown in Fig. 8–50(A). (Deuterium has an I-value of one and C—D couplings lead to the spectral complexity.) The J-resolved 2-D nmr is run by employing gated deuterium (^{2}H) decoupling producing the ^{13}C—^{2}H coupling information on the F_1 axis. The ^{13}C chemical shifts of the four carbons are seen on the F_2 axis. The slices parallel to F_1 and through the different carbons show the single, three, five, and seven peaks for coupling constants of the substituents CH_3, CH_2D, CHD_2, and CD_3. A stick diagram of overlapping resonances can now be readily constructed to interpret the spectra in Fig. 8–50(A).

In spectra that contain both homonuclear and heteronuclear coupling, J spectroscopy can be used to separate and assign the homonuclear and heteronuclear coupling constants, e.g., proton-proton couplings can be distinguished

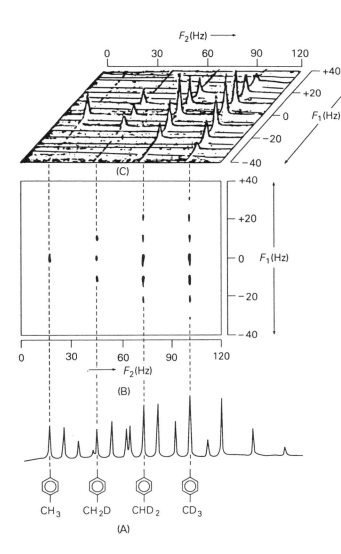

FIGURE 8–50 The proton decoupled, two-dimensional J-resolved ^{13}C spectra of a mixture of methyl deuterated toluenes (^{2}H has an $I = 1$). J coupling is shown on the F_1-axis and the corresponding ^{13}C Shifts on the F_2- axis. The single peak for CH$_3$, triplet for —CH$_2$D, pentet for —CHD$_2$, and septet for CD$_3$ are seen on the F_1 axis for slices through F_2 corresponding to the analogous ^{13}C chemical shift. The one-dimensional ^{13}C spectrum is shown in (A). Dashed lines connect the chemical shifts of the two-dimensional and one-dimensional spectrum.

from phosphorus-proton couplings. The reason for this is that if the proton-proton coupling is modulated during t_1, the phosphorus-proton couplings remain and are present on the v_2-axis. An example is illustrated in Fig. 8–51(A) for the complex Cp—Ru—(CH=CH$_2$)(P(C$_6$H$_5$)$_3$)$_2$. The spectrum of the proton on the vinyl group of the carbon bound to ruthenium is shown in Fig. 8–51(A). This should consist of overlapping triplets from two equivalent phosphorus atoms split into a doublet of doublets from non-equivalent CH$_2$ protons of the vinyl group. The 2-D, J-resolved, ^{1}H nmr spectrum shows the phosphorus splitting on the $F_1(\delta)$ axis, as seen by the triplet in the spectrum in Fig. 8–51(C). The F_2 axis (J) contains the information about the proton-proton coupling. The doublet of doublets is indicated. Using the resolved coupling constants, a stick diagram can be constructed to show how the overlap of peaks leads to the spectra in Fig. 8-51(A).

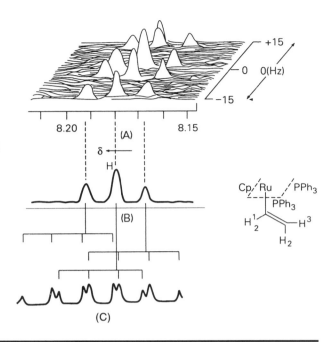

FIGURE 8–51 (A) The proton spectrum of the vinyl CH proton of CpRu(CH=CH$_2$)(Pϕ_3)$_2$. (B) The J-resolved, two-dimensional spectrum. (C) The proton spectra with ^{31}P splittings.

NMR in Solids and Liquid Crystals

8–20 DIRECT DIPOLAR COUPLING

We shall begin this discussion by considering the direct dipole-dipole interaction between two nuclei. This is done by considering *fixed orientations* of two hydrogen atoms in a molecule relative to an external field, as shown in Fig. 8–52. The dashed line is the internuclear axis connecting the two hydrogen atoms, the boldface arrow indicates the orientation of the nuclear moment on b relative to the field, and the curved arrow represents the lines of flux arising from this nuclear moment. We see that for this fixed orientation of the molecule, we would obtain two different peaks in the nmr of the H_a resonance as a consequence of the two different fields from b. This is a through-space effect which, in contrast to the coupling mechanisms discussed earlier, does not involve the electron density in the molecule. The peak separation is indicated by a coupling constant B^{dir} (where dir stands for direct). B can be very large and, for example, is about 120,000 Hz for two protons separated by 1 Å when the H—H internuclear axis is aligned with the field.

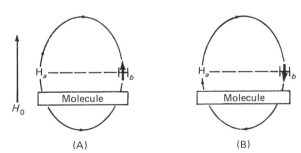

FIGURE 8–52 The direct dipolar interaction of two hydrogens in a molecule.

The magnitude of B will vary with the position of the internuclear axis in relation to the field. The mathematical expression for the magnitude of the direct dipolar interaction between two nuclei p and q, B_{pq}^{dir}, is derived from the expression for the magnetic field due to a point magnetic dipole (*i.e.*, the other nucleus) and the expression for the potential energy of a point dipole (the proton of interest) in a magnetic field. The result is:

$$B_{pq}^{\text{dir}}(\text{Hz}) = -\frac{h}{4\pi^2}\gamma_p\gamma_q\frac{1}{2}\left[\frac{3\cos^2\theta_{pq}-1}{r_{pq}^3}\right] \tag{8-26}$$

where θ_{pq} is the angle that the pq-internuclear axis makes with the external field direction. For a complicated molecule there is one B^{dir} for every pair of magnetic nuclei in the sample.

In a normal solvent, any given internuclear axis is randomly oriented with respect to the external field and rapidly changing its position because the molecule is rotating. The direct dipole-dipole interaction is averaged to zero. Since the indirect coupling constants proceed through the electron density of the molecule, they are not averaged to zero in a solvent. However, the direct dipole-dipole interaction is through space: it is given by equation (8–26) and is averaged out. At very high fields ($\geqslant 14$ T), the effects of imperfectly averaged diamagnetic anisotropy are seen, *e.g.*, broadening of aromatic proton peaks.

8-21 NMR STUDIES OF SOLIDS

In single crystals, the H—H internuclear axes in the molecules have a fixed orientation relative to the applied field and B^{dir} is not averaged out. According to equation (8–26), if one were to study the angular variation of B^{dir} by investigating different orientations of a single crystal, it would be possible to solve the resulting simultaneous equations for $1/r^3$ and to find the magnitude and direction of the H—H internuclear axes. Unfortunately, only rarely are the protons in the crystal few enough in number and far enough apart to permit resolution of the spectral lines and determination of B^{dir}. Generally, the spectra of solids consist of very broad, poorly resolved bands because of the direct dipolar interaction between protons in the molecule and between those from the nearby molecules. However, structural information can be obtained from the broadened resonance line of single crystals or powders by the so-called method of second moments[66]. The second moment is the mean square width $\overline{(\Delta H)^2}$ measured from the center of the resonance line. The center of the resonance line is the average magnetic field, as seen in equation (8–27).

$$H_{\text{av}} = \int_0^\infty Hf(H)\,dH \tag{8-27}$$

where $f(H)$ represents the normalized line shape. The second moment is then given by equation (8–28),

$$\overline{(\Delta H)^2} = \int_0^\infty (H-H_{\text{av}})^2 f(H)\,dH \tag{8-28}$$

which can be evaluated graphically from the observed line shape. Most observed peaks have a Gaussian shape and are described by

$$f(H) = \frac{1}{\Delta H \sqrt{2\pi}} e^{-\{(H - H_{av})^2 / 2(\Delta H)^2\}} \quad (8\text{--}29)$$

The second moment is then determined from the integration of the analytical expression obtained from equations (8–28) and (8–29). The second moment for a single crystal or powder is related to r and θ, and gives structural information. The reader is referred to references 66 through 68, for more details. This technique has been used to show that the infusible white precipitate from the reaction of NH_3 with $HgCl_2$ is NH_2HgCl and not NHg_2ClNH_4Cl or $XHgO(1 - X) HgCl_2 \cdot 2NH_3$.[69] These studies can be carried out on more complex molecules in conjunction with deuteration studies. The magnetic moment of the deuteron is very small, and dipolar interactions that involve deuterium can usually be neglected.

The temperature dependence of the second moment has also been employed to provide information on molecular motion in solids. It has been shown that benzene is fixed in the solid below 90 K, but rotates rapidly about the sixfold axis between 120 and 280 K. The second moment changes gradually from 9.7 gauss^2 below 90 K to 1.6 gauss^2 at 120 K, and remains at this value to 280 K. Rotation of cyclohexane about the S_6 axis has also been demonstrated. This is an extremely sensitive technique, for the rate of rotation required to narrow the resonance is not much higher than the proton resonance frequency in a field of a few gauss, *i.e.*, 10^4 Hz. Second-moment studies demonstrated[70] the rotation of the benzene and cyclopentadiene rings in dibenzene-chromium and ferrocene as well as[71] rotation of benzene in the benzene-silver perchlorate complex. A study[72] of $[Me_3SiNSiMe_2]_2$ demonstrated that at 77 K the methyl groups rotate about the Si—C bond, and this is the only motion in the molecule. At room temperature, methyl groups rotate about the C—Si bond, $(CH_3)_3Si$ groups rotate about the Si—N bond, and the whole molecule rotates about a molecular axis. Thermodynamic data can be obtained for the various motional processes from the temperature dependence of the spectrum.

8–22 NMR STUDIES IN LIQUID CRYSTAL SOLVENTS

Certain materials, *e.g.*,

$$CH_3CH_2O-\langle\bigcirc\rangle-N=N-\langle\bigcirc\rangle-O\overset{O}{\underset{\|}{C}}(CH_2)_5CH_3$$

and

$$CH_3(CH_2)_6-\langle\bigcirc\rangle-\underset{\underset{O}{|}}{N}=N-\langle\bigcirc\rangle-O(CH_2)_6CH_3$$

melt to produce a turbid fluid in which certain domains exist where there is considerable ordering of the molecules. The resulting fluid has some properties of both liquids and crystals. The liquid is strongly anisotropic in many of its properties,[73] but further heating produces an isotropic liquid. This general class of materials is referred to as *thermotropic liquid crystals* (*i.e.*, produced by melting a solid). Three subclassifications are illustrated in Fig. 8–53 by schematically

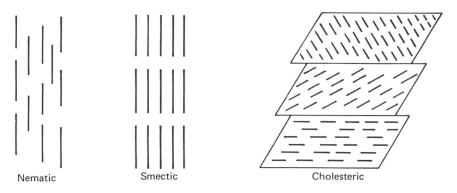

FIGURE 8–53 Types of ordered domains in a liquid crystal.

indicating the type of ordering in the domains. The representations in Fig. 8–53 are of small domains in the whole liquid sample. These domains tend to become aligned in the presence of a magnetic field.

The nmr spectra of the molecules that constitute the liquid crystal are very broad, nondescript resonances that sometimes are so broad as not to be observed. The broadening results because the viscous solvent molecules contain many hydrogens and consequently there are a large number of direct H-H dipole-dipole interactions that are not averaged out in the partially ordered solvent.

If some benzene is dissolved in the liquid crystal as solvent, the spectrum illustrated in Fig. 8–54(A) results.[74] This spectrum of benzene consists of a large number of sharp lines spread out over approximately 2400 Hz. Some of the broad resonances of the solvent are discernible under the sharp benzene peaks. The benzene spectrum is thus somewhere between that obtained in a non-viscous solvent (a single sharp line) and that of a solid (a broad line spread over a wide

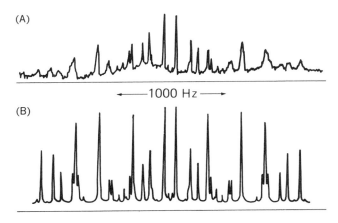

FIGURE 8–54 (A) Liquid crystal nmr spectrum of benzene; (B) simulated spectrum.

field). In an isotropic solvent, the various axes in a solute molecule assume all possible orientations with respect to the magnetic field with equal probability, and B is averaged to zero. As mentioned before, the indirect couplings proceed via the electron density in the molecule and are not averaged to zero. When all orientations of an axis with respect to the field are not equally probable, the contribution to the field at a proton from each surrounding nucleus depends on the orientation of the nuclear moment in the neighbor, and the position of that neighbor nucleus relative to the one being observed. In a solid where the nuclei are fixed and where so many protons affect the field of any one kind of proton, a continuous very broad single peak results. We now must explain why we see such sharp peaks for the solute in the liquid crystal spectrum and why there are so many peaks.

In a liquid crystal, certain orientations of the solute are more favored than others (long axes line up with the solvent long axis) because the magnetic field tends to align the solvent and solute molecules in the field direction. Solute molecules diffuse freely and tumble freely enough so that there is no contribution to B^{dir} from the solvent or other solute molecules. All the couplings are intramolecular. Since the molecule is tumbling rapidly, sharp lines are observed in the nmr. We see many lines in the spectrum because B_{pq}^{dir} [equation (8–26)] makes an observed contribution to the resonance line positions. The anisotropic motion does not average this quantity to zero. Each θ_{pq} of equation (8–26) now becomes some average value for the net orientation of each of the respective H-H axes in this rapidly and anisotropically tumbling molecule.

In benzene, for example, we have six protons and need a matrix of all possible combinations of six spins (i.e., a basis set $\alpha\alpha\alpha\alpha\alpha\alpha$, $\alpha\alpha\alpha\alpha\alpha\beta$, etc.) to describe this system. Thus, we have peaks corresponding to molecules with all these different permutations of spins with B and J values for all pairs of hydrogens. Often, second-order spectra result, further complicating the appearance of the spectrum. The energies are described by the spin Hamiltonian matrix for this system, which is similar to that discussed in treating second-order spectra except that, for every J on the diagonal in the solution problem, we now have a $J + 2B$; and for every J previously on the off diagonal we now have a $J - B$. This problem is solved, *vide infra.* by finding J and B values [$J_{ortho(o)}$, $J_{meta(m)}$, $J_{para(p)}$, B_o, B_m, and B_p] that will reproduce the experimental spectrum. The calculated spectrum is shown in Fig. 8–54(B). For benzene, the resulting values are $B_o = -639.5$ Hz, $B_m = -123.1$ Hz, $B_p = -79.93$ Hz, $J_o = 6.0$ Hz, $J_m = 2.0$ Hz, and $J_p = 1.0$ Hz.

The dipolar couplings are related to the nuclear coordinates by

$$\langle B_{pq} \rangle = \frac{-\gamma_p \gamma_q h}{8\pi^2 \langle \gamma_{pq}^3 \rangle} [S_{zz}(3\cos^2\theta_{pqz} - 1)$$
$$+ (S_{xx} - S_{yy})(\cos^2\theta_{pqx} - \cos^2\theta_{pqy}) + 4S_{xy}\cos\theta_{pqx}\cos\theta_{pqy}$$
$$+ 4S_{xz}\cos\theta_{pqx}\cos\theta_{pqz} + 4S_{yz}\cos\theta_{pqy}\cos\theta_{pqz}] \quad (8\text{--}30)$$

where r_{pq} is the internuclear separation of the coupled nuclei, θ_{pqz} is the angle r_{pq} makes with the z-coordinate, etc. for θ_{pqx} and θ_{pqy}. The axes are fixed within the molecule and S is an averaged-order matrix that relates to the extent of orientation of the molecule by the liquid crystal. If the axis is completely aligned with the field, S_{pq} is 1. If the axis is aligned perpendicular to the field, then $S_{pq} = -(1/2)$. For a random orientation, $S_{pq} = 0$.

If we knew S_{pq}, we could calculate r_{pq} from the experimental B_{pq}^{dir} and get distances and geometries of molecules in solution. In a rigid molecule, the values of S_{pq} for the various axes must be interrelated. This interrelationship between the axes and the orientation of the molecule in three-dimensions can be described with the tensor $\bar{\bar{S}}$. The $\bar{\bar{S}}$ tensor is 3×3, symmetric and traceless, so that only five of the nine elements are independent. As discussed in Chapter 2, we can define a molecular coordinate system that diagonalizes the tensor. This usually corresponds to symmetry axes if the molecule has symmetry; then there will be only two independent elements. (All off-diagonal elements are zero and the diagonal ones are traceless, i.e., their sum equals zero.) For a molecule with a threefold or higher axis, $S_{xx} = S_{yy} = -(1/2)S_{zz}$ and there is only one independent tensor element. Note that the tensor is independent of r. With a threefold or high-order axis, equation (8–30) becomes

$$B_{pq}^{\text{dir}} = \frac{-h}{8\pi^2}\gamma_p\gamma_q \frac{S_{zz}(3\cos^2\theta_{pqz} - 1)}{r_{pq}^3} \qquad (8\text{–}31)$$

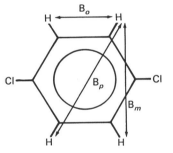

FIGURE 8–55 The B_{pq} interactions in p-$Cl_2C_6H_4$.

We can illustrate the method by considering paradichlorobenzene, Fig. 8-55, as an example. Analysis of the spectra yields three dipolar coupling constants, B_o, B_m, and B_p shown in Fig. 8–55. There are two unknown tensor elements S_{zz} and $(S_{xx} - S_{yy})$ and two distances corresponding to the two sides of the triangle made by B_p, B_m, and H-H in Fig. 8–55. We are thus confronted with three equations and four unknowns. Therefore, it is only possible to get the ratios of all the distances; or, if we can assume one, the others can be calculated. It is often the case in these systems that there is one more unknown than there are knowns. However, even if a particular system has more coupling constants than unknowns, we cannot obtain a unique solution. This can be seen by referring to equation (8–31). For a given S and r that satisfy equation (8–31) we could multiply r by a factor and S by the cube of that factor and get the same B. This corresponds to a uniform expansion of the structure without any change in the ratios of the distances. In spite of this severe limitation, a very considerable amount of information can still be obtained from nmr studies in liquid crystals.

In molecules with a threefold axis or higher, equation (8–31) applies. These systems require only one S element to relate the B_{pq} values to structure. The ratios of the dipolar coupling constants are independent of S and give us the shape of the molecule. The ratio of the dipolar coupling for all proton pair B_{pq} values can be calculated for a regular hexagon by taking the corresponding ratios of equation (8–31) to yield 1 to 0.192 to 0.125. Experimental ratios of 1 to 0.192 to 0.125 are found for benzene, showing that it is a regular hexagon in solution. CpNiNO has two B_q values whose ratio for a regular pentagon is 4.236. A value of 4.211 is found indicating a regular polygon. On the other hand, CpMn(CO)$_3$ produces an experimental ratio of 4.11 indicating that distortion exists.[75,76] In a study[77] of the liquid crystal nmr spectrum of trans-[Cl$_2$Pt(C$_2$H$_4$)C$_5$H$_5$N], the influences that various dynamic processes in solution have on the resulting spectra are discussed. The ordering factor S is very low and severe overlapping of the resonances results. A method was developed in which the spectrum intensities are fitted in the spectral analysis. The ratio of the $r_{\text{gem}}/r_{\text{cis}}$ protons in the coordinated ethylene is consistent with a structure in which the C—H bonds are bent back away from the metal. Except for the dynamic process occurring in solution (the ethylene rotates rapidly about the bond to platinum), the solution structure of the ethylene fragment is similar to that in the solid.

When the chemical shift differences of the protons involved are large and first-order spectra are obtained, the spectrum is readily interpreted by inspection, and information about the symmetry of the molecule can be obtained from the spectrum in a straightforward manner. Consider the results reported[78] for $H_3Ru_3(CO)CH_3$, which led to the proposed structure shown in Fig. 8–56 along with the spectrum observed for this molecule in a liquid crystal. The methyl

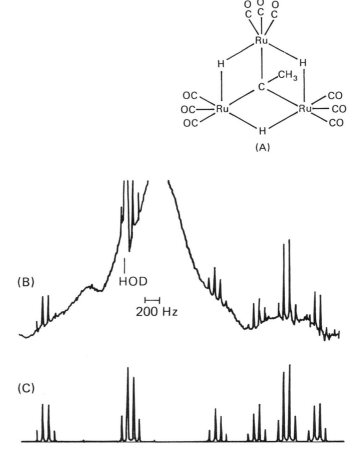

FIGURE 8–56 (A) Structure of $H_3Ru_3(CO)CH_3$. (B) Experimental liquid crystal spectrum (broad absorption is from the liquid crystal). (C) Simulated liquid crystal spectrum. [Reprinted with permission from A. D. Buckingham, *et al.*, J. Amer. Chem. Soc., *95*, 2732 (1973). Copyright by the American Chemical Society.]

group is rapidly rotating about the C—C bond axis. The quartets arise from the three equivalent methyl protons splitting the three bridging protons and vice versa. The three methyl protons are equivalent with respect to the bridging hydrides because of the rapid rotation. In a fixed staggered configuration (*i.e.*, CH_3 versus metal hydrides), the hydride protons would be split by a set of two equivalent protons and one non-equivalent methyl proton, producing six peaks in each triplet component, *i.e.*, 36 peaks total. One of the two groups of triplets arises from direct dipolar coupling of the methyl protons with each other, and the second (more closely spaced) triplet arises from the dipolar coupling of the three bridged hydrides with each other. If we consider one of the protons of the methyl group, it can be split by the other two because $(+\ +), (+\ -), (-\ +)$, and $(-\ -)$ combinations of nuclear moments in different molecules cause this proton to experience different fields, since the molecule is not undergoing isotropic rotation and the direct dipolar coupling is not averaged to zero. The larger coupling constants arise when the protons causing the splitting are closer to each other.

Deuterium nmr in liquid crystals[78,79] leads to a determination of the quadrupole coupling constant q_{zz} (Chapter 14). The interaction of a proton with a deuteron in a partially aligned molecule will give rise to a four-line pattern. The large doublet arises from the quadrupole interaction and the smaller doublet from the dipolar coupling. The value of q_{pzz} is related to the quadrupole coupling constant by

$$q_{pzz} = S_{pq} q_{zz} \qquad (8\text{-}32)$$

FIGURE 8–57 Splitting of the deuterium resonance by a proton.

8-23 HIGH RESOLUTION NMR OF SOLIDS[80]

As described in Section 8–21, dipolar coupling leads to very broad nmr spectra of solids. Also contributing to the broadness is chemical shift and scalar spin-spin coupling (J) anisotropy with both quantities being represented by tensors. The entire range of chemical shifts and couplings spanned by all orientations of crystallites is manifested in the spectra leading to widths of 50–200 ppm in ^{13}C spectra. The first attempts to reduce dipolar coupling in solids involved rapid spinning of the sample about an axis, inclined at the so-called magic angle of 54°44′. All of the moment vectors are pointing in all directions in the static sample. One such moment is illustrated by the vector pq relative to H_o in Fig. 8–58. If we spin rapidly, we average all of these leading to a net moment that points along the spinning axis. This can be seen in Fig. 8–58 where spinning along the axis at 54°44′ sweeps out two circles for opposite ends of pq. The net of all the orientations along the circles is the vector $p'q'$ lying on this spinning axis. At this magic angle for the axis, $3\cos^2\theta - 1$ equals zero. One must spin the sample rapidly compared to the magnitude of the interaction to average the moments. Dipolar coupling of two protons typically is 5×10^4 Hz. Centrifugal force will cause rotors to disintegrate at this speed. Conventional spinning rates are fast enough to remove chemical shift and J coupling anisotropy. Large chemical shift anisotropy produces severe spinning sidebands.

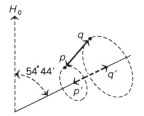

FIGURE 8–58 Averaging of the moment pq to $p'q'$ by spinning about the axis at 54°44′ to the applied H_0 field.

This approach, which will be called magic angle sample spinning, MASS, can remove dipolar coupling in magnetically dilute samples, e.g., ^{13}C. Note in equation (8–26) that the direct coupling is a function of $1/r^3$. Also note (equation (8–26)) that a low γ value for the nucleus of interest also tends to reduce dipolar coupling. Thus, ^{13}C and ^{29}Si give sharp nmr spectra from MASS.

Quadrupolar nuclei also lead to line broadening through the electric quadrupole interaction. For $I = n + 1/2$ nuclei, the center transition ($-1/2 \to 1/2$) is not affected and a sharp line results. The other transitions are generally broadened beyond detection. For nuclei with integer values of I, there is no $-1/2 \to 1/2$ transition and very broad spectra result. Even with $I = n/2$ systems, a second-order quadrupole effect broadens the central transition. This effect becomes less important at high fields.

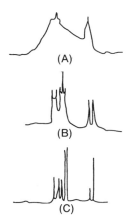

FIGURE 8-59 ^{13}C nmr of polysulfone (—O—C(CH$_3$)$_2$—O—O—O—SO$_2$—O—O—) (A) double resonance; (B) double resonance plus MASS; (C) solution spectrum.

Low values of γ and chemical dilution permits detection of 100% ^{14}N and ^{31}P in solids with broadband proton decoupling. Long T_1's can be shortened by cross polarizing,[81] CP, dilute, and abundant spins. In the CP technique, the proton spin polarization is transferred[82] to the less abundant nuclei, e.g., ^{13}C by applying a proton rf field (H_{1_H}) and then a ^{13}C field (H_{1_C}) such that $\gamma_H H_{1_H} = \gamma_C H_{1_C}$. Efficient transfer of energy occurs between the two spin manifolds generating ^{13}C magnetization along H_{1_C} and increasing the intensity of the transition. Cross polarization depends on the static component of the dipolar interaction and has the greatest effect on neighboring nuclei. It is most efficient for immobile species and ineffective for mobile species. These concepts have been employed[83,84] to study the surface structure and reactivity of silica gel and zeolites. Complementary information about the bulk structure can be obtained using a pulse-delay-observe sequence.

Combining chemical dilution with MASS leads to a further sharpening of the resonances as shown in Fig. 8–59. A pulse sequence referred to as WAHUHA[85] modulates the spin and has reduced line widths from 10^4 to $\sim 10^2$ Hz. These experiments have the potential for determining chemical shift and dipolar anisotropes. Carbon-hydrogen dipolar couplings and with them C–H distances have been obtained on solids by a 2-D pulse sequence.[86] The evolution period corresponds to variable times, τ, for the application of a WAHUHA pulse. The 2-D plot contains the chemical shift as one axis and the dipolar splittings as the other.

REFERENCES CITED

1. F. L. Slejko, R. S. Drago and D. G. Brown, J. Amer. Chem. Soc., *94*, 9210 (1972).
2. S. S. Zumdahl and R. S. Drago, J. Amer. Chem. Soc., *89*, 4319 (1967).
3. S. Alexander, J. Chem. Phys., *37*, 967 (1962); Rev. Mol. Physics, *24*, 74 (1957).
4. C. S. Johnson, Adv. Mag. Res., *1*, 33 (1965).
5. G. Fano, Rev. Mod. Physics, *24*, 74 (1957).
6. S. S. Zumdahl and R. S. Drago, Inorg. Chem., *7*, 2162 (1968).
7. (a) W. D. Perry, R. S. Drago and N. K. Kildahl, J. Coord. Chem., *3*, 203 (1974).
   (b) N. K. Kildahl and R. S. Drago, J. Amer. Chem. Soc., *95*, 6245 (1973).
8. H. S. Gutowsky and C. H. Holm, J. Chem. Phys., *25*, 1228 (1956).
9. M. Rogers and J. C. Woodbrey, J. Phys. Chem., *66*, 540 (1962).
10. C. Looney, W. D. Phillips and E. L. Reilly, J. Amer. Chem. Soc., *79*, 6136 (1957).
11. G. Claeson, G. Androes and M. Calvin, J. Amer. Chem. Soc., *83*, 4357 (1961).
12. E. Grunwald, A. Loewenstein and S. Meiboom, J. Chem. Phys., *27*, 630 (1957).
13. S. Meiboom *et al.*, J. Chem. Phys., *29*, 969 (1958) and references therein.
14. E. L. Muetterties and D. W. Phillips, J. Amer. Chem. Soc., *81*, 1084 (1959).
15. A. Patterson, Jr. and R. Ettinger, Z. Elektrochem., *64*, 98 (1960).
16. W. D. Phillips *et al.*, J. Amer. Chem. Soc., *81*, 4496 (1959).
17. W. N. Lipscomb, Adv. Inorg. Chem. Radiochem., *1*, 132 (1959).
18. B. M. Graybill, J. K. Ruff and M. F. Hawthorne, J. Amer. Chem. Soc., *83*, 2669 (1961).
19. M. J. Bennett, Jr., *et al.*, J. Amer. Chem. Soc., *88*, 4371 (1966); F. A. Cotton, Accts. Chem. Res., *1*, 257 (1968).
20. (a) K. Vrieze and P. W. N. M. van Leeueven, Prog. Inorg. Chem., *14*, 1 (1971).
    (b) "Dynamic Nuclear Magnetic Resonance Spectroscopy," ed. L. M. Jackman and F. A. Cotton, Academic Press, New York, 1975.
21. L. W. Reeves and K. N. Shaw, Can. J. Chem., *48*, 3641 (1970).
22. H. S. Gutowsky and C. J. Hoffman, J. Chem. Phys., *19*, 1259 (1951).

23. J. I. Musher, J. Amer. Chem. Soc., *94*, 5662 (1972) and references therein.
24. (a) W. G. Klemperer, J. Amer. Chem. Soc., *95*, 380 (1973) and references therein.
    (b) J. P. Jesson and P. Meakin, J. Amer. Chem. Soc., *96*, 5760 (1974).
25. G. M. Whitesides and H. L. Mitchell, J. Amer. Chem. Soc., *91*, 5384 (1969).
26. J. R. Shapley and J. A. Osborn, Accts. Chem. Res., *6*, 305 (1973).
27. J. P. Jesson, in "Transition Metal Hydrides," ed. E. L. Muetterties, M. Dekker, New York, 1971.
28. S. S. Eaton, J. R. Hutchison, R. H. Holm and E. L. Muetterties, J. Amer. Chem. Soc., *94*, 6411 (1972).
29. D. J. Duffy and L. H. Pignolet, Inorg. Chem., *11*, 2843 (1972).
30. J. W. Faller, in "Determination of Organic Structures by Physical Methods," Vol. 5, eds. F. Nachod and J. Zuckerman, Academic Press, New York, 1973, and references therein.
31. (a) P. D. Kennewell, J. Chem. Educ., *47*, 278 (1970).
    (b) See, for example, A. Carrington and A. D. McLachlan, "Introduction to Magnetic Resonance," Chapter 13, Harper & Row, New York, 1967.
    (c) J. H. Noggle and R. E. Schermer, "The Nuclear Overhauser Effect: Chemical Applications," Academic Press, New York 1971.
    (d) D. Neuhaus, P. Davis, and M. Williamson, "The Nuclear Overhauser Effect in Stereochemical and Conformational Analysis," VCH Publishers, New York, 1989.
32. G. C. Martin and J. M. Boncella, *Organomet.*, 8, 2969, (1989).
33. (a) G. L. Closs *et al.*, J. Am. Chem. Soc., *92*, 2183, 2185, 2186 (1970).
    (b) S. H. Pine, J. Chem. Educ., *49*, 664 (1972).
34. T. C. Farrar and E. D. Becker, "Introduction to Pulse and Fourier Transform NMR Methods," Academic Press, New York, 1971.
35. C. Brevard and P. Granger, "Handbook of High Resolution Multinuclear NMR," Wiley, New York, 1981.
36. E. D. Becker, J. A. Feretti, M. Gambhir, Anal. Chem. *51*, 1413 (1979).
37. K. L. Kuhlmann, D. M. Grant and R. K. Harris, J. Chem. Phys., *52*, 3439 (1970).
38. T. D. Alger, S. W. Collins and D. M. Grant, J. Chem. Phys., *54*, 2820 (1971).
39. A. Allerhand and D. Doddrell, J. Am. Chem. Soc., *93*, 2777 (1971).
40. R. A. Newmark and J. R. Hill, J. Am. Chem. Soc., *95*, 4435 (1973).
41. (a) O. A. Gansow, A. R. Burke and G. N. LaMar, Chem. Comm., 456 (1972).
    (b) O. A. Gansow *et. al.*, J. Amer. Chem. Soc., *94*, 2550 (1972).
    (c) D. F. S. Natusch, J. Amer. Chem. Soc., *93*, 2566 (1971).
42. D. Doddrell and A. Allerhand, Proc. Natl. Acad. Sci. (USA), *68*, 1083 (1971).
43. O. A. Gansow and W. Schittenhelm, J. Amer. Chem. Soc., *93*, 4294 (1971).
44. N. A. Matwiyoff and D. G. Ott, Science, *181*, 1125 (1973).
45. P. J. Vergamini, N. A. Matwiyoff, R. C. Wohl and T. Bradley, Biochem. Biophys. Res. Comm., *55*, 453 (1973).
46. N. A. Matwiyoff *et al.*, J. Amer. Chem. Soc., *95*, 4429 (1973).
47. R. E. London, C. T. Gregg and N. A. Matwiyoff, Science, *188*, 266 (1975).
48. R. E. London, V. H. Kollman and N. A. Matwiyoff, J. Amer. Chem. Soc., *97*, 3565 (1975).
49. G. A. Olah *et al.*, J. Amer. Chem., Soc., *93*, 4219 (1971).
50. R. G. Parker and J. D. Roberts, J. Amer. Chem. Soc., *92*, 743 (1970).
51. G. M. Bodner *et al.*, Chem. Comm., *1970*, 1530.
52. G. M. Bodner and L. J. Todd, Inorg. Chem., *13*, 360 (1974); *ibid.*, *13*, 1335 (1974).
53. L. J. Todd and J. R. Wilkinson, J. Org. Met. Chem., *77*, 1 (1974).
54. B. E. Mann, Adv. Org. Met. Chem., *12*, 135 (1974).
55. J. Evans and J. R. Norton, Inorg. Chem., *13*, 3043 (1974).
56. R. Ditchfield and P. D. Ellis, in "Topics in Carbon-13 NMR Spectroscopy," Vol. 1, ed. G. C. Levy, Wiley-Interscience, New York, 1974.
57. R. K. Harris and B. E. Mann, "NMR and the Periodic Table," Academic Press, New York, 1978.

58. J. Mason, Chem. Brit., 654 (1983).
59. (a) I. C. P. Smith, G. W. Stockton, A. P. Tulloch, C. F. Polnaszek and K. G. Johnson, J. Colloid Interface Sci., *58*, 439 (1977).
    (b) A. Seelig and J. Seelig, Biochemistry, *13*, 4839 (1974).
    (c) H. H. Mantsch, H. Saitô and I. C. P. Smith, in "Progress in Nuclear Magnetic Resonance Spectroscopy," ed. J. W. Emsley, J. Feeney and L. H. Sutcliffe, Pergamon Press, London, 1977.
60. H. Schmidbaur, O. Gasser, T. E. Fraser and E. A. V. Ebsworth, J. Chem. Soc., Chem. Comm. 34 (1971).
61. A. Derome, "Modern NMR Techniques for Chemistry Research," Pergamon Press, London, 1987.
62. D. M. Dodrell, D. T. Pegg, M. R. Bendall, J. Mag. Reson., *48*, 323 (1982).
63. T. L. Venable, W. C. Hulton and R. N. Grimes, J. Am. Chem. Soc., *106*, 29 (1984).
64. (a) C. Brevard, R. Schimpe, G. Tourne and C. M. Tourne, J. Am. Chem. Soc., *105*, 7059 (1983).
    (b) R. G. Fuke, B. Rapho, R. J. Sexton and P. J. Vomaille, J. Am. Chem. Soc., *108*, 2947 (1986).
65. See, for example, N. W. Alcoch, J. M. Brown, A. E. Derome and A. R. Lucy, J. Chem. Soc., Chem. Comm., 757, (1985).
66. C. P. Slichter, "Principles of Magnetic Resonance," Chapter 3, Harper & Row, New York, 1963.
67. A. Abragam, "The Principles of Nuclear Magnetism," Chapter 4, Clarendon Press, Oxford, 1961.
68. E. R. Andrew, "Nuclear Magnetic Resonance," Chapter 6, Cambridge University Press, New York, 1955; E. R. Andrew and P. S. Allen, J. Chem. Phys., *63*(1), 85–91 (1966).
69. C. M. Deely and R. E. Richards, J. Chem. Soc., 3697 (1954).
70. L. N. Mulay, E. G. Rochow and E. O. Fischer, J. Inorg. Nucl. Chem. *4*, 231 (1957).
71. D. F. R. Gilson and C. A. McDowell, J. Chem. Phys., *40*, 2413 (1964) and references therein.
72. H. Levy and W. E. Grizzle, J. Chem. Phys., *45*, 1954 (1966).
73. P. Diehl, et al., "NMR Basic Principles and Progress," Vol. 1, Springer-Verlag, New York, 1969.
74. L. C. Snyder and E. W. Anderson, J. Am. Chem. Soc., *86*, 5023 (1964).
75. J. W. Emsley and C. S. Lunon, Mol. Phys., *28*, 1373 (1974).
76. C. S. Yannoni, G. P. Ceasar and B. P. Dailey, J. Am. Chem. Soc., *89*, 2833 (1967).
77. D. R. McMillin and R. S. Drago, Inorg. Chem. *13*, 546 (1974).
78. P. Diehl and A. J. Tracey, Molec. Phys., *30*, 1917 (1975).
79. J. W. Emsley *et al.*, J. Chem. Soc., Faraday II, 1365 (1976).
80. R. K. Harris, "Nuclear Magnetic Resonance Spectroscopy," Pitman, London, 1983.
81. A. Pines, M. Gibby and J. S. Waugh, J. Chem. Phys. *59*, 569 (1973) and references therein.
82. S. R. Hartman and E. C. Halsk, Phys. Rev. *128*, 2042 (1962).
83. D. W. Sindorf and G. E. Maciel, J. Am. Chem. Soc. *105*, 1487, 3767 (1983).
84. C. A. Fyfe, G. C. Gobbi, J. Klinowski, J. M. Thomas and S. Ramdas, Nature *296*, 530 (1982); J. Phys. Chem. *86*, 1247 (1982).
85. J. S. Waugh, L. M. Huber and U. Haeberlen, Phys. Rev. Lett., *20*, 180 (1968).
86. J. S. Waugh, Proc. Natl. Acad. Sci., USA, *73*, 1394 (1976); R. K. Hister, J. L. Ackerman, B. Neff and J. S. Waugh, Phys. Rev. Lett. *36*, 1081 (1976).

ADDITIONAL REFERENCES

J. W. Emsley et al., "High Resolution Nuclear Magnetic Resonance Spectroscopy," Vols. 1 and 2, Pergamon Press, New York, 1966.

T. C. Farrar and E. D. Becker, "Introduction to Pulse and Fourier Transform NMR Methods," Academic Press, New York, 1971.

C. P. Slichter, "Principles of Magnetic Resonance," Harper & Row, New York, 1963.

A. Abragam, "The Principles of Nuclear Magnetism," Clarendon Press, Oxford, 1961.

G. C. Levy and G. L. Nelson, "Carbon-13 Nuclear Magnetic Resonance for Organic Chemists," Wiley-Interscience, New York, 1972.

J. B. Stothers, "Carbon-13 NMR Spectroscopy," Academic Press, New York, 1972.

I. Bertini, H. Mollnari and N. Niccolai, "NMR and Biomolecular Structure," VCH Publishers, New York, 1991.

"Nuclear Magnetic Resonance Shift Reagents," ed. R. E. Sievers, Academic Press, New York, 1973.

C. P. Poole, Jr. and H. A. Farach, "The Theory of Magnetic Resonance," Wiley-Interscience, New York, 1972.

E. Breitmaier and W. Voelter, "Carbon-13 NMR Spectroscopy: High-Resolution Methods and Applications in Organic Chemistry and Biochemistry," VCH Publishing, New York, 1987.

A. Derome, "Modern NMR Techniques for Chemistry Research," Pergamon Press, London, 1987.

G. E. Martin and A. S. Zektzer, "Two-Dimensional NMR Methods for Establishing Molecular Connectivity: A Chemist's Guide to Experiment Selection, Performance and Interpretation," VCH Publishing, New York, 1989.

"Two-Dimensional NMR Spectroscopy: Applications for Chemists and Biochemists," eds. W. R. Croasmun and R. M. K. Carlson, VCH Publishing, New York, 1987.

A. E. Tonelli, "NMR Spectroscopy and Polymer Microstructure: The Conformational Connection," VCH Publishing, New York, 1989.

W. S. Brey, ed. "Pulse Methods in 1-D and 2-D Liquid Phase NMR," Academic Press, New York, 1988.

R. Freeman, "A Handbook of Nuclear Magnetic Resonance," Wiley, New York, 1987.

R. R. Ernst, G. Bodenhausen and A. Wokaun, "Principles of Nuclear Magnetic Resonance in One and Two Dimensions," Oxford University Press, Oxford, 1987.

A series of articles by R. W. King and K. R. Williams provides a clear explanation of FT methods:

"The Fourier Transform in Chemistry. Nuclear Magnetic Resonance: Introduction," J. Chem. Educ., 66, A213–A219 (1989).

"The Fourier Transform in Chemistry. Part 2. Nuclear Magnetic Resonance: The Single Pulse Experiment," J. Chem. Educ., 66, A243–A248 (1989).

"The Fourier Transform in Chemistry. Part 3. Nuclear Magnetic Resonance: Multiple Pulse Experiments," J. Chem. Educ., 67, A93–A99 (1990).

"The Fourier Transform in Chemistry. Nuclear Magnetic Resonance: A Glossary of NMR Terms," J. Chem. Educ., 67, A100–A104 (1990).

"The Fourier Transform in Chemistry. Part 4. Nuclear Magnetic Resonance: Two-Dimensional Methods," J. Chem. Educ., 67, A125–A137 (1990).

Series

"Annual Reports on NMR Spectroscopy," Academic Press, New York, 1968–.

"Advances in Magnetic Resonance," ed. by J. S. Waugh, Academic Press, New York, 1965–.

"Advances in Magnetic Resonance, Supplement 1: High Resolution NMR in Solids," U. Haeberlen, Academic Press, New York, 1976.

"NMR Abstracts and Index," Preston Technical Abstracts Co., Evanston, Ill., 1968.

EXERCISES

1. Consider the molecule

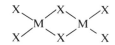

where M and X have $I = \frac{1}{2}$. Sketch the nmr spectra for the following conditions, assuming $\Delta > J$ in all cases (when necessary to assume orders of magnitude for various coupling constants, state your assumption). Ignore coupling if the atoms are not directly bonded.

   a. nmr spectrum of X, no exchange.

   b. nmr spectrum of M, no exchange.

   c. nmr spectrum of X with rapid intermolecular exchange of all X groups.

   d. nmr spectrum of M with rapid $[\tau' < 1/(\nu_A - \nu_B)]$ intermolecular exchange of all X groups.

   e. nmr spectrum of X with rapid intramolecular exchange.

   f. nmr spectrum of M with rapid intramolecular exchange.

2. In the absence of any exchange, two peaks A—H and B—H are separated by 250 Hz. At room temperature, exchange occurs and the peaks are separated by 25 Hz. The spin-lattice relaxation of A—H and B—H is long, and there are equal concentrations (0.2 M) of each. Calculate the lifetime of a proton on A, and from this the rate constant for the exchange (specify units).

3. In a given compound MF_4 (for M, $I = \frac{1}{2}$) the J_{M-F} value is 150 Hz. In the absence of chemical exchange, the F^- and M—F signals are separated by 400 Hz. At room temperature the F^- and MF_4 exchange at a rate such that the fine structure just disappears. Assuming equal concentrations of M—F and F^- species and no stable intermediates, calculate τ' for F^-. What will be the separation of the M—F and F^- peaks under these conditions?

4. The following are ^{31}P nmr spectra, $I_P = I_F = \frac{1}{2}$. No J_{P-H} coupling is ever resolved.

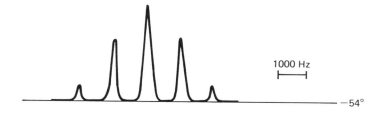

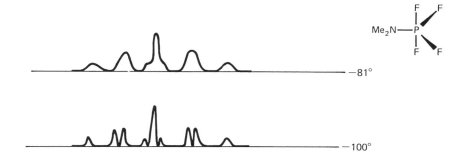

a. Explain the low temperature spectrum.

b. Explain the high temperature spectrum.

c. On the basis of these spectra, what can you say about the mechanism of exchange?

d. What is the relationship between coupling constants in the slow and fast exchange regions?

5. The complex $[(CF_3)Co(CO)_3(PF_3)]$ [J. Amer. Chem. Soc., 91, 526 (1969)] is assumed to be trigonal bipyramidal. The ^{19}F nmr spectra at $+30°C$ and $-70°C$ are given below. Splittings from ^{59}Co are not observed.

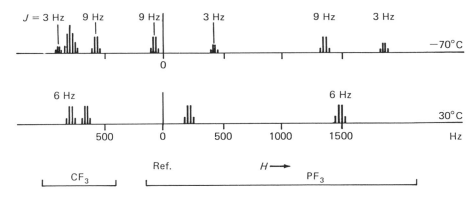

First consider the $-70°C$ spectrum.

a. Explain the reason why four quartets are observed for the fluorines bonded to phosphorus. (No ^{59}Co spin-spin coupling is detected.)

b. Explain how the CF_3 resonances are consistent with your explanation in part a.

c. Describe the reason for the smaller number of PF_3 resonances at $30°C$.

d. What is the significance of the PF_3 resonances being observed as quartets in the $30°C$ spectrum?

6. The line width of the methyl proton resonance of 3-picoline-N-oxide in solutions containing $(3\text{-picoline-}N\text{-oxide})_6 Ni^{2+}$ and excess ligand has been studied as a function of temperature [Inorg. Chem., 10, 1212 (1971)]. The following plot was obtained.

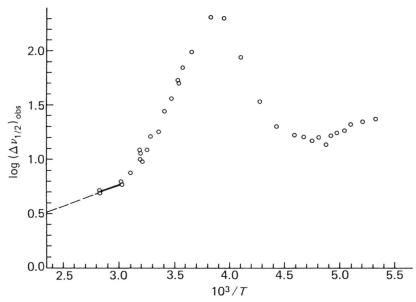

a. Label the fast, near fast, intermediate, and slow exchange regions. Do this by giving approximate boundaries to the region in units of $10^3/T$; e.g., fast: x_1 to x_2 units.

b. These data were analyzed using full line shape analysis from the classical treatment. Why wasn't pyridine-N-oxide used as the ligand?

7. a. When the nmr spectrum of benzonitrile in a liquid crystal is observed, why do solvent resonances not appear?

b. Why are the benzonitrile lines in a liquid crystal much sharper than those in solid benzonitrile?

c. Compared to an nmr spectrum of benzonitrile in CCl_4, why are so many lines observed in the liquid crystal nmr spectrum of benzonitrile?

8. The spectrum and structure of pyrogallol is given below. The OH protons are not

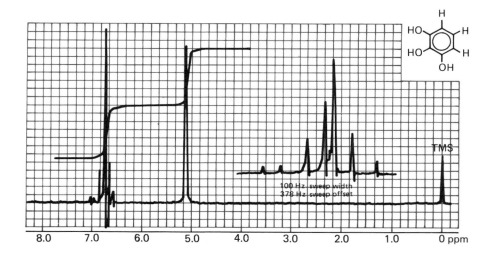

coupled to any other nuclei. Why don't the three phenyl protons give rise to a doublet and a triplet? (The group of peaks centered about 6.7 ppm are expanded in the offset sweep.)

9. ^{57}Fe has an extremely small magnetic moment, and its abundance is only 2.2%. The magnetic moment is not known to very good accuracy, and almost no information exists about chemical shifts of ^{57}Fe nuclei. Even though the resonance signal was expected to cover only a very small frequency range, and thus a slow passage scan through the absorption signal would not take very much longer to observe than would a free induction decay signal, this experiment was not reported until Fourier transform nmr became available.

   a. Why was Fourier transform nmr advantageous for this experiment?

   b. Why wasn't any ^{57}Fe-^{13}C spin-spin splitting observed in the natural-abundance sample?

10. Suppose that a sample has two sites with different chemical shifts. In the continuous wave (CW) experiment, we expect a separate signal for each site.

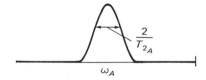
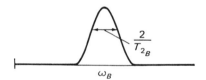

Consider a kinetic process that sets in as the temperature is raised, which exchanges the nuclei between the two sites A and B. (In general, the problem is complex, since the differential equations, one for each site, are coupled by the dynamic process and the individual component magnetizations begin to lose their identities.) Consider the limiting case in which the kinetic process begins to have an observable effect, yet one can still consider the individual M_A magnetization. Argue on a physical basis what effect exchange will have on the width of the A-type resonance. (Hint: Recall the physical process parametrized by T_{2_A}).

11. a. Let the resonance frequency of a water sample be ω_{H_2O}. Show how one can determine the pulse duration τ and a field strength H_1 for an oscillator tuned to ω_{H_2O} that will invert the H_2O magnetization.

   b. Suppose that at a later time a $3\pi/2$ pulse was applied and that no free induction decay curve was observed. What explains this? Of what use is this experiment?

   c. A solute HX is added to the sample, and proton exchange occurs at a rate such that separate but broadened resonances are observed for H_2O and HX. Discuss what effect is expected on solute signal intensity observed in a CW experiment if, while scanning through it, one strongly irradiates the sample of ω_{H_2O}, saturating the H_2O spin system.

12. In the experiment described in the text for measuring T_1, why must the 180° pulse be followed by a 90° pulse?

13. In the slow passage nmr experiment, one frequently observes the following type of pattern when sweeping the field from left to right:

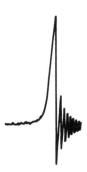

The wiggles to the right of the absorption peak are referred to as "ringing" and are found only when the nucleus observed has a long T_2. Explain ringing in terms of the behavior of the bulk magnetization of the sample in this experiment. Be sure to specify your frame of reference.

14. In the section on the effects of nuclei with quadrupole moments, there was a discussion of the width of the ^{19}F resonance in the spectrum of NF_3 as a function of temperature.

   a. Explain the quadrupole moment relaxation in terms of the concept of spectral density.

   b. Explain the resulting nmr spectrum in terms of spectral density.

15. Consider a frame of reference rotating at a frequency $\omega_1 = 1.0000 \times 10^8$ Hz and a sample of nuclei with a Larmor frequency of $\omega_0 = 1.0001 \times 10^8$ Hz, with $T_1 = 100$ sec and $T_2 = 0.01$ sec. At $t = 0$, a 90° pulse is applied along the u axis using a strong r.f. field of frequency ω_1, in a negligible length of time. Using a uz or vs axis system in each case, draw the net magnetization vector

   a. just before the pulse.

   b. just after the pulse.

   c. at $t = 2.5 \times 10^{-5}$ sec.

   d. at $t = 1$ sec.

   e. at $t = 10^4$ sec.

16. The following ^{13}C T_1's have been determined by Freemen [J. Chem. Phys., 54, 3367 (1971)] on 3,5-dimethylcyclohex-2-ene-1-one:

	T_1 (sec)
C_1	37
C_3	33
3-Me	5.9
C_2	5.4
C_5	5.3
C_6	3.1
C_4	3.1
5-Me	2.7

a. What appears to be the dominant means of spin-lattice relaxation for these ^{13}C atoms?

b. In light of your explanation, why do C_1 and C_3 have such long T_1's?

17. Experiments can be performed to determine the T_1 of a specific atom in a molecule. The most commonly employed technique using FTNMR is to apply a 180° pulse, delay for a time τ, followed by a 90° pulse and immediate acquisition. Tau is usually varied from 0 sec to many times the expected T_1 value.

a. Sketch a generic version of this pulse sequence showing the bulk magnetization after each step in the experiment.

b. Consider the magnetization following the 90° pulse previously stated. Show what the spectral peaks (assume a singlet peak) would look like at the following times: $\tau = 0$, $\tau = $ fraction of T_1, $\tau = T_1$, $\tau \gg T_1$.

c. Would it be possible to perform this experiment to measure T_1 using only a 90° pulse, waiting τ, and then acquiring? Explain.

18. The proton-coupled ^{11}B COSY spectrum of decaborane is shown as follows.

a. Draw the ^{11}B proton-decoupled nmr spectrum.

b. Assuming only directly bound borons are coupled, assign the resonances in the ^{11}B nmr spectrum. Explain.

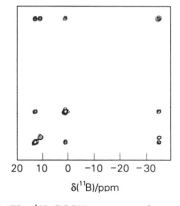
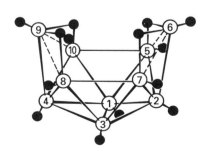

19. The ^{1}H COSY spectrum of a sample is shown below (letters correspond to the same ^{13}C nucleus throughout this problem).

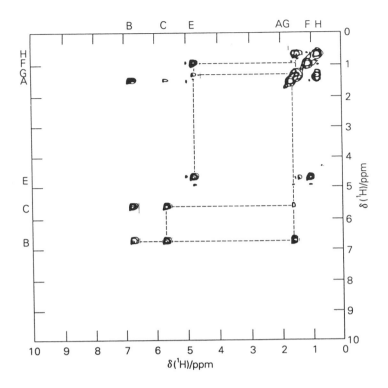

a. Draw the proton spectrum.

b. Using the letters used to label carbons, assign the coupled peaks by letter, *e.g.*, A is coupled to B and C.

The ^{1}H/^{13}C HETCOR spectrum of the sample above is shown next.

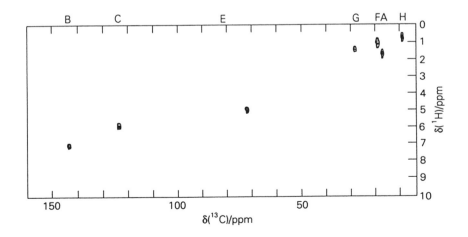

c. Draw the proton-decoupled ^{13}C spectrum.

The heteronuclear ^{13}C 2-D, *J*-spectrum is shown as follows.

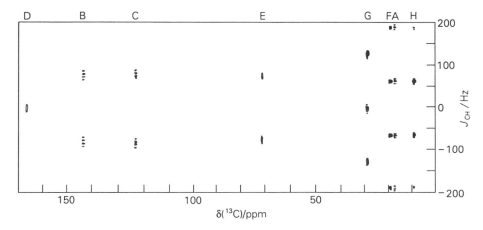

d. Describe how the *J*-spectra aid in peak assignment.

e. The molecule is a partially unsaturated ester with empirical formula $C_8H_{14}O_2$. Draw the structure.

20. The following spectra (acidic conditions) were taken for the molecule whose structure is given below. The *J*-spectrum is given in A and the HETCOR in B. Assign the numbered peaks to the labeled carbons (ignore all unlabeled carbons).

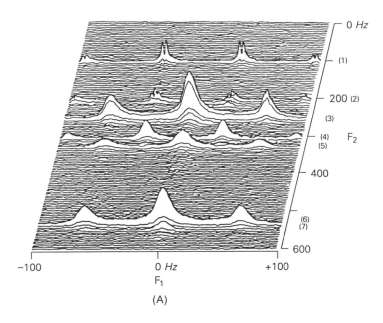

(A)

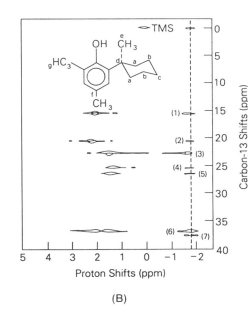

(B)

9 Electron Paramagnetic Resonance Spectroscopy*

Introduction

9-1 PRINCIPLES

Electron paramagnetic resonance is a branch of spectroscopy in which radiation of microwave frequency is absorbed by molecules, ions, or atoms possessing electrons with unpaired spins. This phenomenon has been designated by different names: "electron paramagnetic resonance" (epr), "electron spin resonance" (esr), and "electron magnetic resonance" (emr). These are equivalent and merely emphasize different aspects of the same phenomenon. There are some similarities between nmr and epr spectroscopy that are of help in understanding epr. In nmr spectroscopy, the two different energy states (when $I = \frac{1}{2}$) arise from the alignment of the nuclear magnetic moments relative to the applied field, and a transition between them occurs upon the application of radio-frequency radiation. In epr, different energy states arise from the interaction of the unpaired electron spin moment (given by $m_s = \pm \frac{1}{2}$) with the magnetic field, the so-called electronic Zeeman effect. The Zeeman Hamiltonian for the interaction of an electron with the magnetic field is given by equation (9-1):

$$\hat{H} = g\beta H \hat{S}_z \tag{9-1}$$

where g for a free electron has the value 2.0023193, β is the electron Bohr magneton, $e\hbar/2m_e c$, which has the value $9.274096 \pm (0.000050) \times 10^{-21}$ erg gauss^{-1} $\hat{S}_z$ is the spin operator; and H is the applied field strength. This Hamiltonian operating on the *electron* spin functions α and β corresponding to $m_s = +\frac{1}{2}$ and $-\frac{1}{2}$, respectively, produces the result illustrated in Fig. 9-1. The β spin state has its moment aligned with the field, in contrast to nmr, where the lowest energy state corresponds to $m_I = +1/2(\alpha_N)$. The lowest energy state in epr corresponds to $m_s = -\frac{1}{2}$ because the sign of the charge on the electron is opposite that on the proton. The transition energy is given by equation (9-2):

$$\Delta E = g\beta H \tag{9-2}$$

---

* The Additional References contain reviews of epr. The reference by J. E. Wertz and J. R. Bolton is especially recommended.

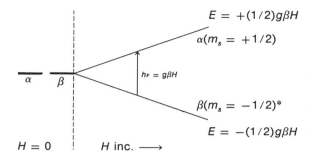

FIGURE 9–1 The removal of the degeneracy of the α and β electron spin states by a magnetic field. *Note the difference in the ground state from nmr.

The energy difference between the α and β spin states in magnetic fields of strengths commonly used in the epr experiment (several thousand gauss) corresponds to frequencies in the microwave region. For a field strength of 10,000 gauss, one can calculate from equation (9–2) that ΔE is 28,026 MHz. This is to be contrasted with the energy for the transition of the proton nuclear moment of 42.58 MHz (*i.e.*, radiation in the radio-frequency region) in a magnetic field of identical strength. The magnetic moment for the electron is -9.2849×10^{-21} erg gauss^{-1} compared to 1.4106×10^{-23} erg guass^{-1} for a proton nuclear moment.

The epr experiment is generally carried out at a fixed frequency. Two common frequencies are in the X-band frequency range (about 9500 MHz or 9.5 gigahertz, GHz,* where a field strength of about 3400 gauss is employed) and the so-called Q-band frequency (35 GHz, where a field strength of about 12,500 gauss is used). Since the sensitivity of the instrument increases roughly as v^2 and better spectral resolution also results, the higher frequency is to be preferred. There are several limitations on the use of the Q-band frequency. Smaller samples are required for Q-band, so the sensitivity is not as much greater as one would predict from v^2. It is more difficult to attain the higher field homogeneity ($\delta H/H$) that is required at higher frequencies. Finally, for aqueous samples, dielectric absorption by the solvent becomes more serious as the frequency increases and this results in decreased sensitivity.

Water, alcohols, and other high dielectric constant solvents are not the solvents of choice for epr because they strongly absorb microwave power. They can be used when the sample has a strong resonance and is contained in a specially designed cell (a very narrow sample tube). Epr measurements on gases, solutions, powders, single crystals, and frozen solutions can be carried out. The best frozen solution results are obtained when the solvent freezes to form a glass. Symmetrical molecules or those that hydrogen bond extensively often do not form good glasses. For example, cyclohexane does not form a good glass, but methylcyclohexane does. Some solvents and mixtures that form good glasses are listed in Table 9–1.

The sample tube employed is also important. If the signal-to-noise ratio is low, a quartz sample tube is preferred, because Pyrex absorbs more of the microwave power and also exhibits an epr signal.

---

*The prefix mega indicates 10^6, and giga indicates 10^9.

TABLE 9–1. Commonly Used Glasses*

Pure Substances

3-Methylpentane	Sulfuric acid	Sugar
Methylcyclopentane	Phosphoric acid	Triethanolamine
Nujol (paraffin oil)	Ethanol	2-Methyltetrahydrofuran
Isopentane	Isopropanol	Di-n-propyl ether
Methylcyclohexane	l-Propanol	cis-trans Decalin
Isooctane	l-Butanol	Triacetin
Boric acid	Glycerol	Toluene

Mixtures

Components	Ratio A/B
3-Methylpentane/isopentane	1/1
Isopentane/methylcyclohexane	1/6
Methylcylopentane/methylcyclohexane	1/1
Pentene-2(*cis*)/pentene-2(*trans*)	
Propane/propene	1/1
Isopropyl benzene/propane/propene	2/9/9
Ethanol/methanol	4/1, 5/2, 1/9
Isopropyl alcohol/isopentane	3/7
Ethanol/isopentane/diethyl ether	2/5/5
Alphanol 79'/mixture of primary alcohols	
Isopentane/n-butyl alcohol	7/3
Isopentane/isopropyl alcohol	8/2
Isopentane/n-propyl alcohol	8/2
Diethyl ether/isooctane/isopropyl alcohol	3/3/1
Diethyl ether/isooctane/ethyl alcohol	3/3/1
Diethyl ether/isopropyl alcohol	3/1
Diethyl ether/ethanol	3/1
Isooctane/methylcyclohexane/isopropyl alcohol	3/3/1
Diethyl ether/toluene/ethanol	2/1/1
Isopropyl alcohol/isopentane	2/5/5
Propanol/diethyl ether	2/5
Butanol/diethyl ether	2/5
Diethyl ether/isopentane/dimethyl formamide/ethanol	12/10/6/1
Water/propylene glycol	1/1
Ethylene glycol/water	2/1
Trimethylamine/isopentane/diethyl ether	2/5/5
Triethylamine/isopentane/diethyl ether	3/1/3
Methylhydrazine/methylamine/trimethylamine	2/4/4
Diethyl ether/isopentane/ethanol/pyridine	12/10/6/1
Di-n-butyl ether/diisopropyl ether/dimethyl ether	3/5/12
Diphenyl ether/1,1-diphenylethane/triphenylmethane	3/3/1
Diethyl ether/isopentane	1/1 to 1/2
Dipropyl ether/isopentane	3/1
Dipropyl ether/methylcylohexane	3/1
Diethyl ether/pentene-2(*cis*)-pentene-2(*trans*)	2/1
Ethyl iodide/isopentane/diethyl ether	1/2/2
Ethylbromide/methylcyclohexane/isopentane/methylcyclopentane	1/4/7/7
Ethanol/methanol/ethyl iodide	16/4/1
Ethanol/methanol/propyl iodide	16/4/1
Ethanol/methanol/propyl chloride	16/4/1
Ethanol/methanol/propyl bromide	16/4/1
Diethyl ether/isopentane/ethanol/1-chloronaphthalene	8/6/2/2
3-Methylpentane/isopentane	1/2
Propyl alcohol/propane/propene	2/9/9
Diisopropylamine/propane/propene	2/9/9
Dipropyl ether/propane/propene	2/4/4

Mixtures	
Components	Ratio A/B
Toluene/methylene chloride	1/1 or excess toluene
Toluene/acetone	1/1 or excess toluene
Toluene/methanol or ethanol	1/1 or excess toluene
Toluene/acetronitrile	1/1 or excess toluene
Toluene/chloroform	1/1 or excess toluene
2-Methyltetrahydrofuran/methanol	2/1
2-Methyltetrahydrofuran/proprionitrile	2/1
2-Methyltetrahydrofuran/methylene chloride	1/1

*Abstracted in part, with permission, from B. Meyer, "Low Temperature Spectroscopy," American Elsevier Publishing Co., New York, 1971

There are many effects that modify the electron energy states in a magnetic field. We shall consider these factors one at a time by discussing the epr spectra of increasingly complex systems.

In the way of introduction, it will simply be mentioned that differences in the energy of the epr transition for different molecules are described by changing the value of g in equation (9–2). This is to be contrasted with nmr, where one customarily holds g_N fixed and introduces the shielding constant to describe the different resonance energies, *i.e.*, for nmr,

$$\Delta E = -g_N \beta_N (1 - \sigma) H \Delta m_I \tag{9–3}$$

As we proceed to more complex systems, we shall discuss the factors that influence the magnitude of g.

9–2 THE HYDROGEN ATOM

Nuclear Hyperfine Splitting

The first contribution to epr transition energies that will be introduced is the electron-nuclear hyperfine interaction. The hydrogen atom (in free space) is a simple system to discuss because, by virtue of its spherical symmetry, anisotropic effects are absent. In the development of epr, we shall employ the Hamiltonian to quantitatively describe the effects being considered. The full interpretation of the esr spectrum of a system is given in terms of an *effective spin Hamiltonian*. This is a Hamiltonian that contains those effects, of the many to be described, which are used to interpret the particular spectrum of the compound studied.

The complete spin Hamiltonian for the hydrogen atom (in free space) is

$$\hat{H} = g\beta \vec{H} \cdot \hat{S} - g_N \beta_N \vec{H} \cdot \hat{I} + a\hat{I} \cdot \hat{S}$$
$$= g\beta(H_x \hat{S}_x + H_y \hat{S}_y + H_z \hat{S}_z) - g_N \beta_N (H_x \hat{I}_x + H_y \hat{I}_y + H_z \hat{I}_z) + a\hat{I}_x \hat{S}_x$$
$$+ a\hat{I}_y \hat{S}_y + a\hat{I}_z \hat{S}_z$$

For a spherical system in a magnetic field that is defined as the z-axis, this simplifies* to

$$\hat{H} = g\beta H \hat{S}_z - g_N \beta_N H \hat{I}_z + a\hat{I} \cdot \hat{S} \qquad (9-4)$$

The first term of this Hamiltonian discussed earlier [equation (9–1)], leads to the energy-field relation shown in Fig. 9–1. The second term of the Hamiltonian is familiar from our discussion of nmr; it describes the interaction of the nuclear moment of the hydrogen atom with a magnetic field. It is of opposite sign (the state with $m_I = +1/2$ is lowest) and smaller in magnitude than the first term. The combined effect of the first two terms in equation (9–4) upon the energies of the spin states of the hydrogen atom in a magnetic field is shown in Fig. 9–2(C). The field strength is fixed in this figure, and the dashed lines simply show the energy changes incurred by adding a new term in the Hamiltonian. In order to determine the energy of the hydrogen atom in a magnetic field, we employ a basis set for this Hamiltonian [equation (9–4)] that consists of the four possible electron and nuclear spin functions. Such a basis set is $\varphi_1 = |\alpha_e \alpha_N\rangle$, $\varphi_2 = |\alpha_e \beta_N\rangle$, $\varphi_3 = |\beta_e \alpha_N\rangle$ and $\varphi_4 = |\beta_e \beta_N\rangle$. Let us begin by calculating the energies arising from the first two terms in the Hamiltonian, $\hat{H}_0$. We must solve the simultaneous equations $\langle\varphi_n|\hat{H}_0|\varphi_m\rangle - E\langle\varphi_n|\varphi_m\rangle = 0$, where n and m may or may not be equal. Thus the 4×4 secular determinant in this basis set contains diagonal terms of the type:

$$\langle\alpha_e\beta_N|g\beta H\hat{S}_z - g_N\beta_N H\hat{I}_z|\alpha_e\beta_N\rangle - E\langle\alpha_e\beta_N|\alpha_e\beta_N\rangle$$
$$= \langle\alpha_e\beta_N|g\beta H\hat{S}_z|\alpha_e\rangle \cdot |\beta_N\rangle - \langle\alpha_e\beta_N|g_N\beta_N H\hat{I}_z|\beta_N\rangle \cdot |\alpha_e\rangle - E = 0$$

Since for this problem the operators are $\hat{I}_z$ and $\hat{S}_z$, α and β are eigenfunctions; i.e., $\hat{S}_z|\alpha_e\rangle = 1/2\alpha_e$, $\hat{S}_z|\beta_e\rangle = -1/2\beta_e$, $\hat{I}_z|\alpha_N\rangle = 1/2\alpha_N$, and $\hat{I}_z|\beta_N\rangle = -1/2\beta_N$. Furthermore, $\hat{S}_z$ does not operate on the nuclear spin function and $\hat{I}_z$ does not operate on the electron spin function, leading to:

$$\langle\alpha_e\beta_N|\hat{H}_0|\alpha_e\beta_N\rangle - E\langle\alpha_e\beta_N|\alpha_e\beta_N\rangle = \frac{1}{2}g\beta H + \frac{1}{2}g_N\beta_N H - E = 0$$

It should now be clear that all the off-diagonal elements will be zero with this Hamiltonian, for all of the off-diagonal elements are of the form $\langle\varphi_n|\hat{H}_0|\varphi_m\rangle - \langle\varphi_n|\varphi_m\rangle$, which equals zero when $n \neq m$. Since the Hamiltonian matrix is diagonal, the determinant is already factored and we get the four energies directly, as shown above for $\alpha_e\beta_N$. They are indicated for E_1, E_2, E_3, and E_4, in Fig 9–2(C). These results can be verified for practice. The usual selection rules in epr are $\Delta m_I = 0$ and $\Delta m_s = \pm 1$. It will be noticed that the two esr transitions ($\Delta m_I = 0$) illustrated in Fig 9–2(C) have the same energy. Considering only the first two terms of the Hamiltonian, the epr spectrum of the hydrogen atom would be the same as that of a free electron, i.e., one line at a field $hv/g\beta$ or $g = 2.0023$.

Next, we must concern ourselves with the $a\hat{I} \cdot \hat{S}$ term in the Hamiltonian. This term describes the coupling of the electron and nuclear spin moments, which

---

* H_x and H_y are zero and $H = H_z$. The effects of S_x, S_y, I_x, and I_y are not necessarily zero.

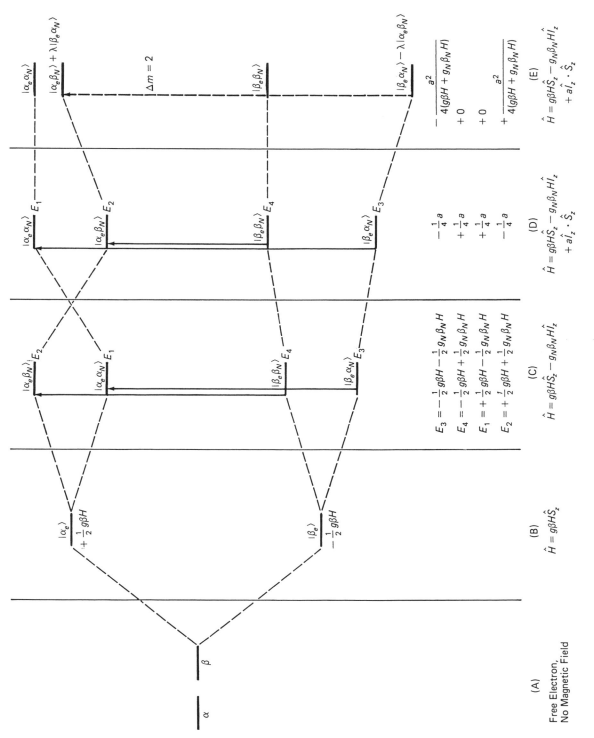

FIGURE 9-2 The influence of various terms in the Hamiltonian on the energy of a hydrogen atom in a magnetic field. $\beta_e \alpha_N$ corresponds to $m_S = -1/2$ and $m_I = +1/2$; $\beta_e \beta_N$ to $m_S = -1/2$ and $m_I = -1/2$; $\alpha_e \beta_N$ to $m_S = +1/2$ and $m_I = -1/2$; and $\alpha_e \alpha_N$ to $m_S = +1/2$ and $m_I = +1/2$.

classically corresponds to the dot product of these two vectors. The quantity a indicates the magnitude of the interaction and has the dimensions of energy. This is referred to as the Fermi contact contribution to the coupling, and its magnitude depends upon the amount of electron density at the nucleus, $\psi_{(0)}^2$, according to:

$$a = \frac{8\pi}{3} g\beta g_N \beta_N |\psi_{(0)}|^2 \tag{9-5}$$

For a hydrogen atom, the Slater orbital function is $\psi_{1s} = (1/\pi a_0^3) \exp(-r/a_0)$ where a_0, the Bohr radius, equals $h^2/me^2 = 0.52918$ Å. Substituting into equation (9–5), the value of ψ for a hydrogen s-orbital at $r = 0$ yields $a/h = 1422.74$ MHz. Since the nuclear hyperfine interaction we are talking about involves the dot product of the nuclear and spin moments, it has x-, y-, and z-components, so

$$a\hat{I} \cdot \hat{S} = a(\hat{I}_x\hat{S}_x + \hat{I}_y\hat{S}_y + \hat{I}_z\hat{S}_z) \tag{9-6}$$

The elements generated by the $\hat{I}_z\hat{S}_z$ term operating on the basis $\alpha_e\alpha_N$, etc., are again only diagonal elements because $\langle \varphi_n | a\hat{I}_z\hat{S}_z | \varphi_m \rangle = 0$ when $m \neq n$; for example, $\langle \alpha_e\beta_N | a\hat{I}_z | \alpha_N \rangle = 0$. The following results are obtained for the diagonal elements:

$$\langle \alpha_e\alpha_N | a\hat{S}_z\hat{I}_z | \alpha_e\alpha_N \rangle = \frac{1}{4}a$$

$$\langle \alpha_e\beta_N | a\hat{S}_z\hat{I}_z | \alpha_e\beta_N \rangle = -\frac{1}{4}a$$

$$\langle \beta_e\alpha_N | a\hat{S}_z\hat{I}_z | \beta_e\alpha_N \rangle = -\frac{1}{4}a$$

$$\langle \beta_e\beta_N | a\hat{S}_z\hat{I}_z | \beta_e\beta_N \rangle = \frac{1}{4}a$$

These have to be added to the energies in Fig. 9–2(C), modifying the energies as shown in Fig. 9–2(D). The contributions of $\pm(^1/_4)a$ to E_1, E_2, etc., from $a\hat{I}_z\hat{S}_z$ are indicated at the bottom of (D). Now we see, looking at the arrows for the two electron spin changes, that the *transition energies are no longer equal*. One transition gives rise to a spectral peak at lower energy than that corresponding to $g = 2.0023$ [see Fig. 9–2(D)] by $(^1/_2)a$, and the other occurs at an energy that is higher by $(^1/_2)a$. The energy separation of the two peaks is a.

To complete the problem, we now have to add the effects of $\hat{I}_x\hat{S}_x$ and $\hat{I}_y\hat{S}_y$. This is best done in terms of the raising and lowering operators, which work in a similar fashion to $\hat{I}_+$ and $\hat{I}_-$ that were discussed earlier. For the electron spin operators, we define:

$$\hat{S}_+ = \hat{S}_x + i\hat{S}_y$$
$$\hat{S}_- = \hat{S}_x - i\hat{S}_y$$

Thus

$$\hat{S}_+\hat{I}_- = (\hat{S}_x\hat{I}_x + \hat{S}_y\hat{I}_y) + i(\hat{S}_y\hat{I}_x - \hat{S}_x\hat{I}_y)$$

and

$$\hat{S}_-\hat{I}_+ = \hat{S}_x\hat{I}_x + \hat{S}_y\hat{I}_y - i(\hat{S}_y\hat{I}_x - \hat{S}_x\hat{I}_y).$$

Combining these equations, we see that:

$$\hat{S}_x\hat{I}_x + \hat{S}_y\hat{I}_y = \frac{1}{2}(\hat{S}_+\hat{I}_- + \hat{S}_-\hat{I}_+)$$

The following results are obtained by analogy to our earlier discussion of $\hat{I}_+$ and $\hat{I}_-$:

$$\hat{S}_+\hat{I}_-|\beta_e\alpha_N\rangle = |\alpha_e\beta_N\rangle$$
$$\hat{S}_-\hat{I}_+|\alpha_e\beta_N\rangle = |\beta_e\alpha_N\rangle$$
$$\hat{S}_-\hat{I}_+|\alpha_e\alpha_N\rangle = 0$$

All other operations of $\hat{S}_-\hat{I}_+$ or $\hat{S}_+\hat{I}_-$ upon the basis set produce zero. Thus, if we consider the 4 × 4 matrix shown in Fig. 9–3, the only non-vanishing matrix elements from $\hat{S}_+\hat{I}_-$ and $\hat{S}_-\hat{I}_+$ are

$$\langle\alpha_e\beta_N|a\hat{S}_+\hat{I}_-|\beta_e\alpha_N\rangle = a$$
$$\langle\beta_e\alpha_N|a\hat{S}_-\hat{I}_+|\alpha_e\beta_N\rangle = a$$

	$\|\alpha_e\alpha_N\rangle$	$\|\alpha_e\beta_N\rangle$	$\|\beta_e\alpha_N\rangle$	$\|\beta_e\beta_N\rangle$
$\|\alpha_e\alpha_N\rangle$	$\frac{1}{2}(g\beta H - g_N\beta_N H) + \frac{1}{4}a - E$	0	0	0
$\|\alpha_e\beta_N\rangle$	0	$\frac{1}{2}(g\beta H + g_N\beta_N H) - \frac{1}{4}a - E$	$\frac{1}{2}a$	0
$\|\beta_e\alpha_N\rangle$	0	$\frac{1}{2}a$	$-\frac{1}{2}(g\beta H + g_N\beta_N H) - \frac{1}{4}a - E$	0
$\|\beta_e\beta_N\rangle$	0	0	0	$-\frac{1}{2}(g\beta H - g_N\beta_N H) + \frac{1}{4}a - E$

FIGURE 9–3 Secular determinant for the field-free hydrogen atom.

The matrix element

$$\langle\alpha_e\beta_N|a\hat{S}_x\hat{I}_x + a\hat{S}_y\hat{I}_y|\beta_e\alpha_N\rangle = \frac{1}{2}a$$

and

$$\langle\beta_e\alpha_N|a\hat{S}_x\hat{I}_x + a\hat{S}_y\hat{I}_y|\alpha_e\beta_N\rangle = \frac{1}{2}a$$

We can summarize this entire section by completing the full determinant for the original spin Hamiltonian equation (9–4), operating on the φ basis set to give energies $\langle \varphi_m | \hat{H} | \varphi_n \rangle = E \langle \varphi_m | \varphi_n \rangle$. The determinant shown in Fig. 9–3 equals zero. Note that it is block diagonal so that two of the energies, E_1 and E_4, are obtained directly. We also see that $\hat{I}_x \hat{S}_x$ and $\hat{I}_y \hat{S}_y$ lead to off-diagonal elements that mix φ_2 and φ_3. A perturbation theory solution* of the resulting 2×2 determinant gives (to second order):

$$E_2 = \frac{1}{2} g\beta H + \frac{1}{2} g_N \beta_N H - \frac{1}{4} a + \frac{a^2}{4(g\beta H + g_N \beta_N H)}$$

$$E_3 = -\left(\frac{1}{2} g\beta H + \frac{1}{2} g_N \beta_N H\right) - \frac{1}{4} a - \frac{a^2}{4(g\beta H + g_N \beta_N H)}$$

We can see in Fig. 9–2(E), where the effects of these off-diagonal elements on the energy levels are illustrated, that the energies of both transitions are increased by the same amount. Since the off-diagonal elements are small compared to the diagonal, effects arising from this part of the Hamiltonian are referred to as second-order effects. Second-order effects thus have no influence on the value of a read off the spectrum, but will change the value read off the spectrum for g. A more interesting contribution to the spectral appearance is that the previously forbidden transition $E_3 \rightarrow E_2$ (the simultaneous electron and nuclear spin flip), now becomes allowed because of the mixing of the basis set.†

9–3 PRESENTATION OF THE SPECTRUM

As in nmr, the epr spectrum can be represented by plotting intensity, I, against the strength of the applied field; but epr spectra are commonly presented as derivative curves, *i.e.*, the first derivative (the slope) of the absorption curve is plotted against the strength of the magnetic field. It is easier to discern features in a derivative presentation if the absorption lines are broad. The two modes of presentation are easily interconverted, and the relationship between the two kinds of spectra is illustrated in Fig. 9–4. In (A), a single absorption peak with no fine structure is represented; (B) is the derivative curve that corresponds to (A). The derivative curve crosses the abscissa at a maximum in the absorption curve, for the slope changes sign at a maximum. Curve (C) is the absorption counterpart of curve (D). Note that the shoulders in (C) never pass through a maximum and as a result the derivative peaks in (D) corresponding to these shoulders do not cross the abscissa. The number of peaks and shoulders in the absorption curve can be determined from the number of minima (marked with an asterisk in (D)) or maxima in the derivative curve.

---

* The exact solution is $E = -(1/4) a \pm (1/2) [(g\beta + g_N \beta_N)^2 H^2 + a^2]^{1/2}$.

† This allowedness is derived in Carrington and McLachlan (see Additional References) on page 401.

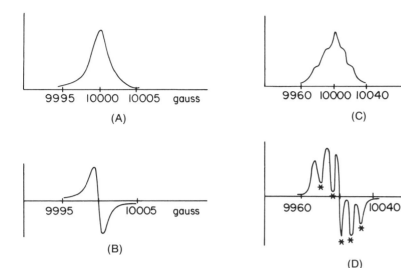

FIGURE 9–4 Comparison of spectral presentation as absorption (A and C) and derivative (B and D) curves.

The epr spectrum of the hydrogen atom is illustrated in Fig. 9–5. To a good approximation, the g-value is measured at x, which is midway between the two solid circles corresponding to absorption peak maxima. The hyperfine splitting, $a/g\beta$, is the separation between the solid circles in gauss. The sign of a generally cannot be obtained directly from the spectrum. The splitting in Fig. 9–2 implies a positive value for a. If a were negative then $m_s = -\frac{1}{2}$ and $m_I = -\frac{1}{2}$; that is $(\beta_e \beta_N)$ would be the low energy state.

The epr spectrometer is designed to operate at a fixed microwave source frequency. The magnetic field is swept, and the horizontal axis in Fig. 9–5 is in units of gauss. One can set the field at any position using the field dial and sweep from that place. For a fingerprint type of identification, greater accuracy is needed than can be obtained using the instrument dials. For this purpose, an external standard, diphenylpicrylhydrazide DPPH, is used together with a microwave frequency counter. *DPPH has a g-value of* 2.0037 ± 0.0002. The field sweep is assumed to be linear, and the g values of other peaks are calculated relative to

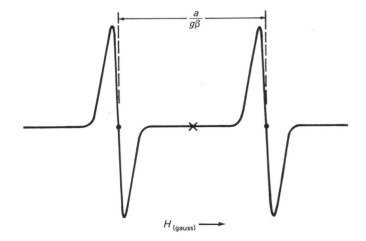

FIGURE 9–5 The epr spectrum of a hydrogen atom (β has units of ergs/gauss).

this standard. The field axis is in units of gauss, and the g value is reported as a dimensionless quantity using

$$g = \frac{h\nu}{\beta H}$$

where ν is the fixed frequency of the probe and H (which is being swept) is obtained from the spectrum. A frequency counter should be used to measure the probe frequency ν.

The value of a is sometimes reported in units of gauss, MHz, or cm^{-1}. It is to be emphasized that the line separation in a spectrum in units of gauss is given by $a/g\beta$ (where a has units of ergs and β has units of ergs/gauss^{-1}). When $g \neq 2$, it is incorrect to report this separation as a in units of gauss. One would have to multiply the line separation by $g\beta$ and divide by $g_e\beta$ (where g_e is the free electron value of 2.0023193) to report a correct value for a in gauss. Since a is an energy, it is best to report its value as an energy. This is simply done by multiplying the line separation in gauss by $g\beta$, with β in units of cm^{-1}/gauss^{-1}. There is no g-value dependence for this unit. The value of a in MHz is obtained by multiplying a (cm^{-1}) by c (3×10^{10} cm sec^{-1}) and dividing by 10^6.

9–4 HYPERFINE SPLITTINGS IN ISOTROPIC SYSTEMS INVOLVING MORE THAN ONE NUCLEUS

The first-order energies of the levels in the hydrogen atom are given by equation (9–7), which ignores the small nuclear Zeeman interaction.

$$E = g\beta H m_s + a m_s m_I \qquad (9\text{–}7)$$

Substituting the values of m_s and m_I into this equation enables one to reproduce the energies given in Fig. 9–2(D). For a nucleus with any nuclear spin, the projection of the nuclear magnetic moment along the effective field direction at the nucleus can take any of the $2I + 1$ values corresponding to the quantum numbers $-I, (-I + 1), \ldots, (I - 1), I$. These orientations give rise to $2I + 1$ different nuclear energy states (one for every value of m_I); and when each of these couples with the electron moment, $2I + 1$ lines result in the epr experiment. Since these energy differences are small, all levels with the same m_s value are equally populated for practical purposes and the epr absorption lines will usually be of equal intensity and equal spacing. For example, three lines are expected for an unpaired electron on ^{14}N, whose $I = 1$.

We shall next consider the effect on the spectrum when the electron interacts with (i.e. is delocalized onto) several nuclei. For simplicity, assume that the species is rotating very rapidly in all directions, and that the g value is close to the free electron value. As an example, the methyl radical will be discussed.[1,2] As illustrated in Fig. 9–6, addition of the nuclear spin angular momentum quantum numbers of the individual protons results in four different values for the total nuclear spin moment, M_I. As indicated in Fig. 9–7, this gives rise to four transitions ($\Delta M_I = 0$, $\Delta m_s = \pm 1$). Since there are three different possible ways to obtain a total of $M_I = +^1/_2$ or $-^1/_2$ (see Fig. 9–6), but only one possible way to obtain

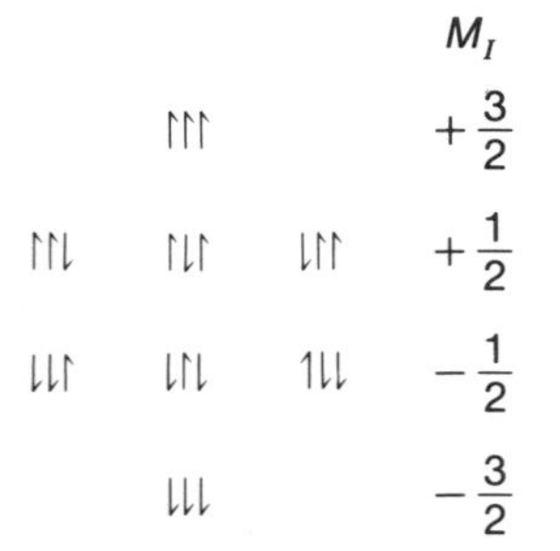

FIGURE 9–6 Possible nuclear spin arrangements of the protons in a methyl radical.

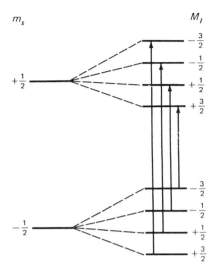

FIGURE 9–7 The four transitions that occur in the epr spectrum of the methyl radical (see Fig. 9–8 for the spectrum). (As with the H atom, the $+m_I$ state is lowest for $m_s = -1/2$ and the $-m_I$ state is lowest for $m_s = +1/2$, from the $\hat{I} \cdot \hat{S}$ term.)

$M_I = +3/2$ or $-3/2$, the former system is three times more probable than the latter and the observed relative intensities for the corresponding transitions (Fig. 9–7) are in the ratio $1:3:3:1$.

In general, when the absorption spectrum is split by n-*equivalent* nuclei of equal spin I_i, the number of lines is given by $2nI_i + 1$. When the splitting is caused by both a set of n-equivalent nuclei of spin I_i and a set of m-equivalent nuclei of spin I_j, the number of lines is given by $(2nI_i + 1)(2mI_j + 1)$. The following specific cases illustrate the use of these general rules.

1. If a radical contains n non-equivalent protons onto which the electron is delocalized, a spectrum consisting of 2^n lines will arise.

2. If the odd electron is delocalized over a number, n, of equivalent protons, a total of $n + 1$ lines, $(2nI + 1)$, will appear in the spectrum. This number is less than the number of lines expected for non-equivalent protons (i.e., 2^n) because several of the possible arrangements of the nuclear spins are degenerate (see Fig. 9–6). The spectrum of the methyl radical illustrated in Fig. 9–8 contains the four peaks expected from these rules.

The spectra expected for differing numbers of equivalent protons can easily be predicted by considering the splitting due to each proton in turn, as illustrated in Fig. 9–9. When the signal is split by two equivalent protons, the total M_I for the three levels can have the values $\Sigma m_I = +1, 0,$ and -1. Since there are two ways in which we can arrange the separate m_I's to give a total $M_I = 0$ (namely $+1/2, -1/2$, and $-1/2, +1/2$), the center level is doubly degenerate. Three peaks are observed in the spectrum ($\Delta M_I = 0, \Delta m_s = \pm 1$), and the intensity ratio is $1:2:1$. The case of three protons (e.g., the methyl radical) was discussed above, and similar considerations are employed for the systems represented in Fig. 9–9 with more than three protons.

The relative intensities of the peaks are given by the coefficients of the binomial expansion. It should be remembered in applying this formula that it is restricted to equivalent protons or other nuclei having $I = 1/2$.

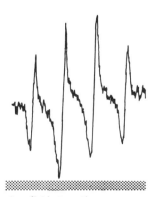

Inc. field strength ⟶

FIGURE 9–8 The derivative spectrum of the methyl radical in a CH_4 matrix at 4.2 K.

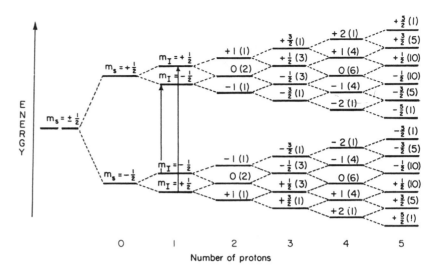

FIGURE 9–9 Hyperfine energy levels resulting from interaction of an unpaired electron with varying numbers of equivalent protons.[2] Each number in parentheses gives the degeneracy of the level to which it refers, and hence the relative peak intensities for the corresponding transitions.

3. If the odd electron is delocalized over two sets of non-equivalent protons, the number of lines expected is the product of the number expected for each set $[(2nI_i + 1)(2mI_j + 1)]$. The naphthalene negative ion, which can be prepared by adding sodium to naphthalene, contains an odd electron that is delocalized over the entire naphthalene ring. Naphthalene contains two different sets of four equivalent protons. A total of $n + 1$, or five peaks, is expected for an electron delocalized on either set of four equivalent protons. In the naphthalene negative ion, the two sets of four equivalent protons should give a total of 25 lines in the epr spectrum. This is found experimentally.

4. If the electron is delocalized on nuclei with spin greater than $^1/_2$, a procedure similar to that for protons can be applied to calculate the number of peaks expected. If the electron is delocalized over several equivalent nuclei that have spins greater than $^1/_2$, the number of peaks expected in the spectrum is predicted from the formula $2nI + 1$. For example, five peaks are expected for an electron delocalized on two equivalent nitrogen atoms. A procedure similar to that in Fig. 9–6 shows that the intensities of the five peaks will be in the ratio $1:2:3:2:1$.

5. If the electron is delocalized over several non-equivalent atoms, the total number of peaks expected is obtained by taking the product of the number expected for each atom. The scheme illustrated in Fig. 9–10 for an electron delocalized onto two non-equivalent nuclei with $I = 1$ is often employed to indicate the splitting expected. The three lines in (A) represent splitting of an epr peak by a nucleus with $I = 1$ and a hyperfine coupling constant a. Each of these lines is split into three components as a result of delocalization of the electron on a second, non-equivalent nucleus with $I = 1$, and a hyperfine coupling constant a', producing in (B) a total of nine lines. In subsequent discussions a scheme similar to that in Fig. 9–10 will be employed for the interpretation of spectra. The shape of the spectrum and the separations of the peaks will depend upon the resonant field, g, and the coupling constants a and a'. Frequently the measured spectrum will not reveal all the lines expected, because the line widths are large compared to $a/g\beta$ and two close lines are not resolved. For example, the spectrum

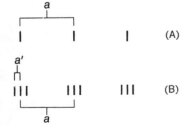

FIGURE 9–10 (A) Three lines expected from an electron on a nucleus with $I = 1$, and (B) nine lines resulting from the splitting by a second non-equivalent nucleus with $I = 1$.

in Fig. 9–11 could result for the hypothetical radical H—Ẋ⁺ ↔ Ḣ—X⁺ where $I = 1$ for X. The two lines in (A) result from the proton splitting. In (B) each line in turn is split into three components owing to interaction with nucleus X; thus we would expect six lines, all of equal intensity. However, it is possible to detect only five lines, if the two innermost components are not resolved. They would give rise to a single peak with twice the area of the other peaks (see Fig. 9–11).

The epr spectrum[3] of bis-salicylaldimine copper(II) in Fig. 9–12 is an interesting example to summarize this discussion of nuclear coupling. This spectrum was obtained on a solid and is not isotropic; this aspect will be discussed shortly. Four main groups of lines result from coupling of the ^{63}Cu nucleus ($I = \frac{3}{2}$) with the electron. The hyperfine structure in each of the four groups consists of eleven peaks of intensity ratio $1:2:3:4:5:6:5:4:3:2:1$. These peaks result from splitting by the two equivalent nitrogens and two hydrogens, H' in Fig. 9–12. The total number of peaks expected is fifteen; $(2n_N I_N + 1)(2n_H I_H + 1) = 5 \times 3 = 15$. The eleven peaks found for each subgroup in the actual spectrum result from overlap of some of the fifteen peaks as indicated in Fig. 9–13. The line for an electron not split by a nucleus is shown in (A). The splittings by the two equivalent nitrogens are indicated in (B) and the subsequent splitting by two equivalent protons is indicated in (C) The two nitrogens split the resonance into five peaks of relative intensity $1:2:3:2:1$. These values are denoted in (B) by $4d$, $4e$, and $4f$ where the intensities correspond to $d = 1$, $e = 2$, and $f = 3$. The splitting by two equivalent protons will give rise to three lines for each line in (B), with an intensity ratio of $1:2:1$. The intensities indicated by letters underneath the lines in (C) result from the summation of the expected intensities. Since the relative intensities are $d = 1$, $e = 2$, and $f = 3$, the ratio of the intensity of the bands in (C) is $1:2:3:4:5:6:5:4:3:2:1$. The experimental spectra agree with this interpretation, which is further substantiated by the following results:

1. Deuteration of the N—H'' groups (see Fig. 9–12) produced a compound which gave an identical spectrum.

2. When the H' hydrogens were replaced by methyl groups, the epr spectrum for this compound consisted of four main groups, each of which consisted of five

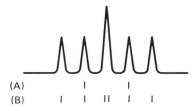

FIGURE 9–11 Hypothetical absorption spectrum for the radical H·X⁺ ($I = 1$ for X).

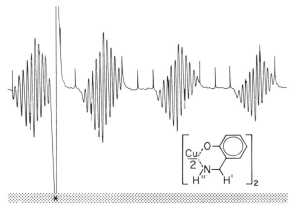

FIGURE 9–12
EPR-derivative spectrum of bis-salicylaldimine copper(II) with isotopically pure ^{63}Cu. Askerisk indicates calibration peak from DPPH. From A. H. Maki and B. R. McGarvey, J. Chem. Phys., *29*, 35 (1958).

Decreasing H ⟶

FIGURE 9–13 Interpretation of the epr spectrum of bissalicylaldimine copper(II). (A) An unsplit transition. (B) Splitting by two equivalent nitrogens, with d, e, and f indicating relative intensities of 1, 2, and 3. (C) Further splitting by two equivalent protons.

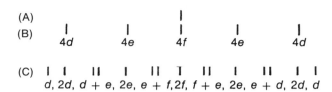

FIGURE 9–14 (A) Basic molecular geometry of the $Co_3(CO)_9Se$ complex. (B) The epr spectrum of a single crystal of $FeCo_2(CO)_9Se$ doped with about 0.5% of paramagnetic $Co_3(CO)_9Se$. This spectrum containing 22 hyperfine components was recorded at 77 K with the molecular threefold axis parallel to the magnetic field direction. [Reprinted with permission from C. E. Strouse and L. F. Dahl, J. Amer. Chem. Soc., *93*, 6032 (1971). Copyright by the American Chemical Society.]

lines resulting from nitrogen splitting only. The hyperfine splitting by the N—H″ proton and that by the protons on the methyl group are too small to be detected.

This spectrum furnishes conclusive proof of the delocalization of the odd electron in this complex onto the ligand. This can be interpreted only as covalence in the metal-ligand interaction, for only by mixing the metal ion and ligand wave functions can we get ligand contributions to the molecular orbital in the complex that contains the unpaired electron.

Another interesting application of epr[4] involves the spectrum of $Co_3(CO)_9Se$, whose structure and epr spectrum are shown in Fig. 9–14. The 22-line spectrum indicates that the one unpaired electron in this system is completely delocalized over the three cobalt atoms ($I_{Co} = {}^7/_2$). This in effect gives rise to an oxidation state of $+{}^2/_3$ for each cobalt atom.

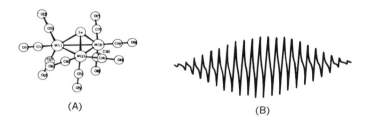

9–5 CONTRIBUTIONS TO THE HYPERFINE COUPLING CONSTANT IN ISOTROPIC SYSTEMS

In Equation (9–5), we saw that:

$$a = \frac{8\pi}{3} g\beta g_N \beta_N |\psi_0|^2$$

In a molecule, the hyperfine splitting from delocalization of unpaired spin density $\rho(r_N)$ onto a hydrogen atom of the molecule is given by equation (9–8):

$$a = \frac{8\pi}{3} g\beta g_N \beta_N \rho_H(r_N) \tag{9-8}$$

where $\rho(r_N)$ can be crudely thought of as the difference in the average numbers of electrons at the nucleus that have spin moments characterized by $m_s = +{}^1/_2$ and $-{}^1/_2$. When there is an excess of $+{}^1/_2$ spin density (*i.e.*, the electron spin moment opposed to the field) at the nucleus, this nucleus is said to experience *negative spin density*. An excess of electron density with $m_s = -\frac{1}{2}$ is called *positive spin density*. Positive spin density is represented by an arrow that is aligned with the external field, and negative spin density is represented by one opposed to the field. To further complicate matters, positive spin density is often referred to as α spin in the literature, even though the wave function for evaluation of the matrix elements for this electron is represented by β_e. Thus, the common convention for labeling spin density is exactly the opposite of that used to label the electron spin wave functions. We will avoid the α and β labels of spin density and use these as symbols for the spin wave functions in this book.

We should also be careful to point out that the amount of unpaired spin density on an atom in the molecule does not correspond directly to the atom contributions in the molecular orbital containing the unpaired electron. We shall refer to the latter effect as *unpaired electron density*. An unpaired electron in an orbital of one atom in a molecule can polarize the paired spins in an orthogonal sigma bond so that one of the electrons is more often in the vicinity of one atom than in the vicinity of the other. This puts unpaired spin density at the nucleus of the atom even though there is no unpaired electron density delocalized onto it. We can make this more specific with the following example.

Some of the first attempts at interpreting hyperfine couplings involved aromatic radicals with the unpaired spin in the π-system, *e.g.*, $C_6H_5NO_2^-$. Hückel calculations were carried out and the squares of the various carbon p_z coefficients in the m.o. containing the unpaired electron were employed to give the amount of unpaired electron density on the various carbon atoms. The hyperfine splittings observed experimentally were from the ring hydrogens, which are orthogonal to the π-system. No unpaired electron density can be delocalized *directly* onto them, but unpaired spin density is felt at the hydrogen nucleus by the so-called *spin-polarization* or *indirect mechanism*. We shall attempt to give a simplified view of this effect using a valence bond formalism. Consider the two resonance forms in Fig. 9–15 for a C—H bond in a system with an unpaired electron in a carbon p_π orbital. In the absence of any interaction between the π and σ systems, the so-called perfect pairing approximation, we can write a valence bond description of the bonding and antibonding sigma orbital wave functions:

$$\psi^\circ = \frac{1}{\sqrt{2}}(\psi_I + \psi_{II}) \quad \text{and} \quad \psi^* = \frac{1}{\sqrt{2}}(\psi_I - \psi_{II})$$

Here ψ_I and ψ_{II} represent wave functions for structures I and II in Fig. 9–15, which we shall not attempt to specify in terms of an a.o. and spin basis set. When interaction of the π and σ systems is considered, we find that I is a more stable structure than II and, accordingly, it contributes to the ground state more than does II, resulting in valence bond functions:

$$\psi^\circ = a\psi_I + b\psi_{II} \quad (a > b)$$

and

$$\psi^* = a'\psi_I - b'\psi_{II} \quad (a' < b')$$

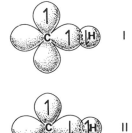

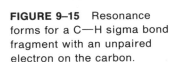

FIGURE 9–15 Resonance forms for a C—H sigma bond fragment with an unpaired electron on the carbon.

This in effect polarizes the electrons in the C—H sigma bond, leaving spin density on the hydrogen that is opposite to the unpaired spin density in the carbon p_z orbital. The stabilizing feature of structure I over structure II is the *electron exchange interaction*. It is the same effect that causes the lowest-energy excited state for helium, $1s^1 2s^1$, to be a triplet instead of a singlet. If we label the two electrons that are mainly on carbon in each of the structures in Fig. 9–15, we see that they can be interchanged in structure I, without changing the m_s value of the unpaired electron:

$$\begin{array}{cc} a\uparrow & b\uparrow \\ C\uparrow\downarrow H & C\uparrow\downarrow H \\ b & a \end{array}$$

This interchange is not possible in structure II. Thus, I is stabilized by a quantum mechanical interaction analogous to resonance.

A more complete molecular orbital description of this effect is presented in Dewar's text.[5] In treating the e^2/r_{ij} interactions in a molecule, one type of integral that is obtained describes a repulsive interaction and has the form

$$J_{mn} = \iint \psi_m^*(i)\psi_m(i) \frac{e^2}{r_{ij}} \psi_n^*(j)\psi_n(j)\, d\tau_i\, d\tau_j \tag{9-9}$$

This represents the Coulomb repulsion of the electron density [$\int \psi_m^*(i)\psi_m(i) d\tau_i$ is the density in i] from electrons i and j in orbitals m and n, where m may or may not equal n. There are other integrals that we shall label K_{mn}, which are zero when the electron spins are paired and are non-zero when the spins are parallel. The K_{mn} integrals are:

$$K_{mn} = \iint \psi_m^*(i)\psi_m(i) \frac{e^2}{r_{ij}} \psi_n^*(j)\psi_n(j)\, d\tau_i\, d\tau_j \tag{9-10}$$

This is called an exchange integral because it corresponds to an exchange of the orbitals containing electrons i and j. When the spins of the two electrons are parallel, the squares of these wave functions show that the two electrons have a drastically reduced probability of being near each other when compared to two electrons with opposite spins. Thus, the Coulomb repulsion is decreased when the spins are parallel because the electrons stay away from each other. The magnitude of the K_{mn} integral depends upon the overlap of the two orbitals m and n. The quantity $\psi_m(j)\psi_n(j) d\tau_j$ was referred to earlier as the differential overlap. The overlap integral of the differential overlap of two orthogonal orbitals is zero; i.e.,

$$\int \psi_m(j)\psi_n^*(j) d\tau_j = 0$$

However, when the differential overlap in the volume element $d\tau_j$ is operated upon by e^2/r_{ij}, multiplied by $\psi_m^*(i)\psi_n(i)$, and integrated over all volume elements $d\tau_i$ and $d\tau_j$, the result is not zero, but is the exchange integral. These exchange effects not only give rise to the spin polarization described above and cause the first excited triplet state of helium to be lower than the first excited singlet, but they also lead to Hund's rule.

The spin density experienced at the hydrogen atom of the C—H bond when there is unpaired electron density in the π-orbital ($2p_z$) is expressed by equation (9–11):

$$a_H = Q\rho_C \tag{9-11}$$

where ρ_C is the *unpaired electron density* in the carbon $2p_z$ orbital and Q is the value of a_H when there is a full electron on the carbon. The spin density at the proton is negative, so a_H is negative and Q must be negative. In systems where ρ_C is known, Q can be calculated and is found experimentally to vary from -22 to -27 gauss. A rough value of -23 gauss for aromatic radicals treated by Hückel theory suffices for most purposes that will concern us. When an extended Hückel calculation is used, the overlap is not set equal to zero and the molecular orbital coefficients are normalized to include overlap. Accordingly, the value of Q used depends on the m.o. calculation used, that is, on the definition of ρ_C.

When there is a node in the m.o. containing the unpaired electron at one of the carbon atoms in a π-system, similar exchange interactions with lower-energy filled pi-molecular orbitals operate to place negative spin density on this carbon. We shall make this more specific in the discussion of the allyl radical below. Resulting exchange interaction of this unpaired spin with the C—H sigma bond places positive spin density on the hydrogen. The Hückel, extended Hückel, or any restricted m.o. calculation (*i.e.*, one where two electrons are fed into each molecular orbital) do not include these exchange interactions. They simply indicate a node at the carbon atom or the hydrogen atom. One attempts to correct for this shortcoming, for example, at a hydrogen directly bonded to a carbon containing unpaired electron density in an orthogonal C_{2p} orbital by employing equation (9–11). Often other polarization effects in a molecule are qualitatively discussed, and the protons or carbons where these effects dominate are ignored when one attempts a quantitative fit of the calculated and experimental coupling constants in a molecule. This discussion can be made more specific by considering the observed[6] proton hyperfine splittings in the allyl radical shown in Fig. 9–16. The radical contains three electrons in the π-system whose wave functions are given by:

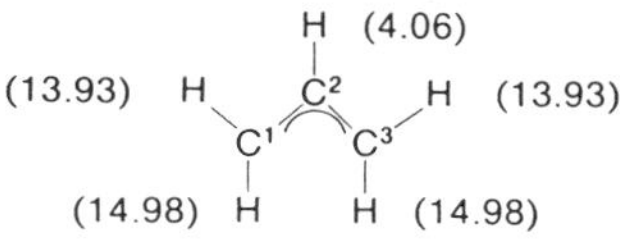

$$\psi_1 = \frac{1}{2}(\varphi_1 + \sqrt{2}\varphi_2 + \varphi_3) \quad \text{(bonding)} \tag{9-12}$$

$$\psi_2 = \frac{1}{\sqrt{2}}(\varphi_1 - \varphi_3) \quad \text{(non-bonding)} \tag{9-13}$$

$$\psi_3 = \frac{1}{2}(\varphi_1 - \sqrt{2}\varphi_2 + \varphi_3) \quad \text{(anti-bonding)} \tag{9-14}$$

FIGURE 9–16 Proton hyperfine splittings in the allyl radical.

The odd electron is placed in ψ_2, so one would predict the unpaired density at C_1 to be $\rho_{C_1} = (1/\sqrt{2})^2 = 0.5$, where $1/\sqrt{2}$ is the C_1 coefficient in the m.o. containing the unpaired electron. Using equation (9–11) with $Q = -23$, one would predict a_H to be equal to -11.5. Furthermore, there would be no unpaired electron density at C_2 and, without some kind of spin polarization involving the carbon π electron density, one would predict a zero coupling constant for this hydrogen or for a ^{13}C at this position. This is not observed. Two polarizations are needed to account for a hydrogen coupling constant from this middle hydrogen. The qualitative explanation involves taking the filled orbital ψ_1 and writing two separate spin orbitals for it, ψ_{1a} and ψ_{1b}. Only one electron is placed in each spin orbital. The wave functions in terms of the foregoing wave functions for allyl become

$$\psi_{1a} = \psi_1 + \lambda\psi_3$$
$$\psi_{1b} = \psi_1 - \lambda\psi_3 \text{ (where } \lambda \ll 1\text{)}$$

For spin aligned with the field in the lower energy ψ_{1a} and opposed to the field in ψ_{1b}, there will be increased spin density aligned with the field on atoms 1 and 3 relative to that on atom 2. In ψ_{1b}, with spin density opposed to the field, there will be more negative spin density on atom 2 than on carbon atoms 1 and 3. By this mechanism, we are not introducing any unpaired electrons into the old ψ_1 orbital, but we simply are influencing the distribution of the paired spins over the three atoms giving rise to negative (opposed to the applied field) spin density on C_2. This negative spin density then undergoes spin polarization with the electron pair in the C—H bond [see the discussion of equation (9–11)] to place spin density on the hydrogen. The exchange interaction of the unpaired electron in ψ_2 (mainly on C_1 and C_3) with the pair in ψ_1 is the effect that lowers the energy of ψ_{1a} relative to ψ_{1b}. The two hydrogens on one of the terminal carbons are not equivalent by symmetry, but our discussion so far has not introduced any effects that would make them non-equivalent from the standpoint of spin distribution. Exchange polarization involving filled sigma molecular orbitals is required.

The phenomena discussed above are all *indirect* mechanisms for placing unpaired spin density on the hydrogen. When the free radical is a sigma radical, *e.g.*, the vinyl radical $H_2C=\dot{C}-H$, the protons in the molecule make a contribution to the sigma molecular orbital containing the unpaired electron. Thus, the unpaired electron is delocalized directly onto the proton and a_H is proportional to ψ^2. Since Hückel calculations are inappropriate for sigma systems of this sort, the initial work in this area utilized extended Hückel molecular orbital calculations. Procedures have been reported[7] for evaluating ψ^2 at the nucleus from the wave function, and a_H is calculated by:

$$a_H \text{ (gauss)} = 1887 \, \psi_{(H)}^2 \qquad (9\text{–}15)$$

Again, spin polarization is foreign to this calculation. Often the majority of the unpaired electron density resides at a given atom in the radical. When this is the case, spin polarization makes a large contribution at protons directly bound to this atom. Poor agreement between calculated and experimental results can be expected for this atom. Generally, for other protons in the molecule (with a few exceptions), spin polarization effects make a relatively insignificant contribution when direct delocalization is appreciable.

Semi-empirical quantitative approaches[8,9] have been reported to incorporate the effects of spin-polarization. An unrestricted molecular orbital calculation, *i.e.*, one utilizing spin orbitals, yields the best results. The most common one at present is the so-called INDO calculation,[9] which has been parametrized to calculate spin densities on hydrogen. It has not been extensively tested on atoms for which spin polarization dominates, but is certainly the method of choice for this calculation at present. The output[10] consists of one-electron orbitals, and all the positive and negative spin densities at an atom in all the filled molecular orbitals are summed to produce the net spin density at the atom.

The application of the results from molecular orbital calculations to the hyperfine splittings from atoms other than hydrogen is considerably more complex. In contrast to protons, which have only the direct and indirect mechanisms described above, ^{13}C hyperfine splittings have contributions from other sources. (1) Unpaired electrons in a $p(\pi)$ orbital can polarize the filled $2s$ and the filled $1s$ orbitals on the same atom. (2) There can be direct delocalization of electron density into the $2s$ orbital in a sigma radical. (3) Spin density on a neighbor carbon, by polarizing the C—C sigma bond, can place spin density into the $2s$ and $2p$ orbitals of the carbon whose resonance is being interpreted. The calculations[10–13] of ^{14}N, ^{33}S, and ^{17}O hyperfine coupling have been more successful than those for ^{13}C. Silicon-containing radicals have also been successfully treated.[13] The effects from spin densities on neighboring atoms are found to be less important for these nuclei than for ^{13}C.

The application of the results from molecular orbital calculations to the assignment of the esr spectrum of an organic radical is an important application of the foregoing discussion. Another application involves determining the geometry of free radicals. For instance, is $CH_3 \cdot$ planar? Does the C—H bond of the vinyl radical lie along the C—C bond axis? When the calculated hyperfine coupling constants are found to vary considerably with geometry (*i.e.*, a whole series of molecular orbital calculations are performed for different geometries), the fit of calculated and experimental results can be used to suggest the actual geometry.[14] In several examples, molecular orbital calculations have provided evidence about the structure of a radical produced in an experiment.[11a, 14] For example, γ irradiation of pyridine produced a radical believed to be the pyridine cation; *i.e.*, one of the lone pair electrons was removed. The results shown in Table 9–2 indicate that the 2-pyridyl radical was actually formed.[11a]

TABLE 9–2. Extended Hückel and Experimental Isotropic Hyperfine Coupling Constants (Gauss)

	Calculated for $C_5H_5N^+$	Calculated for 2-pyridyl C_5H_4N	Experimental
a_N	52.5	33.8	29.7
a_H	27.0 (2,6 H)	5.3 ⎫	
	9.7 (3,5 H)	9.3 ⎬	
	38.8 (4 H)	7.6 ⎭	4.3
		0.0	

These calculations and esr experiments indicate the very extensive amount of electron delocalization that occurs in the sigma system. For example, significant amounts of unpaired electron density are found on the methyl group protons of $CH_3\dot{N}H_2^+$ and $C_2H_5\dot{N}H_2^+$. Thus, the lone pair molecular orbital is not a localized nitrogen lone pair orbital, but is a delocalized molecular orbital. The extent of delocalization onto the protons varies in the different rotamers.[16]

Anisotropic Effects

9–6 ANISOTROPY IN THE g VALUE

The next feature of esr spectroscopy can be introduced by describing the g values for the NO_2 radical trapped[17] in a single crystal of KNO_3. When the crystal is mounted with the field parallel to the z-axis of NO_2 (the twofold rotation axis), a g value of 2.006 is obtained. When the crystal is mounted with the x- or y-axis (the plane of the molecule containing y) parallel to the field, a g value of 1.996 is obtained. The molecule is rapidly rotating about the z-axis in the solid, so the same result is obtained for x or y parallel to the field. The differences in the g values with orientation are even more pronounced in transition metal ion complexes (*vide infra*) and in complexes of the lanthanides and actinides.

The treatment so far has involved so-called isotropic spectra. These are obtained when the radical under consideration has spherical or cubic symmetry. For radicals with lower symmetry, anisotropic effects are manifested in the solid spectra for both the g values and the a values. Usually, for these lower symmetry systems, the solution spectra appear as isotropic spectra because the anisotropic effects are averaged to zero by the rapid rotation of the molecules. Our concern here is how these anisotropic effects arise and how they can be determined. Later (Chapter 13), we shall see how the anisotropy in g and a can be used to provide information about the electronic ground state of transition metal ion complexes.

Anisotropy in g arises from coupling of the spin angular momentum with the orbital angular momentum. The spin angular momentum is oriented with the field, but the orbital angular momentum, which is associated with electrons moving in molecular orbitals, is locked to the molecular wave function. Consider a case where there is an orbital contribution to the moment from an electron in a circular molecular orbit that can precess about the z-axis of the molecule. In Fig. 9–17, two different orientations of this molecular orbital relative to the field are indicated by ellipses. In Fig. 9–17(A), $\bar{\mu}_L$ (the orbital magnetic moment vector*) and $\bar{\mu}_s$ (the spin magnetic moment vector) are in the same direction. In Fig. 9–17(B), a different orientation of the molecule is shown. The electron moment $\bar{\mu}_s$ has the same magnitude as before, but now the net moment, indicated by the boldface arrow, results because $\bar{\mu}_L$ and $\bar{\mu}_s$ do not point in the same direction. If it were not for the orbital contribution, the moment from the electron would be isotropic. *When the effects of the orbital moment are small, they are incorporated into the g-value; and this g-value will be anisotropic.* A tensor, equation (9–16), is needed to describe it. The g-tensor than gives us an effective spin, *i.e.*, $\mathbf{g} \cdot S = S_{\text{eff}}$. For different orientations, the g-tensor lengthens and shortens S_{eff} to incorporate orbital effects. It should be emphasized that even when the ground state of a molecule has no orbital angular momentum associated with it, field-induced mixing in of an excited state that does have orbital angular momentum can lead to anisotropy in g. The information that g-tensor anisotropy provides about the electronic structure of the molecule will be discussed in Chapter 13 on the epr of transition metal ion complexes.

---

*The direction of the orbital angular momentum given by the right-hand rule is opposite to that of the magnetic moment vector of the electron.

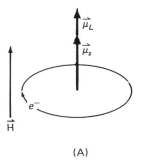 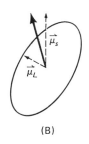

FIGURE 9–17 Coupling of the projections of the spin and orbital angular momentum for two different molecular orientations relative to the applied field H.

(A) (B)

For an isotropic system, we wrote our Hamiltonian to describe the interaction of an electron spin moment with the magnetic field and with a magnetic nucleus in equation (9–4) as:

$$\hat{H} = g\beta H\hat{S}_z - g_N\beta_N H\hat{I}_z + a\hat{I}\cdot\hat{S}$$

Both g and a were scalar quantities. When an anisotropic free radical is investigated in the solid state, both g and a have to be replaced by tensors or matrices. The $g\beta H\hat{S}_z$ term in the Hamiltonian becomes $\beta\hat{S}\cdot\mathbf{g}\cdot H$, which can be represented with matrices as:

$$\beta[S_x \ S_y \ S_z]\begin{bmatrix} g_{xx} & g_{xy} & g_{xz} \\ g_{yx} & g_{yy} & g_{yz} \\ g_{zx} & g_{zy} & g_{zz} \end{bmatrix}\begin{bmatrix} H_x \\ H_y \\ H_z \end{bmatrix} \quad (9\text{-}16)$$

and a in the $aI\cdot S$ term is also replaced by a tensor. Here x, y and z are defined in the laboratory frame; *i.e.*, they are crystal axes. The off-diagonal element g_{zx} gives the contribution to g along the z-axis of the crystal when the field is applied along the x-axis. *This matrix is diagonal when the crystal axes are coincident with the molecular coordinate system that diagonalizes* **g**. When they are not coincident and the crystal is studied along the x, y, and z crystal axes, we get off-diagonal contributions, as we shall show subsequently. The g matrix can be made diagonal by a suitable choice of coordinates.

If one studies the esr of a single crystal with anisotropy in g, the measured g value is a function of the orientation of the crystal with the field because we measure an effective g value oriented along the field. If we define molecular axes X, Y, and Z that diagonalize the g-tensor and pick as an example a case in which they are coincident with the crystal axes, the effective values of g for an arbitrary orientation of the crystal is then given by

$$\begin{aligned}g_{\text{eff}}^2 &= (g^2)_{xx}\cos^2\theta_{Hx} + (g^2)_{yy}\cos^2\theta_{Hy} + (g^2)_{zz}\cos^2\theta_{Hz} \\ &= (g^2)_{xx}l_x^2 + (g^2)_{yy}l_y^2 + (g^2)_{zz}l_z^2 \end{aligned} \quad (9\text{-}17)$$

Here θ_{Hx}, θ_{Hy}, and θ_{Hz} are the angles between the field H and the X-, Y-, and Z-axes, respectively. The symbols l_x, l_y, and l_z are often used to represent the cosines of these angles and are referred to as the *direction cosines*. From trigonometry,

$l_x^2 + l_y^2 + l_z^2 = 1$, so two parameters suffice to specify the direction. The above equation can be indicated in matrix notation as

$$g_{\text{eff}}^2 = [l_x \ l_y \ l_z] \begin{bmatrix} (g^2)_{xx} & 0 & 0 \\ 0 & (g^2)_{yy} & 0 \\ 0 & 0 & (g^2)_{zz} \end{bmatrix} \begin{bmatrix} l_x \\ l_y \\ l_z \end{bmatrix} \quad (9\text{-}18)$$

Since S_x, S_y, and S_z as well as H_x, H_y, and H_z are defined in terms of the molecular coordinate x,y,z system, they can be replaced by the same direction cosines. The molecular coordinate system that diagonalizes the g-tensor may not be coincident with the arbitrary axes associated with the crystal morphology. Since this experiment is carried out using the easily observed axes of the bulk crystal, the above equation has to be rewritten in non-diagonal form as

$$g_{\text{eff}}^2 = [l_x \ l_y \ l_z] \begin{bmatrix} (g^2)_{xx} & (g^2)_{xy} & (g^2)_{xz} \\ (g^2)_{yx} & (g^2)_{yy} & (g^2)_{yz} \\ (g^2)_{zx} & (g^2)_{zy} & (g^2)_{zz} \end{bmatrix} \begin{bmatrix} l_x \\ l_y \\ l_z \end{bmatrix} \quad (9\text{-}19)$$

Equation (9–19) can be used to evaluate all the tensor components. The matrix is symmetric [i.e., $(g^2)_{yx} = (g^2)_{xy}$], so only six independent components need to be evaluated. Since it is most convenient to orient the crystal in the magnetic field relative to the observed crystal axes, the x, y, and z axes are defined in terms of these observed axes of the bulk crystal. S_x, S_y, and S_z, as well as H_x, H_y, and H_z, are defined in terms of these axes. Consider the first case where the crystal is mounted, as shown in Fig. 9–18, with the y-axis perpendicular to the field so that the crystal can be rotated around y with $\vec{H}$ making different angles, θ, to z in the xz-plane. Now, l_z equals $\cos \theta$ and l_x equals $\sin \theta$, where θ is the angle between $\vec{H}$ and the z-axis. Substituting these quantities into equation (9–19) and carrying out the matrix multiplication yields

$$g_{\text{eff}}^2 = (g^2)_{xx} \sin^2 \theta + 2(g^2)_{xz} \sin \theta \cos \theta + (g^2)_{zz} \cos^2 \theta \quad (9\text{-}20)$$

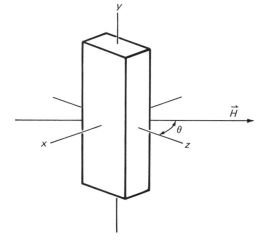

FIGURE 9–18 Mounting of a crystal for rotation in the xz-plane.

For rotations in the yz-plane, we have $l_x = 0$, $l_y = \sin\theta$, and $l_z = \cos\theta$. Substitution into (9–19) and matrix multiplication then yields

$$g_{\text{eff}}^2 = (g^2)_{yy} \sin^2\theta + 2(g^2)_{yz} \sin\theta \cos\theta + (g^2)_{zz} \cos^2\theta \qquad (9\text{–}21)$$

In a similar fashion, rotation in the xy-plane yields

$$g_{\text{eff}}^2 = (g^2)_{xx} \cos^2\theta + 2(g^2)_{xy} \sin\theta \cos\theta + (g^2)_{yy} \sin^2\theta \qquad (9\text{–}22)$$

These equations thus tie our matrix in equation (9–19) to experimental observables, g_{eff}^2. In our experiment, a g-value is obtained, but we do not know its sign. The measured g-value is squared and used in this analysis. For rotations in any one plane, only three measurements of g_{eff}^2 need be made (corresponding to three different θ-values) to solve for the three components of the g^2-tensor in the respective equations. For the xz-plane, one measures $(g^2)_{zz}$ at $\theta = 0$ and $(g^2)_{xx}$ at $\theta = 90°$. With these values and g_{eff}^2 at $\theta = 45°$, one can solve for $(g^2)_{xz}$. In this way, the six independent components of the g^2-tensor can be measured. In practice, many measurements are made and the data are analyzed by the least squares method. One then solves for a transformation matrix that rotates the coordinate system and diagonalizes the g^2-tensor. This produces the molecular coordinate system for diagonalizing the g^2-tensor, and the square roots of the individual diagonal g^2 matrix elements produce g_{xx}, g_{yy}, and g_{zz} in this special coordinate system. In order for this procedure to work as described, it is necessary that all the molecules in the unit cell have the same orientation of their molecular axes relative to the crystal axes. Thus, these measurements are often carried out in conjunction with a single crystal x-ray determination.

9–7 ANISOTROPY IN THE HYPERFINE COUPLING

We introduced Section 6 by describing the anisotropy in g when a single crystal of NO_2 in KNO_3 was examined at different orientations relative to the field. The a-values of this system are also very anisotropic. When the molecular twofold axis is parallel to the applied field, the observed nitrogen hyperfine coupling constant is 176 MHz, while a value of 139 MHz is observed for the orientation in which this axis is perpendicular to the field. In rigid systems, interactions between the electron and nuclear dipoles give rise to anisotropic components in the electron nuclear hyperfine interaction. The classical expression for the interaction of two dipoles was treated in Chapter 8, and the same basic considerations apply here. For the interaction of an electron moment and a nuclear moment, the Hamiltonian is:

$$\hat{H}_{\text{dipolar}} = -g\beta g_N \beta_N \left[\frac{\hat{S}\cdot\hat{I}}{r^3} - \frac{3(\hat{S}\cdot\vec{r})(\hat{I}\cdot\vec{r})}{r^5} \right] \qquad (9\text{–}23)$$

The sign is opposite that employed for the interaction of two nuclear dipoles, which was the problem treated in the section on liquid crystal and solid nmr.

Substituting $\hat{S} = \hat{S}_x + \hat{S}_y + \hat{S}_z$, $\hat{I} = \hat{I}_x + \hat{I}_y + \hat{I}_z$, and $r = x + y + z$, and expanding these vectors, leads to

$$\hat{H}_{\text{dipolar}} = -g\beta g_N \beta_N \left\{ \left[\frac{r^2 - 3x^2}{r^5} \right] \hat{S}_x \hat{I}_x + \left[\frac{r^2 - 3y^2}{r^5} \right] \hat{S}_y \hat{I}_y + \left[\frac{r^2 - 3z^2}{r^5} \right] \hat{S}_z \hat{I}_z \right.$$

$$\left. - \left[\frac{3xy}{r^5} \right] (\hat{S}_x \hat{I}_y + \hat{S}_y \hat{I}_x) - \left[\frac{3xz}{r^5} \right] (\hat{S}_x \hat{I}_z + \hat{S}_z \hat{I}_x) - \left[\frac{3yz}{r^5} \right] (\hat{S}_y \hat{I}_z + \hat{S}_z \hat{I}_y) \right\}$$

(9–24)

When this Hamiltonian is applied to an electron in an orbital, the quantities in brackets must be replaced by average values; we employ angular brackets to refer to the average value over the electronic wave function. In matrix notation, we then have

$$\hat{H}_{\text{dipolar}} =$$

$$-(g\beta g_N \beta_N)[\hat{S}_x \; \hat{S}_y \; \hat{S}_z] \begin{bmatrix} \left\langle \frac{r^2 - 3x^2}{r^5} \right\rangle & \left\langle \frac{-3xy}{r^5} \right\rangle & \left\langle \frac{-3xz}{r^5} \right\rangle \\ \left\langle \frac{-3xy}{r^5} \right\rangle & \left\langle \frac{r^2 - 3y^2}{r^5} \right\rangle & \left\langle \frac{-3yz}{r^5} \right\rangle \\ \left\langle \frac{-3xz}{r^5} \right\rangle & \left\langle \frac{-3yz}{r^5} \right\rangle & \left\langle \frac{r^2 - 3z^2}{r^5} \right\rangle \end{bmatrix} \begin{bmatrix} \hat{I}_x \\ \hat{I}_y \\ \hat{I}_z \end{bmatrix}$$

(9–25)

This equation is abbreviated as

$$\hat{H}_{\text{dipolar}} = h\hat{S} \cdot \mathbf{T} \cdot \hat{I}$$

(9–26)

where **T** is the dipolar interaction tensor (in units of Hz) that gauges the anisotropic nuclear hyperfine interaction. The Hamiltonian now becomes

$$\hat{H} = \beta \hat{S} \cdot \mathbf{g} \cdot H - g_N \beta_N H \cdot \hat{I} + h\hat{S} \cdot \mathbf{A} \cdot \hat{I}$$

(9–27)

where the first term on the right is the electron Zeeman term, the second is the nuclear Zeeman term, and the third is the hyperfine interaction term. The quantity **A** in the third term includes both the isotropic and the anisotropic components of the hyperfine interaction; i.e.,

$$\mathbf{A} = \mathbf{T} + a\,\mathbf{1}$$

(9–28)

In the application of the Hamiltonian given in equation (9–27) to organic free radicals, several simplifying assumptions can be introduced. First, $g_N \beta_N H \cdot \hat{I}$, the nuclear Zeeman effect, usually gives rise to a small energy term compared to the others. (Recall our earlier discussion about the energies of the esr and nmr transitions). Second, g-anisotropy is small, and we shall assume that g is isotropic in treating the hyperfine interaction.* (This would be a particularly bad

---

*If g-tensor anisotropy is comparable to hyperfine anisotropy, this assumption cannot be made. The reader is referred to Chapter 13 and to A. Abragam and B. Bleany, "EPR of Transition Ions," p. 167, Clarendon Press, Oxford, England, 1970 for a discussion of this situation.

assumption for certain transition metal complexes, *vide infra*.) The electron Zeeman term is assumed to be the dominant energy term, so $\hat{S}$ is quantized along $\vec{H}$, which we label as the z-axis. We see in this example, as we shall see over and over again, that *it is often convenient to define the coordinate system to be consistent with the largest energy effect*. Next we have to worry about the orientation of the nuclear moment relative to the z-field. Our discussion is general, but it may help to consider an ethyl radical oriented as shown in Fig. 9–19, with the CH_3 group not undergoing rotation. To make this point, focus attention on nucleus H_a involved in dipolar coupling to the electron. The nuclear moment will not be quantized along z, but along an effective field, $\vec{H}_{eff}$, which is the vector sum of the direct external field $\vec{H}$ (nuclear Zeeman) and the hyperfine field produced by the nearby electron. If the hyperfine interaction is large (~ 100 gauss), the hyperfine field at this nucleus (*i.e.*, the field from the electron magnetic moment felt at the hydrogen nucleus) is about 11,700 gauss. (This is to be contrasted to the field of ~ 3000 gauss from the magnet and the field of ~ 18 gauss felt at the electron from the considerably smaller nuclear moment.) Thus, we may be somewhat justified in ignoring the nuclear Zeeman term, $g_N \beta_N H \cdot \hat{I}$, in equation (9–27). The Hamiltonian for most organic free radicals (where g is isotropic) is then considerably simplified from the form in equation (9–27), and becomes

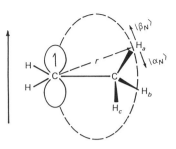

FIGURE 9–19 The orientation of the spin and nuclear moments in an applied field.

$$\hat{H} = g\beta H \hat{S}_z + h\hat{S}_z(A_{zx}\hat{I}_x + A_{zy}\hat{I}_y + A_{zz}\hat{I}_z) \qquad (9\text{–}29)$$

The terms on the far right give the z-component of the electron-nuclear hyperfine interaction with contributions from $\hat{I}_x$ and $\hat{I}_y$ as well as from $\hat{I}_z$, for the z-field does not quantize $\hat{I}$, but does quantize $\hat{S}$. When this Hamiltonian operates on the $|\alpha_e \alpha_N\rangle$ and other wave functions, off-diagonal matrix elements in the secular determinant result. When it is diagonalized and solved for energy, the following results are obtained:

$$\left.\begin{matrix}\alpha_e \alpha_N \\ \alpha_e \beta_N\end{matrix}\right\} E = \frac{1}{2} g\beta H \pm \frac{1}{4} \sqrt{A_{zx}^2 + A_{zy}^2 + A_{zz}^2}$$

$$\left.\begin{matrix}\beta_e \alpha_N \\ \beta_e \beta_N\end{matrix}\right\} E = -\frac{1}{2} g\beta H \pm \frac{1}{4} \sqrt{A_{zx}^2 + A_{zy}^2 + A_{zz}^2}$$

The term containing the square root replaces the $(^1/_4)a$ obtained from the evaluation of the $a\hat{I} \cdot \hat{S}$ term for the hydrogen atom. The energy for the hyperfine coupling is thus given by

$$\Delta E_{hf} = \frac{1}{2} \sqrt{A_{zx}^2 + A_{zy}^2 + A_{zz}^2} \qquad (9\text{–}30)$$

The quantity **A** contains both the isotropic, a, and anisotropic, **T**, components of the hyperfine interaction. Since in solution the anisotropic components are averaged to zero, it becomes a simple matter to take one-third of the trace of **A** to decompose **A** into **T** and a. (This assumes that the solvent or solid lattice has no effect on the electronic structure.)

These expressions apply for any orientation of the molecule relative to the applied field. In a single-crystal experiment, in which the crystal and molecular axes are not aligned, we proceed as in the case of the evaluation of the g-tensor to determine all of the components of the hyperfine tensor. *The coordinate system that diagonalizes the g-tensor need not be the same one that diagonalizes the A-tensor, and neither one of these need be the apparent molecular coordinate system.*[17b] If the molecule has overall symmetry (*i.e.*, the full ligand environment included) such that it possesses an *n*-fold rotation axis, the same axis will be diagonal for *g* and *A*, and it must be coincident with the molecular z-axis.

The angular dependence of the hyperfine interaction for the case where the field from the hyperfine interaction is large, $I = \frac{1}{2}$, and the system has axial symmetry can be expressed by substituting $r \cos\theta$ for z and $r \sin\theta$ for x and y into equation (9–24). We are, in effect, resolving the nuclear moment in Fig. 9–19 into components parallel and perpendicular to the field. The Hamiltonian including the electron Zeeman term ($g\beta H\hat{S}_z$) becomes

$$\hat{H} = g\beta H\hat{S}_z + h\hat{S}_z\{[a + B(3\cos^2\theta - 1)]\hat{I}_z + 3B\cos\theta\sin\theta\hat{I}_x\} \quad (9\text{–}31)$$

The result of this Hamiltonian operating on the basis set produces the energies given by

$$E = g\beta HM_s \pm \frac{hM_s}{2}[(a - B)^2 + 3B(2a + B)\cos^2\theta]^{1/2} \quad (9\text{–}32)$$

where *a* is the isotropic hyperfine coupling constant, *B* is the anisotropic hyperfine coupling constant, and θ is the angle that the z-axis of the molecule makes with the field. The hyperfine coupling constant *A* observed experimentally is the difference between the energies of the appropriate levels and is given (in cm^{-1}) by

$$A = h[(a - B)^2 + 3B(2a + B)\cos^2\theta]^{1/2}$$

One often sees the following equation presented in the literature to describe the anisotropy of *g* and *A* for an axial system:

$$\Delta E = hv = \left(\frac{1}{3}g_\parallel + \frac{2}{3}g_\perp\right)\beta H_0 + am_I + \left[\frac{1}{3}(g_\parallel - g_\perp)\beta H_0 + Bm_I\right](3\cos^2\theta - 1) \quad (9\text{–}33)$$

where θ is the angle between the z-axis and the magnetic field, *a* is the isotropic coupling constant, and *B* is the anisotropic coupling constant. The equation results from equation (9–32) by adding the anisotropy in *g* and by assuming that both the *g*- and *A*-tensors are diagonal in the same axis system.

Since in the analysis of the anisotropy in the hyperfine coupling we deal with $\mathbf{A}^2$ [see equation (9–30)], usually one cannot obtain the sign of the coupling constant from the esr experiment.* However, it can be readily predicted for an organic radical that contains an electron in a *p*-orbital (there is no anisotropic contribution from unpaired electron density in a spherical 2*s* orbital). We shall begin by returning to equation (9–25) to further explore its meaning. Assume

---

*For certain systems, one can determine the sign of the coupling constant by using nmr, because spin aligned with the field causes a downfield shift and spin opposed to the field causes an upfield shift.

that the electron is in a hypothetical orbital that can be represented by a unit vector. When this hypothetical orbital lies along z, we have $z = r$, $x = 0$, and $y = 0$, and all off-diagonal terms are zero. We observe, on substitution into equation (9–25), that $T_{xx} = k\langle 1/r^3 \rangle$, $T_{yy} = k\langle 1/r^3 \rangle$, and $T_{zz} = -k\langle 2/r^3 \rangle$.† [The matrix elements as written in equation (9–25) have the opposite sign, but the whole term is negative to describe the interaction of the positive nuclear moment and the negative electron moment.] Note that the trace is zero.

To make the problem more realistic, we shall next consider the electron to be in a p_z orbital. It is convenient to convert to spherical polar coordinates to solve this problem by substituting $z = r \cos \theta$, $x = r \sin \theta \cos \varphi$, and $y = r \sin \theta \sin \varphi$. The result for an electron at a specific (r, θ) (after consideration of the negative sign) is

$$T_{zz} = g\beta g_N \beta_N \langle (3\cos^2 \theta - 1)/r^3 \rangle$$

$$T_{yy} = -\frac{1}{2} g\beta g_N \beta_N \langle (3 \cos^2 \theta - 1)/r^3 \rangle$$

$$T_{xx} = -\frac{1}{2} g\beta g_N \beta_N \langle (3 \cos^2 \theta - 1)/r^3 \rangle$$

The latter two matrix elements are readily obtained by substituting for x and y in equation (9–25) and substituting $\langle \cos^2 \varphi \rangle = 1/2$ for an axial system. Note that the trace is zero. Now, consider that the electron can be located at any place in the p orbital. Thus, we have to integrate over all possible angles for the radius vector to the electron in this orbital and then over all radii r. In doing so, we get

$$T_{zz} = \frac{4}{5} g\beta g_N \beta_N \langle 1/r^3 \rangle \tag{9–34}$$

where $\langle 1/r^3 \rangle$ is the average value of the quantity $1/r^3$. Abbreviating equation (9–34) as $T_{zz} = 4/5 P_p$, we have

$$T_{xx} = -\frac{2}{5} P_p \quad \text{and} \quad T_{yy} = -\frac{2}{5} P_p \tag{9–35}$$

These considerations are valuable in predicting the signs of the anisotropic components of the hyperfine coupling constant.

It is informative to apply these equations to the anisotropic hyperfine tensor of the ^{13}C nucleus, which depends mainly on the unpaired electron density in the p orbital of this atom. We wish to consider the signs of T_{xx}, T_{yy}, and T_{zz} for this system. The three orientations of the p-orbital in the molecule relative to the applied field are indicated in Fig. 9–20. The dotted lines indicate the regions where the $(3 \cos^2 \theta - 1)$ function is zero. This corresponds to plotting the signs for the various regions of the lines of flux emanating from the nuclear moment. Accordingly, by visual inspection, we can tell whether T_{zz} expressed in equation (9–34) will be positive or negative. For example, as we see in Fig. 9–20(A), when

---

† The proportionality constant is $g\beta_N g_N \beta/h$.

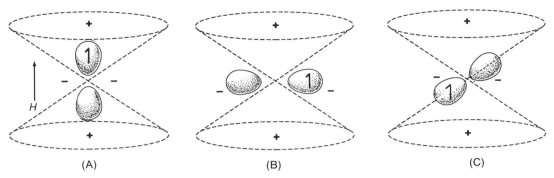

FIGURE 9–20 Visual representation of dipolar averaging of the electron and nuclear moments: (A) p orbital oriented along the field; (B) and (C) p orbital perpendicular to the field.

the p_z orbital is aligned with the field almost the entire averaging of the dipolar interaction of the nuclear moment over the p_z orbital will occur in the positive part of the cone. A large, positive T_{zz}-value is thus expected. For the orientation along the x-axis shown in Fig. 9–20(B), the dipolar interaction, T_{xx}, will be large and negative; the same is true of T_{yy} for the orientation shown in Fig. 9–20(C). Analysis[18,19] of the ^{13}C hyperfine structure of the isotopically enriched malonic acid radical, $H^{13}\dot{C}(COOH)_2$, produces $a_c = 92.6$ MHz, $T_{xx} = -50.4$ MHz, $T_{yy} = -59.8$ MHz, and $T_{zz} = +120.1$ MHz. After the hyperfine tensor is diagonalized, the relative signs of T_{xx}, T_{yy}, and T_{zz} are known (the trace must be zero), but the absolute signs are not; *i.e.*, all those given above could be reversed. However, the arguments based on Fig. 9–20 provide us with good reason to think that the signs presented above are correct.

It is informative to predict the signs of the anisotropic hydrogen hyperfine components of a C—H radical. By analogy to our discussion above, the three orientations of the p_π-orbital of this radical shown in Fig. 9–21 predict that T_{zz} is small, whereas T_{yy} is positive and T_{xx} is negative. Visual averaging of the p orbital with our cone of magnetic nuclear flux also suggests that T_{zz} will be small. Note that the cones representing the nuclear moment lines of flux are drawn at the nucleus whose moment is causing the splitting via the dipolar

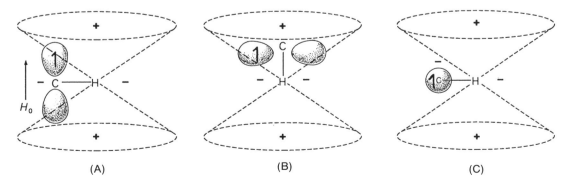

FIGURE 9–21 Visual representation of the dipolar averaging of the nuclear moment on the hydrogen with the electron moment in a p-orbital of carbon. (A) H_0 is parallel to the z-crystal axis; (B) H_0 is parallel to the y-crystal axis; (C) H_0 is parallel to the x-crystal axis, and we are looking down the axis of the p-orbital.

interaction with the electron. If the x, y, and z axes are defined relative to the fixed crystal axes (which are coincident with the molecular axes) as in Fig. 9–21, calculation[20] shows that a full unpaired electron in the carbon p-orbital would lead to an anisotropic hyperfine tensor of

$$\mathbf{T}_H = \begin{bmatrix} -38 & 0 & 0 \\ 0 & +43 & 0 \\ 0 & 0 & -5 \end{bmatrix} \text{ MHz}$$

The experimental proton hyperfine tensor for the α proton of the malonic acid radical was found to be

$$\mathbf{A}_H = \begin{bmatrix} \pm 91 & 0 & 0 \\ 0 & \pm 29 & 0 \\ 0 & 0 & \pm 58 \end{bmatrix} \text{ MHz}$$

Since the isotropic hyperfine coupling constant, a, is one-third the trace of $\mathbf{A}_H$, it equals ± 59 MHz. Accordingly, the anisotropic hyperfine tensor $\mathbf{T}$ must be:

$$\mathbf{T}_H = \begin{bmatrix} -32 & 0 & 0 \\ 0 & +30 & 0 \\ 0 & 0 & +1 \end{bmatrix} \text{ MHz} \quad \text{or} \quad \begin{bmatrix} +32 & 0 & 0 \\ 0 & -30 & 0 \\ 0 & 0 & -1 \end{bmatrix} \text{ MHz}$$

The arguments presented in the discussion of Fig. 9–21 and comparison to the theoretical tensor lead us to predict the tensor on the left to be correct. This matrix arose from $a = -59$ MHz. Since a positive a would have given tensor components that correspond to values greater than that for a full electron, the isotropic hyperfine coupling constant must be negative.

Anisotropic and isotropic hyperfine coupling constants have been measured in several organic and inorganic radicals and have provided considerable information about the molecular orbital containing the unpaired electron. The value of B for one electron in a p orbital of various atoms can be evaluated using an SCF wave function from

$$B = \frac{2}{5} h^{-1} g \beta g_N \beta_N \langle r^{-3} \rangle \tag{9-36}$$

For ^{13}C, the anisotropic hyperfine coupling constants are given by

$$\begin{bmatrix} -B & & \\ & -B & \\ & & +2B \end{bmatrix}$$

where B is calculated from SCF wave functions to be 91 MHz. For $\text{H}\dot{\text{C}}(\text{COOH})_2$, the experimental value of T_{zz} for ^{13}C is found to be $+120.1$ MHz compared to the value of 182 MHz for an electron localized in a C_{2p} orbital. Accordingly, it is concluded that ρ_c is 0.66. From ^{13}C enrichment, it is found that $a_c = +92.6$ MHz. A full electron in an s-orbital has an isotropic hyperfine coupling constant of 3110 MHz. The measured a_c corresponds to a value of 0.03 for C_{2s} spin density. The radical is expected to be nearly planar.

The magnitude of the isotropic ^{13}C hyperfine coupling constant supports a planar[21] CH_3 radical, $a_c = 38.5$ gauss, but the value of $a_c = 271.6$ gauss in the CF_3 radical indicates[21] that it is pyramidal with s-character in the orbital containing the unpaired electron.

The isotropic ^{14}N hyperfine coupling constant in NO_2 is 151 MHz, and the maximum value for the anisotropic hyperfine coupling constant is 12 MHz. With 1540 MHz expected for one electron in a nitrogen 2s orbital and 48 MHz for an electron in a 2p orbital, ρ_s is calculated to be 0.10 and ρ_p is found to be 0.25, for a 2p/2s ratio of 2.5. An sp^2 orbital would have a ratio of 2.0, so this suggests that more p-character is being used in the orbitals to bond oxygen and an angle greater than 120° is predicted. Microwave results gave a value of 134° for NO_2 in the gas phase.

9–8 THE EPR OF TRIPLET STATES

The next complication we shall discuss arises when there is more than one unpaired electron in the molecule. An example is provided by the triplet state that is formed upon u.v. irradiation of naphthalene. The single crystal epr spectrum was studied for a sample doped into durene. The similar shapes of these two molecules allowed the naphthalene to be trapped in the durene lattice; dilution of the naphthalene greatly increases the lifetime of the triplet state. The spectrum consists of three peaks, which changes resonance fields drastically with orientation of the crystal. The changes could not be fitted with the anisotropic g- and a-tensor we have developed. The anisotropy in this system arises from electron-electron spin interaction and is described by the spin Hamiltonian given in equation (9–37). This Hamiltonian is seen to be very similar to that for the dipolar interaction of an electron and nuclear spin [equation (9–23)].

$$\hat{H} = g\beta H \cdot (\hat{S}_1 + \hat{S}_2) + g^2\beta^2 \left\{ \frac{\hat{S}_1 \cdot \hat{S}_2}{r^3} - \frac{3(\hat{S}_1 \cdot \vec{r})(\hat{S}_2 \cdot \vec{r})}{r^5} \right\} \quad (9\text{–}37)$$

where $\vec{r}$ is the vector joining the two electrons labeled 1 and 2.

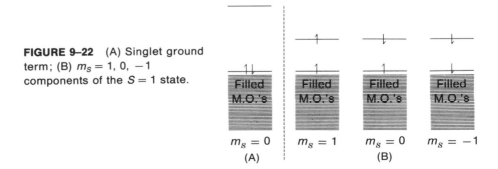

FIGURE 9–22 (A) Singlet ground term; (B) $m_S = 1, 0, -1$ components of the $S = 1$ state.

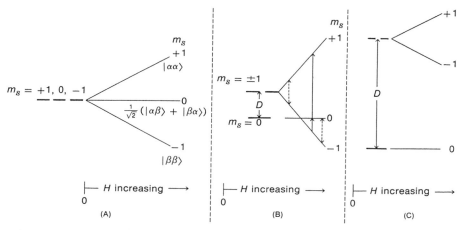

FIGURE 9–23 The effects of zero-field splitting on the expected epr transitions. (A) No zero-field effects. (B) Moderate zero-field splittings. The dashed arrows show the fixed frequency result, and the solid arrows show the fixed field result. (C) Large zero-field effects. The magnetic field is assumed to be parallel to the dipolar axis in the molecule.

The magnitude of the contribution to the epr spectrum from these effects depends upon the extent of the interaction of the two spins. We discussed the Coulomb (J) and exchange (K) integrals earlier in this chapter. When two molecular orbitals are closer in energy than the difference between exchange and repulsive energies, a stable triplet state arises. In the case of naphthalene, the ground state is a singlet, and the excited triplet state has an appreciable lifetime because of the forbidden nature of the triplet-singlet transition. For the ground state $S = 0$ and $m_s = 0$, and this configuration is illustrated in Fig. 9–22(A). For the triplet state we have $S = 1$ and $m_s = 1, 0, -1$. These electronic configurations are illustrated in Fig. 9–22(B). If only exchange and electrostatic interactions existed in the molecule, the three configurations $m_S = 1, 0,$ and -1 would be degenerate in the absence of a magnetic field. The magnetic field would remove this degeneracy as shown in Fig. 9–23(A), and only a single transition would be observed as was the case for $S = 1/2$. However, magnetic dipole-dipole interaction between the two unpaired electrons removes the degeneracy of the m_S components of $S = 1$ *even in the absence of an external field*, as shown in Fig. 9–23(B). This removal of the degeneracy in the absence of the field is called *zero-field splitting*. When a magnetic field is applied, the levels are split so that two $\Delta m_S = \pm 1$ transitions can be detected, as illustrated in Fig. 9–23(B). Earlier we mentioned that three peaks were observed for the epr spectrum of triplet naphthalene. Two of these are the $\Delta m_S = \pm 1$ transitions shown in Fig. 9–23(B); the third is the $\Delta m_S = \pm 2$ (between $m_S = -1$ and $m_S = +1$) transition, which becomes allowed when the zero-field splitting is small compared to the microwave frequency. When this is the case, one cannot assign precise $m_S = +1, 0,$ or -1 values to the states that exist. When the zero-field splitting is very large, as in Fig. 9–23(C), the m_S values become valid quantum numbers and the energies for the $\Delta m_S = \pm 1$ allowed transitions become too large to be observed in the microwave region; accordingly, no spectrum is seen.

Since the electron-electron interaction is dipolar [equation (9–37)], it is expected to be described by a symmetric tensor, the so-called zero-field splitting tensor, **D**.

$$\hat{H}_D = \hat{S} \cdot \mathbf{D} \cdot \hat{S} \tag{9–38}$$

The **D**-tensor elements have the same form as those for **T** given in equation (9–25). This dipolar **D**-tensor accounts for the large anisotropy observed in the spectrum of a bi-radical. In terms of the principal axes that diagonalize the zero-field splitting tensor, we can write

$$\hat{H}_D = -X\hat{S}_x^2 - Y\hat{S}_y^2 - Z\hat{S}_z^2$$

where X is the D_{xx} element in the diagonalized zero-field tensor. Since the tensor is traceless, $(X + Y + Z = 0)$, the zero-field Hamiltonian can be written in terms of two independent constants, D and E:

$$\hat{H}_D = D\left(\hat{S}_z^2 - \frac{1}{3}\hat{S} \cdot \hat{S}\right) + E(\hat{S}_x^2 - \hat{S}_y^2)$$

Since $X = Y$ in a system of axial symmetry, the last term disappears. If the molecule has cubic symmetry, no splitting from this zero-field effect will be observed.

Operating with this Hamiltonian on the triplet state wave functions, one can calculate the energies as a function of field and orientation. The results indicate substantial anisotropy in the spectrum. The spectrum for the naphthalene triplet state is described by g (isotropic) = 2.0030, $D/hc = +0.1012$ cm^{-1}, and $E/hc = -0.0141$ cm^{-1}. The magnitudes of D and E are related to how strongly the two spins interact. These quantities become negligible as the two electrons become localized in parts of the molecule that are very far apart.

In liquids, the traceless tensor **D** averages to zero. The large fluctuating fields arising from the large, rotating anisotropic spin-spin forces in molecules with appreciable zero-field splitting cause effective relaxation. The lines in the spectra are thus usually too broad to be detected. With some exceptions, the epr of triplet states cannot be observed in solution unless the two spins are far apart (*i.e.*, D and E are small).

9–9 NUCLEAR QUADRUPOLE INTERACTION

A nucleus that has a nuclear spin quantum number $I \geq 1$ also has an electric moment, and the unpaired electron interacts with both the nuclear magnetic and electric moments. The electric field gradient at the nucleus can interact with the quadrupole moment as in nqr, and this interaction affects the electron spin energy states via the nuclear-electronic magnetic coupling as a second-order perturbation. The effect of quadrupole interaction is usually complicated because it is accompanied by a much larger magnetic hyperfine interaction. The orientation of the nuclear moment is quantized with respect to both the electric field gradient and the magnetic field axis. When the magnetic field and the crystal axes are parallel, the only quadrupole effect is a small displacement of all the energy levels by a constant amount, which produces no change in the observed transitions. However, when the two axes are not parallel, the effect is a competition between the electric field and the magnetic field. This has two effects on the spacing of the hyperfine lines: (1) a displacement of all energy levels by a constant amount and (2) a change in the separation of the energy levels that causes the spacing between adjacent epr lines to be greater at the ends of the spectrum than in the middle.

This quadrupole effect can easily be distinguished from another second-order effect that produces a gradual increase or decrease in the spacing from one end of the spectrum to the other. The variation in the spacing from this other second-order effect occurs when the magnetic field produced by the nucleus becomes comparable in magnitude to the external field. In this case, the unequal spacing can be eliminated by increasing the applied magnetic field.

A further effect of this competition between the quadrupolar electric field and the magnetic field is the appearance of additional lines that are normally forbidden by the selection rule $\Delta m_I = 0$. Both $\Delta m_I = \pm 1$ and $\Delta m_I = \pm 2$ transitions are sometimes observed.[22] An analysis of the forbidden lines gives the nuclear quadrupole coupling constant.[23] The approach involves a single-crystal epr study of a compound doped into a diamagnetic host. A spectrum containing these transitions is illustrated[24] in Fig. 9–24 for bis(2,4-pentanedionato)copper(II) [^{63}Cu(acac)$_2$], doped into Pd(acac)$_2$. The forbidden transitions are marked in Fig. 9–24(A) with arrows and the other bands are the four allowed transitions [$I(^{63}\text{Cu}) = {}^3/_2$]. The spacing and intensity of the forbidden lines vary considerably with the angle θ. The variation in the spacing is shown in Fig. 9–24(B). By matrix diagonalization, the spectra could be computer-fitted[24] to a spin Hamiltonian:

$$\hat{H} = \beta[g_\parallel H_z \hat{S}_z + g_\perp(H_x \hat{S}_x + H_y \hat{S}_y)] + A\hat{S}_z \hat{I}_z + B(\hat{S}_x \hat{I}_x + \hat{S}_y \hat{I}_y)$$
$$+ Q'\left[\hat{I}_z^2 - \frac{1}{3}I(I+1)\right] - g_N \beta_N H \cdot \hat{I}$$

The term $Q'[I_z^2 - (^1/_3)I(I+1)]$ accounts for the quadrupole effects, and all other symbols have been defined previously. A value of $Q'/hc = (3.4 \pm 0.2) + 10^{-4}$ cm^{-1} fits all the data; and for $I = {}^3/_2$, $4Q'$ is the quantity related to the field gradient e^2Qq (see Chapter 14) at the copper nucleus, i.e., $Q' = 3e^2qQ/4I(2I+1)$. In similar studies of Cu(bzac)$_2$ doped into Pd(bzac)$_2$ and of bis-dithiocarbamato copper(II) [Cu(dtc)$_2$] in diamagnetic Ni(dtc)$_2$, Q'-values of $(3.3 \pm 0.21) + 10^{-4}$ cm^{-1} and $(0.7 \pm 0.1) \times 10^{-4}$ cm^{-1}, respectively, were obtained. A Q'-value of about 16×10^{-4} cm^{-1} is expected for a hole in a $d_{x^2-y^2}$ orbital of a free copper ion, and covalency should lower the value found (Q' measures the electric field gradient in the direction of the symmetry axis). The charge distribution in the sulfur complex is very close to symmetrical and was interpreted to indicate a very considerable amount of covalency in the copper-ligand bond.

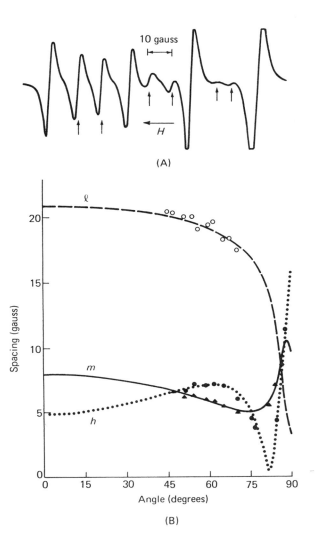

FIGURE 9–24 (A) The Q-band spectrum of Cu(acac)$_2$ at $\theta = 87°$. Forbidden lines are indicated by arrows. (B) Angular dependence of the spacing between each pair of forbidden lines ($\Delta m_I = \pm 1$) in the spectrum in (A). The curves are calculated values using $Q' = 3.4 \times 10^{-4}$ cm^{-1}, and the symbols are experimental values. The letters *l*, *m*, and *h* represent low-field, medium-field, and high-field pairs, respectively. [Reprinted with permission from H. So and R. L. Belford, J. Amer. Chem. Soc., *91*, 2392 (1969). Copyright by the American Chemical Society.]

9–10 LINE WIDTHS IN EPR

In this section, we shall discuss briefly a number of factors that influence the epr line width. Many of these are similar to effects we have discussed in nmr.

Broadening due to spin-lattice relaxation results from the interaction of the paramagnetic ions with the thermal vibrations of the lattice. The variation in spin-lattice relaxation times in different systems is quite large. For some compounds it is sufficiently long to allow the observation of spectra at room temperature, while for others this is not possible. Since relaxation times usually increase as the temperature decreases, many salts of the transition metals need to be cooled to liquid N$_2$, H$_2$, or He temperatures before well-resolved spectra are observed.

Spin-spin interaction results from the small magnetic fields that exist on neighboring paramagnetic ions. As a result of these fields, the total field at the ions is slightly altered and the energy levels are shifted. A distribution of energies results, which produces broadening of the signal. Since this effect varies as $(1/r^3)(1 - 3\cos^2\theta)$, where r is the distance between ions and θ is the angle between the field and the symmetry axis, this kind of broadening will show a marked dependence upon the direction of the field. The effect can be reduced by increasing the distance between paramagnetic ions by diluting the salt with an isomorphous diamagnetic material; for example, small amounts of $CuSO_4$ can be doped into a diamagnetic host $ZnSO_4$ crystal.

As in nmr, rapid chemical processes also influence the spectral line widths and appearance. Resonances that are separate in the stopped exchange limit will broaden as the process rate increases, and they will then coalesce to give a weighted-average single resonance. With one-half the width at half height of an organic free radical being typically ~0.1 gauss, significant line broadening occurs for processes with first order rate constants of 5×10^7 sec^{-1}.

Electron spin exchange processes are very common in free radical systems, and these effects drastically influence line width and spectral appearance. In solution, this is generally a bimolecular process in which two radicals collide and exchange electrons. The effects are similar to those observed in nmr and are illustrated in Fig. 9–25 for solutions of varying concentrations of di-t-butyl nitroxide, $[(CH_3)_3C]_2NO$. As the concentration increases on going from (A) to (C), the rate of bimolecular exchange increases and the resonances broaden. The rate constant for this process has been evaluated in the solvent N,N-dimethyl-formamide and was found to be 7×10^9 l mole^{-1} sec^{-1}, a value corresponding to the diffusion controlled limit. As the solution whose spectrum is shown in Fig. 9–25(C) becomes considerably more concentrated, the single line resonance sharpens as in the very fast exchange limit in nmr. This sharpening is referred to as *exchange narrowing*.

Electron transfer between a radical and a diamagnetic species can also occur at a rate that causes line broadening of the epr spectrum. One of the first systems

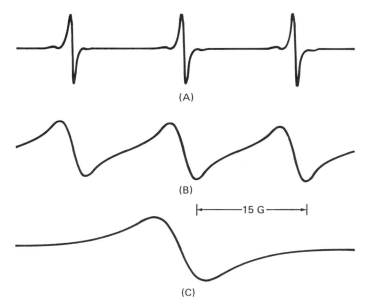

FIGURE 9–25 EPR spectra of the di-t-butyl nitroxide radical in C_2H_5OH at 25° C. (A) 10^{-4} M; (B) 10^{-2} M; (C) 10^{-1} M.

investigated involved the electron exchange between naphthalene and the naphthalene negative ion. A second-order electron transfer rate constant of 6×10^7 l mole^{-1} sec^{-1} was found[25a] in the solvent THF. This is a factor of 100 slower than the diffusion-controlled rate constant; it is thought to be slower because the positive counter-ion of the negative ion radical in the ion pair must also be transferred with the electron.

Many effects cause the line widths of one band to differ from those of another in a given spectrum. We mentioned earlier that the spin density in the CH_3 proton of ethylamine is conformation dependent. The time dependency of this type of process can influence the line widths of different protons in a molecule differently. Rapid interchange between various configurations of an ion pair with an anion or cation radical can also lead to greater line broadening of certain resonances than of others.[25b,26]

9–11 THE SPIN HAMILTONIAN

The spin Hamiltonian operates only on the spin variables and describes the different interactions that exist in systems containing unpaired electrons. It can be thought of as a shorthand way of representing the interactions described above. The epr spin Hamiltonian for an ion in an axially symmetric field (*e.g.*, tetragonal or trigonal) is:

$$\hat{H} = D\left[\hat{S}_z^2 - \frac{1}{3}S(S+1)\right] + g_\parallel \beta H_z \hat{S}_z + g_\perp \beta (H_x \hat{S}_x + H_y \hat{S}_y) \quad (9\text{–}39)$$

$$+ A_\parallel \hat{S}_z \hat{I}_z + A_\perp (\hat{S}_x \hat{I}_x + \hat{S}_y \hat{I}_y) + Q'\left[\hat{I}_z^2 - \frac{1}{3}I(I+1)\right] - g_N \beta_N H_0 \cdot \hat{I}$$

The first term describes the zero-field splitting, the next two terms describe the effect of the magnetic field on the spin degeneracy remaining after zero-field splitting, the terms $A_\parallel$ and $A_\perp$ measure the hyperfine splitting parallel and perpendicular to the unique axis, and Q' measures the changes in the spectrum produced by the quadrupole interaction. All of these effects have been discussed previously. The final term takes into account the fact that the nuclear magnetic moment μ_N can interact directly with the external field $\mu_N H_0 = g_N \beta_N H_0 I$. This interaction can affect the paramagnetic resonance only when the unpaired electrons are coupled to the nucleus by nuclear hyperfine or quadrupole interactions. Even when such coupling occurs, the effect is often negligible in comparison with the other terms.

In the case of a distortion of lower symmetry, the principal g-values become g_x, g_y, and g_z; the hyperfine coupling constants become A_x, A_y, and A_z; and two additional terms need to be included, i.e., $E(\hat{S}_x^2 - \hat{S}_y^2)$ as an additional zero-field splitting and $Q''(\hat{I}_x^2 - \hat{I}_y^2)$ as a further quadrupole interaction. The symbols P and P' are often employed for Q' and Q'', respectively.

The importance of the spin Hamiltonian is that it provides a standard phenomenological way in which the epr spectrum can be described in terms of a small number of constants. Once the values for the constants have been determined from experiment, calculations relating these parameters back to the electronic configurations and the energy states of the ion are possible. It should be pointed out that not all terms in equation (9–39) are of importance for any given system. For a nucleus with no spin, all terms containing I are zero. In the absence of zero-field splitting, the first term is equal to zero.

9–12 MISCELLANEOUS APPLICATIONS

When the epr spectrum for $CuSiF_6 \cdot 6H_2O$, diluted with the corresponding diamagnetic Zn salt, was obtained at 90 K, the spectrum was found to consist of one band with partially resolved hyperfine structure and a nearly isotropic g-value.[27] In a cubic field, the ground state of Cu^{2+} is orbitally doubly degenerate. Although $CuSiF_6 \cdot 6H_2O$ has trigonal rather than cubic symmetry, this orbital degeneracy is not destroyed. Thus, Jahn-Teller distortion will occur. However, there are three distortions with the same energy that will resolve the orbital degeneracy. These are three tetragonal distortions with mutually perpendicular axes (elongation along the three axes connecting *trans* ligands). As a result, three epr transitions are expected, one for each species. Since only one transition was found, it was proposed that the crystal field resonates among the three distortions.[28] When the temperature is lowered, the spectrum becomes anisotropic and consists of three sets of lines corresponding to three different copper ions distorted by three different tetragonal distortions.[29] The transition takes place between 50 and 12 K; the three perpendicular tetragonal axes form the edges of a unit rectangular solid and the trigonal axis is the body diagonal. Other mixed copper salts have been found to undergo similar transitions: $(Cu, Mg)_3La_2(NO_3)_{12} \cdot 24D_2O$ between 33 and 45 K, and $(Zn, Cu)(BrO_3)_2 \cdot 6H_2O$, incomplete below 7 K. The following parameters were reported for $CuSiF_6 \cdot 6H_2O$[30]:

90 K	20 K
$g_\parallel = 2.221 \pm 0.005$	$g_z = 2.46 \pm 0.01$
$g_\perp = 2.230 \pm 0.005$	$g_x = 2.10 \pm 0.01$
$A = 0.0021 \pm 0.0005$ cm^{-1}	$g_y = 2.10 \pm 0.01$
$B = 0.0028 \pm 0.0005$ cm^{-1}	$A_z = 0.0110 \pm 0.0003$ cm^{-1}
	$\left.\begin{array}{c}A_x\\A_y\end{array}\right\} < 0.0030$ cm^{-1}

No quadrupole interaction was resolved. A similar behavior (*i.e.*, a resonating crystal field at elevated temperatures) was detected in the spectra of some tris complexes of copper(II) with 2,2'-dipyridine and 1,10-phenanthroline.[31]

The epr spectrum of the complex $[(NH_3)_5Co-O-O-Co(NH_3)_5]^{5+}$ is an interesting example to demonstrate how structural information can be derived from spin density information and from hyperfine splitting. This complex can be formulated as: (1) two cobalt(III) atoms connected by an O_2^- bridge; (2) cobalt(III) and cobalt(IV) atoms connected by a peroxy, O_2^{2-}, bridge; (3) two equivalent cobalt atoms, owing to equal interaction of one unpaired electron with both cobalt atoms; (4) interaction of the electron with both cobalt atoms, but more with one than, the other.

If (1) were the structure, a single line would result, whereas (2) would give rise to eight lines ($I = ^7/_2$ for Co). Structure (3) would result in 15 lines and (4) in 64. It was found[32] that the spectrum consists of 15 lines, eliminating the unlikely structures (2) and (4) and supporting structures (1) or (3) or a mixture of both. An ^{17}O hyperfine result would be required to determine the importance of structure (1).

A study[33] of the 1:1 adducts of cobalt(II) complexes with dioxygen has led to an internally consistent interpretation of the ^{17}O and ^{59}Co isotropic and anisotropic hyperfine coupling constants. Depending on the ligands attached to cobalt, the adducts are described as consisting of bound O_2 or O_2^-.

One of the advantages of epr is its extreme sensitivity to very small amounts of paramagnetic materials. For example, under favorable conditions a signal for diphenylpicrylhydrazyl (DPPH) radical can be detected if there is 10^{-12} gram of material in the spectrometer. This great sensitivity has been exploited in a study of the radicals formed by heating sulfur. When sulfur is heated, the diamagnetic S_8 ring is cleaved to produce high molecular weight S_x chains that have one unpaired electron at each end. The chains are so long that the concentration of radicals is low, and paramagnetism cannot be detected with a Gouy balance. An epr signal was detected,[34] and the number of unpaired electrons (which is proportional to the area under the absorption curve) was determined by comparing the area of this peak with the area of a peak resulting from a known concentration of added radicals from DPPH. The total number of radicals in the system is thus determined, and since the total amount of sulfur used is also known, the average molecular weight of the species $\cdot SS_xS\cdot$ can be calculated. The radical concentration at 300°C was $1.1 \times 10^{-3} M$, and the average chain length at 171°C was 1.5×10^6 atoms. By studying the radical concentration as a function of temperature, a heat of dissociation of the S—S bond of 33.4 kcal mole^{-1} per bond was obtained.

Copper(II) forms complexes of varying geometries, which have similar electronic spectra and magnetic susceptibilities. Thus, it is often difficult to infer the geometries of these materials in solution or in media other than the solid, where single crystal x-ray studies can be used. A recent ^{17}O study of the five-coordinate adducts formed by various Lewis bases and hexafluoroacetylacetonate copper(II) describes[35] a procedure for determining whether apical or basal isomers of a square pyramid are formed.

This high sensitivity of epr measurements has been of great practical utility in biological systems.[36] Many metalloproteins have been studied in order to determine the metal's oxidation state, the coordination number of the metal, and the kinds of ligands attached. The measurements are generally made on frozen solutions. The interpretation of the results is difficult, and conclusions are based upon analogies between these spectra and those of model compounds. These applications are more appropriately considered in the chapter on the epr spectra of transition metal ions.

There have been several studies reported in which epr has been employed to identify and provide structural information about radicals generated with high energy radiation.[37] The materials O_2^-, ClO, ClO_2, PO_3^{2-}, and ClO_3 are among the many interesting radicals produced by this technique.

Spin traps have been used[38] to detect radicals whose lifetimes or relaxation times are too short to be seen by epr. In a typical experiment, a nitrone or nitroso compound is added to the system in which the radicals are formed. The reactions in equations (9–40) and (9–41) occur to generate a relatively stable nitroxide that can be detected in the epr.

$$\text{(nitrone)} + R\cdot \longrightarrow \text{(nitroxide adduct)} \qquad (9\text{--}40)$$

$$(CH_3)_3C-NO + R\cdot \longrightarrow (CH_3)_3C-N(O\cdot)CH_3 \qquad (9\text{--}41)$$

The epr of the resulting nitroxides show the characteristic 1:1:1 nitrogen hyperfine pattern. In the nitrone, (5,5-dimethyl-1-pyrroline-1-oxide, equation (9–40)) further splitting by the hydrogen labeled H′ in equation (9–40) gives rise to a six-line pattern. Splitting by the methyl group of $(CH_3)_3CN(O)CH_3$ gives an overlapping triplet of quartets. The magnitude of the A_N and A_H values can be used to help identify the radicals trapped. Table 9–3 lists some typical values for spin adducts of 5,5-dimethyl-1-pyrroline-1-oxide.

TABLE 9–3. Hyperfine Splitting (in gauss) for Nitroxides Produced from Radical Trapping by 5,5-Dimethyl-1-pyrroline-1-oxide

R·	a_N	a_H	a_y
1. Methyl	14.31	20.52	
2. Ethyl	14.20	20.49	
3. n-Butyl	14.24	20.41	
4. Benzyl	14.16	20.66	
5. 1-Phenylethyl	14.20	20.49	
6. Phenyl	13.76	19.22	
7. α-Cyanobenzyl	14.39	20.63	
8. Phenoxymethyl	13.79	19.56	
9. 1-Ethoxyethyl	14.20	20.49	
10. Tetrahydrofuranyl	14.12	17.92	
11. Hydroxymethyl	14.66	20.67	
12. 1-Hydroxyethyl	15.03	22.53	
13. 1-Hydroxybutyl	14.89	22.72	
14. 2-Hydroxypropyl	14.58	23.91	
15. Acetyl	14.03	17.87	
16. Benzoyl	13.99	15.57	
17. Aminoformyl	15.23	18.56	
18. Dimethylaminoformyl	14.30	17.37	
19. Trifluoromethyl	13.22	15.54	1.01(3F)
20. Methoxy	13.58	7.61	1.85(1H)
21. Ethoxy	13.22	6.96	1.89(1H)
22. n-Butoxy	13.61	6.83	2.06(1H)
23. t-Butoxy	13.11	7.93	1.97(1H)
24. Benzoyloxy	12.24	9.63	0.87(2H)
25. Superoxide	12.9	6.9	

The couplings are solvent dependent. Oxidation of the nitrone produces a radical with $g = 2.009$ and $A_N = 13$ gauss that can be confused with trapping a radical. Generation and trapping of radicals from solvent oxidation must also be considered. A data base, computer compilation of radicals formed from various traps in a variety of solvents is available.[39]

REFERENCES CITED

1. F. J. Adrian *et al.*, Adv. Chem. Ser., *36*, 50 (1960).
2. A. Carrington, Quart. Rev., *17*, 67 (1963).
3. A. H. Maki and B. R. McGarvey. J. Chem. Phys., *29*, 35 (1958).
4. C. E. Strouse and L. F. Dahl, J. Amer. Chem. Soc., *93*, 6032 (1971).
5. M. J. S. Dewar, "Molecular Orbital Theory of Organic Chemistry," McGraw-Hill, New York, 1969.
6. R. W. Fessenden and R. H. Schuler, J. Chem. Phys., *39*, 2147 (1963).
7. R. S. Drago and H. Petersen, Jr., J. Amer. Chem. Soc., *89*, 3978 (1967).
8. A. D. McLachlan, Mol. Phys., *3*, 233 (1960).
9. J. A. Pople *et al.*, J. Amer. Chem. Soc., *90*, 4201 (1968) and references therein; J. A. Pople and D. L. Beveridge, "Approximate Molecular Orbital Theory," McGraw-Hill, New York, 1970.
10. M. Karplus and G. K. Fraenkel, J. Chem. Phys., *35*, 1312 (1961).
11. a. R. E. Cramer and R. S. Drago, J. Amer. Chem. Soc., *90*, 4790 (1968).
    b. P. D. Sullivan, J. Amer. Chem. Soc., *90*, 3618 (1968).
12. M. Broze, Z. Luz and B. L. Silver, J. Chem. Phys., *46*, 4891 (1967).
13. I. Biddles and A. Hudson, Mol. Phys., *25*, 707 (1973).
14. a. R. S. Drago and H. Petersen, Jr., J. Amer. Chem. Soc., *89*, 5774 (1967).
    b. R. E. Cramer and R. S. Drago, J. Chem. Phys., *51*, 464 (1969).
15. R. W. Fessenden and P. Neta, Chem. Phys. Letters, *18*, 14 (1973).
16. R. Fitzgerald and R. S. Drago, J. Amer. Chem. Soc., *90*, 2523 (1968).
17. a. R. Livingston and H. Zeldes, J. Chem. Phys., *41*, 4011 (1964).
    b. J. C. W. Chien and L. C. Dickerson, Proc. Natl. Acad. Sci. USA, *69*, 2783 (1972).
18. H. M. McConnell and R. W. Fessenden, J. Chem Phys., *31*, 1688 (1959).
19. T. Cole and C. Heller, J. Chem. Phys., *34*, 1085 (1961).
20. H. M. McConnell *et al.*, J. Amer. Chem. Soc., *82*, 766 (1960).
21. R. W. Fessenden and R. H. Schuler, J. Chem. Phys., *43*, 2704 (1965).
22. P. W. Atkins and M. C. R. Symons, J. Chem. Soc., 4794 (1962).
23. N. Hirota, C. A. Hutchison, Jr. and P. Palmer, J. Chem. Phys., *40*, 3717 (1964) and references therein.
24. H. So and R. L. Belford, J. Amer. Chem Soc., *91*, 2392 (1969); R. L. Belford, D. T. Huang and H. So, Chem. Phys. Letters, *14*, 592 (1972) and references therein.
25. a. R. L. Ward and S. I. Weissman, J. Amer. Chem. Soc., *79*, 2086 (1957); T. A. Miller and R. N. Adams, J. Amer. Chem. Soc., *88*, 5713 (1966).
    b. J. R. Bolton and A. Carrington, Mol. Phys., *5*, 161 (1962).
26. J. H. Freed and G. K. Fraenkel, J. Chem. Phys., *37*, 1156 (1962).
27. B. Bleaney and D. J. E. Ingram, Proc. Phys. Soc., *63*, 408 (1950).
28. A. Abragam and M. H. L. Pryce, Proc. Phys. Soc., *63*, 409 (1950).
29. B. Bleaney and K. D. Bowers, Proc. Phys. Soc., *65*, 667 (1952).
30. B. Bleaney, K. D. Bowers and R. S. Trenam, Proc. Roy. Soc. (London) *A228*, 157 (1955).
31. H. C. Allen. Jr., G. F. Kokoszka and R. G. Inskeep, J. Amer. Chem. Soc., *86*, 1023 (1964).
32. E. A. V. Ebeworth and J. A. Weil, J. Phys. Chem., *63*, 1890 (1959).
33. B. Tovrog, D. Kitko and R. S. Drago, J. Amer. Chem. Soc., *98*, 5144 (1976).
34. D. M. Gardner and G. K. Fraenkel, J. Amer. Chem. Soc., *78*, 3279 (1956).
35. D. McMillin, R. S. Drago and J. A. Nusz, J. Amer. Chem. Soc., *98*, 3120 (1976).
36. See, for example, C. S. Yang and F. M. Heuennekens, Biochemistry, *9*, 2127 (1970).
37. P. W. Atkins and M. C. R. Symons, "The Structure of Inorganic Radicals," Elsevier Publishing Co., Amsterdam, 1967.
38. a. E. G. Janzen, Free Radicals Biol., *4*, 115 (1980).
    b. M. J. Perkins, Adv. Phys. Org. Chem., *17*, 1 (1980).
    c. E. Finkelstein, G. M. Rosen and E. J. Rauckman, J. Am. Chem. Soc., *102*, 4994 (1980).
    d. G. M. Rosen and E. J. Rauckman, Proc. Natl. Acad. Sci. U.S.A., *78*, 7346 (1981).
    e. E. G. Janzen, C. A. Evans and J. I.-P. Liu, J. Mag. Res., *9*, 510 (1973).
    f. D. E. Hamilton, R. S. Drago and J. Telser, J. Am. Chem. Soc., *106*, 5353 (1984).

39. A. S. W. Li, A. H. de Haas, L. J. Park, M. S. Watson and C. F. Chignell, "Spin Trap Data Base II (STDBII)" MD1703, LMB, P.O. Box 12233, Research Triangle Park, N.C. 27709.

ADDITIONAL REFERENCES*

J. E. Wertz and J. R. Bolton, "Electron Spin Resonance," McGraw-Hill, New York, 1972.
C. P. Poole and H. A. Farach, "The Theory of Magnetic Resonance," Wiley-Interscience, New York, 1972.
A. Carrington and A. D. McLachlan, "Introduction to Magnetic Resonance," Harper & Row, New York, 1967.
C. P. Poole, Jr., "Electron Spin Resonance—A Comprehensive Treatise on Experimental Techniques," Interscience Publishers, New York, 1967.
B. A. Goodman and J. B. Raynor, Adv. Inorg. Chem. Radiochem., *13*, 135 (1970).
W. Weltner, "Magnetic Atoms in Molecules," Dover Publications, Mineola, N.Y., 1989.
A. Carrington, Quart Rev., *17*, 67 (1963).
N. M. Atherton, "Electron Spin Resonance," Halsted Press, London, 1973.
P. W. Atkins and M. C. R. Symons, "The Structure of Inorganic Radicals," Elsevier Publishing Co., Amsterdam, 1967.
C. P. Slichter, "Principles of Magnetic Resonance," Harper & Row, New York, 1963.
R. M. Golding, "Applied Wave Mechanics," Van Nostrand Reinhold, New York 1964.
A. Abragam, "The Principles of Nuclear Magnetism," Clarendon Press, Oxford, 1961.
G. E. Pake and T. L. Estle, "Paramagnetic Resonance," 2nd ed., W. A. Benjamin, New York, 1973.
H. Fisher, "Magnetic Properties of Free Radicals," Landolt-Bernstein Tables, New Series Group II, Vol. I, Springer-Verlag, Berlin, 1965.

EXERCISES

1. Convert the derivative curves below to absorption curves:

   (A)   (B)

2. a. How many hyperfine peaks would be expected from delocalization of the odd electron in dibenzene chromium cation onto the rings?

   b. Using a procedure similar to that in Fig. 9–6, explain how the number of peaks arises and what the relative intensities would be.

---

*For references heavily oriented toward the epr of transition metal ions, see Chapter 13.

3. **a.** Copper(II) acetate is a dimer, and the two copper atoms are strongly interacting. The epr spectrum consists of seven lines with intensity ratios $1:2:3:4:3:2:1$. Copper nuclei have an I value of $3/2$, and copper acetate consists of a ground state that is a singlet and an excited state that is a triplet. Explain the number and relative intensity of the lines in the spectrum. [For answer, see B. Bleaney and K. D. Bowers, Proc. Roy. Soc. (London), *A214*, 451 (1952).]

   b. What would you expect to happen to the signal intensity as a sample of copper acetate is cooled? Why?

4. Predict the epr spectrum for $(SO_3)_2NO^{2-}$.

5. The mono negative ion 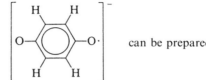 can be prepared.

   a. How many lines are expected in the spectrum, and what would be the relative intensities of these?

   b. What evidence would you employ and what experiments could be carried out to indicate electron delocalization onto the oxygen?

   c. The magnitude of a_H in this material is 2.37 gauss. Compare the spin density on hydrogen in this molecule with that on a hydrogen atom.

   d. How would the sign of the proton hyperfine coupling constant indicate whether the odd electron was in a sigma- or pi-molecular orbital?

   e. Using the value of a_H given above and the fact that the unpaired electron is in the π-system, calculate the spin density on the nearest neighbor carbon.

6. The ^{13}C hyperfine coupling in the methyl radical is 41 gauss, and the proton hyperfine coupling is 23 gauss. Sketch the spectrum expected for $^{13}\cdot CH_3$ radical. [For answer, see T. Cole *et al.*, Mol. Phys., *1*, 406 (1958).]

7. Assume that all hyperfine lines can be resolved and sketch the spectrum for the chlorobenzene anion radical.

8. Assuming all other factors constant, would line broadening be greater for a bimolecular process with a rate constant of 10^7 or with a rate constant of 10^{10}?

9. How many lines would you expect in the epr spectrum of $(CN)_5CoO_2Co(NH_3)_5$? Explain.

10. The spectrum below is obtained for the NH_2 radical:

    a. Convert it to an absorption spectrum.

    b. How could you determine whether the larger or smaller splitting is due to hydrogen?

    c. Assume that the larger splitting is due to nitrogen. Construct a diagram similar to Fig. 9–10 to explain the spectrum.

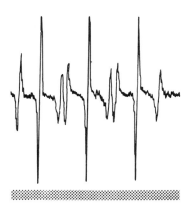

11. a. How many lines would you expect in the spectrum of the hypothetical molecule SCl_3 (I for $S = 0$ and $Cl = {}^3/_2$)?

   b. Using a procedure similar to that in Fig. 9–6 and Fig. 9–7, explain how this number arises and indicate the transitions with arrows. State what the expected relative intensities would be.

12. The epr spectrum of the cyclopentadiene radical ($C_5H_5\cdot$) rapidly rotating in a single crystal of cyclopentadiene is given below.

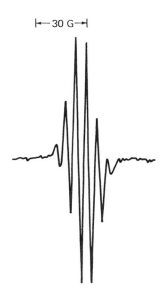

   a. Write the appropriate spin Hamiltonian.

   b. Interpret the spectrum.

13. Interpret the epr spectrum of $\cdot CH_2OH$ given below.

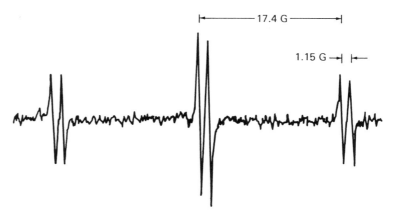

14. The epr spectrum of $C_6H_5Ge(CH_3)_3^-$ is given below. Interpret this spectrum, given the fact that all of the splittings arise from the phenyl ring protons. Calculate the a-values.

15. Given below is the epr spectrum of

Write the spin Hamiltonian, interpret the spectrum, and report the a-values.

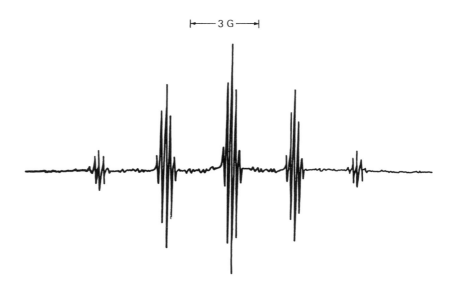

16. a. Interpret the epr spectrum given below and calculate the *a*-value(s) for the substituted nitrosyl nitroxide,

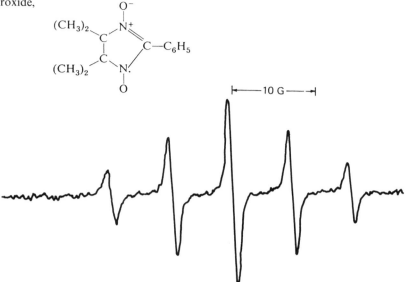

b. What can you conclude about the delocalization of the unpaired electron?

17. The epr spectrum of the potassium salt of the biphenyl anion,

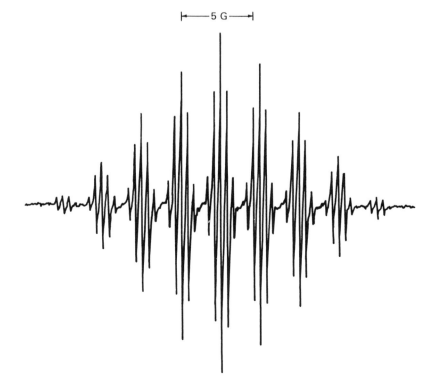

is given as follows. Interpret the spectrum and calculate the *a*-values.

406 Chapter 9 Electron Paramagnetic Resonance Spectroscopy

18. The epr spectrum of the pyrazine anion, 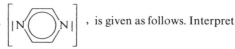, is given as follows. Interpret the spectrum and calculate the *a*-value(s).

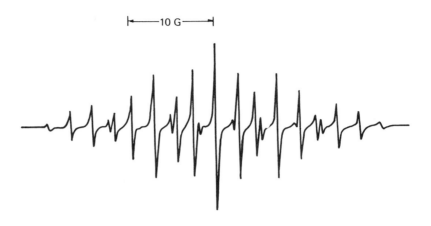

19. Below is the epr spectrum of a sample of S_2^- that has 40% ^{32}S ($I = 0$) nuclei and 60% ^{33}S ($I = 3/2$) nuclei. Interpret the spectrum and determine *a* for ^{33}S.

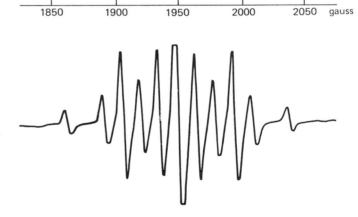

20. McConnell's relation allows a rough prediction of the magnitude of proton hyperfine coupling constants on conjugated organic systems by performing a Hückel m.o. calculation on the system. The hyperfine constant for the *i*th proton, a_i, is given by $a_i = Q\rho_i$, where $\rho_i = C_{ji}^2$. C_{ji} is the coefficient of the various carbon $2p_z$-atomic orbitals in the molecular orbital containing the unpaired electron.

   a. The carbon $2p_z$ atomic orbitals that make up the π-system are orthogonal to the C—H sp^2-sigma bond. Why then, does any unpaired electron density reside on the proton?

   b. The molecular orbital scheme for benzene is:

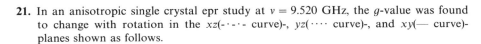

In the benzene anion, the unpaired electron can be in either ψ_4 or ψ_5. A Hückel m.o. calculation for these wave functions gives

$$\psi_4 = \frac{1}{2}(\varphi_2 - \varphi_3 + \varphi_5 - \varphi_6)$$

$$\psi_5 = \frac{1}{\sqrt{12}}(2\varphi_1 - \varphi_2 - \varphi_3 + 2\varphi_4 - \varphi_5 - \varphi_6)$$

In *p*-xylene, the degeneracy of these two m.o.'s is lifted, with ψ_4 lower in energy. Use McConnell's relation and calculate the proton hyperfine coupling constants for the *p*-xylene anion. Draw the epr spectrum.

21. In an anisotropic single crystal epr study at $v = 9.520$ GHz, the *g*-value was found to change with rotation in the xz(-·-·- curve)-, yz(···· curve)-, and xy(— curve)- planes shown as follows.

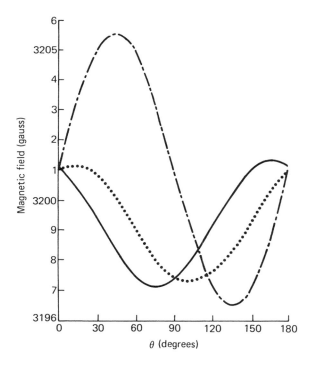

The field position for resonance, H, is given. This is converted to a *g*-value using $\Delta E = h\nu$ and equation (9–2). One obtains $g_{\text{eff}} = \dfrac{h\nu}{\beta H} = \dfrac{6801.9}{H}$ for a 9.520-GHz microwave frequency.

a. Interpolate from the plot and evaluate all of the elements of the g^2-tensor.

b. What would you learn by diagonalizing the g^2-tensor?

c. What steps are required to diagonalize the g^2-tensor?

d. What steps are required to obtain the direction cosine matrix?

e. Write the spin Hamiltonian.

The Electronic Structure and Spectra of Transition Metal Ions

10

Introduction

The subject of this chapter has also been the topic of several textbooks.[1-12] Here we shall present an overview of the electronic structure of transition metal ions. In so doing, we will develop some important ideas for the understanding of the spectroscopy of transition metal ion complexes—our main objective. Transition metal ion systems are covered separately in this book because their unpaired electrons introduce several complications. As is so often the case, these complicating factors, when understood, provide a wealth of information. The complications involve electron-electron interactions, spin-orbit coupling, and the influence that a magnetic field has on systems with unpaired electrons. We have discussed many of these topics earlier, but their full implication is best demonstrated with examples from transition metal ion chemistry.

Free Ion Electronic States

10–1 ELECTRON-ELECTRON INTERACTIONS AND TERM SYMBOLS

There are numerous ways in which one or more electrons can be arranged in the five d orbitals of a gaseous metal ion. We can indicate the energy differences arising from different interelectronic repulsions and different orbital angular momenta for these various arrangements with *term symbols*. Any one term symbol groups together all of the degenerate arrangements in the gaseous ion. The simplest case to consider first is d^1. There are five ways to arrange an electron with $m_s = +^1/_2$ in the five d orbitals. Each arrangement is called a *microstate configuration*.

In the absence of external electric and magnetic fields, the five microstates are degenerate, and there are five others that are also degenerate with these, corresponding to $m_s = -^1/_2$. *These ten microstates comprise the tenfold degeneracy of the so-called ^2D term (vide infra).*

The ground state term for any d^n configuration can be deduced by arranging the electrons in the d orbitals, filling those with the largest m_ℓ values first and not pairing up any electrons until each orbital has at least one; i.e., Hund's rules are obeyed. The m_ℓ values of the orbitals containing electrons may be algebraically summed to produce the L value for the term. More completely, the m_ℓ quantum number for an individual electron is related to a vector with component $m_\ell(h/2\pi)$ in the direction of an applied field. The M_L value is the sum of the one-electron m_ℓ values. Vector coupling rules demand that M_L have values of $L, (L-1), \ldots, -L$, so we deduce that the maximum M_L value is given by the value of L. The following letters are used to indicate the L values: S, P, D, F, G, H, I corresponding to $L = 0, 1, 2, 3, 4, 5, 6$, respectively. The $2L + 1$ arrangements refer to the orbital degeneracy and are described by M_L with values of $L, (L-1), \ldots, -L$. As mentioned earlier, the ^2D term symbol describes the d^1 case. It is tenfold degenerate with a fivefold orbital degeneracy corresponding to M_L values of 2, 1, 0, -1, -2. In the d^1 ion, the ground ^2D term is the only one arising from the $3d$ orbitals.

The spin multiplicity of a state is defined by $2S + 1$ (S, in analogy with L, is the largest possible M_S, where $M_S = \Sigma m_s$) and is indicated by the superscript to the upper left of the term symbol. The spin multiplicity refers to the number of possible projections of $\mathbf{S}$ along a magnetic field; e.g., when $S = 1$, the multiplicity of three refers to $M_S = 1, 0, -1$ (giving the z-component of spin angular momentum aligned with, perpendicular to, and opposed to the field). *The total degeneracy of a term is given by $(2L+1)(2S+1)$.* The value of the S quantum number for the term (or state) is given by the maximum M_S, which equals the sum of the m_s values of all unpaired electrons. *Complete subshells contribute nothing to L or S, because the sum of the m_s and the m_ℓ values is zero.*

Next, consider the d^2 configuration. There are 45 ways to arrange two electrons with $m_s = \pm 1/2$ in the five d orbitals. Using the procedures described above for the microstate $\boxed{1\ |\ 1\ |\ \ |\ \ |\ \ }$, we see that $L = 3$ and $S = 1$. This leads to a ^3F ground term that is 21-fold degenerate in the absence of spin-orbit coupling. The other 24 microstates comprise higher energy (excited) states; i.e., electron-electron repulsions are larger for these states. All of the terms for a d^n configuration can be found by constructing a table like that shown in Table 10–1 for a d^2 ion. To be systematic, we can begin construction of the table with the row $M_L = 4$. This M_L value can be obtained only by having two electrons in $m_\ell = 2$, and this can be done only if the spins are paired up. The resulting microstate is abbreviated as 2^+2^-. The microstate is placed in the $M_S = 0$ column. Next, ways to obtain $M_L = 3$ are shown. The possibilities are $2^+1^+, 2^+1^-, 2^-1^+$, and 2^-1^-, corresponding to M_S values of $+1, 0, 0$, and -1, respectively. The procedure is repeated for M_L values of 2, 1, and 0. The microstates corresponding to negative M_L values are not indicated in the table. They are obtained by multiplying the M_L values above for the positive M_L microstates by -1; e.g., for $M_L = -3$, the possibilities are $-2^+ - 1^+, -2^+ - 1^-, -2^- - 1^+$, and $-2^- - 1^-$. Note that, for example, 2^+1^- is not distinct from 1^-2^+ since the order in which we list the electrons is irrelevant; but it is distinct from 2^-1^+ because in this latter microstate the electron with $m_\ell = 2$ no longer has $m_s = +1/2$.

Starting with the highest M_L value, we can conclude that there must be a ^1G term or state and that it has M_L components of 4, 3, 2, 1, 0, -1, -2, -3, -4. A box is used to set these configurations apart. The choice of the $M_L = 3$

TABLE 10–1. Microstates for a d^2 Ion for Positive M_L Values

M_L \ M_S	(+ +) +1 (↑↑)	(+ −) 0 (↑↓)	(− −) −1 (↓↓)
4		$\boxed{2^+2^-}$	
3	(2^+1^+)	$\boxed{2^+1^-}$ (2^-1^+)	(2^-1^-)
2	(2^+0^+)	$\boxed{2^+0^-}$ (2^-0^+) $\left(1^+1^-\right)$	(2^-0^-)
1	$(2^+ -1^+)$ $[1^+0^+]$	$\boxed{2^+ -1^-}$ $(2^- -1^+)$ $\left(1^+0^-\right)$ $[1^-0^+]$	$(2^- -1^-)[1^-0^-]$
0	$(2^+ -2^+)[1^+ -1^+]$	$\boxed{2^+ -2^-}$ $(2^- -2^+)$ $\left(1^+ -1^-\right)$ $[1^- -1^+]$ 0^-0^+	$(2^- -2^-)[1^- -1^-]$

configuration for the ^1G term is arbitrary because we are only bookkeeping with this procedure. The actual wave function for this component of the ^1G term is a linear combination* of the two microstates indicated for $M_L = 3$. Raising and lowering operators can be employed[1] to produce the wave function for given values of L, M_L, S, and M_S. The same is true whenever there is a choice of microstates. Now, we proceed to the next highest M_L value that remains, namely, $M_L = 3$. With $M_S = +1, 0, -1$ components, it can be deduced that there must be a ^3F term. This term will be 21-fold degenerate; the 12 microstates, with non-negative M_L values, arbitrarily assigned to it are enclosed with parentheses in Table 10–1. Next, we come to a state with $L = 2$, which must be a singlet (i.e., $S = 0$). The microstates arbitrarily assigned to this ^1D term are circled. Next, we enclose with brackets those microstates of the ^3P term. The remaining term is ^1S. Each of these terms constitutes a degenerate set of states, and each term differs in energy from any other.†

---

*Each of the microstates is really an abbreviation for a determinantal wave function, e.g.,

$$2^+1^- = \frac{1}{\sqrt{2}} \begin{vmatrix} 2^+(1) & 1^-(1) \\ 2^+(2) & 1^-(2) \end{vmatrix}$$

†One often finds the individual eigenfunctions of a term referred to as states. The entire term is also referred to as a state. The difference is usually obvious. If it is not, we shall use *term* or *level* to describe the entire collection of degenerate states and *component states* for the individual states.

The energies for all of these terms can be calculated[1] and expressed with the Condon-Shortley parameters F_0, F_2, and F_4. These parameters are abbreviations for the various electron repulsion integrals of the ion. The energy expression for any term as a function of these parameters is independent of the metal ion. The magnitude of the parameters, on the other hand, varies with the metal ion. For example, $E(^3F) = F_0 - 8F_2 - 9F_4$ and $E(^3P) = F_0 + 7F_2 - 84F_4$. The transition energy $^3P - {}^3F$ is the difference between the energies of these two terms of $15F_2 - 75F_4$. Similar expressions exist for all the other transitions involving terms of the gaseous ion. The entire spectrum can be fit with the parameters F_2 and F_4. This is true for any d^2 ion.

In the V(III) ion, the ^3F-^3P transition occurs at 13,000 cm^{-1} and the ^3F-^1D transition occurs at 10,600 cm^{-1}. Solving the two simultaneous equations

$$E(^3P - {}^3F) = 15F_2 - 75F_4 = 13,000 \text{ cm}^{-1} \quad (10\text{–}1)$$

$$E(^1D - {}^3F) = 5F_2 + 45F_4 = 10,600 \text{ cm}^{-1} \quad (10\text{–}2)$$

one obtains $F_2 = 1310$ cm^{-1} and $F_4 = 90$ cm^{-1}.

The *Racah* parameter redefined the empirical Condon-Shortley parameters so that the separation between states having the maximum multiplicity is a function of only a single parameter, B:

$$B = F_2 - 5F_4 \quad (10\text{–}3)$$

A second parameter, C, is needed to express the energy difference between terms of different multiplicity:

$$C = 35F_4 \quad (10\text{–}4)$$

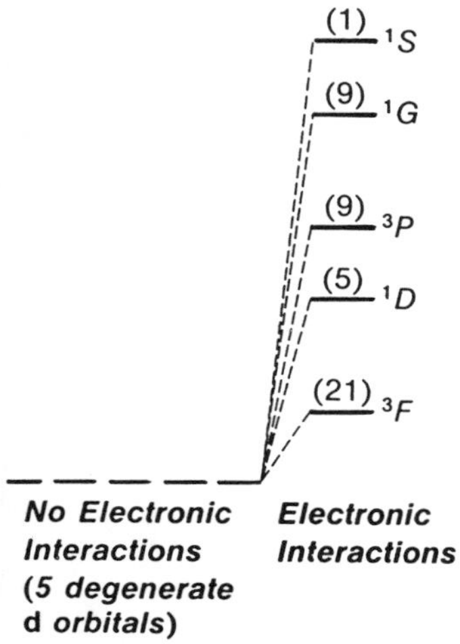

Rearranging, we obtain $F_2 = B + C/7$ and $F_4 = C/35$. Substituting these into equations (10–1) and (10–2), we obtain:

$$E(^3P - {}^3F) = 15B$$

$$E(^1D - {}^3F) = 5B + 2C$$

For V(III), we find $B = 866$ cm^{-1} and $C/B = 3.6$.

In summary, the results of the electron-electron interactions in a d^2 ion give rise to a ^3F ground term and the excited states arising from the d orbitals shown in Fig. 10–1. The degeneracy of each term is indicated in parentheses.

FIGURE 10–1 The terms arising from the electron-electron interactions in a d^2 gaseous ion. The number in parentheses indicates the degeneracy of each level (excluding any spin-orbit coupling).

No Electronic Interactions (5 degenerate d orbitals)

Electronic Interactions:
- (1) ^1S
- (9) ^1G
- (9) ^3P
- (5) ^1D
- (21) ^3F

TABLE 10–2. Free Ion Terms for Various d^n Ions

n		Terms
d^1 d^9	^2D	
d^2 d^8	^3F ^3P ^1G ^1D ^1S	
d^3 d^7	^4F ^4P ^2H ^2G ^2F ^2D ^2D ^2P	
d^4 d^6	^5D ^3H ^3G ^3F ^3F ^3D ^3P ^3P ^1I ^1G ^1G ^1F ^1D ^1D ^1S ^1S	
d^5	^6S ^4G ^4F ^4D ^4P ^2I ^2H ^2G ^2G ^2F ^2F ^2D ^2D ^2D ^2P ^2S	

By using procedures similar to those employed in Table 10–1, we can determine the terms arising from various d^n ions. The results for $n = 1$ to $n = 9$ are presented in Table 10–2.

The d^9 configuration is for many purposes considered to be equivalent to the d^1 case, if we think in terms of the degenerate states that would arise from degeneracies associated with the positive hole that exists in the d^9 case. It may help to think of d^9 as being a d^{10} case with a positron that can annihilate any one of the 10 electrons. This concept is referred to as the *hole formalism*. By the same token, the following equivalences arise:

$$d^2 \approx d^8$$

$$d^3 \approx d^7$$

$$d^4 \approx d^6$$

10–2 SPIN-ORBIT COUPLING IN FREE IONS

As discussed in Chapter 9 (Fig. 9–18), the coupling of the magnetic dipole from the electron spin moment with the orbital moment, $\vec{L} \cdot \vec{S}$, is spin-orbit coupling. Variations in the amount of spin-orbit coupling in the different electronic configurations also lead to splitting of the terms derived so far. Two schemes are widely used to deal with this effect: the so-called *Russell-Saunders* or $L \cdot S$ *coupling* scheme, and the $j \cdot j$ *coupling* scheme. When the electron-electron interactions give rise to large energy splittings of the terms compared to the splittings from spin-orbit coupling, the former scheme is used. With the $L \cdot S$ scheme, we essentially treat the effects of spin-orbit coupling as a perturbation on the individual term energies. On the other hand, the $j \cdot j$ coupling scheme is used when a large splitting results from spin-orbit coupling and the electron-electron interactions are sufficiently small to be treated as a perturbation on the spin-orbit levels. The $j \cdot j$ scheme is applied to the rare earth elements as well as the third row transition metal ions. Briefly, in the $j \cdot j$ scheme the spin angular momentum of an individual electron couples with its orbital momentum to give a resultant angular momentum, $\vec{j}$, for that electron. The indivdual $\vec{j}$'s are coupled to produce the resultant vector $\vec{J}$ for the system, labeling the overall angular momentum for the atom. The $L \cdot S$ coupling scheme is applicable to most first row transition metal ions, and we shall discuss this scheme in more detail. We previously mentioned that the individual orbital angular momenta of the electrons, m_ℓ, couple to produce a resultant angular momentum indicated by $\vec{L}$. The spin moments couple to give $\vec{S}$. The resultant angular momentum including spin-orbit coupling is given by $\vec{J}$, and *the corresponding quantum number J can take on all consecutive integer values ranging from the absolute values of $|L - S|$ to $|L + S|$. For the ground term, the minimum value of J refers to the lowest energy state of the manifold if the subshell (e.g., the d orbitals) is less than half filled; and the maximum value of J refers to the lowest energy state when the subshell is more than half filled.* If the shell is half filled, there is only one J value because $L = 0$.

Our discussion in this section can be made clearer by working out some examples. The box diagram for the ground state of the carbon atom is:

1s	2s	2p +1 0 −1
↑↓	↑↓	↑ ↑

The value of the L quantum number, obtained by adding the m_l values for all the electrons in incomplete orbitals, is 1 for carbon: $L = +1 + 0 = 1$. This L value corresponds to a P state. The sum of the spin quantum numbers ($m_s = \pm{}^1/_2$) for all unpaired electrons, is 1 for carbon: $S = {}^1/_2 + {}^1/_2 = 1$. The multiplicity is three, and the term symbol for the ground state is ^3P. The values for J (given by $|L − S|, \ldots, |L + S|$) are $|L − S| = 1 − 1 = 0, |L + S| = 1 + 1 = 2$, so $J = 0$, 1, 2 (one being the only integer needed to complete the series). The subshell involved is less than half filled, so the state with minimum J has the lowest energy. The term symbol for the ground state of carbon is $^3P_0$, with the zero subscript referring to the J value.

The box diagram for the ground state of V^{3+} is

+2	+1	0	−1	−2
↑	↑			

with term symbol $^3F_2$ ($L = 3, S = 1, J = 4, 3, 2$). An excited state for this species is represented by | ↑↓ | | | | | ; this microstate belongs to the term with term symbol $^1G_4$ ($L = 4, S = 0, J = 4$). For nitrogen with a box diagram | ↑ | ↑ | ↑ |, $L = 0, S = {}^3/_2$, and $J = {}^3/_2$ so the term symbol $^4S_{3/2}$ results. Note that with $L = 0$ there is only one J value because $|L + S| = |L − S| = {}^3/_2$.

For practice, one can determine the following term symbols for the ground states of the elements in parentheses: $^3P_2$ (S), $^2P_{3/2}$ (Cl), $^3F_2$ (Ti), $^5D_0$ (Cr), $^3F_4$ (Ni), $^3P_0$ (Si), $^4S_{3/2}$ (As), and $^4I_{9/2}$ (Pr).

Two parameters, ξ and λ, are commonly used to describe the magnitude of the energy of the spin-orbit coupling interaction. The parameter ξ *is used to describe the spin-orbit coupling energies for a single electron*. It measures the strength of the interaction between the spin and orbital angular momenta of a single electron of a particular microstate, and is thus a property of the microstate and not of the term. The operator is $\xi \vec{l} \cdot \vec{s}$. The value of ξ is given by

$$\xi = \frac{Z_{\text{eff}} e^2}{2m^2 c^2} \langle r^{-3} \rangle \qquad (10\text{--}5)$$

where $\langle r^{-3} \rangle$ is the average value of r^{-3}, m is the mass of the electron, c is the speed of light, and Z_{eff} is the effective nuclear charge.

When the parameter λ *is used to describe the term*, the operator now becomes $\lambda \vec{L} \cdot \vec{S}$. The values of λ and ξ are related by

$$\lambda = \pm \xi / 2S \qquad (10\text{--}6)$$

The parameter ξ is fundamentally a positive quantity. If the shell is less than half filled, the sign of λ is positive; if it is more than half filled, λ is negative. This makes sense if we think in terms of positive holes requiring the sign of equation (10–5) to change for the configurations of more than half-filled shells. In summary, then, for a shell that is less than half filled, the lowest value of J corresponds to the lowest energy and λ is positive.

An equivalent operator form for $\hat{L}\cdot\hat{S}$ is given by $1/2(\hat{J}^2 - \hat{L}^2 - \hat{S}^2)$ when the states can be characterized by quantum numbers L, S, and J. The spin-orbit contribution to the energy of any level is then given by:

$$\tfrac{1}{2}\lambda[J(J+1) - L(L+1) - S(S+1)] \tag{10–7a}$$

The energy difference between two adjacent spin-orbit states in a term is given by

$$\Delta E_{J,J+1} = \lambda(J+1) \tag{10–7b}$$

For example, the energy separation between the $J = 3$ and $J = 4$ states of a term is 4λ. Furthermore, in the $L \cdot S$ scheme, spin-orbit splitting occurs so as to preserve the center of gravity of the energy of the term, i.e., the average energy remains the same. The ground state for a d^2 system, ^3F, has J values of 4, 3, and 2, with 2 lowest since the shell is less than half filled. The complete ground state term symbol is $^3F_2$. The ^1D excited state has only one possible J state, equal to 2.

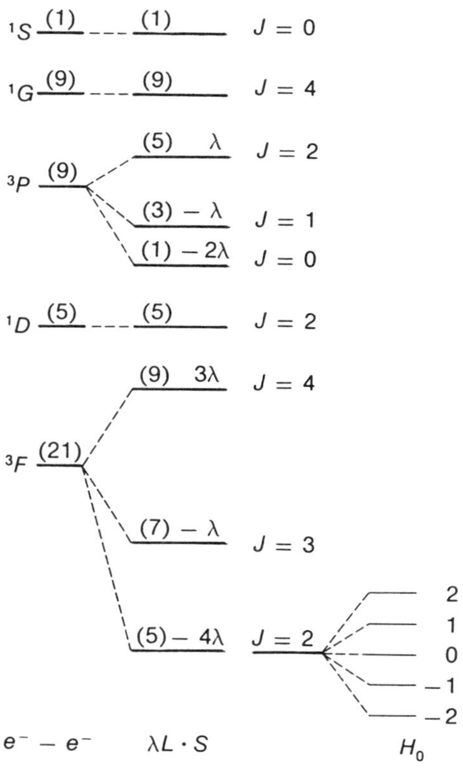

FIGURE 10–2 Spin-orbit states from a d^2 configuration. The splitting of the $^3F_2$ state by a magnetic field H_0 is indicated on the far lower right.

For the excited state, ^3P, we have $J = 0$, 1, and 2, whereas ^1G has only $J = 4$ and ^1S has only $J = 0$.

Now, using equation (10–7a), we can calculate the spin-orbit contribution to the energies of all the J states. For the ground level of ^3F where $J = 2$, we obtain $(1/2)\lambda[2(2 + 1) - 3(3 + 1) - 1(1 + 1)] = -4\lambda$. This result is summarized in Fig. 10–2 along with the results of similar calculations of the effects of $\lambda L \cdot S$ on all the states of a d^2 system. Not all the degeneracy is removed by spin-orbit coupling, and the remaining degeneracy, corresponding to integer values of M_J from J to $-J$, is indicated in parentheses over each level. Note that equation (10–7b) is obeyed and the center of gravity is preserved. *For example, in the ^3P term, the degeneracy times the energy change gives $5\lambda - 3\lambda - (1)(2)\lambda = 0$. The degeneracy of the individual J states is removed by a magnetic field.* The splitting into the M_J states is indicated only for the ground $J = 2$ term in Fig. 10–2.

Crystal Fields

10–3 EFFECTS OF LIGANDS ON THE d ORBITAL ENERGIES

We usually do not work with gaseous ions but with transition metal ions in complexes. There are two crystal field type approaches to determine the effects that these ligands in a transition metal ion complex have on the energies of the d orbitals. The metal ion electrons in a complex undergo interelectronic repulsions and are also repelled by the electron density of the Lewis base (ligand). When the repulsions between the metal electrons and the electron density of the ligands is small compared to interelectronic repulsions, the so-called *weak field approach* is employed. When the ligands are strong Lewis bases, the ligand electron–metal electron repulsions are larger than the interelectron repulsions and the *strong field approach* is employed.

The basis set used in these problems can be the orbitals represented by complex wave functions whose angular dependences are given by the spherical harmonics

$$Y_2^0 = (5/8)^{1/2}(3\cos^2\theta - 1) \cdot (2\pi)^{-1/2}$$
$$Y_2^{\pm 1} = (15/4)^{1/2} \sin\theta \cos\theta \cdot (2\pi)^{-1/2} e^{\pm i\varphi}$$
$$Y_2^{\pm 2} = (15/16)^{1/2} \sin^2\theta \cdot (2\pi)^{-1/2} e^{\pm 2i\varphi}$$

Alternatively, the real trigonometric wave functions, which are linear combinations of the complex orbitals taken to eliminate i, can be employed. These are given by:

$$d_{z^2} = |0\rangle \qquad (d_{z^2} \text{ is really } d_{(z^2 - r^2/3)})$$
$$d_{yz} = (i/\sqrt{2})[|-1\rangle + |1\rangle]$$
$$d_{xz} = (1/\sqrt{2})[|-1\rangle - |1\rangle]$$
$$d_{xy} = -(i/\sqrt{2})[|2\rangle - |-2\rangle]$$
$$d_{(x^2 - y^2)} = (1/\sqrt{2})[|2\rangle + |-2\rangle]$$

In the weak field approach, the free-ion term state eigenfunctions (which take into account the interelectronic repulsions in the d-manifold) are employed as the basis set. As an example, for the ^3F term, the wave functions corresponding to $M_L = \pm 3, \pm 2, \pm 1, 0$ are used. They are abbreviated as $|3\rangle$, $|2\rangle$, etc. The Hamiltonian is given as:

$$\hat{H} = \hat{H}_0 + \hat{V}$$

where $\hat{H}_0$ is the free-ion Hamiltonian and $\hat{V}$ is taken as a perturbation from the ligand electron density on $\hat{H}_0$. The perturbation, $\hat{V}$, has a drastically simplified form incorporating only the electrostatic repulsion from the ligands, which are represented simply as point charges. For an octahedral complex, the perturbation is given by:

$$\hat{V} = \sum_{i=1}^{6} eZ_i/r_{ij} \qquad (10\text{–}8)$$

where e is the charge on the electron, Z_i is the effective charge on the ith ligand, and r_{ij} is the distance from the d-electron (this is a d^1 problem) to the ith charge. This is to be compared to the full Hamiltonian given in Chapter 3. Using the simplified Hamiltonian [equation (10–8)] leads to *crystal field theory*. It is to be emphasized that this formulation of the problem simply describes the electrostatic repulsion between the d-electrons and the ligand electron density, and as such can tell us directly only about *relative* energies of the d orbitals.

In order to evaluate the integrals $\langle M_L | \hat{V} | M_L' \rangle$, $\hat{V}$ is written in a form that facilitates integration.[2,3] When this is done, many quantities related to the radial part of the matrix elements appear in the secular determinant with the form $\frac{1}{6}(Ze^2\overline{r_2^{-4}}a^{-5})$. Here, $\overline{r_2^{-4}}$ corresponds to the mean fourth-power radius of the d-electrons of the central ion, a is the metal-ligand distance, and Ze has the same units as e. This radial quantity is referred to as *10Dq* and has units of energy.

It is informative to write the secular determinant for an octahedral complex with this Hamiltonian acting upon a d^1 configuration. Employing the complex d-orbital basis set, we obtain

$$\begin{vmatrix} & |2\rangle & |1\rangle & |0\rangle & |-1\rangle & |-2\rangle \\ |2\rangle & Dq - E & & & & 5Dq \\ |1\rangle & & -4Dq - E & & & \\ |0\rangle & & & 6Dq - E & & \\ |-1\rangle & & & & -4Dq - E & \\ |-2\rangle & 5Dq & & & & Dq - E \end{vmatrix} = 0$$

This gives roots

$$E(|1\rangle) = -4Dq$$
$$E(|-1\rangle) = -4Dq$$
$$E(|0\rangle) = 6Dq$$

and the determinantal equation

$$\begin{array}{c|cc} & |2\rangle & |-2\rangle \\ \hline |2\rangle & Dq - E & 5Dq \\ |-2\rangle & 5Dq & Dq - E \end{array} = 0$$

This determinant is solved to produce two energies: one at $-4Dq$ and one at $6Dq$.

As in a Hückel calculation, the energies can be substituted into the secular equations written from the secular determinant, and the wave functions thus obtained. The results are

$$\psi_4 = -i2^{-1/2}(|2\rangle - |-2\rangle)$$

and

$$\psi_5 = 2^{-1/2}(|2\rangle + |-2\rangle)$$

These are the wave functions for the d_{xy} and $d_{x^2-y^2}$ orbitals, the latter pointing at the ligands and the former in between the ligands. *Note that the octahedral crystal field mixes the $|2\rangle$ and $|-2\rangle$ wave functions and makes it more convenient to employ the real d_{xy} and $d_{x^2-y^2}$ orbitals in the description of the complex.* Since the degeneracy of $|1\rangle$ and $|-1\rangle$ is not removed, the wave functions ψ_1 to ψ_3 can be considered as the real or imaginary combination as is convenient. These results are summarized in Fig. 10–3, where it can be seen that two degenerate sets of orbitals result which are separated by $10Dq$. Thus, we expect one d-d electronic transition for a d^1 system with an energy corresponding to $10Dq$. In O_h symmetry, the three degenerate d_{xz}, d_{yz}, and d_{xy} orbitals transform as t_{2g} and lead to the $^2T_{2g}$ ground state, while d_{z^2} and $d_{x^2-y^2}$ lead to the ^2E_g excited state. The following information is conveyed by these symbols: (1) the symbol T indicates that the state is orbitally triply degenerate and E doubly degenerate; (2) the superscript 2 indicates a spin multiplicity of two, i.e., one unpaired electron; and (3) the g indicates a gerade or symmetric state.

Next, we shall treat a weak field, octahedral d^2 complex. Our initial concern will be with the ^3F term described by the basis set $|3\rangle \ldots |0\rangle \ldots |-3\rangle$. Since our basis set for ^3F consists of all $M_S = 1$ functions, this has been dropped from the symbol and only M_L is indicated; i.e., $|3, 1\rangle$, etc., would be a more complete description. We shall not present the wave functions for these basis functions. Procedures for obtaining these wave functions using the raising and lowering operators are presented by Ballhausen.[1] The secular determinant is given below:

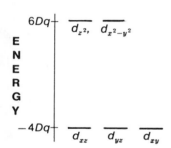

FIGURE 10–3 Splitting of the one-electron d orbitals by an O_h crystal field.

$$\begin{array}{c|ccccccc} & |3\rangle & |2\rangle & |1\rangle & |0\rangle & |-1\rangle & |-2\rangle & |-3\rangle \\ \hline \langle 3| & -3Dq - E & & & & 15^{1/2}Dq & & \\ \langle 2| & & 7Dq - E & & & & 5Dq & \\ \langle 1| & & & -Dq - E & & & & 15^{1/2}Dq \\ \langle 0| & & & & -6Dq - E & & & \\ \langle -1| & 15^{1/2}Dq & & & & -Dq - E & & \\ \langle -2| & & 5Dq & & & & 7Dq - E & \\ \langle -3| & & & 15^{1/2}Dq & & & & -3Dq - E \end{array} = 0$$

The solutions are

$$E_1 = -6Dq \quad E_5 = 2Dq$$
$$E_2 = -6Dq \quad E_6 = 2Dq$$
$$E_3 = -6Dq \quad E_7 = 12Dq$$
$$E_4 = 2Dq$$

The energies* and wave functions for the respective levels are indicated in Fig. 10–4.

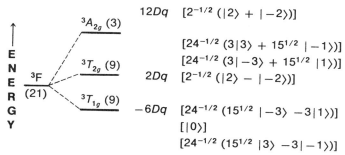

FIGURE 10–4 Splitting of the ^3F term of a d^2 ion by an O_h crystal field.

In this analysis, we have ignored any covalency in the metal-ligand bond. As a result, if we were to attempt a quantitative calculation of Dq, it would differ considerably from that found experimentally. *Ligand field theory* admits to covalency in the bond and treats Dq (and other parameters to be discussed shortly) as an empirical parameter that is obtained from the electronic spectrum. The formulation of the problem in all other respects is identical.

10–4 SYMMETRY ASPECTS OF THE d-ORBITAL SPLITTING BY LIGANDS

As is usually the case when a vastly simplified Hamiltonian is employed, the correct predictions from the method are symmetry determined. For example, we mentioned in Chapter 2 that appropriate combinations of the binary products of the x, y, and z vectors gave the irreducible representations for the d orbitals and their degeneracies. We can use principles already covered (Chapter 2) to illustrate a procedure for deriving all of the states arising from one-electron levels. This procedure can then be extended in a straightforward way to derive all of the states arising for various multielectron configurations in various geometries.

We begin by examining the effect of an octahedral field on the total representation for which the set of d wave functions forms the basis. To determine this total representation, we must find the elements of matrices that express the

---

*Dq is defined for a one-electron system. For a polyelectron system, one should employ a different Dq value for each state. (C. J. Ballhausen and H. B. Gray, "Chemistry of Coordination Compounds," Volume I, ed. A. E. Martell, Van Nostrand Reinhold, Princeton, N.J., 1960.) This is seldom done in practice because of the crude nature of this model.

effect upon our basis set of d orbitals of each of the symmetry operations in the group. The characters of these matrices will comprise the representation we seek. Since all of the d orbitals are gerade, *i.e.* symmetric to inversion, no new information will result as a consequence of the inversion symmetry operation. Thus, we can work with the simpler pure rotational subgroup O instead of O_h. If you need to convince yourself of this, note that in any group containing i (*e.g.*, D_{4h} or C_{3h}), the corresponding rotation group (*e.g.*, D_4 or C_3) has the same irreducible representation for the binary products except for the u and g subscripts in the former group. Recall that the d wave functions consist of radial, spin, and angular (θ and φ) parts. The radial part is neglected, for it is nondirectional and hence unchanged by any symmetry operation. We shall assume that the spin part is independent of the orbital part and ignore it for now. The angle θ is defined relative to the principal axis (*i.e.*, the rotation axis), so it is unchanged by any rotation and can be ignored. Only φ will change; the form of this part of the wave function is given by $e^{im_\ell \varphi}$. (For the d orbitals, $\ell = 2$ and m_ℓ has values 2, 1, 0, -1, -2.) To work out the effects of a rotation by α on $e^{im_\ell \varphi}$, we note that such a rotation causes the following changes:

$$\begin{bmatrix} e^{2i\varphi} \\ e^{i\varphi} \\ e^0 \\ e^{-i\varphi} \\ e^{-2i\varphi} \end{bmatrix} \xrightarrow[\text{by } \alpha]{\text{rotate}} \begin{bmatrix} e^{2i(\varphi+\alpha)} \\ e^{i(\varphi+\alpha)} \\ e^0 \\ e^{-i(\varphi+\alpha)} \\ e^{-2i(\varphi+\alpha)} \end{bmatrix}$$

The matrix that operates on our d-orbital basis set is

$$\begin{bmatrix} e^{2i\alpha} & 0 & 0 & 0 & 0 \\ 0 & e^{i\alpha} & 0 & 0 & 0 \\ 0 & 0 & e^0 & 0 & 0 \\ 0 & 0 & 0 & e^{-i\alpha} & 0 \\ 0 & 0 & 0 & 0 & e^{-2i\alpha} \end{bmatrix}$$

A general form of this matrix for the rotation of any set of orbitals is given by:

$$\begin{bmatrix} e^{\ell i \alpha} & 0 & \cdots & 0 & 0 \\ 0 & e^{(\ell-1)i\alpha} & \cdots & 0 & 0 \\ \vdots & \vdots & \vdots & \vdots & \vdots \\ 0 & 0 & \cdots & e^{(1-\ell)i\alpha} & 0 \\ 0 & 0 & \cdots & 0 & e^{-\ell i \alpha} \end{bmatrix}$$

When $\alpha = 0$, each element is obviously one. We give the trace of this latter matrix without proof* as:

$$\chi(\alpha) = \frac{\sin\left(\ell + \frac{1}{2}\right)\alpha}{\sin\left(\frac{\alpha}{2}\right)} \tag{10–9}$$

---

*The quantities summed form a geometric progression.

where $\alpha \neq 0$. From the trace determined with this formula, we have the character of the representation for any rotation operation. Substituting directly into this formula, we find that the total character for the C_3 rotation of the five d orbitals is given by

$$\chi(C_3) = \frac{\sin\left[\left(2 + \frac{1}{2}\right)\left(\frac{2\pi}{3}\right)\right]}{\sin\left(\frac{2\pi}{3 \times 2}\right)} = \frac{\sin\frac{5\pi}{3}}{\sin\frac{\pi}{3}} = \frac{-\sin\frac{\pi}{3}}{\sin\frac{\pi}{3}} = -1$$

Characters for other rotations can be worked out in a similar way. To obtain the characters when $\alpha = 0$ and $\alpha = 2\pi$, one must evaluate the limit of an indeterminate form, e.g.,

$$\frac{\sin\left(\ell + \frac{1}{2}\right)(2\pi)}{\sin \pi} \to \frac{0}{0}$$

Using l'Hopital's rule, one obtains

$$\chi(0) = 2\ell + 1$$
$$\chi(2\pi) = 2\ell + 1 \text{ for integer } \ell \text{ or}$$
$$-(2\ell + 1) \text{ for half-integer } \ell$$

Thus, for the identity, $\alpha = 0$ and $\chi(E)$ is given by $2\ell + 1$. Referring to the character table for the O point group and using the above formula to determine the characters for the various operations on the five d orbitals, we have:

	E	$6C_4$	$3C_2(=C_4{}^2)$	$8C_3$	$6C_2$
χ_T	5	-1	1	-1	1

Using the decomposition formula, we obtain the result $\chi_T = E + T_2$. Since the d orbitals are gerade, we can write this as

$$\chi_T = E_g + T_{2g}$$

This was the result of our crystal field analysis. By a similar symmetry analysis, the results summarized in Table 10–3 can be obtained.

We could work out the effect of other symmetries on the one-electron levels in a similar way. Alternatively, one can use a correlation table that shows how the representations of the group O_h are changed or decomposed into those of its sub-groups when the symmetry is altered. Table 10–4 contains such information for some of the symmetries commonly encountered in transition metal ion complexes.

With the irreducible representations given in Table 10–3 for various atomic orbitals in O_h symmetry and with the correlation table given in Table 10–4, we can ascertain the irreducible representations of the various orbitals in different symmetry environments. The results for single electrons in various orbitals apply

TABLE 10-3. Representations for Various Orbitals in O_h Symmetry

Type of Orbital	ℓ Value	Irreducible Representation[a]
s	0	a_{1g}
p	1	t_{1u}
d	2	$e_g + t_{2g}$
f	3	$a_{2u} + t_{1u} + t_{2u}$
g	4	$a_{1g} + e_g + t_{1g} + t_{2g}$
h	5	$e_u + 2t_{1u} + t_{2u}$
i	6	$a_{1g} + a_{2g} + e_g + t_{1g} + 2t_{2g}$

[a] The subscripts g and u are determined by the g or u nature of the atomic orbitals involved. When ℓ is even, the orbital is gerade; when ℓ is odd, the orbital is ungerade.

TABLE 10-4. Correlation Table for the O_h Point Group

O_h	O	T_d	D_{4h}	D_{2d}	C_{4v}	C_{2v}	D_{3d}	D_3	C_{2h}
A_{1g}	A_1	A_1	A_{1g}	A_1	A_1	A_1	A_{1g}	A_1	A_g
A_{2g}	A_2	A_2	B_{1g}	B_1	B_1	A_2	A_{2g}	A_2	B_g
E_g	E	E	$A_{1g} + B_{1g}$	$A_1 + B_1$	$A_1 + B_1$	$A_1 + A_2$	E_g	E	$A_g + B_g$
T_{1g}	T_1	T_1	$A_{2g} + E_g$	$A_2 + E$	$A_2 + E$	$A_2 + B_1 + B_2$	$A_{2g} + E_g$	$A_2 + E$	$A_g + 2B_g$
T_{2g}	T_2	T_2	$B_{2g} + E_g$	$B_2 + E$	$B_2 + E$	$A_1 + B_1 + B_2$	$A_{1g} + E_g$	$A_1 + E$	$2A_g + B_g$
A_{1u}	A_1	A_2	A_{1u}	B_1	A_2	A_2	A_{1u}	A_1	A_u
A_{2u}	A_2	A_1	B_{1u}	A_1	B_2	A_1	A_{2u}	A_2	B_u
E_u	E	E	$A_{1u} + B_{1u}$	$A_1 + B_1$	$A_2 + B_2$	$A_1 + A_2$	E_u	E	$A_u + B_u$
T_{1u}	T_1	T_2	$A_{2u} + E_u$	$B_2 + E$	$A_1 + E$	$A_1 + B_1 + B_2$	$A_{2u} + E_u$	$A_2 + E$	$A_u + 2B_u$
T_{2u}	T_2	T_1	$B_{2u} + E_u$	$A_2 + E$	$B_1 + E$	$A_2 + B_1 + B_2$	$A_{1u} + E_u$	$A_1 + E$	$2A_u + B_u$

also to the terms arising from multielectron systems. For example, we can take the ^3F, ^3P, ^1G, ^1D, and ^1S terms of the d^2 configuration and treat them like f, p, g, d, and s orbitals. The g or u subscripts given in Table 10-3 will not apply, but will depend upon the g or u nature of the atomic orbitals involved. Thus, Table 10-3 applies to terms as well as orbitals. The D term, for example, is fivefold degenerate like the five d orbitals, the former being described by a wave function for each of the five M_L values. These wave functions have a Φ part given by $e^{iM_L\varphi}$. Combining Tables 10-3 and 10-4, we can see that the D state of the free ion splits into $E_g + T_{2g}$ states in an octahedral field and into $A_{1g} + B_{1g} + E_g + B_{2g}$ states in a tetragonal D_{4h} field. Similarly, the ^3F term gives rise to $A_{2g} + T_{1g} + T_{2g}$ in an octahedral field and $B_1 + A_2 + 2E + B_2$ in a C_{4v} field.

Next, we shall consider the states that arise from a d^2 configuration when the crystal field is large compared to the interelectronic repulsions.* The interelectronic repulsions are treated as perturbations on the strong field d-electron configurations. In other words, the various crystal field states are identified, eigenfunctions are constructed, and e^2/r_{ij} is used as a perturbation. The terms are readily written for the various d^n configurations. For d^1, the two terms are $^2T_{2g}$ and ^2E_g. The terms that arise from a configuration with an additional electron, d^{n+1}, are given by the direct product of the d^n term symmetries and that of the added electron. For d^2, we have t_{2g}^2, $t_{2g}^1 e_g^1$, and e_g^2 arrangements. Accordingly, for t_{2g}^2, we have

$$t_{2g} \times t_{2g} \text{ which leads to } T_{1g} + T_{2g} + E_g + A_{1g}$$

For $t_{2g}^1 e_g^1$, we have

$$t_{2g} \times e_g \text{ which leads to } T_{1g} + T_{2g}$$

and for e_g^2, we have

$$e_g \times e_g \text{ which leads to } E_g + A_{1g} + A_{2g}$$

Next we have to determine the multiplicity of these terms and show the connection of these strong terms to those of the gaseous ions. The multiplicities are determined by the *method of descending symmetries*.[1,2] By considering the e_g^2 configuration, we can show that when we lower the symmetry from O_h to D_{4h} to remove the degeneracy, the determination of the spin degeneracy becomes straightforward. The correlation table (Table 10–4) shows

$$e_g \begin{array}{c} \nearrow \underline{\qquad} a_{1g} \\ \searrow \underline{\qquad} b_{1g} \end{array}$$
$$O_h \qquad D_{4h}$$

---

*The total degeneracy that exists for a system of q electrons filling a subshell with z-degenerate orbitals (each occupied by two electrons with opposite spin) is given by:

$$\frac{(2z)!}{q!(2z-q)!}$$

For a t_{2g}^2 configuration, we have:

$$\frac{(2 \times 3)!}{2!(6-2)!} = 15$$

For $t_{2g}^2 e_g^1$, we have

$$\left[\frac{6!}{2!(6-2)!}\right]\left[\frac{4!}{1!(4-1)!}\right] = [15] \cdot [4] = 60$$

In addition to the orbital correlation shown above, it is important to remember that states behave in the same way. In D_{4h}, a_{1g}^2 must be singlet because of the Pauli principle. For example, a_{1g}^2 leads to $a_{1g} \times a_{1g}$ which leads to $^1A_{1g}$, while b_{1g}^2 leads to $b_{1g} \times b_{1g}$ which leads to $^1A_{1g}$, and $a_{1g}^1 b_{1g}^1$ leads to $b_{1g} \times a_{1g}$ which leads to $^1B_{1g}$ or $^3B_{1g}$. The states in octahedral symmetry (A_{1g}, A_{2g}, and E_g) must change to those in D_{4h} symmetry as shown in the group correlation table (Table 10–4) and as summarized below:

O_h	D_{4h}
A_{1g}	$\longrightarrow A_{1g}$
A_{2g}	$\longrightarrow B_{1g}$
E_g	$\longrightarrow A_{1g}$
	$\searrow B_{1g}$

We have determined the multiplicity of the D_{4h} states; since lowering the symmetry cannot change the spin degeneracies, we can work backwards to determine the spin degeneracy of the O_h states. The $^1A_{1g}$ state in D_{4h} must correspond to a singlet $^1A_{1g}$ in O_h. The other $^1A_{1g}$ must arise from E_g, requiring that this be ^1E_g. The B_{1g} state in D_{4h} arising from ^1E_g must also be a singlet. We are left with the fact that $^3B_{1g}$ in D_{4h} is associated with A_{2g}, requiring this to be $^3A_{2g}$.

D_{4h}	O_h
$^1A_{1g}$	$\longrightarrow A_{1g}(^1A_{1g})$
$^3B_{1g}$	$\longrightarrow A_{2g}(^3A_{2g})$
$^1A_{1g}$	$\longrightarrow E_g(^1E_g)$
$^1B_{1g}$	$\nearrow$

The t_{2g}^2 configuration gives rise to $A_{1g} + E_g + T_{1g} + T_{2g}$ states. We must examine the correlation table for these states in O_h symmetry and find a lower symmetry that converts these states to one-dimensional representations or a sum of one-dimensional representations. Table 10–4 shows that C_{2h} and C_{2v} satisfy this requirement. The results for C_{2h} are summarized as

O_h	C_{2h}
A_{1g}	$\longrightarrow A_g$
E_g	$\longrightarrow A_g + B_g$
T_{1g}	$\longrightarrow A_g + B_g + B_g$
T_{2g}	$\longrightarrow A_g + A_g + B_g$

Since t_{2g} in O_h gives rise to $a_g + a_g + b_g$ orbitals in C_{2h}, the possible configurations are a_{g1}^2, $a_{g1}^1 a_{g2}^1$, $a_{g1}^1 b_g^1$, a_{g2}^2, $a_{g2}^1 b_g^1$, b_g^2. As previously:

$$a_{g1} \times a_{g1} \text{ leads to } A_g$$
$$a_{g1} \times a_{g2} \text{ leads to } A_g$$
$$a_{g1} \times b_g \text{ leads to } B_g$$
$$a_{g2} \times a_{g2} \text{ leads to } A_g$$
$$a_{g2} \times b_g \text{ leads to } B_g$$
$$b_g \times b_g \text{ leads to } A_g$$

The subscripts 1 and 2 on the a_g orbital are used to distinguish the two different a_g representations that result from t_{2g} in this lower symmetry. Since the first, fourth and sixth binary products listed correspond to both electrons occupying the same orbital* (a_{g1}, a_{g2}, and b_g, respectively), these must be singlet states: ^1A_g, ^1A_g, and ^1A_g, respectively. The second, third, and fifth binary products correspond to electrons in different orbitals and give rise to singlet and triplet states: $^1A_g + {}^3A_g$ and $2\,{}^1B_g + 2\,{}^3B_g$. The result in C_{2h} is summarized in the left column of Table 10–5. In the right-hand column, we connect up the correlating states in O_h from Table 10–4. Since we have three triplet states in C_{2h} of B_g, B_g, and A_g symmetry, these must arise from $^3T_{1g}$. All of the other states from t_{2g}^2 are singlets. No other possible correspondence exists. One can work with the $t_{2g}^1 e_g^1$ configuration in a similar manner.

Summarizing the above procedure, we note that we have used the method of descent in symmetry as follows:

1. The orbitals in the higher symmetry are correlated with the orbitals in the lower symmetry.
2. The required number of electrons are added to these lower symmetry orbitals.
3. The electronic states resulting from the electron configurations in the lower symmetry are determined.
4. The lower symmetry electronic states, including spin multiplicity, are correlated with the electronic states of the higher symmetry case. (This is, in effect, an ascent in symmetry.)

The terms of the d^3 strong field configuration are given by taking the direct product of the terms for d^2 with t_{2g} and e_g for the added electron.

Next, we have to show how the terms that arise in a strong field are related to those in a weak field, for there must be a continuous change from one to the other as a function of the ligand field strength in a series of complexes that a given metal forms. We can illustrate this with a d^2 ion by employing two principles:

1. A one-to-one correspondence exists between the states of the same symmetry and spin multiplicity at the weak field and strong field extremes.

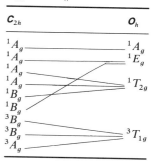

TABLE 10–5. Relation of the Multiplicities in C_{2h} and O_h

---

*If we label the orbitals 1, 2, and 3, we have 1×1, 1×2, 1×3, 2×2, 2×3, and 3×3. For 1×1, 2×2, and 3×3, we have two electrons in the same orbital.

2. States of the same symmetry and spin degeneracy cannot cross as the ligand field strength is varied.

Figure 10–5 contains the free ion and corresponding weak field states on the left and the strong field states on the right. The configurations in an infinitely strong field are indicated on the far right.

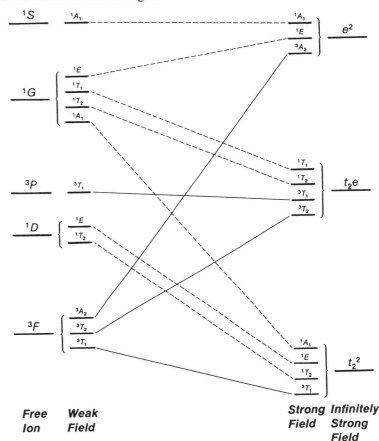

FIGURE 10–5 Correlation of the strong and weak field states of a d^2 ion.

If we begin with the infinitely strong field e^2 configuration, we note a ^1A state. There is also a $^1A_1$ state in t_2^2. Only if the connections are made as shown can we avoid crossing. The same is true for ^1E. Since there is only one $^3A_2$ state in the weak field, this connection is straightforward. Proceeding to the other states from t_2e and t_2^2, only the connections shown will lead to non-crossing of states with the same symmetry and multiplicity.

The results of the evaluation of energies of the various levels in going from weak field to strong field cases are presented in graphical form in the *Tanabe and Sugano diagrams*[13] contained in Appendix C for various d^n configurations. The energies are plotted as E/B versus Dq/B, where B is the Racah parameter. The quantities E/B and Dq/B are plotted because they enable one to convert the equation for the energy into a convenient form for plotting. Since states of different multiplicities are involved, they are a function of the Racah parameter C as well as B. Therefore, the diagram can be constructed only for a given ratio of C/B. The lowest term is taken as the zero of energy in these diagrams.

The *Orgel diagrams* are used to present some of the information in the more complete Tanabe and Sugano diagrams. Orgel diagrams contain only those terms that have the same multiplicity as the ground state. Accordingly, they suffice for the interpretation of the electronic spectra of multiplicity-allowed transitions and will be employed often in the rest of the chapter (*e.g.*, Fig. 10–9).

10–5 DOUBLE GROUPS

We have previously shown how to use the character tables to find the character of the representation for which the p and d orbitals form a basis in various symmetries. In the preceding section, we showed that for any symmetry operation corresponding to a rotation by an angle α on an orbital or state wave function having an angular momentum quantum number ℓ, or L, the character $\chi(\alpha)$ for which this forms a basis is given by equation (10–9):

$$\chi(\alpha) = \frac{\sin\left(\ell + \frac{1}{2}\right)\alpha}{\sin\left(\frac{\alpha}{2}\right)}$$

This equation can also be applied to those states characterized by the total angular momentum J (where $J = L + S$) by simply substituting J for ℓ. When there are an even number of electrons and J is an integer value, the total representation in any symmetry can be decomposed into the irreducible representations of the point group as done in the previous section. However, when J is half-integral (*i.e.*, S is odd), a rotation by 2π (which is the identity operation) does not produce the identity for the character:

$$\chi(\alpha + 2\pi) = \frac{\sin\left(J + \frac{1}{2}\right)(\alpha + 2\pi)}{\sin\left[\frac{(\alpha + 2\pi)}{2}\right]} = \frac{\sin\left[\left(J + \frac{1}{2}\right)\alpha + \left(J + \frac{1}{2}\right)2\pi\right]}{\sin\left(\frac{\alpha}{2} + \pi\right)}$$

$$= \frac{\sin\left(J + \frac{1}{2}\right)\alpha}{-\sin\left(\frac{\alpha}{2}\right)} = -\chi(\alpha)$$

It can be shown that rotation by 4π is needed to produce the identity. To avoid this difficulty, rotation by 2π in this instance is treated as a symmetry operation that we shall label R. The ordinary rotation group is expanded by taking the product of R with all existing rotations. The new group is called a *double group*. Using equation (10–9), the characters for the rotations can be worked out. The characters of E and R (*i.e.*, $\alpha = 0$ and $\alpha = 2\pi$) require evaluation of the limit of the indeterminate form

$$\frac{\sin\left(J + \frac{1}{2}\right)2\pi}{\sin \pi} \to \frac{0}{0}$$

leading to $\chi(0) = 2J + 1$ and $\chi(2\pi) = 2J + 1$ for integer J or $-(2J + 1)$ for half-integer J, as mentioned before. The character tables for the rotation double groups D_4' and O' are given in Appendix B. Those for other groups have been reported.* Two commonly encountered systems for labeling the irreducible representations are given. One uses a serially indexed set of Γ_i's, and the other uses primed symbols similar to those we have been employing. The direct products of representations of double groups can be taken as before and reduced to sums of irreducible representations.

The above discussion can be clarified by considering some examples. We need to employ double groups to determine the effects of spin-orbit coupling when J is half-integral. Since spin-orbit effects arise from coupling of spin and orbital momenta of electrons, we are concerned with the direct product representation of these two effects. As an example, we shall work out the effects of an octahedral field and spin-orbit coupling on the ^4F free ion state of a d^7 ion. As in the previous section, we can work out the total representation in the O point group and factor it to obtain

$$\chi_T(\ell = 3) = A_2 + T_1 + T_2$$

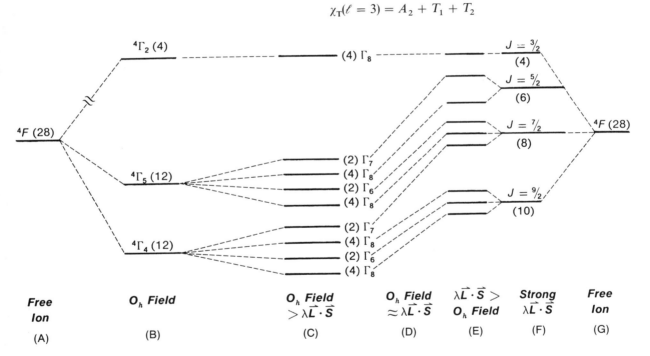

FIGURE 10–6 (A) The gaseous ion, (B) split by a strong O_h field, (C) followed by smaller $\lambda \vec{L} \cdot \vec{S}$. On the right, (G) free ion, (F) split by large spin-orbit coupling, (E) followed by a weaker ligand field. Part (D) indicates the correlation of states in the intermediate region. For convenience, none of the states are shown to cross. States of different double-group symmetries may cross. States of the same double-group symmetry will undergo configuration interaction.

---

*S. Sugano, Y. Tanabe, and H. Kamimura, "Multiplets of Transition Metal Ions in Crystals," Academic Press, New York, 1970.

A d^7 ion in a weak O_h field leads, as shown in the Tanabe and Sugano diagram, to a $^4T_{1g}$ ground state and $^4T_{2g}$ and $^4A_{2g}$ excited states. In the O' double group, these correspond to $T_1'(\Gamma_4)$, $T_2'(\Gamma_5)$, and $A_2'(\Gamma_2)$, respectively. Using $S = {}^3\!/_2$ and substituting S for ℓ in equation (10–9), we generate in the O' point group an irreducible representation of $^1G(\Gamma_8)$, i.e., one of the new irreducible representations of the double group. Now, we take the direct products of the spin and orbital parts and decompose them as before, leading to

$$\Gamma_2 \times \Gamma_8 = \Gamma_8$$
$$\Gamma_4 \times \Gamma_8 = \Gamma_6 + \Gamma_7 + 2\Gamma_8$$
$$\Gamma_5 \times \Gamma_8 = \Gamma_6 + \Gamma_7 + 2\Gamma_8$$

As we see, spin-orbit effects do not split Γ_2, but they split Γ_4 into four states and Γ_5 into four states. We could have converted L and S to J and employed equation (10–9) on J values of ${}^9\!/_2$, ${}^7\!/_2$, ${}^5\!/_2$, and ${}^3\!/_2$ to obtain the double group representations. This procedure would have been followed if spin-orbit coupling were comparable to or greater than the crystal field. Using the approach employed above, we have assumed a large crystal field and a small spin-orbit perturbation on it. We can summarize the results with the diagram in Fig. 10–6.

It is important to remember that whenever one is concerned with the effects of spin-orbit coupling (as we shall often be in subsequent chapters) *in a system with half-integral J values, the double group should be employed.*

10–6 THE JAHN-TELLER EFFECT

There is one other effect that influences the electronic structure of a complex, which we should consider before discussing electronic spectra. Consider a molecule with an unpaired electron in a doubly degenerate orbital, e.g., an octahedral Cu(II) system. Note that by distorting the molecule from its most symmetrical geometry (O_h) to, say, D_{4h}, it is possible to lower its energy:

$$\begin{array}{cccc}
e_g & \uparrow\downarrow \ \uparrow & & \uparrow \quad b_{1g} \\
 & & \rightarrow & \uparrow\downarrow \quad a_{1g} \\
t_{2g} & \uparrow\downarrow \ \uparrow\downarrow \ \uparrow\downarrow & & \uparrow\downarrow \quad b_{2g} \\
 & & & \uparrow\downarrow \ \uparrow\downarrow \quad e_g \\
 & O_h & \rightarrow & D_{4h}
\end{array}$$

Here the e_g set splits into b_{1g} and a_{1g} components. Since two electrons are in the stabilized a_{1g} orbital and only one is in the destabilized b_{1g} orbital, the molecule as a whole is stabilized. It is easy to see how this happens by remembering the basis of simple electrostatic crystal field theory: an orbital pointing at a ligand is destabilized. The closer the ligand is, the higher the energy. A tetragonal elongation (a lengthening of two M—L bonds on the z-axis and shortening of the other four on the x- and y-axes) destabilizes the $d_{x^2-y^2}$ (b_{1g}) orbital and stabilizes the d_{z^2} (a_{1g}) orbital. Similarly, a tetragonal compression would raise d_{z^2} and lower $d_{x^2-y^2}$. Jahn and Teller first pointed out that, for a non-linear molecule, when such a distortion can occur to lower the energy, it will. We thus expect that there will be a Jahn-Teller distortion any time we have an orbitally degenerate (E or T) state and when a proper symmetry vibrational mode exists which enables the molecule to move from one geometry to the other. One unpaired electron in a doubly degenerate pair of e orbitals gives rise to an E state and one or two unpaired electrons in three triply degenerate t orbitals gives rise to a T state.

Note that this criterion is very similar to the criterion for spin-orbit coupling. A simple one-electron picture can be used to predict when orbital angular momentum contributions are expected and when they are not. To obtain orbital angular momentum the electron must be in a set of degenerate orbitals that permit it to move freely from one degenerate orbital to the next and, in so doing, circulate around an axis. Consider, for example, the d_{xz} and d_{yz} orbitals of a metallocene. Degeneracy of this pair permits circulation and angular momentum about the z-axis. All E and T states will have spin-orbit coupling, except for E states in the O_h and T_d point groups. In these latter cases, the E states are composed of $d_{x^2-y^2}$ and d_{z^2} so degeneracy does not allow circulation about an axis.

Some people, including Teller, argue that, if a state is split for any reason, there is no Jahn-Teller effect. Others talk about a Jahn-Teller distortion combining with other factors that remove degeneracy. This latter approach brings up an interesting dilemma: When a degenerate state splits, is this due to a Jahn-Teller distortion, distortion from lower symmetry components in the structure, or spin-orbit coupling? Since the magnitude of these effects is often comparable (200 to 2000 cm^{-1}), it may only be possible to say that the splitting is due to some unspecified combination of these effects. One guideline is that Jahn-Teller distortions are generally larger in E states than in T states, so spin-orbit coupling is generally the dominant effect in T states.

10–7 MAGNETIC COUPLING IN METAL ION CLUSTERS

In metal clusters containing two or more paramagnetic metal centers, the electron spins of the individual metal ions $S_1, S_2 \ldots S_n$, interact. The interaction between a pair of spins i and j is described by the Hamiltonian operator:

$$\hat{H} = J\hat{S}_i \cdot \hat{S}_j \qquad (10\text{–}10)$$

where J in energy units is called the exchange coupling constant or exchange parameter. It is to be distinguished from the J quantum number. With this Hamiltonian, the exchange parameter J is negative for an antiferromagnetic interaction, which leads to a pairing of electrons, and is positive for a ferromagnetic interaction. The literature is confusing in this area, for the above Hamiltonian has been written with J replaced by $2J$, $-J$ or $-2J$.

Dimeric copper(II) acetate dihydrate is the classic example of this type of system. The structure of this molecule is shown in Fig. 10–7, and the metal-metal axis is taken as the z-axis. The copper(II) ions have a d^9 configuration. At low temperatures the compound is found to be diamagnetic, and at elevated temperatures it is paramagnetic. We can view this system as one in which two molecular orbitals exist that are largely metal $d_{x^2-y^2}$ (with a significant contribution from the bridging acetate). A simplified representation of this part of the

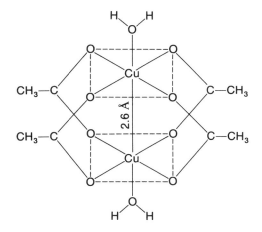

FIGURE 10–7 The structure of $Cu_2(CH_3CO_2)_4 \cdot 2H_2O$. The Cu—Cu distance is 2.6 Å.

molecular orbital diagram of the complex is shown in Fig. 10–8.* (If necessary, the reader is referred back to the chapter on epr to review the discussion on the triplet state and exchange interactions.) With Δ as the energy separation between the bonding and antibonding molecular orbitals (Fig. 10–8) and K as the spin pairing energy, we obtain, when $\Delta < K$, a ferromagnetic system. When $\Delta > K$, an antiferromagnetic coupled system results. The copper(II) acetate dimer is an example of the latter. In $Cu_2(CH_3CO_2)_4 \cdot 2H_2O$, the ground state is diamagnetic, but the excited triplet state is close by in energy and is thermally populated at elevated temperatures. The value of J is found[14] from magnetic susceptibility studies (*vide infra*) to be 284 cm^{-1}.

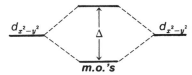

FIGURE 10–8 A simplified representation of the interaction of two metal $d_{x^2-y^2}$ orbitals in $Cu_2 \cdot (CH_3CO)_4 \cdot 2H_2O$ to produce two nondegenerate levels.

In the discussion of the epr spectra of diradicals, new states with $S = 0$ and $S = 1$ were shown to arise (see Fig. 9–22). The general expression for the new states arising from the exchange coupling of S_1 and S_2 centers is given by:

$$(S_1 + S_2), \ldots, (S_1 - S_2) \qquad (10\text{–}11)$$

When $S_1 = S_2 = 1/2$ we obtain $S = 0$ and $S = 1$. When $S_1 = 3/2$ and $S_2 = 1/2$ we obtain $S = 4, 3, 2$ and 1.

We define a total spin quantum number for the system, $\vec{S}$, as

$$\vec{S} = \vec{S}_1 + \vec{S}_2$$

Accordingly

$$\vec{S}^2 = \vec{S}_1^2 + \vec{S}_2^2 + 2\vec{S}_1 \cdot \vec{S}_2$$

which upon rearranging defines $\vec{S}_1 \cdot \vec{S}_2$ as

$$\vec{S}_1 \cdot \vec{S}_2 = \tfrac{1}{2}[\vec{S}^2 - \vec{S}_1^2 - \vec{S}_2^2] \qquad (10\text{–}12)$$

---

*This is referred to as a super exchange pathway involving the bridging acetates. There is controversy regarding the amount of a direct exchange contribution, which involves a direct overlap of the two orbitals on each copper.

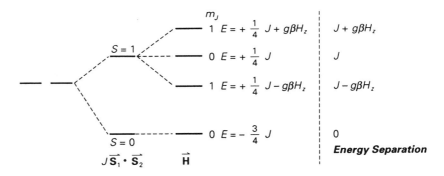

FIGURE 10–9 Energies of the levels in a dimeric d^9 system.

Recalling that $\hat{S}^2\psi = S(S+1)\psi$, we have the following result:

$$E(S) = J\hat{S}_1 \cdot \hat{S}_2 \psi = \frac{J}{2}[S(S+1) - S_1(S_1+1) - S_2(S_2+1)]\psi \quad (10\text{–}13)$$

For dimeric copper(II) acetate dihydrate, we have

$$S_1 = S_2 = \tfrac{1}{2} \quad \text{and} \quad S = 0 \quad \text{or} \quad S = 1$$

For $S = 0$, we obtain

$$\frac{J}{2}\left[0 - \frac{3}{4} - \frac{3}{4}\right] = -\frac{3}{4}J$$

whereas for $S = 1$, we obtain

$$\frac{J}{2}\left[2 - \frac{3}{4} - \frac{3}{4}\right] = +\frac{1}{4}J$$

By adding $g\beta\hat{S}_z\hat{H}_z$ to equation (10–10) the Hamiltonian becomes $\hat{H} = g\beta\hat{S}_z H_z + J\hat{S}_1 \cdot \hat{S}_2$ which produces the results shown in Fig. 10–9. Calling the ground state energy zero, we calculate the energy separation to the excited states by subtracting the ground state energy.

In a similar fashion the reader is left to show that when a $S = {}^5/_2$ center is magnetically coupled to an $S = {}^3/_2$ center, equation (10–11) leads to states 4, 3, 2, and 1 with energies from equation (10–14) of $9J$, $5J$, $2J$, and 0. In a magnetic field, each state will be split into the M_s components given by $S, (S-1) \ldots -S$. The wave functions $|S, M_s\rangle$ are expressed as linear combinations of the two centers $|S_1, S_2, M_s, M_{s/2}\rangle$ weighted by their appropriate coefficients $C_{1,2}$.

10–8 SURVEY OF THE ELECTRONIC SPECTRA OF O_h COMPLEXES

Applications

The electronic spectra of transition metal complexes can be interpreted with the aid of crystal field theory. In our discussion of O_h complexes in this section, our concern will be with systems in which the local symmetry is O_h, though the overall molecular symmetry may not be. Throughout the remainder of this chapter, we shall use the symmetry terms very loosely to describe the type and arrangements of donor atoms directly bonded to the metal, with the rest of the ligand atoms being ignored. It should be realized that this assumption is not always justified. Upon completion of this section, we shall be in a position to assign and predict the electronic spectrum as well as rationalize the magnitudes of the d-orbital splittings observed. The treatment here will not be encyclopedic; selected topics will be covered. The aim is to give an appreciation for a very powerful tool in coordination chemistry: the utilization of electronic spectra in the solution of structural problems. More advanced treatments containing references to the spectra of many complexes are available.[1,2,4,5,9,10,12]

The discussion in Chapter 5 of selection rules for electronic transitions should be reviewed if necessary. Here we shall apply these rules to some transition metal ion systems. We begin by discussing high spin, octahedral complexes of Mn(II), a d^5 case, where there are no spin-allowed d-d transitions. All d-d transitions in this case are both multiplicity and Laporte forbidden. If it were not for vibronic coupling and charge transfer transitions, Mn(II) complexes would be colorless. Hexaquomanganese(II) ion is very pale pink, with all absorption peaks in the visible region being of very low intensity.

The fact that multiplicity-allowed transitions are usually broad, while multiplicity-forbidden transitions are usually sharp, aids in making band assignments. Multiplicity-allowed $t_{2g} \to e_g$ transitions lead to an excited state in which the equilibrium internuclear distance between the metal ion and ligand is larger than in the ground state. In the course of the electronic transition no change in distance can occur (Franck-Condon principle), so the electronically excited molecules are in vibrationally excited states with bond distances corresponding to the configuration of the ground state. The interaction of an excited state with solvent molecules not in the primary coordination sphere is variable because neighboring solvent molecules are various distances away when the excited molecule is produced. Since the solvent cannot rearrange in the transition time, a given excited vibrational state in different molecules will undergo interactions with solvent molecules located at varying distances. Varying solvation energies produce a range of variable energy, vibrationally excited states and a broad band results.

In some spin-forbidden transitions, rearrangement occurs in a given level. For example, in Cr(III) complexes a transition occurs from a ground state containing three unpaired electrons in t_{2g} to an excited state in which t_{2g} has two paired electrons and one unpaired electron. In these multiplicity-forbidden transitions there is often little difference in the equilibrium internuclear distances of the excited and ground electronic states. Sharp lines result from these transitions to a low energy vibrational level of an excited state whose potential energy curve is similar in both shape and in equilibrium internuclear distance to that of the ground state.

As discussed earlier, there is no center of symmetry in a tetrahedral molecule, so somewhat more intense absorptions ($\varepsilon = 100$ to 1000) than those in octahedral complexes are often obtained for d-d transitions in T_d complexes.

d^1 and d^9 Complexes

The simplest case with which to illustrate the relation between Dq and the color of a transition metal ion complex resulting from a d-d transition is d^1 e.g., Ti(III) in an octahedral field. The ground state of the free ion is described by the term symbol ^2D and, as indicated earlier, the degenerate d levels are split in the presence of an octahedral field into a triply degenerate $^2T_{2g}$ and doubly degenerate ^2E_g set. The splitting is equal to $10Dq$. This is represented graphically in Fig. 10-10.

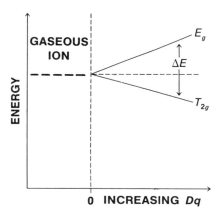

FIGURE 10–10 Splitting of d levels for a d^1 case, O_h field.

As Dq increases, ΔE, the energy (hence the frequency) of the transition increases. The slope of the T_{2g} line is $-4Dq$ and that of E_g is $+6Dq$. The value of Δ (in units of cm^{-1}) can be obtained directly from the frequency of the absorption peak. For example, Ti(H$_2$O)$_6^{3+}$ has an absorption maximum at about 5000 Å (20,000 cm^{-1}). The Δ value for water attached to Ti^{3+} is about 20,000 cm^{-1} (Dq is 2000 cm^{-1}). Since this transition occurs with the absorption of the yellow-green component of visible light, the color transmitted is purple (blue + red). As the ligand is changed, Dq varies and the color of the complex changes. The color of the solution is the complement of the color or colors absorbed, because the transmitted bands determine the color. Caution should be exercised in inferring absorption bands from visual observations; e.g., violet and purple are often confused.

For a d^9 complex in an *octahedral* field, the energy level diagram is obtained by inverting that of the d^1 complex (see Fig. 10–11). The inversion applies because the ground state of a d^9 configuration is doubly degenerate [$t_{2g}^6 e_g^3$ can be $t_{2g}^6(d_{x^2-y^2})^2(d_{z^2})^1$ or $t_{2g}^6(d_{x^2-y^2})^1(d_{z^2})^2$] and the excited state is triply degenerate [$t_{2g}^5 e_g^4$ can be $(d_{xy})^2(d_{yz})^2(d_{xz})^1(e_g)^4$ or $(d_{xy})^2(d_{yz})^1(d_{xz})^2(e_g)^4$ or $(d_{xy})^1(d_{yz})^2(d_{xz})^2(e_g)^4$]. Therefore, the transition is $^2E_g \to {}^2T_{2g}$. In effect, the electronic transition causes the motion of a positive hole from the e_g level in the ground state to the t_{2g} level in the excited state, and the appropriate energy

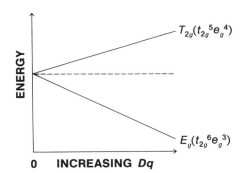

FIGURE 10–11 Splitting of the d levels in a d^9 complex, O_h field.

diagram results by inverting that for the electronic transition for a d^1 case. In order to preserve the center of gravity, the slopes of the lines in Fig. 10–11 must be $-6Dq$ for E and $+4Dq$ for T_2.

The results described above are often summarized by an Orgel diagram as in Fig. 10–12. For d^1, the tetrahedral splitting is just the opposite of that for octahedral splitting, so d^1 tetrahedral and d^9 octahedral complexes have similar Orgel diagrams, as indicated in Fig. 10–12. The splitting of the states as a function of Dq for octahedral complexes with electron configurations d^1 and d^6 and for tetrahedral complexes with d^4 and d^9 electron configurations is described by the right half of Fig. 10–12. Only one band arises in the spectra from d-d transitions, and this is assigned as $^2T_{2g} \rightarrow {}^2E_g$. The left-hand side of the Orgel diagram applies to octahedral d^4 and d^9 as well as tetrahedral d^1 and d^6 complexes. The single d-d transition that occurs is assigned as $^2E \rightarrow {}^2T_2$.

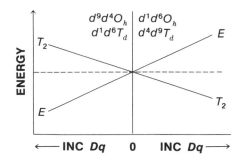

FIGURE 10–12 Orgel diagram for high spin d^1, d^4, d^6, and d^9 complexes. (The g subscript would be added to the symmetry designation for an O_h complex.)

d^2, d^7, d^3, and d^8 Configurations

The two triplet states for a d^2 gaseous ion were earlier shown to be ^3F and ^3P. Furthermore, we showed that an octahedral field split the ^3F terms into the triplet states $^3T_{1g}$, $^3T_{2g}$, and $^3A_{2g}$; the $^3A_{2g}$ state arises from e_g^2, the $^3T_{1g}$ states arise from t_{2g}^2, and (though not worked out earlier) another $^3T_{2g}$ state arises from $t_{2g}^1 e_g^1$. In understanding the electronic spectra, only the triplet states need be considered, for the ground state is triplet. A simplified, one-electron orbital description of these triplet states will be presented. The following degenerate arrangements are possible for the ground state of the octahedral d^2 complex: $d_{xy}^1, d_{xz}^1, d_{yz}^0$; $d_{xy}^0, d_{xz}^1, d_{yz}^1$; $d_{xy}^1, d_{xz}^0, d_{yz}^1$. The ground state is orbitally triply degenerate and the symbol $^3T_{1g}(F)$ is used to describe this state; the (F) indicates that the state arose from the gaseous ion F term. In addition to the $^3T_{1g}(F)$ ground state, an excited state exists corresponding to the configuration in which the two electrons are paired in the t_{2g} level. Transitions to these states are multiplicity forbidden, but sometimes weak absorption bands assigned to these transitions are observed.

The triplet excited state $t_{2g}{}^1e_g{}^1$ will be considered next. If an electron is excited out of d_{xz} or d_{yz} so that the remaining electron is in d_{xy}, the excited electron will encounter less electron-electron repulsion from the electron in d_{xy} if it is placed in d_{z^2}. The $d_{x^2-y^2}$ orbital is less favorable because of the proximity of an electron in this orbital to the electron remaining in d_{xy}. This gives rise to arrangement (A) in Fig. 10–13. Similarly, if the electron is excited out of d_{xy} it

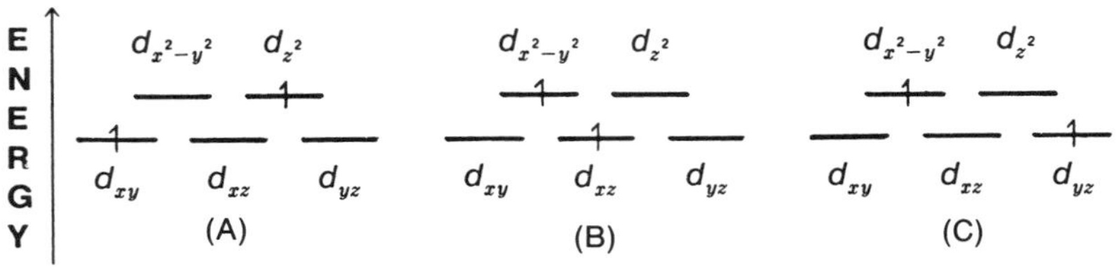

FIGURE 10–13 Possible electron arrangements for $t_{2g}{}^1e_g{}^1$.

will be most stable in $d_{x^2-y^2}$. The remaining electron can be in either d_{xz} or d_{yz}, giving rise to (B) and (C). This set [Fig. 10–13(A, B, and C)] gives rise to the $^3T_{2g}$ state, which is orbitally triply degenerate and has a spin multiplicity of three. The arrangements $(d_{xy}{}^1, d_{x^2-y^2}{}^1; d_{xz}{}^1, d_{z^2}{}^1; d_{yz}{}^1, d_{z^2}{}^1)$ are higher in energy and also produce an orbitally triply degenerate state, $^3T_{1g}(P)$. Other possible arrangements corresponding to $t_{2g}{}^1e_g{}^1$ involve reversing one of the electron spins to produce states with singlet multiplicity. Transitions to these states from the triplet ground state are multiplicity forbidden. Finally, a two-electron transition producing the excited state $e_g{}^2$ or $d_{z^2}{}^1, d_{x^2-y^2}{}^1$ gives rise to a singly degenerate $^3A_{2g}$ state. It is instructive to indicate how these states relate to the gaseous ion. As illustrated in the section on term symbols (Section 10–1), the ground state for the gaseous ion V(III) (a d^2 ion) is ^3F. The ligand field in the complex removes the sevenfold orbital degeneracy of this state (i.e., $M_L = 3, 2, 1, 0, -1, -2, -3$) into two threefold degenerate states, $^3T_{1g}(F)$ and $^3T_{2g}$, and one non-degenerate state, $^3A_{2g}$. This is indicated in the Orgel diagram (Fig. 10–14) for a d^2, O_h complex.

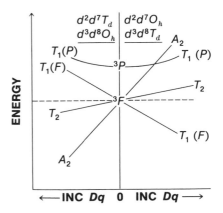

FIGURE 10–14 Orgel diagram for high spin d^2, d^3, d^7, and d^8 complexes.

TABLE 10–6. Absorption Maxima of Octahedral Ni^{II} Complexes (ν_{max} in cm^{-1})

Ligand	$^3A_{2g} \to {}^3T_{2g}$	$^3A_{2g} \to {}^3T_{1g}(F)$	$^3A_{2g} \to {}^3T_{1g}(P)$
H$_2$O	8500	15,400	26,000
NH$_3$	10,750	17,500	28,200
(CH$_3$)$_2$SO	7730	12,970	24,040
HC(O)N(CH$_3$)$_2$	8500	13,605 (14,900)	25,000
CH$_3$C(O)N(CH$_3$)$_2$	7575	12,740 (14,285)	23,810

For zero Dq (i.e., the gaseous ion), only two triplet states, ^3F and ^3P, exist. As Dq increases, ^3F is split into the $^3T_{1g}(F)$, $^3T_{2g}$, and $^3A_{2g}$ states. The degeneracy of the ^3P state is not removed by the ligand field, and this state becomes the triplet $^3T_{1g}(P)$ state in an octahedral complex. The (P) indicates that this state arises from the gaseous ion ^3P state. The energies of these states as a function of Dq are presented in the Orgel diagrams as well as the Tanabe and Sugano diagrams[13] (Appendix D). Use of the Orgel diagrams in predicting spectra and making assignments will be demonstrated by considering V(III) and Ni(II) complexes.

For V(III), three transitions involving the states shown in Fig. 10–14 could occur: $^3T_{1g}(F) \to {}^3T_{2g}$, $^3T_{1g}(F) \to {}^3T_{1g}(P)$, and $^3T_{1g}(F) \to {}^3A_{2g}$. The transition to $^3A_{2g}$ in V(III) is a two-electron transition. Such transitions are relatively improbable, and hence have low intensities. This transition has not been observed experimentally. The spectra obtained for octahedral V(III) complexes consist of two absorption bands assigned to $^3T_{1g}(f) \to {}^3T_{2g}(F)$ and $^3T_{1g}(F) \to {}^3T_{1g}(P)$. In V(H$_2$O)$_6{}^{3+}$ these occur at about 17,000 and 24,000 cm^{-1}, respectively.

For octahedral nickel(II) complexes, the Orgel diagram (left-hand side of Fig. 10–14, d^8) indicates three expected transitions: $^3A_{2g} \to {}^3T_{2g}$, $^3A_{2g} \to {}^3T_{1g}(F)$, and $^3A_{2g} \to {}^3T_{1g}(P)$. (A similar result is obtained from the use of the Tanabe and Sugano diagram in Appendix D.) Experimental absorption maxima corresponding to these transitions are summarized in Table 10–6 for octahedral Ni(II) complexes. (Numbers in parentheses correspond to shoulders on the main band.) Spectra of the octahedral NH$_3$, HC(O)N(CH$_3$)$_2$, and CH$_3$C(O)N(CH$_3$)$_2$ complexes are given[14] in Fig. 10–15. These complexes are colored purple, green, and yellow, respectively.

10–9 CALCULATION OF Dq AND β FOR O_h Ni(II) COMPLEXES

The graphical information contained in the Orgel diagrams is more accurately represented by the series of equations that relates the energies of these various states to the Dq value of the ligand. These energies were derived in Section 10–3. For Ni(II) in an octahedral field, the energies, E, of the states relative to the spherical field are given by the following equations.

for $^3T_{2g}$: $\quad\quad\quad\quad E = -2Dq \quad\quad\quad\quad$ (10–14a)

for $^3A_{2g}$: $\quad\quad\quad\quad E = -12Dq \quad\quad\quad\quad$ (10–14b)

for $^3T_{1g}(F)$ and $^3T_{1g}(P)$:

$$[6Dqp - 16(Dq)^2] + [-6Dq - p]E + E^2 = 0 \quad\quad (10\text{–}14c)$$

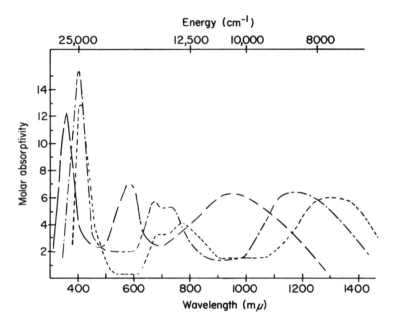

FIGURE 10–15 Molar absorptivity, ϵ, for some nickel(II) complexes in CH_3NO_2 solution ———, $Ni(NH_3)(ClO)_2$; ---, $Ni[HC(O)N(CH_3)_2]_6(ClO_4)_2$; —·—, $Ni[CH_3C(O)N(CH_3)_2]_6(ClO_4)_2$.

where p is the energy of the ^3P state. There are two roots to the last equation corresponding to the energies of the states $^3T_{1g}(F)$ and $^3T_{1g}(P)$.

From the equations it is seen that the energies of both $^3T_{2g}$ and $^3A_{2g}$ are linear functions of Dq. For any ligand that produces a spin-free octahedral nickel complex, the difference in energy between the $^3T_{2g}$ state and the $^3A_{2g}$ state in the complex is $10Dq$. As can be seen from the Orgel or Tanabe and Sugano diagrams, the lowest energy transition is $^3A_{2g} \rightarrow {}^3T_{2g}$. Since this transition is a direct measure of the energy difference of these states, Δ (or $10Dq$) can be equated to the transition energy, i.e., the frequency of this band (cm^{-1}).

Equation (10–14c) can be solved for the energies of the other states. However, the above equations have been derived by assuming that the ligands are point charges or point dipoles and that there is no covalence in the metal-ligand bond. If this were true, the value for Dq just determined could be substituted into equation (10–14c), the energy of ^3P obtained from the atomic spectrum of the gaseous ion,[10] and the energy of the other two levels in the complex calculated from equation (10–14c). The frequencies of the expected spectral transitions are calculated for one band corresponding to the difference between the energies of the levels $^3T_{1g}(F) - {}^3A_{2g}$ and for the other band from the energy difference $^3T_{1g}(P) - {}^3A_{2g}$. The experimental energies obtained from the spectra are almost always lower than the values calculated in this way. The deviation is attributed to covalency in the bonding.

The effect of covalency is to reduce the positive charge on the metal ion, as a consequence of electron donation by the ligands. With reduced positive charge, the radial extension of the d orbitals increases; this decreases the electron-electron repulsions, lowering the energy of the ^3P state. Covalency is foreign to the crystal field approach and is incorporated into the ligand field approach by providing an additional parameter, as we shall discuss next.

The difference in energy between the ^3P and ^3F states in the complex relative to that in the gaseous ion is decreased by covalency and, as a result, the gas phase value cannot be used for p [in equation (10–14c)]; rather, p must be experimentally evaluated for each complex. Equation (10–14c) can be employed for this calculation by using the Dq value from the $^3A_{2g} \rightarrow {}^3T_{2g}$ transition and the experimental energy, ΔE, for the $^3A_{2g} \rightarrow {}^3T_{1g}(P)$ transition. The only unknown remaining in equation (10–14c) is p. The lowering of ^3P is a measure of covalency, among other effects. It is referred to as the *nephelauxetic effect* and is sometimes expressed by a parameter $\beta°$, a percentage lowering of the energy of the ^3P state in the complex compared to the energy of ^3P in the free, gaseous ion.[13] It is calculated by using the equation:

$$\beta° = [(B - B')/B] \times 100$$

where B is the Racah parameter discussed earlier for the free gaseous ion and B' is the same parameter for the complex. It should be noted that p of equation (10–14c) is proportional to B. In the case of nickel(II), the energy of ^3P in the complex can be substituted along with Dq into equation (10–14c) and the other root calculated. The difference in the energy between this root and the energy of $^3A_{2g}$ gives the frequency of the middle band $[^3A_{2g} \rightarrow {}^3T_{1g}(E)]$. The agreement of the calculated and experimental values for this band is good evidence for O_h symmetry. The above discussion will be made clearer by referring to Appendix E, where a sample calculation of Dq, $\beta°$, and the frequency of the $^3A_{2g} \rightarrow {}^3T_{1g}(E)$ transition is presented for $Ni[(CH_3)_2SO]_6(ClO_4)_2$.

Most often the quantity β is used instead of $\beta°$, where β is defined as:

$$\beta = \frac{B'}{B} \qquad (10\text{–}15)$$

The two quantities are easily related if equation (10–13) is rewritten as $\beta° = (1 - \beta) \times 100$.

With many other ions the spectral data cannot be solved easily for Dq and β because of complications introduced by spin-orbit coupling. The consequences of this effect on a d^1 ion are illustrated in Fig. 10–16. The triply degenerate T_{2g} state is split by spin-orbit (s.o.) coupling as indicated in Fig. 10–16(B). Coupling lowers the energy of the ground state and the extent of lowering depends upon the magnitude of the coupling. When the ground state is lowered by spin-orbit coupling, the energies of all the bands in the spectrum have contributions from this lowering, $\Delta_{s.o.}$. When the contribution to the total energy from $\Delta_{s.o.}$ cannot be determined, the evaluation of Δ and β is not very accurate. Spin-orbit coupling in an excited state is not as serious a problem because transitions to both of the split levels often occur and the energies can be averaged. When the ground state is split, only the lower level is populated. Thus accurate values for Dq and β without corrections for spin-orbit coupling can be obtained only for ions in which

FIGURE 10–16 Contribution to Δ from spin-orbit coupling. (A) d-Level splitting with no spin-orbit coupling. (B) Splitting of T_{2g} level by spin-orbit coupling.

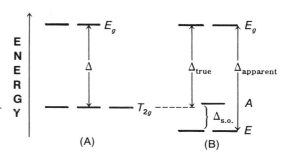

the ground state is A or E (*e.g.*, Ni(II)). A further complication is introduced by the effect that Jahn-Teller distortions have on the energies of the levels.

Ni(II), Mn(II) (weak field), Co(III) (strong field), and Cr(III) form many octahedral complexes whose spectra permit accurate calculation of Dq and β without significant complications from spin-orbit coupling or Jahn-Teller distortions. Ti(III) has only minor contributions from these complicating effects. In the case of tetrahedral complexes, the magnitude of the splitting by spin-orbit interactions more nearly approaches that of crystal field splitting (Dq', the splitting in a tetrahedral field, is about $^4/_9 Dq$). As a result, spin-orbit coupling makes appreciable contributions to the energies of the observed bands. A procedure has been described[14] that permits evaluation of Dq and β for tetrahedral Co(III). A sample calculation is contained in Appendix E.

Both σ and π bonding of the ligand with the metal ion contribute to the quantity Dq. When π bonding occurs, the metal ion t_{2g} orbitals will be involved, for they have the proper directional and symmetry properties. If π bonding occurs with empty ligand orbitals (*e.g.*, d in Et_2S or p in CN^-), Dq will be larger than in the absence of this effect. If π bonding occurs between filled ligand orbitals and filled t_{2g} orbitals (as in the case with the ligand OH^- and Co(III)), the net result of this interaction is antibonding and Dq is decreased. These effects are illustrated with the aid of Fig. 10–17. In Fig. 10–17(A), the d electrons in t_{2g} interact with the empty ligand orbitals, lowering the energy of t_{2g} and raising the energy of the mainly ligand π orbitals in the complex. An empty π^* orbital of the ligand could be involved in this type of interaction. Since t_{2g} is lowered in energy and e_g is not affected (the e_g orbitals point toward σ electron pairs on the ligands), Dq will increase. In the second case [Fig. 10–17(B)], filled ligand π orbitals interact with higher energy filled metal d orbitals, raising the energy of t_{2g} and lowering Dq.

It is informative to relate the energies of the observed d-d transitions to the energy levels associated with the molecular orbital description of octahedral complexes. The scheme for an O_h complex is illustrated in Fig. 10–18 (which neglects π bonding). The difference between T_{2g} and E_g^* is $10Dq$. As metal-ligand σ bond strength increases, E_g is lowered, E_g^* is raised by the same amount, and Dq increases. If T_{2g} metal electrons form π bonds with empty p or d orbitals of the ligand, the energy of the T_{2g} level in the complex is lowered and Dq is increased. Electron-electron repulsions of the T_{2g} electrons and the metal nonbonding electrons raise the energy of the T_{2g} set and decrease Δ. The above ideas have been employed in the interpretation of the spectra of some transition metal acetylacetonates.[15,16,17]

10–9 Calculation of Dq and β for O_h Ni(II) Complexes 441

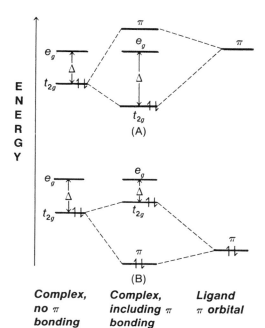

FIGURE 10–17 Effect of π bonding on the energy of a t_{2g} orbital and on Dq. (A) Filled metal orbitals, empty ligand orbitals. (B) Filled metal orbitals, filled ligand orbitals.

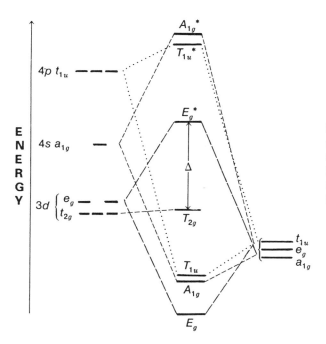

FIGURE 10–18 Molecular orbital description of an octahedral complex (π bonding effects and electrons are not included).

The magnitude of Dq is determined by many factors: interactions from an electrostatic perturbation, the metal-ligand σ bond, the metal to ligand π-bond, the ligand to metal π bond, and metal electron-ligand electron repulsions. Additional references[9,10,18] on the material presented in this section are available.

Much useful information regarding the metal ion-ligand interaction can be obtained from an evaluation of Dq and β. For a series of amides of the type

$R_1C(O)N\begin{smallmatrix}R_2\\R_3\end{smallmatrix}$ it was found that whenever R_1 and R_2 are both alkyl groups, lower values for Dq and β result for the six-coordinate nickel complexes than when either R_1 or both R_2 and R_3 are hydrogens. This is not in agreement with the observation that toward phenol and iodine the donor strengths of these amides are found to increase with the number of alkyl groups. It was proposed that a steric effect exists between neighboring coordinated amide molecules[14] in the metal complexes. A study of the nickel(II) complexes of some primary alkyl amines indicated that even though water replaces the amines in the complexes, the amines interact more strongly with nickel than does water, and almost as strongly as ammonia.[19] A large Dq is also reported for the nickel complex of ethyleneimine.[20] These results are interpreted and an explanation involving solvation energies is proposed for the instability of the alkylamine complexes in water.[19]

The magnitude of $10Dq$ for various metal ions generally varies in the following order:

$$\text{Mn(II)} < \text{Ni(II)} < \text{Co(II)} < \text{Fe(II)} < \text{V(II)} < \text{Fe(III)} < \text{Cr(III)} < \text{V(III)}$$
$$< \text{Co(III)} < \text{Mn(IV)} < \text{Mo(III)} < \text{Rh(III)} < \text{Pd(IV)} < \text{Ir(III)} < \text{Re(IV)}$$
$$< \text{Pt(IV)}$$

Representative ligands give rise to the following order, referred to as the spectrochemical series:

$$\text{I}^- < \text{Br}^- < -\text{SCN}^- < \text{F}^- < \text{urea} < \text{OH}^- < \text{CH}_3\text{CO}_2^- < \text{C}_2\text{O}_4^{2-} <$$
$$\text{H}_2\text{O} < -\text{NCS}^- < \text{glycine} < \text{C}_5\text{H}_5\text{N} \sim \text{NH}_3 < \text{ethylenediamine} <$$
$$\text{SO}_3^{2-} < o\text{-phenanthroline} < \text{NO}_2^- < \text{CN}^-$$

TABLE 10–7. Empirical Parameters for Predicting $10Dq$ and B with Equations (10–16a) and (10–16b)[a]

Ligands	f	h	Metal Ions	$g(10^3 \text{ cm}^{-1})$	k
6F⁻	0.9	0.8	V(II)	12.3	0.08
6H₂O	1.00	1.0	Cr(III)	17.4	0.21
6 urea	0.91	1.2	Mn(II)	8.0	0.07
6NH₃	1.25	1.4	Mn(IV)	23	0.5
3 en	1.28	1.5	Fe(III)	14.0	0.24
3ox²⁻	0.98	1.5	Co(III)	19.0	0.35
6Cl⁻	0.80	2.0	Ni(II)	8.9	0.12
6CN⁻	1.7	2.0	Mo(III)	24	0.15
6Br⁻	0.76	2.3	Rh(III)	27	0.30
3dtp⁻	0.86	2.8	Re(IV)	35	0.2
C₅H₅N	1.25	—	Ir(III)	32	0.3
			Pt(IV)	36	0.5

[a] From C. K. Jørgensen, "Absorption Spectra and Chemical Bonding in Complexes," Pergamon Press, New York, 1962.

Jørgensen[4a,b] has reported a remarkable set of parameters that enable one to predict $10Dq$ and β for various transition metal ion complexes. When the empirical parameters given in Table 10–7 are substituted into equations (10–16a) and (10–16b), the values of $10Dq$ and B for the complex result.

$$10Dq = fg \ (\text{cm}^{-1} \times 10^{-3})$$

$$B = B_0(1 - hk)$$

B_0 is the free ion interelectronic repulsion parameter.

10–10 EFFECT OF DISTORTIONS ON THE d-ORBITAL ENERGY LEVELS

Since octahedral, square planar, and tetrahedral crystal fields cause different splittings of the five d orbitals, geometry will have a pronounced effect upon the $d \to d$ transitions in a metal ion complex. Spectral data for these transitions should provide information about the structure of complexes. Our initial concern will be with how structure affects the energies of the various states in a metal ion. This information will then be applied to determine the structures of various complexes.

The structures of six-coordinate complexes can be classified as cubic, axial, or rhombic, if the equivalences of the ligands along the x, y, and z axes are represented by $x = y = z$, $x = y \neq z$, or $x \neq y \neq z$, respectively. Tetragonal and trigonal distortions (*i.e.*, an elongation or compression along the threefold axis) are commom axial examples. The splitting of the states for a d^1 case is represented in Fig. 10–19. Since d-d electron repulsions are not present in the d^1 case, the

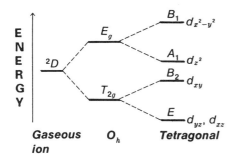

FIGURE 10–19 Orbital splitting for d^1 complexes in octahedral and tetragonal fields.

states can be correlated with the d orbitals as indicated. For the splitting in a tetragonal complex, *trans*-TiA$_4$B$_2{}^{3+}$, to arise as indicated in Fig. 10–18, ligand A must occupy a higher position in the spectrochemical series than B. Metal electron–ligand electron repulsions are less for states consisting of electrons in orbitals directed toward the ligands, B, located on the z-axis. As a result, d_{z^2} is lower in energy than $d_{x^2-y^2}$, and d_{xz} and d_{yz} are lower than d_{xy}. More bands will be observed in the spectrum of the tetragonal complex than in the spectrum of the octahedral complex.

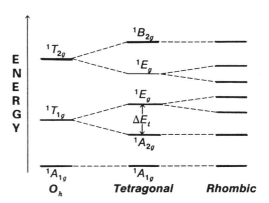

FIGURE 10–20 Splitting of the various states for cobalt(III) in octahedral, tetragonal, and rhombic fields.

The energies and splittings of the various states for a spin-paired Co(III) complex are indicated in Fig. 10–20. Since this is an ion with more than one d electron, we must concern ourselves with states and not orbitals. The $^1T_{1g}$ excited state of the octahedral complex splits into $^1A_{2g}$ and ^1E_g states in a tetragonal field, while $^1T_{2g}$ splits into ^1E_g and $^1B_{2g}$. The splitting that occurs in a rhombic field is also indicated. Hydrated tris(glycinato)cobalt(III) exists as a violet α isomer and a red β isomer. One isomer must be cubic (where $x = y = z$), and the other isomer must be rhombic. The spectrum[18] of the β isomer consists of two bands. The α isomer also gives rise to two bands, one of which is asymmetric and must consist of two or more absorption bands that are not resolved. Therefore, the α isomer must be the rhombic isomer and the β isomer the cubic isomer.

In the *cis*- and *trans*-CoA₄B₂⁺ complexes, the difference in energy between the ^1A2g and ^1Eg states (ΔE_t in Fig. 10-20) is usually about twice as large in *trans* complexes as the corresponding transition in *cis* complexes[21,22] [$\Delta E_t(cis) = -C(\Delta_A - \Delta_B)$ and $\Delta E_t(trans) = 2C(\Delta_A - \Delta_B)$, where C is a constant usually less than one, Δ_A and Δ_B represent the crystal field splittings of ligands A and B (*i.e.*, their positions in the spectrochemical series), and the minus sign accounts for the fact that the energies of ^1E_g and $^1A_{2g}$ are interchanged in *cis* and *trans* complexes]. Usually, when Δ_A and Δ_B differ appreciably, the $^1A_{2g}$ and ^1E_g splitting gives rise to a doublet for the $^1A_{1g} \rightarrow {}^1T_{1g}$ peak in the *trans* compound, while this band is simply broadened in the *cis* compound.[22] It also found that *cis* isomers often have larger molar absorptivity values for $d \rightarrow d$ transitions than *trans* isomers. Typical spectra[23] are illustrated in Fig. 10–21. When Δ_A and Δ_B are similar, this criterion cannot be employed. The ultraviolet charge transfer band can also be used to distinguish between *cis*- and *trans*-cobalt(III) complexes when both isomers are available. The frequency of the band is usually higher in the *cis* than the *trans* compound. The benzoylacetonates of Co(III) and Cr(III) are examples in which Δ_A and Δ_B are nearly equal and the meridional (*trans*) complex has a larger ε value than the facial (*cis*) complex.[24]

If certain assumptions are made concerning the metal-ligand distance, the dipole moment of the ligand, and the effective charge of the nickel nucleus, it is possible to calculate[25] the change in energy that occurs for the various states as the ligand arrangement is varied from O_h to D_{4h}. This corresponds to lengthening the metal-ligand distance along d_{z^2} to infinity. The change in energy for the various levels as a function of the distortion is illustrated in Fig. 10–22. The percentages on the abscissa indicate progressive weakening of two *trans*

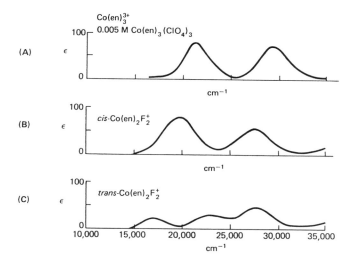

FIGURE 10–21 Spectra of (A) octahedral trisethylenediamine Co(III), (B) the *cis* difluoro compound, and (C) the *trans* difluoro compound. All spectra are taken in water.

metal-ligand bonds (*e.g.*, 100% weakening represents a square planar complex). It is thus easy to see why the spectrum of a square planar or tetragonally distorted nickel complex should differ from that of a regular octahedral complex.

As indicated in Fig. 10–22, there will be a continuous change in spectral properties as the amount of tetragonal distortion increases. Eventually, for highly distorted tetragonal complexes, the spectra will resemble those of square planar complexes. With large distortion, the multiplicity of the lowest energy state for Ni(II) becomes singlet and a diamagnetic complex results. The diamagnetic

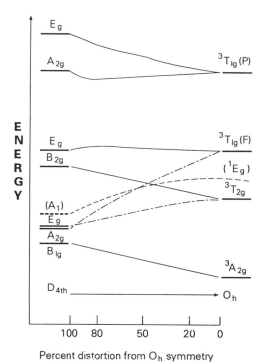

FIGURE 10–22 Effect of tetragonal distortion on the energy levels of nickel(II). [From C. Furlani and G. Sartori, J. Inorg. Nucl. Chem., *8*, 126 (1958).]

tetragonal or square planar complexes have high intensity absorption bands ($\varepsilon = 100$ to 350) with maxima in the 14,000 to 18,000 cm^{-1} region. The spectra may contain one, two, or three peaks,[26,27] and band assignments are often difficult. However, by using spectral and magnetic data, square planar or highly distorted tetragonal nickel(II) complexes can be easily distinguished from nearly octahedral or tetrahedral complexes.

Just as distortion from O_h and D_{4h} changes the energies and properties of the various levels, so does distortion from D_{4h} (planar) to D_{2d} to T_d (Fig. 10–23).

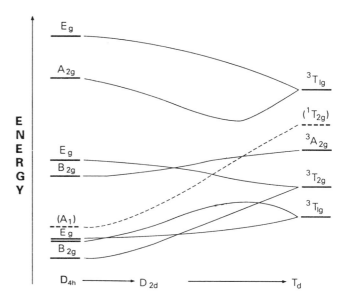

FIGURE 10–23 Energy levels for T_d, D_{4h}, and D_{2d} complexes. [From C. Furlani and G. Sartori, J. Inorg. Nucl. Chem., 8, 126 (1958).]

Figure 10–24 contains spectra for O_h, D_{4h}, D_{2d}, and T_d complexes.[25,28–30] The transitions of the T_d nickel complex NiCl$_4^{2-}$ have large ε values because the complex does not have a center of symmetry. In this case, the d and p orbitals can mix in a molecular orbital description. The p orbital contribution of the ground and excited states gives some allowed $d \rightarrow p$ character to the transition, and the intensity increases. Mixing in non-centrosymmetric ligand molecular orbitals will also enhance the intensity. Accordingly, $\varepsilon/5$ is plotted in Fig. 10–24. As can be seen, different spectra are obtained for different structures. The spectrum for a T_d complex is expected (see Fig. 10–23) to contain three bands v_1, v_2, and v_3 corresponding to the three-spin-allowed transitions: $^3T_1(F) \rightarrow {}^3T_2$, v_1; $^3T_1(F) \rightarrow {}^3A_2$, v_2; and $^3T_1(F) \rightarrow {}^3T_1(P)$, v_3 (see Fig. 10–24). The v_1 band occurs in the range between 3000 and 5000 cm^{-1} and is often masked by absorption by either the organic part of the molecule or the solvent. It has been observed for Ni(II) in silicate glasses and in NiCl$_4^{2-}$. The v_2 band occurs in the 6500 to 10,000 cm^{-1} region and has small molar absorptivity ($\varepsilon = 15$ to 50). The v_3 band is found in the visible region (12,000 to 17,000 cm^{-1}) and shows moderate absorption ($\varepsilon = 100$ to 200).

It is proposed[28] that the complex Ni[OP(C$_6$H$_5$)$_3$]$_4$(ClO$_4$)$_2$ has a D_{2d} configuration. Absorption peaks occur in the spectrum at 24,300, 14,800, and 13,100 cm^{-1} with ε values of approximately 24, 8, and 9, respectively.

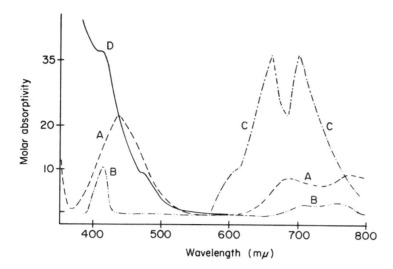

FIGURE 10–24 Electronic absorption spectra of some nickel complexes. (A) Ni(ϕ_3PO)$_4$(ClO$_4$)$_2$ in CH$_3$NO$_2$ (D_{2d}); (B) Ni[(CH$_3$)$_2$SO]$_6$(ClO$_4$)$_2$ in (CH$_3$)$_2$SO (O_h); (C) NiCl$_4^{2-}$ in CH$_3$NO$_2$(T_d); (D), Ni(II) (dimethylglyoxinate)$_2$ in CHCl$_3$ (D_{4h}). Curve C is a plot of $\varepsilon/5$.

10–11 STRUCTURAL EVIDENCE FROM THE ELECTRONIC SPECTRUM

The electronic spectrum can often provide quick and reliable information about the ligand arrangement in transition metal ion complexes. Tetrahedral complexes are often readily distinguished from six-coordinate ones on the basis of the intensity of the bands. The spectra of nickel(II) and cobalt(II) are particularly informative. The complex Ni{OP[N(CH$_3$)$_2$]$_3$}$_4$Cl$_2$ could have tetrahedral, D_{2d}, square planar, tetragonal, or other distorted octahedral geometries. The similarity of the electronic spectrum of this complex to that of NiCl$_4^{2-}$ [see Fig. 10–24(C)] implied[30] that this was the first cationic, tetrahedral nickel(II) complex ever prepared. Further confirmation of the structure comes from the similarity of the x-ray powder diffraction patterns of the nickel(II) and zinc(II) complexes. The latter, with a $3d^{10}$ configuration, is expected to be tetrahedral. The Orgel or Tanabe-Sugano diagram for a T_d, d^8 complex is the same as that for octahedral cobalt(II) ($O_h d^7$) with a low Dq value. Accordingly, the high energy visible band is assigned to $^4T_1(F) \rightarrow {}^4T_1(P)$ and the low energy band to $^4T_1(F) \rightarrow {}^4A_2$. The commonly observed splitting of the visible band is attributed to spin-orbit coupling, which lifts the degeneracy of $^4T_1(P)$ state. It is recommended that any discussion of band assignments in this section be accompanied by reference to the Tanabe and Sugano (or Orgel) diagrams.

In another complex, Ni(NO$_3$)$_4^{2-}$, it was shown[31] that the electronic spectrum of the nickel is that of a six-coordinate complex. Some of the nitrate groups must be bidentate. The color of a transition metal ion complex is often a very poor indicator of structure. Octahedral nickel(II) complexes usually have three absorption bands in the regions from 8000 to 13,000 cm^{-1}, from 15,000 to 19,000 cm^{-1}, and from 25,000 to 29,000 cm^{-1}. The exact position will depend upon the quantities Δ and β. The molar absorptivities of these bands are generally below 20. The ligand field fit of the calculated and experimental frequencies of the middle peak has been proposed as confirmatory evidence for the existence of an O_h complex.

Spin-free tetragonal nickel(II) complexes, in which the two ligands occupying either *cis* or *trans* positions have Dq values that are similar to the other four, will give spectra that will be very much like those of the O_h complexes. In general, molar absorptivities will be higher for tetragonal than for octahedral complexes. A rule of *average environment* relates the band maxima in these slightly distorted tetragonal complexes to the Dq values of the ligands. The band position is determined by a Dq value that is an average of all the surrounding ligands.[4,32]

Nickel(II) forms a large number of five-coordinate complexes.[33] Geometries based on both the trigonal bipyramid and the tetragonal (square) pyramid are known. Many of the complexes are distorted significantly from this geometry.[35] The electronic spectra have been analyzed in detail by Ciampolini,[35] and the interested reader is referred to his account. It is often difficult to detect the difference between tetrahedral and certain five-coordinate geometries on the basis of their electronic spectra.

The electronic spectra of cobalt(II) complexes can often provide reliable structural information. Most six-coordinate complexes are high spin. The Orgel diagram is given in Fig. 10–14. The ground state is $^4T_{1g}$ and a substantial amount of spin-orbit coupling is expected. Three transitions are predicted, $^4T_{1g}(F) \rightarrow {}^4T_{2g}$, $^4T_{1g}(F) \rightarrow {}^4A_{2g}$, and $^4T_{1g}(F) \rightarrow {}^4T_{1g}(P)$. The $^4T_{1g}(F) \rightarrow {}^4A_{2g}$ transition is a two-electron transition and is not observed. The electronic spectrum of octahedral $Co(H_2O)_6^{2+}$ and tetrahedral $CoCl_4^{2-}$ are shown in Fig. 10–25. The band in the octahedral complex at $\sim 20,000$ cm^{-1} is assigned as the $^4T_{1g}(F) \rightarrow {}^4T_{1g}(P)$ transition. The shoulder results because spin-orbit coupling in the excited $^4T_{1g}(P)$ state causes the degeneracy to be lifted. The other absorption band, at 8350 cm^{-1}, is assigned to $^4T_{1g}(F) \rightarrow {}^4T_{2g}$.

Tetrahedral complexes of Co(II) have an energy level diagram like that for Cr(III). The complexes will always be high spin (see the Tanabe and Sugano diagrams in Appendix D). The absorption band at $\sim 15,000$ cm^{-1} is assigned to $^4A_2 \rightarrow {}^4T_1(P)$. The fine structure is attributed to spin-orbit coupling of the T state. The existence of spin-orbit coupling also allows some quartet $\rightarrow$ doublet spin transitions to occur. The other band shown is assigned to $^4A_2 \rightarrow {}^4T_1(F)$. The expected $^4A_2 \rightarrow {}^4T_2$ transition is predicted to occur at 3000 to 4500 cm^{-1}; this is outside the range of most u.v.-visible instruments and is often overlapped by ligand infrared bands. Several five-coordinate complexes of cobalt(II) have been prepared and their spectra reported and interpreted.[35a]

Copper(II) complexes[35b] take on a wide range of geometries, often with low symmetry. One generally finds, for most geometries, one very broad band with a maximum around $15,000 \pm 5,000$ cm^{-1}, which is thought to contain all of the expected transitions. It is possible, however, that the highest-energy transition occurs farther out in the ultraviolet region under charge transfer bands. Thus, the electronic spectrum of copper(II) is of little value in structure assignment. The band position can be correlated roughly with the ligand field strength of the bonding groups.

Spectral data are available to enable similar conclusions to be drawn concerning the structures of other transition metal ions. The salient differences between spectra for various structures have been summarized.[10] Infrared and magnetic data should be used in conjunction with electronic spectral data to aid

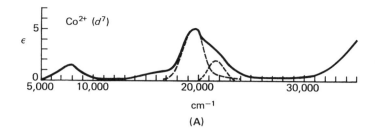

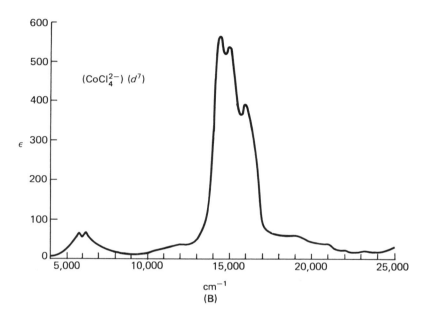

FIGURE 10–25 The electronic spectra of (A) Co(H$_2$O)$_6^{2+}$ [0.021 M Co(BF$_4$)$_2$ in H$_2$O]; (B) CoCl$_4^{2-}$ [0.001 M CoCl$_2$ in 10 M HCl]. The dotted line in (A) gives the resolution of the two bands contributing to the observed spectrum.

in assigning structures to complexes.[36] Use of magnetic data will be described in the next chapter.

The above examples are only a few of the very many cases that indicate the utility of near infrared, visible, and ultraviolet spectroscopy in providing information about the structures of complexes. The number of bands, their frequencies, and their molar absorptivities should all be considered. Solution spectra should be checked against the solid-state spectra (reflectance or mulls) to be sure that changes in structure are not occurring in solution. These changes could involve ligand rearrangement, ligand replacement by solvent, or expansion of the coordination number by solvation.

Other kinds of structural applications of visible spectroscopy have been reported. The Dq values for the nitrite ion are different for the nitro (—NO$_2$) and nitrito (—ONO) isomers. As a result of the difference in average Dq, [Co(NH$_3$)$_5$ONO]$^{2+}$ is red, while [Co(NH$_3$)$_5$NO$_2$]$^{2+}$ is yellow. Limitations of this kind of application have been reported.[37]

The use of electronic spectra to provide structural information is nicely illustrated in a study of the electronic structure of the vanadyl ion.[38] Spectra of the vanadyl ion, VO^{2+}, are interpreted to indicate that there is considerable oxygen-to-metal π bonding in the V—O bond. The similarity in the charge transfer spectra of solids known by x-ray analysis to contain the VO^{2+} group and of solutions is presented as evidence that aqueous solutions contain the species $VO(H_2O)_5^{2+}$ and not $V(H_2O)_6^{4+}$ or $V(H_2O)_4(OH)_2^{2+}$. Protonation of VO^{2+} would have a pronounced effect on the charge transfer spectrum. It is proposed that the oxygen is not protonated because its basicity is weakened by π bonding with vanadium. A complete molecular orbital scheme for $VO(H_2O)_5^{2+}$ is presented,[38] and assignments are made for the spectrum of $VOSO_4 \cdot 5H_2O$ in aqueous solution. Similar studies on other oxy-cations provide evidence for considerable metal-oxygen π bonding[39] and aid in elucidating the electronic structures of these species.

The electronic spectrum has been particularly valuable in determining the coordination number and ligand arrangement in metallo-enzymes. When zinc(II) is replaced by cobalt(II) in carbonic anhydrase, the electronic spectrum indicates that the metal ion is in a distorted tetrahedral site.[40] In such applications, one must be particularly careful to ascertain that the structure of the enzyme has not been changed by metal substitution. If the enzyme is still active, one can have some confidence that the structure is the same. In this example, subsequent x-ray structure determination confirmed the distorted tetrahedral ligand arrangement around zinc(II). The band position in the visible spectrum of a copper(II) protein, erythrocuprein, was interpreted[40] as indicative of coordination by at least four nitrogen donor ligands.

The blue copper(II) protein stellacyanin has been converted to a cobalt(II) derivative.[41] Copper(II) and cobalt(II) were shown to compete for the same site in the protein. Since cobalt(II) spectra are more readily interpreted than those of copper(II), the authors were able to conclude that the cobalt was in either a distorted tetrahedral or a five-coordinate environment. A strong charge transfer band indicated the existence of a Co-SR linkage. All of the bands in the native copper protein were assigned by analogy. The existence of porphyrin complexes in an enzyme system can be detected by the characteristic *Soret band* around 25,000 cm^{-1}. This is a ligand-based $\pi \rightarrow \pi^*$ charge transfer type of transition, discussed in Chapter 5. Two other lower intensity bands are also found in the electronic spectra of these complexes. The existence of these bands and their shifts upon placing substituents on the rings are understood in terms of results from molecular orbital calculations.[42] The positions of these bands have been employed to classify a whole host of cytochromes.

Bonding Parameters from Spectra

10-12 σ AND π BONDING PARAMETERS FROM THE SPECTRA OF TETRAGONAL COMPLEXES

As mentioned earlier, when the symmetry of a complex is lowered from a local O_h or T_d environment additional bands appear in the spectrum. This is illustrated in Fig. 10-25, where the spectra of $Co(NH_3)_6^{3+}$ and $Co(NH_3)_5Cl^{2+}$ are compared. As indicated by the Tanabe and Sugano diagram (Appendix D), for a strong field d^6 complex the spin-allowed transitions are $^1A_{1g} \rightarrow {}^1T_{1g}$ and $^1A_{1g} \rightarrow {}^1T_{2g}$ and the bands are assigned as indicated. When one of the ammonia molecules in the complex is replaced by a different group, which we arbitrarily locate on d_{z^2}, that group interacts differently with cobalt than does ammonia. If this group were chloride, the σ interaction would be weaker and the π interaction greater. The relative energies of d_{z^2}, d_{xz}, and d_{yz} will be affected differently in $Co(NH_3)_5Cl^{2+}$ and $Co(NH_3)_6^{3+}$; the stronger π interaction of Cl^- raises the energy of d_{xz} and d_{yz} and a weaker σ interaction from Cl^- lowers that of d_{z^2}. The degeneracy of the triplet state is removed and $^1T_1$ is split into ^1E and ^1A states, producing the spectrum indicated in Fig. 10–26(B).

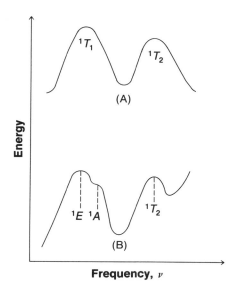

FIGURE 10–26 Spectra of (A) $Co(NH_3)_6^{3+}$ and (B) $Co(NH_3)_5Cl^{2+}$.

The splitting of the $^1T_2$ band is predicted by theory[22] to be too small to be observed. With the additional bands, there is need for additional parameters (besides $10Dq$ and B) to describe the spectrum. The calculation of values for these parameters depends very much upon the assignment of the observed electronic transitions.

Several lower symmetry arrangements are treated by Gerloch and Slade.[3] Here, we shall briefly consider tetragonally distorted six-coordinate complexes. The total potential is now considered as

$$V(D_{4h}) = V_{oct} + V_{tetr} \qquad (10\text{–}17)$$

When the matrix elements are evaluated, we obtain quantities that we earlier labeled Dq, plus two other radial quantities that arise from V_{tetr}, which we shall label Ds and Dt. Ds involves $\overline{r^{-2}}$ terms and Dt involves $\overline{r^{-4}}$ terms, as shown in equations (10–18) and (10–19):

$$Ds = \frac{2}{7} Ze^2 \overline{r^{-2}} \left(\frac{1}{a_{(xy)}^3} - \frac{1}{b_{(z)}^3} \right) \qquad (10\text{–}18)$$

$$Dt = \frac{2}{7} Ze^2 \overline{r^{-4}} \left(\frac{1}{a_{(xy)}^5} - \frac{1}{b_{(z)}^5} \right) \qquad (10\text{–}19)$$

Here a and b refer to the different metal-ligand distances (a refers to xy-ligands and b refers to z-ligands). Dt, being an $\overline{r^{-4}}$ function, can be related to Dq. (Recall that Dq is a function of r^{-4}.) It is a measure of the difference in Dq between the axial and equatorial sites, as shown in equation (10–20):

$$Dt = \frac{4}{7}[Dq(xy) - Dq(z)] \qquad (10\text{–}20)$$

It is to be emphasized that neither $Dq(xy)$ nor $Dq(z)$ is the same as that for the ligand in an octahedral complex.[43] Assuming that the symmetry of the complex is such that the metal e_g orbitals are only σ-antibonding and that the metal t_{2g} orbitals are only π-antibonding, McClure[22] has reported parameters $\delta\sigma$ and $\delta\pi$, defined as:

$$\delta\sigma = \sigma_z - \sigma_{xy} = -\frac{12}{8}Ds - \frac{15}{8}Dt \qquad (10\text{–}21)$$

$$\delta\pi = \pi_z - \pi_{xy} = -\frac{3}{2}Ds + \frac{5}{2}Dt \qquad (10\text{–}22)$$

The σ and π parameters reportedly indicate the relative σ and π antibonding properties of the ligands. Values for various ligands have been reported and interpreted.[43–45] The following order of σ bond interaction with the metal ion results from spectral studies on these complexes:

$$NH_3 > H_2O > F^- > Cl^- > Br^- > I^-$$

The order of π repulsion is:

$$I^- > Br^- > Cl^- > F^- > NH_3$$

10–13 THE ANGULAR OVERLAP MODEL

An alternative approach (with several advantages) to the parameterization of the spectra of transition metal complexes is the angular overlap model.[3,46] This model has its origins in a crude molecular orbital treatment of the energies of the transition metal compounds. A simple case, which we will consider first, is that of a monocoordinated complex ML:

$$M\text{—}L$$

If the metal is a transition metal, we are most interested, from a spectroscopic point of view, in the energies of the d orbitals in the complex. The five d orbitals in the $C_{\infty v}$ symmetry of the complex span the σ, π, and δ representations; i.e., $d(z^2)$ is σ, $d(xz)$ and $d(yz)$ are π, and $d(xy)$ and $d(x^2 - y^2)$ are δ. Considering, for instance, the σ interaction, we can write the secular equations:

$$\begin{vmatrix} H_M{}^\sigma - E & H_{ML}{}^\sigma - S_{ML}{}^\sigma E \\ H_{ML}{}^\sigma - S_{ML}{}^\sigma E & H_L{}^\sigma - E \end{vmatrix} = 0 \qquad (10\text{–}23)$$

where H_M^σ and H_L^σ are the energies of the appropriate metal and ligand orbitals respectively; H_{ML}^σ describes the exchange integral between the metal and ligand orbitals, and S_{ML}^σ is the overlap integral. In general, it is found that $H_M^\sigma \gg H_L^\sigma$ and that the diatomic overlap integral S_{ML}^σ is small, so that one of the roots, say E_1, will be quite close in energy to H_M^σ, and the other, E_2, will be quite close to H_L^σ. Invoking this assumption enables us to write two approximate determinants:

$$\begin{vmatrix} H_M^\sigma - E_1 & H_{ML}^\sigma - S_{ML}^\sigma H_M^\sigma \\ H_{ML}^\sigma - S_{ML}^\sigma H_M^\sigma & H_L^\sigma - H_M^\sigma \end{vmatrix} = 0$$

$$\begin{vmatrix} H_M^\sigma - H_L^\sigma & H_{ML}^\sigma - S_{ML}^\sigma H_L^\sigma \\ H_{ML}^\sigma - S_{ML}^\sigma H_L^\sigma & H_L^\sigma - E_2 \end{vmatrix} = 0 \quad (10\text{--}24)$$

The values of the roots can be written explicitly as

$$E_1 = H_M^\sigma + \frac{(H_{ML}^\sigma)^2 - 2H_{ML}^\sigma H_M^\sigma S_{ML}^\sigma + (S_{ML}^\sigma H_M^\sigma)^2}{H_M^\sigma - H_L^\sigma} \quad (10\text{--}25)$$

$$E_2 = H_L^\sigma - \frac{(H_{ML}^\sigma)^2 - 2H_{ML}^\sigma H_L^\sigma S_{ML}^\sigma + (S_{ML}^\sigma H_L^\sigma)^2}{H_M^\sigma - H_L^\sigma} \quad (10\text{--}26)$$

Using the Wolfsberg-Helmholz approximation for H_{ML}^σ in the form

$$H_{ML}^\sigma = S_{ML}^\sigma (H_M^\sigma + H_L^\sigma) \quad (10\text{--}27)$$

equations (10–25) and (10–26) can be rewritten as

$$E_1 - H_M^\sigma = E_\sigma^* = \frac{(H_M^\sigma + H_L^\sigma)^2}{H_M^\sigma - H_L^\sigma} (S_{ML}^\sigma)^2 \quad (10\text{--}28)$$

$$E_2 - H_L^\sigma = E_\sigma = -\frac{(H_M^\sigma + H_L^\sigma)^2}{H_M^\sigma - H_L^\sigma} (S_{ML}^\sigma)^2 \quad (10\text{--}29)$$

E_σ^* is positive, and therefore represents the destabilizing effect on the metal orbital energies, whereas E_σ is negative and represents the stabilizing effect on the ligand orbitals. As is shown by equation (10–28), the destabilizing effect on a particular metal orbital is proportional to $(S_{ML}^\sigma)^2$. This effect has to be small, when $H_M^\sigma - H_L^\sigma$ is large and S_{ML}^σ is small. E_σ^* might in principle be calculated using (10–28), but in practice it is more profitable to express it parametrically. For this purpose, the overlap integral S_{ML}^σ can be factored into a radial and an angular product:

$$S_{ML}^\sigma = S_{ML} F_\sigma^d \quad (10\text{--}30)$$

where S_{ML} is the integral of the radial functions of the metal and of the ligand orbitals, and F_σ^d refers to the angular part. S_{ML} depends on the particular metal and ligand orbitals considered and on the metal-ligand distance, whereas F_σ^d depends only on the geometrical dispositions of the metal and the ligand. Once the geometry is known, F_σ^d can be easily calculated. We can demonstrate this factoring with a simple example of a ligand L overlapping a p_z orbital on a metal, M, in an M—L fragment. In Fig. 10–27(A), the ligand is located on the z-axis

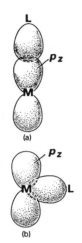

FIGURE 10–27 Orbital overlap with a p_z orbital in M—L fragments.

and the M—L bond is coincident with this axis. If we define F_σ^p (sigma ligand overlap with a p orbital) as one, the overlap is given by the magnitude of S_{ML}. When the M—L bond is the same length but not coincident with the z-axis, the net overlap will be decreased. The radial part S_{ML} stays the same, so the decrease is accomplished by decreasing F_σ^p, the angular part. In Fig. 10–27(B), the M—L bond is along the x-axis, the ligand σ orbital is orthogonal to p_z (overlap is zero) and, since the radial part doesn't change, this is accomplished by having the angular term, F_σ^p, become zero.

Substituting equation (10–30) into (10–28), one gets

$$E_\sigma^* = \frac{(H_M^\sigma + H_L^\sigma)^2}{H_M^\sigma - H_L^\sigma} S_{ML}^2 (F_\sigma^d)^2 \tag{10-31}$$

Now letting

$$e_\sigma = \frac{(H_M^\sigma + H_L^\sigma)^2}{H_M^\sigma - H_L^\sigma} S_{ML}^2 \tag{10-32}$$

equation (10–31) becomes

$$E_\sigma^* = e_\sigma (F_\sigma^d)^2 \tag{10-33}$$

Equation (10–33) shows that the σ-antibonding effect on a particular d orbital can be expressed by a parameter, e_σ, and a number (F_σ^d), which can be obtained from standard tables (*vide infra*).

In the case of π-bonding, the ligand can interact either through ligand low energy filled orbitals, so that the same conditions hold as for the σ case, or through ligand high energy empty orbitals, so that $H_M^\pi \ll H_L^\pi$, and the correction on the energies of the metal orbitals is of the bonding type. In both cases, however, e_π can be defined as in equation (10–32), the only difference being in the sign (e_π is positive for antibonding, and negative for π–back bonding into empty ligand orbitals). The energies of the metal d orbitals in a monocoordinated M—L complex are as follows:

$$\begin{aligned} E(d_{z^2}) &= e_\sigma \\ E(d_{xz}) &= E(d_{yz}) = e_\pi \\ E(d_{xy}) &= E(d_{x^2-y^2}) = e_\delta \end{aligned} \tag{10-34}$$

A transition metal ion complex differs from this simple M—L example, for we are concerned with overlap of d orbitals and we have many ligands. For an octahedral complex, the coordinate system, shown in Fig. 10–28(A), fixes the location of the real d orbitals. We can now employ a local coordinate system on each ligand L, such that the metal-ligand bond is called the z'-axis. The x'-axis is in the plane formed by z and z'. This local coordinate system is shown in Fig. 10–28(B) for ligand L_2. The polar coordinates of the ligand can be used to express the relation between the coordinates in the primed coordinate system and those in the unprimed coordinate system. Our concern is how to describe a d orbital, whose position in the unprimed system is known, with the variables of the primed

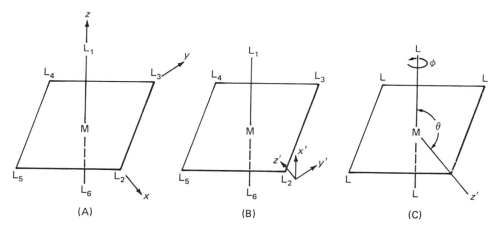

FIGURE 10–28 (A) A coordinate system for a six-coordinate complex; (B) a local coordinate system; (C) definition of θ and φ in Table 10–8.

coordinate system. These relationships are worked out and the results summarized in Table 10–8. The results can be used for a complex of any geometry.

We illustrate the use of Table 10–8 by determining the relation between the d orbitals in the primed and unprimed set for L_2. The ligands in an octahedral complex, labeled as in Fig. 10–28, have angular polar coordinates that can be expressed as follows:

Ligand	1	2	3	4	5	6
θ	0	90	90	90	90	180
φ	0	0	90	180	270	0

Considering L_2, we have $\theta = 90°$ and $\varphi = 0$. Substituting these values into row 1 of Table 10–8, we find what combination of d orbitals we must take in the

TABLE 10–8. Relationships of the d Orbitals in the Primed and Unprimed Coordinate Systems of Fig. 10–25

	z'^2	$y'z'$	$x'z'$	$x'y'$	$x'^2 - y'^2$
z^2	$\frac{1}{4}(1 + 3\cos 2\theta)$	0	$-\frac{\sqrt{3}}{2}\sin 2\theta$	0	$\frac{\sqrt{3}}{4}(1 - \cos 2\theta)$
yz	$\frac{\sqrt{3}}{2}\sin \varphi \sin 2\theta$	$\cos \varphi \cos \theta$	$\sin \varphi \cos 2\theta$	$-\cos \varphi \sin \theta$	$-\frac{1}{2}\sin \varphi \sin 2\theta$
xz	$\frac{\sqrt{3}}{2}\cos \varphi \sin 2\theta$	$-\sin \varphi \cos \theta$	$\cos \varphi \cos 2\theta$	$\sin \varphi \sin \theta$	$-\frac{1}{2}\cos \varphi \sin 2\theta$
xy	$\frac{\sqrt{2}}{4}\sin 2\varphi (1 - \cos 2\theta)$	$\cos 2\varphi \sin \theta$	$\frac{1}{2}\sin 2\varphi \sin 2\theta$	$\cos 2\varphi \cos \theta$	$\frac{1}{4}\sin 2\varphi (3 + \cos 2\theta)$
$x^2 - y^2$	$\frac{\sqrt{3}}{4}\cos 2\varphi (1 - \cos 2\theta)$	$-\sin 2\varphi \sin \theta$	$\frac{1}{2}\cos 2\varphi \sin 2\theta$	$-\sin 2\varphi \cos \theta$	$\frac{1}{4}\cos 2\varphi (3 + \cos 2\theta)$

primed system (listed across the top) to be equivalent to d_{z^2}. The result is obtained by substituting these values for θ and φ into the first row:

$$d_{z^2} = \frac{1}{4}(1 + 3\cos 2\theta)(z'^2) + 0(y'z') - \frac{\sqrt{3}}{2}\sin 2\theta(x'z')$$

$$+ 0(x'y') + \frac{\sqrt{3}}{4}(1 - \cos 2\theta)(x'^2 - y'^2)$$

$$= -\frac{1}{2}z'^2 + \frac{\sqrt{3}}{2}(x'^2 - y'^2) \qquad (10\text{--}35)$$

The interaction energies of the orbitals z'^2 and $x'^2 - y'^2$ with the ligand will be given by e_σ and e_δ, where the subscripts indicate that the ligand-d_{z^2} interaction is σ and that with $d_{x^2-y^2}$ is δ. The coefficients $-1/2$ and $+\sqrt{3}/2$ can be considered as the overlap of the d_{z^2} orbital with $d_{z'^2}$ and $d_{x'^2-y'^2}$. Since the energy is proportional to the overlap squared [equation (10–32)], we can write the energy of the ligand-d_{z^2} interaction as

$$E(d_{z^2}) = \frac{1}{4}e_\sigma + \frac{3}{4}e_\delta \qquad (10\text{--}36)$$

We now must evaluate the effect of L_2 on the energies of all the d orbitals. We accomplish this by substituting $\theta = 90°$ and $\varphi = 0°$ into all the expressions in Table 10–8. We can express the result in the form of a matrix as

$$\begin{bmatrix} -\frac{1}{2} & 0 & 0 & 0 & \frac{\sqrt{3}}{2} \\ 0 & 0 & 0 & -1 & 0 \\ 0 & 0 & -1 & 0 & 0 \\ 0 & 1 & 0 & 0 & 0 \\ \frac{\sqrt{3}}{2} & 0 & 0 & 0 & \frac{1}{2} \end{bmatrix}$$

Since the columns are in the primed coordinate system, they correspond to σ, π, π, δ, δ interactions. These coefficients in the primed set produce the energies by squaring, leading to:

$$E(d_{z^2}) = \frac{1}{4}e_\sigma + \frac{3}{4}e_\delta$$

$$E(d_{yz}) = e_\delta$$

$$E(d_{xz}) = e_\pi \qquad (10\text{--}37)$$

$$E(d_{xy}) = e_\pi$$

$$E(d_{x^2-y^2}) = \frac{3}{4}e_\sigma + \frac{1}{4}e_\delta$$

Substituting values of zero for θ and φ of ligand L_1 into Table 10–8 yields the matrix

$$\begin{bmatrix} 1 & 0 & 0 & 0 & 0 \\ 0 & 1 & 0 & 0 & 0 \\ 0 & 0 & 1 & 0 & 0 \\ 0 & 0 & 0 & 1 & 0 \\ 0 & 0 & 0 & 0 & 1 \end{bmatrix}$$

Squaring the coefficients, we obtain an energy of interaction of 1 for ligand L_1 with each orbital, *i.e.*, the result given earlier by equation (10–34).

Repeating this procedure for the other ligands, we obtain:

	Ligand 1	Ligand 2	Ligand 3
$E(d_{z^2})$	e_σ	$\frac{1}{4}e_\sigma + \frac{3}{4}e_\delta$	$\frac{1}{4}e_\sigma + \frac{3}{4}e_\delta$
$E(d_{yz})$	e_π	e_δ	e_π
$E(d_{xz})$	e_π	e_π	e_δ
$E(d_{xy})$	e_δ	e_π	e_π
$E(d_{x^2-y^2})$	e_δ	$\frac{3}{4}e_\sigma + \frac{1}{4}e_\delta$	$\frac{3}{4}e_\sigma + \frac{1}{4}e_\delta$

	Ligand 4	Ligand 5	Ligand 6
$E(d_{z^2})$	$\frac{1}{4}e_\sigma + \frac{3}{4}e_\delta$	$\frac{1}{4}e_\sigma + \frac{4}{3}e_\delta$	e_σ
$E(d_{yz})$	e_δ	e_π	e_π
$E(d_{xz})$	e_π	e_δ	e_π
$E(d_{xy})$	e_π	e_π	e_δ
$E(d_{x^2-y^2})$	$\frac{3}{4}e_\sigma + \frac{1}{4}e_\delta$	$\frac{3}{4}e_\sigma + \frac{1}{4}e_\delta$	e_δ

You can check your result by summing each column and thereby noting that each ligand contributes one σ, two π, and two δ types of interactions. Summing up the contributions of the individual ligands, we obtain finally the expressions of the energies of the d orbitals for octahedral compounds as:

$$E(d_{z^2}) = 3e_\sigma + 3e_\delta$$
$$E(d_{yz}) = 4e_\pi + 2e_\delta$$
$$E(d_{xz}) = 4e_\pi + 2e_\delta \qquad (10\text{–}38)$$
$$E(d_{xy}) = 4e_\pi + 2e_\delta$$
$$E(d_{x^2-y^2}) = 3e_\sigma + 3e_\delta$$

Of course, the three t_{2g} orbitals have the same energy and so do the two e_g orbitals. If the e_δ contribution is neglected, the difference between the e_g and t_{2g} orbitals is $3e_\sigma - 4e_\pi$, which corresponds to the Δ of the crystal field theory. In a T_d complex, the e orbitals have energies of $^8/_3 e_\pi + ^4/_3 e_\delta$, whereas the t_2 orbitals have energies of $^4/_3 e_\sigma + ^8/_9 e_\pi + ^{16}/_9 e_\delta$. Note that with these parameters, $\Delta_{T_d} = ^4/_9 \Delta_{O_h}$. In complexes with lower symmetry, the values of all of the ligands are added and the d-orbital energies are calculated. Numerical values for the e_σ, e_π, and e_δ parameters are obtained from octahedral complexes. The e_σ values employed in the different complexes are scaled according to the overlap integral. The value of the model is that one set of parameters results for a particular ligand and metal, which can be scaled for geometry and overlap to account for the spectra of many transition metal ion complexes. The relationships of Dq, Ds, Dt, $\delta\sigma$, and $\delta\pi$ with e_σ and e_π have been presented.[47]

In closing this section, it should be emphasized that for these studies, and other applications dependent upon band assignments, the assignments should be verified by polarized single crystal results (see Chapter 5).

Miscellaneous Topics Involving Electronic Transitions

10–14 ELECTRONIC SPECTRA OF OXO-BRIDGED DINUCLEAR IRON CENTERS

The electronic spectra of high spin octahedral and tetrahedral iron(III) compounds are as expected from the Tanabe and Sugano diagrams. Three transitions are found: $^6A_{1g} \to {}^4T_2$, $^6A_{1g} \to {}^4T_1$, and $^6A_{1g} \to {}^4A_{1g}$ [four transitions are found when $^4E(D)$ is low enough in energy]; and since Dq is larger for octahedral complexes than for tetrahedral ones, the $^4T_1$ and $^4T_2(G)$ transitions occur at higher energy in the former complexes. All of the d-d transitions are multiplicity forbidden and are weak. However, when the electronic spectrum of the six-coordinate oxo-bridged dimer (HEDTA Fe)$_2$O (where HEDTA is hydroxyethylethylenediaminetriacetate) is examined,[48,49] the surprising result shown in Fig. 10–29 is obtained.

The bands labeled **a** through **d** are in the correct place to be assigned to the four $d \to d$ transitions, $^6A_1 \to {}^4T_1$, $^6A_1 \to {}^4T_2(G)$, $^6A_1 \to A_1{}^4E_1$, and $^6A_1 \to {}^4E(D)$, of an octahedral complex. These bands are two orders of magnitude more intense than those of a typical iron(III) complex. *This intensity enhancement of spin forbidden bands is common to spin-coupled systems because the coupling partially relaxes the spin selection rule.*

The four intense (**e–h**) u.v. bands are too intense for a d-d transition on a single metal ion. The bands have been attributed[48] to simultaneous d-d transitions on the two iron(III) centers. They are coupled so that the pair excitation is spin allowed.[49] The band labeled **e** at 29.2×10^3 cm^{-1} is assigned to the simultaneous transition of **a** on one center and **b** on the other ($\bar{v}_a + \bar{v}_b = 29.4 \times 10^3$ cm^{-1}); band **f** at 32.5×10^3 cm^{-1} to **a** + **c** (32.2×10^3 cm^{-1}); band **g** at 36.8×10^3 cm^{-1} to **b** + **b**; and band **h** at 42.6×10^3 cm^{-1} to **b** and **d**. There is considerable interest in this spectral feature because of the presence of the Fe-O-Fe group in a wide variety of iron proteins including hemerythrin and methane monooxygenase. Resonance Raman[50] and other studies[51] of a series of hemerythrin derivatives and model compounds has led to the conclusion that

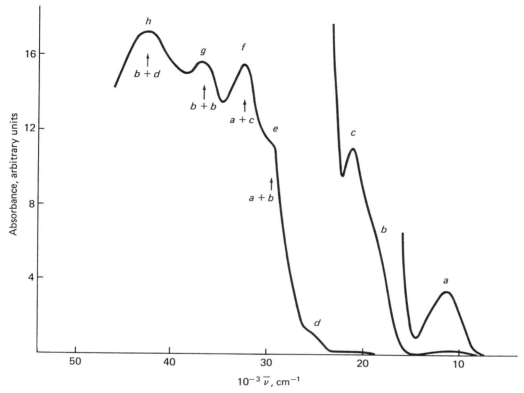

FIGURE 10–29 Absorption spectrum of 0.20 F aqueous solution of [(HEDTA Fe)$_2$O]·6H$_2$O at 296 K. Calculated positions of SPE excitations are indicated by arrows.

the intense near-u.v. features common to all oxo-bridged dinuclear iron(III) compounds should be designated as oxo to iron(III) charge transfer transitions. Similar effects have been observed on other systems[52,53] and have been thoroughly studied[53] for (NH$_3$)$_5$CrOCr(NH$_3$)$_5^{4+}$. *It is concluded that the intensity enhancement arises from a vibronic, exchange-induced, electric dipole mechanism.*

10–15 INTERVALENCE ELECTRON TRANSFER BANDS

Metallomers, molecules with two or more metals, can be prepared in which the bridging group permits the metals to exist in two different oxidation states, *e.g.*,

$$[L_5M^{III}-X-M^{II}L_5]^{+n}$$

These "mixed valence" compounds have long been of interest because of the intense colors they possess. For example, KFe[Fe(CN)$_6$], Prussian blue, has a deep color that is absent in K$_3$Fe(CN)$_6$ and K$_4$Fe(CN)$_6$. This color has been attributed to an *intervalence transfer transition*. In order to understand this phenomenon, let us consider as an example the ruthenium pyrazine dimer[54]:

$$(NH_3)_5Ru(II)\text{—}N\bigcirc N\text{—}Ru(III)(NH_3)_5{}^{5+} \xrightarrow{h\nu}$$

$$(NH_3)_5Ru(III)\text{—}N\bigcirc N\text{—}Ru(II)(NH_3)_5{}^{5+}$$

In order to clarify several points to be made about this system, we shall simplify our discussion by ignoring any ligands in the coordination sphere of Ru(II) and Ru(III) except the bridging pyrazine. Furthermore, we shall label the two ruthenium atoms as $Ru_a{}^{III}\text{—}N\bigcirc N\text{—}Ru_b{}^{II}$. We shall draw a potential energy curve[55] for the molecule as a whole as we vary the $Ru_a{}^{III}\text{—}N$ distance, X, in the molecule in such a way that the $Ru_a\text{—}N$ plus $Ru_b\text{—}N$ distance is constant. Curve a of Fig. 10–30 results when a harmonic potential function is assumed, i.e., $V = (^1/_2)kX^2$.

Next we shall construct a similar curve for the $Ru_a{}^{II}\text{—}N$ distance for the state $Ru_a{}^{II}\text{—}N\bigcirc N\text{—}Ru_b{}^{III}$. The equilibrium $Ru_a{}^{II}\text{—}N$ distance is longer than the equilibrium $Ru_b{}^{III}\text{—}N$ distance. The curve for the $Ru_a{}^{II}\text{—}N$ state is indicated by curve b. We can view the two curves as defining our total system in which we now have provision for the electron jumping from one center to the other. E_{th} is the activation energy for the jump, and it corresponds in this system to the molecule arriving at a place where the $Ru_a\text{—}N$ and $Ru_b\text{—}N$ bond lengths are identical. The thermal energy required for the electron to surpass the barrier between metals, E_{th}, is given[55] by

$$E_{\text{th}} = \frac{1}{2}k\left(\frac{1}{2}X_0\right)^2 = \frac{1}{4}\left(\frac{1}{2}kX_0{}^2\right) = \frac{1}{4}E_{\text{IT}} \qquad (10\text{–}39)$$

By the Franck-Condon principle, electronic absorption occurs without change in nuclear coordinates, so the energy of the intervalence transfer band in the absorption spectrum (i.e., a charge transfer transition from one Ru atom to the other) is represented by the vertical line, E_{IT}.

The actual situation is more complex, since the coordinate X is a complicated coordinate involving all the ligands and electrons in the complex. Depending upon the proximity of the two metal centers and the overlap of various orbitals, the two orbitals (one on each ruthenium) that can contain the odd electron may mix to form a bonding and antibonding combination. This would produce the situations depicted in Figs. 10–30(B) or (C). The quantity labeled $2H_{\text{res}}$ in Fig. 10–30(B) is two times the resonance integral, i.e., the off-diagonal element between the two d-orbitals which can hold the odd electron (one on each ruthenium) in the secular determinant.[56] In Fig. 10–30(B), we have shown $E_{\text{th}} < (^1/_4)E_{\text{IT}}$. In Fig. 10–30(C), the odd electron is completely delocalized on all time scales, and the electron absorption cannot properly be termed an intervalence transfer.

Robin and Day[57] have pointed out that in Fig. 10–30(B), one of the energy minima corresponds to an electronic wave function in which the odd electron is mainly on one metal but, to a small extent, is delocalized onto the other metal. They have proposed that mixed valence compounds be classified as Class I, II, or III, depending on whether none, some, or half of the unpaired electron density is delocalized from one metal center onto the other at any one instant.

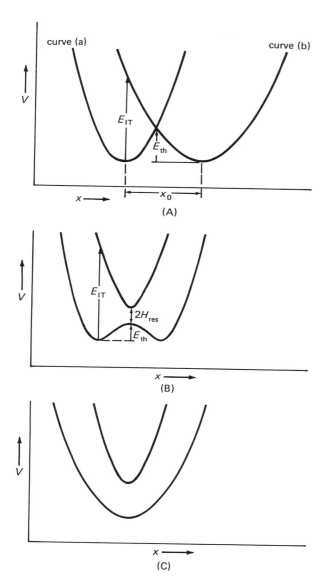

FIGURE 10-30 Potential energy curves for various classes of mixed valence compounds. The systems vary from completely localized systems (A) to intermediate (B) to completely delocalized (C). X is a coordinate expressing the Ru_a^{III}—N distance as the sum of Ru_a—N and Ru_b—N is held constant.

The ruthenium pyrazine dimer[54] shows an intense band in the near infrared at 1570 mμ, assigned to the electronic transition shown in Fig. 10-30(A) and labeled E_{IT}. From the energy of this transition, a thermal rate of electron exchange of 3×10^9 sec^{-1} has been calculated employing a crude model and equation (10-39). The electronic transition is described as a [2, 3] → [3, 2] transition.

The existence of high intensity bands in the d-d region in other complexes has been used as evidence in support of mixed valence in molecules containing more than one metal ion. The assignment of the band to a mixed valence species can be supported by following the intensity of the transition as a function of the extent of oxidation or reduction of the [n, n] oxidation state complex to form [n, m]. A maximum is observed when [n, m] is formed, and it disappears on formation of [m, m].

10–16 PHOTOREACTIONS

The principles discussed is this chapter and Chapter 5 are of importance to chemists for applications other than structure determination. Fluorescence, phosphorescence, and photochemistry all have to do with electronic transitions. In photochemical reactions, the reactant is a molecule in a reactive excited state. Understanding of the photochemical reaction requires an understanding of the structure and reactivity of the excited state. In some cases, molecules that are singlets in the ground state become reactive radicals by being excited to a triplet state containing unpaired electrons. Often, two molecules that do not form a complex in the ground state form a complex (called an exciplex) when one of the molecules is in an excited state. The principles of electronic transitions studied here and in Chapter 5 are thus important in many areas.

REFERENCES CITED

1. C. J. Ballhausen, "Introduction to Ligand Field Theory," McGraw-Hill, New York, 1962.
2. B. N. Figgis, "Introduction to Ligand Fields," Interscience, New York, 1966.
3. M. Gerloch and R. C. Slade, "Ligand Field Parameters," Cambridge University Press, New York, 1973.
4. a. C. K. Jørgensen, "Absorption Spectra and Chemical Bonding in Complexes," Pergamon Press, New York, 1962.
   b. C. K. Jørgensen, "Modern Aspects of Ligand Field Theory," North Holland Publishing Co., Amsterdam, 1971.
5. H. L. Schläfer and G. Gliemann, "Basic Principles of Ligand Field Theory," Interscience, New York, 1969.
6. C. E. Moore, "Atomic Energy Levels Circular 467," Vol. 2, National Bureau of Standards, 1952. Volumes 1 and 3 contain similar data for other gaseous ions.
7. E. U. Condon and G. H. Shortley, "Theory of Atomic Spectra," Cambridge University Press, New York, 1957.
8. J. S. Griffith, "The Theory of Transition Metal Ions," Cambridge University Press, New York, 1961.
9. D. S. McClure, "Electronic Spectra of Molecules and Ions or Crystals," Academic Press, New York, 1959.
10. A. B. P. Lever, "Inorganic Electronic Spectroscopy," 2nd Ed. Elsevier, New York, 1984.
11. "Spectroscopic Properties of Inorganic and Organometallic Compounds," Vols. I–IV, The Chemical Society, London, 1968–1971.
12. T. M. Dunn, "Modern Coordination Chemistry," eds. J. Lewis and R. G. Wilkins, Interscience, New York, 1960.
13. Y. Tanabe and S. Sugano, J. Phys. Soc. Japan, 9, 753, 766 (1954).
14. R. S. Drago, D. W, Meek, M. D. Joesten, and L. LaRoche, Inorg. Chem., 2, 124 (1963).
15. F. A. Cotton and M. Goodgame, J. Amer. Chem. Soc., 83, 1777 (1961).
16. D. W. Barnum, J. Inorg. Nucl. Chem., 21, 221 (1961).
17. T. S. Piper and R. L. Carlin, J. Chem. Phys., 36, 3330 (1962).
18. A. D. Liehr and C. J. Ballhausen, Phys. Rev., 106, 1161 (1957).
19. R. S. Drago, D. W. Meek, R. Longhi, and M. D. Joesten, Inorg. Chem., 2, 1056 (1963).
20. R. W. Kiser and T. W. Lapp, Inorg. Chem., 1, 401 (1962).
21. F. Basolo, C. J. Ballhausen, and J. Bjerrum, Acta Chem. Scand., 9, 810 (1955).
22. D. S. McClure, "Advances in the Chemistry of Coordination Compounds," ed. S. Kirshner, p. 498, Macmillan, New York, 1961 and references therein.

23. F. Basolo, J. Amer. Chem. Soc., *72*, 4393 (1950); Y. J. Shimura, J. Amer. Chem. Soc., *73*, 5079 (1951); K. Nakamoto, J. Fujita, M. Kobayashi, and R. Tsuchida, J. Chem. Phys., *27*, 439 (1957); and references therein.
24. R. C. Fay and T. S. Piper, J. Amer. Chem. Soc., *84*, 2303 (1962).
25. C. Furelani and G. Sartori, J. Inorg. Nucl. Chem., *8*, 126 (1958).
26. W. Manch and W. Fernelius, J. Chem. Educ., *38*, 192 (1961).
27. G. Maki, J. Chem. Phys., *29*, 162 (1958); *ibid*., *29*, 1129 (1958).
28. F. A. Cotton and E. Bannister, J. Chem. Soc., *1960*, 1873.
29. N. S. Gill and R. S. Nyholm, J. Chem. Soc., *1959*, 3997.
30. J. T. Donoghue and R. S. Drago, Inorg. Chem., *1*, 866 (1962).
31. D. K. Straub, R. S. Drago, and J. T. Donoghue, Inorg. Chem., *1*, 848 (1962).
32. C. K. Jørgensen, "Energy Levels of Complexes and Gaseous Ions," Gjellerups, Copenhagen, 1957.
33. L. Sacconi, Pure Appl. Chem., *17*, 95 (1968); Transition Metal Chem., *4*, 199 (1968); R. Morassi, I. Bertini, and L. Sacconi, Coord. Chem. Rev., *11*, 343 (1973).
34. C. Furlani, Coord. Chem. Rev., *3*, 141 (1968).
35. a. M. Ciampolini, Struct. Bonding, *6*, 52 (1969).
    b. B. J. Hathaway and D. E. Billing, Coord. Chem. Rev., *5*, 143 (1970).
36. B. N. Figgis and J. Lewis, "Modern Coordination Chemistry," eds. J. Lewis and R. G. Wilkins, Interscience, New York, 1960.
37. M. Linhard, H. Siebert, and M. Weigel, Z. Anorg. Allgem. Chem., *278*, 287 (1955).
38. C. J. Ballhausen and H. B. Gray, Inorg. Chem., *1*, 111 (1962).
39. H. B. Gray and C. R. Hare, Inorg. Chem., *1*, 363 (1962); C. R. Hare, I. Bernal, and H. B. Gray, Inorg. Chem., *1*, 831 (1962).
40. A. S. Brill, B. R. Martin, and R. J. P. Williams, "Electronic Aspects of Biochemistry," ed. B. Pullman, Academic Press, New York, 1964.
41. D. R. McMillin, R. A. Holwerda, and H. B. Gray, Proc. Nat. Acad. Sci., *71*, 1339 (1974).
42. W. S. Caughey *et al.*, J. Mol. Spect., *16*, 451 (1965) and references therein.
43. D. A. Rowley and R. S. Drago, Inorg. Chem., *7*, 795 (1968); *ibid*., *6*, 1092 (1967).
44. a. A. B. P. Lever, Coord. Chem. Rev., *3*, 119 (1968).
    b. R. L. Chiang and R. S. Drago, Inorg. Chem., *10*, 453 (1971).
45. C. D. Burbridge, D. M. L. Goodgame, and M. Goodgame, J. Chem. Soc. (A), *1967*, 349.
46. a. C. E. Schäffer and C. K. Jørgensen, Mol. Phys., *9*, 401 (1965).
    b. E. Larsen and G. N. LaMar, J. Chem. Educ., *51*, 633 (1974).
47. I. Bertini, D. Gatteschi, and A. Scozzafara, Inorg. Chem., *15*, 203 (1976).
48. H. J. Schugar, G. R. Rossman, J. Thibeault, and H. B. Gray, Chem. Phys. Letters, *6*, 26 (1970).
49. L. Lohr, Coord. Chem. Rev., *8*, 241 (1972).
50. I. Sandere-Loehr *et al*. J. Am. Chem. Soc., *111*, 8084 (1989).
51. R. C. Deem *et al*. J. Am. Chem. Soc., *111*, 4688 (1989).
52. A. E. Hansen and C. J. Ballhausen, Trans. Faraday Soc., *61*, 631 (1965).
53. H. U. Güdel and L. Dubicki, Chem. Phys., *6*, 272 (1974) and references therein.
54. C. Creutz and H. Taube, J. Amer. Chem. Soc., *91*, 3988 (1969).
55. N. S. Hush, Prog. Inorg. Chem., *8*, 391 (1967) and references therein.
56. B. Mayoh and P. Day, J. Chem. Soc., Dalton, *1974*, 846.
57. M. B. Robin and P. Day, Adv. Inorg. Chem. Radiochem., *10*, 248 (1967).

EXERCISES

1. Refer to the Tanabe and Sugano diagrams in Appendix D. For octahedral Cr(III) and a ligand with Dq/B of 1, how many bands should occur in the spectrum? Label these transitions and list them in order of increasing wavelength.

2. In complexes with weak field ligands ($Dq/B = 0.7$), octahedral Co^{2+} exhibits a spectrum with three well-separated bands. Make a tentative assignment using the Tanabe and Sugano diagrams and list the assignments in order of decreasing frequency. Would the spectrum of a strong field complex be any different? Describe the spectrum you would expect for a strong field complex.

3. A nickel complex NiR_4Cl_2 has an absorption spectrum with peaks that have ε values of around 150. R and Cl occupy similar positions in the spectrochemical series. Are the chlorines coordinated?

4. Two different isomers of $Co(NH_3)_4(SCN)_2{}^+$ were separated. How could you determine whether the SCN groups in both were bonded through the sulfur? If both isomers were coordinated through sulfur, how would you determine which is *cis* and which is *trans*? (Hint: —SCN is near Cl^- in the spectrochemical series, while —NCS$^-$ creates a stronger field; $Co(NH_3)_4Cl_2{}^+$ is easily prepared.)

5. Using the Tanabe and Sugano diagrams, assign the following spectra of six-coordinate aquo species [except for (A)].

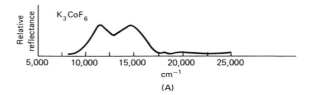

(A)

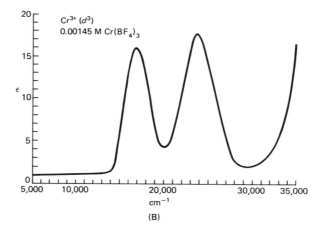

(B)

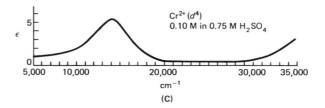

(C)

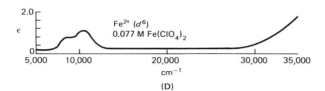

(D)

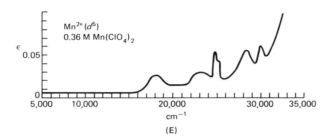

(E)

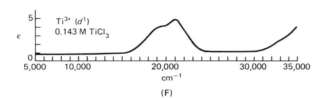

(F)

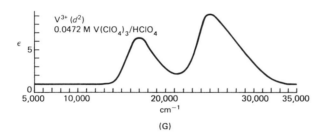

(G)

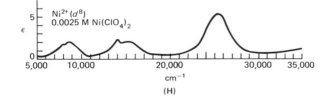

(H)

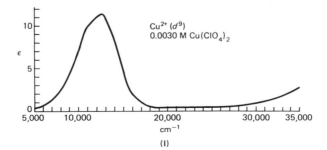

(I)

6. Using the spectrum in problem 5(H), calculate the value of $10Dq$ and B for H_2O.

7. Given the following information for d^3 [see, e.g., A. B. P. Lever, J. Chem. Ed., 45, 711 (1968)]: $^4A_{2g} \rightarrow {}^4T_{2g} = 10Dq$ and $^4A_{2g} \rightarrow {}^4T_{1g}(F) = 7.5B' + 15Dq - \frac{1}{2}(225B'^2 + 100Dq^2 - 180B'Dq)^{1/2}$; calculate the value of $10Dq$ and β (B'/B where $B = 1030$ cm^{-1}) for:

   a. H_2O, using the spectrum in problem 5(B).

   b. $C_2O_4^{2-}$, using the following spectrum:

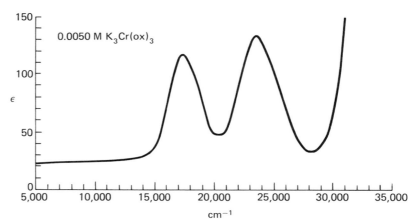

   c. The value of $10Dq$ for H_2O toward Ni^{2+} is 8500 cm^{-1}, and $\beta = 0.88$. Compare your results in part (a) with these values and offer an explanation. Also compare your results toward Cr(III) for water with those for oxalate.

   d. Given that $^4A_{2g} \rightarrow {}^4T_{1g}(P)$ is given by $7.5B' + 15Dq + \frac{1}{2}(225B'^2 + 100Dq^2 - 180B'Dq)^{1/2}$, calculate the frequency of this band in $Cr(C_2O_4)_3^{3-}$

8. The ion $[Ni(pyridine)_4(H_2O)_2]^{2+}$ has d-d absorption bands at 27,000, 16,500, and 10,150 cm^{-1}. No low symmetry splitting is observed. Treating it as an octahedral complex, determine $10Dq$. Compare this value with the average (rule of average environment) of the values predicted from the two six-coordinate complexes. The Dq values for the six-coordinate complexes can be predicted from Table 10–6.

9. Why are octahedral Mn^{2+} complexes (weak field) much less intensely colored than those of Cr^{3+}?

10. The electronic spectrum of trisoxalatochromium(III) doped into a host lattice of $NaMgAl(C_2O_4)_3 \cdot 9H_2O$ has been reported. The ground state is $^4A_{2g}$ if octahedral symmetry is assumed. The lowest excited states (not m.o.'s) for octahedral symmetries are then ^2E_g, $^2T_{1g}$, $^2T_{2g}$, and $^4T_{2g}$. The observed bands and extinction coefficients are:

17,500 cm^{-1}	$\varepsilon = 40$
23,700 cm^{-1}	$\varepsilon = 67$
14,500 cm^{-1}	$\varepsilon = 2.6$
15,300 cm^{-1}	$\varepsilon = 2.0$
20,700 cm^{-1}	$\varepsilon = 0.3$

a. The above transitions have been shown to be electronically allowed. Why is this spectrum inconsistent with octahedral symmetry? (Multiplicity-forbidden transitions have low intensities; generally, ε is less than 5.)

b. Actually, the spectrum is consistent with the true symmetry of the molecule, D_3. Lowering the symmetry causes the following:

O_h	D_3
A_{2g}	A_2
T_{1g}	$A_2 + E$
T_{2g}	$A_1 + E$
E_g	E

If the spectrum is taken with light polarized perpendicular and parallel to the trigonal axis, all the absorptions except the one at 17,500 cm^{-1} occur with perpendicularly polarized light, but only the bands at 17,500 and 15,300 cm^{-1} are present with parallel polarized light. Given that splittings of the doublet octahedral states in D_3 symmetry are unresolved and that ^2E_g is lower in energy than $^2T_{1g}$, assign the transitions using the D_3 excited states. You must explain your choices. (Remember that some doublet bands represent unresolved multiplets and will consequently correspond to more than one transition.)

11. In $Re_2Cl_8^{2-}$ (see problem 6, Chapter 3) the transition of an electron from the b_{2g} orbital to the b_{1u} orbital is a d-d transition in a molecule with a center of inversion. Is it allowed? Explain.

12. The following is the splitting of the state energies of a high spin $3d^2$ ion in an O_h field:

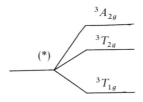

a. Calculate the free ion ground term symbol (*). (Note the T and A states written above will have the same *spin* multiplicity as the free ion.)

b. What would this diagram look like if we considered spin-orbital ($\lambda L \cdot S$) effects? (Do only the T_{1g} state.) Show the splitting and label each level with its J value and its degeneracy.

c. Show the splitting pattern for a d^2 case for all J states above and label with appropriate M_J values. Show the effect of a magnetic field on the levels.

13. a. Of the states arising from an ion with a configuration of two $2p$ electrons, only one is a spin triplet. What is the term symbol of this triplet? Show all work.

b. Now consider spin-orbit coupling. What J states will this triplet give rise to? Which will have the lowest energy?

14. a. Determine the irreducible representations for the terms arising from the splitting of a gaseous ion ^4F state by an octahedral ligand field.

b. Using the method of descent in symmetry, determine the spin multiplicity of the states in part a.

15. Determine the irreducible representations of the states into which the $^2T_{2g}$ level of $Fe(CN)_6^{3-}$ is split by spin-orbit coupling.

16. a. List the operations of the double group D_3' that are obtained by combining the operations of D_3 with the operation R. Make sure your list has the properties of a group.

b. How many classes are in D_3'? What are the operations in each class? You may wish to utilize the fact that equivalent symmetry operations are members of the same class.

17. Indicate the S quantum numbers and calculate the energies of the states arising from the magnetic coupling of $S_1 = 2$ and $S_2 = 5/2$ metal centers.

Magnetism 11

11–1 INTRODUCTION

In this chapter we will consider certain aspects of magnetism that are critical to an understanding of the nmr and esr spectra of transition metal ion complexes. Magnetic effects arise mainly from the electrons in a molecule because the magnetic moment of an electron is about 10^3 times that of the proton. In the chapters on nmr, we have discussed the electron circulations of paired electrons that give rise to diamagnetic effects. When there are unpaired electrons in the system, we observe magnetic behavior that is related to the number and orbital arrangement of the unpaired electrons. The magnetic behavior is determined by measuring (*vide infra*) the magnetic polarization of a substance by a magnetic field. Various types of behavior are illustrated in Fig. 11–1. It is convenient to

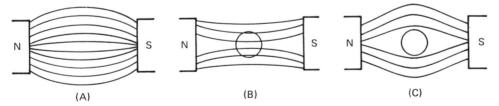

FIGURE 11–1 (A) Magnetic field lines of flux (*i.e.*, contour lines of constant field values) in vacuum; (B) the lines of flux for a paramagnetic substance in a field; (C) the lines of flux for a diamagnetic substance in a field.

define a quantity called magnetic induction, $\vec{B}$, in order to describe the behavior of substances in a field.

$$\vec{B} = \vec{H}_0 + 4\pi \vec{M} \qquad (11\text{–}1)$$

Here $\vec{H}_0$ is the applied field strength and $\vec{M}$ is the magnetization, *i.e.*, the intensity of magnetization per unit volume. When we divide* by $\vec{H}_0$, equation (11–2) results:

---

*We assume in an isotropic system that the directions of $\vec{H}_0$ and $\vec{M}$ are coincident. Thus, though we cannot divide a vector by a vector, we can factor out the directional property and perform the division, leaving an equation in which only the magnitudes appear.

$$\frac{B}{H_0} = 1 + 4\pi \frac{M'}{H_0} = 1 + 4\pi\chi_v \tag{11-2}$$

Here M'/H_0 is given the symbol χ_v and is referred to as the *magnetic susceptibility per unit volume*. The volume susceptibility is thus related to the magnetization by

$$\chi_v = \frac{M'}{H_0} \text{ (dimensionless)}$$

B/H_0 is the permeability of the medium and is the magnetic counterpart of the dielectric constant. Dividing χ_v by the density of the substance, d, produces the gram susceptibility, χ_g:

$$\frac{\chi_v}{d} = \chi_g \text{ (cm}^3\text{/gram)} \tag{11-3}$$

Multiplying χ_g by the molecular weight produces a molar susceptibility, χ:

$$\chi_g \times \text{MW} = \chi \text{ (cm}^3\text{/mole)} \tag{11-4}$$

The value of χ is negative for a diamagnetic substance and positive for a paramagnetic one. In an ordered crystal, the susceptibility may be anisotropic, *i.e.*, represented by a tensor with several components. We shall discuss five types of magnetic behavior: diamagnetism, paramagnetism, superparamagnetism, ferromagnetism, and antiferromagnetism. The behaviors corresponding to various classifications are described in Table 11–1 and superparamagnetism will be discussed in a later section. The latter two types of behavior can be checked by studying the field dependence of χ. The behavior of the susceptibility as a function of temperature is also quite characteristic for these different substances. This is illustrated in Fig. 11–2.

TABLE 11–1. Various Types of Magnetic Behavior

Type	Sign	Magnitude	Field Dependence of χ	Origin
Diamagnetism	−	10^{-6} emu units	Independent	Field induced, paired electron circulations
Paramagnetism	+	0 to 10^{-4} emu units	Independent	Angular momentum of the electron
Ferromagnetism	+	10^{-4} to 10^{-2} emu units	Dependent	Spin alignment from dipole-dipole interaction of moments on adjacent atoms, ↑↑
Antiferromagnetism	+	0 to 10^{-4} emu units	Dependent	Spin pairing, ↑↓, from dipole-dipole interactions

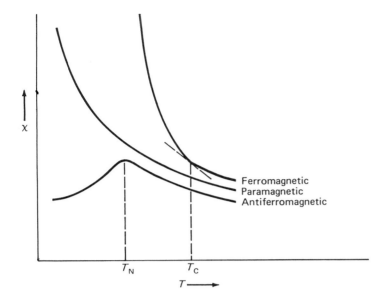

FIGURE 11–2 Temperature dependence of ferromagnetic, paramagnetic, and antiferromagnetic behavior.

The temperature at which the maximum occurs in the plot of antiferromagnetic behavior is referred to as the *Néel temperature*. The temperature at which the break occurs in the ferromagnetic plot is called the *Curie temperature*. Many compounds which, in the solid state, exhibit paramagnetic behavior around room temperature exhibit slight ferromagnetic or antiferromagnetic behavior below liquid helium (4.2 K) temperature.

11–2 TYPES OF MAGNETIC BEHAVIOR

Diamagnetism

As mentioned earlier, diamagnetism arises from field-induced electron circulations of paired electrons, which generate a field opposed to the applied field. Thus, all molecules have contributions from diamagnetic effects. The diamagnetic susceptibility of an atom is proportional to the number of electrons, n, and the sum of the squared values of the average orbital radius of the ith electron $\bar{r}_i$:

$$\chi_A = -\frac{Ne^2}{6mc^2} \sum_i^n \bar{r}_i^2 = -2.83 \times 10^{10} \sum_i^n \bar{r}_i^2 \tag{11-5}$$

Larger atoms with more electrons have greater diamagnetic susceptibilities than smaller atoms with fewer electrons. The molar diamagnetic susceptibility of a molecule or complex ion, χ, can be obtained to a good approximation by summing the diamagnetic contributions from all of the atoms, χ_A, and from all of the bonds in functional groups, χ_B.

$$\chi = \sum_i \chi_{A_i} + \sum_j \chi_{B_j} \tag{11-6}$$

472 Chapter 11 Magnetism

TABLE 11–2. Pascal's Constants

	Atoms, χ_A				Bonds, χ_B	
Atom	χ_A ($\times 10^{-6}$ cm^3 mole^{-1})	Atom	χ_A ($\times 10^{-6}$ cm^3 mole^{-1})	Bond	χ_B ($\times 10^{-6}$ cm^3 mole^{-1})	
H	−2.93	F	−63	C=C	+5.5	
C	−6.00	Cl	−20.1	C≡C	+0.8	
C (aromatic)	−6.24	Br	−30.6	C=N	+8.2	
N	−5.57	I	−44.6	C≡N	+0.8	
N (aromatic)	−4.61	Mg^{2+}	−5	N=N	+1.8	
N (monamide)	−1.54	Zn^{2+}	−15	N=O	+1.7	
N (diamide, imide)	−2.11	Pb^{2+}	−32.0	C=O	+6.3	
O	−4.61	Ca^{2+}	−10.4			
O_2 (carboxylate)	−7.95	Fe^{2+}	−12.8			
S	−15.0	Cu^{2+}	−12.8			
P	−26.3	Co^{2+}	−12.8			
		Ni^{2+}	−12.8			

The values of χ_A and χ_B are referred to as *Pascal's constants*, and some common ones are listed in Table 11–2. The calculation of χ is illustrated here for pyridine and acetone.

C$_5$H$_5$N

SUM OF CONTRIBUTIONS TO χ ($\times 10^{-6}$ CM^3 MOLE^{-1})

$$5 \times C \text{ (ring)} = -31.2$$

$$5 \times H = -14.6$$

$$1 \times N \text{ (ring)} = -4.6$$

$$\chi = \sum_i \chi_{A_i} + \sum_j \chi_{B_j} = -50.4 \times 10^{-6} \text{ cm}^3 \text{ mole}^{-1}$$

The functional groups are accounted for by using the ring values for carbon and nitrogen, so $\sum_j \chi_{B_j}$ equals zero.

(CH$_3$)$_2$C=O

SUM OF ATOM CONTRIBUTIONS ($\times 10^{-6}$ CM^3 MOLE^{-1})

$$3 \times C = -18.0$$

$$6 \times H = -17.6$$

$$1 \times O = -4.6$$

SUM OF BOND CONTRIBUTIONS ($\times 10^{-6}$ CM^3 MOLE^{-1})

$$1 \times \text{C=O} = +6.3$$

$$\sum \chi_A + \sum \chi_B = -33.9 \times 10^{-6} \text{ cm}^3 \text{ mole}^{-1}$$

For a transition metal complex, one can measure only the net magnetism, χ_{MEAS}, which is the sum of the paramagnetic, χ_{PARA}, and diamagnetic, χ_{DIA}, contributions.

$$\chi_{PARA} = \chi_{MEAS} - \chi_{DIA} \qquad (11\text{–}7)$$

Thus, to obtain the paramagnetic susceptibility, the diamagnetic susceptibility must be subtracted from the net susceptibility. This can be accomplished by: (1) using the values for Pascal's constants reported in Table 11–2 to estimate χ_{DIA}; (2) by measuring the diamagnetic susceptibility of the ligand and adding that of the metal from Table 11–2 to obtain χ_{DIA}; or (3) by making an analogous diamagnetic metal complex and using its value as an estimate of χ_{DIA}.

Paramagnetism in Simple System where $S = \frac{1}{2}$

The paramagnetic contribution to the susceptibility arises from the spin and orbital angular momenta of the electrons interacting with the field. First, we shall consider a system that is spherical, contains only one electron, and has no orbital contribution to the moment. The magnetic moment, $\vec{\mu}$, associated with such a system is a vector quantity given by equation (11–8):

$$\vec{\mu} = -g\beta\vec{S} \qquad (11\text{–}8)$$

where $\vec{S}$ is the spin angular momentum operator, g is the electron g-factor discussed in Chapter 9, and β is the Bohr magneton of the electron, also discussed in Chapter 9 ($\beta = 0.93 \times 10^{-20}$ erg gauss^{-1}).

The Hamiltonian describing the interaction of this moment with the applied field, $\vec{H}$, is given by

$$\hat{H} = -\vec{\mu}\cdot\vec{H} = g\beta\vec{S}\cdot\vec{H} \qquad (11\text{–}9)$$

This Hamiltonian, operating on the spin wave functions, has two eigenvalues (see Fig. 9–1) with energies given by

$$E = m_s g\beta H \qquad \text{with } m_s = \pm\tfrac{1}{2} \qquad (11\text{–}10)$$

with an energy difference

$$\Delta E = g\beta H \qquad (11\text{–}11)$$

When H is about 25 kilogauss, ΔE for a free electron with $g = 2.0023$ is about 2.3 cm^{-1}, which is small enough compared to kT (205 cm^{-1} at room temperature) that both states are populated at room temperature, with a slight excess in the ground state.

The magnitude of the projection along the field direction of the magnetic moment, μ_n, of an electron in a quantum state n is given by the partial derivative of the energy of that state, E_n, with respect to the field, H, as shown in equation (11–12):

$$\mu_n = -\frac{\partial E_n}{\partial H} = -m_s g\beta \qquad (11\text{–}12)$$

In order to determine the bulk magnetic moment of a sample of any material, we must take a sum of the individual moments of the states weighted by their Boltzmann populations.

The Boltzmann factor for calculating the probability, P_n, for populating discrete states having energy levels E_n at thermal equilibrium is given by equation (11–13):

$$P_n = \frac{N_n}{N} = \frac{\exp\left(\frac{-E_n}{kT}\right)}{\sum_n \exp\left(\frac{-E_n}{kT}\right)} \tag{11-13}$$

Here N_n refers to the population of the state n, while N refers to the total population of all existing states. We have a wave function for each state, and we use the term 'level' to indicate all of the states that have the same energy. The population-weighted sum of magnetic moments over the individual states, which is the macroscopic magnetic moment, M, then, is given for a mole of material by equation (11–14):

$$M = \underline{N} \sum_{m_s} \mu_n P_n \tag{11-14}$$

where $\underline{N}$ is Avogadro's number. Substituting equation (11–13) for P_n in equation (11–14) produces (11–15) for an $S = \frac{1}{2}$ system.

$$M = \frac{\underline{N} \sum_{m_s=-1/2}^{+1/2} \mu_n \exp\left(\frac{-E_n}{kT}\right)}{\sum_{m_s=-1/2}^{+1/2} \exp\left(\frac{-E_n}{kT}\right)} \tag{11-15}$$

Substituting equation (11–12) for μ_n and equation (11–10) for E_n, and summing over $m_s = \pm \frac{1}{2}$, produces:

$$M = \frac{\underline{N} g\beta}{2} \left[\frac{\exp\left(\frac{g\beta H}{2kT}\right) - \exp\left(\frac{-g\beta H}{2kT}\right)}{\exp\left(\frac{g\beta H}{2kT}\right) + \exp\left(\frac{-g\beta H}{2kT}\right)} \right] \tag{11-16}$$

When $(g\beta H/kT) \ll 1$ ($g\beta H$ equals ~ 1 cm^{-1} for $g = 2.0$ and common fields of 5000 to 10,000 gauss, and kT is 205 cm^{-1} at room temperature), we can introduce the following approximation:

$$\exp\left(\frac{\pm g\beta H}{2kT}\right) \approx \left(1 \pm \frac{g\beta H}{2kT}\right) \tag{11-17}$$

Substituting equation (11–17) into (11–16) and simplifying leads to

$$M = \frac{\underline{N} g^2 \beta^2 H}{4kT} \tag{11-18}$$

Since the molar susceptibility is related to the moment by

$$\chi = \frac{M}{H} \tag{11-19}$$

we can write

$$\chi = \frac{Ng^2\beta^2}{4kT} \tag{11-20}$$

Equation (11–20) is the so-called *Curie law*, and it predicts a straight-line relation between the susceptibility and the reciprocal of temperature, giving a zero intercept; *i.e.*, $\chi \to 0$ as T approaches infinity. This type of behavior is usually not observed experimentally. Straight-line plots are obtained for many systems, but the intercept is non-zero:

$$\chi = \frac{C}{T - \theta} \tag{11-21}$$

Equation (11–21), where $C = Ng^2\beta^2/4k$ and θ corrects the temperature for the non-zero intercept, describes the so-called *Curie-Weiss* behavior. It is common to have a non-zero intercept in systems that are not magnetically dilute (*i.e.*, pure solid paramagnetic material). In these systems, interionic or intermolecular interactions[1] cause neighboring magnetic moments to become aligned and contribute to the value of the intercept.

For the case of molecules with no orbital angular momentum, one commonly sees equation (11–20) written as

$$\chi = \frac{Ng^2\beta^2}{3kT} S(S + 1) \tag{11-22}$$

with units of Bohr magneton, BM.

This reduces to equation (11–20) for $S = \frac{1}{2}$ and yields the so-called spin-only magnetic susceptibilities of complexes containing different numbers of unpaired electrons.

Both χ and M are macroscopic properties. In describing the magnetic properties of transition metal complexes, it is common to employ a microscopic quantity called the effective magnetic moment μ_{eff}. It is *defined* as follows:

$$\mu_{\text{eff}} = \left(\frac{3k}{N\beta^2}\right)^{1/2} (\chi T)^{1/2} = 2.828(\chi T)^{1/2} \text{ (BM)} \tag{11-23}$$

The susceptibility employed in this equation is χ_{PARA} described above. Equation (11–23) can be obtained by replacing $g^2 S(S + 1)$ in equation (11–22) by μ_{eff}^2 and solving for μ_{eff}. Thus, any effects that tend to make S *not a good quantum number* become incorporated into the *g*-value (Recall the variability of the *g*-value in epr,

TABLE 11–3. Spin-Only Magnetic Moments for Various Numbers of Unpaired Electrons

Number of Unpaired Electrons	S	μ_{eff}(spin-only)(BM)
1	1/2	1.73
2	1	2.83
3	3/2	3.87
4	2	4.90
5	5/2	5.92
6	3	6.93
7	7/2	7.94

Chapter 9.) Equation (11–24) can be used to calculate the spin-only magnetic moments for various values of S:

$$\mu_{\text{eff}} \text{ (spin-only)} = g[S(S+1)]^{1/2} \text{ (BM)} \tag{11-24}$$

where g equals 2.0 for an electron with no orbital angular momentum.

The spin-only results in Table 11–3 are obtained for various numbers of unpaired electrons. In many transition metal ion complexes, values close to those predicted by the spin-only formula are observed.[1-9] However, in many other complexes, the moments and temperature dependence of the susceptibility are at variance with these predictions. Other effects are operative, and a more complete analysis is in order.

11–3 VAN VLECK'S EQUATION

General Basis of the Derivation

In this analysis, we will introduce orbital contributions and also anisotropy in the magnetic susceptibility for low symmetry molecules. Defining the principal molecular axis as the z-axis, we can write the necessary part of the Hamiltonian including these additional effects as:

$$\hat{H} = \lambda \hat{L} \cdot \hat{S} + \beta(\hat{L} + g_e \hat{S}) \cdot H \tag{11-25}$$

where $\hat{L}$ and $\hat{S}$ are operators with x, y, and z components. In this chapter, g_e will be used when referring to the free electron g-value. The first term on the right of the equality sign describes the spin-orbit coupling (λ is the spin-orbit coupling constant) and is seen to be field independent. The other term sums the spin and orbital contributions to the electron moment. [Note its resemblance to equation (11–9).] In using this Hamiltonian, we have to worry about what basis set to employ. In a free ion, for a d^1 case with $\vec{L} \cdot \vec{S}$ coupling ignored, the complex functions $|+2\rangle$, $|+1\rangle$, etc., are a good choice for they are already eigenfunctions of $\hat{H}$. Hence, when the full matrix with elements $\langle \varphi_n | L_z + g_e S_z | \varphi_m \rangle \beta H$ is evaluated, there will be no off-diagonal elements.

In a d^1 complex, with the ligands defining the x, y, and z axes, the orbitals are a convenient basis set. Now we have, for example:

$$d_{xy} = \frac{1}{i\sqrt{2}}(d_{+2} - d_{-2})$$

and

$$d_{x^2-y^2} = \frac{1}{\sqrt{2}}(d_{+2} + d_{-2})$$

In this basis set, the off-diagonal elements will be non-zero; for example,

$$\langle d_{xy}|\hat{L}_z|d_{x^2-y^2}\rangle = \frac{1}{2i}[\langle d_{+2}|\hat{L}_z|d_{+2}\rangle + \langle d_{+2}|\hat{L}_z|d_{-2}\rangle - \langle d_{-2}|\hat{L}_z|d_{+2}\rangle$$
$$- \langle d_{-2}|\hat{L}_z|d_{-2}\rangle]$$
$$= \frac{1}{2i}[2 + 0 + 0 - (-2)] = -2i$$

The $\hat{S}_z$ contribution to this matrix element will be zero, since

$$\langle d_{xy}{}^+|\hat{S}_z|d_{x^2-y^2}{}^+\rangle = \langle d_{xy}|d_{x^2-y^2}\rangle\langle +|\hat{S}_z|+\rangle = (0)\left(\frac{1}{2}\right) = 0$$

The non-zero off-diagonal elements account for a distortion of the ground state wave function by the applied field [we worked out the above matrix elements for $\hat{L}_z$ and $\hat{S}_z$, but the full Hamiltonian is $\beta(\hat{L} + g_e\hat{S}) \cdot \vec{H}$]. This distortion is accomplished by mixing in appropriate excited states. The diagonal elements are called the first-order Zeeman terms, and the off-diagonal elements give rise to second-order Zeeman terms. If there were no off-diagonal terms, all of the diagonal matrix elements would be to the first power in H and the resulting energies would be first order in H.

Off-diagonal elements connecting states of very different energies are generally small compared to the energy difference, so this problem is generally treated with perturbation theory. We saw in the discussion of the Ramsey equation (Chapter 8) that this approach gave rise to terms of the general form

$$\sum_{m \neq n} \frac{\langle \psi_n|\widehat{Op}|\psi_m\rangle^2}{E_m - E_n}$$

where $\widehat{Op}$ is an abbreviation for an unspecified operator. The field-induced mixing of excited states was used as an interpretation of the paramagnetic contribution to the chemical shift. In the present case, we obtain from perturbation theory a term that looks similar:

$$\sum_{j \neq i} \frac{[\langle \psi_i|\hat{L}_z + g_e\hat{S}_z|\psi_j\rangle \beta H]^2}{E_i - E_j} \qquad (11-26)$$

When the ion configuration becomes other than d^1, the advantage of the perturbation treatment is seen, for the full matrix would be large.

For a weak complex, the basis set for a full matrix evaluation would involve using the wave functions that resulted after a weak field approximation in the crystal field analysis of the electron-electron repulsions. For a strong field complex, the real d orbitals provide a good basis set for the complex. Thus, we see that the relative magnitudes of the factors influencing the d orbital energies are important in determining the best basis set. We can list some common, rough orders of magnitude as follows, using C.F. to abbreviate the crystal field:

(A) Weak field, first row transition metal ion:

$$\begin{array}{cccc} e^2/r_{ij} & \geqslant \quad \text{C.F.} & > \text{low-symmetry C.F. perturbation} > & \lambda \vec{L}\cdot\vec{S} \\ 10^4 \text{ cm}^{-1} & 10^4 \text{ cm}^{-1} & 10^2 \text{ to } 10^3 \text{ cm}^{-1} & 10^2 \text{ cm}^{-1} \end{array}$$

(B) Strong field, first row:

$$\begin{array}{cccc} \text{C.F.} & > & e^2/r_{ij} & \gtrsim \text{C.F. perturbation} > & \lambda \vec{L}\cdot\vec{S} \\ 5\times 10^4 \text{ cm}^{-1} & & 3\times 10^3 \text{ cm}^{-1} & 10^3 \text{ cm}^{-1} & 10^2 \text{ cm}^{-1} \end{array}$$

(C) Third row:

$$\begin{array}{ccc} \text{C.F.} & > \quad \lambda \vec{L}\cdot\vec{S} \geqslant & e^2/r_{ij} \\ 10^4 \text{ cm}^{-1} & 10^3 \text{ cm}^{-1} & 10^3 \text{ cm}^{-1} \end{array}$$

(D) Lanthanides:

$$\begin{array}{ccc} \lambda\vec{L}\cdot\vec{S} & \geqslant \quad e^2/r_{ij} & > \quad \text{C.F.} \\ 5\times 10^3 \text{ cm}^{-1} & 5\times 10^3 \text{ cm}^{-1} & 10^3 \text{ cm}^{-1} \end{array}$$

The magnetic field effect is about 1 cm^{-1}.

In our analysis so far, we have not taken spin-orbit coupling (the $\lambda\vec{L}\cdot\vec{S}$ term) into account. For first row transition metal ions, this is accomplished by adding the effects of $\vec{L}\cdot\vec{S}$ to the energies as a perturbation on their magnitude. This is a good approximation only when $\lambda\vec{L}\cdot\vec{S}$ is small compared to electron-electron repulsions and crystal field effects. The diagonal $\vec{L}\cdot\vec{S}$ matrix elements are evaluated in the real orbital basis set and added to the energies as corrections. When spin-orbit coupling is large, this perturbation approach is not appropriate. For example, d_2^- and d_1^+ (signs refer to the electron m_s value) have the same m_J value ($^3/_2$) and are mixed by $\vec{L}\cdot\vec{S}$.

Derivation of the Van Vleck Equation

This very general discussion of how to proceed in a crystal field evaluation of the effects of the Hamiltonian in equation (11–25) on the molecule or ion of interest is sufficient for our purposes. We shall now return to a discussion of the influence of these factors on the magnetic moment. When we list the contributions to the energy of a given state, n, from the factors discussed in the earlier section for $S = ^1/_2$ systems, in terms of the field dependence of the effects, equation (11–27) results:

$$E_n = E_n^{(0)} + HE_n^{(1)} + H^2 E_n^{(2)} \qquad (11\text{--}27)$$

$\lambda \text{L}\cdot\text{S}$ first-order Zeeman   second-order Zeeman
                   (diagonal terms)   (off-diagonal terms)

Recalling that the projection of the magnetic moment in the field direction is given by $-\partial E_n/\partial H$ [equation (11–12)], we see that the first term, $E_n^{(0)}$, makes no contribution to the moment of a given state; the second term makes a contribution that is independent of the field strength; and the third term makes a field-dependent contribution. The $E_n^{(1)}$ term in equation (11–27) is the same term that we had in the Curie law derivation, except that the orbital momentum is now included. The magnitude of the second-order contribution will depend upon $E_i - E_j$. It can be very large when the electronic excited state is close in energy to the ground state and has correct symmetry.

In order to determine the influence of these effects on the susceptibility, we return to the earlier Curie law derivation and rewrite equation (11–15) by replacing $\exp(-E_n/kT)$ with

$$\exp\left(\frac{-E_n^{(0)} - HE_n^{(1)} - H^2 E_n^{(2)} + \cdots}{kT}\right) \cong \left(1 - \frac{HE_n^{(1)}}{kT}\right) \exp\left(\frac{-E_n^{(0)}}{kT}\right) \tag{11–28}$$

Also, we let

$$\mu_n = \frac{-\partial E_n}{\partial H} = -E_n^{(1)} - 2HE_n^{(2)} \tag{11–29}$$

Making these substitutions into equation (11–15) leads to:

$$M = N \frac{\sum_n (-E_n^{(1)} - 2HE_n^{(2)}) \left(1 - \frac{HE_n^{(1)}}{kT}\right) \exp\left(\frac{-E_n^{(0)}}{kT}\right)}{\sum_n \exp\left(\frac{-E_n^{(0)}}{kT}\right)\left(1 - \frac{HE_n^{(1)}}{kT}\right)} \tag{11–30}$$

Limiting this derivation to paramagnetic substances, this equation must yield $M = 0$ at $H = 0$ and, in order for this to happen, the following must be true:

$$-\sum_n E_n^{(1)} \exp\left(\frac{-E_n^{(0)}}{kT}\right) = 0 \tag{11–31}$$

Expanding the numerator, and neglecting terms higher than $E_n^{(2)}$ as well as the $E_n^{(2)} E_n^{(1)}$ product in equation (11–30), and recalling that $\chi = M/H$ [equation (11–19)], we obtain from equations (11–30) and (11–31):

$$\chi = N \frac{\sum_n \left[\frac{(E_n^{(1)})^2}{kT} - 2E_n^{(2)}\right] \exp\left(\frac{-E_n^{(0)}}{kT}\right)}{\sum_n \exp\left(\frac{-E_n^{(0)}}{kT}\right)} \tag{11–32}$$

where $E_n^{(0)}$ has contributions from $\lambda \vec{L} \cdot \vec{S}$, etc. The $E_n^{(0)}$ term is always zero for the ground level; for a higher energy state, the quantity giving rise to this energy term in the absence of a field is substituted for $E_n^{(0)}$. The term $E_n^{(1)}$ contains the $m_s g \beta H$ and other first-order contributions, and $E_n^{(2)}$ has the contributions from the second-order Zeeman term.

Thus, the susceptibility is determined by taking a population-weighted average of the susceptibility of the level. *An r-fold degenerate level has r component states, each of which must be included in the summations of equation (11–32).* Its use will become more clear by working out some examples.

Application of the Van Vleck Equation

We shall demonstrate the use of the Van Vleck equation [equation (11–32)] by applying it to the *ground state* of a free metal ion with quantum number J (Russell-Saunders coupling applies). In all of the examples worked out in this section, it is important to appreciate that all we are doing is taking a population-weighted average of the individual moments of the levels. The $2J + 1$ degeneracy is removed by a magnetic field, and the relative energies of the resulting levels are given by $m_J g \beta H$. We are considering only the ground level $E_n^{(0)}$ and $E_n^{(2)}$, which are taken as zero. (In doing a Boltzmann population analysis, the zero of energy is arbitrary; we set the energy of the ground level in the absence of H [i.e., $E_n^{(0)}$] at zero for convenience.) Equation (11–32) becomes

$$\chi = N \sum_{\substack{m_J = -J \\ n = 0}}^{+J} \frac{m_J^2 g^2 \beta^2}{kT(2J+1)}$$

$$= \frac{Ng^2\beta^2}{kT}\left(\frac{J^2 + (J-1)^2 + \cdots + (-J+1)^2 + (-J)^2}{2J+1}\right)$$

$$= \frac{Ng^2\beta^2}{kT}\left(\frac{J(J+1)(2J+1)}{3(2J+1)}\right) = \frac{Ng^2\beta^2}{3kT}J(J+1) \qquad (11\text{–}33)$$

Written in terms of μ_{eff}, we obtain for the free ion

$$\mu_{\text{eff}} = g[J(J+1)]^{1/2} \text{ (BM)} \qquad (11\text{–}34)$$

For a free ion, following the Russell-Saunders coupling scheme, we give without derivation[2] the expression for the g-value as:

$$g = 1 + \frac{S(S+1) - L(L+1) + J(J+1)}{2J(J+1)} \qquad (11\text{–}35)$$

We see from this equation that, in the free ion, contributions to μ_{eff} arise from both the spin and the orbital angular momenta. Furthermore, when $L = 0$, then $J = S$. Then $g = 2.00$ and equation (11–34) reduces to the spin-only formula given in equation (11–24).

The next example selected to illustrate the use of equation (11–32) is a $Ti^{3+}(d^1)$ complex.[3] The splitting of the gaseous ion terms by the crystal field, spin-orbit coupling, and the magnetic field is illustrated in Fig. 11–3. The expressions for the energies given in the figure are obtained from the wave functions resulting from a weak crystal field analysis by operating on them with the $\lambda \hat{L} \cdot \hat{S}$ and $\beta(\hat{L} + g_e\hat{S}) \cdot \vec{H}$ operators of equation (11–25). Since $10Dq$ is generally large in an octahedral complex, we can ignore the ^2E state in evaluating the susceptibility with equation (11–32). We shall discuss this entire problem by starting with the ground $^2T_2$ level and numbering the states 1 to 4 in order of

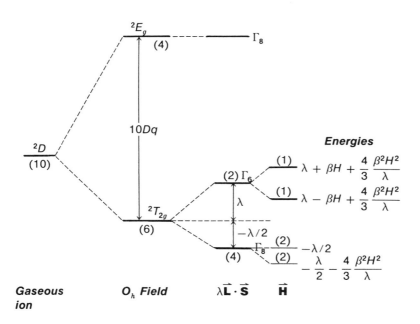

FIGURE 11-3 The splitting of the gaseous ion ^2D state by an O_h field, by $\lambda \hat{\mathbf{L}} \cdot \hat{\mathbf{S}}$ and by a magnetic field. The degeneracies of the levels are indicated in parentheses, and the energies are listed on the right.

increasing energy. The $E_n^{(0)}$ terms for states 1 to 4 are $-\lambda/2$, $-\lambda/2$, λ, and λ, respectively (see Chapter 10). The $E_n^{(1)}$ terms are 0, 0, $-\beta H$, and $+\beta H$, respectively, while the $E_n^{(2)}$ terms are $-\frac{4}{3}(\beta^2 H^2/\lambda)$, 0, $+\frac{4}{3}(\beta^2 H^2/\lambda)$, and $+\frac{4}{3}(\beta^2 H^2/\lambda)$, respectively. Substituting these values into equation (11-32) and multiplying each term by the degeneracy of the corresponding level produces:

$$\frac{\chi}{N} = \left\{ 2\left[\left(\frac{0^2}{kT}\right) + (2)\left(\frac{4}{3}\right)\left(\frac{\beta^2}{\lambda}\right)\exp\left(\frac{\lambda}{2kT}\right)\right] + 2\left[\left(\frac{0^2}{kT}\right) - (2)(0)\exp\left(\frac{\lambda}{2kT}\right)\right] \right.$$
$$\left. + \left[\left(\frac{\beta^2}{kT}\right) - (2)\left(\frac{4}{3}\right)\left(\frac{\beta^2}{\lambda}\right)\right]\exp\left(\frac{-\lambda}{kT}\right) + \left[\left(\frac{\beta^2}{kT}\right) - (2)\left(\frac{4}{3}\right)\left(\frac{\beta^2}{\lambda}\right)\right]\exp\left(\frac{-\lambda}{kT}\right)\right\}$$
$$\div 2\left[\exp\left(\frac{\lambda}{2kT}\right) + \exp\left(\frac{\lambda}{2kT}\right) + \exp\left(\frac{-\lambda}{kT}\right)\right]$$

Recalling that $\chi = (N\beta^2/3kT)\mu_{\text{eff}}^2$, we obtain:

$$\mu_{\text{eff}}^2 = \frac{8 + \left[\frac{3\lambda}{kT} - 8\right]\exp\left(\frac{-3\lambda}{2kT}\right)}{\frac{\lambda}{kT}\left[2 + \exp\left(\frac{-3\lambda}{2kT}\right)\right]} \beta^2$$

where β is the Bohr magneton. Note that our analysis predicts that the Curie law will not hold. As T approaches infinity, μ_{eff}^2 approaches zero, As T becomes small, μ_{eff}^2 becomes small; and as T approaches zero, the equations no longer apply because $g\beta H \approx kT$. As λ approaches zero, μ_{eff}^2 approaches 3. Finally, as T approaches zero, we have the very interesting result that a system with one unpaired electron has zero susceptibility. The result arises because the spin and orbital contributions cancel. These predictions are confirmed by experiment.

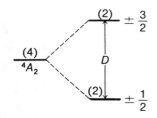

FIGURE 11-4 The splitting of the $^4A_2$ state by a tetragonal field D. (D is the tetragonal splitting or zero-field splitting parameter.)

In our analysis of this problem, we have ignored any contributions from the ^2E_g excited level. However, the above approximation is valid for many magnetic applications. In the more sensitive epr technique, one can detect the contribution from the excited state to the g-value (see Chapter 12). With spin-orbit coupling, the ground level (Γ_8) and the Γ_8 excited level from ^2E_g ($\Gamma_3 \times \Gamma_6 = \Gamma_8$) can mix,[4] changing the g-value from $4\beta H/3\lambda$ to $4\lambda/\Delta + 4\beta H/3\lambda$, where Δ is $10Dq$. The second-order Zeeman term mixes the ground level with the excited level, and the extent of mixing depends upon Δ.

The next system [9] that we shall consider is chromium(III) (d^3). An octahedral field gives rise to a $^4A_2$ ground state and $^4T_2$ and $^4T_1$ excited levels, as shown for the quartet states in the Tanabe and Sugano diagrams. Since $^4T_2$ is about 18,000 cm^{-1} higher in energy than $^4A_2$, its contribution to the susceptibility can be ignored. Since the ground state is orbitally singlet (A), there is no orbital contribution to the magnetic susceptibility (*vide infra*). The magnetism is predicted with the spin-only formula and $S = {}^3/_2$. Next, we shall consider the effect of a tetragonal distortion on chromium(III). This removes the degeneracy of the $m_s = \pm {}^1/_2$ and the $m_s = \pm {}^3/_2$ states as shown in Fig. 11-4. The splitting by the tetragonal component is described by the parameter D. Since this splitting exists in the absence of a field, it is one of the many effects that are referred to as a *zero-field splitting*. For the case of an axial zero-field splitting, one can represent this with the Hamiltonian, $D\hat{S}_z^2$. The susceptibility for this system when the applied field is parallel to the principal molecular axis is obtained by inserting values for $E_n^{(1)}$, which equals $({}^1/_2)g_z\beta$ and $({}^3/_2)g_z\beta$, respectively, for the two levels shown in Fig. 11-4. With $E_n^{(0)}$ given a value of zero for the lower energy level and a value of D for the higher one, we have:

$$\frac{\chi_z}{N} = \left[\frac{2\left(\frac{1}{2}g_z\beta\right)^2}{kT}\exp(0) + \frac{2\left(\frac{3}{2}g_z\beta\right)^2}{kT}\exp\left(\frac{-D}{kT}\right) \right] \bigg/ \left[2\exp(0) + 2\exp\left(\frac{-D}{kT}\right) \right]$$

$$= \frac{g_z^2\beta^2}{4kT}\frac{\left[1 + 9\exp\left(\frac{-D}{kT}\right)\right]}{\left[1 + \exp\left(\frac{-D}{kT}\right)\right]}$$

Thus we see that when $(D/kT) \ll 1$, which is true for a very small distortion or at a very high temperature, the expression for χ_z reduces to $({}^5/_4)Ng_z^2\beta^2/kT$, while the spin-only formula for $S = {}^1/_2$ results as T approaches zero or D becomes very large.* To calculate the powder average susceptibility, χ_x and χ_y must be evaluated using the x and y components of $\hat{L}$ and $\hat{S}$ in equation (11–25) with the $D\hat{S}_z^2$ term added. The anisotropy in χ can be calculated this way.

As a final example, we shall consider a nickel(II) complex with a small tetragonal distortion. The splitting is shown in Fig. 11–5.

---

*Experimentally, D is approximately 0.1 cm^{-1} for pseudooctahedral Cr^{3+} from a spin-orbital mixing in of excited states.

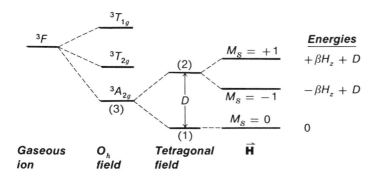

FIGURE 11–5 Splitting of a nickel(II) ion in a tetragonal field.

$$\frac{\chi_\parallel}{N} = \frac{(0)\exp(-0) + \dfrac{(-g_z\beta)^2}{kT}\exp\left(\dfrac{-D}{kT}\right) + \dfrac{(g_z\beta)^2}{kT}\exp\left(\dfrac{-D}{kT}\right)}{1 + 2\exp\left(\dfrac{-D}{kT}\right)}$$

$$\chi_\parallel = \frac{2Ng_z^2\beta^2}{kT}\frac{\exp\left(\dfrac{-D}{kT}\right)}{1 + 2\exp\left(\dfrac{-D}{kT}\right)}$$

When $D \ll kT$, the following expression results:

$$\chi_\parallel = \frac{2Ng_z^2\beta^2\left(1 - \dfrac{D}{kT}\right)}{kT\left(1 + 2 - \dfrac{2D}{kT}\right)} \approx \frac{2Ng_z^2\beta^2}{3kT}\left(1 - \dfrac{D}{kT}\right)$$

For $(NH_4)_2Ni(SO_4)_2 \cdot 6H_2O$, the experimental value[4] of g_z is 2.25 and D is -2.24 cm^{-1}, giving a value of $\chi_\parallel$ of 4260×10^{-6} cm^3 mole^{-1}. The experimental value is 4230×10^{-6} cm^3 mole^{-1}, whereas the spin-only formula predicts a value of 3359×10^{-6} cm^3 mole^{-1}.

There is one additional point to be made in this section. If electrons are delocalized onto the ligands as a result of covalency in the metal ligand bond, the matrix elements corresponding to orbital angular momentum are reduced below the value calculated by using the metal-centered operator $\hat{L}$. To compensate for this, one can use $k\hat{L}$ (where k, *the orbital reduction factor*, is a constant less than one) to correct for delocalization of electron density onto the ligand where it will have reduced orbital angular momentum.

11–4 APPLICATIONS OF SUSCEPTIBILITY MEASUREMENTS

Spin-Orbit Coupling

When equation (11–34) is applied to complexes of the rare earth ions, excellent agreement between calculated and observed susceptibilities results, as shown for some trivalent ions in Table 11–4.

TABLE 11–4. Calculated and Experimental Magnetic Moments for Some Trivalent Rare Earth Ions

Element	Config.	Term	μ_{eff}(calc)	μ_{eff}(exp)
Ce^{3+}	$4f^1$	$^2F_{5/2}$	2.54	2.4
Pr^{3+}	$4f^2$	$^3H_4$	3.58	3.5
Nd^{3+}	$4f^3$	$^4I_{9/2}$	3.62	3.5
Pm^{3+}	$4f^4$	$^5I_4$	2.68	
Sm^{3+}	$4f^5$	$^6H_{5/2}$	0.84	1.5
Eu^{3+}	$4f^6$	$^7F_0$	0	3.4
Gd^{3+}	$4f^7$	$^8S_{7/2}$	7.94	8.0
Tb^{3+}	$4f^8$	$^7F_6$	9.72	9.5
Dy^{3+}	$4f^9$	$^6H_{15/2}$	10.63	
Ho^{3+}	$4f^{10}$	$^5I_8$	10.60	10.4
Er^{3+}	$4f^{11}$	$^4I_{15/2}$	9.59	9.5
Tm^{3+}	$4f^{12}$	$^3H_6$	7.57	7.3
Yb^{3+}	$4f^{13}$	$^2F_{7/2}$	4.54	4.5

TABLE 11–5. Calculated and Observed Magnetic Moments for Complexes of the 3d Ions
(Calculated results are presented using equation (11–34) and the spin-only formula.)

Ion	Config.	Term	$g[J(J+1)]^{1/2}$	$2[S(S+1)]^{1/2}$	μ_{eff}(exp)
Ti^{3+}, V^{4+}	$3d^1$	$^2D_{3/2}$	1.55	1.73	1.7–1.8
V^{3+}	$3d^2$	$^3F_2$	1.63	2.83	2.6–2.8
Cr^{3+}, V^{2+}	$3d^3$	$^4F_{3/2}$	0.77	3.87	~3.8
Mn^{3+}, Cr^{2+}	$3d^4$	$^5D_0$	0	4.90	~4.9
Fe^{3+}, Mn^{2+}	$3d^5$	$^6S_{5/2}$	5.92	5.92	~5.9
Fe^{2+}	$3d^6$	$^5D_4$	6.70	4.90	5.1–5.5
Co^{2+}	$3d^7$	$^4F_{9/2}$	6.63	3.87	4.1–5.2
Ni^{2+}	$3d^8$	$^3F_4$	5.59	2.83	2.8–4.0
Cu^{2+}	$3d^9$	$^2D_{5/2}$	3.55	1.73	1.7–2.2

This excellent agreement is obtained because the crystal field from the ligands does not effectively quench the orbital angular momentum of the electrons in the inner 4f orbitals. A very much different result is obtained with the 3d transition series where, as can be seen in Table 11–5, the spin-only formula comes much closer to predicting the observed results.

Thus, in many of the complexes, the orbital contribution is largely quenched by the crystal field. There is a very simple model that enables one to predict when the orbital moment will not be completely quenched. If an electron can occupy degenerate orbitals that permit circulation of the electron about an axis, orbital angular momentum can result.* There must not be an electron of the same spin in the orbital into which the electron must move.

In an octahedral d^1 complex, for example, the electron can occupy d_{xz} and d_{yz} to circulate about the z-axis, and the complex possesses orbital angular momentum. In an octahedral, d^3 complex, there are electrons with the same spin quantum number in both d_{xz} and d_{yz}, so this ion does not have orbital angular

---

*This rule is often expressed by requiring that a rotation axis exists that enables one to rotate one of the degenerate orbitals containing the electron into another vacant one.

momentum. Using this crude model, we would predict that the following octahedral complexes would have effectively all of the orbital contribution to the moment quenched:

High-Spin:

$$t_{2g}^{3},\ t_{2g}^{3}e_{g}^{1},\ t_{2g}^{3}e_{g}^{2},\ t_{2g}^{6}e_{g}^{2},\ t_{2g}^{6}e_{g}^{3}$$

Low-Spin:

$$t_{2g}^{6}\ \text{and}\ t_{2g}^{6}e_{g}^{1}$$

The one electron in e_g could only occupy d_{z^2} and $d_{x^2-y^2}$ so circulation about an axis cannot result. If an E state in an appropriate ligand field consisted of an electron in d_{xy} and $d_{x^2-y^2}$ orbitals, an orbital contribution would be expected.

For tetrahedral complexes, for which high spin complexes result, the orbital contribution is quenched in e^1, e^2, $e^2t_2^{3}$, $e^3t_2^{3}$, and $e^4t_2^{3}$.

Many molecules with A_{2g} and E_g ground states have moments that differ from the spin-only value. This variation results from two sources: (1) mixing in of excited states that have some contributions from spin-orbit coupling, and (2) second-order Zeeman effects (temperature-independent paramagnetism). For example:

$$\mu_{\text{eff}}(A_{2g}) = \mu(\text{spin only})\left(1 - \frac{4\lambda}{10Dq}\right) + \frac{8N\beta^2}{10Dq} \quad (11\text{--}36)$$

The last term is the temperature independent paramagnetism which is field induced. The $4\lambda/10Dq$ term arises from the mixing in of an excited state via spin-orbit coupling. In lower symmetry complexes, the states are split and more mixing becomes possible. For an E_g ground state, the moment from mixing in an excited state is given by:

$$\mu_{\text{eff}}(E_g) = \mu(\text{spin only})\left(1 - \frac{2\lambda}{10Dq}\right) + \frac{4N\beta^2}{10Dq} \quad (11\text{--}37)$$

As we can see from the above discussion, the magnetic moments of transition metal ion complexes are often quite characteristic of the electronic ground state and structure of the complex. There have been many reported examples of this kind of application. A few nickel(II) and cobalt(II) complexes will be discussed here to illustrate this application.

In an octahedral field, nickel(II) has an orbitally nondegenerate ground state, $^3A_2(t_{2g}^{6}e_g^{2})$, and no contribution from spin-orbit coupling is expected. The measured moments are in the range from 2.8 to 3.3 BM, very close to the spin-only value of 2.83 BM. Values for octahedral complexes slightly above the spin-only value arise from slight mixing with a multiplet excited state in which spin-orbit coupling is appreciable. Tetrahedral nickel(II) has a $^3T_1$ ground state that is essentially $(e_g^{4}t_{2g}^{4})$, and a large orbital contribution to the moment is expected. As a result, even though both octahedral and tetrahedral nickel(II) complexes contain two unpaired electrons, tetrahedral complexes have magnetic moments around 4 BM compared to 3.3 BM or less for octahedral complexes. Experimentally, it is found[8,10] that $NiCl_4^{2-}$, $Ni(HMPA)_4^{2+}$ [10a] (HMPA = hexamethylphosphoramide), and [10b] $NiX_2 \cdot 2(C_6H_5)_3AsO$ (X = halogen) have moments in excess of 4 BM. An inverse relationship exists between the magnitude of the moment and the distortion of nickel(II) complexes from tetrahedral symmetry. The complex $NiX_2 \cdot 2(C_6H_5)_3P$, which is known[11] to be seriously distorted, has a moment of about 3 BM. The structures of $NiX_2 \cdot 2HMPA$ and $CoX_2 \cdot 2HMPA$ (where X = Cl^-, Br^-, I^-, NO_3^-) were deduced from combined magnetic and spectroscopic studies.[12] This work provides a good illustration of the use of these techniques.

In octahedral cobalt(II) complexes, the ground state is $^4T_{1g}$ and a large orbital contribution to the moment is expected. Mixing in of an excited state lowers the moment somewhat but a value in excess of 5 BM is usually found [μ(spin-only) = 3.87 BM]. The ground state for tetrahedral cobalt(II) complexes is $^4A_2$ and a low moment approaching the spin-only value might be expected. However, an excited magnetic state is comparatively low in energy in the tetrahedral complexes and can be mixed with the ground state. Moments in the range from 4 to 5 BM have been predicted[13] and are found experimentally. An inverse relationship exists for tetrahedral cobalt(II)[13] complexes between the magnitude of the moment of a complex and the value of Dq as predicted by equation (11–36).

11–5 INTRAMOLECULAR EFFECTS

In our treatment so far, we have assumed that there is no interaction between the electron spins on the individual metal ions in the solid. Next we wish to consider molecules containing more than one metal ion with unpaired spins. Consider the planar $M_2L_4X_2$ system:

$$\begin{array}{c} L \\ L \end{array} \! \! >\!\! M \!\! <\!\! \begin{array}{c} X \\ X \end{array} \!\! >\!\! M \!\! <\!\! \begin{array}{c} L \\ L \end{array}$$

The influence of one metal, i, on the other, j, can be treated as arising from pairwise interactions of metal d orbitals on centers i and j leading dimeric m.o.'s, ψ_i and ψ_j. When two d orbitals, each containing one electron, are orthogonal, the two m.o.'s are degenerate, a triplet state arises, and the coupling is said to be ferromagnetic. When the two a.o.'s undergo a bonding interaction, ψ_i and ψ_j differ in energy. If the energy difference is greater than the pairing energy, a singlet state arises and the metals are said to be antiferromagnetically coupled. The Hamiltonian for the interaction between a pair of spins on atoms i and j is given in equation (11–38)

$$\hat{H} = J\hat{S}_i \cdot \hat{S}_j \tag{11-38}$$

In the literature, one finds equation (11–38) written with $-J$, $2J$, or $-2J$ instead of J. With the convention in equation (11–38), J is positive for an antiferromagnetic interaction and negative for a ferromagnetic system.

Substituting the energies for $S_1 = S_2 = \frac{1}{2}$ from Fig. 10–8 into equation (11–32) produces

$$\frac{\chi}{N} = \frac{\dfrac{2(g\beta)^2}{kT}\exp\left(\dfrac{-J}{4kT}\right)}{3\exp\left(\dfrac{-J}{4kT}\right) + \exp\left(\dfrac{+3J}{4kT}\right)} = \frac{\dfrac{2g^2\beta^2}{kT}}{3 + \exp\left(\dfrac{J}{kT}\right)}$$

Rearranging, we find

$$\chi = \frac{2Ng^2\beta^2}{3kT}\frac{1}{1 + \dfrac{\exp(J/kT)}{3}}$$

Since $N\beta^2/3k$ is about $\frac{1}{8}$ we obtain when $g = 2$

$$\chi \cong \frac{1}{T}\frac{1}{1 + \dfrac{\exp(J/kT)}{3}} \tag{11-39}$$

From this expression we find that χ approaches zero as T approaches infinity and becomes small as T becomes small. As a result χ must have a maximum, which is given by setting $\partial \ln \chi/\partial T$ equal to zero. From this we find

$$\frac{J}{kT_N} \cong \frac{8}{5} \tag{11-40}$$

i.e., χ has a maximum value at

$$T_N = \frac{5}{2}\frac{J}{k}$$

When $J \ll kT$ or when $T \gg T_N$, the susceptibility follows the Curie-Weiss law, i.e.,

$$\chi = \frac{3}{4}\frac{1}{T + \theta} \tag{11-41}$$

where $\theta = J/4k$. We also see that as the quantity J approaches positive infinity, χ approaches zero.

Magnetic susceptibility measurements have provided information[14] concerning the influence of electronic and structural change on the magnitude of J, the energy separation of the triplet and singlet states. A molecular orbital analysis[15] provides a model to explain these influences on the magnitude of J. To a first approximation, the energy separation of the singlet E_S and triplet E_T states is given by

$$J = E_T - E_S = -2K_{ab} + \frac{(\varepsilon_1 - \varepsilon_2)^2}{H_{aa} - H_{ab}} \tag{11-42}$$

where H refers to a two-electron Coulomb integral, K an exchange integral, and ε_1 and ε_2 the m.o. (ψ_1 and ψ_2) energies. When the two m.o.'s are degenerate, $\varepsilon_1 = \varepsilon_2$, $J = -2K_{ab}$ and a ferromagnetic system results. When the energies differ, the second term dominates and the splitting leads to a singlet ground state. The value of J, which corresponds to the energy difference of the singlet and triplet states, has contributions from a direct overlap of the metal orbitals and from an indirect mechanism (referred to as superexchange) in which the orbitals interact through the intermediacy of lone pairs or π-electrons of a bridging ligand.[14] Orbital interactions involving orbitals on the bridging atom that can lead to antiferromagnetic or ferromagnetic interactions are shown in Fig. 11–6(A) and (B), respectively,[16]

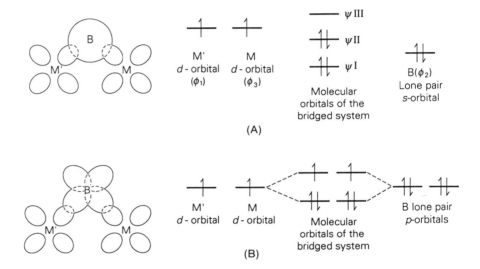

FIGURE 11–6 Orbital overlap and molecular orbitals leading to (A) an antiferromagnetic interaction and (B) a ferromagnetic interaction. M represents the same metal with one electron in its d orbital and B the bridging group. In (A) B has a pair of electrons in an s orbital and in (B) pairs of electrons in two orthogonal p orbitals.

In complexes in which the M—B—M angle is 90°, the situation is close to that in Fig. 11–6(A) with the low energy $2s$ orbital of chlorine making only small contributions to the bonding. As the angle changes to 180°, more s character is employed in the chlorine hybrids used to bridge the two atoms and the situation described in Fig. 11–6(A) results. Experimentally, larger values of J are found as the bridge angle increases to 180°. The J values of a large number of di-μ-oxo-bridged copper(II) complexes whose structures are known from x-ray crystallography have been compiled.[17] For square planar and square pyramidal geometries the unpaired electron resides in $d_{x^2-y^2}$ orbitals. The J values for the dihydroxo-bridged complexes vary linearly with the bridge angle being ferromagnetically coupled when the angle is less than 96°. At angles of 105° the J value is greater than 500 cm^{-1}. A general trend of increasing J with increasing angle is also observed in this system for bridging alkoxo, phenoxo, keto, and N-oxo groups. Additional studies involving a binuclear Schiff base ligand system shows that J decreases from 660 cm^{-1} for the Cu—Cu pair to 150 cm^{-1} for a Cu—Ni

---

*The combination $-S_1(S_1 + 1) - S_2(S_2 + 1)$ is constant for each level of the dimer and is frequently dropped; i.e., the zero of energy is redefined.

pair. Ferromagnetic systems arise for Cu—V(IV)O and Cu—Cr(III), which have electrons in orthogonal $d_{x^2-y^2}$ and d_{xy} orbitals. In higher symmetry complexes, with triply degenerate T ground states, the problem becomes more complicated with the spin Hamiltonian formalism breaking down and more than one parameter being needed to describe the exchange interaction.

A bimetallic doubly bridged oxo system formed by coordinating *bis*(hexafluoroacetylacetonato) copper(II) to the oxygens of a Schiff base copper(II) complex was found[18] to have a J of only 20 cm^{-1}. The ligand distortions in the adduct, were determined by x-ray diffraction allowing an assignment of the location of the $d_{x^2-y^2}$ orbital containing the unpaired electron on the two centers. Of the two bridging oxo groups only one is common to the two $d_{x^2-y^2}$ orbitals. As a result, this complex does not obey the above correlations for doubly bridged oxo dimers. The reader is referred to extensive literature[19-21] for additional studies of the influence of geometry and orbital energies on the values of J.

11–6 HIGH SPIN–LOW SPIN EQUILIBRIA

If one examines the Tanabe and Sugano diagrams for d^4, d^5, d^6, and d^7 octahedral complexes, it can be seen that for certain values of Dq/B the ground state changes from a high spin to a low spin complex. For d^4 the ^5E_g and $^3T_{1g}$ states are involved, whereas d^5 involves $^6A_{1g}$ and $^2T_{2g}$, d^6 involves $^5T_{2g}$ and $^1A_{1g}$, and d^7 involves $^4T_{1g}$ and ^2E_g. When the ligand field is such that the two states are close in energy, the excited state can be thermally populated and the system will consist of an equilibrium mixture of the two forms. There have been many reported studies of this type of behavior.[22a] A typical system, which is selected for discussion because it has been studied[22b] in the solid state and in solution, involves iron(II) complexes of ligands that are Schiff base type condensation products of tren [N(CH$_2$CH$_2$NH$_2$)$_3$] and 2-pyridinecarboxaldehyde. The resulting complex is shown in Fig. 11–7, where only one of the pyridine aldimines has been drawn in for clarity of presentation. The whole series of compounds in which R, R', and R'' are —H or —CH$_3$ were prepared.

The symbol (I) will be used to abbreviate the complex [Fe(Py)$_3$tren]$^{2+}$ where R = R' = R'' = H; (II) symbolizes [Fe(6MePy)(Py)$_2$tren]$^{2+}$ where R = R' = H and R'' = CH$_3$; (III) symbolizes [Fe(6MePy)$_2$(Py)tren]$^{2+}$ where R = H and R' = R'' = CH$_3$; and (IV) symbolizes [Fe(6MePy)$_3$tren]$^{2+}$ where R = R' = R'' = CH$_3$.

An equilibrium involving the low spin $^1A_{1g}(t_{2g}{}^6)$ and high spin $^5T_{2g}(t_{2g}{}^4 e_g{}^2)$ states was found both in the solid state and in solution for several of these complexes. Complex (I) is fully low spin at and below room temperature, while (II) and (III) undergo spin equilibrium both in the solid state and in solution. In solution, complex (IV) is essentially high spin at all temperatures above 180 K. In the solid state, a spin equilibrium exists that is very anion dependent. Thermodynamic data for the interconversion can be determined from the change in susceptibility with temperature, and are reported to be +4.6 and +2.8 kcal mole^{-1}, respectively, for complexes (II) and (III) in solution. An x-ray crystal structure determination indicated that the methyl substituents on the pyridine ring interact with the adjacent pyridine moiety. Thus, the ligand field is weakened in compound (IV) to the extent that a high spin compound results, while compound (I) is low spin. In the case of compound (IV), the average metal nitrogen

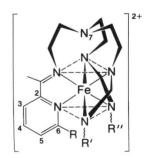

FIGURE 11–7 Structural formula of tris{4-[(6-R)-2-pyridyl]-3-aza-3-butenyl} amine iron(II) complex.

distance is found to decrease by about 0.12 Å in going from the high spin to the low spin complex.

11-7 MEASUREMENT OF MAGNETIC SUSCEPTIBILITIES

In this section, the measurement of bulk magnetic susceptibility will be briefly covered[23] in the course of presenting the pertinent references for a more complete discussion. The Gouy method[23b] employs a long, uniform glass tube packed with the solid material or solution, which is suspended in a homogeneous magnetic field. The sample is weighed in and out of the field, and the weight difference is related to the susceptibility and field strength. If a standard of known susceptibility is used, the field strength need not be known. Evans[23c] has reported a very clever and inexpensive device for routine measurement of the magnetic susceptibility by the Gouy method.

The Faraday method[23a] uses a small amount of sample that is suspended in an inhomogeneous field such that $H(\partial H/\partial X)$ is a constant over the entire volume of the sample. The method is very sensitive, so small samples can be used and studies can be made on solutions.

Susceptibilities can also be determined conveniently over a wide temperature range, down to liquid helium temperatures, with a vibrating sample magnetometer.[23d] The change in the inductance of a coil upon insertion of a sample can be related to the sample susceptibility. A mutual inductance bridge has been described and used for susceptibility determination.[23e,f] An ultrasensitive, superconducting quantum magnetometer with a Josephson junction element has also been described.[24]

When one studies the susceptibility of single crystals, the anisotropy in the susceptibility can be determined.[25-28] This information has several important applications, as we shall see in our study of the nmr and epr of transition metal ions. The Krishnan critical torque method has been commonly used.[29,30]

An nmr method has been reported[31] for the measurement of the magnetic susceptibility of materials in solution. In this method, a solution of the paramagnetic complex containing an internal standard is added to the inner of two concentric tubes. A solution of the same inert standard, dissolved in the same solvent containing the dissolved complex, is placed in the outer of the two concentric tubes. Two separate nmr lines corresponding to the standard will be observed, with the line from the paramagnetic solution lying at higher frequency.

The contribution to the absolute chemical shift of a nucleus from the bulk magnetic susceptibility of the sample $(\Delta v/v_0)_\chi$, is given by[32]:

$$\left(\frac{\Delta v}{v_0}\right)_\chi = m\left(\frac{1}{3} - \alpha\right)\left[\chi - \chi_0\left(1 + \frac{\rho_0 - \rho_s}{m}\right)\right] \qquad (11\text{-}43)$$

where Δv is the shift in H_z, v_0 the spectrometer frequency, m the mass of solute per unit volume, α a demagnetization factor (*vide infra*), χ and χ_0 the mass susceptibility of the solute and solvent, respectively, and ρ_o and ρ_s the densities of the solvent and solution respectively. χ has contributions from a paramagnetic χ^P and diamagnetic term χ^D as follows:

$$\chi = \chi^P - \chi^D \qquad (11\text{-}44)$$

For a spherical sample $\alpha = 1/3$ and the contribution to the shift from bulk susceptibility vanishes. For a cylindrical sample whose cylinder axis is perpendicular to the magnetic field direction, $\alpha = 1/2$. In most superconducting magnets the magnetic field direction is parallel to the cylinders axis, and $\alpha = 0$. Thus, $[(1/3) - \alpha]$ equals $-1/6$ in the former case and $1/3$ in the latter.

The shift of the standard in the paramagnetic solution is measured as just described. The measurement is repeated for an analogous diamagnetic complex (*e.g.*, the zinc complex). If the concentration of paramagnetic and diamagnetic complex used is the same, the difference in the two shifts $\Delta(\Delta v/v_o)$ is related to the mass susceptibility χ_g^P by:

$$\Delta\left(\frac{\Delta v}{v_0}\right) = m\left(\frac{1}{3} - \alpha\right)\chi^P \qquad (11\text{--}45)$$

The paramagnetic contribution to the molar susceptibility χ^P, which is related to μ_{eff} (equation (11–23)), is given by:

$$\Delta\left(\frac{\Delta v}{v_0}\right) = 1000\, M\left(\frac{1}{3} - \alpha\right)\chi^P \qquad (11\text{--}46)$$

Errors can arise if the reference material undergoes a chemical interaction with the paramagnetic center. It is best to repeat the experiment with different types of references.

11–8 SUPERPARAMAGNETISM*

Superparamagnetism is observed in very small particles of transition metals and their compounds, particularly their oxides. It can be used to characterize fine dispersions of metal(s), alloy(s), and their oxides and has applications in several areas including catalysis.

It should be noted that parameters such as the magnetization per cubic centimeter (M) or magnetization per gram (σ) are better parameters than the susceptibility (χ) for describing superparamagnetism and magnetically ordered ferro- and ferrimagnetic materials. Magnetization (M) is expressed in units of "gauss"; this unit in reality stands for "gauss per cm^3." Quite often "emu/cm^3" is employed. Dividing gauss by the density (g/cm^3) gives "gauss cm^3 per g," which is an awkward unit for the magnetization per gram (σ). This situation is often circumvented by naming it "emu/g." It should be stressed that "emu" is an abbreviation for electromagnetic system of units in the cgs system and "emu" itself is not really a unit. The applied magnetic field (H) is expressed in oersteds (Oe). Furthermore, the magnetic moment in ferro-, ferri-, and superparamagnetism

---

*This contribution provided by L. N. Mulay is an abbreviated, modified form of his Chapter 42 in "Catalyst Characterization Science," Ed. M. L. Deviney and J. L. Gland, A. C. S. Symposium Series No. 288, Washington, D.C. (1985). This material is reproduced by permission of the author and the American Chemical Society.

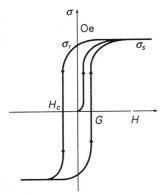

FIGURE 11–8 Schematic diagram of a hysteresis curve for a typical ferromagnet showing magnetization (σ) as a function of the applied magnetic field (H). Saturation magnetization is indicated by σ_s.

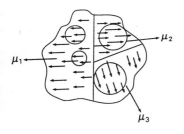

FIGURE 11–9 Schematic representation of a multidomain ferromagnetic particle in the unmagnetized state. Each of the three domains with net moments μ_1, μ_2, μ_3 remain randomized in this state. Circles of varying sizes represent the subdomain superparamagnetic clusters.

refers to the saturation moment (μ or μ_s) and not the effective moment (μ_{eff}) which is important in paramagnetism;[33,34]

$$\mu_{\text{eff}} = \sqrt{n(n+2)}$$

where n is the number of unpaired spins associated with discrete species such as ions and free radicals. The saturation moment is given by

$$\mu_s = n \quad (11\text{–}47)$$

This situation arises because n, the number of spins in a cluster, is very large—of the order of 10^3. A discussion of various systems of units, with factors for conversion to SI units is given by Mulay and Boudreaux[33] and by Mulay and Mulay[34], who discuss many applications to catalysis.

Superparamagnetism is best understood by comparing it with paramagnetism and ferromagnetism. Ferromagnetic metals, such as Fe, Co, Ni, and insulators such as $\gamma\text{-Fe}_2\text{O}_3$ and Fe_3O_4, show the well-known hysteresis curve (Fig. 11–8), which stems from their domain structure. In their unmagnetized state, the unpaired electrons associated with each atom (or structural unit) have a net (spontaneous) magnetic moment (μ_s) or magnetization (σ_s) which is the vector sum of all unpaired electrons in that domain; these are shown by the large arrows in Fig. 11–9. Saturation is shown by σ_s, the remanence, the remaining magnetic induction by σ_r, and the reverse field necessary to bring σ_r to 0 is given by H_c, the coercive force. Between the domains, a "Bloch wall" is formed, with spins curled up in a helical fashion. The magnetization present within a ferromagnetic domain at zero degrees Kelvin and with zero applied field is called the spontaneous, intrinsic, or technical magnetization (I_{sp}). The *subdomain* clusters with varying sizes represent superparamagnetic clusters (Fig. 11–9), with no interaction between neighbors. These clusters are thermally unstable, that is, their magnetic moments (represented by moment vectors) experience thermal fluctuations with great ease, as is the case with paramagnetic species due to the lack of long-range ordering.

It is well known that when ferromagnetic and ferrimagnetic materials are heated above a critical Curie temperature (T_c), they change over to paramagnetic behavior; thus the hysteresis disappears. In contrast, ideal superparamagnetic systems, when sufficiently cooled below a critical blocking temperature (T_B) will experience a very slow relaxation time. Their net magnetic moment will align parallel to the applied field (H) and appear to behave as if they had an apparent "bulk-like" ferromagnetic behavior. This aspect will result in hysteresis of "apparent" ferromagnetic behavior. Conversely, above the T_B, the hysteresis will disappear and the clusters will show a unique curve with no hysteresis (Fig. 11–10).

Each ferromagnetic domain consists of myriads of spins. One can imagine small clusters within each domain having different volumes ($V_1, V_2, \ldots, V_i$). These *subdomain* particles, which consist of several thousand spins are the superparamagnetic clusters. They will have large magnetic moments of several thousand BM (Bohr magnetons: recall one BM $= eh/4\pi mc$). When such superparamagnetic clusters are well-dispersed on (or "within") a substrate (e.g., SiO_2, carbon, zeolites, etc.), there is no magnetic interaction between their moments. Hence, the

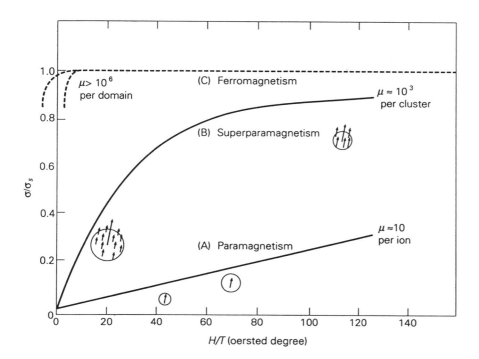

FIGURE 11–10 Plot of the relative magnetization σ/σ_s, as a function of H/T. (A) A paramagnetic system is characterized by an effective moment with a maximum Bohr Magneton number of ~ 10 per ion and by the absence of hysteresis. The very small slope of the straight line is exaggerated for clarity. Paramagnetic saturation occurs at a very high ratio $H/T = 10^4$ Oe/K. (B) Langevin curve for superparamagnetic clusters. Large clusters with relatively high moments are shown to reach saturation easily at lower values of H/T. Similarly small clusters with a net small moment are shown to reach saturation with difficulty, at high values of H/T. (C) Top part of a hysteresis curve observed in a ferromagnetic curve.

superparamagnetic clusters behave in the same manner as magnetically dilute paramagnetic ions. For paramagnetic ions, the susceptibility ($\chi = \sigma/H$) and the μ_{eff} are good parameters for the interpretation of magnetic results. However, for the superparamagnetic clusters with a giant BM number of $\sim 10^4$, the per gram magnetization (σ) is a good parameter. The σ increases with increasing H, and with decreasing temperature (T). Therefore, it is customary to measure σ at constant field strength and varying temperature, and vice versa. When the relative magnetization σ/σ_s (where σ_s is the saturation magnetization) is plotted as a function of H/T, one obtains an excellent superposition of data points, and there is no coercive force ($H_c = 0$). Therefore, hysteresis is not observed in Fig. 11–8 in contrast to the bulk ferromagnetic material. Hence, superparamagnetic behavior may be regarded as a phenomenon intermediate between paramagnetism and ferromagnetism. Note that for pure ferromagnetic materials, without mechanical strain, the H_c can be as small as 0.5 oersteds.

Since "superparamagnetism," as the name implies, is similar to paramagnetism, one can apply the classical Langevin function, first derived for the non-interacting paramagnetic spins, to the non-interacting, ideal S.O. clusters. Thus,

$$\sigma/\sigma_s = \coth(\mu_c H/kT) - (kT/\mu_c(H)) \qquad (11\text{--}48)$$

Here, μ_c stands for the (giant) magnetic moment of the cluster which replaces the μ for single, isolated paramagnetic spins; k is the Boltzmann constant.

In the above equation, all quantities except μ_c can be measured. The μ_c can be derived for (ideal) superparamagnetic clusters by a curve-fitting procedure (*cf.* Ref. 3). The gist of obtaining the average volume (v) of a superparamagnetic cluster lies in the basic definition of σ_s, which for a superparamagnetic system can be written as $\sigma_s = \mu_c/v$. It should be noted that at the atomic level, the molar saturation magnetization (χ_M) is given by

$$\chi_M = N\beta\mu_s \quad \text{emu per mole}$$

here N is the Avogadro's number of species, each carrying a moment $\mu_s = n$ (the number of unpaired electrons per species), and β is the Bohr magneton. (A convenient number, $N\beta = 5585$).

From the low-field (LF) and high-field (HF) approximations[35,36] of the Langevin function stated below, one can calculate the v_{LF} for large clusters (which magnetically saturate easily at low values of H/T) and v_{HF} for small clusters (which saturate with difficulty at high values of H/T). From the v_{LF} and v_{HF}, the average volume v of clusters can be estimated by taking the arithmetic mean.[35,36]

$$\bar{d}_{LF}^3 = \frac{18k}{\pi I_s}(\sigma/\sigma_s)/(H/T) \qquad (11\text{-}49)$$

$$\bar{d}_{HF}^3 = \frac{6k}{\pi I_s}(1 - \sigma/\sigma_s)^{-1}/(H/T) \qquad (11\text{-}50)$$

Here I_s is the spontaneous magnetization for the bulk material. For elemental Fe, I_s is 1707 gauss (or oersted) at room temperature. Cluster diameters as small as 2 nm and ranging up to 20 nm have been deduced from magnetic results for various catalysts such as Fe, Fe—Co, Fe-Ru, and Ni, supported on high surface area alumina, silica and carbon.[34–36]

REFERENCES CITED

1. A. P. Ginsberg and M. E. Lines, Inorg. Chem., *11*, 2289 (1972).
2. G. Herzberg, "Atomic Spectra and Atomic Structure," 2nd ed., Dover Publications, New York 1944, 109.
3. M. Kotani, J. Phys. Soc. Japan, *4*, 293 (1949).
4. B. Bleaney and K. W. H. Stevens, Rep. Prog. Phys., *16*, 108 (1953).
5. F. E. Mabbs and D. J. Machin, "Magnetism and Transition Metal Complexes," John Wiley, New York, 1973.
6. a. B. N. Figgis and J. Lewis, "The Magnetic Properties of Transition Metal Complexes," in "Progress in Inorganic Chemistry," Volume 6, ed. F. A. Cotton, Interscience, New York, 1964.
   b. B. N. Figgis and J. Lewis, "The Magnetochemistry of Complex Compounds," in Modern Coordination Chemistry," eds. J. Lewis and R. G. Wilkins, Interscience, New York, 1960.

7. R. S. Nyholm, J. Inorg. Nuclear Chem., *8*, 401 (1958).
8. N. S. Gill and R. S. Nyholm, J. Chem. Soc., *1959*, 3997.
9. R. L. Carlin, J. Chem. Educ., *43*, 521 (1966).
10. **a.** J. T. Donoghue and R. S. Drago, Inorg, Chem., *1*, 866 (1962) and references therein.
    b. F. A. Cotton et al., J. Amer Chem. Soc., *83*, 4161 (1961) and references therein.
11. L. M Venanzi, J. Chem. Soc., *1958*, 719.
12. J. T. Donoghue and R. S. Drago, Inorg. Chem., *2*, 572 (1963).
13. F. A. Cotton et. al., J. Chem. Soc., *1960*, 1873.
14. G. F. Kokoszka and G. Gordon, Trans. Metal Chem., *5*, 181 (1969).
15. P. J. Hay, J. C. Thibeault and R. Hoffmann, J. Am. Chem. Soc. *97*, 4884 (1975).
16. R. S. Drago, Coord. Chem. Rev., *32*, 97 (1980).
17. D. Gatteschi, "The Coordination Chemistry of Metalloenzymes," ed. I. Bertini, R. S. Drago, and C. Luchinat, D. Reidel Dordrecht, Holland, 1982.
18. K. A. Leslie, R. S. Drago, G. D. Stucky, D. J. Kitko and J. A. Breese, Inorg Chem. *18*, 1885 (1979).
19. **a.** B. N. Figgis and R. L. Martin, J. Chem. Soc., *1956*, 3837.
    b. E. Kokot and R. L. Martin, Inorg. Chem., *3*, 1306 (1964) and references therein.
    c. W. E. Hatfield in "Theory and Applications of Molecular Paramagnetism" ed. E. A. Boudreaux and C. N. Mulay, Wiley, New York, 1976.
    d. D. J. Hodgson, Prog. Inorg. Chem., *19*, 173 (1975); "Extended Interactions between Metal Ions in Transition Metal Complexes, ed." L. V. Interrante, ACS Symposium Series No. S, ACS, Washington, D.C., 1974.
20. J. A. Bertrand, J. H. Smith and P. G. Eller, Inorg. Chem. *13*, 1649 (1974).
21. B. K. Teo, M. B. Hall, R. F. Fenske and L. F. Dahl, J. Organomet-Chem., *70*, 413 (1979).
22. **a.** R. L. Martin and A. H. White, Trans. Metal Chem., *4*, 113 (1968).
    b M. A. Hoselton, L. J. Wilson and R. S. Drago, J. Amer. Chem. Soc., *97*, 1722 (1975).
23. **a.** See, for example, L. N. Mulay and I. L. Mulay, Anal. Chem., *44*, 324R (1972).
    b. B. Figgis and J. Lewis, "Techniques of Inorganic Chemistry," eds. H. Jonassen and A. Weissberger, Volume IV, p. 137, Interscience, New York, 1965.
    c. D. F. Evans, J. Physics E; Sci. Instr., *7*, 247 (1974).
    d. S. Foner, Rev. Sci. Instr., *30*, 548 (1959).
    e. W. L. Pillinger, P. S. Jastram and J. G. Daunt, Rev. Sci. Instr., *29*, 159 (1958).
    f. J. N. McElearney, G. E. Shankle, R. W. Schwartz and R. L. Carlin, J. Chem. Phys., *56*, 3755 (1972).
24. J. W. Dawson et al., Biochemistry, *11*, 461 (1972) and references therein.
25. S. Mitra, Trans. Metal Chem., *7*, 183 (1972).
26. M. Gerloch et al., J. Chem. Soc., Dalton, *1972*, 1559; ibid., *1972*, 980 and references therein.
27. W. D. Horrocks, Jr. and J. P. Sipe, III, Science, *177*, 944 (1972).
28. B. N. Figgis, L. G. B. Wadley and M. Gerloch, J. Chem. Soc., Dalton, *1973*, 238.
29. K. S. Krishnan et al., Phil. Trans. Roy. Soc. A, *231*, 235 (1933).
30. D. A. Gordon, Rev. Sci. Instr., *29*, 929 (1958).
31. **a.** D. F. Evans, J. Chem. Soc., *1959*, 2005.
    b. D. Ostfeld and I. A. Cohen, J. Chem. Educ., *49*, 829 (1972).
32. W. C. Dickinson, Phys. Rev., *81*, 717 (1951).
33. L. N. Mulay and E. A. Boudreaux (Eds.), "Theory and Applications of Molecular Paramagnetism" and "Theory and Applications of Molecular Diamagnetism," Wiley Interscience, New York 1976.
34. L. N. Mulay and Indu Mulay, "Static Magnetic Techniques and Applications," Chapter 3 in "Physical Methods of Chemistry," Vol. III B, ed. B. W. Rossiter and J. F. Hamilton, Wiley Interscience Publishers, New York, 1989.
35. H. Yamamura and L. N. Mulay, J. Appl. Phys., *1979*, 50, 7795 (1979).
36. P. W. Selwood, "Chemisorption and Magnetization," Academic Press, New York, 1975.

EXERCISES

1. Consider the following system of energy levels, which could arise from zero-field splitting of an $S = 1$ state:

$$E(|0\rangle) = -\frac{2}{3}D - \frac{g^2\beta^2 H^2}{3D}$$

$$E(|-1\rangle) = \frac{1}{3}D - g\beta H + \frac{g^2\beta^2 H^2}{6D}$$

$$E(|1\rangle) = \frac{1}{3}D + g\beta H + \frac{g^2\beta^2 H^2}{6D}$$

Here D is the zero-field splitting parameter and all other symbols have their usual meanings. Determine the molar paramagnetic susceptibility of this system.

2. The p^1 configuration of a free ion has a ^2P ground state. Consider the ion in a magnetic field directed along the z-axis.

   a. What is the degeneracy of this state?

   b. Write the wave functions for this state in Dirac notation, i.e., $|M_L, M_S\rangle$.

   c. The major perturbation on this state is spin-orbit coupling:

   $$\hat{H} = \lambda \vec{L} \cdot \vec{S}$$

   Consider the major contribution in a strong z field leading to:

   $$\hat{H} \cong \lambda \hat{L}_z \hat{S}_z$$

   Does this give diagonal elements, off-diagonal elements, or both, in the basis set of part b? Evaluate the energies of these wave functions using $\hat{H}$.

   d. The Zeeman Hamiltonian is $\hat{H} = \beta \vec{H} \cdot (\vec{L} + 2\vec{S}) = \beta H_z(\hat{L}_z + 2\hat{S}_z)$. Evaluate the energies of wave functions of part b using this Hamiltonian.

   e. Use the energies obtained in parts c and d, and the Van Vleck equation, to calculate χ.

3. The following represents the approach taken to the magnetic susceptibility data of a complex of the form shown as follows.
   Consider octahedral Fe(II). In many systems, the low spin $^1A_{1g}$ state is less than 1000 cm^{-1} below the high spin $^5T_{2g}$ state, so that both are appreciably populated at room temperature. The ^1A state is non-degenerate and the ^5T is 15-fold degenerate (why?). Mathematically, a T state in an octahedral field is equivalent to a P state in a free ion; i.e., it may be considered to have $L = 1$ and $M_L = -1, 0, +1$. Consider the ^1A state to be unshifted by all perturbations.

   a. Write the 15 basis functions for the ^5T state in Dirac notation, using T-P equivalence. Hint: The wave function having $M_L = 1$ and $M_S = -2$ is denoted $|1, -2\rangle$. In this formulation, M_L and M_S are often referred to as "good quantum numbers."

[Structural diagram showing a tripodal ligand with three CH₂CH₂-N=C-pyridine arms (labeled R', R'', R''') reacting with Fe(II) to form an octahedral complex with M at the center coordinated by six N atoms.]

b. Now, let the complex undergo a trigonal distortion, *i.e.*, one for which the octahedron's threefold axis is retained. The effect of this distortion is described phenomenologically by the Hamiltonian

$$\hat{H} = \delta\left(\frac{2}{3} - \hat{L}_z^2\right)$$

where δ is the trigonal distortion or zero-field splitting parameter and $\hat{L}_z$ is the z-component of the equivalent orbital angular momentum. As is often the case, chemical factors influencing the magnitude of δ are poorly understood, and values obtained in experiments such as this may help elucidate the factors involved. The Zeeman Hamitonian for a magnetic field oriented along z is:

$$\hat{H} = \beta H_z(\hat{L}_z + g_e \hat{S}_z)$$

Here we are considering only the z-direction (parallel to the threefold axis), so our final expression will be for $\chi_\parallel$.

The following illustrate the use of the Hamiltonian and wave functions:

$$\left(\frac{2}{3} - \hat{L}_z^2\right)|1,-2\rangle = \left(\frac{2}{3} - M_L^2\right)|1,-2\rangle = \left(\frac{2}{3} - 1\right)|1,-2\rangle = -\frac{1}{3}|1,-2\rangle$$

$$\hat{S}_z|1,-2\rangle = M_S|1,-2\rangle = -2|1,-2\rangle$$

Apply the two terms of the Hamiltonian discussed above to the ^5T wave functions derived in part a. Show that all off-diagonal elements of the 15 × 15 matrix must be zero. (This means that all $E^{(2)}$ terms of the Van Vleck equation will be zero.) Hence, determine the energies of the 15 wave functions. Confirm that, as the total Hamiltonian has been constructed, the center of gravity is maintained;

i.e., that $\hat{H} = \delta\hat{L}_z^2$, though giving the correct splitting pattern would not maintain a center of gravity.

c. Let the difference between the ^1A and ^5T states be parametrized by E. The energy level diagram should look like this:

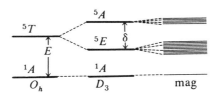

Use the Van Vleck equation to determine $\chi_\parallel$ as a function of E, δ, and T.

In the actual experiment, spin-orbit coupling was included, introducing off-diagonal elements that complicated the analysis. The experimental χ vs. T relation was computer fitted to the theoretical expression to yield best values for λ, E, and δ. It is interesting to note that a poor fit was obtained unless E was allowed to vary with temperature.

4. The 300 K molar magnetic susceptibility for a solid sample of

[structure: Cu(hfac)$_2$—O—N bonded to a tetramethylpiperidine-type ring with H$_3$C, H$_3$C substituents]

was determined to be $\chi = -186 \times 10^{-6}$ cm^3 mole^{-1}. Calculate the molar paramagnetic susceptibility by correcting for the diamagnetism of the complex. What is μ_{eff}? How can you explain this μ_{eff}?

hfac$^-$ = [structure: CF$_3$—C(O)—CH—C(O)—CF$_3$ with negative charge]

5. a. Co(N$_2$H$_4$)$_2$Cl$_2$ has a magnetic moment of 3.9 BM. Is hydrazine bidentate? Propose a structure.

b. How could electronic spectroscopy be employed to support the conclusion in part a?

6. In which of the following tetrahedral complexes would you expect contributions from spin-orbit coupling? V^{3+}, Cr^{3+}, Cu^{2+}, Co^{2+}, Fe^{2+}, Mn^{2+}.

7. In which of the following low spin square planar complexes would you expect orbital contributions? d^2, d^3, d^4, d^5, d^6.

8. Why is Fe$_2$(CO)$_9$ (with three bridging and six terminal carbonyls) diamagnetic?

9. Explain why mixing of a D_{4h} component in with a T_d ground state lowers the magnetic moment in nickel(II) complexes.

10. What is the expected magnetic moment for Er^{3+}?

11. In the figure below, effective magnetic moment *vs.* temperature curves are shown for two similar tris-bidentate Fe(III) compounds:

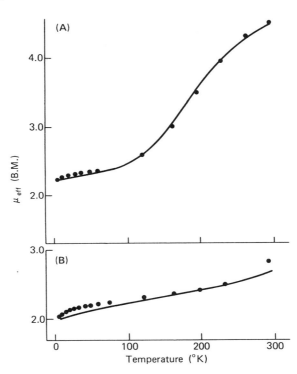

A: $R = n$-propyl
B: $R =$ isopropyl

a. Provide an explanation in terms of the electronic structure of an Fe(III) complex that accounts for the changes in μ_{eff} over the given temperature range for each of the two complexes. In other words, why are the two curves so different while the two complexes *appear* to be so similar?

b. In the above plots the points represent the experimental data and the lines represent least-squares fits to theoretical equations. With the use of an energy level diagram, describe what parameters might be used in such a theoretical treatment.

12 Nuclear Magnetic Resonance of Paramagnetic Substances in Solution

IVANO BERTINI AND CLAUDIO LUCHINAT

12–1 INTRODUCTION

In the early days of nmr, there was a widespread belief that one could not detect the nmr spectrum of a paramagnetic complex because the electron spin moment was so large that it would cause rapid nuclear relaxation leading to a short nuclear T_1 and a broad nmr line. This is the case for complexes of certain metal ions [*e.g.*, those of Mn(II)], but it is not the case for many others. For instance, in Fig. 12–1, the proton nmr spectrum[1] of the paramagnetic complex $Ni(CH_3NH_2)_6^{2+}$ is presented (B) and compared to that of CH_3NH_2 (A).

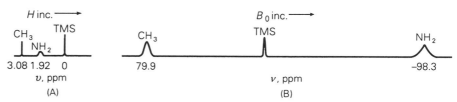

FIGURE 12–1 Proton nmr spectra (simulated) of solutions of (A) CH_3NH_2 and (B) $Ni(CH_3NH_2)_6^{2+}$. Note that the relative scales in (A) and (B) differ.

These spectra raise several questions:

1. Why do we see a spectrum for the paramagnetic complex?
2. Why are the observed shifts from tms in the complex so large relative to those of the uncoordinated ligand? The normal range of proton shifts for most organic compounds is about 10 to 15 ppm and the shifts in the complex are well outside this range.
3. Why does the NH proton resonance shift upfield, whereas that for the CH_3 protons shifts downfield?

We shall answer all of these questions in the course of logically developing this topic and in the process introduce a powerful structural tool.

12-2 PROPERTIES OF PARAMAGNETIC COMPOUNDS

We shall briefly review and expand upon those properties of paramagnetic compounds important to understanding their nmr. Unpaired electrons have magnetic moments whose intensity is given by:

$$\mu = g_e \, [S(S+1)]^{1/2} \mu_B \qquad (12-1)$$

where g_e is the free electron g factor, S is the total spin quantum number that is equal to one-half times the number of unpaired electrons, and μ_B is the Bohr magneton*, equal to $e\hbar/2m_e$ (9.274096×10^{-24} J·T^{-1}). Unpaired electrons reside in molecular orbitals and are delocalized over the entire molecule. The unpaired electron in one molecular orbital can also spin polarize an electron pair in a doubly occupied molecular orbital as discussed in accounting for the proton hyperfine in the epr of the methyl radical. At every point in the molecular frame there are contributions from the direct (except at a node) and spin polarization mechanisms. Thus, a nucleus in a molecule, containing one or more unpaired electrons, may experience spin density at its point in space and/or feel spin density that is nearby. Referring to atomic orbitals as composing molecular orbitals, net spin density exists at the nucleus of the atom under investigation only through its atomic s orbitals since only the wave functions for s orbitals have a finite value at the nucleus. A node exists at the nucleus in all other type orbitals. In addition, there are fractional unpaired electrons in orbitals other than s on the atom of interest and on neighboring atoms. Finally, a large fraction of the unpaired electrons in the molecule is always located in the paramagnetic metal orbitals.

Magnetic nuclei in a paramagnetic molecule interact with the unpaired electrons just as a magnetic dipole interacts with another magnetic dipole. The interaction is formally divided in two parts, one with the spin density at the resonating nucleus and another with the spin density cloud in the rest of the molecule. The former is called *Fermi contact coupling*, whereas the latter is *through-space dipolar coupling*.[2] The Fermi contact coupling can arise from either the direct delocalization mechanism or spin polarization. The calaculation of the dipolar interaction requires evaluation of an integral over all space, along with a detailed knowledge of the spin density distribution. It is customary to further divide the dipolar interaction into a so-called *metal-centered term* and a *ligand-centered term*. The former considers the unpaired electrons localized on the metal center and the interaction is evaluated using a point dipole approximation. The latter considers unpaired electron density delocalized on the ligand and is often restricted to one or a few atomic orbitals of the atom in the ligand whose nucleus is being studied. Indeed, spin density near the resonating nucleus is always relatively small providing a significant contribution only at small distances.

In the absence of an external magnetic field, the coupling between a magnetic nucleus and an electron gives rise to new energy levels. In the case of $I = \frac{1}{2}$ and $S = \frac{1}{2}$, the coupling partially removes the degeneracy of the four functions shown in Fig. 12–2(A) leading to Fig. 12–2(B).

---

* In this chapter we shall use μ_B instead of β for the Bohr magneton to avoid confusion with the β spin state. SI units are used throughout except sec is used for seconds.

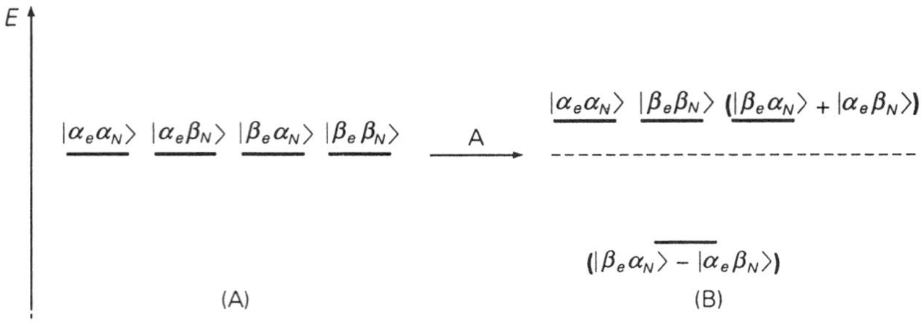

FIGURE 12–2 Energies of the spin wave functions for an $S = 1/2$, $I = 1/2$ system in the absence (A) and in the presence (B) of hyperfine coupling between the electron and nuclear spins. The lower-level and one of the upper-level functions are linear combinations of the non-interacting $|M_S, M_I\rangle$ functions. If the coupling, A, is negative, the splitting would be reversed, with the single degenerate level higher in energy.

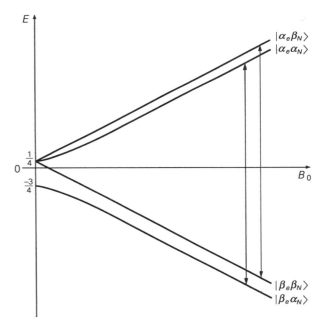

FIGURE 12–3 Effect of the application of a magnetic field B_0 of increasing strength on the spin energy levels of an $S = 1/2$, $I = 1/2$ system. The labeling of the spin functions in the high field limit is shown. The figure is drawn for a positive value of A. The electron transitions are indicated by arrows.

The application of an external magnetic field* splits the levels according to the orientation of the electronic or nuclear dipoles with respect to the orientation of the magnetic field (Fig. 12–3; also see Fig. 9–2). When we consider the nucleus–electron coupling from the point of view of the nucleus, we should keep in mind that the orientation of the magnetic dipole of the electron changes very fast compared to that of the nucleus. Nuclei relax with lifetimes of the order of seconds whereas electron relaxation leads to lifetimes of 10^{-5} to 10^{-6} sec for radicals and 10^{-7} to 10^{-13} sec for metal ions.[2] Therefore, the nucleus sees an oscillating magnetic dipole corresponding to the electron exchanging between the positions depicted in Fig. 12–3. To a first approximation the average coupling

---

*In this chapter the magnetic induction, B_0, will be used instead of the magnetic field, H. The SI unit for B is the tesla (T) and all equations are presented in this unit. In the discussion, magnetic induction will be called the magnetic field.

energy is zero because the two electronic orientations provide equal moments with different signs. However, in a magnetic field, the different populations of the two m_s levels lead to a time average different from zero, as shown in Fig. 12–4.

The nuclear coupling with an average electronic spin gives rise to a chemical shift contribution in the nmr spectrum whose magnitude is proportional to the nature and extent of the coupling. The magnitude of the shift may be anisotropic depending on the orientation of the molecule with respect to the external magnetic field. In solution, rapid molecular rotation produces an average value of the shift anisotropy. Consequently, this shift is called the isotropic shift or isotropic hyperfine shift and can be determined experimentally by measuring the difference in shift of a given atom in a paramagnetic compound and that of the same atom in an analogous diamagnetic compound. In the next section on the isotropic shift, we will address the questions concerning the large shifts seen in paramagnetic proton nmr as well as why some shifts are upfield and others downfield.

Electron relaxation provides mechanisms for nuclear relaxation because the nucleus senses a fluctuating magnetic field. In some cases, the effect is moderate and reasonably sharp nmr resonances result. Both the shifts and nuclear relaxation times can be analyzed to produce valuable structural and dynamic information. In other cases, the efficient nuclear relaxation mechanisms lead to broad nmr lines, often to the point where the spectral line escapes detection. However, even very broad lines can be exploited. For example, when the protons of water coordinated to a paramagnetic metal ion are too broad to be detected, the spectrum can be studied in water as a solvent. When the exchange of the coordinated water molecules with solvent water is fast compared to the difference in chemical shift between the bound and free water, a single line results whose line width is a linear function of the mole fraction of the bound and solvent water. The smaller the T_2 of the bound water proton, the greater the possibility of detecting line broadening of the average line in dilute aqueous solution. Thus, information about the structure and dynamics of the systems can be obtained by understanding the effect of hyperfine coupling on nuclear relaxation.

When the discussion of relaxation is completed, we will understand why the nmr of some paramagnetic complexes can be detected while others are too broad to be seen. Finally, examples of the analysis of shift and relaxation data to chemical problems will be presented along with the use of nmr to investigate the electronic properties of paramagnetic polymetallic clusters. We will see that answering the simple questions raised at the start of this chapter provides us with a powerful tool for studying the structure and dynamics of paramagnetic compounds.

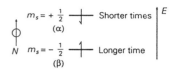

FIGURE 12–4 The nucleus, N, senses a weighted time average of the magnetic field produced by the unpaired electron in the two energy states. The time average of the electron magnetic field is non-zero and along the direction of B_0 because the electron spin spends slightly more time in the lower level.

12–3 CONSIDERATIONS CONCERNING ELECTRON SPIN

So far the discussion has involved pure electron spins. In contrast to the nuclear spin, the electron moves around the nucleus giving rise to an orbital angular momentum that is small but different from zero.[3] We discussed the coupling of the spin and angular momentum in Fig. 9–17. The orbital contribution should be kept in mind when, for the sake of simplicity, we refer to electron spin. The electron in a magnetic field has two different orientations whose energies are $-(1/2)g_e\mu_B B_0$ and $(1/2)g_e\mu_B B_0$. If, however, the total magnetic moment is larger or smaller than that of the free electron because of spin orbit effects, the Zeeman energies will be changed accordingly as modeled in ligand field theory. We account for these changes by using a g-value that differs from the free electron value, Fig. 12–5. Therefore, we will refer to $g_e = 2.0023$ for the free electron and to $g \neq 2.0023$ when spin orbit coupling is taken into consideration. Information about the actual value of g can be obtained directly or indirectly with epr spectroscopy (Chapter 13).

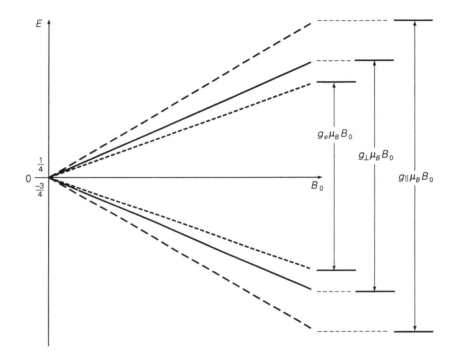

FIGURE 12–5 In the presence of spin orbit coupling, the electron spin transition energy differs from $g_e\mu_B B_0$ and depends on the orientation of the molecule in the magnetic field, B_0. The difference can be expressed in terms of changes in the value of the g-factor. In this example, both the extreme values $g_\perp$ and $g_\parallel$ are taken to be larger than g_e $(g_\parallel > g_\perp > g_e)$.

The Expectation Value of S_z, $\langle S_z \rangle$

When a bulk sample of a paramagnetic molecule containing one unpaired electron is placed in a magnetic field, the two m_s spin states are not equally populated. The rate of electron exchange between the two levels is described by the quantity T_2^{-1}. The picture is equivalent to that of a single molecule ($S = 1/2$) in a magnetic field spending more time in one level than the other. The population fraction of the two spin levels is described by N_α (the fraction of spins with $m_s = -1/2$ or α) and N_β (the fraction of spins with $m_s = +1/2$ or β). One may speak of the probability, P_α and P_β, that a single spin has an m_s value of $-1/2$ or $+1/2$. The average value of the spin along the field direction, z, is referred to as the *expectation value of S_z*, $\langle S_z \rangle$. This numerical value is proportional to the modulus of the vector component along the z-direction of either a single spin or of an assembly of spins normalized to one spin. A single nuclear spin feels this component of the electron spin and appropriately the remainder of this section will deal with the evaluation of an expression for $\langle S_z \rangle$. Recall that

$$\boldsymbol{\mu} = -g\mu_B \mathbf{S} \tag{12-2}$$

The Hamiltonian for the interaction of the electron magnetic moment with the field was given by

$$\hat{H} = -\boldsymbol{\mu} \cdot \mathbf{B}_0 = g\mu_B \mathbf{S} \cdot \mathbf{B}_0 \tag{12-3}$$

For an $S = \frac{1}{2}$ system, applying Boltzmann statistics, we obtain

$$\frac{N_\alpha}{N_\beta} = \exp(-\Delta E/kT) = \exp(-g\mu_B B_0/kT) \tag{12-4}$$

Solving for the total number of electron spins in the β state, we get

$$N_\beta = N_\alpha \exp\left(\frac{g\mu_B B_0}{kT}\right) \tag{12-5}$$

Since $(g\mu_B B_0/kT) \ll 1$,

$$\exp(\pm g\mu_B B_0/kT) \simeq 1 \pm (g\mu_B B_0/kT) \tag{12-6}$$

and equation (12–5) becomes

$$N_\beta \simeq N_\alpha(1 + g\mu_B B_0/kT) = (N - N_\beta)(1 + g\mu_B B_0/kT) \tag{12-7}$$

where $N = N_\alpha + N_\beta$ (i.e., the total number of spins). Rearranging this expression by solving for N_β, we obtain

$$N_\beta = N \frac{1 + (g\mu_B B_0/kT)}{2 + (g\mu_B B_0/kT)} = \frac{N}{2} \frac{1 + (g\mu_B B_0/kT)}{1 + (g\mu_B B_0/2kT)} \tag{12-8}$$

The numerator in parentheses in equation (12–8) is of the form $1 + nx$ where $n = 2$ and $x = g\mu_B B_0/2kT$. When $x \ll 1$, we may use the approximation $(1 + x)^n \simeq 1 + nx$ to obtain

$$N_\beta \simeq \frac{N}{2} \frac{(1 + g\mu_B B_0/2kT)^2}{(1 + g\mu_B B_0/2kT)} = \frac{N}{2}(1 + g\mu_B B_0/2kT) \tag{12-9}$$

Similarly, we can show that

$$N_\alpha \simeq (N/2)(1 - g\mu_B B_0/2kT) \tag{12-10}$$

The difference between N_β and N_α gives the excess spin along the field direction and is related to the modulus of the resultant vector of the up and down spin vectors. It represents the net magnetic effect induced in the system's unpaired electrons by the external field. The magnitude of the expectation value of S_z is the quantity of interest to us in the nmr experiment. This is given by a population-weighted average of the $+\frac{1}{2}$ and $-\frac{1}{2}$ states:

$$\langle S_z \rangle = \frac{(+\frac{1}{2})N_\alpha + (-\frac{1}{2})N_\beta}{N} = \frac{-g\mu_B B_0}{4kT} \tag{12-11}$$

where $\frac{1}{2}$ and $-\frac{1}{2}$ are the m_s-components along B_0 of α and β spins, respectively.

There is a direct relationship between $\langle S_z \rangle$ and the bulk magnetic moment M discussed in Chapter 11. We have seen that the number of excess spins in the lowest level is twice the expectation value of S_z times the total number of electrons N. This number of excess spins must be multiplied by the component along the field direction of the magnetic moment of one spin in the lowest state, $(\frac{1}{2})g\mu_B$, in order to find the average magnetic moment, that is,

$$M = (\tfrac{1}{2})g\mu_B(2N\langle S_z \rangle) = Ng^2\mu_B^2 B_0/4kT \tag{12-12}$$

This is the Curie law expression, described in Chapter 11 and arrived at by population weighting of the various levels.

For the general case of more than one spin, $\langle S_z \rangle$ is given by

$$\langle S_z \rangle = \frac{\sum_{M_s=-S}^{S} S \exp(-E_{M_s}/kT)}{\sum_{M_s=-S}^{S} \exp(-E_{M_s}/kT)} \tag{12-13}$$

Here E_{M_s} is the energy of the state with quantum number M_s and according to equation (12–1) is equal to $g\mu_B M_S B_0$. Thus, equation (12–13) is consistent with the equation for one electron spin ($S = \frac{1}{2}$). In equation (12–13), a state with a particular value of the spin magnetic quantum number S is weighted according to its equilibrium population. The weighted values are summed over all the energy levels and divided by the total number of levels to give the expectation value of S_z. If $\Delta E \ll kT$, then $\exp(-E/kT) = 1 - E/kT$, and after some algebraic manipulations, equation (12–14) results:

$$\langle S_z \rangle = \frac{-g\mu_B B_0 S(S+1)}{3kT} \tag{12-14}$$

For a free electron, the energy separation between two M_S levels is $g\mu_B B_0$ with $g = 2.0023$. The contribution from the orbital magnetic moment (in the spin Hamiltonian formalism) is accommodated by using values for g that are different from 2.0023 and different for different orientations of the molecule with respect to the molecular frame. For a molecule or ion rotating rapidly in solution, the g value of Equation (12–14) is substituted by an average value of g

$$\langle S_z \rangle = -\frac{g_{av}\mu_B B_0 S(S+1)}{3kT} \tag{12-15}$$

Now that we have an expression for $\langle S_z \rangle$, we shall proceed to evaluate the quantitative expression for the nmr contact shift assuming equation (12–15). This holds everytime a single S multiplet is populated. However, when two or three S multiplets are degenerate (as in the cases of E and T ground states) or close in energy with respect to kT, the energy separation between M_s states is different from $g_{av}\mu_B B_0$, and equation (12–15) becomes a rough approximation. More thorough calculations are needed for these cases.[4]

12–4 THE CONTACT SHIFT

The contact shift arises from the presence of unpaired spin density at the resonating nucleus. It is proportional to the amount of spin density. It is also called the Fermi contact shift. Fermi provided the first theoretical justification for electron density at the nucleus in accounting for some details in the electronic spectrum of the hydrogen atom.[5] The contact shift is also referred to as the scalar shift because it is independent of molecular rotation.

The spin density at the nucleus of an atom in a ligand is a small fraction of an electron which, in the presence of an external magnetic field, gives rise to a permanent, time-averaged, additional magnetic field that adds to the applied magnetic field. Therefore, the nucleus experiences a further shift in resonance frequency. In other words, the nuclear energy levels are affected by the presence of spin density at the nucleus as shown in Fig. 12–2(B). If we had a full unpaired electron at the nucleus, this nucleus would sense a magnetic moment equal to $g\mu_B \langle S_z \rangle$ aligned with the external magnetic field, with $\langle S_z \rangle$ given by equation (12–15). We do not have a full electron at the nucleus whose nmr is studied, but, in general, only a small fraction of an electron, resulting from direct delocalization and spin polarization. For simplicity let us consider the case of a single electron in a molecular orbital Ψ that is directly delocalized on the resonating nucleus without any spin polarization. The spin density at the nucleus is $\Psi(0)^2$ where 0 means at 0 distance from the nucleus. The shift from paramagnetism comes from the magnetic field B' arising from the electron density at the nucleus:

$$B' = g\mu_B \langle S_z \rangle \Psi(0)^2 = g^2 \mu_B^2 B_0 S(S+1) \Psi(0)^2 / 3kT \qquad (12–16)$$

For more than one unpaired electron, a normalization term $(1/2)S$ must also be included. The expression for the shift is:

$$\frac{\Delta v}{v} = -\frac{\Delta B}{B_0} = \frac{1}{2S} g^2 \mu_B^2 \frac{S(S+1)}{3kT} \Sigma \Psi(0)^2 \qquad (12–17)$$

where $\Psi(0)^2$ is summed over all molecular orbitals containing unpaired electrons.

In quantum mechanical terms, the contact hyperfine term is

$$\hat{H} = A_c \mathbf{I} \cdot \mathbf{S}. \qquad (12–18)$$

where A_c is the contact hyperfine coupling constant and is the same as a discussed in Chapter 13. Capital A indicates that there may be more than one unpaired electron in the molecule or ion. A_c is defined as:

$$A_c = \frac{\mu_0}{3S} g_N \mu_N g \mu_B \Psi(0)^2 \qquad (12–19)$$

If μ_B, μ_N (the nuclear Bohr magneton), k, T, Ψ, and μ_0 are expressed in SI units, A_c is given in $J(A^{-2}J^3T^{-2}m^2)$. However, it is customary to report the A_c value in terms of $A_c/(\hbar/\text{rad sec}^{-1})$ or A_c/h (sec^{-1} or Hz). If we substitute the equation for A_c into equation (12–17), we obtain the following expression for the contact shift:

$$\frac{\Delta v}{v} = -\frac{\Delta B}{B_0} = A_c \frac{g\mu_B S(S+1)}{g_N \mu_N 3kT} = -\frac{A_c}{g_N \mu_N B_0} \langle S_z \rangle \qquad (12\text{–}20)$$

We can now see that a very small A_c value can give rise to a huge isotropic shift. For example, an A_c value of 12^{-27} J gives a contact shift of 40 ppm. The reader is encouraged to convert the isotropic frequency shifts in Fig. 12–1 (at room temperature) to the equivalent field shifts in tesla. The equality $-\Delta B/B_0 = \Delta v/v$, where v is the fixed probe frequency, follows directly from the fact that $hv = g_N \mu_N b_0$ for nuclear spins.

If one measures the temperature dependence of Δv, a plot of Δv vs. $1/T$ should produce a straight line with a slope proportional to A_c for systems exhibiting Curie law behavior. For systems with an orbitally non-degenerate ground state, such as octahedral nickel(II) and tetrahedral cobalt(II) complexes, the application of equation (12–20) is valid; otherwise, equation (12–20) represents an approximation. Equation (12–20) holds independently of the orientation of the molecular frame as long as $g = g_e$. It is also valid in solution when an average value of g is considered.

12–5 THE PSEUDOCONTACT SHIFT

Magnetic moments aligned along the external magnetic field from spin density at every point all over the molecule (except that at the resonating nucleus) can also couple with the nuclear magnetic moment. The coupling is dipolar in origin. The relevant Hamiltonian for the general case of coupled magnetic moment vectors $\boldsymbol{\mu}_1 = g_N \mu_N \mathbf{I}$ and $\boldsymbol{\mu}_2 = g\mu_B \mathbf{S}$ was given in equation (9–23).

$$\hat{H}_{\text{dip}} = -g\mu_B g_N \mu_N \left(\frac{\mathbf{S} \cdot \mathbf{I}}{r^3} - \frac{3(\mathbf{S} \cdot \mathbf{r})(\mathbf{I} \cdot \mathbf{r})}{r^5} \right) \qquad (12\text{–}21)$$

When both I and S are aligned along the external magnetic field (as is the case when g is isotropic) the energy of interaction is:

$$E_{\text{dip}} = \frac{\mu_1 \mu_2}{r^3} (3 \cos^2 \gamma - 1) \qquad (12\text{–}22)$$

where μ_1 is the nuclear magnetic moment, μ_2 is the electron magnetic moment, and r is the distance between the origins of the two vectors. The meaning of the angle γ is illustrated in Fig. 12–6.

The electron magnetic moment arises from the excess population of the Zeeman levels times the spin density at a given point and r is the distance between the nucleus and the given point. In order to evaluate the interaction energy over

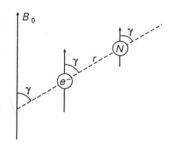

FIGURE 12–6 The dipolar interaction energy between the electron and nuclear magnetic dipoles in an isotropic situation is a function of the angle γ between the electron-nucleus vector, r, and the external magnetic field, B_0.

all space, integration is needed. This dipolar contribution to the electron–nucleus hyperfine coupling can be measured in the solid state or a frozen solution, with epr and ENDOR. In solution, where random and isotropic rotation occurs, the energy described above averages to zero for an isotropic g because the integral for $\cos \gamma$ ranging from 1 to -1 of equation (12–22) is zero. For this averaging to occur, the rotation rate in radians/sec^{-1}, has to be faster than $E_{dip}/\hbar$. This condition is easily satisfied in solution even for macromolecules.

Now, let us suppose that μ_2 is not isotropic but changes its magnitude as the orientation of the molecule changes with respect to the external magnetic field. Under these circumstances, μ_2 is anisotropic and the integration of equation (12–22) over $\cos \gamma$ is not zero. Thus, an added contribution to the hyperfine coupling results, that is, an additional magnetic field is sensed by the nucleus giving rise to a contribution to the chemical shift. The value of μ_2 will be anisotropic when there is an orbital contribution to the magnetic moment which, by its nature, changes in different directions of the molecular frame (see Fig. 9–17). In this case, g also becomes anisotropic and therefore orientation dependent. Thus, the same effects that give rise to anisotropy in g also give rise to a shift contribution. This shift contribution is isotropic because the dipolar effect is averaged by rotation in solution. Because it is isotropic, like the contact contribution, it is called a pseudocontact shift and is also referred to in the literature as an isotropic dipolar shift. The through space nature of the pseudocontact effect is comparable to the neighbor anisotropic contribution discussed in Chapter 8, which was seen to be dependent upon differences in χ_{DIA} for different orientations. In a similar fashion, anisotropy for χ_{para} leads to the pseudocontact shift.

The complete evaluation of the pseudocontact shift requires knowledge of the spin density distribution over the entire molecule. This information is usually not available. If the unpaired electrons are assumed to be localized only on the metal ion, a point dipole model leads to equations for evaluating the pseudocontact contribution to the isotropic shift. The general equation in SI units* is[7,8]:

$$\frac{\Delta v_{i(pc)}}{v} = -\frac{1}{4\pi}(^1/_3 N_A)[\chi_{zz} - ^1/_2(\chi_{xx} + \chi_{yy})]\frac{3\cos^2\theta_i - 1}{r_i^3}$$

$$-\frac{1}{4\pi}(^1/_2 N_A)(\chi_{xx} - \chi_{yy})\frac{\sin^2\theta_i \cos 2\phi_i}{r_i^3} \quad (12\text{–}23)$$

Here, N_A is Avogadro's constant, χ_{xx}, χ_{yy}, and χ_{zz} are the susceptibility components, and the angles θ and ϕ are defined in Fig. 12–7. The coordinates x, y, and z are defined by the principal directions of the magnetic susceptibility tensor.

For cases where the magnetic susceptibility along a given coordinate k can be approximated by equation (12–24), equation (12–23) can then be rewritten in terms of g values.

$$\chi_{kk} = \mu_0 \frac{g_{kk}^2 \mu_B^2 S(S+1)}{3kT} \quad (12\text{–}24)$$

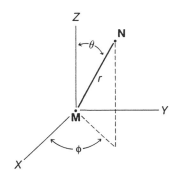

FIGURE 12–7 Definition of the quantities in equation (12–23).

---

*Note that $\chi(SI) = 4\pi \times 10^{-6} \chi$ (cgs emu).

The anisotropy in the g tensor is more easily obtained than that in χ, especially for $S = \frac{1}{2}$ systems.

Rewriting equation (12–23) in terms of Δg for an axially symmetric system, we obtain the following equation for the pseudocontact contribution:

$$\frac{\Delta v_{i(pc)}}{v} = \frac{\mu_0}{4\pi} \frac{\mu_B^2 S(S+1)}{9kT} \frac{3\cos^2\theta_i - 1}{r_i^3} (g_\parallel^2 - g_\perp^2) \qquad (12\text{–}25)$$

An example of how these terms are defined for a typical molecule with local axial symmetry is given in Fig. 12–8.

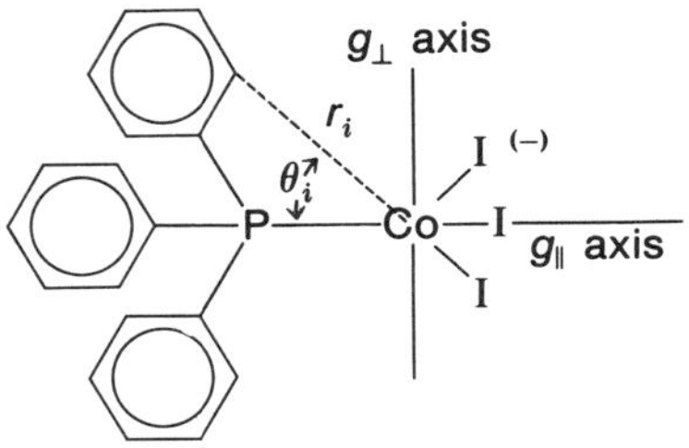

FIGURE 12–8 Definition of terms in equation (12–25). The parallel and perpendicular axes are shown. The radius vector from the metal center to the nucleus being investigated is labeled r_i, and the angle that it makes with the highest-fold rotation axis is θ_i. For a proton shift, r_i would be drawn to the hydrogen of the phenyl ring.

In view of the simplicity of equation (12–25), many studies have been performed on $S = \frac{1}{2}$ systems. Equation (12–25) is valid when the nucleus studied feels the unpaired electron as a point dipole (*i.e.*, when delocalization can be neglected and the nucleus is relatively far from the metal).[9] The significant delocalization of unpaired electrons, at least within the coordination cage, may prevent the quantitative use of equations (12–23) and (12–25) for donor atom nuclei or for nuclei close to the donor atoms.[10]

The foregoing discussion should be made more specific by considering different types of nuclei. Protons and deuterons only have spin density in an s orbital and since this spin density is isotropic, a pseudocontact shift does not result. If the protons are attached to a carbon or nitrogen atom, for example, bearing spin density in a p_π orbital, a pseudocontact shift can arise at the proton that is then called ligand centered. In this case, ligand-centered effects cannot be neglected. Combined information on the carbon or nitrogen and an attached hydrogen atom is needed for a complete treatment. It is also sometimes useful to obtain independent information from the relaxation parameters (*vide infra*). The pseudocontact shift equations are particularly useful for nuclei in metalloproteins which are within ~10 Å of the metal but are removed from the metal by many sigma bonds. In this case, the spin density and the contact shift is zero as is spin density on neighboring atoms.

12–6 LANTHANIDES

Lanthanides are characterized by large spin-orbit effects so that both the spin and orbital angular momenta are sizable and strongly coupled. This means that for each electron, the various projections of the orbital angular momentum m_l will interact differently with the various projections of the spin angular momentum, m_s, giving a total angular momentum component equal to $m_l + m_s = m_j$. The m_j values identify a j multiplet, just as m_l and M_s were related to the l and

TABLE 12-1. Some Electronic Properties of Lanthanide Ions

Ion	Configuration	$^{2S+1}L_j$ of Ground State (multiplicity)	g_j^a	$\langle S_z \rangle_j^b$	$\langle S_z \rangle_j^c$	$\left(\dfrac{\Delta v}{v_0}\right)^{dip}$ (ppm)^d
Ce^{3+}	$4f^1$	$^2F_{5/2}(6)$	$6/7$	1.07	0.98	+1.6
Pr^{3+}	$4f^2$	$^3H_4(9)$	$4/5$	3.20	2.97	+2.7
Nd^{3+}	$4f^3$	$^4I_{9/2}(10)$	$8/11$	4.91	4.49	+1.05
Pm^{3+}	$4f^4$	$^5I_4(9)$	$3/5$	4.80	4.01	−0.6
Sm^{3+}	$4f^5$	$^6H_{5/2}(6)$	$2/3$	1.79	0.06	+0.17^e
Eu^{3+} (Sm^{2+})	$4f^6$	$^7F_0(1)$	—	—	−10.68	−1.0^e
Gd^{3+} (Eu^{2+})	$4f^7$	$^8S_{7/2}(8)$	2	−31.5	−31.50	0
Tb^{3+}	$4f^8$	$^7F_6(13)$	$3/2$	−31.5	−31.82	+20.7
Dy^{3+}	$4f^9$	$^6H_{15/2}(16)$	$4/3$	−28.30	−28.54	+23.8
Ho^{3+}	$4f^{10}$	$^5I_8(17)$	$5/4$	−22.50	−22.63	+9.4
Er^{3+}	$4f^{11}$	$^4I_{15/2}(16)$	$6/5$	−15.30	−15.37	−7.7
Tm^{3+}	$4f^{12}$	$^3H_6(13)$	$7/6$	−8.17	−8.21	−12.7
Yb^{3+}	$4f^{13}$	$^2F_{7/2}(8)$	$8/7$	−2.57	−2.59	−5.2

^{a} Calculated from equation (12–27): the formula does not hold for f^6 ions.
^{b} Calculated according to equation (12–26).
^{c} Calculated by inclusion of the excited states.[12]
^{d} Dipolar shift predicted from Equation (12–28) for $r = 300$ pm, $3\cos^2\theta - 1 = 1$, $T = 300$ K, axial symmetry, and D_z values for each lanthanide estimated from reference 11.
^{e} Including contributions from excited J manifolds.[11] For f^6 ions, the contribution of the ground state manifold is zero.

s manifolds. When electrons are allowed to interact, then J terms are obtained. The ground states for lanthanides in the oxidation state +3 are reported in Table 12-1.[11,12] They are characterized by S, L, and J. For example, cerium(III) has one electron in an f-orbital; $S = 1/2$ combines with $L = 3$ to give both $J = 3 + 1/2 = 7/2$ and $3 - 1/2 = 5/2$. The latter term is the ground state. Although it is not completely correct, we will consider only the ground state in our analysis. This is an even poorer approximation for actinides, which have larger spin orbit coupling and excited states of different J multiplicity closer in energy than the lanthanides.

The f orbitals are often assumed not to be involved in covalent bonding with the donor atoms. Little electron delocalization on the ligand molecules is assumed to result and the contact contribution is often neglected. The validity of this assumption depends on the number and type of bonds between the metal ion and the resonating nucleus. It is not justified for nuclei directly coordinated to the metal ion. We should next consider the spin density at the resonating nucleus and $\langle S_z \rangle$, which is now labeled $\langle S_z \rangle_J$. It is generally assumed that crystalline field effects are much less than kT and completely neglected in the evaluation of $\langle S_z \rangle_J$:[12]

$$\langle S_z \rangle_J = -g_J(g_J - 1)J(J+1)\frac{\mu_B B_0}{3kT} \quad (12-26)$$

where

$$g_J = 1 + \frac{J(J+1) - L(L+1) + S(S+1)}{2J(J+1)} \quad (12-27)$$

The numerical values are reported in Table 12–1.

The values of $\langle S_z \rangle$ are shown in Fig. 12–9.[11,12] Note that whereas $\langle S_z \rangle$ values for unpaired electrons in d orbitals are negative, they can be either negative or positive in lanthanides. A positive value results when the orbital contribution to the energy level is larger than and different in sign from the spin contribution.

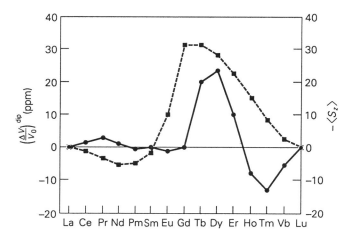

FIGURE 12–9 Calculated patterns of (●) pseudocontact shifts[11] and (■) $-\langle S_z \rangle$ values[12] for lanthanide ions. The shifts are for a nucleus 3Å from the metal with 3 $\cos^2 \theta - 1$ equal to 1 at 300 K.

Due to the small contact terms, the point dipole approximation for the pseudocontact contribution is a better approximation for lanthanides that for first row transition metal ions. The ligand field splitting causes the magnetic anisotropy. The D_z, D_x, and D_y parameters take the ligand field splitting into account. As long as D is larger than the Zeeman energy and smaller than kT, the expression for the pseudocontact term is:[11]

$$\frac{\Delta v_{i(pc)}}{v} = -\frac{\mu_0}{4\pi} \frac{g_J^2 \mu_B^2 J(J+1)(2J-1)(2J+3)}{60(kT)^2}$$

$$\times \frac{D_z(3\cos^2\theta_i - 1) + (D_x - D_y)\sin^2\theta_i \cos 2\Omega_i}{r_i^3} \quad (12\text{–}28)$$

The shifts at 300 K for a nucleus i at 3Å with $3\cos^2\theta_i - 1 = 1$ and $\Omega = 0$ are reported in Fig. 12–9.

12–7 FACTORING THE CONTACT AND PSEUDOCONTACT SHIFTS

In principle, the observed isotropic hyperfine shifts contain contact and pseudocontact contributions. The latter can be further divided into a metal-centered and ligand-centered contributions. In lanthanides, the metal-centered pseudocontact term prevails at least for nuclei a few angstroms away from the metal. Furthermore, since equations are available for both contact and pseudocontact terms, attempts have been made to factor out the two contributions.[12,13] If we consider a series of complexes differing in the metal ion, with the same nucleus and same contact hyperfine constant, the factorization is possible.[13,14]

TABLE 12–2. Contributions to the Hyperfine Shift of Carbon Atoms in a Porphyrin Ring of a Low Spin Iron(III) Complex[17]

Contribution	α Pyrrole $^{13}C^a$	β Pyrrole $^{13}C^a$	Meso ^{13}C	Pyrrole $^{13}CH_3$
$(\Delta v/v_0)^{iso}$	101.5	36.0	73.2	46.8
$(\Delta v/v_0)^{MC}_{dip}$	31	11	22	5.0
$(\Delta v/v_0)^{LC}_{dip}$	36	78	−10	
$(\Delta v/v_0)^{con}$	34	−53	61	47

^{a} The values for α and β carbons refer to TPP; those for meso and pyrrole CH_3 carbons refer to etioporphyrin.

Let us consider two different metal complexes, i and j, of similar geometry. The measured shifts for a given proton are given for axial symmetry by

$$\left(\frac{\Delta v}{v}\right)_i = A\langle S_z \rangle_i + D_{zi} \cdot G$$

$$\left(\frac{\Delta v}{v}\right)_j = A\langle S_z \rangle_j + D_{zj} \cdot G$$

Here, the unknowns are A and G for each nucleus, whereas $\langle S_z \rangle$ and D_z depend only on the lanthanide, and reasonable estimates of their values are available.[11,12,15] By using these estimates for $\langle S_z \rangle$ and D_z, we are left with two equations and two unknowns. If more than two metal complexes are studied, further equations can be used without introducing more unknowns. More complicated methods that do not involve the assumption of axial symmetry have been proposed.[16]

For first row transition metal ions factoring is more complex. Ligand-centered effects complicate the interpretation of nuclei other than protons and deuterons. In Table 12–2, the relative contributions to the hyperfine shifts of carbon atoms in a porphyrin ring containing low spin iron(III) are shown. The contact and the two pseudocontact contributions are similar in magnitude. For protons and deuterons, the metal-centered pseudocontact contribution can be of the same magnitude as the contact term.

For orbitally non-degenerate systems with excited states far removed in energy, the orbital contribution is small and so is the magnetic anisotropy. Accordingly, the pseudocontact shifts are small. Examples include octahedral nickel(II) and tetrahedral cobalt(II). Some anisotropy in the magnetic susceptibility arises from zero field splitting as in the case of high spin iron(III) complexes.[18] Orbitally quasi-degenerate ground states provide large magnetic anisotropies and large pseudocontact contributions (*e.g.*, octahedral cobalt(II)[19] and pseudotetrahedral nickel(II)[20] systems). The knowledge of the magnetic susceptibility tensor provides a direct tool for the evaluation of pseudocontact contributions. Some examples are reported in Table 12–3 where the magnetic susceptibility information obtained on single crystals is transferred to the molecule in solution.

When equation (12–24) is valid (*e.g.*, $S = {}^1/_2$ systems), single-crystal epr studies provide g values and an estimate of the pseudocontact shifts through equation (12–25).[27]

TABLE 12–3. Isotropic Shifts and Pseudocontact Contributions from Known Magnetic Susceptibility Tensors

(A) ^{1}H and ^{13}C Isotropic Shifts (ppm) of the Pyridine Signals in the *bis*-Pyridine Adduct of Cobalt(II) *bis*-Acetylacetonate Using Single Crystal Magnetic Susceptibility Data[21,22]

Atom	$\left(\dfrac{\Delta v}{v_0}\right)^{iso}$	$\left(\dfrac{\Delta v}{v_0}\right)^{dip}$
α—H	+32.9	−39.5
β—H	+5.0	−18.1
γ—H	−9.4	−15.6
α—C	−199	−92.5
β—C	+229	−35.7
γ—C	−73.8	−28.3

(B) The Aromatic Protons in Nickel(II) Bis-Salicyladiminate Using Single Crystal Magnetic Susceptibility Data[20,23,24]

Proton	Structure	$\left(\dfrac{\Delta v}{v_0}\right)^{iso}$	$\left(\dfrac{\Delta v}{v_0}\right)^{dip}$
3		−23.7	+1.4
4		+19.0	−2.3
5		−23.7	−4.5
6		+3.7	−8.4

(C) Phenyl Protons in Dichlorobis(triphenylphosphine)Cobalt(II) and Nickel(II)[25,26]

Metal/Proton		$\left(\dfrac{\Delta v}{v_0}\right)^{iso}$	$\left(\dfrac{\Delta v}{v_0}\right)^{dip}$
Co	ortho	−10.35	−2.13
	meta	+8.08	−0.54
	para	−11.51	−0.53
Ni	ortho	−6.38	+0.56
	meta	+8.03	+0.14
	para	−11.16	+0.14

12–8 THE CONTACT SHIFT AND SPIN DENSITY

As mentioned, if the contact contribution is dominating the isotropic shifts or if we have factored out the pseudocontact shift, a measure of the contact hyperfine coupling constant is available. The contact contribution is then related to the actual spin density on the resonating nucleus through equation (12–19). Spin density arises from a direct delocalization mechanism and from spin polarization of an orbital occupied by two electrons. In principle, both contributions can be calculated through a molecular orbital treatment. If K is the hyperfine coupling due to a full electron on an atom (listed inside the back cover), then the direct delocalization contribution is obtained from equation (12–29)

$$\frac{A_d}{h} = \frac{K\rho}{2S} \qquad (12\text{–}29)$$

where A_d is the contribution from direct delocalization and $\rho/2S$ is the spin density normalized to one electron.[28]

In the 1970s Drago *et al.* pioneered the calculations of direct delocalization on hydrogen nuclei through extended Hückel calculations.[29] Unpaired spin density delocalizes through the highest occupied orbital and spin density is evaluated by calculating $\psi^2_{(0)}$ at the hydrogen nucleus. The orientation of the electron in the magnetic field is such that the m_s ground state is $-1/2$. Direct delocalization from the metal implies that the same sign is maintained; then $\langle S_z \rangle$ is negative and from equation (12–20) the shift in Hz is positive or downfield. If we maintain the sign of $\langle S_z \rangle$ negative, we have to change the sign of the hyperfine coupling constant when spin polarization leads to spin density with an $m_s = 1/2$ ground state. The aliphatic protons of octahedral amino complexes of nickel(II) and cobalt(II) experience direct delocalization and shift downfield. The amino protons, on the contrary, experience an upfield (negative figures in Hz) shift because the large spin density on the nitrogen atom makes spin polarization dominant at the proton. In these cases, the unpaired electrons are in the e_g metal orbitals, which have the correct symmetry to form a σ bond to the nitrogen lone pair. When the unpaired electrons are present only in t_{2g} metal orbitals in octahedral complexes, they cannot contribute to the direct σ delocalization. Spin polarization of the σ molecular orbitals is needed to convey unpaired spin density onto the ligand. This is the reason that early transition metal ions give rise to opposite shift patterns compared to those on the right hand side of the series. In the metal–ligand molecular orbital scheme, the antibonding orbital containing the unpaired electron (Fig. 12–10) is partly constituted of the ligand highest occupied molecular orbital (HOMO) when the metal unpaired electron is in an e_g orbital. (Usually the HOMO contains the lone pair.) The coefficients of the atomic orbital in that molecular orbital determine the spin density at nuclei. Spin delocalization through the highest occupied molecular orbital decays rapidly with the increasing number of σ bonds from the metal to the resonating nucleus. Some examples are reported in Table 12–4. The treatment for carbon nuclei differs because direct spin delocalization on a p orbital induces spin density contributions at the carbon nucleus from 1s and 2s orbitals through spin polarization. This complicates the problem.[30]

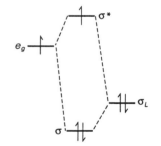

FIGURE 12–10 Unpaired spin density transferred onto the ligand by sigma bonding between the e_g metal orbital containing one electron and the ligand donor lone pair σ_L.

TABLE 12–4. Estimated A_c/h Values (MHz) for ^{1}H, ^{13}C, and ^{14}N Nuclei of *n*-Ethyl Amine Coordinated to *bis*-Acetylacetonate Nickel(II)[30]

^{14}N	N^{1}H$_2$	^{13}CH$_2$	C^{1}H$_2$	^{13}CH$_3$	C^{1}H$_3$
14.4	−2.64	−0.70	1.02	1.80	0.22

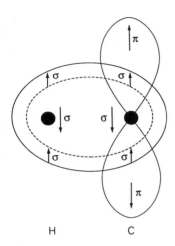

FIGURE 12–11 Positive p_π spin density on an sp^2 carbon induces unbalancing of the spin density on the doubly occupied C—H bonding σ orbital. The sign of the induced spin density is positive in the outer region and negative in the inner region of the σ orbital.

Spin polarization effects are, in general, hard to handle with theoretical tools. Some results have been obtained in the case of aromatic moieties since independent tests come from the investigation of radicals through epr. Assume some spin density is present in a p_π orbital of a carbon atom that is hybridized sp^2. To first order, it does not contribute to contact contributions on either the carbon or the attached hydrogen nuclei. However, spin polarization (Fig. 12–11) induces spin density at the carbon nucleus through the 1s and 2s orbitals as well as to the nuclei of attached atoms through the σ covalent bond. An attached proton therefore experiences spin density at the nucleus. The larger the spin density on the p_π orbital, the larger the spin density at the hydrogen atom. The proportionality constant for this relationship is constant for various systems as long as the nature of the orbitals involved is the same.

$$\frac{A_p}{h} = \frac{Q\rho_C^\pi}{2S} \qquad (12\text{–}30)$$

where A_p is the hyperfine constant due to spin polarization, ρ_C^π is the π spin density on the carbon atom, and Q is the proportionality constant, which is about -70 MHz.[31] The negative sign is due to the fact that spin polarization induces spin density on the hydrogen atom of different sign than the spin density in the p_π orbital. If the carbon bearing the spin density in the p_π orbital is attached to a CH_3 moiety, the 1s orbital of the CH_3 protons can directly overlap with the p_π orbital and then directly transfer spin density with the same sign onto the hydrogen nuclei. The proportionality constant $Q(C-CH_3)$ is now positive and is predicted to be around 75 MHz when rotation around the C—C axis is fast enough to provide a space-averaged hyperfine coupling constant.[32] Experimentally, the value of Q is found to vary considerably from one system to another. Furthermore, delocalization through the π molecular orbital leads to large values of contact shifts for nuclei at a large distance from the paramagnetic center.

In the case of nuclei of the second row of the periodic table, there are several contributions to spin density at the nucleus. As previously mentioned, the unpaired spin on a p_π orbital induces spin density through 1s and 2s orbitals of the same atom. Furthermore, the p_π spin density of the nearby atom contributes, via spin polarization of the σ-bond electrons, to the spin density at the resonating nucleus. For an sp^2 carbon atom, the contributions to spin density through the 2s orbital are given by[33]:

$$\frac{A_p}{h} = \sum_{i=1}^{3} Q_{CX_i}^C \rho_C^\pi + \sum_{i=1}^{3} Q_{X_iC}^C \rho_{X_i}^\pi \qquad (12\text{–}31)$$

where X_i are the three atoms bound to the carbon under consideration, $Q_{CX_i}^C$ is a constant describing the polarization effect of the spin density in the p_π carbon on the two electrons of the C—X_i bond, and Q_{X_iC} is the constant relative to the polarization effect of the spin density in the p_π orbital of X_i on the same C—X_i bond. Some of these Q values are reported[33–35] in Table 12–5. To this value of A_p, a contribution from the polarization of the 1s orbital should be added.

$$\frac{A_p}{h} = \text{const } \rho_c^\pi \qquad (12\text{–}32)$$

where the const is -35.5 MHz.[33]

TABLE 12–5. Q Values (MHz) for Fragments Involved in π Electron Spin Delocalization Systems[33-35]

Q_{CH}^C	Q_{CH}^H	$Q_{CC'}^C$	$Q_{C'C}^C$	$Q_{CCH_3}^H$	Q_{CF}^F	Q_{FC}^F
+54.6	−65.8	+40.3	−39.0			
	−75.6	+53.2				
				+14 ÷ +115		
					−410	+2370

Spin density on p_π orbitals are present when there is π spin delocalization from the metal.[36] For example, vanadium(III) is d^2 with two unpaired electrons in t_{2g} orbitals in octahedral complexes. These orbitals have the correct symmetry to overlap with π ligand molecular orbitals. It should be noted, however, that the π ligand molecular orbitals can be either empty (Fig. 12–12(A)) or full (Fig. 12–12(B)). In the latter case, the description is similar to that of the σ bonds. The

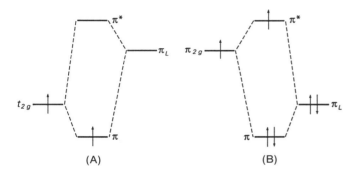

FIGURE 12–12 Unpaired spin density can be transferred onto the ligand by π-bonding between a t_{2g} metal orbital containing one unpaired electron and an empty (A) or full (B) ligand π orbital.

unpaired electron resides in an orbital that is partly constituted of the ligand highest occupied molecular orbital. This mechanism is called ligand-to-metal charge transfer. Electron delocalization from t_{2g} directly into the ligand π system occurs when the π orbital is unoccupied. π delocalization is improbable in ligands having only σ bonds. The lowest unoccupied molecular orbital of the ligand is far removed in energy and we should also consider whether the d electron has the correct symmetry to interact with the ligand orbital or not. In the latter case, spin density on the ligand occurs through spin polarization.

All the ligands of the type

where the X donor group can be $-NR_2$, $-N=CHR$, or O^-,[37-39] experience unpaired spin density in the π system that arises from non-orthogonality between the coordination bond and the π system. The ortho signals are shifted upfield (*i.e.*, the spin density is negative (excess of spin = $^1/_2$) on the hydrogen atom and

TABLE 12–6. ^{1}H nmr Contact Shift Patterns (ppm) on Some Aromatic Ligands

L		2	3	4	5	6	Ref. No.
[N-methylaniline]	a	−4.8	1.7	−5.2	—	—	37
[pyridine-N-oxide]	b	−12.8	10.7	−17.1	—	—	40, 41
[triphenylphosphine]	c	−8.2	8.6	−11.0	—	—	25, 42
[aniline]	d	−6.9	7.9	−11.3	—	—	
[benzylamine]	e	−1.5	1.4	−1.6	—	—	51
[pyridine]	f	72.4	23.1	6.2	—	—	21, 43, 52
[N-methylimidazole]	g	87.7	26.6	8.0	—	—	
[salicylaldimine]	h	55.0	12.0(CH$_3$)	52.0	71.0	—	47
[salicylaldimine]	i	—	−25.1	21.3	−19.2	12.1	20, 24, 39

^{a} In Ni(2,4 pentanedionate)$_2$L$_2$.
^{b} In NiL$_6^{2+}$.
^{c} In CoL$_2$Cl$_2$.
^{d} In NiL$_2$Cl$_2$.
^{e} In NiL$_6^{2+}$.
^{f} In Co(2,4 pentanedionate)$_2$L$_2$.
^{g} In Ni(2,4 pentanedionate)$_2$L$_2$.
^{h} In CoL$_6^{2+}$.
^{i} In NiL$_2$ (R = i-propyl).

positive on the p_π orbital of the carbon atom). The para proton experiences a similar shift both in magnitude and sign. The meta protons experience downfield shifts. This alternation of spin density on six-membered aromatic rings is characteristic of π spin delocalization and arises from a node in the π system leading to spin polarization of the meta carbon.

The same pattern is observed for pyridine-N-oxide[29,40,41] and for triphenylphosphines.[25,42] Pyridine[43–46] and imidazole[47,48] experience σ spin density delocalization, although spin polarization introduces some spin density on the π system. When the aromatic ring is three bonds from the metal as in the case of triphenylphosphineoxide,[49] or benzaldiminates,[50] or benzamides,[40] the shifts still alternate but are about one order of magnitude smaller than in the case of phenylamine type of ligands. In the case of salicylaldiminates, two pathways sum up: one typical of the phenate moiety and one typical of benzaldiminates. The latter gives rise to a smaller shift and lead to a system that is not alternating[39] (Table 12–6). The porphyrins are another interesting ligand system that will be treated later.

12–9 FACTORS AFFECTING NUCLEAR RELAXATION IN PARAMAGNETIC SYSTEMS

As seen in Chapter 7, nuclear relaxation is determined by fluctuating magnetic fields, and the larger the coupling between the nuclear magnetic moment and the fluctuating magnetic fields, the faster the relaxation. In general, we can write

$$T_{1,2}^{-1} = (E^2/\hbar^2) f(\tau_c, \omega) \quad (12\text{–}33)$$

where $E^2/\hbar^2$ is the square of the aforementioned coupling energy in frequency units and $f(\tau_c, \omega)$ is a function of the correlation time, τ_c, and of the Larmor frequency, ω. The correlation time, discussed in Chapter 7, is a time constant typical of the system under investigation. Its reciprocal, τ_c^{-1}, is the average rate of change of the reciprocal position or orientation between the resonating nucleus and any other particle whose movement relative to the resonating nucleus causes the fluctuating magnetic field. The correlation function is of the form

$$f(\tau_c, \omega) = \frac{\tau_c}{1 + \omega^2 \tau^2} \quad (12\text{–}34)$$

and provides the density of frequencies present in the system from the motion described by the correlation time, τ_c. The frequency ω is that needed to induce the nuclear transition.

The difference in a paramagnetic system compared to those in the previous discussion of nuclear relaxation arises from the contributions from the fluctuating magnetic fields caused by the unpaired electrons.

The unpaired electrons cause fluctuating magnetic fields by three main mechanisms[53]: (1) the electron relaxation itself, (2) the rotation of the frame which contains the electron and the nuclear spins both aligned along the external magnetic field (Fig. 12–13), and (3) chemical exchange involving bound and free ligands (Fig. 12–14).

The first mechanism is self-explanatory. Electron relaxation occurs with $T_{1,2}$ of the order of 10^{-5} to 10^{-7} sec for radicals and 10^{-7} to 10^{-13} sec for metal ions. From time to time, in an absolutely random manner, the electron spin changes its orientation. In a large ensemble of electrons, the process of changing orientation is a first-order process having a time constant $T_{1,2}$ where T_1 and T_2 are electron relaxation times. Here $T_{1,2}$ represents the correlation time for the nucleus as far as this mechanism is concerned. The second mechanism leads to a rotation of the electron magnetic moment directly around the resonating

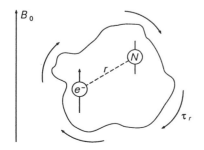

FIGURE 12–13 Rotation of the molecular frame holding the electron and the nucleus together causes nuclear relaxation. τ_r is the rotational correlation time.

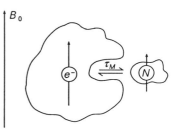

FIGURE 12–14 Binding and detachment of the molecular frame, holding the nucleus to the molecular frame, holding the electron, causes relaxation. τ_M is the chemical exchange time.

nucleus. This causes a fluctuation of the magnetic field at the nucleus; the correlation time is the rotational correlation time. The third mechanism is typically produced by chemical exchange. A given molecule containing the resonating nucleus binds to a molecule containing the unpaired electron and then dissociates from it. The correlation time for this mechanism is the residence time in the bound state.

As discussed in Section 12–2, we have spin density both on the resonating nucleus and outside it that fluctuates with a time constant τ_s due to spin reorientation. This is a fraction of an unpaired electron, equally distributed between the Zeeman levels, which relaxes between the Zeeman levels just like a full electron. Consideration of the existence of an excess population between the Zeeman levels does not appreciably change the picture under these conditions owing to the small extent of such an excess. (This is at variance with the isotropic shift that arises only because of the excess Zeeman population.) When there also is chemical exchange, spin density appears and disappears at the nucleus with a time constant τ_m. Thus, spin density causes fluctuating magnetic fields due to the electron relaxation and in certain cases, to chemical exchange. The energy of the coupling is the contact hyperfine coupling constant, $A_c/\hbar$, in frequency units.

Spin density outside the resonating nucleus, including the large spin density at the metal ion where a large fraction of the unpaired electrons reside, is a source of nuclear relaxation through dipolar coupling. The coupling energy now results from an integral over all space of terms of the form of equation (12–22) summed over the spin density distribution.[2,53,54] Upon rotation the integral averages to zero; however, its square [see equation (12–33)] does not average to zero upon rotation because $(3\cos^2\gamma - 1)^2$ averages to $4/5$. As with the isotropic shift, a metal-centered contribution and a ligand-centered contribution (plus a cross term) is considered.[2,55] The presence of spin density in a p_π orbital of a carbon atom affects the relaxation of a proton attached to the carbon.[56,57] The r^{-6} dependence of the average squared energy enhances the effect of the small fraction of unpaired electrons on the nucleus when it is close to the resonating nucleus. The ligand-centered contribution is often treated as a parameter.[48] The relevant nuclear relaxation mechanisms, arising from the dipolar coupling with spin densities and unpaired electrons, are the electronic relaxation, the rotation of the molecular frame, and chemical exchange.

When more than one effect is operative, the fastest one is dominant. Recall the fastest process corresponds to the smallest correlation time. If the reorientation due to rotation occurs for example with a time constant, $(1/\tau)$, of 10^{-7} sec and that due to intrinsic electron relaxation occurs with a time constant of 10^{-13} sec, the electron spin reorientation occurs on average one million times for a single rotational reorientation and electron spin relaxation dominates.

We can define a correlation time, τ_c^{-1} as

$$\tau_c^{-1} = \tau_r^{-1} + \tau_s^{-1} + \tau_m^{-1} \tag{12–35}$$

where τ_r is the rotational correlation time, τ_s the electronic correlation time, and τ_m the residence time. In a more precise way, we should define τ_{s1} and τ_{s2} depending on whether the correlation time is the longitudinal or transverse electron relaxation time, but this distinction can be overlooked without serious errors.[2,53] For different situations, each of the three mechanisms can dominate the dipolar relaxation. In the case of certain copper(II) proteins $\tau_s^{-1} > \tau_r^{-1}$, $\tau_r^{-1} \approx 10^7$ sec^{-1} and τ_s^{-1} dominates τ_c^{-1}. In smaller molecules τ_r^{-1} approaches 10^{11} sec^{-1} and dominates τ_c^{-1}. The time-average coupling energy is zero in every case, whereas its squared value is different from zero.

In the case of contact relaxation, molecular tumbling is ineffective for nuclear relaxation and the correlation time is:

$$\tau_c^{-1} = \tau_s^{-1} + \tau_m^{-1} \tag{12–36}$$

We have a further source of relaxation that is relevant under certain conditions. Let us return to the consideration of dipolar coupling in the absence of chemical exchange. Whatever the correlation time, we have assumed that the population of the Zeeman levels is equal. Correction for the Boltzmann population difference does not change our conclusions regarding the mechanisms previously discussed. However, this effect introduces a value of $\langle S_z \rangle$ different from zero giving rise to a permanent magnetic moment according to equation (12–13). This small magnetic moment is seen by the nucleus as fluctuating with a correlation time equal to τ_r. If the correlation time for dipolar nuclear relaxation is also τ_r, the effect of the excess population is negligible. When the correlation time for dipolar relaxation is dominated by τ_s, τ_r is much longer, and the unpaired electron (or a fraction of it) is reorienting with a time constant τ_s. On the other hand, the permanent magnetic moment due to $\langle S_z \rangle$ fluctuates, around the nucleus, with a time constant τ_r (Fig. 12–15). The quantity $\langle S_z \rangle$ is proportional to B_0 [Equation (12–15)] and when τ_r is large, as with macromolecules in solution at high magnetic fields where the excess Boltzmann population of the aligned state increases, the latter rotation leads to an efficient relaxation mechanism. The relaxation mechanism is also called Curie relaxation.

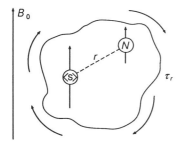

FIGURE 12–15 Rotation of the molecular frame holding the permanent magnetic moment $<S_z>$ and the nucleus together causes nuclear relaxation. τ_r is the rotational correlation time (cf. Fig. 12–13).

Equations for Contact Relaxation

Nuclear relaxation in the case of contact relaxation can be simply visualized in the absence of chemical exchange. The transition between Zeeman levels of the electron (*i.e.*, the spin change) may cause nuclear reorientation. The transition frequency required by the coupled electron–nucleus system is $\omega_I - \omega_s$. This frequency must be provided by the lattice. The equation for the longitudinal relaxation[58] is:

$$T_{1M}^{-1} = \frac{2}{3} S(S+1) \left(\frac{A_c}{\hbar}\right)^2 \frac{\tau_c}{1 + (\omega_I - \omega_s)^2 \tau_c^2} \tag{12-37}$$

where the subscript M indicates the paramagnetic contribution, A_c is proportional to the interaction energy involving the spin density, equation (12–19), $S(S+1)$ introduces the square of the intensity of the electron spin angular momentum, ω_I is the nuclear Larmor frequency, ω_s is the electron Larmor frequency, τ_c is given by equation (12–36), and $(\omega_I - \omega_s)^2$ can be approximated by ω_s^2.

The transverse relaxation is given[58] by:

$$T_{2M}^{-1} = \frac{1}{3} S(S+1) \left(\frac{A_c}{\hbar}\right)^2 \left(\tau_c + \frac{\tau_c}{1 + (\omega_I - \omega_s)^2 \tau_c^2}\right) \tag{12-38}$$

In contrast to T_1^{-1}, T_2^{-1} contains a term differing from equation (12–34) in that it contains $\omega = 0$ in the denominator because near zero frequencies affect T_2.

Equations for Dipolar Relaxation

For electrons localized on the metal ion and a resonating nucleus at a distance r, the energy given in equation (12–22) averages to zero with time under any modulating mechanism that is dipolar in nature. As mentioned before, E^2 does not average to zero. The correlation time τ_c is given by equation (12–35). The equations for relaxation are:

$$T_{1M}^{-1} = \frac{2}{15} \left(\frac{\mu_0}{4\pi}\right)^2 \frac{g_N^2 \mu_N^2 g^2 \mu_B^2 S(S+1)}{\hbar^2 r^6} \left(\frac{7\tau_c}{1 + \omega_s^2 \tau_c^2} + \frac{3\tau_c}{1 + \omega_I^2 \tau_c^2}\right) \tag{12-39}$$

$$T_{2M}^{-1} = \frac{1}{15} \left(\frac{\mu_0}{4\pi}\right)^2 \frac{g_N^2 \mu_N^2 g^2 \mu_B^2 S(S+1)}{\hbar^2 r^6} \left(\frac{13\tau_c}{1 + \omega_s^2 \tau_c^2} + \frac{3\tau_c}{1 + \omega_I^2 \tau_c^2} + 4\tau_c\right) \tag{12-40}$$

where $g_N \mu_N / \hbar$ is μ_1 of equation (12–22) and $g\mu_B/\hbar$ is μ_2. The frequency ω_I corresponds to the nuclear Larmor frequency and ω_s is the electron Larmor frequency. The latter is an approximation of $\omega_I - \omega_s$ (as in the contact term) and of $\omega_I + \omega_s$. The difference in frequency corresponds to the simultaneous flipping of both spins, for instance, one decreasing by one unit in m_s and the other increasing by one unit in m_I. The transition is called a zero-quantum transition and corresponds to W_0 of Section 8–6. The sum in frequency corresponds to an equal change in m_I and m_s. It is a double-quantum transition and corresponds to W_2. As usual, T_2^{-1} contains a non-dispersive term without ω and corresponds to near zero frequency oscillations.

Similar equations can be derived for spin densities at specific atoms in the molecule other than the metal ion.[55-57] These would be the equations for ligand-centered relaxation, which would require knowledge of the spin density distribution in the molecule and the introduction of cross terms with the metal-centered relaxation equations.

Equations for Curie Relaxation

We have already seen that a paramagnetic molecule has a magnetic moment given by equation (12–13). Upon rotation, this magnetic moment causes a fluctuating magnetic field at the resonating nucleus and causes relaxation (Fig. 12–15). This mechanism differs from those above because it depends on $\langle S_z \rangle$. Its effect depends on the square of the magnetic moment[59] and according to equation (12–15) on the square of the applied magnetic field.[60,61] The correlation times are

$$\tau_c^{-1} = \tau_m^{-1} + \tau_r^{-1} \qquad (12\text{–}41)$$

for the dipolar coupling and

$$\tau_c^{-1} = \tau_m^{-1} \qquad (12\text{–}42)$$

for contact coupling. These equations differ from equations (12–36) and (12–37) in that the term τ_s is absent. Otherwise the same considerations hold. For usual fields and small molecules, the effect is negligible. The rotational correlation time can be predicted for a spherical molecule through the Stokes-Einstein equation[2]:

$$\tau_r = \frac{4\pi \eta a^3}{3kT} \qquad (12\text{–}43)$$

where η is the viscosity of the solvent and a the radius of the molecule. Consistent with this formula, τ_r for a hexaaqua complex is about 3×10^{-11} sec and that for a protein of molecular weight 30,000 in solution about 10^{-8} sec. The expression for T_2^{-1} for the dipolar coupling between the permanent magnetic moment and the resonating nucleus (dipolar Curie relaxation) is[60,61]:

$$T_{2M}^{-1} = \frac{1}{5}\left(\frac{\mu_0}{4\pi}\right)^2 \frac{\omega_I^2 g^4 \mu_B^4 S^2(S+1)^2}{(3kT)^2 r^6}\left(4\tau_r + \frac{3\tau_r}{1+\omega^2\tau_r^2}\right) \qquad (12\text{–}44)$$

where ω_I is $g_N \mu_N B_0 / \hbar$. Since the line width is proportional to T_2^{-1},

$$T_2^{-1} = \pi \, \Delta v(\tfrac{1}{2}) \qquad (12\text{–}45)$$

where $\Delta v(\tfrac{1}{2})$ is the line width at half height, the Curie relaxation can affect the line width. At 600 MHz, the protons within 10 Å of a spin $S = \tfrac{3}{2}$ metal ion and with a τ_r of 10^{-8} sec are broadened. They are broadened beyond detection when r is smaller than 5 Å.

The expression for T_2^{-1} [equation (12–14)] contains in the $f(\tau_c, \omega)$ part, a term in τ_c. Relaxation mechanisms whose correlation time is long largely affect

T_2^{-1}. For T_1^{-1}, there are only terms of the type $\tau/(1 + \omega^2\tau^2)$ and for large τ values they tend to zero.[53] In both cases, if τ_c is small, the effect is small.

General Comments Concerning Relaxation

All the foregoing equations have been derived with the assumption that the S manifold is not split at zero magnetic field but is only split by the magnetic field into the $2S + 1$ levels of the S multiplet. However, we know from epr spectroscopy and from magnetic susceptibility measurements that this is rarely the case. Zero field splitting is usually present when $S > \frac{1}{2}$. Furthermore, hyperfine coupling causes splitting of the S manifold at zero magnetic field (see, *e.g.*, Fig. 12–2) and the splitting is sizable when the coupling occurs with magnetic nuclei which bear the unpaired electrons. These splittings do not have an effect on nuclear relaxation as long as the splitting energy E_{sp} is smaller[53] than $\hbar\tau_c^{-1}$. For example, for $\tau_c^{-1} = 10^{-11}$ sec, E_{sp} has to be smaller than 1 cm^{-1}. When this is not the case, the foregoing equations provide qualitative predictions, and when quantitative treatments are needed the reader is referred to available literature.[2,53,62-71]

As in the case of the isotropic shift, the factorization among the various contributions is the first step in any quantitative exploitation of the relaxation parameters. Although the nature of the problem is complicated, we are fortunate in this case because the various contributions may differ by orders of magnitude. Therefore, in general, only one contribution is dominant and simple consideration and sample calculations can show which mechanism dominates. For example, T_1 of nondirectly coordinated hydrogen atoms in paramagnetic metal complexes is determined by dipolar contributions, which sometimes may be approximated by a metal-centered model. Nuclei of atoms directly coordinated to paramagnetic centers may have large contact contributions to T_1 and T_2. In the case of the proton nmr of copper complexes, T_1 is determined by the dipolar contribution, whereas T_2, which contains the non-dispersive τ_c term, is determined by the contact contribution.[72] In fact, the correlation time for contact relaxation is τ_s, which is $\sim 10^{-9}$ sec, whereas the correlation time for T_1 is τ_r, which can be as short as 10^{-11} sec.

12–10 RELAXOMETRY

The equations for nuclear relaxation by unpaired electrons always contain either or both the nuclear and electronic Larmor frequencies. Such terms make the nuclear relaxation magnetic field dependent. A single relaxation measurement does not provide both the correlation time and the electron–nuclear coupling energy. In the ideal situation, enough experimental data would be available at different magnetic fields to describe the curve for the entire field dependence of T_{1M}^{-1} or T_{2M}^{-1}.

The expected plots of T_1^{-1} and T_2^{-1} as a function of the applied magnetic field are reported in Fig. 12–16 for the contact and in Fig. 12–17 for the dipolar relaxation.[2,53,58,59] Experimentally, at the time of writing, we are limited on the high field side to 600 MHz, whereas on the low field side the limit is the sensitivity which is proportional to the strength of the applied magnetic field, to the nuclear magnetic moment, and to the concentration of nuclei. Devices exist for measuring

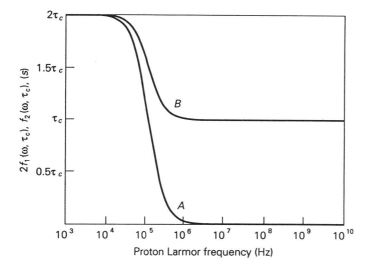

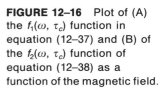

FIGURE 12–16 Plot of (A) the $f_1(\omega, \tau_c)$ function in equation (12–37) and (B) of the $f_2(\omega, \tau_c)$ function of equation (12–38) as a function of the magnetic field.

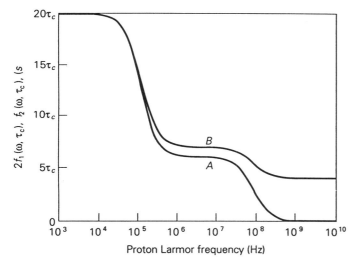

FIGURE 12–17 Plot of (A) the $f_1(\omega, \tau_c)$ in equation (12–39) and (B) the $f_2(\omega, \tau_c)$ function of equation (12–40) as a function of the magnetic field.

T_1, and sometimes T_2, of solvent nuclei, particularly water hydrogens, at very low magnetic fields.[2,53,73,74]

The problem of chemical exchange as a source of nuclear relaxation has been neglected in this chapter except when τ_m becomes the correlation time. When τ_m^{-1} is larger than the separation in Hz (sec^{-1}) between the signal in the A site and that in the B site, then only one signal is observed whose shift is the average, weighted for the occupancy of the two sites. The treatment is analogous to that in diamagnetic systems. If τ_m^{-1} is larger than the difference in both T_1^{-1} and T_2^{-1}, then the resulting relaxation values are also a weighted average. When τ_m^{-1} is of the order of or smaller than the relaxation rates and the chemical shift difference, then more specialized literature should be consulted.[2,75–78] Here we assume that chemical exchange is fast and is averaged by the mole fraction relation.

We will treat coordination compounds in aqueous solution having a water molecule in the coordination sphere. The water proton T_1 values are measured in a range of magnetic fields between 0.01 and 100 MHz. First, we shall consider

FIGURE 12–18 T_2^{-1} (○) and T_1^{-1} (●) of water protons in 6.65×10^{-2} M Cu^{2+} solutions as a function of the proton Larmor frequency at 25° C. The continuous lines represent a fitting of the data according to the Solomon equations (12–40) and (12–39), respectively.[79]

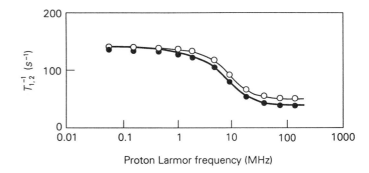

the complex $Cu(OH_2)_6^{2+}$. The experimental data reported[79] in Fig. 12–18 are a part of the curve reported in Fig. 12–17(A) for dipolar relaxation. By fitting the curve with equations (12–39) and (12–40), a value of τ_c of 3.0×10^{-11} sec is obtained at 25° C. This is a typical time constant for rotation of a hexaquacomplex. Since the correlation time is the rotational time, the nuclear relaxation must depend on dipolar coupling for only this effect depends on rotation. Assuming that the measured effect is entirely due to the presence of water in the first coordination sphere, and that there are 12 equivalent protons, the metal–proton distance is estimated to be 2.8 Å, in agreement with the expected 2.1 Å metal–oxygen distance.

These conclusions show that the proton–metal coupling is a metal-centered coupling for which equation (12–39) holds. The validity of the metal-centered point dipole approximation has been shown to be purely accidental as a result of cancellation of terms due to the delocalization of the electron within the Cu-O_6 volume and that due to spin density on the oxygen atom.[54] Proton contact contributions in all the water–copper(II) interactions are negligible.[2,53]

When the solution is made more viscous by adding ethyleneglycol or when the copper ion is bound to a macromolecule, τ_r increases. For a protein of molecular weight 32,000, such as copper–zinc superoxide dismutase (SOD hereafter), τ_r is expected to be around 3×10^{-8} sec. The rotational mechanism does not determine τ_c any more because τ_r is now longer than either or both τ_s or τ_m. Furthermore, there is a splitting of the S manifold at zero magnetic field due to the coupling of the electron with the magnetic nucleus of copper. Such splitting is of the order of 10^{-2} cm^{-1} and is smaller than $\hbar\tau_c^{-1}$ when τ_c is 3×10^{-11} sec (equal to 0.2 cm^{-1}). The splitting is larger than $\hbar\tau_c^{-1}$ when τ_c is of the order of 10^{-9} sec. The experimental data for an aqueous solution of SOD are reported[80] in Fig. 12–19. Their fitting to the dipolar coupling model, including the coupling between the electron and the copper nucleus, has provided a τ_c of 3×10^{-9} sec a copper oxygen distance of 2.4 Å for a single water molecule in the coordination sphere.[65] The value of τ_c could correspond to the water exchange rate, τ_m, or to the electronic relaxation time, τ_s. It is reported that τ_c corresponds to τ_s.[65] Relaxometry is, therefore, a tool to determine the electronic relaxation times, provided that rotation is slower than electronic relaxation.

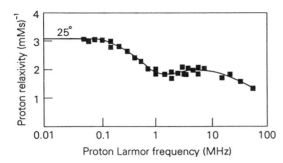

FIGURE 12–19 Water proton relaxation data for superoxide dismutase solutions.[80] The solid line is the best-fit curve obtained with the inclusion of the effect of hyperfine coupling with the metal nucleus[65].

12–11 ELECTRONIC RELAXATION TIMES

The electronic relaxation times estimated with relaxometry are reported in Table 12–7.[2,53] They are referred to a magnetic field of 2.35 T, (100 MHz ^{1}H resonance frequency) because electronic relaxation has been also found to be field dependent in some cases. These relaxation times are very relevant for understanding the nmr of paramagnetic molecules. In the cases of dipolar relaxation dominating

TABLE 12–7. Electronic Relaxation Rates for Various Metal Ions and Nuclear Line Broadening Due to Dipolar Relaxation[a]

Metal Ion	τ_s^{-1} (sec^{-1})	Nuclear Line Broadening (Hz)[a]
Ti^{3+}	$10^9 - 10^{10}$	3000–500
VO^{2+}	$10^8 - 10^9$	20,000–3000
V^{3+}	2×10^{11}	100
V^{2+}	2×10^9	9000
Cr^{3+}	2×10^9	9000
Cr^{2+}	10^{11}	300
Mn^{3+}	$10^{10} - 10^{11}$	3000–300
Mn^{2+}	$10^8 - 10^9$	200,000–40,000
Fe^{3+} (H.S.)	$10^{10} - 10^{11}$	5000–400
Fe^{3+} (L.S.)	$10^{11} - 10^{12}$	40–10
Fe^{2+} (H.S.)	10^{12}	70
Co^{2+} (H.S.)	$10^{11} - 10^{12}$	200–50
Co^{2+} (L.S.)	$10^9 - 10^{10}$	3000–500
Ni^{2+}	$10^{10} - 10^{12}$	1000–25
Cu^{2+}	$3 \times 10^8 - 10^9$	9000–3000
Ru^{3+}	$10^{11} - 10^{12}$	10–40
Re^{3+}	10^{11}	100
Gd^{3+}	$10^8 - 10^9$	400,000–60,000
Ln^{3+}	10^{12}	30–100

[a] For ^{1}H, $r = 500$ ppm, $B_0 = 2.35$ T.

and τ_s being the correlation time, the line width of a proton at a distance of 5 Å from the metal is predicted from equation (12–40) and reported in Table 12–7.

From inspection of the table, it is seen that some metal ions, including most of the lanthanides, give rise to little nmr line broadening. Low spin iron(III), high spin octahedral cobalt(II), and high spin tetrahedral nickel(II) fall in this category. Manganese(II), gadolinium(III), copper(II), and vanadium(IV) give rise to very broad signals. The first class of metal ions are called shift reagents because they provide hyperfine shifts with little broadening, whereas the latter class is called relaxation reagents because they relax the nuclei. In excess ligand and under fast exchange conditions, the line broadening, or T_1^{-1} enhancement, is reduced according to the mole fraction of bound ligand. Under these conditions, a ligand nucleus can be observed in the presence of a relaxing agent yielding fruitful information from the relaxation parameters.

Knowing the values of τ_s is a starting point to attempt an understanding of the factors that determine electronic relaxation in solution. Every time τ_r is shorter than τ_s, we can propose that fluctuating magnetic fields associated with rotation cause electronic relaxation. In addition to the intensity of the magnetic moments, the electron spin differs from the nuclear spin because the electron spin angular momentum is coupled with the orbital angular momentum. The angular momentum is orientation dependent. This leads to g values and zero field splitting values that are orientation dependent. We say that rotation modulates the g anisotropy and the zero field splitting. Such modulations are associated with modulations of the orbital magnetic moment and cause electron spin relaxation. When rotation is slower than electron relaxation, rotation cannot be the mechanism that leads to electron relaxation. In this case, it is instructive to refer to solid-state mechanisms. Indeed, lanthanides have comparable electron relaxation rates in solution and in the solid state, which are much larger than rotation rates. It may be assumed that on the time scale of electron relaxation the surrounding solvent of a metal ion in solution can be visualized as a small crystal. In this context, we shall discuss some results in solids.

In solids, there are crystal vibrations of low energy whose excited levels are populated according to the temperature.[81] Changes in the excited vibrational levels cause modulation of the metal–donor distances, which are reflected in changes of the orbital angular momentum, which in turn lead to electron relaxation. Differences in crystal vibration frequencies due to population or depopulation of a given level are frequencies called phonons.[82] In crystals there are continuous emissions or absorptions of phonons through exchange with other properties of the solid (*e.g.*, with the electron spin levels of a paramagnetic substance). Around room temperature there are many phonons with frequencies of the order of 100–1000 cm^{-1}. Therefore, if the substance has electronic spin levels with energy separations of this order of magnitude, electron relaxation will be very efficient.[83] This is the case of lanthanides where the ground levels reported in Table 12–1 are split by the above order of magnitude from the action of the crystalline field.[84] The same situation holds for high spin octahedral cobalt(II) and tetrahedral nickel(II) complexes.[85] The ground state is a $^4T_{2g}$ and a $^3T_1$ respectively (*i.e.*, it is orbitally triply degenerate). Low symmetry components and spin orbit coupling give rise to six Kramers' doublets in the former case and up to nine levels in the latter. Among these, energy separations of the order of the phonon energies exist at room temperature. Octahedral nickel(II) and tetrahedral cobalt(II) have an orbitally non-degenerate ground state so the low symmetry component and spin orbit coupling provide smaller values of the zero field splitting. It follows that electron relaxation is slower because there are fewer phonons of that energy.

The general idea is that, whenever there are low-lying excited states, phonons can provide the energy for transitions involving a change in m_s (Fig. 12–20(A)). This mechanism is called the Orbach process. When there are no excited states at energies around 100 cm^{-1}, the transitions between Zeeman levels occur through a combination of two phonons whose energy difference matches the energy required for electron relaxation. This process, called the Raman process[84,86] [(Fig. 12–20(B)], is far less efficient and accounts for the long electron relaxation times of copper(II), vanadium(IV), high spin manganese(II), etc. In the case of high spin iron(III), (second order) spin orbit coupling is rather efficient in splitting the ^6S. When there are strong anisotropies as in the case of some porphyrin complexes, the splitting is of the order of several tens of wavenumbers and the ^1H nmr signal can be observed. In general, the electron relaxation rates of high spin iron(III) complexes are difficult to predict because zero field splitting is variable.

In summary, electrons can relax through the rotational mechanisms for compounds with slow electron relaxation rates in solution. When rotation cannot account for the observed values of electron relaxation, we should refer to the

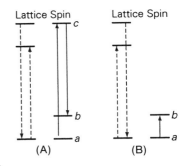

FIGURE 12–20 (A) Orbach and (B) Raman processes. In an Orbach process, the electron spin is excited to level c by absorption of a lattice phonon and then emits another phonon to level b. In a Raman process, the electron spin is excited directly to an excited level by simultaneous absorption and emission (scattering) of a single phonon.

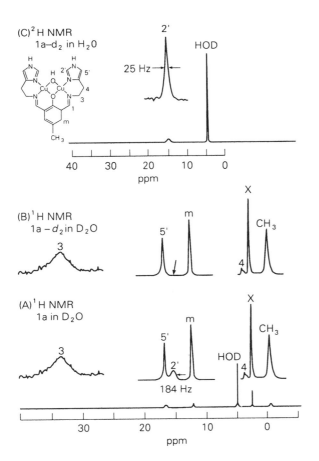

FIGURE 12–21 (A) ^1H nmr spectra at 23° C of the compound shown and (B) its partially deuterated form in D_2O (shaded arrow denoted the resonance position of the 2-H signal) and (C) ^2H nmr spectrum at 23° C in H_2O.

type of mechanisms observed in the solid state. The presence of unpaired electrons always causes line broadening. Equations (12–38) and (12–40) provide an estimate of the line broadening as a function of τ_c; the latter may or may not be equal to the electronic relaxation times. In any case, the electronic relaxation times represent an upper limit for the broadening. For example, a proton at 5 Å from copper(II) would experience a broadening of $\simeq 5000$ Hz at 100 MHz (Table 12–7), if the correlation time were determined by τ_s. This holds for macromolecules that rotate slowly. In small complexes, however, τ_c is determined by a short τ_r and therefore the line broadening is smaller. Finally, a proper choice of the magnetic field should be made by taking into account equation (12–44) (Curie relaxation).

A comment is also appropriate regarding the choice of the resonating nucleus. The metal nuclei bearing the unpaired electrons cannot be observed because the coupling is too strong and the line width too large. Nuclei with small magnetic moments are broadened less because the coupling energy is small (although the magnitude of the hyperfine shift decreases when expressed in Hz). It is common to use a deuteron instead of a proton when the linewidths of the latter are too large to be detected. In Fig. 12–21, the ^{1}H and ^{2}H spectra of copper(II) complexes are shown.[87]

12–12 CONTRAST AGENTS

NMR imaging is a new clinical technique to obtain an image that is used particularly for the head and spine. The technique is sensitive to the protons of tissue water and fat in both soft tissue and bone marrow, and is insensitive to bone itself. Proton density varies only slightly among tissues. The change is not large enough for images of proton density alone to be clinically useful. However, the relaxation rates of tissue protons vary significantly from one tissue to another, so that contrast in the nmr image depends mainly on the relaxation rates, T_1 and T_2. The contrast can be optimized case by case with the appropriate weighting of the relaxation rates.[88]

It is the aim of researchers in the field to improve contrast and resolution for a given image in order to obtain a better image or a comparable image in less time. A reasonable way involves altering relaxation rates by introducing a paramagnetic compound into the tissue, which acts as contrast-enhancing agent. From the inorganic point of view, a contrast agent is a complex with a capability of dramatically increasing the proton T_1^{-1} and T_2^{-1}. According to equations (12–37) to (12–40), the complex should have a large S and a long τ_s. These properties are met in Gd^{3+} ($S = 7/2$), Mn^{2+} and Fe^{3+} ($S = 5/2$). Furthermore, water molecules should be in fast chemical exchange conditions with the paramagnetic center as is usually the case for water molecules coordinated to Mn^{2+} and Gd^{3+}. The correct approach requires measurement of T_1 and T_2 of a water solution of the complex at various magnetic fields in order to check the relaxivity (solvent relaxation rate enhancement per mM solute) at the imager frequency.[89] Then the complex should be tested in tissues and the water proton relaxation rates remeasured directly at different times after the injection. Tissues behave essentially as liquids.[88] Experiments in tissues can be made in vivo in test animals or after surgical asportation of pathological samples.

The organic part of the ligand is responsible for targeting a certain tissue when the contrast agent is administered as a drug. At present, the contrast agent

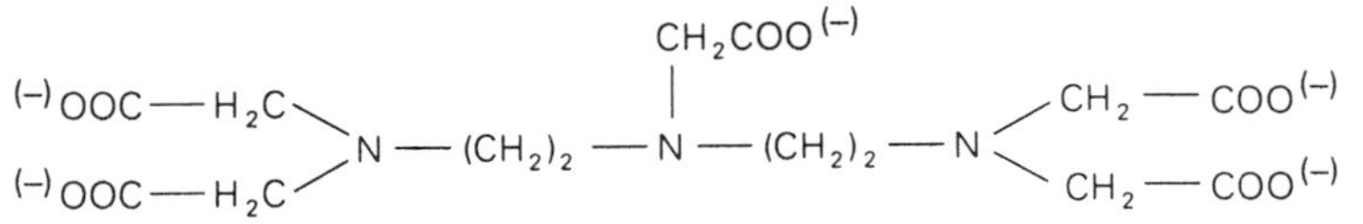

FIGURE 12–22 The DTPA^{5-} anion.

gadolinium (diethylenetriamine pentaacetate)$^{2-}$ (GdDTPA^{2-}, with DTPA^{5-} shown in Fig. 12–22) is commercially available. Other gadolinium complexes are being investigated.[90]

12–13 TRENDS IN THE DEVELOPMENT OF PARAMAGNETIC NMR

One main goal in nmr is the assignment of peaks. T_1^{-1} and T_2^{-1} are often dominated by dipolar coupling and, therefore, depend at least qualitatively on the sixth power of the nucleus–metal distance. These measurements, together with the analysis of the shifts and some experience, can lead to an assignment. The relative intensities of the signals and substitution of a nucleus, often a proton, with another group or isotope help in the assignment. NMR in diamagnetic systems has gone far in establishing criteria for assignment and in relating nuclear properties with internuclear distances. The trends in nmr of paramagnetic molecules parallel the applications of the well-established techniques in diamagnetic systems, with the limitations imposed on the paramagnetic system from fast nuclear relaxation.

The Nuclear Overhauser Effect

The nuclear Overhauser effect (nOe), in its more classical form, is the fractional variation of a signal intensity when another signal is saturated for a long time.[91,92] This is called steady-state nOe (see the diamagnetic nmr chapter, Section 8–6). If the nuclei corresponding to two signals interact through dipolar coupling, then a steady-state nOe is observed which, as already discussed, is given by the equation

$$\eta = \frac{W_2 - W_0}{2W_1 + W_2 + W_0} = \frac{\sigma}{\rho} \qquad (12\text{–}46)$$

where σ is the dipolar cross relaxation and ρ can be taken as T_1^{-1}. Here, σ corresponds to the net magnetization transfer from the saturated nucleus to its neighbor under observation and is given by

$$\sigma \propto \frac{\mu_1^2 \mu_2^2}{r^6} \left(\frac{6\tau_c}{1 + 4\omega^2 \tau_c^2} - \tau_c \right) \qquad (12\text{–}47)$$

where μ_1 and μ_2 are the nuclear magnetic moments, r is their distance, and τ_c is the correlation time for their reciprocal reorientation.

Paramagnetic molecules are characterized by large ρ values and therefore η is small. For protons, in the ortho position of aromatic rings or geminal in a CH_2 moiety, the distance is small and even if ρ is of the order of 10^2 sec^{-1}, η

can be of the order of 1 to 5%. Therefore, nOe can be measured under high signal-to-noise conditions. The conditions for observing nOe are best if the magnetic field is high and τ_c is large. Therefore, applications are successfully made in the study of paramagnetic metalloproteins where τ_c is large corresponding to the rotational correlation time, equation (12–33). Some examples of nOe experiments in different systems are shown in Fig. 12–23.[93,94]

Steady-state nOe in diamagnetic molecules result in the observation of secondary nOe through spin diffusion. The effect is greatly reduced in paramagnetic systems where nuclear relaxation is so fast that secondary nOe do not have time to build up.[95]

Transient nOe is the fractional variation of a signal intensity when a related signal is inverted with a 180° pulse. Since the inverted signal returns to equilibrium with a time constant T_1^{-1}, which is large in a paramagnetic system, magnetization transfer is small. In general, transient nOe experiments give poorer results than steady-state nOe experiments.[95] Nuclear Overhauser effect measurements provide information on internuclear distances and, therefore, provide a unique criterion to perform a scientifically correct assignment.

The Effect of Fast Nuclear Relaxation on 2-D Spectra

The 2-D experiments, COSY and NOESY (see Section 8–19), are also possible for paramagnetic systems. In this case, the limit is that the magnetization transfer

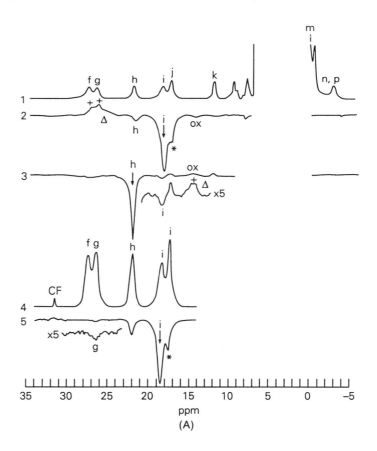

FIGURE 12–23 Some examples of nOe difference spectra of paramagnetic metalloproteins. (A) Reduced ferredoxin from *P. umbilicalis* in D_2O.[93] (B) N_3^- adduct of copper(II)-cobalt(II) superoxide dismutase in H_2O. The nOe's are typically of a few percent.

or redistribution responsible for cross peaks should occur before the spin system returns to the starting equilibrium situation. The return to equilibrium is governed by the relaxation times T_1 and T_2, which, as we have said, may be very short. Sometimes, T_1 and T_2 are so short that there is not enough time to alter the magnetization of the coupled spins. Recent experiments have shown that for T_1 or T_2 of the order of 50 msec, 2-D experiments are feasible with millimolar solutions. The results are useful if other conditions, like a long τ_r for NOESY experiments or large J values for COSY experiments, are satisfied.[96-99] One example is reported in the following section.

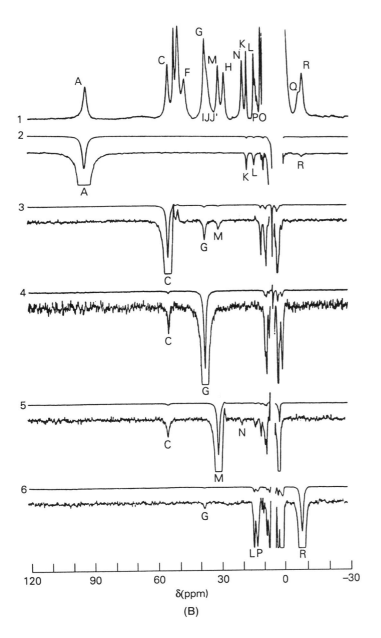

FIGURE 12–23 (continued)

12-14 SOME APPLICATIONS

Planar-Tetrahedral Equilibria in Nickel

One of the early applications of ^{1}H nmr to paramagnetic complexes involved *bis* N-substituted aminotroponeiminate nickel(II) complexes.[100] A chemical equilibrium exists between planar, diamagnetic, and tetrahedral, paramagnetic species.[101] Peak assignments were performed on the basis of chemical substitution and on the analysis of the first-order J splitting that is observed when the diamagnetic species is dominant. The shifts on the ring are alternating, as expected for a π delocalization pattern.[102] A large variety of substituents, R, on nitrogen were investigated. When R is aliphatic, the shifts decrease with the distance from the paramagnetic center, whereas in the case of aromatic substituents large shifts are also observed for protons very far away (Table 12–8).[103]

Since the planar-tetrahedral equilibrium is fast on the nmr time scale, the isotropic shift depends on the molar fraction of the paramagnetic species. The equilibrium constant is expressed as

$$K = \frac{[P]}{[D]} \frac{\alpha}{1-\alpha}$$

TABLE 12–8. ^{1}H Contact Shifts or R Groups in N-substituted Nickel Aminotroponeiminates^{a}[103]

R		δ/δ_β
CH$_3$		3.36
CH$_2$—CH$_3$	CH$_2$	3.24
	CH$_3$	0.21
CH$_2$—CH$_2$CH$_3$	CH$_2$	3.44
	CH$_2$	0.23
	CH$_3$	0.12
(phenyl)	o	−0.36
	m	0.38
	p	−0.51
(p-tolyl)	o	−0.37
	m	0.37
	CH$_3$	0.63
(biphenyl)	o	−0.37
	m	0.38
	o'	−0.037
	m'	0.051
	p'	−0.059
(diphenyl ether)	o	−0.39
	m	0.34
	o'	−0.01
	m'	~0
	p'	−0.01

^{a} The shifts are relative to the shift of the β proton of the troponeiminate ring, δ_β. Such shifts are in the 20 to 30 ppm range.

where [P] is the concentration of the tetrahedral paramagnetic species and [D] is the concentration of the planar diamagnetic species.[36] The molar fraction of paramagnetic species is given by α:

$$\alpha = \frac{(\Delta v)_{\text{observed}}}{(\Delta v)_{\text{fully paramagnetic}}}$$

When Δv of the fully paramagnetic species is not available, one should rely on the validity of equation (12–17) for estimating such a value.[104]

Diasteroisomerism

Bis-N-substituted salicylaldiminates also undergo planar $\rightleftharpoons$ tetrahedral equilibria. Studies similar to those of the aminotroponeiminates have been performed.[105] The *bis*-chelate complexes are dissymmetric and can be Λ or Δ. As the R = sec-butyl group bears a center of asymmetry, the *bis* N-sec-butylsalicyladiminate complexes can exist as two diastereoisomers $\Lambda+$ and $\Lambda-$, which are pairwise equivalent to $\Delta-$ and $\Delta+$, respectively.[106] The two diastereoisomers are detected thanks to sharp lines and to the two different isotropic shifts (Fig. 12–24).

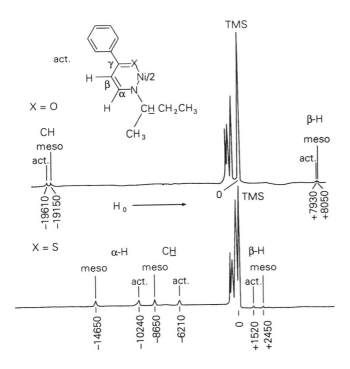

FIGURE 12–24 100 MHz ^{1}H nmr spectra of mixtures of active and meso isomers of Ni(sBuPhHH)$_2$(X = O) and Ni(sBu-SPhHH)$_2$ (X = S) in CDCl$_3$ solution at $-30°$. Chemical shifts are given in Hz relative to TMS.[106]

Diastereoisomerism and Diastereotopism in Cobalt(II)

N-alkyl pyridin-2-aldimines give rise to tris-chelate cobalt(II) complexes. Due to the asymmetric nature of the ligand, *cis* and *trans* isomers are expected and observed (Fig. 12–25).[107] The contact and pseudocontact contributions to the shifts lead to a large difference in shifts for the various isomers. When R is

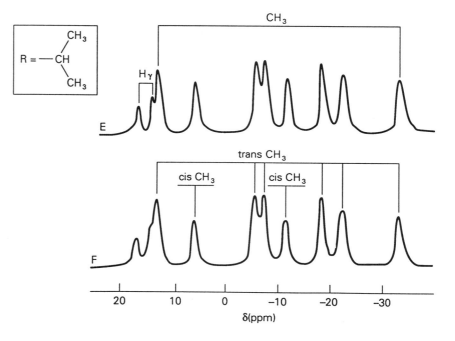

FIGURE 12–25 ^{1}H nmr spectra at 26° of Co(N-isopropyl pyridin-2-aldimine)$_3^{2+}$ in d_3-acetonitrile (top) and d_6-acetone (bottom).[107]

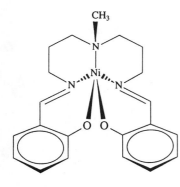

isopropyl, the two CH_3 groups are inequivalent because they sense an asymmetric center that is the metal. Two signals are expected for the *cis* isomer and six for the *trans* isomer. Indeed, eight well-separated signals are observed. In every case the intensity of the *cis* signal is minor corresponding to a less stable isomer (Fig. 12–25).[107]

Group Inequivalence in NiSALMeDPT

The five coordinated, *bis* (3,3'-salicylideneiminatopropyl) methyl amino nickel(II) complex is asymmetric, with the CH_3 group making the two salicylaldiminate moieties inequivalent when coordinated. The signals of the two salicylideneiminatopropyl groups will differ (as seen in Fig. 12–26). This system was first studied by LaMar and Sacconi,[108] who related the difference in shift to the existence of a more square pyramidal (almost equivalent groups) or more trigonal bypyramidal (more inequivalent groups) isomers. The generality of the inequivalence in one isomer was then shown for the *bis* thio analog.[109] The assignment of the aromatic signals was performed by varying substituents. Next, the spectral width was increased to include all the signals.[2] Finally, T_1 measurements and 2-D spectra have led to a complete assignment.[96]

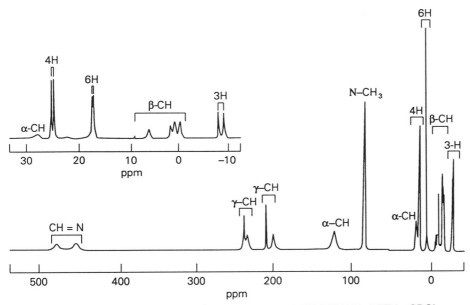

FIGURE 12–26 Room temperature ^{1}H nmr spectra of 5Cl-NiSALMeDPT in CDCl$_3$. Each proton in the molecule (except the methyl protons) gives rise to a separate signal.

Ion Pairing

Ion pairs involving a paramagnetic anion and a diamagnetic cation have been investigated through ^{1}H nmr.[110,111] The protons of the tetraalkylammonium cation experience an isotropic shift, which is probably contact in nature (Fig. 12–27). When the pair is formed, it appears that spin density is transferred onto the diamagnetic cation. The approach of the two anions occurs from every possible direction, and the free $\rightleftharpoons$ paired equilibrium is fast on the nmr time scale. Therefore, all the equivalent cation protons remain equivalent upon pairing. These systems were investigated through the shifts[110,111] and T_1 values.[111] The formation of ion pairs is enough to provide orbital overlap and transfer of spin density. Cases can be conceived where a moiety can approach a paramagnetic system at van der Waals distance (*i.e.*, the spin transfer is negligible, but the nucleus still senses the unpaired electron through space). In this case, the correlation time for the interaction can be the diffusion correlation time τ_D and the relaxation equations have a form of the type[112,113]

$$T_1^{-1} = (32\pi/405)\gamma_I^2\gamma_S^2\hbar^2 S(S+1)(N_A/1000)([M]/(d/D)) \cdot [J_2(\omega_S - \omega_I)$$
$$+ 3J_1(\omega_1) + 6J_2(\omega_S + \omega_I)] \qquad (12\text{–}48)$$

$$T_2^{-1} = (16\pi/405)\gamma_I^2\gamma_S^2\hbar^2 S(S+1)(N_A/1000)([M]/(dD)) \cdot [4J_1(0)$$
$$+ J_2(\omega_S - \omega_I) + 3J_1(\omega_I) + 6J_2(\omega_S + \omega_I) + 6J_2(\omega_S)] \qquad (12\text{–}49)$$

where N_A is Avogadro's constant, d is the distance of closest approach, D is the sum of the diffusion coefficients for the two molecules, and [M] is the

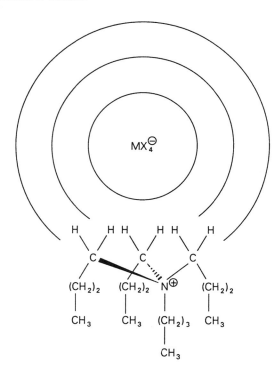

FIGURE 12–27 A contact interaction occurs in $(NR_4)^+(MX_4)^-$ ion pairs involving the α-CH_2 groups of the cation.[110,111]

concentration of the molecule bearing the paramagnetic center. Under a wide range of conditions the spectral density functions are given by

$$J_1 = J_2 = (1 + 5z/8 + z^2/8)/(1 + z + z^2/2 + z^3/6 + 4z^4/81 + z^5/81 + z^6/648) \quad (12\text{–}50)$$

with

$$z = (2\omega\tau_D)^{1/2} \quad (12\text{–}51)$$

In turn, τ_D is given by

$$\tau_D = \frac{2d^2}{D} \quad (12\text{–}52)$$

A model of this type has been used to describe the interaction of water with $Cr(en)_3^{3+}$.[114]

Hemin-Imidazole-Cyanide

The spectrum of a heme complex (*i.e.*, hemin-imidazole-cyanide) is an example of a low spin ($S = \frac{1}{2}$) iron(III) heme system. Heme complexes are present in many proteins and studies have led to understanding of these important complex systems. The spectrum of a model is shown in Fig. 12–28).[115] The four intense signals are assigned to the four ring methyls. Their precise assignment can be

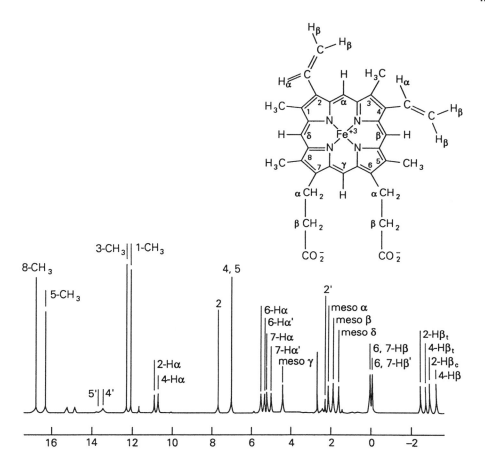

FIGURE 12–28 Room temperature 600 MHz ^{1}H nmr spectrum of hemin-imidazole-cyanide. The hemin scheme is reported in the inset.

made by using selectively deuterated samples. Alternatively, 2-D NOESY and COSY spectra can be employed. In the NOESY spectra, cross peaks occur between adjacent protons. In the COSY spectra, the cross peaks occur between signals related by J coupling. Vinyl and propionyl group correlations are found both in NOESY and COSY spectra, as shown in Fig. 12–29.[115] Methyl groups are distinguished because 1 and 3 give a nOe with vinyl groups. Finally, methyls 1 and 8 give a nOe to one another. The methyl signals, once assigned, give a nOe with the meso protons and, therefore, the spectrum is fully assigned. As far as the imidazole is concerned, the 1'-H signal disappears when the spectrum is recorded in D$_2$O and 2'-H as well as 4'-H have the shortest T_1. All these spectra were recorded at 40 mM heme concentration.[115] In the case of proteins, the concentration is about one order of magnitude lower and, as a result, some cross peaks may be lost.

Spin Delocalization in Iron Porphyrins

Iron porphyrins are important molecules especially for their biological relevance. Iron(III) can be high spin or low spin, although in some cases spin admixtures are proposed. The electronic levels in the two cases are shown in the following

540 Chapter 12 Nuclear Magnetic Resonance of Paramagnetic Substances in Solution

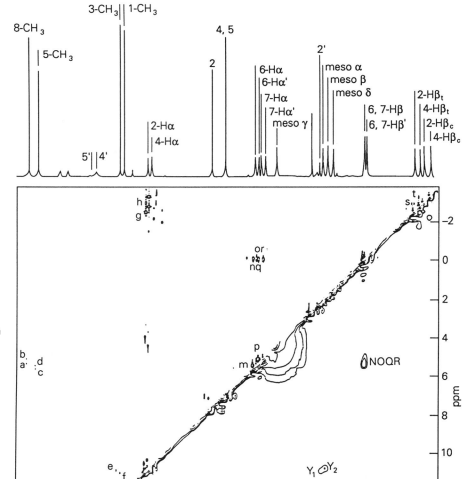

FIGURE 12–29
Two-dimensional ¹H COSY (upper left) and NOESY (lower right) spectra of hemin-imidozole-cyanide. The labeling on the spectrum (lowercase = COSY, uppercase = NOESY) indicates the following correspondences to the scheme in Fig. 12–28.

a/A:	8-CH₃/7-Hα	h:	2-Hα/2-Hβ_c	p:	7-Hα/7-Hα'	X_1:	3-CH₃/meso α	
b/B:	8-CH₃/7-Hα'	i:	4-Hα/4-Hβ_t	q/Q:	7-Hα/7-Hβ,β'	X_2:	1-CH₃/meso δ	
c/C:	5-CH₃/6-Hα	j:	4-Hα/4-Hβ_c	r/R:	7-Hα'/7-Hβ,β'	Y_1:	2-Hα/meso α	
d/D:	5-CH₃/6-Hα'	l:	2/4,5	s:	2-Hβ_t/2-Hβ_c	Y_2:	4-Hα/meso β	
e:	3-CH₃/4-Hα	m:	6-Hα/6-Hα'	t:	4-Hβ_t/4-Hβ_c	Z_1:	3-CH₃/4-Hβ_t	
f:	1-CH₃/2-Hα	n/N:	6-Hα/6-Hβ,β	U:	8-CH₃/meso δ	Z_2:	1-Ch₃/2-Hβ_t	
g:	2-H_α/2-Hβ_t	o/O:	6-Hα'/6-Hβ	V:	5-CH₃/meso β			

scheme. High spin complexes have unpaired electrons in σ-type orbitals.[116] Therefore, a dominant σ delocalization and large downfield shifts are expected. The pyrrole protons of meso-tetraphenylporphyrin (TPP) chloride complexes are far downfield (Fig. 12–30).[2,116]

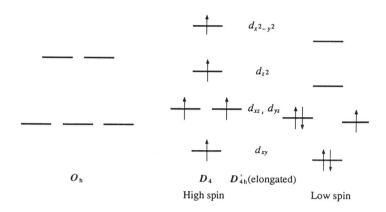

Similar considerations hold for high spin iron(II). In contrast, the unpaired electron is in a π-type orbital in the low spin iron(III) TPP *bis*-imidazole complex.[116] The shifts are smaller and the pyrrole proton is shifted upfield (Fig. 12–30). Since low spin iron(III) complexes have short electronic relaxation times,

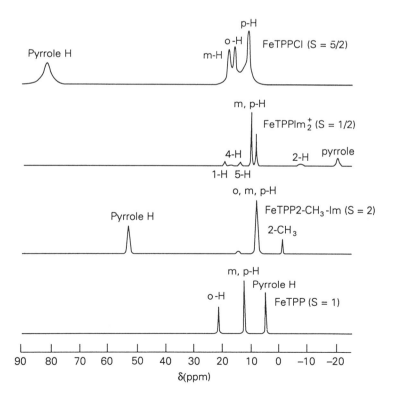

FIGURE 12–30 Simulated ^{1}H nmr spectra of meso-tetraphenylporphyrin (TPP) iron complexes.[2] Chemical shifts are in ppm from TMS.

the nmr signals are quite sharp. The ^{1}H nmr spectra of iron(III) hemin-imidazole-cyanide,[115] shown in Fig. 12–28, provides a typical example of this class of compounds.

Cobalt-substituted Carbonic Anhydrase

Carbonic anhydrase (CA) is a zinc protein of MW 30,000. Zinc can be removed and substituted by cobalt(II) without loss of activity. Cobalt(II) in CoCA is high spin. Three histidine imidazoles and a water molecule are coordinated. One of

the imidazoles is coordinated to the metal through its Nδ1 nitrogen, whereas the other two are coordinated through their Nϵ2 nitrogens. If NO_3^- is added to the CoCA solution, it binds cobalt giving rise to a five-coordinate derivative where one oxygen of nitrate adds as a fifth ligand. The spectrum is shown in Fig. 12–31 as an example of a spectrum of a metalloprotein.[117]

Four signals are observed far downfield. Three of them disappear when the spectrum is recorded in D_2O and these are assigned to the exchangeable NH protons of the three coordinated histidine rings. The non-exchangeable signal at 71 ppm is of similar line width, and is thus assigned to the Hδ2 proton of His-119, the only one not adjacent to a coordinating nitrogen. This signal gives an nOe with the NH signal at 80 ppm, which is therefore its neighbor (Hϵ2) on the imidazole ring of His-119.[118] The other signals between +30 and −20 ppm belong to protons that are all close to cobalt and the shifts are all dipolar in origin. Their assignment[118] is a challenge with implications of paramount importance for further investigation of proteins.

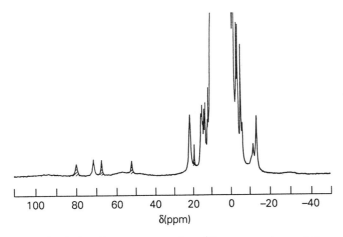

FIGURE 12–31 Room temperature ^{1}H nmr spectrum of the nitrate adduct of cobalt(II) substituted carbonic anhydrase.[117,118] The shaded signals disappear in D_2O.

12–15 THE INVESTIGATION OF BIMETALLIC SYSTEMS

We now consider the consequences of magnetic coupling in bimetallic systems (see Chapter 11) on nmr. For simplicity and without loss of generality, consider that one nucleus senses only one metal ion of the couple. If this were not the case, the effects of each metal ion, calculated separately, would be added. Our working scheme is therefore

$$N \text{———} M_1 \text{———} M_2$$

where N is the nucleus and M_1 and M_2 two metal ions. Also consider that magnetic coupling between M_1 and M_2 is a weak perturbation with respect to the energy and distribution of electrons of the moieties to which M_1 and M_2 belong in the absence of magnetic coupling. This means that spin density on nucleus N is not affected by the existence of magnetic coupling. Furthermore, we can also consider that the electronic relaxation time of metal 1 changes its value in a perturbative fashion when interacting with the electron of metal 2. This consideration is not fully justified, since it is arbitrary to refer to only one electronic relaxation time of a metal ion.* In principle, there is an electronic relaxation time for each level arising from magnetic coupling. Nevertheless, this reasoning leads to the conclusion that for homodimeric systems there is no change

---

*Electronic relaxation time is treated as a single transition; when there are several transitions as in the case of zero field split levels, it is appropriate to assign a relaxation time for each transition.

of electronic relaxation times, whereas in heterodimetallic systems one metal ion will have larger electronic relaxation times than the other. In the latter case, the slow relaxing metal ion will take advantage of magnetic coupling, and its electrons will relax with the mechanism of the fast-relaxing metal ion. Thus, magnetic coupling causes a shortening of the electronic relaxation time of the slow-relaxing metal ion. These conclusions are consistent with observations. Equations and discussions are available in the literature that relate the extent and nature of J with the values of electron relaxation times.[53,119]

The ^{1}H nmr of Dimeric Complexes

The following ligand

indicated as PMK, provides a nice frame for bimetallic systems where antiferromagnetic coupling is established between the two metal ions. The systems studied include CoCo, NiNi, CuCo, and CuNi.[120] With Zn being diamagnetic CuZn, CoZn, and NiZn represent the blank.[70,121] The J value is relatively small (*i.e.*, $\ll kT$ at room temperature). Consequently, all the energy levels arising from magnetic coupling are almost equally populated. The isotropic shift experienced by each proton is the sum of the contributions of the two metal ions (*i.e.*, the

TABLE 12–9. 200 MHz ^{1}H nmr Shift and T_1 Values at 300 K for the Isotropically Shifted Signals in the Cu_2Co_2SOD Derivative Together with Their Assignment (signal labeling refers to Figs. 12–33 and 12–34)

Signal	δ, ppm	T_1, ms	Assignment
A	66.2	1.5	His-61 Hδ2
B	56.5	7.8	His-118 Hδ1
C	50.3	4.2	His-44 Hε2
D	49.4	3.8	His-69 Hδ2
E	48.8	4.6	His-78 Hδ2
F	46.7	2.1	His-78 Hε2 (His-69 Hε2)
G	40.6	3.5	His-44 Hδ2
H	39.0	1.8	His-118 Hε1
I	37.4	1.7	Asp-81 Hβ1 (Asp-81 Hβ2)
J'	35.6	1.7	Asp-81 Hβ2 (Asp-81 Hβ1)
J	35.4	d	His-69 Hε2 (His-78 Hε2)
K	34.5	8.0	His-46 Hδ1
L	28.4	4.3	His-46 Hδ2
M	25.3	2.7	His-44 Hε1
N	24.1	2.9	His-118 Hδ2
O	19.6	1.9	His-46 Hε1
P	18.7	1.6	His-44 Hβ1
Q(R)	−6.2	2.4	His-69 Hβ2
R(Q)	−6.2	2.4	His-44 Hβ2

sum of the shift of a given proton in the M_1Zn and the ZnM_2 complexes). As far as the line widths are concerned, those of the CuZn complex are very large for those protons sensing copper because $\tau_s \simeq 10^{-9}$ sec and τ_c is given by τ_r, which is $\simeq 10^{-11}$ sec.[120] Such a relatively long value of τ_s provides a short nuclear T_2 [equation (12–38)]. In the CuCo system, cobalt(II) has an electronic relaxation time of $\simeq 10^{-12}$ sec (see Table 12–9) and the τ_s of copper approaches this value, becoming much shorter than the value of the uncoupled system. The ^{1}H nmr signals are now quite sharp[121] (Fig. 12–32).

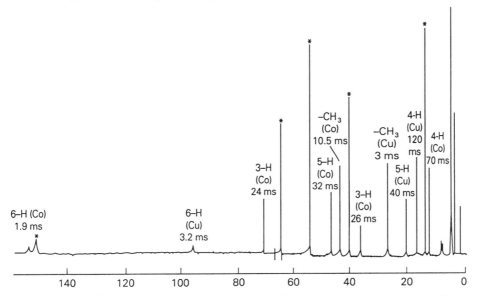

FIGURE 12–32 ^{1}H nmr spectrum of CuCo (PMK)$_3^{4+}$ in D$_2$O. Peaks of Co$_2$(PMK)$_3^{4+}$ are indicated with an asterisk.[121]

Cu$_2$Co$_2$ Superoxide Dismutase

The enzyme Cu$_2$Co$_2$SOD, where SOD refers to superoxide dismutase, is analogous to the bimetallic systems. SOD catalyzes the dismutation of the superoxide ion. One common enzyme is a dimeric protein of MW equal to 32,000 with a MW of 16,000 in each subunit. In each subunit, zinc is bound to three histidine nitrogens and one oxygen of an aspartate[122] (Fig. 12–33). Copper is bound to four histidine nitrogens and is exposed to solvent (see Figure 12–33). One histidine is in common to the two ions, bridging them. The protein rotates with $\tau_r \simeq 10^{-8}$ sec, whereas τ_s of copper(II) is 2×10^{-9} sec.[65] The protons sensing copper(II) have such a broad nmr line that they escape detection. The protein can be manipulated to substitute zinc with cobalt. The copper and cobalt ions experience antiferromagnetic coupling through the bridging imidazolate with $J = 17$ cm^{-1}.[123] The electronic relaxation times of copper(II) approach those of cobalt(II), which are of the order of 10^{-12} sec. The ^{1}H nmr signals of the ligands of both metal ions can now be observed since they are relatively sharp.[48,124] Through nOe it has been possible to propose a full assignment[125] since several interproton distances could be calculated and compared to those of the x-ray structure (see Table 12–9 and Fig. 12–34).

546 Chapter 12 Nuclear Magnetic Resonance of Paramagnetic Substances in Solution

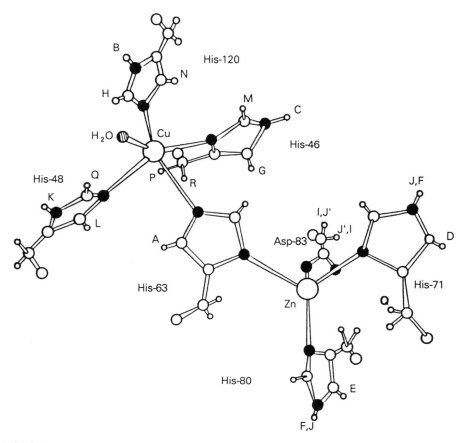

FIGURE 12–33 Scheme of the metal sites in Cu_2Zn_2SOD.[122]

FIGURE 12–34 ^1H nmr spectrum at 300 K and pH 5.6 acetate buffer of the Cu_2Co_2SOD derivative.

^1H nmr Spectra of Fe_2S_2 Ferredoxins

The Fe_2S_2 ferredoxins are electron transfer proteins of low molecular weight ($\simeq 10{,}000$) containing the following bimetallic moiety.

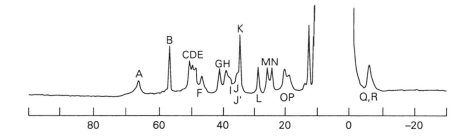

The oxidized form of the protein contains two iron(III) ions, whereas the reduced form contains one iron(III) and one iron(II). In the oxidized form, the two iron(III) ions are antiferromagnetically coupled with $J = 400$ cm^{-1}, giving an $S' = 0$ ground state. The energy levels, calculated according to equation (11–40) are shown in Fig. 12–35. Since $J \simeq kT$, only part of the levels are occupied and the magnetic susceptibility of the system is reduced with respect to that expected for an uncoupled system. The electronic relaxation times of iron(III) are long and the ^{1}H nmr signals of the β-CH$_2$ (see the foregoing scheme) are expected to be very broad. Although still very broad, they are observed in the dimer because of the reduced magnetic susceptibility. All eight protons fall inside a broad envelope (Fig. 12–36).

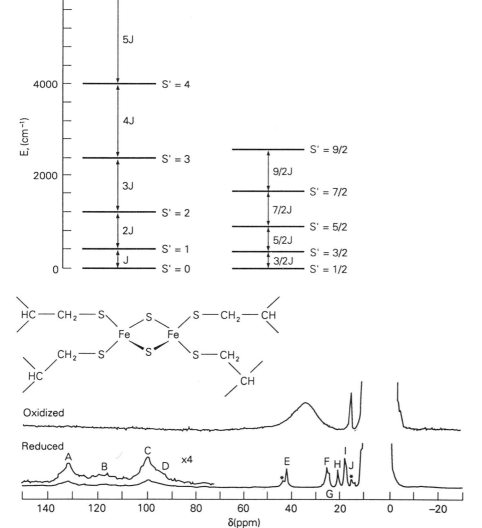

FIGURE 12–35 Energy levels for Fe$_2$S$_2$ dimetallic centers calculated for $J = 400$ cm^{-1} in the oxidized form and $J = 200$ cm^{-1} in the reduced form.[126]

FIGURE 12–36 ^{1}H nmr spectra of oxidized and reduced forms of Fe$_2$S$_2$ ferredoxin from spinach.[126,127] The signals marked with asterisks belong to impurities.

In the reduced case, iron(II) has short electronic relaxation times and the whole system becomes more suitable for nmr investigation. The spectrum is shown in Fig. 12–36.[127] The J value is now 200 cm^{-1} and the energy levels are reported in Fig. 12–35. The ground state is given by $S' = 1/2$ (*i.e.*, by $S' = S_1 - S_2$ where S_1 refers to iron(III) and S_2 to iron(II)). If only this level were populated, the isotropic shifts of iron(III) would be far downfield (because $\langle S_{1z} \rangle$ is negative according to equation (12–11) and the spin density for σ delocalization is positive [see equations (12–15) and (12–20)]. The isotropic shifts of iron(II) would be upfield because of a negative S_2 leading to a positive $\langle S_{2z} \rangle$. The four signals far downfield are assigned to the two β-CH$_2$ protons of ligands bound to iron(III). Figure 12–37. Signals F, G, H, and I are assigned to the β-CH$_2$ of ligands bound to iron(II). They are slightly downfield because of population of excited S' levels. The assignment is based on the temperature dependence of the shifts. The signals assigned to iron(III) are expected to follow a Curie dependence (*i.e.*, their shifts decrease with increasing temperature because the contribution of iron(III) to the magnetic susceptibility of the system decreases with increasing temperature as seen qualitatively with the Curie law [see equation (12–12)]. In the case of iron(II), the temperature increase causes an increase in the population of the states with positive S_z, since the highest S' level is given by $S_1 + S_2$. Accordingly, the downfield isotropic shift increases with increasing temperature. In Fig. 12–37, the experimental temperature dependence of the isotropic shifts is shown together with expectations from considering the energy levels of Fig. 12–35.[126,127] These results show that upon reduction of the oxidized protein only one localized iron of the pair is reduced and a mixed valent system does not result.[126] NOE studies have suggested that the reducible iron is the one closer to the surface of the protein.[128]

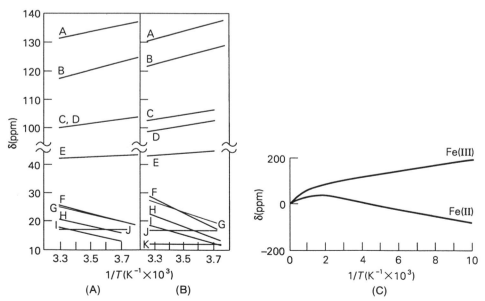

FIGURE 12–37 Temperature dependence of the isotropic shifts of the ^{1}H nmr signals in reduced spinach (A) and *P. umbilicalis* (B) ferredoxin. The predicted behavior is shown in (C).[126]

Perspectives in Cluster Investigations

Paramagnetic polymetallic systems are becoming more readily available from coordination chemistry and metalloproteins. Extension of the theory of dimetallic systems[126] allows the investigation of more complex systems (*e.g.*, Fe_4S_4, Fe_3S_4, Co_4S_4 clusters). The reader is referred to specialized literature.[126,129-133]

Shift Reagents

Rare earth tris-chelates of β-diketonate derivatives are Lewis acids and cause large pseudocontact shifts in Lewis bases that are added to solutions containing these complexes. The resulting resonances for several of these ions (*e.g.*, europium and praseodymium complexes) are relatively sharp. In Fig. 12-38, the 100-MHz proton magnetic resonance spectra of $CDCl_3$ solutions of *cis*-4-butylcyclohexanol containing varying amounts of $Eu(dpm)_3$ [dpm is dipivaloylmethanato, $(CH_3)_3CC(O)CHC(O)C(CH_3)_3$] are shown[134] to illustrate this behavior. The complex pattern obtained for the pure alcohol is shown in the bottom spectrum. A first-order spectrum with all resonances well separated is obtained in curve D. Note that the spin-spin couplings are still intact.

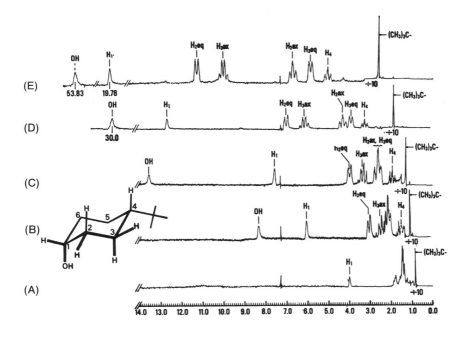

FIGURE 12-38 100-MHz pmr spectra of *cis*-4-*tert*-butylcyclohexanol (20 mg, 1.28×10^{-4} mol) in $CDCl_3$ (0.4 ml) containing various amounts of $Eu(dpm)_3$: (A) 0.0 mg; (B) 10.3 mg; (C) 16.0 mg; (D) 33.1 mg; (E) 60.2 mg. [Reprinted with permission from P. V. Demarco *et al.*, J. Am. Chem. Soc., *92*, 5734 (1970). Copyright by the American Chemical Society.]

Rare earth complexes that exhibit this behavior are called[135] shift reagents, S.R. As is usually the case with these systems, the S.R. complex is in fast exchange with an excess of the Lewis base. The lanthanide complexes of the fluorinated chelate 1,1,1,2,2,3,3-heptafluoro-7,7-dimethyl-4,6-octanedione, $(CH_3)_3CC(O)CHC(O)CF_2CF_2CF_3$, abbreviated as fod, are stronger Lewis acids and produce larger shifts.[136] Since these early reports there have been hundreds of papers dealing with applications of shift reagents.[137] High field spectrometers have caused a decrease in interest in shift reagents. Here, we shall deal with examples of some different types of applications and discuss some of the potential complications encountered.

One use of a shift reagent involves simplifying second-order (*i.e.*, $J \approx \Delta$) spectra. Different protons in the molecule are shifted by different amounts upon complexation. Thus, as the shift reagent concentration is increased, Δ becomes greater than J. This behavior is seen in Fig. 12–38.

A second potentially exciting application of shift reagents involves their use in determining geometries of molecules in solution.[138] This experiment is usually done in the fast exchange region. The proton nmr spectral shifts induced by shift reagents are assumed to be almost exclusively dipolar in origin. Otherwise, factorization is needed. Ideally, a structure would be assumed for the molecule and equation (12–23) would be used to calculate the expected dipolar shifts at a large number of various nuclei in the molecule whose structure is to be determined. The assumed structure of the molecule would be varied to produce a best fit of the experimental shifts. Since the structure of the molecule being investigated and that of the complex in solution are not known, nor are the magnitude and position of the magnetic dipole of the metal center in the complex, there are eight unknowns in the system. These unknowns are best seen by looking at Fig. 12–39, where the example of a rigid ligand such as pyridine complexed to a S.R. is illustrated. Four parameters are needed to fix the orientation of the molecule being studied relative to the shift reagent: (1) the metal–donor atom distance, r; (2) the ligand atom–donor atom–metal bond angle, α; (3) β, the angle between the ligand's molecular plane and the magnetic xy-plane of the metal; and (4) γ, the angle

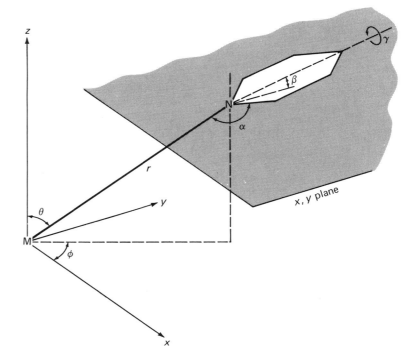

FIGURE 12–39 Definition of the variables in a shift reagent determination of molecular structure.

measuring the twist of the ligand's molecular plane around the nitrogen-para-carbon axis. In addition, two angles are needed to define the orientation of the magnetic axis with respect to the metal-donor bond. Two additional unknowns are needed to account for the anisotropy in χ. To complete our discussion of a rigorous solution of this problem, we should consider the possibility of having significant amounts of both 2:1 and 1:1 complexes in solution. (Most shift reagents are potentially diacidic, having two acid sites.) If neither the 2:1 nor the 1:1 complex is the predominant species in solution, both K_1 and K_2 must be known as well as two sets of the unknowns mentioned before, one set for each of the two complexes. However, it is obvious from the preceding discussion that it is not simple to determine geometries of molecules in solution in a rigorous manner. Thus, to gain any information about geometries in solution, more independent information is needed, such as the relaxation rates, and some simplifying assumptions must be made.

If threefold or higher axial symmetry and the presence of only one predominant species are assumed, the problem becomes tractable. With axial symmetry, equation (12–25) can be used instead of equation (12–25). By taking the ratio of all the observed nmr shifts with respect to the largest observed shift, Δv_j, equation (12–53) can be derived from equation (12–25):

$$\frac{\Delta v_i}{\Delta v_j} = \left(\frac{1 - 3\cos^2\theta_i}{r_i^3}\right) \bigg/ \left(\frac{1 - 3\cos^2\theta_j}{r_j^3}\right) \qquad (12\text{--}53)$$

and the magnitude of the magnetic anisotropy in the complex is eliminated from the problem. (This is possible only if one ignores the chelate ligand protons and works only with the coordinated Lewis base.)

The complications to obtain structures in solution include:

1. The stoichiometry of the adduct in solution must be determined but is often assumed.

2. The treatment assumes that there is only a single geometrical isomer of the adduct in solution. Many are possible. Furthermore, lanthanides have a high affinity for water, which if present in the system can compete for Lewis acid sites with the substrate, producing even more complexes in the system.

3. When the substrate is involved in a dynamic, intramolecular process, the average conformation of the molecule is determined, and this might represent a relatively unstable one.

4. The complex is generally assumed to be axially symmetric, so that the dipolar shifts can be evaluated with a $\langle(1 - 3\cos^2\theta)/r^3\rangle$ term. Those systems studied to date do not possess axial symmetry (however, see reference 139).*

5. The location of the principal magnetic axis in the complex is often not known relative to the substrate ligand.

6. When the substrate is a complex one, it is necessary to assume that there is only one binding site.

---

*Some more accurate studies were possible in proteins where many experimental data are available, and in this case no assumption of this type is needed.[2,139]

In spite of these complications, the method is capable of distinguishing gross structural features such as those that exist in many different isomeric compounds. For example, the compounds

are readily distinguished,[134] as are [141]

Shift reagents have been used to differentiate between internal and external phospholipid layers in membranes.[142] They have also been employed in conformational studies of nucleotides in solution[143] and in other metalloprotein systems.[144] one must always be concerned with the problem that coordination of the shift reagent may affect the conformation of the biomolecule.

Another interesting application involves the use of optically active shift reagents to determine optical purity. The idea is similar to that discussed in Chapter 8, in which optically active solvents were employed. Here, different stability constants for forming the different diastereoisomeric adducts exist, leading to different mole fraction averaged shifts for the enantiomeric bases. Several reports of different optically active rare earth complexes that can be used for this application have appeared.[145–147]

REFERENCES CITED

1. R. J. Fitzgerald and R. S. Drago, J. Am. Chem. Soc., *90*, 2523 (1968).
2. I. Bertini and C. Luchinat, "NMR of Paramagnetic Molecules in Biological Systems," Benjamin/Cummings, Menlo Park, CA, 1986.
3. J. S. Griffith, "The Theory of Transition Metal Ions," Cambridge University Press, London, 1971.
4. R. M. Golding, "Applied Wave Mechanics," Van Nostrand, London, 1969.
5. E. Fermi, Z. Phys., *60*, 320 (1930).
6. H. M. McConnell, Proc. Natl. Acad. Sci. USA, *69*, 335 (1972).
7. R. J. Kurland and B. R. McGarvey, J. Magn. Reson., *2*, 286 (1970).

8. W. deW. Horrocks, Jr. and D. D. Hall, Inorg. Chem., *10*, 2368 (1971).
9. R. M. Golding, R. O. Pascual, and B. R. McGarvey, J. Magn. Reson., *46*, 30 (1982).
10. R. M. Golding, R. O. Pascual, and J. Vrbancich, Mol. Phys., *31*, 731 (1976).
11. B. Bleaney, J. Magn. Reson., *8*, 91 (1972).
12. R. M. Golding and M. P. Halton, Aust. J. Chem., *25*, 2577 (1972).
13. C. N. Reilley, B. W. Good, and J. F. Desreux, Anal. Chem., *47*, 2110 (1975).
14. J. Reuben and G. A. Elgavish, J. Magn. Reson., *39*, 421 (1980).
15. B. Bleaney, C. M. Dobson, B. A. Levine, R. B. Martin, R. J. P. Williams, and A. V. Xavier, J. Chem. Soc. Chem. Commun., 791 (1972).
16. J. G. Shelling, M. E. Bjornson, R. S. Hodges, A. K. Taneja, and B. D. Sykes, J. Magn. Reson., *57*, 99 (1984).
17. H. M. Goff, J. Am. Chem. Soc., *103*, 3714 (1981).
18. G. N. LaMar, G. R. Eaton, R. H. Holm, and F. A. Walker, J. Am. Chem. Soc., *95*, 63 (1973).
19. W. DeW. Horrocks, Inorg. Chem., *9*, 690 (1970).
20. C. Benelli, I. Bertini, and D. Gatteschi, J. Chem. Soc. Dalton, 661 (1972).
21. D. Doddrell and J. D. Roberts, J. Am. Chem. Soc., *92*, 6839 (1970).
22. W. DeW. Horrocks and D. D. Hall, Inorg. Chem., *10*, 2368 (1971).
23. M. Gerloch and R. C. Slade, J. Chem. Soc., *A*, 1022 (1969).
24. R. H. Holm, G. W. Everett, and A. Chakravorty, Progr. Inorg. Chem., *7*, 83 (1966).
25. W. DeW. Horrocks and E. S. Greenberg, Inorg. Chem., *10*, 2190 (1971).
26. E. A. Lalancette and D. R. Eaton, J. Am. Chem. Soc., *86*, 5145 (1964).
27. J. P. Jesson, J. Chem. Phys., *45*, 1049 (1966).
28. F. Keffer, T. Oguchi, W. O'Sullivan, and J. Yamashita, Phys. Rev., *115*, 1553 (1959).
29. W. D. Perry, R. S. Drago, D. W. Herlocker, G. K. Pagenkopf, and K. Czworniak, Inorg. Chem., *10*, 1087 (1971).
30. J. P. Quaegebeur, C. Chachaty, and T. Yasukawa, Mol. Phys., *37*, 409 (1979).
31. H. M. McConnell, Proc. Natl. Acad. Sci. USA, *69*, 335 (1972).
32. E. W. Stone and A. H. Maki, J. Chem. Phys., *37*, 1326 (1962).
33. M. Karplus and G. K. Fraenkel, J. Chem. Phys., *35*, 1312 (1961).
34. E. T. Strom, G. R. Underwood, and D. Jurkowitz, Mol. Phys., *24*, 901 (1972); W. Derbishire, Mol. Phys., *5*, 225 (1962).
35. D. R. Eaton, A. D. Josey, W. D. Phillips, and R. E. Benson, Mol. Phys., *5*, 407 (1962).
36. "NMR of Paramagnetic Molecules," eds. G. N. LaMar, W. DeW. Horrocks, Jr., and R. H. Holm, Academic Press, New York, 1973.
37. R. W. Kluiber and W. DeW. Horrocks, Inorg. Chem., *6*, 430 (1967).
38. I. Bertini and L. J. Wilson, J. Coord. Chem., *1*, 237 (1971).
39. R. H. Holm, A. Chakravorty, and G. O. Dudek, J. Am. Chem. Soc., *86*, 379 (1964).
40. a. M. Wicholas and R. S. Drago, J. Am. Chem. Soc., *91*, 5463 (1969).
    b. I. Bertini, D. Gatteschi, and A. Scozzafava, Inorg. Chim. Acta, *6*, 185 (1972).
41. I. Bertini, C. Luchinat, and A. Scozzafava, Inorg. Chim. Acta, *19*, 201 (1976).
42. G. N. LaMar, W. DeW. Horrocks, and L. C. Allen, J. Chem. Phys., *41*, 2126 (1964).
43. J. A. Happe and R. L. Ward, J. Chem. Phys., *39*, 1211 (1963).
44. R. W. Kluiber and W. DeW. Horrocks, Inorg. Chem., *6*, 166 (1967).
45. R. H. Holm, G. W. Everett, and W. DeW. Horrocks, J. Am. Chem. Soc., *88*, 1071 (1966).
46. a. G. M. Zhidomirov, P. V. Schastnev, and N. D. Chuvylkin, J. Struct. Chem., *11*, 458 (1970).
    b. R. E. Cramer and R. S. Drago, J. Am. Chem. Soc., *92*, 66 (1970).
47. I. Bertini, G. Canti, C. Luchinat, and F. Mani, J. Am. Chem. Soc., *103*, 7784 (1981).
48. L. Banci, I. Bertini, C. Luchinat, and A. Scozzafava, J. Am. Chem. Soc., *109*, 2328 (1987).
49. I. Bertini, C. Luchinat, and A. Scozzafava, Inorg. Nucl. Chem. Lett., *15*, 89 (1979).
50. I. Bertini, D. L. Johnston, and W. DeW. Horrocks, Inorg. Chem., *9*, 693 (1970); ibid., p. 698.
51. R. J. Fitzgerald and R. S. Drago, J. Am. Chem. Soc., *89*, 287 (1967).

52. I. Morishima, T. Yonezawa, and K. Goto, J. Am. Chem. Soc., *92*, 6651 (1970).
53. L. Banci, I. Bertini, and C. Luchinat, "Electron and Nuclear Relaxation," VCH, Heidelberg, 1991.
54. L. Nordenskjold, A. Laaksonen, and J. Kowalewski, J. Am. Chem. Soc., *104*, 379 (1982).
55. D. Waysbort and G. Navon, J. Chem. Phys., *68*, 3704 (1978).
56. H. P. W. Gottlieb, M. Barfield, and D. M. Doddrell, J. Chem. Phys., *67*, 3785 (1977).
57. D. M. Doddrell, P. C. Healy, and M. R. Bendall, J. Magn. Reson., *29*, 163 (1978).
58. N. Bloembergen, J. Chem. Phys., *27*, 575 (1957).
59. I. Solomon, Phys. Rev., *99*, 559 (1955).
60. M. Gueron, J. Magn. Reson., *19*, 58 (1975).
61. A. J. Vega and D. Fiat, Mol. Phys., *31*, 347 (1976).
62. H. Sternlicht, J. Chem. Phys., *42*, 2250 (1965).
63. I. Bertini, C. Luchinat, M. Mancini, and G. Spina, in "Magneto-Structural Correlations in Exchange Coupled Systems", eds. D. Gatteschi, O. Kahn, and R. D. Willett, Reidel, Dordrecht, 1985, p. 421.
64. I. Bertini, C. Luchinat, M. Mancini, and G. Spina, J. Magn. Reson., *59*, 213 (1984).
65. I. Bertini, F. Briganti, C. Luchinat, M. Mancini, and G. Spina, J. Magn. Reson., *63*, 41 (1985).
66. I. Bertini, G. Lanini, C. Luchinat, M. Mancini, and G. Spina, J. Magn. Reson., *63*, 56 (1985).
67. L. Banci, I. Bertini, and C. Luchinat, Inorg. Chim. Acta, *100*, 173 (1985).
68. I. Bertini, C. Luchinat, and J. Kowalewski, J. Magn. Reson., *62*, 235 (1985).
69. L. Banci, I. Bertini, F. Briganti, and C. Luchinat, J. Magn. Reson., *66*, 58 (1986).
70. C. Owens, R. S. Drago, I. Bertini, C. Luchinat, and L. Banci, J. Am. Chem. Soc., *108*, 3298 (1986).
71. I. Bertini, C. Luchinat, C. Owens, and R. S. Drago, J. Am. Chem. Soc., *109*, 5208 (1987).
72. W. G. Espersen and R. B. Martin, J. Am. Chem. Soc., *98*, 40 (1976).
73. A. S. Anderson and A. G. Redfield, Phys. Rev., *116*, 583 (1959).
74. S. H. Koenig and R. D. Brown, III, in "NMR Spectroscopy of Cells and Organisms," CRC Press, Boca Raton, FL 1987.
75. H. S. Gutowsky, D. M. McCall, and C. P. Slichter, J. Chem. Phys., *21*, 279 (1953).
76. H. S. Gutowsky and C. H. Holm, J. Chem. Phys., *25*, 1228 (1956).
77. T. J. Swift and R. E. Connick, J. Chem. Phys., *37*, 307 (1962).
78. R. A. Dwek, "Nuclear Magnetic Resonance in Biochemistry," Oxford University Press, Oxford, 1973, Chapter 11.
79. R. Hausser and F. Noack, Z. Physik, *182*, 93 (1964).
80. B. P. Gaber, R. D. Brown, S. H. Koenig, and J. A. Fee, Biochim. Biophys. Acta, *271*, 1 (1972).
81. L. T. Muus and P. W. Atkins, eds., "Electron Spin Relaxation in Liquids," Plenum Press, New York and London, 1972.
82. N. W. Ashcroft and N. D. Mermin, "Solid State Physics," Saunders, Philadelphia, 1976.
83. A. A. Manenkov and R. Orbach, "Spin Lattice Relaxation in Ionic Solids," Harper & Row, New York, 1966.
84. R. Orbach, Proc. R. Soc. A, *264*, 458 (1961); ibid., p. 485.
85. W. B. Lewis and L. O. Morgan, Trans. Met. Chem., *4*, 33 (1968).
86. J. H. Van Vleck, Phys. Rev., *57*, 426 (1940).
87. M. Maekawa, S. Kitagawa, M. Munkata, and H. Masuda, Inorg. Chem., *28*, 1904 (1989).
88. S. H. Koenig and R. D. Brown, III, in "Metal Ions in Biological Systems", eds. H. Sigel and A. Sigel, Vol. 21, Dekker, New York, 1987.
89. I. Bertinin, F. Capozzi, and C. Luchinat, Magn. Reson. Imaging, *9*, (1991); S. H. Koenig and R. D. Brown III, Prog NMR Spectr *22*, 487 (1990).

90. G. Hernandez, M. F. Tweedle, and R. G. Bryant, Inorg. Chem., *29*, 5109 (1990).
91. J. H. Noggle and R. E. Schirmer, "The Nuclear Overhauser Effect," Academic Press, New York, 1971.
92. D. Neuhaus and M. Williamson, "The Nuclear Overhauser Effect in Structural and Conformational Analysis," VCH, New York, 1989.
93. L. B. Dugad, G. N. LaMar, L. Banci, and I. Bertini, Biochemistry, *29*, 2263 (1990).
94. L. Banci, A. Bencini, I. Bertini, C. Luchinat, and M. Piccioli, Inorg. Chem., *29*, 4867 (1990).
95. I. Bertini, L. Banci, C. Luchinat, and M. Piccioli, FEBS Lett., *272*, 175 (1990).
96. C. Luchinat, S. Steuernagel, and P. Turano, Inorg. Chem., *29*, 4351 (1990).
97. S. D. Emerson and G. N. LaMar, Biochemistry, *29*, 1545 (1990).
98. Y. Fengo and S. W. Englander, Biochemistry, *29*, 3505 (1990).
99. N. C. Veitch and R. J. P. Williams, Eur. J. Biochem., *189*, 351 (1990).
100. W. D. Phillips and R. E. Benson, J. Chem. Phys., *33*, 607 (1960); R. E. Benson, D. R. Eaton, A. D. Josey, and W. D. Phillips, J. Am. Chem. Soc., *83*, 3714 (1961).
101. D. R. Eaton and W. R. McClellan, Inorg. Chem., *6*, 2134 (1967).
102. D. R. Eaton, A. D. Josey, and R. E. Benson, J. Am. Chem. Soc., *89*, 4040 (1967).
103. D. R. Eaton, A. D. Josey, W. D. Phillips, and R. E. Benson, J. Chem. Phys, *37*, 347 (1962).
104. W. DeW. Horrocks, Inorg. Chem., *9*, 690 (1970).
105. R. E. Ernst, M. J. O'Connor, and R. H. Holm, J. Am. Chem. Soc., *90*, 5735 (1968).
106. D. H. Gerloch and R. H. Holm, J. Am. Chem. Soc., *91*, 3457 (1969).
107. L. J. Wilson and I. Bertini, J. Coord. Chem., *1*, 237 (1971).
108. G. N. LaMar and L. Sacconi, J. Am. Chem. Soc., *89*, 2282 (1967).
109. I. Bertini, L. Sacconi, and P. Speroni, Inorg. Chem., *11*, 1323 (1972).
110. a. I. M. Walker and R. S. Drago, J. Am. Chem. Soc., *90*, 6851 (1968).
  b. D. G. Brown and R. S. Drago, J. Am. Chem. Soc., *92*, 1871 (1970).
  c. Y. Y. Lim and R. S. Drago, J. Am. Chem. Soc., *94*, 84 (1972).
111. I. Bertini, C. Luchinat, and E. Borghi, Inorg. Chem., *20*, 303 (1981).
112. L.-P. Hwang and J. H. Freed, J. Chem. Phys., *63*, 4017 (1975); J. H. Freed, J. Chem. Phys., *68*, 4034 (1978).
113. C. F. Polnaszek and R. G. Bryant, J. Chem. Phys., *81*, 4038 (1984).
114. C. C. Lester and R. G. Bryant, J. Phys. Chem., *94*, 3654 (1990).
115. I. Bertini, F. Capozzi, C. Luchinat, and P. Turano, J. Magn. Reson., (1991), in press.
116. G. N. LaMar and F. A. Walker, in "The Porphyrins," ed. D. Dolphin, Academic Press, New York, 1979, Vol. 4, p. 61.
117. I. Bertini, G. Canti, C. Luchinat, and F. Mani, J. Am. Chem. Soc., *103*, 7784 (1981).
118. Unpublished results from our laboratory.
119. I. Bertini, G. Lanini, C. Luchinat, M. Mancini, and G. Spina, J. Magn. Reson., *63*, 56 (1985).
120. C. Benelli, A. Dei, and D. Gatteschi, Inorg. Chem., *21*, 1284 (1982).
121. I. Bertini, C. Owens, C. Luchinat, and R. S. Drago, J. Am. Chem. Soc., *109*, 5208 (1987).
122. J. A. Tainer, E. D. Getzoff, K. M. Beem, J. S. Richardson, and D. C. Richardson, J. Mol. Biol., *160*, 181 (1982).
123. I. Morgenstern-Badarau, D. Cocco, A. Desideri, G. Rotilio, J. Jordanov, and N. Dupre, J. Am. Chem. Soc., *108*, 300 (1986).
124. I. Bertini, G. Lanini, C. Luchinat, L. Messori, R. Monnanni, and A. Scozzafava, J. Am. Chem. Soc., *107*, 4391 (1985).
125. L. Banci, I. Bertini, C. Luchinat, M. Piccioli, A. Scozzafava, and P. Turano, Inorg. Chem., *28*, 4650 (1989).
126. L. Banci, I. Bertini, and C. Luchinat, Struct. Bonding, *72*, 113 (1990).
127. I. Bertini, G. Lanini, and C. Luchinat, Inorg. Chem., *23*, 2729 (1984).
128. L. B. Dugad, G. N. LaMar, L. Banci, and I. Bertini, Biochemistry, *29*, 2263 (1990).
129. I. Bertini, F. Briganti, C. Luchinat, A. Scozzafava, and M. Sola, J. AM. Chem. Soc., *113*, 1237 (1991).

130. I. Bertini, F. Briganti, and C. Luchinat, Inorg. Chim. Acta, *175*, 9 (1990).
131. I. Bertini, C. Luchinat, L. Messori, and M. Vasak, J. Am. Chem. Soc., *111*, 7296 (1988); ibid, p. 7300.
132. L. Banci, I. Bertini, F. Briganti, and C. Luchinat, New J. Chem., *15*, 467 (1991).
133. L. Banci, I. Bertini, F. Briganti, C. Luchinat, A. Scozzafava, and M. Vicens Oliver, Inorg. Chem., (1991), in press.
134. P. V. Demarco *et al.*, J. Am. Chem. Soc., *92*, 5734 (1970).
135. C. C. Hinckley, J. Am. Chem. Soc., *91*, 5160 (1969).
136. R. E. Rondeau and R. E. Sievers, J. Am. Chem. Soc., *93*, 1522 (1971).
137. G. E. Hawkes *et al.*, "Nuclear Magnetic Shift Reagents," ed. R. E. Sievers, Academic Press, New York (1973).
138. M. F. Rettig and R. S. Drago, J. Am. Chem. Soc., *88*, 2966 (1966).
139. G. N. LaMar and E. A. Metz, J. Am. Chem. Soc., *96*, 5611 (1974).
140. L. Lee and B. D. Sykes, Biochemistry, *22*, 4366 (1983).
141. T. Akutani *et al.*, Tetrahedron Lett., 1115 (1971).
142. S. B. Andrews *et al.*, Proc. Nat. Acad. Sci. USA, *70*, 1814 (1973) and references therein.
143. C. D. Barry *et al.*, Biochim. Biophys. Acta, *262*, 101 (1972) and references therein.
144. R. A. Dwek *et al.*, J. Biochem., *21*, 204 (1971).
145. G. M. Whitesides and D. W. Lewis, J. Amer. Chem. Soc., *92*, 6979 (1970); *93*, 5914 (1971).
146. H. L. Goering *et al.*, J. Am. Chem. Soc., *93*, 5913 (1971).
147. R. R. Fraser *et al.*, Chem. Commun., 1450 (1971).

EXERCISES

1. For each of the following, state whether the contact contribution to the isotropic shift in the proton nmr is close to zero. Also state whether the pseudocontact contribution is zero. Explain all answers.

    a. $Ni(:N\text{-}C_6H_5)_6^{2+}$ (octahedral about Ni).

    b. The *n*-butanol adduct of $Eu(dpm)_3$, where dpm is

    $$(CH_3)_3C-\underset{\underset{H}{|}}{C}(\overset{O^-}{\underset{\|}{C}})_2 C(CH_3)_3$$

2. The following Δv(iso) values are reported for $Ni(C_6H_5CH_2NH_2)_6^{2+}$: NH_2, -105 CH_2, 34.7; phenyl protons: *o*, -1.3; *m*, 1.6; *p*, -1.4 (ppm). Interpret these shifts in terms of delocalization mechanisms.

3. The stable free radical illustrated gives a sharp three-line esr signal in benzene solution at room temperature with a concentration of about 10^{-4} M. The nmr signal is too broad to be detected. On increasing radical concentration, the esr signal broadens and is barely discernible at the 1 M level, though now the following nmr signal is seen at 60 MHz.

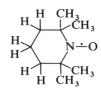

a. Explain the low concentration esr signal and the high concentration nmr signals.

b. Why does the esr signal disappear with increasing concentration while the nmr signal grows with concentration?

4. Four-coordinate complexes of Ni(II) may be (1) low spin ($S = 0$) square planar, (2) high spin ($S = 1$) tetrahedral, (3) an intermediate D_{2d} geometry with singlet and triplet state close in energy, or (4) situations (1) and (2) in equilibrium and interchanging with a rate constant on the order of 10^7 sec^{-1}.

a. Describe what result you would expect to obtain for each of the above cases if the magnetism were studied as a function of temperature (*i.e.*, Curie-Weiss behavior or not, and why).
b. Could cases (3) and (4) be distinguished by employing infrared, nmr, or esr spectroscopy? (Indicate what is expected, and why, for each of those techniques.)
c. Describe what you would expect to see from a study of the nmr spectrum and its temperature dependence for each of the above four cases.

5. Water has one σ MO capable of π-bonding with the t_{2g} orbitals in octahedral hexaqua complexes. Predict the order of ^{1}H hyperfine shifts in the d^1–d^9 series.

6. Oxygen-17 nmr has been observed in aqueous solutions of paramagnetic ions. Water that is coordinated to a paramagnetic ion has an ^{17}O resonance that is shifted far up- or downfield in relation to uncoordinated water. An aqueous solution containing the paramagnetic ion Dy^{3+} has only one ^{17}O resonance. Aqueous solutions of the diamagnetic ion Al^{3+} exhibit two ^{17}O resonances. See diagram of spectra.

a. Briefly explain why one ^{17}O resonance is observed in aqueous Dy^{3+} solutions, while two are observed in Al^{3+} solutions.

b. A solution containing 0.1 mole of Dy^{3+} and 10.0 moles of water has an ^{17}O resonance 190 ppm upfield from uncoordinated water. Addition of 0.2 mole of anhydrous Al^{3+} causes this peak to shift to 216 ppm. Explain.

c. Calculate the number of water molecules in the coordination sphere of Al^{3+}.

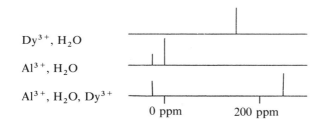

558   Chapter 12   Nuclear Magnetic Resonance of Paramagnetic Substances in Solution

7. Calculate the contact shift of a proton in a metal complex of axial symmetry if (a) the isotropic shift is 103 ppm, (b) the metal coordinates are $x = 0$, $y = 0$, $z = 0$, and the proton coordinates are $x = 3.51$, $y = 2.76$, $z = 1.12$ Å, (c) the magnetic susceptibility anisotropy is 2.3×10^{-8} m^3 mol^{-1}, (d) the direction cosines of the z-axis of the χ-tensor are $a = 0.128$, $b = 0.692$, $c = 0.710$.

8. Are T_1 and T_2 of a proton at 300 MHz mainly contact, mainly dipolar, or both contact and dipolar in origin if the contact shift is $+173$ ppm, the metal–proton distance is 4.20 Å, τ_r is 7×10^{-11} sec and the electron relaxation time is

   a. 10^{-12} sec?

   b. 10^{-8} sec?

9. Determine the magnetic field at which the line width, expressed both in Hz and ppm, is sharpest for a proton at 5.2 Å from a metal with five unpaired electrons in a complex having $\tau_r = 3 \times 10^{-8}$ sec and $\tau_s = 10^{-11}$ sec. Neglect the contact contribution.

10. Predict and explain the order of the electron relaxation times for the following metal ions: Cu^{2+}, Co^{2+} (H.S.), Yb^{3+}.

11. Predict the relative ease of observing COSY and NOESY cross peaks at 200 and 600 MHz between two signals in the paramagnetic complex of exercise 9. Remember that during the buildup of the cross correlation, the COSY intensities decrease with T_2 and the NOESY intensities decrease with T_1.

12. Calculate the distance between two protons A and B in a paramagnetic complex with $\tau_r = 10^{-9}$ sec knowing that, by irradiating A in a steady-state experiment at 500 MHz, a nOe of 2.5% on B is obtained and the T_1 of B is 25 msec.

13. Explain why ^{1}H nmr signals from the metal ligands in a antiferromagnetically coupled nickel(II)-copper(II) system are sharper than in the analogous zinc(II)-copper(II) system.

14. Are the nmr lines of a copper(II) complex broadest

   a. in the presence of antiferromagnetic coupling with a high spin cobalt(II) complex?

   b. in the presence of ferromagnetic coupling of equal magnitude?

   c. in the absence of magnetic coupling?

Electron Paramagnetic Resonance Spectra of Transition Metal Ion Complexes

13

13-1 INTRODUCTION

This chapter is an extension of Chapter 9; the principles covered there, as well as those in Chapters 10 and 11, should be well understood before this chapter is begun. It is for this reason that the presentation of the subject matter in this chapter did not follow Chapter 9 directly.

The epr spectra of transition metal ion complexes contain a wealth of information about their electronic structures. The additional information and accompanying complications that are characteristic of transition metal ion systems arise because of the approximate degeneracy of the *d*-orbitals and because many of the molecules contain more than one unpaired electron. These properties give rise to orbital contributions and zero-field effects. As a result of appreciable orbital angular moments, the *g*-values for many metal complexes are very anisotropic. Spin-orbit coupling also gives rise to large zero-field splittings (of 10 cm^{-1} or more) by mixing ground and excited states.

An important theorem that summarizes the properties of multielectron systems is *Kramers' rule*.[1] This rule states that if an ion has an odd number of electrons, the degeneracy of every level must remain at least twofold in the absence of a magnetic field. With an odd number of electrons, m_J quantum numbers will be given by $\pm^1/_2$ to $\pm J$. Therefore, any ion with an odd number of electrons must always have as its lowest level at least a doublet, called a *Kramers' doublet*. This degeneracy can then be removed by a magnetic field, and an epr spectrum should be observed. On the other hand, for a system with an even number of electrons, $m_J = 0, \pm 1, \ldots \pm J$. The degeneracy may be completely removed by a low symmetry crystal field, so only singlet levels remain that could be separated by energies so large that an epr transition would not be observed in the microwave region. This discussion is illustrated by the energy level splittings in Fig. 13–1. For the even electron system, the ground state is non-degenerate and the $J = 0$ to 1 transition energy is quite frequently outside the microwave region.

A few additional comments relative to some experimental procedures (others are given in Chapter 9) are in order in the way of introduction to transition metal systems. A number of factors, other than those that are instrumental, affect the epr line width. As in nmr, spin-lattice, spin-spin, and exchange interactions are important.

Broadening due to spin-lattice relaxation results from the interaction of the

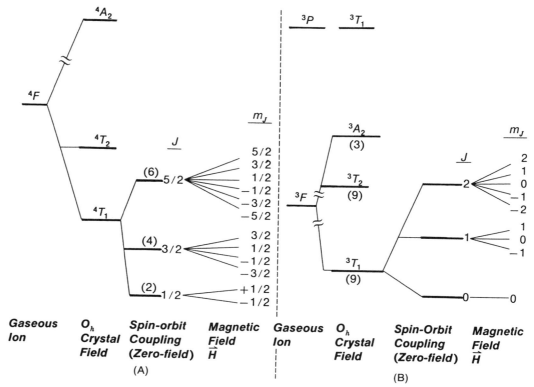

FIGURE 13–1 The splitting of the gaseous ion degeneracy of (A) Co^{2+} and (B) V^{3+} by the crystal field, spin-orbit coupling, and a magnetic field. The T state is regarded as having an effective L, called L', equal to 1; for an A state, $L' = 0$. Then $J = L' + S, \ldots, L' - S$. Only the zero-field and magnetic field splittings of the ground state are shown.

paramagnetic ions with the thermal vibrations of the lattice. The variation in spin-lattice relaxation times in different systems is quite large. For some compounds the lifetime is sufficiently long to allow the observation of spectra at room temperature, while in others this is not possible. Since these relaxation times generally increase as the temperature decreases, many of the transition metal compounds need to be cooled to liquid N_2 or liquid He temperatures before well-resolved spectra are observed.

Spin-spin interaction results from the magnetic fields that originate in neighboring paramagnetic ions. As a result of these fields, the total field at each ion is slightly altered and the energy levels are shifted. A distribution of energies results, which produces a broad signal. This effect varies as $(1/r^3) \cdot (1 - 3 \cos^2 \theta)$, where r is the distance between ions and θ is the angle between the field and the symmetry axis. Since this effect is reduced by increasing the distance between paramagnetic ions, it is often convenient to examine transition metal ion systems by diluting them in an isomorphous diamagnetic host. For example, a copper complex could be studied as a powder or single crystal by diluting it in a host lattice of an analagous zinc complex or by examining it in a frozen solution. Dilution of the solid isolates the electron spin of a given complex from that of another paramagnetic molecule, and the spin lifetime is lengthened. (Recall our discussion of the spectra in Fig. 9–25.) If a frozen solution is used, it must form a good glass; otherwise, paramagnetic aggregates form, which lead to a spectrum with broadened lines. It is often necessary to remove O_2 from the solvent because this can lead to a broadening of the resonance. Even in a well-formed glass, one cannot usually detect hyperfine splittings smaller than 3 or 4 gauss.

Line widths are altered considerably by chemical exchange processes. This effect can also be reduced by dilution. If the exchange occurs between equivalent paramagnetic species, the lines broaden at the base and become narrower at the center. When exchange involves dissimilar ions, the resonances of the separate lines merge to produce a single line, which may be broad or narrow depending upon the exchange rate. Such an effect is observed for $CuSO_4 \cdot 5H_2O$, which has two distinct copper sites per unit cell.[2]

In single crystals the anisotropy in the esr parameters can be obtained. Information about the anisotropy in the system can also often obtained in powders and glasses, because the resulting spectra are not those of a motionally averaged system. We will next consider why anisotropic information is obtained from the spectra even though the molecules in a glass or powder sample exist in an extremely large number of possible orientations relative to the applied field. Consider a molecule with a threefold or higher symmetry axis, which can be described by a $g_\parallel$ and a $g_\perp$. As can be seen from Fig. 13–2, there are many axes

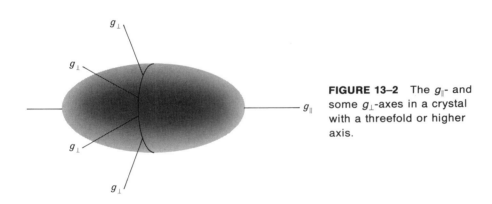

FIGURE 13–2 The $g_\parallel$- and some $g_\perp$-axes in a crystal with a threefold or higher axis.

that could be labeled $g_\perp$. In a bulk sample containing many orientations for the assemblage of crystallites, there are more possible orientations that have the $g_\perp$ axes aligned with the applied field than there are orientations that have the $g_\parallel$ axis aligned. The g-value for any orientation is given by:

$$g^2 = g_\perp^2 \sin^2 \theta + g_\parallel^2 \cos^2 \theta \qquad (13\text{–}1)$$

where θ is the angle that the principal axis (*i.e.*, the $g_\parallel$ axis) makes with the applied field. Since all orientations of the crystallites in a solid are equally probable, absorption will occur at all fields between that associated with $g_\parallel$ and that associated with $g_\perp$. Since many more crystallites have $g_\perp$ aligned than $g_\parallel$, the most intense absorption will correspond to $g_\perp$. When one considers the probabilities of the various orientations and the transition probability corre-

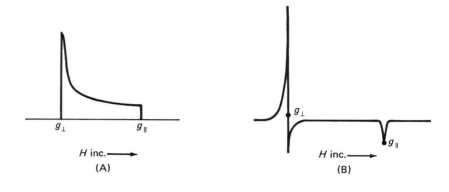

FIGURE 13–3 idealized absorption (A) and derivative (B) spectra for an unoriented system with $S = {}^1/_2$, axial symmetry, and no hyperfine interaction ($g_\perp > g_\parallel$).

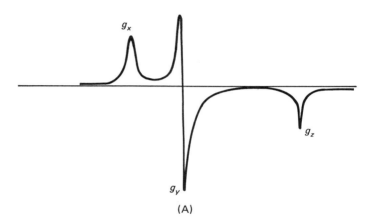

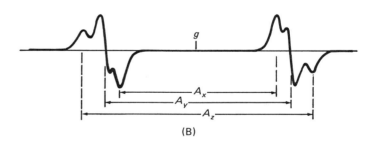

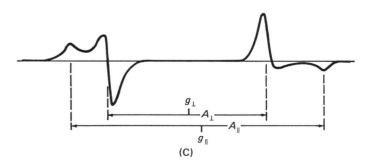

FIGURE 13–4 Powder epr spectra of $S = {}^1/_2$ systems. (A) An orthorhombic system with $I = 0$; (B) isotropic g with $I = {}^1/_2$, $g_\perp > g_\parallel$, and $A_\parallel > A_\perp$. In the latter cases, accurate g- and A-values are available only from computer simulation.

sponding to each of these, the absorption spectrum shown in Fig. 13–3(A) is predicted. This is converted to the derivative spectrum in Fig. 13–3(B). This is an idealized example, and often one finds that the overlapping features generated by $g_\parallel$ and $g_\perp$ make it difficult to obtain their values.

When the system is orthorhombic and $g_x > g_y > g_z$, the powder spectrum obtained for $I = 0$ is like that in Fig. 13–4(A). When $I = \frac{1}{2}$ and the system has nearly isotropic g-values, but $A_z > A_y > A_x$, the spectrum in Fig. 13–4(B) is expected. The spectrum for a complex with axial symmetry and $I = \frac{1}{2}$, in which $g_\perp > g_\parallel$ and $A_\parallel > A_\perp$, is illustrated in Fig. 13–4(C). Other systems become quite complex, and the possibility for misassignments becomes very large. Only in the relatively simple cases can the g- and A-values be determined with confidence. Computer programs are available to simulate powder epr spectra for simple systems.

Liquid crystal nematic phases can also be used[3] to orient a molecule for epr work. The molecule to be studied, which is the solute, cannot be spherical; as an example, consider the molecule Co(Meacacen) in Fig. 13–5(A). The liquid crystal solution of this low spin Co(II) complex is placed in a magnetic field to orient the liquid crystal molecules (and, in turn, the solute molecules) and is then cooled. This is schematically illustrated in Fig. 13–5(B). The epr spectrum[4a] in

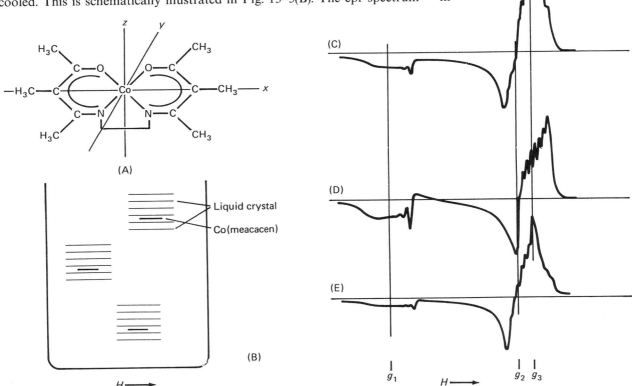

FIGURE 13–5 The epr spectra [4] of Co(Meacacen) at 77° K. (A) Structural formula; (B) orientation of the molecule in a frozen oriented liquid crystal; (C) unoriented frozen solution; (D) frozen liquid crystal oriented as in (B); (E) frozen liquid crystal reoriented 90° from (B). The phasing in this spectrum is inverted relative to what one normally employs. [Reprinted with permission from B. M. Hoffman, F. Basolo, and D. L. Diemente, J. Amer. Chem. Soc. *95*, 6497 (1973). Copyright by the American chemical Society.]

(D) is for the sample oriented relative to the magnetic field as shown in (B), while in spectrum (E) the sample is rotated 90° around the z-axis (*i.e.*, y parallel to the field) relative to the magnetic field. Upon rotation, the portion of the spectrum corresponding to g_2 is enhanced, but that for g_1 is not. One could easily make the mistake of assuming that this is an axial system with g_1 assigned to the z-axis (*i.e.*, $g_\|$, the axis perpendicular to the plane) and g_2 and g_3 assigned to $g_\perp$, where g_x and g_y are similar. However, with the molecular coordinate system as defined in Fig. 13–5(A), g_z must be assigned to g_3, g_x to g_1, and g_y to g_2. These assignments have subsequently been confirmed by a single crystal epr study.[4b] Difficulties can arise in this application if care is not taken to demonstrate that the liquid crystal is not coordinating the complex being studied.

13–2 INTERPRETATION OF THE *g*-VALUES

Introduction

In contrast to organic free radicals, the *g*-values of transition metal ions can differ appreciably from the free electron value of 2.0023. Such deviations provide considerable information about the electronic structure of the complex. Different g-values arise because spin-orbit coupling is much greater in many transition metal ion complexes than in organic free radicals (*vide infra*). Thus, spin-orbit effects become essential to an understanding of esr.

The value of *g* for an unpaired electron in a gaseous atom or ion, for which Russell-Saunders coupling is applicable, was given earlier by the expression

$$g = 1 + \frac{J(J+1) + S(S+1) - L(L+1)}{2J(J+1)} \tag{13-2}$$

In condensed phases, first row transition metal ion systems not only do not have *g*-values in accord with this expression, but they often deviate from the spin-only value. In condensed phases, the orbital motion of the electron is strongly perturbed and the orbital degeneracy, if it existed before application of the chemical environment, is partly removed or "quenched." If the electron has orbital angular momentum, the angular momentum tends to be bolstered by being weakly coupled to the spin. There is therefore a competition between the quenching effect of the ligands—the "crystal field"—and the sustaining effect of the spin-orbit coupling. Were it not for spin-orbit coupling, we should always observe an isotropic *g*-value of 2.0023.

These effects can be illustrated by considering the influence of a crystalline field on a d^1 ion as shown for O_h and D_{4h} (z-axis compression) in Fig. 13–6. Equation (13–2) would describe the ^2D gaseous ion. The octahedral crystal field splits ^2D into $^2T_{2g}$ and ^2E_g states. The degenerate T_{2g} state may be further split by distortion (*e.g.*, Jahn-Teller effects) or by a tetragonal ligand field into E and B_2 levels. Spin-orbit coupling, on the other hand, tends to preserve a small amount of orbital angular momentum, so, in the tetragonal complex, the orbital angular momentum is not completely quenched. One generally refers to this as mixing in of a nearby excited state by spin-orbit coupling. When the amount of spin-orbit coupling is small compared to the tetragonal distortion (*i.e.*, in the case of large distortions), the mixing can be treated by perturbation theory. In an octahedral

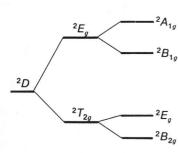

FIGURE 13–6 Splitting of the ^2D state by O_h and D_{4h} fields (z-axis compression in D_{4h}).

complex, spin-orbit coupling is present in the ground $^2T_{2g}$ state; in order to obtain the accuracy needed to understand the epr spectrum, this situation cannot be adequately treated with perturbation theory. Recall that such a treatment was employed in Chapter 11 on magnetism.

$S = \frac{1}{2}$ Systems with Orbitally Non-degenerate Ground States

The full Hamiltonian for our system with spin-orbit coupling in a magnetic field is given by

$$\hat{H} = \hat{H}(\text{Zeeman}) + \hat{H}(\text{SO}) = \beta \vec{H} \cdot (\hat{L} + g_e \hat{S}) + \lambda \hat{L} \cdot \hat{S} \qquad (13\text{--}3)$$

One of the effects of spin-orbit coupling is to modify the simple one-electron d orbital wave functions. This is described by the $\lambda \hat{L} \cdot \hat{S}$ term in the Hamiltonian. For example, the spin wave function for the ground state $^2B_{2g}$ of a d^1 ion in a tetragonal complex is modified by the spin-orbit interaction $\lambda \hat{L} \cdot \hat{S}$. From first-order perturbation theory, the wave function for the Kramers' doublet $|\pm\rangle$ including spin-orbit effects is given by:

$$|\pm\rangle = N|0\rangle + (1-N^2)^{-1/2} \sum_{M_L M_S} \frac{\langle n|\lambda \hat{L} \cdot \hat{S}|0\rangle}{E(0) - E(n)} |n\rangle \qquad (13\text{--}4)$$

The term $|0\rangle$ is the $^2B_{2g}$ ground state before spin-orbit effects are considered (*i.e.*, for d^1 with a tetragonal compression, this one electron is in the d_{xy} orbital), while the summation indicates the contribution made by spin-orbit admixture of the excited states. In this example, the ΔE term in the denominator indicates that the ^2E state will make the largest contribution of all the states that mix in. We can see from this expression that when there is no orbital angular momentum mixed into the ground state, $|\pm\rangle = |0\rangle$. Evaluation of the matrix elements in equation (13–4) gives the coefficients necessary to write the appropriate wave functions. These functions are then used with the Zeeman Hamiltonian in equation (13–3), *i.e.*,

$$\hat{H}' = \beta \hat{L} \cdot \vec{H} + g_e \beta \hat{S} \cdot \vec{H}$$

to set up the 2 × 2 matrix involving $|+\rangle$ and $|-\rangle$. Note that we have worked with the full Hamiltonian in equation (13–3), using the two parts separately. The $\lambda \hat{L} \cdot \hat{S}$ term modified the wave function on which we are now operating with the Zeeman Hamiltonian. The problem is solved by using the raising and lowering operators. Energies are obtained which are expressed as $g\beta H m_s$, where g is the effective g-factor in the direction of the field component H_i ($i = x, y, z$). In this way, using $\hat{L}_z$, $\hat{S}_z$, and the Zeeman Hamiltonian, we obtain

$$g_z = 2.0023 + 2\lambda \sum_n \left(\frac{\langle 0|\hat{L}_z|n\rangle \langle n|\hat{L}_z|0\rangle}{E(0) - E(n)} \right) \qquad (13\text{--}5)$$

where $|0\rangle$ is the ground state wave function and $|n\rangle$ is that of one of the n excited states. Since the ground and excited states corresponding to electronic transitions in an $S = \frac{1}{2}$ complex with small spin-orbit coupling are adequately described by real orbital occupations, the state and orbital designations can be used interchangeably. The g-value is thus seen to be very dependent upon the mixing in of the excited state by spin-orbit coupling.

The matrix elements $\langle 0|\hat{L}_z|n\rangle$ are non-zero only when the m_ℓ value of $|0\rangle$ equals the m_ℓ value of $|n\rangle$. In a real orbital basis set with

$$d_{x^2-y^2} = \left(\frac{1}{\sqrt{2}}\right)(|+2\rangle - |-2\rangle)$$

and

$$d_{xy} = \left(\frac{-i}{\sqrt{2}}\right)(|+2\rangle + |-2\rangle)$$

we will have non-zero matrix elements for $\langle d_{x^2-y^2}|\hat{L}_z|d_{xy}\rangle$. (This can be seen from the triple product $\Gamma_{x^2-y^2} \times \Gamma_{L_z} \times \Gamma_{xy}$.) For a d^1 complex, where we can use one-electron oribitals instead of states, an evaluation of all the matrix elements leads to

$$g_z = 2.0023 + \frac{8\xi}{E(0) - E(n)} \tag{13-6}$$

where $\lambda = +\xi/2S$ for a shell that is less than half-filled. The symbol ξ represents the one-electron spin-orbit coupling. When the shell is more than half-filled, we generally think in terms of formation of the positive holes that accompany the electronic transitions. The sign of the second term, $8\xi/[E(0) - E(n)]$, is changed by changing the denominator to $E(n) - E(0)$. When the shell is less than half-filled, spin-orbit effects reduce g from 2.0023; but when the shell is more than half-filled, spin-orbit effects increase g above g_e. $E(0)$ and $E(n)$ are the energies of the ground and excited states, respectively.

In a tetragonal complex we have

$$g_x = 2.0023 + \frac{2\lambda \sum_n \langle 0|\hat{L}_x|n\rangle\langle n|\hat{L}_x|0\rangle}{E(0) - E(n)} \tag{13-7}$$

where $g_x = g_y$. These matrix elements are evaluated by using the raising and lowering operators, so matrix elements are non-zero only when the ground and excited state m_ℓ values differ by ± 1. For a d^1 tetragonal complex (Fig. 13–6) we have:

$$g_x = 2.0023 + \frac{2\xi}{E_{xy} - E_{xz}} \tag{13-8}$$

The results of the evaluation of matrix elements by this procedure can be summarized by writing the following general expression for the g-values of $S = \frac{1}{2}$ systems:

$$g = 2.0023 + \frac{\mathbf{n}\xi}{E(0) - E(n)} \tag{13-9}$$

The values of **n** are obtained from the so-called magic pentagon shown in Fig. 13–7, which summarizes the results of the evaluation of matrix elements $\langle 0|\hat{L}|n\rangle$.

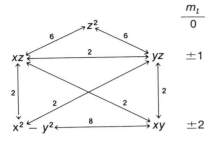

FIGURE 13-7 The "magic pentagon" for evaluating n of equation (13-9).

We will repeat the problem discussed earlier (the d^1 tetragonal field) with the use of this pentagon. We first determine that, for an electron in an xy orbital (*i.e.*, an xy ground state), only electron circulation into the $x^2 - y^2$ orbital could give orbital angular momentum along the z-axis. The quantity g_z then has contributions only from xy and $x^2 - y^2$, which is seen in Fig. 13-7 to have the value **n** = 8. Equation (13-9) then becomes:

$$g_z = 2.0023 + \frac{8\xi}{E_{xy} - E_{x^2-y^2}} \quad (13\text{-}10)$$

Since g_x has non-zero matrix elements only when $|0\rangle$ and $|n\rangle$ differ by $\Delta m_\ell = 1$, we obtain

$$g_x = 2.0023 + \frac{2\xi}{E_{xy} - E_{xz}} \quad (13\text{-}11)$$

and

$$g_y = 2.0023 + \frac{2\xi}{E_{xy} - E_{yz}} \quad (13\text{-}12)$$

These formulas in effect tell us what orbitals permit electron circulations about the respective axes; that is, $x^2 - y^2$ and xy about z, xy and xz about x, or xy and yz about y. The "magic pentagon" is easily constructed. The three rows represent orbitals possessing different m_ℓ values; the top row corresponds to $m_\ell = 0$, the second row to $m_\ell = \pm 1$, and the third row to $m_\ell = \pm 2$. It should be emphasized that this whole treatment arises from a perturbation assumption and is valid only when $\mathbf{n}\xi/[E(0) - E(n)]$ is small compared to the diagonal Zeeman elements. When second-order perturbation theory is pertinent, a term proportional to $\xi^2/\Delta E^2$ is added.

Before concluding this section, an application of the use of g-values will be presented. In Fig. 13-8, the crystal field splittings are indicated for tetragonal copper(II) complexes with a metal-ligand distance on the fourfold (z)-axis that is compressed [part (A)] and elongated ϵpart (B)] along this direction. In (A), the unpaired electron is in d_{z^2} with $m_\ell = 0$. There is no other orbital with an $m_\ell = 0$ component, so $g_z = 2.0023$. The value of $g_x = g_y$ is given by

$$g_x = 2.0023 + \frac{6\xi}{E_{yz} - E_{z^2}} \quad (13\text{-}13)$$

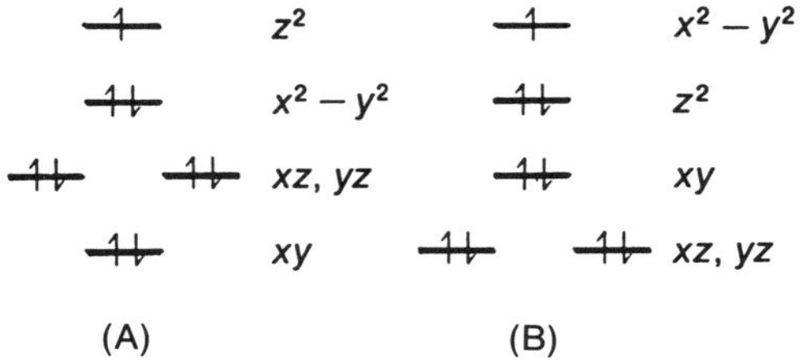

FIGURE 13–8 The d-level splittings for a tetragonally compressed complex (A) and a tetragonally elongated complex (B).

(A)
- ─┼─ z^2
- ─╫─ $x^2 - y^2$
- ─╫─ ─╫─ xz, yz
- ─╫─ xy

(B)
- ─┼─ $x^2 - y^2$
- ─╫─ z^2
- ─╫─ xy
- ─╫─ ─╫─ xz, yz

Note that we have switched the order of the energies in the denominator because the shell is more than half-filled. On the other hand, for a tetragonal elongation [Fig. 13–8(B)], with the unpaired electron in $d_{x^2-y^2}$, we obtain

$$g_z = 2.0023 + \frac{8\xi}{E_{xy} - E_{x^2-y^2}} \tag{13–14}$$

The value of $g_x = g_y$ is given by

$$g_x = 2.0023 + \frac{2\zeta}{E_{yz} - E_{x^2-y^2}} \tag{13–15}$$

Thus, the g-values can be used to distinguish the two structures. Complexes with a tetragonal compression are very rare, but elongated ones are common. Copper(II) porphyrin[5] complexes have been reported with g-values of $g_z = 2.70$ and $g_x = 2.04$.

The g-values in some six-coordinate copper(II) complexes exhibit interesting behavior. As mentioned in Chapter 9, when the single crystal epr spectrum for [Cu(H$_2$O)$_6$]SiF$_6$, diluted with the corresponding diamagnetic zinc salt, was obtained at 90 K, the spectrum was found to consist of one band with partly resolved hyperfine structure and a nearly isotropic g-value.[6] Jahn-Teller distortion is expected, but there are three distortions with the same energy that will resolve the orbital degeneracy. These are three mutually perpendicular tetragonal distortions (elongation or compression along the three axes connecting *trans* ligands). As a result, three distinguishable epr spectra are expected, one for each species. Since only one transition was found, it was proposed that the crystal field resonates among the three distortions (a so-called dynamic Jahn-Teller distortion). When the temperature is lowered, the spectrum becomes anisotropic and consists of three sets of lines corresponding to three different copper ion environments distorted by three different tetragonal distortions.[8] The following parameters were reported for CuSiF$_6 \cdot$ 6H$_2$O:[9]

90 K	20 K
$g_\parallel = 2.221 \pm 0.005$	$g_z = 2.46 \pm 0.01$
$g_\perp = 2.230 \pm 0.005$	$g_x = 2.10 \pm 0.01$
$A = 0.0021 \pm 0.0005$ cm^{-1}	$g_y = 2.10 \pm 0.01$
$B = 0.0028 \pm 0.0005$ cm^{-1}	$A_z = 0.0110 \pm 0.0003$ cm^{-1}
	$\left.\begin{array}{c} A_x \\ A_y \end{array}\right\} < 0.0030$ cm^{-1}

Other mixed copper salts have been found to undergo similar transitions: $(Cu, Mg)_3La_2(NO_3)_{12} \cdot 24D_2O$ between 33 and 45 K, and $(Zn, Cu)(BrO_3)_2 \cdot 6H_2O$, incomplete below 7 K.

A similar behavior (*i.e.*, a resonating crystal field at elevated temperatures) was detected in,[10] the spectra of some tris complexes of copper(II) with 2,2'-dipyridine and 1,10-phenanthroline[11] as well as with trisoctamethyl-phosphoramide.[12] This latter ligand, $[(CH_3)_2N]_2P(O)OP(O)[N(CH_3)_2]_2$, is a bidentate chelate in which the phosphoryl oxygens coordinate. The complex can be studied without having to dilute it in a diamagnetic host. A complete single crystal epr study and an x-ray diffraction study are reported on this system.[12] At 90 K, the spectra indicated that at least three different magnetic sites exist, each of which is described by a g-tensor. Analysis of the spectra indicates that these sites correspond to distortions along the x-, y-, and z-axes, respectively, in effect locking in the various extremes in the dynamic Jahn-Teller vibrations occurring at room temperature. The interesting problem of determining the chirality of the molecule in a single crystal epr spectral analysis has also been discussed.[12]

In single crystal epr studies on transition metal ion systems, it is common[13–15] to find complexes in which the g- and A-tensors are not diagonal in the obvious crystal field coordinate system. An axis that is perpendicular to a reflection plane or lies on a rotation axis must be one of the three principal axes of the molecule. The g-tensor of the molecule and the A-tensor for any atom lying on the axis must have principal values along this coordinate. If a molecule contains only a single axis that meets the above requirements, the *other two axes used as the basis for the crystal field analysis will not necessarily be the principal axes of the corresponding tensors of g and A*; *i.e.*, selecting these axes may not lead to a diagonal tensor. For example, *bis*(diselenocarbamate) copper(II) has C_{2h} symmetry.[13,14] The twofold rotation axis is one of the axes that diagonalize the corresponding g- and A-tensor components, but the other two components are not diagonal in an axis system corresponding to the axes of the crystal field. Had the molecule possessed D_{2h} symmetry, the three twofold rotation axes of this point group would have been the principal axes for both the A- and g-tensors. Thus, epr studies can provide us with information about the symmetry of the molecule. For a molecule with no symmetry, none of the molecular axes need be coincident with the axes that diagonalize the g-tensor or A-tensor.* As a matter of fact, the axis system that diagonalizes A may not diagonalize g. In vitamin B_{12}, for example,[15] the angle between the principal xy-axis system that leads to a diagonal A-tensor and those that lead to a diagonal g-tensor is 50°.

Systems in which Spin-Orbit Coupling Is Large

When spin-orbit coupling is large, perturbation theory cannot be used to give the appropriate wave functions; *i.e.*, equation (13–4) does not apply. An octahedral d^1 complex with a $^2T_{2g}$ ground state is a case in which spin-orbit coupling is

---

* For low symmetry molecules, the g- and A-tensors may in fact be asymmetric (*i.e.*, $a_{xy} \neq a_{yx}$). The g^2-tensor will always be symmetric.

large. When the spin-orbit operator $\xi \hat{L} \cdot \hat{S}$ operates on the sixfold degenerate real orbital basis set

$$\psi_1 = (1/\sqrt{2})\left(\left|2, \tfrac{1}{2}\right\rangle - \left|-2, \tfrac{1}{2}\right\rangle\right) \qquad \psi_4 = \left|1, -\tfrac{1}{2}\right\rangle$$

$$\psi_2 = (1/\sqrt{2})\left(\left|1, -\tfrac{1}{2}\right\rangle - \left|-2, -\tfrac{1}{2}\right\rangle\right) \qquad \psi_5 = \left|-1, \tfrac{1}{2}\right\rangle$$

$$\psi_3 = \left|1, \tfrac{1}{2}\right\rangle \qquad \psi_6 = \left|-1, -\tfrac{1}{2}\right\rangle$$

(the wavefunctions are expressed as $|M_L, M_S\rangle$), the result is a new set of orbitals, appropriate for d^1 systems with a large amount of spin-orbit coupling. These are obtained by evaluating the 6×6 determinant containing matrix elements $\langle \psi_i | \xi \hat{L} \cdot \hat{S} | \psi_j \rangle$ to obtain the energies. The energies are substituted back into the secularlike equations to obtain the eigenfunctions φ_i listed below.

$$\varphi_1 = (1/\sqrt{3})(\sqrt{2}\psi_1 + \psi_6)$$
$$\varphi_2 = (1/\sqrt{3})(-\sqrt{2}\psi_2 + \psi_3)$$
$$\varphi_3 = \psi_4$$
$$\varphi_4 = \psi_5$$
$$\varphi_5 = (1/\sqrt{3})(\psi_2 + \sqrt{2}\psi_3)$$
$$\varphi_6 = (1/\sqrt{3})(\psi_1 - \sqrt{2}\psi_6)$$

The corresponding energies are $E_1 = -\xi/2$, $E_2 = -\xi/2$, $E_3 = -\xi/2$, $E_4 = -\xi/2$, $E_5 = \xi$, and $E_6 = \xi$. We see from this analysis that spin-orbit coupling has removed the sixfold orbital degeneracy of the T state, giving a twofold set of levels and a lower-energy fourfold set, corresponding to Γ_7 and Γ_8 in the O' double group.* Next, we need to determine the effect of the magnetic field. Since systems with O_h symmetry are magnetically isotropic (x, y, and z), it is only necessary to work out the effect of H_z. The Hamiltonian operator, $\hat{H}$ (parallel to z) is $\beta(\hat{L}_z + g_e\hat{S}_z)H_z$. The resulting energies were derived in Chapter 11 and are indicated in Fig. 13–9. The splitting in the low-energy set is very small, being second order in H (i.e., H^2). When one solves for $g(\Delta E = g\beta H)$, the result $g\beta H = 4\beta^2 H/3\xi$ or $g = 4\beta H/3\xi \cong 0$ is obtained for the lowest level.† With appreciable separation of Γ_8 and Γ_7 (e.g., $\xi = 154$ cm^{-1} for Ti^{3+}), epr will not be detected unless the Γ_7 state is populated. Solving the expression for $g\beta H$ for these levels as above results in a g-value of 2.0. This state will be populated at room temperature, but the large amount of spin-orbit coupling gives rise to a short τ_e and no spectra are observed. Liquid helium temperatures are needed to lengthen τ_e, but then only the Γ_8 states are populated. Thus, the inability to observe epr spectra on these systems is a direct prediction of crystal field theory.

---

* The Γ_7 and Γ_8 states correspond to $J = \tfrac{1}{2}$ and $J = \tfrac{3}{2}$, respectively, for the free ion. These designations are derived by factoring the product of T_2' (orbital part) and E_2' (Γ_6 for spin = $\tfrac{1}{2}$) in the O' double group.

† In some systems, transitions with very small g-values have been detected.

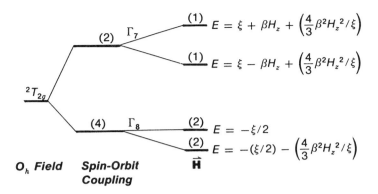

FIGURE 13-9 The influence of spin-orbit coupling and an applied magnetic field on a T state.

13-3 HYPERFINE COUPLINGS AND ZERO FIELD SPLITTINGS

Hyperfine and Zero-Field Effects on the Spectral Appearance

Transition metal systems are rich in information arising from metal hyperfine coupling and zero-field splitting. Figure 9-14 illustrates the rich cobalt hyperfine interaction in $Co_3(CO)_9Se$. Before continuing with this discussion, the reader should review the section on "Anisotropy in the Hyperfine Coupling" in Chapter 9, as well as that on "The epr of Triplet States," if necessary. The spin Hamiltonian for a single nucleus with spin I and a single effective electronic spin S can be written to include these extra effects as

$$\hat{H} = \beta \vec{H} \cdot \mathbf{g} \cdot \hat{S} + \hat{S} \cdot \mathbf{D} \cdot \hat{S} + h\hat{S} \cdot \mathbf{A} \cdot \hat{I} - g_N \beta_N \vec{H} \cdot \hat{I} \qquad (13-16)$$

In the previous section, we discussed the $\beta \vec{H} \cdot \mathbf{g} \cdot \hat{S}$ term and the complications introduced by orbital contributions. The next term incorporates the zero-field effects previously described by the dipolar tensor, $\mathbf{D}$, which has a zero trace. The last two terms arise when $I \neq 0$.

In Chapter 9, we discussed zero-field effects that arose from the dipolar interaction of the two or more electron spin moments. In transition metal ion systems, this term is employed to describe any effect that removes the spin degeneracy, including dipolar interactions and spin-orbital splitting. A low symmetry crystal field often gives rise to a large zero-field effect.

In an axially symmetric field (*i.e.*, tetragonal or trigonal), the epr spin Hamiltonian that can be used to fit the observed spectra for effective spin systems lower than quartet takes the form

$$\hat{H}_{spin} = D\left[\hat{S}_z^2 - \frac{1}{3}S(S+1)\right] + g_\parallel \beta H_z \hat{S}_z + g_\perp \beta(H_x \hat{S}_x + H_y \hat{S}_y) +$$
$$A_\parallel \hat{S}_z \hat{I}_z + A_\perp(\hat{S}_x \hat{I}_x + \hat{S}_y \hat{I}_y) + Q'\left[\hat{I}_z^2 - \frac{1}{3}I(I+1)\right] - \gamma \beta_N \vec{H}_0 \cdot \hat{I}$$

The first term describes the zero-field splitting, the next two terms describe the effect of the magnetic field on the spin multiplicity remaining after zero-field splitting, the terms in $A_\parallel$ and $A_\perp$ measure the hyperfine splitting parallel and perpendicular to the main axis, and Q' measures the small changes in the spectrum produced by the nuclear quadrupole interaction. All of these effects have been discussed previously (see Chapter 9). The final term takes into account the fact that the nuclear magnetic moment μ_N can interact directly with the external field $\mu_N H_0 = \gamma \beta_N \vec{H}_0 \cdot \hat{I}$, where γ is the nuclear magnetogyric ratio and β_N is the nuclear Bohr magneton. This is the nuclear Zeeman effect, which gave rise to transitions in nmr. This interaction can affect the paramagnetic resonance spectrum only when the unpaired electrons are coupled to the nucleus by nuclear hyperfine or quadrupole interactions. Even when such coupling occurs, the effect is usually negligible in comparison with the other terms.

In the case of a distortion of lower symmetry, there are three different components g_x, g_y, and g_z, and three different hyperfine interaction constants A_x, A_y, and A_z. Two additional terms need to be included: $E(\hat{S}_x^2 - \hat{S}_y^2)$ as an additional zero-field splitting and $Q''(\hat{I}_x^2 - \hat{I}_y^2)$ as a further quadrupole interaction. The symbols P and P' are often used in place of Q' and Q'', respectively.

The importance of the spin Hamiltonian is that it provides a standard phenomenological way in which the epr spectrum can be described in terms of a small number of constants. Once values for the constants have been determined from experiment, calculations relating these parameters to the electronic configurations and the energy states of the ion are often possible.

The splitting of the ^6S state of an octahedral manganese(II) complex is illustrated in Fig. 13–10(A). Here we have the interesting case in which O_h Mn^{2+} has a $^6A_{1g}$ ground state, which is split by zero-field effects. Spin-orbit coupling mixes into the ground state excited $^4T_2$ states that are split by the crystal field, and this mixing gives rise to a small zero-field splitting in Mn^{2+}. The dipolar interaction of the electron spins is small in comparison to the higher state mixing in this complex. The orbital effects are very interesting in this example because the ground state is ^6S, and thus the excited $^4T_2$ state can only be mixed in by second order spin-orbit effects. Thus, the zero-field splitting is relatively small, for example, of the order of 0.5 cm^{-1} in certain manganese(II) porphyrins.[16a] As indicated in Fig. 13–10, the zero-field splitting produces three doubly degenerate spin states, $M_s = \pm^5/_2, \pm^3/_2, \pm^1/_2$, (Kramers' degeneracy). Each of these is split into two singlets by the applied field, producing six levels. As a result of this splitting, five transitions ($-^5/_2 \to -^3/_2$, $-^3/_2 \to -^1/_2$, $-^1/_2 \to ^1/_2$, $^1/_2 \to ^3/_2$, $^3/_2 \to ^5/_2$) are expected. The spectrum is further split by the nuclear hyperfine interaction with the manganese nucleus ($I = ^5/_2$). This would give rise to thirty peaks in the spectrum.

In contrast to hyperfine splitting, the term *fine splitting* is used when an absorption band is split because of non-degeneracy arising from zero-field splitting. Components of fine splitting have varying intensities: the intensity is greatest for the central lines and smallest for the outermost lines. In simple cases, the separation between lines varies as $3\cos^2\theta - 1$, where θ is again the angle between the direction of the field and the molecular z-axis.

In Fig. 13–11, the influence of zero-field splitting on an $S = 1$ system is indicated for a fixed molecular orientation. (Recall that there is no Kramers' degeneracy.) In the absence of zero-field effects [Fig. 13–11(A)], the two $|\Delta m_s| = 1$ transitions are degenerate and only one peak would result.

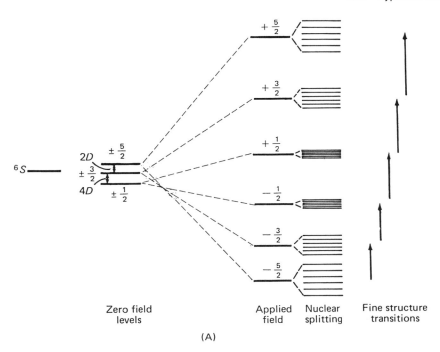

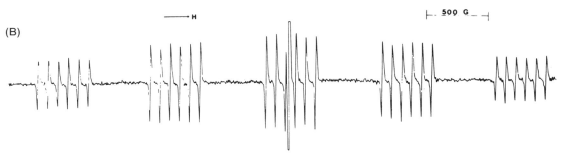

FIGURE 13–10 (A) Splitting of the levels in an octahedral Mn(II) spectrum. (B) Spectrum of a single crystal of Mn^{2+} doped into MgV_2O_6, showing the five allowed transitions (fine structure), each split by the magnanese nucleus ($I = {}^5/_2$) (hyperfine structure). At 300 K, $g_x = 2.0042 \pm 0.0005$, $g_y = 2.0092 \pm 0.001$, and $g_z = 2.0005 \pm 0.0005$; $A_z = A_y = A_z = -78 \pm 5$ G; and $D_x = 218 \pm 5$ G, $D_y = -87 \pm 5$ Gm abd$D_z = -306 \pm 20$ g. [Modified from H. N. Ng and C. Calvo, Can. J. Chem., *50*, 3619 (1972). Reproduced by permission of the National Research Council of Canada.]

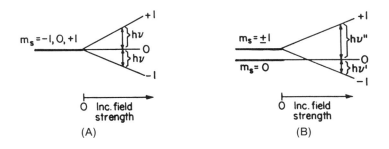

FIGURE 13–11 The energy level diagram and transitions for a molecule or ion with $S = 1$, (A) in the absence of and (B) in the presence of zero-field splitting. The system in (B) is aligned with the z-axis, for the effect is very anisotropic.

For the splitting shown in Fig. 13–11(B), two transitions would be observed in the spectrum. A specific example of this type of system is the $^3A_{2g}$ ground state of nickel(II) in an O_h field. Spin-orbit coupling mixes in excited states, which split the $^3A_{2g}$ configuration. *Recall that zero-field splitting is very anisotropic, providing a relaxation mechanism for the electron spin state.* Accordingly, epr spectra of Ni(II) O_h complexes are difficult to detect, and when they are studied, liquid nitrogen or helium temperatures must generally be employed. At room temperature, nmr spectra can be measured. In some systems, sharp double quantum transitions ($|\Delta m_s| = 2$) can be seen in the epr spectrum.[16b]

As mentioned in Chapter 9, the zero-field splitting can be so large that the $\Delta m_s = \pm 1$ transitions are not observed. For example, in V^{3+} (d^2), $\Delta m_s = \pm 1$ transitions are not detected, but one observes a weak $\Delta m_s = 2$ transition ($-1 \rightarrow +1$) split into eight lines by the nuclear spin of ^{51}V ($I = 7/2$). The mechanism whereby the $\Delta m_s = 2$ transition becomes allowed is the same as that discussed for organic triplet states in Chapter 9.

The next topic to be discussed is the definition of the term *effective spin, S'*. We have already been using this idea, but now formally define it to describe how some of the effects that we have discussed are incorporated into the spin Hamiltonian. When a cubic crystal field leaves an orbitally degenerate ground state (*e.g.*, a *T* state), the effect of lower symmetry fields and spin-orbit coupling will remove this degeneracy as well as the spin degeneracy. In the case of an odd number of unpaired electrons, Kramers' degeneracy leaves the lowest spin state doubly degenerate. If the splitting is large, this doublet will be well isolated from higher-lying doublets. Transitions will then be observed only in the low-lying doublet, which behaves like a simpler system having $S = 1/2$. We then say that the system has an effective spin S' of only $1/2$ ($S' = 1/2$). An example is Co^{2+}. The cubic field leaves a ^4F ground state which, as a result of lower symmetry fields and spin-orbit coupling, gives rise to six doublets. When the lowest doublet is separated from the next by appreciably more than kT, the effective spin has a value of $1/2$ ($S' = 1/2$). instead of $3/2$. A spin Hamiltonian can be written in terms of S' rather than S.

It should be clear that all of the effects discussed above can have a pronounced influence on the spectral appearance. The qualitative interpretation of the epr spectra of transition metal ions by inspection is thus not trivial. Proficiency is obtained by looking at many spectra and drawing analogies to known systems when dealing with new systems. Practice will be afforded in a later section where the representative spectra of different d^n systems will be considered, and also in the exercises at the end of the chapter.

Contributions to A

The hyperfine coupling interaction has contributions to it from Fermi contact, dipolar nuclear spin–electron spin, and nuclear spin–electron orbit mechanisms. These effects have been discussed in Chapter 9. The reader is referred to equation (9–25) and the subsequent discussion of it for a treatment of the dipolar contribution. *The trace of the tensor in equation (9–25) is zero, so information about the dipolar contribution can be obtained only from ordered or partially ordered systems.* As mentioned in equations (9–34) and (9–35), where the traceless tensor components of A were indicated by T, the contribution to dipolar hyperfine coupling for an electron in a p_z orbital is given by

$$T_{zz} = \left(\frac{4}{5}\right)P_p \quad T_{xx} = -\left(\frac{2}{5}\right)P_p \quad T_{yy} = -\left(\frac{2}{5}\right)P_p$$

with

$$P_p = g_e \beta g_N \beta_N \left\langle \frac{1}{r^3} \right\rangle_p \tag{13–17}$$

TABLE 13–1. Contributions to the Dipolar Hyperfine Coupling from Electrons in d orbitals

Orbital	$T_{zz}(P_d)$	$T_{xx}(P_d)$	$T_{yy}(P_d)$
d_{z^2}	$4/7$	$-2/7$	$-2/7$
$d_{x^2-y^2}$	$-4/7$	$2/7$	$2/7$
d_{yz}	$2/7$	$-4/7$	$2/7$
d_{xz}	$2/7$	$2/7$	$-4/7$
d_{xy}	$-4/7$	$2/7$	$2/7$

For an electron in a p_z orbital, we have

$$T_{zz} = -\left(\frac{2}{5}\right)P_p \qquad T_{yy} = -\left(\frac{2}{5}\right)P_p \qquad T_{xx} = \left(\frac{4}{5}\right)P_p$$

Similar expressions can be derived for an electron in one of the d orbitals. The quantitites in Table 13–1 must be multiplied by

$$P_d = g_e \beta g_N \beta_N \left\langle \frac{1}{r^3} \right\rangle_d \qquad (13\text{–}18)$$

to obtain the dipolar contribution. The signs and magnitudes can be easily remembered. The orbital is located in the $3\cos^2\theta - 1$ plot of the lines of flux from the nuclear moment, as shown in Fig. 13–12 for the d_{xz} orbital. In this figure, the z-axis of the molecule is aligned with the z-axis of the field to give T_{zz} as a small positive number. Next rotate the orbital counterclockwise 90°, without rotating the cone, so that the x-axis of the molecule is aligned parallel to the field (which is still along the z-axis shown in Fig. 13–12) to give a small positive T_{xx}. Next, start with Fig. 13–12 and rotate the molecule so that the z-axis of the molecule is perpendicular to the page and the y-axis of the molecule is parallel to the field (*i.e.*, the z-axis in Fig. 13–12). Now the lobes of d_{xz} are in the negative region of the cone and a large negative T_{yy} is expected. These rotations correspond

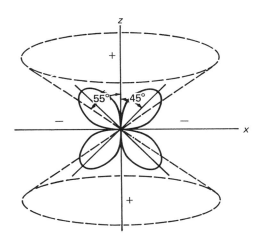

FIGURE 13–12 Orientation of the d_{xz} orbital relative to the $3\cos^2\theta - 1$ cone.

to the three mutually perpendicular orientations of the molecule relative to the field. With information on the signs and magnitudes of the components of the hyperfine tensor, one can obtain information about the atomic orbitals in a complex that make the principal contribution to the molecular orbital containing the unpaired electron.

The Fermi contact contribution has been discussed in detail in Chapters 9 and 12. Unpaired spin density is felt at the nucleus by direct admixture of the *s* orbitals into the m.o. containing the unpaired electron and by spin polarization of filled inner *s* orbitals by unpaired electron density in *d* orbitals. When the 4*s* orbital of the metal is empty, it can mix into the largely metal *d*-antibonding orbital; and if this m.o. contains unpaired spin, the electron partly occupies the metal 4*s* orbital.

Spin polarization can have two different results. When an inner 1*s*- or 2*s* orbital is spin-polarized by α-spin in a 3*d* orbital, an excess of β-spin results at the nucleus. When the filled 3*s* orbital is spin-polarized by an electron with α-spin in a 3*d* orbital, α-spin results at the nucleus. The effect of direct delocalization into the 4*s*-orbital can be shown by comparing Fermi contact hyperfine values ($A_{F.C.}$) determined for various cobalt(II) complexes. It has been found[17a] that $A_{F.C.}$ for sixfold coordination falls in the -30 to -45 G region, for fivefold coordination in the -5 to -25 G region, and for square planar fourfold coordination in the 0-G region. One can write an equation summarizing contributions to $A_{F.C.}$:

$$A_{F.C.} = x(A_{4s}) + (1 - x)(A_{3d}) \tag{13-19}$$

where A_{4s} is the direct hyperfine interaction of one unpaired electron in a 4*s* orbital, 1320 G × $g\beta$; and A_{3d} is the hyperfine interaction arising from spin-polarization of filled *s* orbitals by an unpaired electron in a 3*d* orbital, -90 G × $g\beta$. Applying this equation to the sixfold, fivefold, and fourfold cobalt(II) complexes, one obtains $x = 3$ to 4%, $x = 4.5$ to 6%, and $x = 6.5\%$ respectively, for admixture of 4*s* into the m.o. containing the unpaired electron.

The nuclear spin-electron orbit contribution to the coupling constant is related to the pseudocontact contribution discussed in Chapter 12. The Hamiltonian is

$$\hat{H} = \left(\frac{P_d}{7}\right)[-(\vec{L}\cdot\vec{S})(\vec{L}\cdot\vec{I}) - (\vec{L}\cdot\vec{I})(\vec{L}\cdot\vec{S})] \tag{13-20}$$

Here the nuclear moment is interacting not only with the spin moment (discussed earlier), but also with the orbital moment. The nuclear interaction with the spin moment is a traceless tensor, but the interaction with the orbital moment is not. Since it is not, a pseudocontact contribution is observed in the nmr spectrum in solution.

13–4 LIGAND HYPERFINE COUPLINGS

Hyperfine splittings from ligand atoms have contributions from a Fermi contact term, F.C.; dipolar contributions from the metal, DIP; dipolar contributions from electron density in *p* orbitals of the ligands, LDP; and the metal pseudocontact contribution at the ligand, LPC, which results from the interaction of the orbital angular momentum of the unpaired electron with the ligand nuclear spin. When ligand hyperfine structure is resolved, this latter term is generally small compared to the other contributions. When there is extensive spin-orbit coupling, a large pseudocontact contribution would be expected, but relaxation effects lead to difficulty in observing the epr spectrum and consequently the ligand hyperfine splittings. The $A_\|$ and $A_\perp$ values are given by equations (13–21) and (13–22):

$$A_\| = A_{\text{F.C.}} + 2(A_{\text{DIP}} + A_{\text{LDP}}) \qquad (13\text{–}21)$$

$$A_\perp = A_{\text{F.C.}} - (A_{\text{DIP}} + A_{\text{LDP}}) \qquad (13\text{–}22)$$

These equations refer to the parallel and perpendicular components of the ligand hyperfine tensor. When an investigator measures the ligand hyperfine interaction (A_l) from the fine structure of a metal hyperfine peak (e.g., nitrogen superhyperfine structure on a cobalt hyperfine peak), difficulty will arise if the two hyperfine coupling tensors are not diagonal in the same coordinate system. Single-crystal x-ray diffraction and single-crystal epr studies are necessary for complete understanding of these systems. If one carries out a solution epr study with the ligand hyperfine structure resolved, $A_{\text{F.C.}}$ can be measured directly. Equations (13–21) and (13–22) cannot be solved for A_{DIP} and A_{LDP}. However, a reasonable value for A_{DIP} can be calculated from a knowledge of the structure, by use of equation (13–23):

$$A_{\text{DIP}} = \frac{g_e g_N \beta \beta_N}{a^3} \, (\text{cm}^{-1}) \qquad (13\text{–}23)$$

where a is the metal-ligand distance. Equation (13–23) is derived for circumstances in which the metal-ligand distance is large compared to the metal nucleus–electron distance. Where this approximation is poor, other equations have been reported.[18] The $A_{\text{F.C.}}$ and A_{LDP} values are related to the s orbital contribution α_s and p orbital contribution α_p to the molecular orbital by the following equations:

$$A_{\text{F.C.}} = \frac{16\pi}{3} \gamma \beta \beta_N |\psi(0)|^2 \alpha_s^2 \qquad (13\text{–}24)$$

$$A_{\text{LDP}} = \frac{2}{5} \gamma \beta \beta_N \left\langle \frac{1}{r^3} \right\rangle_p \alpha_p^2 \qquad (13\text{–}25)$$

Values of $A_{\text{F.C.}}$ and A_{LDP} are reported[19] for one electron in s and p orbitals from which the %s and %p character can be calculated. From these equations, the ratio %s/%p can be determined to indicate the "hybridization of the ligand" in the complex. This treatment tacitly assumes that delocalization occurs mainly by a direct mechanism; i.e., spin polarization contributions are ignored.

The application of these ideas will be illustrated with a few brief examples. The epr spectrum[20] of

[structure: Fe center with NO ligand on top, coordinated by four S atoms from two dithiolate-type ligands with C=O groups; overall charge 2−]

indicates $A_{\parallel}(^{14}\text{N})/g\beta = 15.6$ G, $A_{\perp}(^{14}\text{N})/g\beta = 12.7$ G, and $A_{\text{iso}} = 13.7$ G. Values of A reported for a full electron in s and p orbitals are $A_p°/g\beta = 34.1$ G and $A_s°/g\beta = 550$ G. This leads to a ratio of $\%p/\%s$ of about 1, which indicates an sp hybrid and a linear Fe—N—O structure. On the other hand, the ratio of p/s for the nitrosyl nitrogen[21] of $\text{Fe(CN)}_5\text{NO}^{3-}$ is 1.6, consistent with a bent metal-nitrosyl structure (an sp^2 hybrid would have had a ratio of 2).

Recently, phosphorus hyperfine couplings were observed[22] in the epr spectra of $(C_6H_5)_3P$ and PF_3 adducts of tetraphenylporphyrin cobalt(II). The p/s ratio was reported as 2.7 in the former case and 0.47 in the latter.

13–5 SURVEY OF THE EPR SPECTRA OF FIRST-ROW TRANSITION METAL ION COMPLEXES

In this section, we shall briefly survey some of the results of epr studies on various d^n complexes. For a more complete discussion, the reader is referred to references 23 and 24. Before beginning this survey, we should mention that *spin-orbit coupling provides the dominant mechanism for electron relaxation in these systems.* In your reading, you will find statements like "zero-field splitting causes rapid relaxation" or "g-value anisotropy leads to a short electron spin lifetime," etc. It is to be emphasized that these are all manifestations of the effects of spin-orbit coupling in the molecule. We have previously discussed the relationship of spin-orbit coupling to these effects. Second and third row transition metal complexes become increasingly more difficult to study by the epr technique because the spin-orbit coupling constants are much larger.

d^1

The results of studies on ions with the electronic configuration d^1 can be fit with the spin Hamiltonian:

$$\hat{H} = \beta[g_x H_x \hat{S}_x + g_y H_y \hat{S}_y + g_z H_z \hat{S}_z] + A_x \hat{S}_x \hat{I}_x + A_y \hat{S}_y \hat{I}_y + A_z \hat{S}_z \hat{I}_z \quad (13\text{–}26)$$

In an octahedral ligand field, the ground state is $^2T_{2g}$ and there is considerable spin-orbit coupling in this ground state. All of the Kramers' doublets from the $^2T_{2g}$ term are close in energy and extensively mixed by spin-orbit coupling. This leads to a short τ_e. In a tetrahedral ligand field, an ^2E ground state ($x^2 - y^2, z^2$) with no first-order spin-orbit coupling results. In this geometry, admixture of the nearby $^2T_{2g}$ excited states into the ground state by second-order* spin-orbit coupling provides a mechanism leading to a short spin relaxation time for the electron and broad absorption lines. The complexes usually must be studied at temperatures approaching that of liquid helium. The ^2T excited state is split by spin-orbit coupling. When the ligand field is distorted (e.g., in VO^{2+}), the ground state becomes orbitally singlet and the excited states are well removed. Sharp epr spectral lines result even at higher temperatures.

Equations have been derived[24] using the d^1 wave functions and the appropriate spin Hamiltonian, to relate the g-value in a trigonally distorted complex to the amount of distortion. The distortion is expressed in terms of δ (cm^{-1}), the splitting of the ^2T state. A large distortion with $\delta = 2000$ to 4000 cm^{-1} is found in tris(acetylacetonato) titanium(III). As a result of this splitting, the electron spin lifetime is increased and one is able to detect the epr spectrum at room temperature.

d^2

Very few examples of epr spectra of these ions in octahedral complexes have been reported because of the extensive spin-orbit coupling in the $^3T_1$ ground state. Tetrahedral complexes have an $^3A_2$ ground state, so we would expect longer relaxation times and more readily observed epr spectra. The spectra of these systems can be fitted with $S = 1$ and the spin Hamiltonian

$$\hat{H} = g_z \beta H_z \hat{S}_z + g_x \beta H_x \hat{S}_x + g_y \beta H_y \hat{S}_y + D\left[\hat{S}_z^2 - \frac{2}{3}\right]$$
$$+ E[\hat{S}_x^2 - \hat{S}_y^2] + A\hat{S}_z\hat{I}_z + B[\hat{S}_x\hat{I}_x + \hat{S}_y\hat{I}_y] \tag{13-27}$$

V^{3+} in an octahedral environment[25] in an Al_2O_3 lattice gave $g_\parallel = 1.92$, $g_\perp = 1.63$, $D = +7.85$, and $A = 102$. An example of a T_d complex that has been studied[23] is V^{3+} in a CdS lattice. At liquid N_2 temperature, a g-value of 1.93 and D of 1 cm^{-1} are observed.

d^3

Octahedral d^3 complexes have an $^4A_2$ ground state, which must have a Kramers' doublet lowest in energy. When the zero-field splitting is small, as shown in Fig. 13–13(A) three transitions can sometimes be detected and the zero-field parameter can be obtained from the two affected transitions. When the zero-field splitting is large compared to the spectrometer frequency, only one line will be observed, as shown in Fig. 13–13(B). In general, the spectra can be fitted to the spin Hamiltonian as follows:

$$\hat{H} = g_z \beta H_z \hat{S}_z + g_x \beta H_x \hat{S}_x + g_y \beta H_y \hat{S}_y + D\left[\hat{S}_z^2 - \frac{5}{4}\right] + E[\hat{S}_x^2 - \hat{S}_y^2]$$
$$+ A_\parallel \hat{S}_z\hat{I}_z + A_\perp[\hat{S}_x\hat{I}_x + \hat{S}_y\hat{I}_y] \tag{13-28}$$

---

* In second-order perturbation theory, the E-component of this split state is mixed into the ground state.

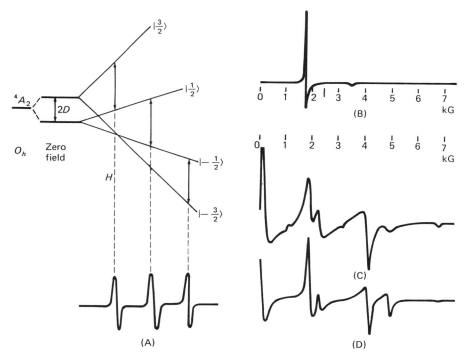

FIGURE 13–13 (A) Small zero-field and magnetic field splitting of the $^4A_2$ ground state (field along z) for a d^3 case and the resulting spectrum.
(B) *Trans*-[Cr(C$_5$H$_5$N)$_4$I$_2$]$^+$ in DMF, H$_2$O, CH$_2$O, CH$_3$OH glass$^{(33)}$ at 9.3 GHz. $D > 0.4$ cm^{-1}, $E < 0.01$.
(C) *Trans*-[Cr(C$_5$H$_5$N)$_4$Cl$_2$]$^+$ in DMF, H$_2$O, CH$_3$OH glass$^{(33)}$ at 9.211 GHz.
(D) Computer simulation$^{(33)}$ of (C) with $g_\parallel = g_\perp = 1.99$, $D = 0.164$ cm^{-1}, $E = 0$.
[Reprinted with permission from E. Pedersen and H. Toftlund, Inorg. Chem., *13*, 1603 (1974). Copyright by the American Chemical Society.]

It is difficult to recognize "typical patterns" for Cr^{3+} in some systems.$^{(26)}$ The spectrum shown in Fig. 13–13(D) was calculated by a computer, using an isotropic g, $D = 0.164$ cm^{-1}, and $E = 0$ cm^{-1}. Reference 26 contains many spectra of tetragonal Cr(III) complexes and a detailed analysis of them.

The d^3 system has been very extensively studied, particularly Cr^{3+}. In octahedral complexes, the metal electrons are in t_{2g} orbitals, so ligand hyperfine couplings are usually small. The g-value for this system is given, according to crystal field theory, by

$$g = 2.0023 - \frac{8\lambda}{\Delta E(^4T_{2g} - {}^4A_{2g})} \tag{13–29}$$

The ground state, being $^4A_{2g}$, has no spin-orbit coupling and a small amount is mixed in *via* the $^4T_{2g}$ state. Equation (13–29) differs from those presented earlier in two ways. The spin-orbit coupling is described by λ (which can be positive or negative) and characterizes a state. With more than one unpaired electron, the energy differences also must be expressed in terms of the energy differences of the appropriate electronic states. Calculating g for V(H$_2$O)$_6{}^{2+}$ using $\Delta E = 11,800$ cm^{-1} and $\lambda = 56$ cm^{-1} gives a value $g = 1.964$, which is close to the observed$^{(27)}$ value of 1.972. For Cr(H$_2$O)$_6{}^{3+}$, $\Delta E = 17,400$ cm^{-1}, $\lambda = 91$ cm^{-1}, and the predicted g-value is smaller than the experimental value[28] of 1.977. In the case of Mn^{4+}, the discrepancy is even larger, with a calculated $g = 1.955$ and an experimental value of 1.994. This is in keeping with the fact that the crystal field approximations are poorer and covalency becomes more important as the charge of the central ion increases.

d^4

There are very few epr spectra reported for this d-electron configuration. The ground state in this system, in a weak crystalline O_h field, which is ^5E, has no orbital angular momentum, so S is a good quantum number. Zero-field splitting of the ± 2, ± 1, and 0 levels leads to four transitions when the splitting is small, as shown in Fig. 13–14, and none when the splitting is large. Jahn-Teller distortions and the accompanying large zero-field splittings that are expected often make it impossible to see a spectrum.

For low spin d^4 complexes, the reader is referred to the d^2 section (recall the hole formalism).

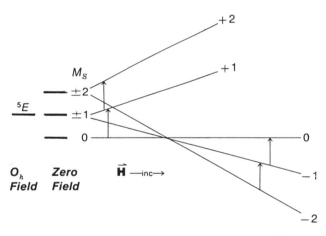

FIGURE 13–14 Zero-field and magnetic field splitting of the ^5E ground state (field along z) for a d^4 case.

d^5 Low Spin, $S = \frac{1}{2}$

In a strong ligand field of octahedral symmetry, the ground state is $^2T_2$. Spin-orbit coupling splits this term into three closely spaced Kramers' doublets; however, epr spectra can be seen only at temperatures close to those of liquid helium because of the large amount of spin-orbit coupling present. Since there are five d-electrons in these systems, the situation is analogous to d^1, except that in this case we are working with a positive hole. Jahn-Teller forces tend to distort systems such as MX_6^{n-}, so the g-values contained in equation (13–9) are rarely observed. The splitting of the free ion doublet state by an O_h field, a D_3 distortion, spin-orbit coupling, and a magnetic field are shown in Fig. 13–15. Since we have non-integral spin, the double group representations are employed when spin-orbit coupling is considered, and primed symbols are employed for the representations. If one

582 Chapter 13 Electron Paramagnetic Resonance Spectra of Transition Metal Ion Complexes

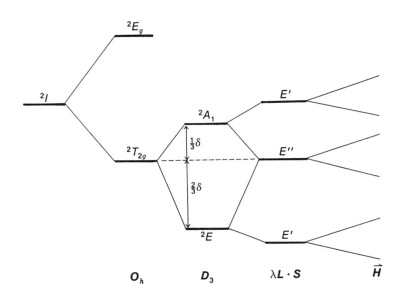

FIGURE 13–15 Energy level diagram for a low spin d^5 system in an O_h field and a D_3 field, followed by spin-orbit coupling and the magnetic field-induced splittings. With spin-orbit effects included, the double group D_3' representations are employed (primed values).

defines a distortion parameter, δ, as in Fig. 13–15, equations for the g-values can be derived[29]:

$$g_\parallel = \frac{3(\xi + 2\delta)}{[(\xi + 2\delta)^2 + 8\xi^2]^{1/2}} - 1 \qquad (13\text{–}30)$$

$$g_\perp = \frac{(2\delta - 3\xi)}{[(\xi + 2\delta)^2 + 8\xi^2]^{1/2}} + 1 \qquad (13\text{–}31)$$

Note that as $\delta \to 0$, both $g_\parallel$ and $g_\perp \to 0$, in accord with the equations in Fig. 13–9. For example, Fe^{3+} in $K_3Co(CN)_6$ exhibits[30] an epr spectrum with $g_\parallel = 0.915$ and $g_\perp \approx 2.2$. Substituting into equation (13–30) and employing $\xi = -103$ cm^{-1} for the free ion produces a value of $\delta \approx 200$ cm^{-1}.

Large deviations from octahedral symmetry cause an orbitally singlet state to lie lowest in energy, well removed from orbitally non-degenerate excited states. Longer electron relaxation times result, and epr spectra can be observed at higher temperatures. Examples of such a system are low spin derivatives of ferric hemoglobin[31] (Fig. 13–16), which possesses a large tetragonal distortion as a result of the heme plane. Examples of bases, B, that produce a low spin environment are N_3^-, CN^-, and OH^-. Experimental g-values for the N_3^- species are $g_x = 1.72$, $g_y = 2.22$, and $g_z = 2.80$. The large anisotropy in g_x and g_y is thought to arise from interaction of a specific iron d orbital with a nitrogen π orbital from the coordinated histidine group that ties the globin to the heme unit. EPR spectra[32] of bipyridyl and phenanthrolene complexes of iron(III), ruthenium(III), and osmium(III) are analyzed in terms of an energy diagram similar to Fig. 13–15, but one in which the distortion produces the ^2A state lower in energy than ^2E; i.e., δ is negative. Spin delocalization onto the ligand was studied as a function of the metal in this series.

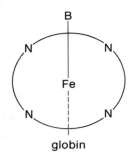

FIGURE 13–16 Schematic formula for hemoglobin.

d^5 High Spin

This d-electron configuration has been very thoroughly studied. The high spin complexes have ^6S ground states, and there are no other sextet states. The $^4T_1$ is the closest other term, and second-order spin-orbit coupling effects are needed to mix in this configuration, so the contributions are small. Thus, the electron spin lifetime is long and epr spectra are easily detected at room temperature in all symmetry crystal fields. Furthermore, with an odd number of electrons, Kramers' degeneracy exists even when there is large zero-field splitting. The results for high spin complexes are fit by:

$$\hat{H} = g\beta \vec{\mathbf{H}} \cdot \hat{S} + D\left[\hat{S}_z^2 - \frac{35}{12}\right] + E[\hat{S}_x^2 - \hat{S}_y^2] + A\hat{S} \cdot \hat{I}$$
$$+ \frac{1}{6}a\left[\hat{S}_x^4 + \hat{S}_y^4 + \hat{S}_z^4 - \frac{707}{16}\right] + \frac{1}{180}F\left[35\hat{S}_z^4 - \frac{475}{2}\hat{S}_z^2 + \frac{3255}{16}\right] \quad (13\text{--}32)$$

The higher-power terms in $\hat{S}$ arise because the octahedral crystal field operator couples states with M_s values differing by ± 4, leading to a more complex basis set and more non-zero off-diagonal matrix elements.

The splitting of the energy levels and the spectrum expected for an undistorted octahedral iron(III) complex are shown in Fig. 13–17. Zero field leads to the degenerate $^1\!/_2$ and $^5\!/_2$ states and a non-degenerate $^3\!/_2$ state at lower energy by $3a'$. The magnetic field further removes the degeneracy giving rise to the transitions and spectra shown in Fig. 13–17. This spectrum is observed for iron(III) doped into $SrTiO_3$. The iron is coordinated with an undistorted octahedron of oxygen atoms.

In iron(III) complexes with small tetragonal distortion, $D \ll h\nu$ and $E = 0$. The energy levels and expected spectrum are illustrated in Fig. 13–18(A). Observed g-values are very close to 2.00 because of the extremely small amount of spin-orbit coupling. This fact also allows easy observation of epr spectra at room temperature. In zero field, the $^1\!/_2$ state is lowest with $^3\!/_2$, higher by $2D$, and the $^5\!/_2$ state above the $^3\!/_2$ state by $5D$. With small distortions relative to $h\nu$ all levels are populated and five transitions are seen with relative intensities of $5:8:9:8:5$.

If $D \gg h\nu$, the situation shown in Fig. 13–18(B) exists, and only one peak is seen in the epr corresponding to the transition between $|+^1\!/_2\rangle$ and $|-^1\!/_2\rangle$. Even if the higher levels are populated, $\Delta M_s \neq 1$ for the possible transition and no spectral bands are observed. The g-values can be calculated, using only $|^5\!/_2, ^1\!/_2\rangle$ and $|^5\!/_2, -^1\!/_2\rangle$ as a basis set and employing the Zeeman Hamiltonian, $\hat{H} = g_\parallel'\beta H_z\hat{S}_z + g_\perp'\beta(H_x\hat{S}_x + H_y\hat{S}_y)$. When H is parallel to z, we have

$$\hat{H} = \begin{array}{c} \\ \langle ^5\!/_2, ^1\!/_2 | \\ \langle ^5\!/_2, -^1\!/_2 | \end{array} \begin{array}{c} |^5\!/_2, ^1\!/_2\rangle \quad |^5\!/_2, -^1\!/_2\rangle \\ \left| \begin{array}{cc} ^1\!/_2 g_e\beta H_z & 0 \\ 0 & -^1\!/_2 g_e\beta H_z \end{array} \right| \end{array} \quad (13\text{--}33)$$

Solving equation (13–33) leads to $\Delta E = g_e \beta H_z$ and $g_\parallel = g_e$.

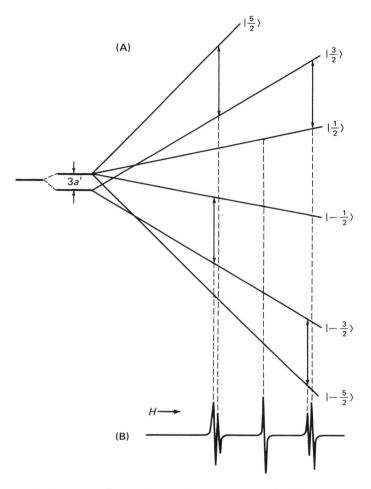

FIGURE 13–17 The splitting of the energy levels (A) and the spectrum (B) expected for an octahedral iron (III) complex (H parallel to a principal axis of the octahedron).

For H parallel to x, after using $S^{\pm} = S_x \pm iS_y$ we obtain

$$\hat{H} = \begin{array}{c} \\ \langle {}^5/_2, {}^1/_2| \\ \langle {}^5/_2, -{}^1/_2| \end{array} \begin{array}{c} |{}^5/_2, {}^1/_2\rangle |{}^5/_2, -{}^1/_2\rangle \\ \left| \begin{array}{cc} 0 & {}^3/_2 g_e \beta H_x \\ {}^3/_2 g_e \beta H_x & 0 \end{array} \right| \end{array} \quad (13\text{–}34)$$

Diagonalization of equation (13–34) leads to $\Delta E = 3g_e \beta H_z$ and $g_\perp = 3g_e \approx 6.0$.

Such a situation is well represented by Fig. 13–16, where B is a weak field ligand such as F^- or H_2O, which lead to a high spin complex. The zero-field

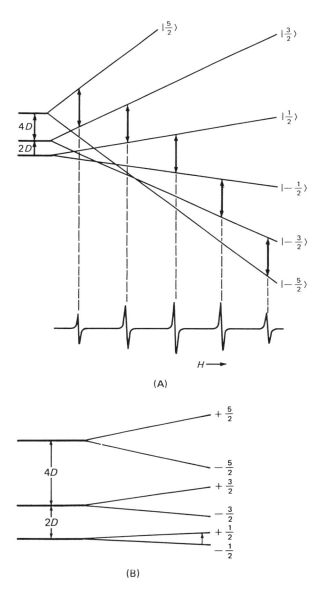

FIGURE 13–18 Energy levels and expected spectrum for a d^5 ion in a weak (A) and strong (B) tetragonal field (H parallel to the tetragonal axis). [From G. F. Kokoszka and R. W. Duerst, Coord. Chem. Rev., 5, 209 (1970).]

splitting parameter, D, has been measured for several systems of this type by examining the far infrared spectrum in a magnetic field. Values ranging from 5 to 20 cm^{-1} are found for various complexes.[33] A sample epr spectrum of this system is that of a frozen aqueous solution of the acid form of ferrimyoglobin shown in Fig. 13–19.

The final case to be considered is one in which a geometric distortion occurs in a complex that removes the axial character. In this case, the rhombic complex is described with the parameters D and E not being zero. If $E/D \simeq 0$ ($E \sim 0$) represents axial symmetry, an increase in E/D represents a departure toward rhombic symmetry. The maximum possible distortion is $E/D = 0.33$, leading to three equally spaced Kramers' doublets. This Hamiltonian again produces three Kramers' doublets, as shown in Fig. 13–20. Solving this Hamiltonian matrix,

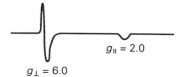

FIGURE 13–19 The X-band epr spectrum of a frozen aqueous solution (77 K) of the acid form of ferrimyoglobin. A similar spectrum results for cytochrome C peroxidase (pH = 7.0 at 4.2 K).

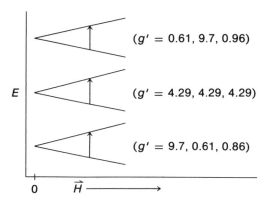

FIGURE 13-20 Kramers' doublets in rhombic symmetry (D and E not equal to zero) for high spin iron(III). The three principal components are listed in parentheses.

using the wave functions that diagonalize it, we find each of the three Kramers' doublets to be a linear combination of $|{}^5/_2, \pm {}^5/_2\rangle$, $|{}^5/_2, \pm {}^3/_2\rangle$, and $|{}^5/_2, \pm {}^1/_2\rangle$. Thus, transitions within each Kramers' doublet are allowed; the corresponding g-values are indicated in Fig. 13–20. The separation between the Kramers' doublets is large enough that transitions between different ones are not observed, but at most temperatures all three are significantly populated, and many resonances are observed. An example[34a] of this situation is Na[Fe(edta)]·4H$_2$O (where edta is ethylenediamine tetraacetate) diluted in a single crystal of the analogous Co(III) complex. The spectrum, Fig. 13–21, shows one almost isotropic transition at $g = 4.27$ and two very anisotropic ones with principal g-values of 9.64 and 1.10, respectively.

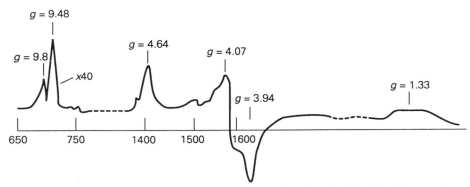

FIGURE 13-21 EPR powder spectrum at 77 K of Fe(III) in Co(III) DTA. The analysis of these spectra for D and E provides information on the extent of distortion. Cases in which the distortions are small produce spectral appearances different from those shown here. The reader is referred to reference 34 for more details.

d^6

This system has not been studied extensively. Low spin complexes are diamagnetic and high spin O_h complexes are similar to d^4. High spin iron(II) has a g-value of 3.49 at 4.2 K and a line width of 500 gauss even at this low temperature. Spin-orbit coupling in the ground state is very large and there are nearby excited states that can be mixed in. With a $J = 1$ ground state, two transitions would be observed if zero-field effects were small. In a distorted octahedral field, zero-field effects are large and no epr spectrum is observed. Deoxyhemoglobin is such a species, and no one has obtained an epr spectrum on this system.

d^7

The ground state for an O_h high spin d^7 complex is $^4T_{1g}(F)$. With extensive spin-orbit coupling, epr measurements are possible only at low temperatures. With $S = {}^3/_2$ and three orbital components in T, a total of 12 low-lying spin states result. Low temperatures are needed to observe the epr spectra, because of spin relaxation problems. Only the low-lying doublet is populated, giving a single peak, with a g-value of 4.33 from an effective $S' = {}^1/_2$. Spectra of these systems have been reviewed.[35]

In tetrahedral symmetry, cobalt(II) complexes with $^4A_2$ ground states are similar to tetrahedral d^3 except that the $^4T_2$ excited state is closer in energy to the $^4A_2$ state in cobalt. With more spin-orbit coupling, broader lines are found for cobalt(II).

In square planar, square pyramidal, and tetragonal complexes, the ground state becomes a doublet, $S = {}^1/_2$ state. With little spin-orbit coupling and no nearby states with spin-orbit coupling, electron spin lifetimes are long, allowing observation of the epr at room temperature. The X-band spectra of Co(DPGB)$_2 \cdot$CH$_3$OH (—C$_2$H$_2$—)$_2$S at 25°C in CH$_2$Cl$_2$/toluene and in a CH$_2$Cl$_2$/toluene glass at $-180°$ C is shown[37] in Fig. 13–22. The DPGB is an abbreviation for a BF$_2$ capped *bis*-diphenylglyoximato ligand system. An isotropic spectrum results at 25° producing an average g of 2.129. At low temperatures, an anisotropic spectrum results in the frozen glass. The cobalt hyperfine is nicely resolved in the parallel region.

The spin Hamiltonian for the low spin d^7 system is usually given as:

$$\hat{H} = \beta[g_x H_x \hat{S}_x + g_y H_y \hat{S}_z + g_z H_z \hat{S}_z] + A_x \hat{S}_x \hat{I}_x + A_y \hat{S}_y \hat{I}_y + A_z \hat{S}_z \hat{I}_z \quad (13\text{–}35)$$

In the cases of the five- and six-coordinate tetragonal d^7 systems, the unpaired electron resides in the d_{z^2} orbital. For this electronic configuration, assuming axial symmetry,

$$g_\parallel = 2.0023 \qquad (13\text{–}36)$$

$$g_\perp = 2.0023 - \frac{6\xi}{\Delta E(z^2 - xz,\ yz)} \qquad (13\text{–}37)$$

This fact has been used[25] to study the adduct formation of coordinatively unsaturated cobalt complexes with varieties of axially coordinated bases, B. Good overlap between the donor lone pair and d_{z^2} causes readily observed hyperfine splitting from bases coordinating *via* nitrogen or phosphorus atoms. Wayland has utilized the large gyromagnetic ratio (and hence large hyperfine coupling) of ^{31}P to obtain hybridization ratios for varieties of PX$_3$ donors with Co(tetraphenylporphyrin)[22] and Co(salen).[36]

As shown in Fig. 13–23, the epr spectra of adducts containing nitrogen donor ligand orbitals that can mix with the d_{z^2} orbital of cobalt(II) (containing the unpaired electron) possess nitrogen hyperfine. Three-line hyperfine is seen in the

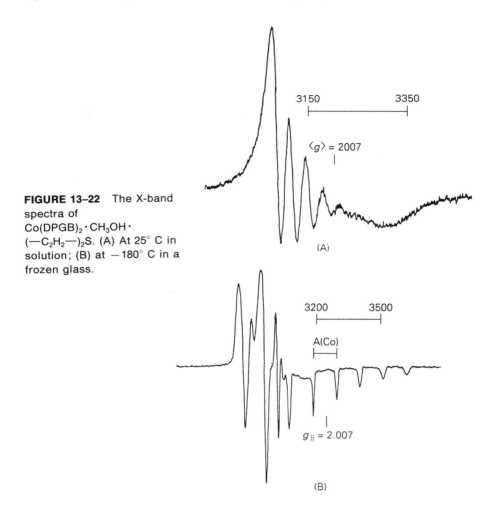

FIGURE 13–22 The X-band spectra of Co(DPGB)$_2$·CH$_3$OH·(—C$_2$H$_2$—)$_2$S. (A) At 25° C in solution; (B) at −180° C in a frozen glass.

complex in Fig. 13–23(A), which contains one acetonitrile and one methanol on the z-axis and five lines (2Σ I + 1) are seen in the complex in Fig. 13–23(B) containing two pyridine ligands on the z-axis. These spectral differences have been used to probe site isolation in polymers containing basic nitrogen functional groups by adding a solution of Co(DPGB)$_2$·2CH$_3$OH to the polymer.[38] A three-line nitrogen hyperfine indicates that two nitrogens coordinate and the complex binds to one site. A five-line pattern indicates that the donor sites are not isolated. Polymer mobility can permit even a dilute concentration of sites to behave in a non-isolated manner. The complex Co(DPGB)$_2$·2CH$_3$CN can be used to study basic sites of nuclei with $I = 0$.

The ground state of the planar, low-spin, four coordinate complexes is more complex[39,40] with ground states of d_{z^2} reported[40] for symmetrical porphyrins and d_{yz} for unsymmetrical Schiff bases.[40]

Many five-coordinate cobalt(II) complexes reversibly bind O$_2$ forming an end on the O$_2$-bound adduct. The unpaired electron in the mainly d_{z^2} orbital of the square pyramidal cobalt(II) complex spin pairs with one of the unpaired

13-5 Survey of the EPR Spectra of First-Row Transition Metal Ion Complexes

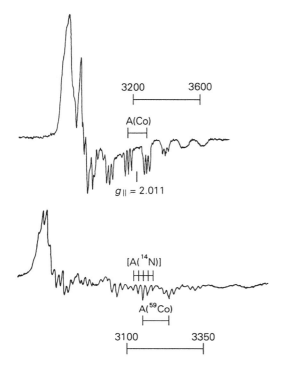

FIGURE 13-23 Frozen glass, X-band epr of (top) Co(DPGB)$_2 \cdot$ CH$_3$OH $\cdot$ CH$_3$CN; (bottom) Co (DPGB)$_2 \cdot$ 2C$_5$H$_5$N.

electrons of O$_2$ forming a sigma bond. This bonding description is referred to as the spin-pairing model.[41] The resulting adduct has one unpaired electron in an adduct molecular orbital that is composed essentially of oxygen a.o.'s. In the O$_2$ adducts, $g_\| > g_\perp$ as seen in Fig. 13-24, making adduct formation easy to recognize. This type of spectrum with $g_\| > g_\perp$ will arise independently of the charge on O$_2$. The charge is determined by the coefficients in the bonding m.o., which contains a pair of electrons. The cobalt hyperfine arises mainly from spin polarization of the pair of electrons in the Co—O$_2$ sigma bonding molecular orbital. Using an approach similar to that described for spin polarization of organic radicals (Chapter 9), an analysis[41] of a series of O$_2$ adducts indicates that the partial negative charge on the bound O$_2$ varies over a range of ~ 0.1 to

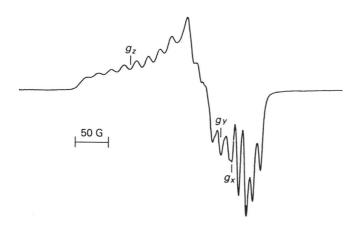

FIGURE 13-24 X-band epr spectrum of a cobalt(II) dioxygen adduct in a frozen glass.

~0.8. A detailed analysis[42] of conflicting interpretations of the cobalt hyperfine in these systems supports the spin-pairing model.

d^8 High Spin

The ground state of the gaseous ion is ^3F with an orbital singlet state lowest in an octahedral field. The d-shell is more than half-filled, so spin-orbit coupling leads to g-values greater than the free electron value. The zero-field splitting makes it difficult to detect epr spectra except at low temperatures. The g-values found are usually close to isotropic.

d^9

The d^9 configuration has been very extensively studied. In an octahedral field, the ground state is ^2E_g. A large Jahn-Teller effect is expected, making observation of the epr spectrum at room temperature possible. In tetragonal complexes, the ground state is $d_{x^2-y^2}$ (x- and y-axes pointing at the ligands) and sharp lines result. Note that the quadrupolar interaction of the copper nucleus (see Chapter 9) can be determined from this experiment. The epr results fit the spin Hamiltonian

$$\hat{H} = \beta[g_z H_z \hat{S}_z + g_x H_x \hat{S}_x + g_y H_y \hat{S}_y] + A_z \hat{S}_z \hat{I}_z + A_x \hat{S}_x \hat{I}_x$$
$$+ A_y \hat{S}_y \hat{I}_y + Q'\left[\hat{I}_z^2 - \frac{1}{3}I(I+1)\right] - g_N \beta_N \vec{H} \cdot \hat{I} \quad (13\text{-}38)$$

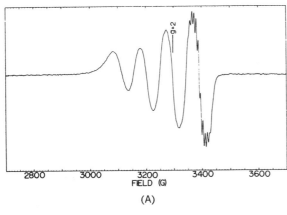

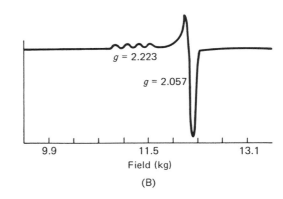

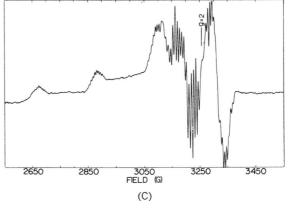

FIGURE 13–25 "Typical" epr spectra for copper(II) complexes. (A) A typical solution spectrum for a square planar Schiff base ligand. [From E. Hasty, T. J. Colburn, and D. N. Hendrickson, Inorg. Chem., 12, 2414 (1973).] (B) Glass or powder sample run at Q-band frequencies on an axial complex. (C) Glass or doped powder spectrum of the complex in (A) run at X-band frequencies.

Some typical copper(II) spectra are shown in Fig. 13–25. In (A), an isotropic solution spectrum is shown. Both nitrogen and proton ligand hyperfine structures are seen on the high field peak, but not on the low field peaks. This is attributed to differences in the relaxation times for the transition, which depend upon the m_I-value associated with the transition.[43] The solvent employed influences the molecular correlation time, which in turn also influences the spectral appearance.[43]

In (B), an anisotropic spectrum is shown at Q-band frequencies. Such spectra are observed in glass or powder samples of copper complexes diluted in hosts. The low field $g_\parallel$ and high field $g_\perp$ peaks are well separated. With the higher microwave energy, the individual peaks are broader so the super-hyperfine splitting is not detected on the $g_\parallel$ peak. In the spectrum in (C), at X-band frequencies, the $g_\parallel$ and $g_\perp$ transitions overlap, but much more ligand hyperfine structure is detected. As mentioned earlier, the temperature dependence of the spectra of many copper(II) systems has been interpreted in terms of Jahn-Teller effects.

13–6 THE EPR OF METAL CLUSTERS

EPR studies of molecules containing two or more metal ions have provided considerable information on indirect (superexchange) exchange mechanisms. The spin Hamiltonian is given by[44]

$$H = \hat{S}_1 \cdot J \cdot \hat{S}_2 \tag{13-39}$$

where J is a matrix connecting the two spin operators $\hat{S}_1$ and $\hat{S}_2$ of metal atoms 1 and 2. In the common case of a weak bonding interaction involving isotropic exchange interactions of the same transition metal ions we can write

$$\hat{H} = J\hat{S}_1 \cdot \hat{S}_2 \tag{13-40}$$

As mentioned in Chapter 8, $J\hat{S}_1 \cdot \hat{S}_2$ groups the energy levels according to

$$|S_1 - S_2| \cdots S \cdots |S_1 + S_2| \tag{13-41}$$

with energies

$$E(S) = (J/2)|S(S+1) - S_1(S_1+1) - S_2(S_2+1)| \tag{13-42}$$

The coefficients indicating the contributions of the spin Hamiltonian parameters for g and A are calculated[45] by projecting the spin Hamiltonian parameters of the individual ions onto the S manifold. For example, g_c in the coupled system as well as A_{c_1} and A_{c_2}, the hyperfine coupling of metals 1 and 2 in the coupled system, is given by[46]

$$g_c = c_1 g_1 + c_2 g_2 \tag{13-43}$$

$$A_{c_1} = C_1 A_1 \tag{13-44}$$

$$A_{c_2} = C_2 A_2 \tag{13-45}$$

where g_1, g_2, A_1, and A_2 refer to the g- and A-values of the metals 1 and 2 in an analogous complex containing one paramagnetic center, i.e., an M_1-Zn(II) or Zn(II)-M_2 complex. The coefficients C_1 and C_2 are given by

$$C_1 = |S(S+1) + S_1(S_1+1) - S_2(S_2+1)|/2S(S+1) \quad (13\text{-}46)$$

$$C_2 = |S(S+1) + S_2(S_2+1) - S_1(S_1+1)|/2S(S+1) \quad (13\text{-}47)$$

Expressions for the zero–field splitting contributions have been reported.[40] For two $S = 1/2$ metal ions interacting ($S_1 = 1/2$, $S_2 = 1/2$, $S = 1$), values of $C_1 = 1/2$ and $C_2 = 1/2$ result. For an $S_1 = 1/2$ atom interacting with an $S_2 = 1$ atom, two S states result, $S = 3/2$ and $S = 1/2$. For the former, the coefficients are $C_1 = 1/2$ and $C_2 = 2/3$ and for $S = 1/2$, they are $C_1 = -1/3$ and $C_2 = 2/3$.

For dimers, in which the two metal ions are identical, e.g., Cu-Cu, the g-values are identical to the values for a single ion, e.g., Cu-Zn. For practice, one can calculate the g-values and A_{Cu} for a Cu-Mn ($S = 2$) pair using $g_z = 2.32$, $g_x = 2.08$, $g_y = 2.06$, and $A_2 = 139$ for a Cu-Zn pair in a Cu-Zn(py-O)Cl$_4$(H$_2$O)$_2$ complex[46,47] assuming $g = 2$ for the manganese ion. Values of 1.95, 1.99, 1.94, and 23 should result. Note the greatly reduced copper hyperfine.

The spectrum of a polycrystalline sample of a molecule containing two copper atoms is illustrated in Fig. 13–26. The main peaks at 2500 and 3700 gauss are assigned to two $g_\perp$ components. The seven copper hyperfine components [two Cu ($I = 3/2$) nuclei] are seen in the low field $g_\parallel$ peak. The other high field $g_\perp$ peak is not seen, for it is out of the spectrometer range. The peak at $\sim$3200 G is assigned to a $\Delta m_s = 2$ transition. The origin of the doublet will be understood after discussion of the spin Hamiltonian. The single crystal spectra[48] of molecules containing two copper(II) atoms with $S = 1$ are fitted to the spin Hamiltonian

$$\hat{H} = \beta[g_z H_z \hat{S}_z + g_x H_x \hat{S}_x + g_y H_y \hat{S}_y] + D\left[\hat{S}_z^2 - \frac{2}{3}\right]$$
$$+ E[\hat{S}_x^2 - \hat{S}_y^2] + A_x \hat{S}_x \hat{I}_x + A_y \hat{S}_y \hat{I}_y + A_z \hat{S}_z \hat{I}_z \quad (13\text{-}48)$$

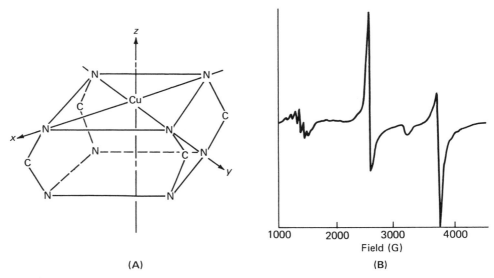

FIGURE 13–26 (A) A schematic representation of a dimeric copper adenine complex; (B) the polycrystalline epr of this complex.

The results for dimeric copper(II) acetate are[49] $g_z = 2.344$, $g_x = 2.053$, $g_y = 2.093$, $D = 0.345$ cm^{-1}, $E = 0.007$ cm^{-1}, $A_z = 0.008$ cm^{-1}, and A_x, $A_y < 0.001$ cm^{-1}. The influence that the J, D, and E parameters have on the epr spectrum is illustrated in Fig. 13–27. With $|J| = 260$ cm^{-1} in Cu$_2$(AOc)$_4$, transitions from the $S = 0$ to $S = 1$ state are not observed in the epr. The exchange interaction gives rise to a lower energy $S = 0$ state, so the intensity of the signals decreases with decreased temperature. This temperature dependence indicates a J of -260 cm^{-1}, corresponding to a separation of the $S = 0$ and $S = 1$ states of 2J or 520 cm^{-1}. In the powder spectrum discussed earlier, split $g_\parallel$ and $g_\perp$ peaks arise from the two $\Delta M_S = \pm 1$ transitions averaged over the orientations. In the relatively rare situation where the exchange parameter J is smaller than the available microwave energy of the epr experiment, it is possible to see epr transitions between the $S = 1$ and $S = 0$ electronic states of a copper dimer. The first example involved a copper dimer and was reported[50] for the "outer-sphere"

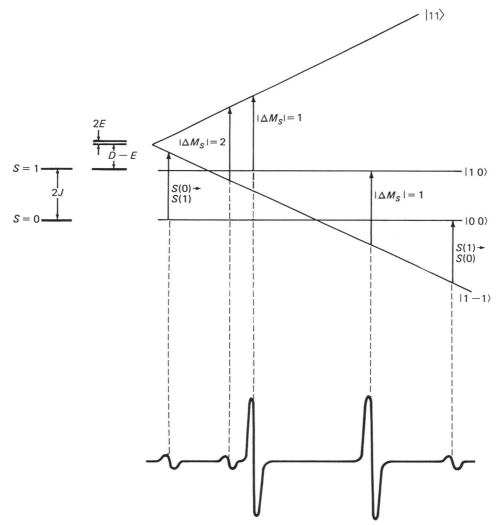

FIGURE 13–27 The influence of J interactions and zero-field effects on the energy levels and single-crystal epr spectrum of a molecule containing two d^9 copper atoms.

dimers in $[Cu_2(tren)_2X_2](BPh_4)_2$, where $X = NCO^-$ and NCS^- and tren is 2,2′,2″-triaminotriethylamine. The "outer-sphere" dimeric association occurs between two Cu(II) trigonal bipyramids by virtue of hydrogen bonding between the uncoordinated O(S) end of $OCN^-(SCN^-)$ nitrogen bonded to one copper and an N—H proton of the coordinated tren. In the case of X-band epr, J-values of ~ 0.15 cm^{-1} to ~ 0.05 cm^{-1} can be gauged by the observation of singlet-to-triplet transitions as illustrated in Fig. 13–27.

13–7 DOUBLE RESONANCE AND FOURIER TRANSFORM EPR TECHNIQUES

Consider an experiment in which the epr transition is broadened by coupling to an ^{17}O nucleus. If one sweeps the nmr ^{17}O frequencies in a decoupling experiment and monitors the epr intensity, a plot of this intensity produces the ^{17}O nmr spectrum. This is referred to as an ENDOR experiment[51] (electron-nucleus double resonance) and combines the inherent sensitivity of the epr experiment with the resolution of the nmr. One generally performs this experiment at several different epr frequencies (i.e., g-values) to maximize the information obtained. A recent report[52] involved iron-sulfur cluster interactions with a ^{17}O-labeled substrate. A broad epr peak at $g = 1.88$ is monitored as the ^{17}O frequency was swept over ~ 10 MHz. Ten peaks corresponding to two five-line patterns from non-equivalent ^{17}O ligands ($I = ^5/_2$) are seen in the ENDOR spectrum even though no ^{17}O hyperfine is observed in the epr.

ELDOR is a double-resonance experiment in which the epr is examined while another electron spin transition is saturated. This application is similar in concept to the nmr decoupling experiment. Intensity increases from the nuclear Overhauser effect can result.

Fourier transform epr provides some of the advantages of FTNMR. The measurement of electron spin lifetimes is possible. The irradiation time must be significantly shorter than T_1 and T_2. Using available pulses as short as 10 nanoseconds coupled with a dead time allowance of 50 to 100 nanoseconds permit the measurement of T_1-values of 10^{-6} to 10^{-7} seconds.[53]

A second application of FTEPR is electron spin echo envelope modulation spectroscopy.[54] One selects a field (i.e., g-value) in the epr spectrum and, with a sequence of microwave pulses, generates a spin echo. The intensity of the echo is modulated as a function of the delay between pulses due to the magnetic interaction of nuclear spins coupled to electron spins. Fourier transform leads to an ENDOR-like spectrum containing nuclear hyperfine and quadrupole splittings. In a three-pulse[54c] sequence it is possible to suppress one nuclear modulation frequency enabling one to eliminate undesired hyperfine couplings and focus on those of interest. For example, one can see deuterium modulation while suppressing proton modulation when both nuclei are present in a sample.

REFERENCES CITED

1. See M. Tinkham, "Group Theory and Quantum Mechanics," p. 143, McGraw-Hill, New York (1964).
2. D. M. S. Bagguley and J. H. E. Griffiths, Nature, *162*, 538 (1948).
3. J. P. Fackler and J. A. Smith, J. Amer. Chem. Soc., *92*, 5787 (1970); J. P. Fackler, J. D. Levy, and J. A. Smith, *ibid.*, *94*, 2436 (1972).

4. **a.** B. M. Hoffman, F. Basolo, and D. L. Diemente, J. Amer. Chem. Soc., *95*, 6497 (1973).
   b. F. Cariati, F. Morazzoni, C. Busetto, D. Del Piero, and A. Zazzetta, J. Chem. Soc. Dalton, Trans., 556 (1975).
5. P. T. Manohanon and M. T. Rogers in "Electron Spin Resonance of Metal Complexes," ed. Teh Fu Yen, Plenum Press, New York, 1969.
6. B. Bleaney and D. J. Ingram, Proc. Phys. Soc., *63*, 408 (1950).
7. A. Abragam and M. H. L. Pryce, Proc. Phys. Soc., *63*, 409 (1950).
8. B. Bleaney and K. D. Bowers, Proc. Phys. Soc., *65*, 667 (1952).
9. B. Bleaney, K. D. Bowers, and R. S. Trenam, Proc. Roy. Soc. (London), *A228*, 157 (1955).
10. H. C. Allen, Jr., G. F. Kokoszka, and R. G. Inskeep, J. Amer. Chem. Soc., *86*, 1023 (1964).
11. For other examples, see F. S. Ham in "Electron Paramagnetic Resonance," ed. S. Geschwind, Plenum Press, New York, 1972.
12. R. C. Koch, M. D. Joesten, and J. H. Venable, Jr., J. Chem. Phys. *59*, 6312 (1973).
13. R. Kirmse *et al.* J. Chem. Phys., *56*, 5273 (1972).
14. E. Buluggiu and A. Vera, J. Chem. Phys., *59*, 2886 (1973).
15. J. R. Pilbrow and M. E. Winfield, Mol. Phys., *25*, 1073 (1973).
16. **a.** Yonetoni *et al.*, J. Biol. Chem., *245*, 2998 (1970).
    b. J. W. Orton *et al.*, Proc. Phys. Soc., *78*, 554 (1961).
17. **a.** M. C. R. Symons and J. G. Wilkerson, J. Chem. Soc. (A), 2069 (1971).
    b. A. H. Maki, N. Edelstein, A. Davison, and R. H. Holm, J. Amer. Chem. Soc., *86*, 4580 (1964).
18. B. A. Goodman and J. B. Raynor, Adv. Inorg. Chem. Radiochem., *13*, 135 (1970).
19. P. W. Atkins and M. C. R. Symons, "The Structure of Inorganic Radicals," Elsevier, New York, 1967.
20. W. V. Sweeney and R. E. Coffman, J. Phys. Chem., *76*, 49 (1972).
21. D. A. C. McNeil, J. B. Raynor, and M. C. R. Symons, Proc. Chem. Soc., 364 (1964).
22. B. B. Wayland and M. E. Abd-Elmageed, J. Amer. Chem. Soc., *96*, 4809 (1974).
23. R. M. Golding, "Applied Wave Mechanics," Van Nostrand, New York (1969).
24. B. R. McGarvey, "Electron Spin Resonance of Transition Metal Complexes," in "Transition Metal Chemistry," Vol. 3, pp. 89–201, ed. R. L. Carlin, Marcel Dekker, New York (1967).
25. S. Foner and W. Low, Phys. Rev., *120*, 1585 (1960).
26. E. Pedersen and H. Toftlund, Inorg. Chem., *13*, 1603 (1974).
27. R. H. Borcherts and C. Kikuchi, J. Chem. Phys., *40*, 2270 (1964).
28. C. K. Jørgensen, Acta Chem. Scand., *8*, 1686 (1957).
29. A. Carrington and A. D. McLachlan, "Introduction to Magnetic Resonance," Harper & Row, New York, 1967.
30. W. C. Lin, C. A. McDowell, and D. J. Ward, J. Chem. Phys., *49*, 2883 (1968).
31. D. J. E. Ingram, "Biological and Biochemical Applications of Electron Spin Resonance," Plenum Press, New York, 1969,
32. R. E. DeSimone and R. S. Drago, J. Amer. Chem. Soc., *92*, 2343 (1970).
33. G. C. Brackett, P. L. Richards, and W. S. Caughey, J. Chem. Phys., *54*, 4383 (1971).
34. **a.** R. Aasa and T. Vanngard, Arkiv Kemi, *24*, 331 (1965).
    b. R. Aasa, J. Chem. Phys., *52*, 3919 (1980).
    c. R. D. Dowsing and J. F. Gibson, J. Chem. Phys., *50*, 294 (1968).
    d. J. R. Pilbrow, J. Mag. Res., *19*, 308 (1975).
    e. J. Peisach "Frontiers of Biological Science," ed. P. L. Dutton et al., Vol. 2, Academic Press, New York, 1978.
    f. M. Sato and H. Kon, Chem. Phys., *12*, 199 (1976).
35. F. S. Kenedy *et al.*, Biochem. Biophys. Res. Comm., *48*, 1533 (1972).
36. B. B. Wayland, M. E. Abd-Elmageed, and L. F. Mehne, submitted.
37. B. Tovrog, D. J. Kitko, and R. S. Drago, J. Amer. Chem. Soc., *98*, 5144 (1976).
38. R. S. Drago and J. H. Gaul, Inog. Chem., *18*, 2019 (1979).
39. F. Urbach, J. Amer. Chem. Soc., *96*, 5063 (1974).

40. **a.** B. R. McGarvey, Can. J. Chem., *53*, 2498 (1975).
    b. V. Malatesta and B. R. McGarvey Can. J. Chem., *53*, 3791 (1975).
    c. A. Reuveni, V. Malatesta, and B. R. McGarvey Can. J. Chem., *55*, 70 (1977).
41. R. S. Drago and B. B. Corden, Accts. Chem. Res., *13*, 353 (1980) and references therein.
42. T. D. Smith and J. R. Pilbrow, Coord. Chem. Rev., *39*, 295 (1981).
43. W. B. Lewis and L. O. Morgan, "Transition Metal Chemistry," Vol. 4, p. 33, ed. R. L. Carlin, Dekker, New York, 1968.
44. T. Maryia in "Magnetism" eds. G. T. Rado and H. Suhl, Vol. 1, Academic Press, New York, 1963.
45. J. Scaringe, D. J. Hodgson, and W. E. Hatfield, Mol. Phys., *35*, 701 (1978).
46. **a.** D. Gatteschi "The Coordination Chemistry of Metalloenzymes," eds. I. Bertini, R. S. Drago, and C. Luchinat, D. Reidel, Holland, 1983.
    b. L. Banci, A. Bencini, and D. Gatteschi, Inorg. Chem., *20*, 2734 (1981).
47. J. A. Paulson, D. A. Knost, G. L. McPherson, R. D. Rogers, J. L. Atwood, Inorg. Chem., *19*, 2519 (1980).
48. G. F. Kokoszka and R. W. Duerst, Coord. Chem. Rev., *5*, 209 (1970).
49. B. Bleaney and K. D. Bowers, Proc. Roy. Soc. (London), *A214*, 451 (1952).
50. D. M. Duggan and D. N. Hendrickson, Inorg. Chem., *13*, 2929 (1974).
51. J. S. Hyde, R. C. Sneed, and G. H. Rist, J. Chem. Phys., *51*, 1404 (1969) and references therein.
52. J. Telser, M. H. Emptage, H. Merkle, M. C. Kennedy, H. Beinert, and B. M. Hoffman, J. Biol. Chem., *261*, 4840 (1986).
53. **a.** R. Orbach and H. J. Stapleton, "Electron Spin-Lattice Relaxation," in "Electron Paramagnetic Resonance" ed. by S. Geschwind, Plenum Press, New York 1972.
    b. H. J. Stapleton, Mag. Reson. Rev., *1*, 65 (1972).
54. **a.** L. Kevan in "Time Domain Electron Spin Resonance," eds. L. Kevan and R. N. Schwartz, Wiley-Interscience, New York, 1979.
    b. L. Kevan, J. Phys. Chem., *85*, 1628 (1981).
    c. L. Kevan, Accts Chem. Res., *20*, 1 (1987).
    d. W. B. Mims and J. Peisach, J. Biol. Mag. Reson., Vol. 3, eds. J. Berliner and J. Ruben, Plenum Press, New York, 1981.

ADDITIONAL REFERENCES

J. E. Wertz and J. R. Bolton, "Electron Spin Resonance," McGraw-Hill, New York, 1972.

A. Abragam and B. Bleaney, "EPR of Transition Ions," Clarendon Press, Oxford, England, 1970.

J. W. Orton, "Electron Paramagnetic Resonance," Gordon and Breach, New York, 1968.

R. M. Golding, "Applied Wave Mechanics," Van Nostrand, New York, 1969.

J. S. Griffith, "The Theory of Transition-Metal Ions," Cambridge University Press, New York, 1961.

A. Carrington and A. D. McLachlan, "Introduction to Magnetic Resonance," Harper & Row, New York, 1967.

C. P. Poole, "Electron Spin Resonance," Interscience, New York, 1967.

B. R. McGarvey, "Electron Spin Resonance of Transition Metal Complexes," in "Transition Metal Chemistry," Vol. 3, pp. 89–201, ed. R. L. Carlin, Dekker, New York, 1967.

G. Kokoszka and G. Gordon, "Technique of Inorganic Chemistry," Vol 7, pp. 151–271, ed. H. B. Jonassen and A. Weissberger, Interscience, New York, 1968.

B. A. Goodman and J. B. Raynor, Adv. Inorg. Chem. Radiochem., *13*, 135 (1970).

R. Bersohn, "Determination of Organic Structures by Physical Methods," Vol. 2, eds. F. C. Nachod and W. D. Phillips, Academic Press, New York, 1961.

A. Carrington, Quart. Rev., *17*, 67 (1963).

A. Carrington and H. C. Longuet-Higgins, Quart. Rev., *14*, 427 (1960).

EXERCISES

1. Below is a typical Cu^{2+} solution spectrum. It can be characterized by $g = (1/3)g_{\parallel} + 2/3 g_{\perp}$ and $A = 1/3 A_{\parallel} + 2/3 A_{\perp}$.

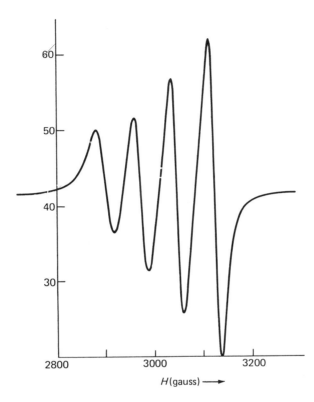

a. Determine g and A.

b. This is an X-band spectrum (9.4×10^9 Hz). Why should A be expressed in Hertz?

2. Readers often find it difficult to understand why the epr spectrum of a powder or frozen solution should yield (in an axially symmetric compound, for example) $g_{\parallel}$, $g_{\perp}$, $A_{\parallel}$, and $A_{\perp}$. "After all," the argument goes, "you are looking at a superposition of all possible orientations. At intermediate orientations you have an average g and A, e.g., $g^2 = g_{\parallel}^2 \cos^2 \theta + g_{\perp}^2 \sin^2 \theta$. Thus, the spectrum should be very broad with few features,." This argument is correct as far as it goes, but it ignores the fact that we are looking at a first derivative. For example, suppose that a compound has $g_{\parallel} = 2.0$ and $g_{\perp} = 2.1$. Recalling that there are many more perpendicular directions than there are parallel ones, the absorption spectrum should appear in the ideal case as in (A):

Because of line broadening, the real absorption curve resembles that in (B). The first derivative will then be

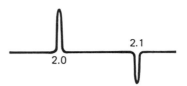

The argument is further complicated, but essentially unchanged, by including hyperfine coupling.

Below is the powder spectrum of copper diethyldithiophosphinate (for ^{63}Cu, $I = 3/2$). Explain as many features as possible. Hint: $A_\parallel(\text{Cu}) \approx 5A_\perp(\text{Cu})$. What other nuclei might give hyperfine splitting?

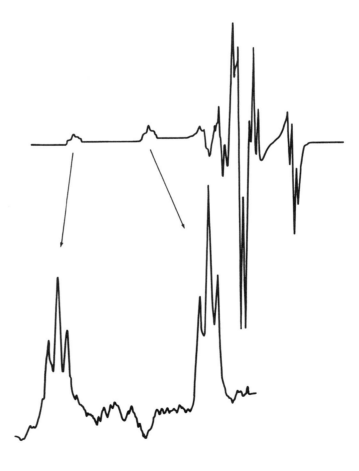

3. The solution spectrum below is typical for a vanadyl complex (for V, $I = 7/2$). Explain the splitting pattern.

[VO(glycolate)$_2$]$^{2-}$ in H$_2$O — DPPH

100 G   H→

4. Below are the frozen solution (top) and room temperature (bottom) spectra of a vanadyl complex of *d,l*-tartrate (only one species is present). What can you deduce about the structure of this complex? (Tartaric acid has the formula

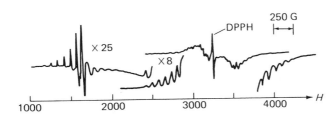
)

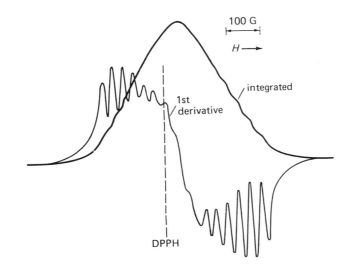

5. Diphenylpicrylhydrazyl (DPPH) is a common reference in epr spectroscopy. Its structure is

In powder or concentrated xylene solution, it gives a sharp one-line pattern shown at top left. In a 10^{-3}-M xylene solution, it gives the pattern shown at the left. How does this splitting pattern arise? Why is it present only in dilute solution?

6. Copper(II) tetraphenylporphyrin [CuTPP; see part (D) below.] has a square planar D_{4h} structure with the copper lying in the plane of the four equivalent porphyrin nitrogens. EPR spectra of a sample of CuTPP doped into the free ligand are shown below. This sample is 100% ^{63}Cu. Natural abundance copper is 69% ^{63}Cu and 31% ^{65}Cu, and $I = 3/2$ for both isotopes.

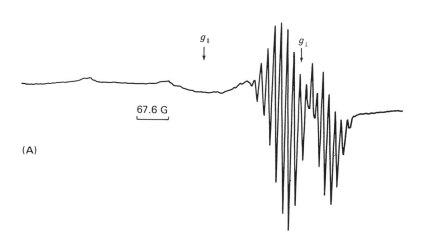

(A)

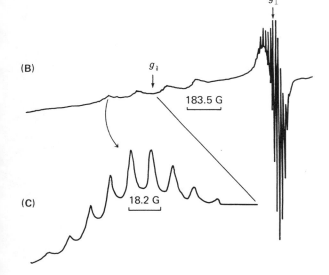

(B)

(C)

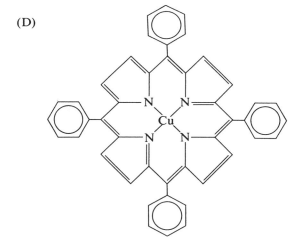

(D)

a. Spectrum (A) is a powder spectrum obtained at a frequency of 9.4 GHz (X-band), while spectrum (B) was obtained at 35 GHz (Q-band). Why are $g_\parallel$ and $g_\perp$ well separated in (B) compared with those in (A)?

b. Explain the source of the number of peaks observed in the $g_\parallel$ region in spectrum (B). Also explain the splitting pattern found for each peak [see insert (C)].

c. What is the crystal field splitting diagram for a square planar complex? How is observation of the splittings in insert (C) consistent with the orbital occupancy predicted for a Cu(II) complex by your crystal field diagram?

d. Qualitatively, if this copper compound did not have a threefold or higher axis of symmetry, what would the powder spectrum look like?

e. If natural abundance copper were used, what effect would this have on the appearance of insert (C)?

Note that insert (C) is obtained by increasing the gain for the set of low-field lines.

7. Below is a spectrum of cobalt phthalocyanine (for Co, $I = 7/2$), a D_{4h} Co(II) porphyrin similar to the copper porphyrin in problem 6.

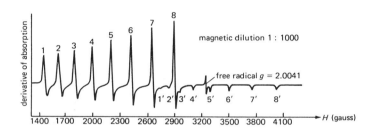

a. What are the values of $g_\parallel$, $g_\perp$, $A_\parallel$, and $A_\perp$? What is the origin of the observed hyperfine structure?

b. Dissolving cobalt phthalocyanine in 4-methylpyridine produces a 1:1 adduct. The frozen solution spectrum at 77 K is shown below. Utilizing the D_{4h} splitting diagram for low spin Co(II), state what orbital the unpaired electron occupies. Why is each of the eight upfield components split into three lines?

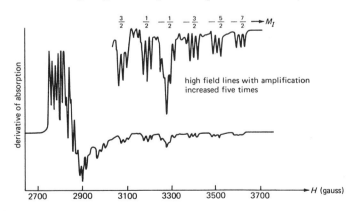

c. Why do you see super-hyperfine interaction from the pyridine nitrogen but not from the four phthalocyanine donor nitrogen atoms?

8. Predict the number of spectral lines for

a. $Co(H_2O)_6^{2+}$

b. $Cr(H_2O)_6^{3+}$

Indicate how zero-field splitting and Kramers' degeneracy applies in these examples.

9. The following epr data on a series of high-spin octahedral metal hexafluoride complexes are taken from Proc. Roy. Soc. (London), *236*, 535 (1956):

Complex	g	A_{metal} $\times 10^4$ cm^{-1}	A_F $\times 10^4$ cm^{-1}	I_{metal}	Temp.
MnF_6^{4-}	$g_x = g_y = g_z = 2.00$	96	17	$^{55}Mn = ^5/_2$	300 °K
CoF_6^{4-}	$g_x = 2.6$ $g_y = 6.05$ $g_z = 4.1$	$A_x = 43$ $A_y = 217$ $A_z = 67$	$A_x = 20$ $A_y = 32$ $A_z = 21$	$^{59}Co = ^7/_2$	20 °K
CrF_6^{3-}	$g = 2.00$ $g = 1.98$	$A = 16.2$ $A = 16.9$	$A = 3$ $A = 1$	$^{53}Cr = ^3/_2$	300 °K

a. Why do CrF_6^{3-} and MnF_6^{4-} give sharp epr at room temperature while CoF_6^{4-} does so only at 20 K? What effect or effects cause these differences in ability to observe the epr? Which of these complexes would be best for a room temperature nmr study?

b. Why are the CrF_6^{3-} and MnF_6^{4-} g-value fairly isotropic and close to 2.0 while the CoF_6^{4-} values are anisotropic and deviate from 2.0?

c. Why do CoF_6^{4-} and MnF_6^{4-} have larger A_F values than CrF_6^{3-}?

10. The November, 1973, issue of *Inorganic Chemistry* reports the liquid and solid solution epr spectra of some 10^{-3} M vanadyl dithiophosphinate complexes:

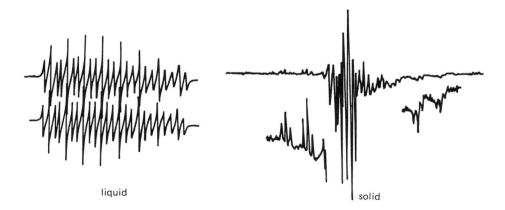

a. Explain the liquid spectrum, for which $R = CH_3$. (The lower one is a computer simulation.) What parameters are needed to characterize it? V has $I = ^7/_2$, P has $I = ^1/_2$, and H has $I = ^1/_2$.

b. Explain the solid solution spectrum, for which R = phenyl. What parameters are needed to characterize it?

c. What should be the mathematical relationship between the parameters in part a and those in part b?

11. The spectrum below is that of an axially symmetric Cu^{2+} complex in a frozen solution. It is a d^9 system. Assume 100% abundance for ^{63}Cu ($I = 3/2$). The following constants will be needed: $\beta = 9.27 \times 10^{-21}$ erg/gauss; $h = 6.67 \times 10^{-27}$ erg sec; $\nu = 9.12 \times 10^9$ Hz for the spectrometer on which the spectrum was obtained.

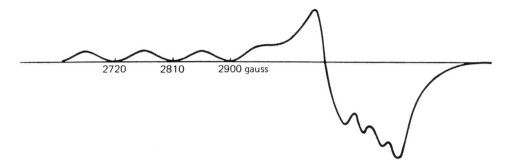

a. How many parameters are required to explain this spectrum? What features of the spectrum suggest these parameters?

b. Suggest why the high-field half of the spectrum ($\perp$ region) is more intense.

c. What is the absolute value of $g_\parallel$? Justify the terms in the equation you use.

12. Assume that the square pyramidal complex $Cu(hfac)_2 P(C_6H_5)_3$ (for ^{31}P, $I = 1/2$; for ^{63}Cu, $I = 3/2$) can exist as either of the following isomers [Inorg. Chem., *13*, 2517 (1974)]:

What might be learned to aid in distinguishing the isomers by looking at

a. the electronic spectrum?

b. the infrared and Raman spectra (answer in a general way; i.e., don't work out the total representation)?

c. the epr spectrum?

d. an ^{17}O-labeled epr spectrum (for ^{17}O, $I = 5/2$)?

e. the nmr spectrum?

14 Nuclear Quadrupole Resonance Spectroscopy, NQR*

14–1 INTRODUCTION

When a nucleus with an electric quadrupole moment (nuclear spin $I \geq 1$; see the second paragraph of Chapter 7 and Fig. 7–1) is surrounded by a non-spherical electron distribution, *the quadrupolar nucleus will interact with the electric field gradient* from the asymmetric electron cloud to an extent that is different for the various possible orientations of the elliptical quadrupolar nucleus. Since the quadrupole moment arises from an unsymmetric distribution of electric charge in the nucleus, it is an electric quadrupole moment rather than a magnetic moment that concerns us. The allowed nuclear orientations are quantized with $2I + 1$ orientations, described by the nuclear magnetic quantum number m, where m has values $+I$ to $-I$ differing by integer values. The quadrupole energy level that is lowest in energy corresponds to the orientation in which the greatest amount of positive nuclear charge is closest to the greatest density of negative charge in the electron environment. The energy differences of the various nuclear orientations are not very great, and at room temperature a distribution of orientations is populated in a group of molecules.

If the electron environment around the quadrupolar nucleus is spherical (as in free Cl^-), all nuclear orientations are equivalent and the corresponding quadrupole energy states are degenerate. If the nucleus is spherical ($I = 0$ or $1/2$), there are no quadrupole energy states even with a non-spherical electron cloud. In nqr spectroscopy, we study the energy differences of the non-degenerate nuclear orientations. These energy differences generally correspond to the radio frequency region of the spectrum, *i.e.*, ~ 0.1 to 700 MHz.

It is helpful to consider the interaction of charges, dipoles, and quadrupoles with negative electron density in order to define some terms important for nqr (and Mössbauer spectroscopy, Chapter 15). In Fig. 14–1(A), we illustrate the interaction of a positive charge on the z-axis with negative electron density. The energy is given by $-e^2/r$ or $-eV$, where $V = -e/r$ is the electron potential felt by the positive charge located at the point r. In Fig. 14–1(B), we represent a dipole moment in the field of electronic charge. Now the energy associated

---

*Several of the principles covered in Sections 14–1 to 14–4 are common to both nqr and Mössbauer spectroscopy, so these sections should be read before Chapter 15.

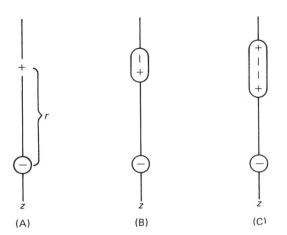

FIGURE 14–1 The interaction of (A) a positive charge, (B) a dipole, and (C) a quadrupole with the z-component of the electric field arising from a unit negative charge ⊖.

with the orientation of the dipole depends upon how the potential energy changes over the dipole. Thus, we are interested in how the electrostatic potential changes over the dipole or $\partial V/\partial z$. This is referred to as the z-component of the electric field, E_z. In Fig. 14–1(C), we illustrate the interaction of a quadrupolar distribution with the electric field. Now we have the electric field from the electron in effect interacting with two dipoles whose configuration relative to each other is fixed, *i.e.*, a quadrupole. The energy will depend upon the rate of change (or gradient) of the electric field over the quadrupole. Thus, we are concerned with the "change in the change" of the potential from the electron, or the second derivative of V with respect to z, that is, $\partial^2 V/\partial z^2$. This quantity, which is also the change in the electric field component, $\partial E_z/\partial z$, is called the *electric field gradient*.

In our molecule we have a nucleus imbedded in a charge cloud of electron density. The electric field gradient is defined in terms of a time-averaged electric potential from an electron. Furthermore, the electron field gradient is described by a 3 × 3 tensor **V**, which is symmetric and has a zero trace. The *nuclear quadrupole moment* is also described by a 3 × 3 tensor **Q**. The nuclear quadrupole coupling energy E_Q is given by

$$E_Q = -\frac{e}{6} \mathbf{Q}_{ij} \mathbf{V}_{ij}$$

where $\mathbf{Q}_{ij}$ is the nuclear quadrupole moment tensor and $\mathbf{V}_{ij}$ is the electrostatic field gradient tensor arising from the electron distribution. The product will depend upon the mutual orientation of the two axis systems. For **Q**, it is convenient to select an axis system that coincides with that of the spin system. When this is done, the cylindrical symmetry of the nucleus permits definition of the tensor in terms of one parameter, the nuclear quadrupole moment Q. The sign of this quantity must be known to obtain the sign of the electric field gradient. An axis system can also be chosen in which the electrostatic field gradient tensor is diagonal. This is called the *principal axis system of the field* *gradient tensor*, and the only non-zero elements are the diagonal elements whose magnitudes produce a traceless tensor:

$$\frac{\partial^2 V}{\partial x^2} + \frac{\partial^2 V}{\partial y^2} + \frac{\partial^2 V}{\partial z^2} = 0$$

When comparing a given atom in different molecules, it is necessary to know the orientation of the field gradient principal axis system in the molecular framework axis system. Three Eulerian angles (α, β, and γ) are required.

The asymmetry of a molecule and the direction of the z-axis of the field gradient, q_{zz}, relative to the crystal axes can be investigated by studying the nqr spectrum of a single crystal in a magnetic field. The Zeeman splitting is a function of orientation, and detailed analysis of the spectra for different orientations enables one to determine the direction of the z-axis of the field gradient, q_{zz}. This axis can be compared to the crystal axes.

When the principal axes of the coordinate system of the molecule are principal axes of the electric field gradient tensor, the potential energy E_Q for interaction of the quadrupole moment with the electric field gradient at the nucleus is given by

$$E_Q = \frac{e}{6}(V_{xx}Q_{xx} + V_{yy}Q_{yy} + V_{zz}Q_{zz}) \tag{14-1}$$

We define the electric field gradient V_{zz} as eq_{zz}, where e is the electron charge (4.8×10^{-10} esu). Since the trace of the electric field gradient tensor is zero, we need define only one more quantity to specify the field gradient, and this is done in equation (14–2):

$$\eta = \frac{V_{xx} - V_{yy}}{V_{zz}} \tag{14-2}$$

The quantity η is called the *asymmetry parameter*. The quantities V_{xx}, V_{yy}, and V_{zz} are often written as eq_{xx}, eq_{yy}, and eq_{zz}. By convention, $|q_{zz}| > |q_{xx}| > |q_{yy}|$, so η ranges from 0 to 1 as a result. With a zero trace, the field gradient is completely defined by eq and η.

Substituting these definitions of q_{zz} and η into equation (14–1) and using the fact that $q_{zz} + q_{xx} + q_{yy} = 0$, we obtain

$$E_Q = \frac{e^2 q}{6}\left[\frac{1}{2}(\eta - 1)Q_{xx} - \frac{1}{2}(\eta + 1)Q_{yy} + Q_{zz}\right] \tag{14-3}$$

When axial symmetry pertains, η equals zero and E_Q becomes equal to $(1/4)e^2qQ_{zz}$. (Note that $Q_{zz} + Q_{xx} + Q_{yy} = 0$.)

The classical considerations given above are readily expressed by a quantum mechanical operator:

$$\hat{H}_Q = \left(\frac{e^2}{6}\right)\sum_{i,j} q_{ij}Q_{ij} \tag{14-4}$$

where the summation is over the components of the nuclear quadrupole moments Q_{ij} and the electric field gradients q_{ij}. In the principal axis system with η defined as above, the most common form of the Hamiltonian is given by:

$$\hat{H}_Q = \frac{e^2 Qq}{4I(2I-1)}\left[3\hat{I}^2 - I(I+1) - \frac{\eta}{2}(\hat{I}_+^2 + \hat{I}_-^2)\right] \tag{14-5}$$

The product e^2Qq or e^2Qq/h (often written as eQq_{zz} or eQq_{zz}/h) is called the quadrupole coupling constant. The operator $\hat{H}_Q$ operates on the nuclear wave functions. When $\eta = 0$, the terms involving the raising and lowering operators drop out. We shall not be concerned with the explicit evaluation of matrix elements; the interested reader can consult references 1–3. Suffice it to say that a series of secular equations can be written and solved to give the energies of the nuclear spin states in the electric field gradient resulting from the distribution of the electron density of the molecule.

14–2 ENERGIES OF THE QUADRUPOLE TRANSITIONS

In an axially symmetric field ($\eta = 0$), the energies of the various quadrupolar nuclear states are summarized by the following equation:

$$E_m = \frac{e^2Qq[3m^2 - I(I+1)]}{4I(2I-1)} \tag{14-6}$$

where I is the nuclear spin quantum number, and m is the nuclear magnetic quantum number. For a nuclear spin of $I = {}^3/_2$, m can have values of ${}^3/_2$, ${}^1/_2$, $-{}^1/_2$, $-{}^3/_2$. For $m = {}^3/_2$, substitution into equation (14–6) produces the result $E_{3/2} = +e^2Qq/4$. Since m is squared, the value for $m = -{}^3/_2$ will be identical to that for $m = +{}^3/_2$, and a doubly degenerate set of quadrupole energy states results. Similarly, the state from $m = \pm{}^1/_2$ will be double degenerate. The transition energy, ΔE, indicated by the arrow in Fig. 14–2, corresponds to $e^2Qq/4 - (-e^2Qq/4) = e^2Qq/2$. Thus, for a nucleus with a spin $I = {}^3/_2$ in an axially symmetric field, a single transition is expected, and the quantity e^2Qq expressed in energy units can be calculated directly from the frequency of absorption $e^2Qq = 2\Delta E = 2h\nu$. The quantity e^2Qq is often expressed as a fre-

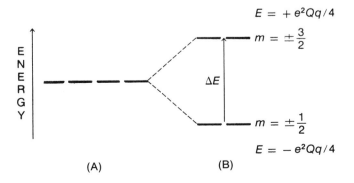

FIGURE 14–2 Quadrupole energy levels in a spherical field (A) and an axially symmetric field (B).

quency in MHz, although strictly speaking this should be e^2Qq/h. For the above case, e^2Qq would be twice the frequency of the nqr transition.

The number of transitions and the relationship of the frequency of the transition to e^2Qq can be calculated in a similar manner for other nuclei with different I values in axially symmetric fields by using equation (14–6). For $I = {}^7/_2$, four energy levels ($E_{\pm 1/2}, E_{\pm 3/2}, E_{\pm 5/2}$, and $E_{\pm 7/2}$) and three transitions result. The selection rule for these transitions is $\Delta m = \pm 1$, so the observed transitions are $E_{\pm 1/2} \to E_{\pm 3/2}$, $E_{\pm 3/2} \to E_{\pm 5/2}$, and $E_{\pm 5/2} \to E_{\pm 7/2}$. (Recall that all levels are populated under ordinary conditions). Substitution of I and m into equation (14–6) produces the result that the energy of the $E_{\pm 3/2} \to E_{\pm 5/2}$ transition is twice that of the $E_{\pm 1/2} \to E_{\pm 3/2}$ transition. The energies of these levels and the influence of the asymmetry parameter η on these energies are illustrated in Fig. 14–3. In measured spectra, deviations from the frequencies predicted when $\eta = 0$ are attributed to deviations from axial symmetry in the sample, and, as will be seen shortly, this deviation can be used as a measure of asymmetry.

FIGURE 14–3 Nuclear quadrupole energy level diagram for $I = {}^7/_2$.

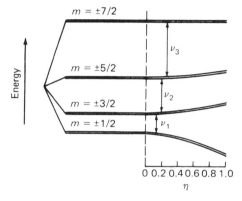

In a nuclear quadrupole resonance (nqr)* experiment, radiation in the radio frequency region is employed to effect transitions among the various orientations of a quadrupolar nucleus in a non-spherical field. The experiment is generally carried out on a powder. Different orientations of the small crystals relative to the r.f. frequency direction affect only the intensities of the transitions but not their energies. Structural information about a compound can be obtained by considering how different structural and electronic effects influence the asymmetry of the electron environment. One set of resonances is expected for each chemically or crystallographically inequivalent quadrupolar nucleus. Crystallographic splittings are often small compared to splittings from chemical nonequivalence.

Two types of oscillators have been commonly used in nqr, the superregenerative and the marginal oscillator. The superregenerative oscillator is most common because it allows broad band scanning in searching for resonances and is not complicated to operate. It has the disadvantage of producing a multiplet of lines for each resonance, as shown in Fig. 14–4(A), because of its particular

---

* For reviews of nqr, see the General References.

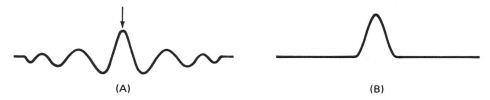

FIGURE 14–4 (A) Multiplet of peaks for a single resonance from a superregenerative oscillator; (B) single peak from a marginal oscillator for the same resonance as in (A).

operational characteristics. The true resonance frequency is the center line of the multiplet. The marginal oscillator gives a single peak for each absorption, but it requires constant adjustment and is tedious to operate. Details regarding the instrumentation have been published.[4]

The ^{35}Cl nqr spectrum of Cl_3BOPCl_3 is shown in Fig. 14–5. Resonances for three non-equivalent chlorines ($Cl_3POBCl_2{}^+Cl^-$) are found. Two of the resonance centers are indicated by × marks on the spectrum at 30.880 MHz and 31.280 MHz. The center of the third resonance around 30.950 MHz is difficult to determine accurately because of overlap with the resonance at 31.280 MHz.

In addition to this direct measurement of the quadrupolar energy level difference by absorption of radio frequency radiation, the same information may also be obtained from the fine structure in the pure rotation (microwave)

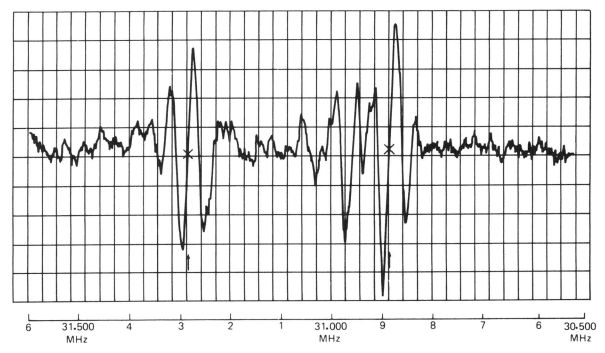

FIGURE 14–5 The ^{35}Cl nqr spectrum of Cl_3BOPCl_3 at 77 K with 25-kHz markers.

spectrum of a gas. The different nuclear orientations give slightly different moments of inertia, resulting in fine structure in the microwave spectrum. The direct measurement by absorption of radio frequency radiation must be carried out on a solid. In a liquid or even in some solids (especially near the melting point), collisions and vibrations modulate the electric field gradient to such an extent that the lifetime of a quadrupole state becomes very short. This leads to uncertainty broadening, and the line is often not detected.

The energy difference between the various levels and, hence, the frequency of the transition will depend upon both the field gradient, q, produced by the valence electrons and the quadrupole moment of the nucleus. The quadrupole moment, eQ, is a measure of the deviation of the electric charge distribution *of the nucleus* from spherical symmetry. For a given isotope, eQ is a constant, and values for many isotopes can be obtained from several sources.[5,6] They can be measured with atomic beam experiments. The units of eQ are charge times distance squared, but it is common to express the moment simply as Q in units of cm². For example, ^{35}Cl with a nuclear spin $I = {}^3/_2$ has a quadrupole moment Q of -0.08×10^{-24} cm², the negative sign indicating that the charge distribution is flattened relative to the spin axis (see Fig. 7–1).

The second factor determining the extent of splitting of the quadrupole energy levels is the *field gradient, q,* at the nucleus produced by the *electron distribution* in the molecule. The splitting of a quadrupole level will be related to the product e^2Qq. For a molecule with axial symmetry, q often lies along the highest-fold symmetry axis, and when eQ is known, one can obtain the value of q. In a non-symmetric environment, the energies of the various quadrupole levels are no longer given by equation (14–6), because the full Hamiltonian in equation (14–5) must be used. For $I = {}^3/_2$, the following equations can be derived[2] for the energies of the two states:

$$E_{\pm 3/2} = \frac{3e^2Qq\sqrt{1 + \eta^2/3}}{4I(2I - 1)} \tag{14–7}$$

$$E_{\pm 1/2} = \frac{-3e^2Qq\sqrt{1 + \eta^2/3}}{4I(2I - 1)} \tag{14–8}$$

From the difference of equations (14–7) and (14–8) and $E = h\nu$, it is seen that one transition with frequency $\nu = (e^2Qq/2h)\sqrt{1 + \eta^2/3}$ is expected for $I = {}^3/_2$. Since there are two unknowns, η and q, the value for e^2Qq cannot be obtained from a measured frequency. As will be seen shortly, this problem can be solved with results from nqr experiments on a sample in a weak magnetic field.

The equations for nuclei with I values other than ${}^3/_2$ have been reported, and in many instances where more than one line is observed in the spectrum, both the asymmetry parameter, η, and e^2Qq can be obtained from the spectrum. For $I = 1$, the equations are:

$$E_0 = \frac{-2e^2Qq}{4I(2I - 1)} \tag{14–9}$$

and

$$E_{\pm 1} = \frac{e^2Qq(1 \pm \eta)}{4I(2I-1)} \quad (14\text{--}10)$$

The corresponding energy levels are indicated in Fig. 14–6(B), where K stands for $e^2Qq/(4I(2I-1))$. The perturbation of these levels by an applied magnetic field is indicated in Fig 14–6(C). This effect will be discussed in the next section. As can be seen from Fig. 14–6(B), there are three transitions for $I = 1$ and $\eta \neq 0$ (labeled v_+, v_-, and v_0), so the two unknowns, e^2Qq and η, can be determined directly from the spectrum. The energies of the levels in Fig. 14–6 are given by equations (14–9) and (14–10). The energies of v_+, v_-, and v_0 are found by the differences in the energies of the levels. For any two transitions, the resulting equations can be solved for the two unknowns e^2Qq and η. It is interesting to point out that for a nucleus with $I = 1$ in an axially symmetric field, only one line is expected in the spectrum [see Fig. 14–6(A)].

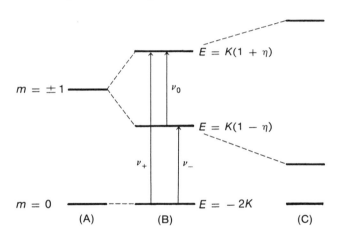

FIGURE 14–6 Energy levels for $I = 1$ under different conditions. (A) $\eta = 0$ and applied magnetic field $H_0 = 0$. (B) $\eta \neq 0$ and $H_0 = 0$. (C) $\eta \neq 0$ and $H_0 \neq 0$ but constant.

The energies of the quadrupole levels as a function of η have been calculated[2] for cases other than $I = 1$ or $^3/_2$. Tables have also been compiled[3] that permit calculations of η and e^2Qq from spectral data for nuclei with $I = ^5/_2$, $^7/_2$, and $^9/_2$. When η is appreciable, the selection rule $\Delta m = \pm 1$ breaks down, and spectra containing $\Delta m = \pm 2$ bands are often obtained.

14–3 EFFECT OF A MAGNETIC FIELD ON THE SPECTRA

When an nqr experiment is carried out on a sample placed in a static magnetic field, the Hamiltonian $\hat{H}_M$ describing the influence of the magnetic field on the nuclear magnetic dipole moment must be added to the nqr Hamiltonian:

$$\hat{H}_M = -g_N\beta_N \mathbf{H} \cdot \hat{I} \quad (14\text{--}11)$$

In weak magnetic fields ($g_N\beta_N H \ll e^2Qq$) the magnetic field acts as a perturbation on $\hat{H}_Q$. In general, the influence of this term on the energies is to *shift the energy of a non-degenerate quadrupole level and split a doubly degenerate level* [i.e., one having $m \neq 0$; see equation (14–6)]. This change in energy of the non-degenerate quadrupolar levels is indicated in Fig. 14–6(B) and (C) for a nucleus in which $I = 1$ and $\eta \neq 0$.

For a nucleus with $I = 1$, a quadrupole spectrum with two lines can arise from either of two distinct situations: for $\eta \neq 0$ as noted above (Fig. 14–6(B)), or for nuclei with $\eta = 0$ located in two non-equivalent lattice sites. Examination of the spectrum of the sample in an external field allows a distinction to be made between these two possibilities. In the former case ($\eta \neq 0$), two lines would again be observed, but with different energies than those obtained in the absence of the field; in the latter case, each doubly degenerate level would be split, giving a spectrum with four lines.

As mentioned earlier, the levels $E_{\pm 1/2}$ and $E_{\pm 3/2}$ are each doubly degenerate for $I = {}^3/_2$ in a non-symmetric field. As a result, e^2Qq and η cannot be determined directly. This degeneracy is removed by a magnetic field, with four levels resulting. Four transitions are observed in the spectrum: $+{}^1/_2 \to +{}^3/_2$, $+{}^1/_2 \to -{}^3/_2$, $-{}^1/_2 \to +{}^3/_2$, and $-{}^1/_2 \to -{}^3/_2$ ($\Delta|m| = +1$). The energy differences corresponding to these transitions are functions of H, e^2Qq, and η [see equations (14–7) and (14–8)], so q and η can be evaluated[7] for this system from the spectra of the sample taken both with and without an applied magnetic field.

14–4 RELATIONSHIP BETWEEN ELECTRIC FIELD GRADIENT AND MOLECULAR STRUCTURE

Our next concern is how we obtain information about the electronic structure of a molecule from the values of q and η. The field gradient at atom A in a molecule, $q_{mol}{}^A$, and the electronic wave function are related by equation (14–2):

$$q_{mol}{}^A = e \left\{ \sum_{B \neq A} [Z_B(3\cos^2\theta_{AB} - 1)/R_{AB}{}^3] - \int \psi^* \sum_n [(3\cos^2\theta_{A_n} - 1)/r_{A_n}{}^3] \psi \, d\tau \right\} \quad (14-12)$$

The molecular field gradiant q_{mol} is seen to be a sensitive measure of the electronic charge density in the immediate vicinity of the nucleus because equation (14–12) involves the expectation value $\langle 1/r^3 \rangle$. The first term in the equation is a summation over all nuclei external to the quadrupolar nucleus, and the second term is a summation over all electrons. If the molecular structure is known, the first term is readily evaluated. Z_B is the nuclear charge of any atom in the molecule other than A, the one whose field gradient is being investigated; θ_{AB} is the angle between the bond axis or highest-fold rotation axis for A and the radius vector from A to B, R_{AB}. The second term represents the contribution to the field gradient in the molecule from the electron density, and it is referred to as the electric field gradient q_{el}. Finally, ψ is the ground state wave function, and θ_{A_n} is the angle between the bond or principal axis and the radius vector r_{A_n} to the nth electron. This integral is difficult to evaluate. In the LCAO approximation, we can write

$$q_{el}{}^A = -e2 \sum_u^{occ} \sum_i \sum_j C_{iu} C_{ju} \int \varphi_i^* \hat{q}_A \varphi_j \, d\tau \quad (14-13)$$

where $\hat{q}_A = (3\cos^2\theta - 1)/r^3$; u is the index over the molecular orbitals; and i and j are indices for atomic orbitals. C_{iu} and C_{ju} represent the LCAO coefficients for the atomic orbitals φ_i and φ_j in the u molecular orbital. The integral may involve one, two, or three centers. Obviously, good wave functions are required, and an involved evaluation of equation (14–13) is needed to interpret q. For certain atoms (*e.g.*, N and Cl), the three-center contribution is small and can be ignored. Separating the one- and two-center terms in equation (14–13) and abbreviating the intergrals $\int \varphi_i \hat{q}_A \varphi_j \, d\tau$ as $q_A{}^{ij}$, etc., we can write equation (14–13) as (14–14):

$$q_{el}{}^A = -e\left[2\sum_u\sum_i^A C_{iu}{}^2 q_A{}^{ii} + 4\sum_u^A\sum_i^{B\neq A}\sum_j C_{iu}C_{ju}q_A{}^{ij} + 2\sum_u^{B\neq A}\sum_j C_{ju}{}^2 q_A{}^{jj}\right] \quad (14\text{--}14)$$

There have been several semi-empirical methods proposed for evaluating electric field gradients. Cotton and Harris[8] assumed that the nuclear term of equation (14–12) (*i.e.*, the first term) was cancelled by the part of $q_{el}{}^A$ arising from the gross atomic populations on the neighbor atom B; *i.e.*,

$$e\sum_{B\neq A}\left[\frac{Z_B(3\cos^2\theta_{AB} - 1)}{R_{AB}{}^3}\right] \cong e\left[2\sum_u^{B\neq A}\sum_j C_{ju}{}^2 q_A{}^{jj} + 2\sum_u^A\sum_i^{B\neq A}\sum_j C_{iu}C_{ju}q_A{}^{ij}\right] \quad (14\text{--}15)$$

This leads to

$$q_{mol}{}^A = -e\left[2\sum_u\sum_i^A C_{iu}{}^2 q_A{}^{ii} + 2\sum_u^A\sum_i^{B\neq A}\sum_j C_{iu}C_{ju}q_A{}^{ij}\right] \quad (14\text{--}16)$$

Further, by assuming that the two-center integral can be formulated as proportional to the overlap integral

$$q_A{}^{ij} = S_{ij}q_A{}^{ii} \quad (14\text{--}17)$$

we can write:

$$q_{mol}{}^A = -e\sum_i^A q_A{}^{ii}\left(2\sum_u C_{iu}{}^2 + 2\sum_u\sum_j^{B\neq A} C_{iu}C_{ju}S_{ij}\right) \quad (14\text{--}18)$$

We see then that the field gradient is given by multiplying $q_A{}^{ij}$ by the gross atomic population, P [equation (3–31)], or:

$$q_{mol}{}^A = -eP\sum_i^A \varphi_i\left|\frac{3\cos^2\theta - 1}{r^3}\right|\varphi_i \quad (14\text{--}19)$$

For an atom A with valence s and p orbitals, the above summation is over four orbitals for which q_A for the s orbital is zero, and

$$q = -eq_{at}\left(P_z - \frac{1}{2}P_x - \frac{1}{2}P_y\right)$$

where q_{at} is the field gradient for one electron in a p orbital. The quantity in parentheses in equation (14–18) is just the molecular-orbital expression for the gross atomic population P of atomic orbital φ_i. (See Chapter 3.) Depending on the relative populations of p_z, p_x, and p_y, q can be positive or negative. If the quadrupole coupling constant is expressed in megahertz, we have

$$\frac{e^2Qq}{h} = -\left(\frac{e^2Qq_{at}}{h}\right)\left(P_z - \frac{1}{2}P_x - \frac{1}{2}P_y\right) \tag{14–20}$$

where e^2Qq/h is the quadrupole coupling constant, and e^2Qq_{at}/h is the coupling constant for a single electron in the p orbital. The p_z orbital should be coincident with q_{zz} in order to apply these equations with much accuracy.

Equation (14–20) is the molecular orbital analog to the valence-bond expression of Townes and Dailey,[9] which had been reported earlier. This approach is based on the following arguments. Since the s orbital is spherically symmetric, electron density in this orbital will not give rise to a field gradient. As long as the atom being studied is not the least electronegative in the bonds to other atoms in the molecule, the maximum field gradient at this atom, q_{mol}, is the atomic field gradient, q_{at}, for a single electron in a p_z orbital of the isolated atom. When the atom being investigated is more electronegative than the atom to which it is bonded, the quadrupolar atom has greater electron density around it in the molecule than in the isolated atom. The relationship between the electron "occupation" of the p orbitals of a quadrupolar atom in a molecule, e^2Qq_{at}, and the quantity e^2Qq_{mol} (which is determined from the nqr spectrum of the molecule under consideration) is given by:

$$e^2Qq_{mol} = [1 - s + d - i(1 - s - d)]e^2Qq_{at} \tag{14–21}$$

where e^2Qq_{at} is the quadrupole coupling constant for occupancy of the p orbital by a single electron, s is the fraction of s character employed by the atom in the bond to its neighbor, d is the fraction of d character in this bond, and i is the fraction of ionic character in the bond (for a molecule A—B, $\psi^0 = c\psi_A + d\psi_B$ and $i = c^2 - d^2$). When the atom being studied is electropositive, i changes sign. Values of e^2Qq_{at} and q_{at} have been tabulated for several atoms.[4,5,10,11] When π bonding is possible, this effect must also be included. A modified form of equation (14–21) has also been employed:[12]

$$e^2Qq_{mol} = (1 - s + d - i - \pi)e^2Qq_{at} \tag{14–22}$$

where π is the extent of π bonding, and all other quantities are the same as before.

As the amount of ionic character in a bond increases, the electronic environment approaches spherical symmetry (where $q_{mol} = 0$) and e^2Qq_{mol} decreases. Hybridization of the p orbital with an s orbital also decreases e^2Qq_{mol}, as indicated by equation (14–22). Mixing of the s orbital with the p orbital decreases the field gradient, because the s orbital is spherically symmetric. In a covalent molecule, d orbital contribution to the bonding increases the field gradient.

There are several problems associated with these approaches to the interpretation of field gradients. First, even if the above approach were correct for the system of interest, there are four unknowns in equation (14–22) and we have only one measurable quantity, e^2Qq_{mol}. Investigators attempting to interpret this quantity are thus forced to assume the answer and provide a reasonable interpretation of the data based on these models. Second, O'Konski and Ha[13] have shown that the assumption of Cotton and Harris[8] indicated in equation (14–15) is not generally correct. When this equation does not apply, equations (14–20), (14–21), and (14–22) are also not correct.

A semiempirical approach to the interpretation of field gradients has been reported,[14] which does not make the suspect approximations described above. The reader is referred to the original literature for details of this method. Electrostatic, E, covalent, C, and transfer, T, parameters derived from bond energies are reported to provide an interpretation of chlorine field gradients.[14b]

One additional factor complicating the quantum mechanical calculation and interpretation of nqr and Mössbauer (*vide infra*) parameters is the *Sternheimer effect*.[15] This effect is the polarization of the originally spherically symmetric inner shell electrons by the valence electrons. When the core electrons lose their spherical symmetry, they contribute to the field gradient at the nucleus. This effect, like spin polarization (discussed in Chapter 12), is an artifact of not performing a full molecular orbital calculation on the whole crystal. There are two contributions to Sternheimer shielding, and these can be illustrated if we consider the ligands or ions external to the electron density of the atom as point charges. The spherical expansion of the core electron density is illustrated for positive point charges in Fig. 14–7(A) and (B). The asymmetry induced is shown in Fig. 14–7(C). We have artificially broken up the polarization to illustrate these two contributions to the Sternheimer shielding.

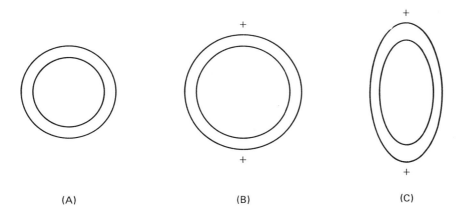

FIGURE 14–7 Schematic illustration of the Sternheimer effect. (A) A spherical shell of electron density; (B) the radial expansion resulting from two external positive charges; (C) the elliptical polarization by two positive charges.

The shielding for closed shell systems is usually expressed by the equation:

$$e^2Qq_{obs} = (1 - \gamma_{ri})e^2Qq_0$$

where e^2Qq_0 is the field gradient calculated in the absence of any Sternheimer effects and γ_{ri} is the shielding parameter for the ith charge located at a distance r from the nucleus. The sign of γ_{ri} is a function of the distance. For charges external to the core electron density of the central atom, γ_{ri} will generally be negative and is said to give rise to an *antishielding* contribution. For charges inside the valence orbitals of the central atom, the sign of γ_{ri} is generally positive and gives rise to a *shielding* contribution. Shielding constants have been calculated for charges external to a large number of ions and are designated by the symbol γ_∞. The value of γ_{ri} used to evaluate the Sternheimer effect in molecules is difficult to determine but appears to be considerably below that of γ_∞. Most workers in the field assume that this effect is constant in a similar series of complexes.

14–5 APPLICATIONS

NQR spectra of a number of molecules containing the following nuclei have been reported: ^{27}Al, ^{75}As, ^{197}Au, ^{10}B, ^{11}B, ^{135}Ba, ^{137}Ba, ^{209}Bi, ^{79}Br, ^{81}Br, ^{43}Ca, ^{35}Cl, ^{37}Cl, ^{59}Co, ^{63}Cu, ^{69}Ga, ^{71}Ga, ^{2}H, ^{201}Hg, ^{127}I, ^{115}In, ^{25}Mg, ^{14}N, ^{55}Mn, ^{23}Na, ^{93}Nb, ^{17}O, ^{185}Re, ^{187}Re, ^{33}S, ^{121}Sb, ^{123}Sb, ^{181}Ta.

The interpretation of e^2Qq Data

As mentioned above, it is not possible to interpret e^2Qq values rigorously in terms of ionic character, π bonding, and s and d hybridization. For compounds in which there are large differences in ionic character, the effect can be clearly seen, as indicated by the data in Table 14–1. More positive e^2Qq_{mol} values are obtained for ionic compounds. Attempts have been made to "explain" e^2Qq data for a large number of halogen compounds by assuming that d hybridization is not important and estimating ionic character or s hybridization.[12,16]

When the differences in ionic character are large enough to be predictable, the e^2Qq_{mol} values manifest the proper trends. However, for systems in which the differences in ionic character are not obvious from electronegativity and other considerations, the interpretation of e^2Qq differences is usually ambiguous in terms of the relative importance of the effects in equation (14–22). One of the more successful studies involves a series of substituted chlorobenzenes. A linear relation is found between the Hammett σ constant of the substituent and the quadrupole resonance frequency.[17] The data obeyed the equation v (MHz) $= 34.826 + 1.024\sigma$. The electron-releasing substituents have negative σ values and give rise to smaller e^2Qq values, because the increased ionic character in the C—Cl bond makes the chlorine more negatively charged.

TABLE 14–1. Values of e^2Qq_{mol} for Some Diatomic Halides

Molecule[a]	e^2Qq_{mol}	Molecule[a]	e^2Qq_{mol}
F<u>Cl</u>	−146.0	<u>Br</u>Cl	876.8
<u>Br</u>Cl	−103.6	Li<u>Br</u>	37.2
I<u>Cl</u>	−82.5	Na<u>Br</u>	58
Tl<u>Cl</u>	−15.8	D<u>Br</u>	533
K<u>Cl</u>	0.04	F<u>Br</u>	1089
Rb<u>Cl</u>	0.774	D<u>I</u>	−1827
Cs<u>Cl</u>	3	Na<u>I</u>	−259.87

[a] Values for e^2Qq_{at} are ^{35}Cl $= -109.74$, ^{79}Br $= 769.76$, and ^{127}I $= -2292.84$. e^2Qq_{mol} applies to the underlined atom.

The general problem of covalency in metal-ligand bonds in a series of complexes has been investigated by direct measurement of the quadrupole resonance.[18] By comparing the trends in force constants with the changes in chlorine quadrupole coupling constants, a consistent interpretation of the data for a whole series of MCl_6^{n-} compounds has been offered.[19] A rigorous interpretation of the data (especially where small differences are involved) is again hampered by the lack of information regarding the four variables in equation (14–22).

Two transitions are observed in the nqr spectra of ^{55}Mn ($I = \frac{5}{2}$), so both q_{zz} and η can be obtained. Measurements on a series of organometallic manganese carbonyls have been reported.[20] Substitution of methyl groups in the series $(CH_3)_nC_6H_{6-n}Mn(CO)_3^+$ causes only minor changes in the asymmetry parameter. A summary of cobalt-59 nqr investigations has appeared.[21] The magnitudes of the field gradients are indicative of *cis* or *trans* octahedal geometry. The e^2Qq value of [*cis*-Co(en)$_2$Cl$_2$]Cl is 33.71, whereas that for [*trans*-Co(en)$_2$Cl$_2$]Cl is 60.63. A rationalization of this difference will be provided in the next chapter, in the section on partial field gradient parameters.

Thermochemical bond energies have been estimated and compared with the results from chlorine-35 nqr for the $GeCl_6^{2-}$, $SnCl_6^{2-}$ and $PbCl_6^{2-}$ ions. The authors[22] support an earlier conclusion[23] that these results are not consistent with assigning a higher electronegativity to lead than to germanium or tin.

This coverage of quadrupole investigations concerned with elucidating covalency and other characteristics of chemical bonds is by no means complete. The presentation is brief because of the ambiguity that exists in the interpretation of the results. More information than just the field gradient is needed to sort out the variables in equation (14–22).

Effects of the Crystal Lattice on the Magnitude of e^2Qq

A further complication and limitation of the application of nqr spectroscopy results from the fact that direct measurement of nuclear quadrupole transitions can be obtained only on solids. For very complex molecules, this is the only source of nuclear quadrupole information because of the complexity of the microwave spectrum. Measurements on solids introduce the complexities of lattice effects. For those molecules which have been studied by both methods (direct measurement and microwave), it is found that e^2Qq is usually 10 to 15% lower in the solid state. It has been proposed[5] that the decrease is due to increased ionicity in the solid. There are several examples of molecules being extensively associated in the solid but not in the gas phase or in solution (*e.g.*, I_2 and CNCl). In these cases, considerable care must be exercised in deducing molecular properties from e^2Qq values. The e^2Qq values for cobalt in selected cobalt complexes illustrates[21] the problem. The values for *trans*-[Co(NH$_3$)$_4$Cl$_2$]Cl and *trans*-[Co(en)$_2$Cl$_2$]Cl are 59.23 MHz and 60.63 MHz; *i.e.*, there is a difference of 1.40 MHz. On the other hand, *trans*-[Co(en)$_2$Cl$_2$]NO$_3$ has an e^2Qq(Co) value of 62.78 MHz. Changing the anion causes a difference of 2.15 MHz. The e^2Qq(Cl) values of K$_2$SnCl$_6$ and [(CH$_3$)$_4$N]$_2$SnCl$_6$ are 15.063 MHz and 16.674 MHz, respectively. Potential causes for these differences have been offered.[22b,22c] Examples in which the crystal lattice affects the number of lines observed in a spectrum will be discussed in the next section.

Structural Information from NQR Spectra

Since different field gradients will exist for non-equivalent nuclei in a molecule, we should expect to obtain a different line (or set of lines, depending on I) for each type of nuclear environment. In general, the environment of an atom as determined by nqr studies is in agreement with results obtained from x-ray studies. Only one line is found[18] in the halogen nqr spectrum of each of the following: K_2SeCl_6, Cs_2SeBr_6, $(NH_4)_2TeCl_6$, $(NH_4)_2SnBr_6$, and K_2PtCl_6. This is consistent with O_h structures for these anions.

As mentioned above, the following effects can give rise to multiple lines in the nqr spectrum:

1. Chemically non-equivalent atoms in the molecule.

2. Chemically equivalent atoms in a molecule occupying non-equivalent positions in the crystal lattice of the solid.

3. Splitting of the degeneracy of quadrupole energy levels by the asymmetry of the field gradient. Splittings of the quadrupole levels by other magnetic nuclei in the molecule, similar to spin-spin splittings in nmr, are often not detected in nqr spectra, but are being found with improved instrumentation. Usually these splittings are less than, or of the same order of magnitude as, the line widths. One case in which such splitting has been reported[24] is in the spectrum of HIO_3. The ^{127}I nucleus is split by the proton.

The non-equivalence of lattice positions (2) is illustrated in the bromine nqr spectrum of K_2SeBr_6, which gives a single line at room temperature and two lines at dry ice temperature. A crystalline phase change accounts for the difference. NQR is one of the most powerful techniques for detecting phase changes and obtaining structural information about the phase transitions. Although K_2PtI_6, K_2SnBr_6, and K_2TeBr_6 contain "octahedal" anions, the halides are not equivalent in the solid lattice, and the halide nqr spectra all consist of three lines, indicating at least three different halide environments.

Four resonance lines were found in the chlorine quadrupole spectra of each of $TiCl_4$,[25a] $SiCl_4$,[25b] and $SnCl_4$.[25b] It was concluded that the crystal structures of these materials are similar.

The problem of distinguishing non-equivalent positions in the lattice from chemical non-equivalence in the molecular configuration (which is related to the structure of the molecule in the gas phase or in a non-coordinating solvent) is difficult in some cases. In general, the frequency difference for lines resulting from non-equivalent lattice positions is small compared to the differences encountered for chemically non-equivalent nuclei in a molecule. When only slight separations between the spectral lines are observed, it is difficult to determine from this technique alone which of the two effects is operative. Another difficulty encountered in interpretation is illustrated by the chlorine spectrum of M_2Cl_{10} species.[26] Even though the Nb_2Cl_{10} molecule has a structure containing both terminal and bridging chlorines, only a single chlorine resonance assigned[26] to the bridging atom is observed in the spectrum. In Re_2Cl_{10} and W_2Cl_{10}, a large number of resonances were obtained.[26] Consequently, arguments based on assigning the number of different types of nuclei in the molecular structure to the number of resonances observed can be of doubtful validity.

NQR studies of ^{14}N ($I = 1$) are difficult to carry out but produce very interesting results. Since $I = 1$, both e^2Qq and the asymmetry parameter η can be evaluated from the nqr spectrum. In BrCN, the ^{14}N nqr resonance is a doublet. This could result from two non-equivalent nitrogen atoms in the crystal lattice or from a splitting of the nitrogen resonance because of an asymmetric field gradient. The former explanation was eliminated by a single crystal x-ray study.[27] The structure of solid BrCN was found to consist of linear chains of the type:

$$\cdots Br-C\equiv N|\cdots Br-C\equiv N|\cdots Br-C\equiv N|$$

The nitrogen has axial symmetry here, and only one line is expected. However, it is proposed that interactions between chains reduce the symmetry at the nitrogen and lead to the two lines. Various resonance forms can be written for BrCN, and the e^2Qq values indicate that the bromine has a formal positive charge. An appreciable increase in e^2Qq for bromine is observed in the solid relative to the gaseous state spectrum of BrCN. This could be due to increased contributions to the ground state from the structure Br^+CN^- in the solid because of stabilization of Br^+ by coordination. If the $N\cdots Br-C$ bond is described as a pd hybrid, the increased d contribution in the bromine carbon bond will also increase e^2Qq.

The ^{14}N quadrupole transitions in pyridine and various coordination compounds of pyridine show[28] large changes in the transition energies. As discussed in conjunction with Fig. 14–6, three transitions (v_+, v_0, and v_-) are expected when $\eta \neq 0$, and e^2Qq and η can be determined from the data. Typical results are summarized in Table 14–2.

The iodine nqr spectrum of solid iodine indicates a large asymmetry parameter, η.[29] Since the iodine atom in an iodine molecule is axially symmetric, the large asymmetry is taken as indication of intermolecular bonding in the solid. A large η in the iodine nqr spectrum of the molecule HIO_3 supports the structure $IO_2(OH)$ instead of HIO_3. The structure HIO_3 has a C_3 axis, so $q_{xx} = q_{yy}$. References 1 and 2 contain many additional examples of studies of this kind.

Information regarding π bonding can be obtained from the asymmetry parameter η. Methods for evaluating η for various nuclei have been discussed. A single σ bond to a halogen should give rise to an axially symmetric field gradient. Double bonding leads to asymmetry, and the extent of π bonding is

TABLE 14–2. ^{14}N Quadrupole Transitions and Field Gradient Parameters in Pyridine and Coordinated Pyridine (All Frequencies in kHz, Temp. = 77 K). (Crystallographic Non-Equivalences, Where Present, Are Resolved.)

Compound	v_+	v_-	v_0	$\dfrac{e^2Qq}{h}$	η
Pyridine (Py)	3892	2984	908	4584	0.396
Pyridinium nitrate	1000	580	420	1053	0.798
Py$_2$ZnCl$_2$	2387	2078	309	2977	0.207
	2332	2038	294	2913	0.202
Py$_2$Zn(NO$_3$)$_2$	2124	1884	240	2672	0.180
	2097	1877	220	2649	0.166
Py$_2$CdCl$_2$	2850	2298	552	3432	0.320

related to η. It was concluded that there is appreciable ($\sim 5\%$) carbon-halogen π bonding in vinyl chloride, vinyl bromide, and vinyl iodide.[30] Bersohn[31] has made a complete study of the problem of quantitatively estimating the extent of carbon-halogen π bonding from η.

The values of η obtained from the nqr spectra of SiI_4, GeI_4, and SnI_4 were interpreted to indicate a very small degree (about 1%) of double bond character[32] in the halogen bond to the central atom. This could be due to a solid state effect.

The ^{77}As, $^{121,123}Sb$, ^{209}Bi, $^{35,37}Cl$, and $^{79,81}Br$ nqr spectra of compounds with general formula R_3MX_2 have been studied[33] for R = CH_3, $CH_2C_6H_5$, and C_6H_5, for X = F, Cl, and Br, and for M = As, Sb, and Bi. Very small asymmetry parameters were found, indicating a threefold axis in all compounds. The results suggest that most of the compounds are trigonal bipyramidal. However, the arsenic compound $[(CH_3)_3AsBr^+]Br^-$ is not, but probably has a cation with C_{3v} symmetry.

The nqr spectra (^{37}Cl, ^{35}Cl, ^{121}Sb, ^{127}I) of $2ICl \cdot AlCl_3$ and $2ICl \cdot SbCl_5$ indicate[34] that these materials should be formulated as $ICl_2^+AlCl_4^-$ and $ICl_2^+SbCl_6^-$. The nqr spectra of the v-shaped cations ICl_2^+, I_3^+, and I_2Cl^+ have been studied in several different compounds. The ^{35}Cl nqr spectra of the chloroaluminate group in a wide variety of $M_n(AlCl_4)_m$ compounds have been studied.[35] The transition energies can be used to indicate whether the relatively free ion $AlCl_4^-$ exists or whether this anion is strongly coordinated to the cation. "Ionic" $AlCl_4^-$ transitions are found in the 10.6 to 11.3 MHz range. Strong coordination of $AlCl_4^-$ results in an elongation of the Al—Cl bridge bonds and an appreciable increase in the range and average frequency of the chlorine transitions.

In the next chapter, on Mössbauer spectroscopy, we shall discuss an additive, *partial field gradient* (pfg), model for correlating field gradients at central atoms. It is useful for deducing structures of molecules from experimental Mössbauer data. This model can also be used to infer structures from nqr data. Since most of the data used to derive and test the pfg model are Mössbauer results, we shall treat this topic in the next chapter.

14–6 DOUBLE RESONANCE TECHNIQUES

Nuclear double resonance techniques have been reported[36–38] for observing quadrupolar transitions. These methods greatly increase sensitivity over previous techniques and also permit one to observe transitions with very low frequencies; *e.g.*, in ^2H spectra, transitions are observed in the 100 to 160 kHz region.[39]

The adiabatic demagnetization double resonance experiment is a novel technique in this general category. Consider a sample that has a quadrupolar nucleus in a molecule containing several protons. When this sample is placed in a magnetic field and we wait long enough for equilibrium to be obtained, there will be an excess of proton nuclear moments aligned with the field that undergo Larmor precession and give rise to a net magnetization, as discussed in the chapters on nmr. When the sample is removed from the field, the net magnetization is reduced to zero as the sample is removed because the individual moments now become aligned with their local fields. A random orientation of these local fields in the absence of an external field produces a zero net magnetization. This is illustrated in Fig. 14–8 on the left-hand side, in the region labeled "sample removed."

If T_1 is longer than the time required to remove the sample, the excess population of $+1/2$ spins will remain, but they now precess about a net local field felt at the nucleus from spin-spin interaction with neighboring protons. Over the full sample, the magnetization is zero; but when this sample is reinserted into a magnetic field, magnetization is simultaneously induced into the sample without having to wait the required time for the T_1 process. This is illustrated in the region of Fig. 14–8 labeled "sample reinserted." The intensity of the magnetization can be measured by employing a 90° pulse immediately after reinserting the sample in the magnetic field and measuring the FID curve (see Fig. 14–8). If the time between the removal of the sample from the magnetic field and reinserting it is long compared to T_1, the magnetization will decrease as the spins become randomized.

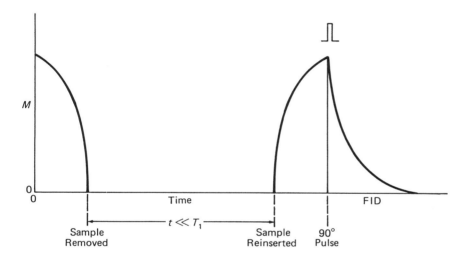

FIGURE 14–8 Plot of the magnetization when a sample is removed from a magnetic field, reinserted, subjected to a 90° pulse, and then allowed to undergo free induction decay.

Now consider an experiment in which the sample is irradiated with an rf frequency corresponding to the quadrupolar nucleus B transition after the sample has been removed from the field. Furthermore, we shall assume that the time between removal and reinsertion is small compared to T_1 for the protons. The effect of this rf field is to randomize the B nucleus by inducing quadrupolar transitions in the B spin system. Provided that the appropriate conditions are met, in terms of the amplitude of the applied rf field in relationship to the local field experienced at the protons, the randomization of the B spin system influences that of the proton system. This occurs by the following process. When the sample is removed from the field, the energy difference between the $m = +1/2$ and $-1/2$ states (*i.e.*, the transition energy of the H nucleus) decreases toward zero. In the process, there is a time at which the energy difference for the hydrogen nuclei matches the energy difference of the quadrupolar states of the B nucleus. A resonance energy exchange occurs, tending to randomize the proton nuclei. The process of randomization is often referred to as an increase in the *spin temperature* of the system. As a result of the randomization of the proton system from transitions in B, the magnetization that is recovered when the sample is returned to the magnetic field is less than it would otherwise be. As a result, the

FID measured is less than it would otherwise be. When the frequency used for the quadrupole transition is not appropriate for resonance, the B nucleus is not randomized and a large amount of the magnetization is recovered. By systematically incrementing the frequency of the B transmitter, the "spectra" of the quadrupole transitions of the B system are mapped out in terms of their effect on the FID of the abundant proton spin system.

By using a spin echo double resonance experiment,[39–41] much of the inhomogeneous dipolar broadening (crystal imperfections, etc.) that leads to very broad lines in the direct nqr experiment can be eliminated. The nqr of Al_2Br_6 has been determined with this technique.[40]

If one has two quadrupolar nuclei surrounded by nuclei with $I = 0$ [e.g., as in $D-Mn(Co)_5$], the dipolar coupling of the manganese and deuterium nuclei can be observed.[42] As discussed in the nmr chapters, the bond distance can be obtained from the magnitude of the dipolar coupling. A Mn—D bond distance of 1.61 Å is calculated from the ^{55}Mn nqr spectrum, in excellent agreement with the neutron diffraction result.[43]

REFERENCES CITED

1. C. P. Slichter, "Principles of Magnetic Resonance," Harper & Row, New York, 1963, Sec. 6.3.
2. R. Bersohn, J. Chem. Phys., *20*, 1505 (1952).
3. R. Livingston and H. Zeldes, "Tables of Eigenvalues for Pure Quadrupole Spectra," Oak Ridge Natl. Lab. Rept. ORNL-1913 (1955).
4. a. T. P. Das and E. L. Hahn, "Nuclear Quadrupole Coupling Spectroscopy," Academic Press, New York, 1958.
   b. E. A. C. Lucken, "Nuclear Quarupole Coupling Constants," Academic Press, New York, 1969.
5. C. T. O'Konski, "Determination of Organic Structures by Physical Methods," Vol. 2, ed. F. C. Nachod and W. D. Phillips, Academic Press, New York, 1962.
6. J. A. Pople, W. G. Schneider, and H. J. Bernstein, "High Resolution Nuclear Magnetic Resonance," McGraw-Hill, New York 1959.
7. C. Dean, Phys. Rev., *86*, 607A (1952).
8. F. A. Cotton and C. B. Harris, Proc. Natl. Acad. Sci. U.S., *56*, 12 (1966); Inorg. Chem., *6*, 376 (1967).
9. C. H. Townes and B. P. Dailey, J. Chem. Phys., *17*, 782 (1949).
10. M. H. Cohen and F. Reif, "Solid State Physics," Vol. 5, ed. F. Seitz and D. Turnbull, Academic Press, New York, 1957.
11. W. J. Orville-Thomas, Quart. Rev., *11*, 162 (1957).
12. B. P. Dailey, J. Chem. Phys., *33*, 1641 (1960).
13. C. T. O'Konski and T. K. Ha, J. Chem. Phys., *49*, 5354 (1968).
14. a. W. D. White and R. S. Drago, J. Chem. Phys., *52*, 4717 (1970).
    b. R. S. Drago, N. Wong, and D. Ferris, J. Amer. Chem. Soc., *113*, 1970 (1991).
15. R. M. Sternheimer, Phys. Rev., *130*, 1423 (1963); *ibid.*, 146, 140 (1966) and references therein.
16. C. H. Townes and B. P. Dailey, J. Chem. Phys., *23*, 118 (1955); W. Gordy, Disc. Faraday Soc., 19, 14 (1955); M. A. Whitehead and H. H. Jaffe, Trans. Faraday Soc., *57*, 1854 (1961).

17. H. C. Meal, J. Amer. Chem. Soc., *74*, 6121 (1952); P. J. Bray and R. G. Barnes, J. Chem. Phys., *27*, 551 (1957) and references therein.
18. D. Nakamura, K. Ito, and M. Kubo, Inorg. Chem., 1, 592 (1962); *ibid.*, *2*, 61 (1963); J. Amer. Chem. Soc., 82 5783 (1960); *ibid.*, *83*, 4526 (1961); *ibid.*, *84*, 163 (1962); M. Kubo and D. Nakamura, Adv. Inorg. Chem. Radiochem., *8*, 257 (1966).
19. T. L. Brown, W. G. McDugle, and L. G. Kent, J. Amer. Chem. Soc., *92*, 3645 (1970).
20. T. B. Brill and A. J. Kotlar, Inorg. Chem., *13*, 470 (1974).
21. T. L. Brown, Accts. Chem. Res., *408*, (1974) and references therein.
22. a. W. A. Welsh, T. B. Brill, *et al.*, Inorg. Chem., 13, 1797 (1974).
    b. T. B. Brill, R. C. Gearhart, and W. A. Welsh, J. Mag. Res., *13*, 27 (1974).
    c. T. B. Brill, J. Chem. Phys., *61*, 424 (1974).
23. R. S. Drago and N. A. Matwiyoff, J. Organometal. Chem., *3*, 62 (1965).
24. R. Livingston and H. Zeldes, J. Chem. Phys., *26*, 351 (1957).
25. a. A. H. Reddoch, J. Chem. Phys., *35*, 1085 (1961).
    b. A. L. Schawlow, J. Chem. Phys., *22*, 1211 (1954).
26. P. A. Edwards and R. E. McCarley, Inorg. Chem., *12*, 900 (1973).
27. S. Geller and A. L. Schawlow, J. Chem. Phys., *23*, 779 (1955).
28. Y. N. Hsieh, G. R. Rubenacker, C. P. Cheng, and T. L. Brown, J. Amer. Chem. Soc., *99*, 1384 (1977).
29. H. G. Dehmelt, Naturwissenschaften, *37*, 398 (1950).
30. J. A. Howe and J. H. Goldstein, J. Chem. Phys., *26*, 7 (1957); *27*, 831 (1957) and references therein.
31. R. Bersohn, J. Chem. Phys., *22*, 2078 (1954).
32. H. Robinson, H. G. Dehmelt, and W. Gordy, J. Chem. Phys., *22*, 511 (1954); S. Kojima *et al.*, J. Phys. Soc. Japan, *9*, 805 (1954).
33. T. B. Brill and G. G. Long, Inorg. Chem., *13*, 1980 (1970).
34. D. J. Merryman and J. D. Corbett, Inorg. Chem., *13*, 1258 (1974)
35. D. J. Merryman, P. A. Edwards, J. D. Corbett, and R. E. McCarley, Inorg. Chem., *13*, 1471 (1974).
36. S R. Hartmann and E. L. Hahn, Phys. Rev., *128*, 2042 (1962).
37. R. E. Slusher and E. L. Hahn, Phys. Rev., *166*, 332 (1968).
38. D. T. Edmonds *et al.*, "Advances in Quadrupole Resonance," Vol. 1, p. 145, Heyden, London, 1974; Rev. Pure Appl. Chem., *40*, 193 (1974).
39. J. L. Ragle and K. L. Sherk, J. Chem. Phys., *50*, 3553 (1969).
40. M. Emshwiller, E. L. Hahn, and D. Kaplan, Phys. Rev., *118*, 414 (1960).
41. J. L. Ragle *et al.*, J. Chem. Phys., *61*, 429, 3184 (1974).
42. a. P. S. Ireland, L. W. Olson, and T. L. Brown, J. Amer. Chem. Soc., *97*, 3548 (1975).
    b. P. S. Ireland and T. L. Brown, J. Mag. Res., *20*, 300 (1975).
43. S. J. LaPlaca *et al.*, Inorg. Chem., *8*, 1928 (1969).

ADDITIONAL REFERENCES

M. H. Cohen and F. Reif, "Solid State Physics," Vol. 5, ed. F. Seitz and D. Turnbull, Academic Press, New York, 1957.

T. P. Das and E. L. Hahn, "Nuclear Quadrupole Resonance Spectroscopy," Academic Press, New York, 1958.

E. A. C. Lucken, "Nuclear Quadrupole Coupling Constants," Academic Press, New York, 1969.

J. A. S. Smith, ed., "Advances in Nuclear Quadrupole Resonance," Heyden and Sons, London: Vol. 1, 1974; Vol. 2, 1975; Vol. 3, 1977.

EXERCISES

1. **a.** Calculate the energies of all the quadrupolar energy states for a nucleus with $I = 2$. Express the energies as a function of e^2Qq.

   b. How many transitions are expected, and what is the relationship between the energy of the transitions and e^2Qq?

2. **a.** Using the equations presented in this chapter for the energies of the 0 and ± 1 levels of a nucleus with $I = 1$ in an asymmetric field, calculate the frequency in terms of e^2Qq and η for the $0 \to +1$ and $0 \to -1$ transitions.

   b. Express the energy difference between the two transitions in part a in terms of η and e^2Qq.

   c. Show how η and q can be determined from this information.

3. Describe an nqr experiment that would give information regarding the extent of π bonding in the phosphorus-sulfur bonds in $PSCl_3$ and $(C_6H_5)_3PS$. Can you determine whether the sulfur is hybridized sp^2 and utilizes a p orbital in bonding or whether the p_x and p_y orbitals of sulfur participate equally in bonding with nqr experiments? (Note: for ^{33}S, $I = ^3/_2$.)

4. It has been reported that the ^{127}I quadrupole resonance in AsI_3 is a singlet but has a very large asymmetry parameter.[28] A single crystal x-ray study indicates that the As is nearly octahedral. Explain the large asymmetry parameter.

5. Indicate the number of resonance lines expected for the following nuclei under the conditions given:

   a. ^{127}I $(I = ^5/_2)$; $\eta = 0$; $H_0 = 0$.

   b. ^{14}N $(I = 1)$; $\eta = 0$; $H_0 = 0$.

   c. ^{75}As $(I = ^3/_2)$; $\eta = 0$; $H_0 = 0$.

   d. ^{121}Sb $(I = ^5/_2)$; $\eta = 1$; $H_0 = 0$.

   e. ^{14}N $(I = 1)$; $\eta = 1$; $H_0 = 0$.

   f. ^{14}N $(I = 1)$; $\eta = 1$; $H_0 \neq 0$.

6. The quadrupolar energy of a nucleus is given by

$$E_m = \frac{e^2Qq[3m^2 - I(I+1)]}{4I(2I-1)}$$

^{59}Co has $I = ^7/_2$ and a natural abundance of 100%.

   a. How many cobalt nqr transitions will be observed for $K_3Co(CN)_6$? What are the transition energies in terms of e^2Qq?

   b. Repeat part a for $K_3Co(CN)_5Br$.

7. Consider the nitrogen nqr of each of the following systems. How many lines would you expect with and without a magnetic field? (For ^{14}N, $I = 1$.)

   a. NH_3

   b. NH_4^+

   c.

8. The ^{59}Co ($I = 7/2$) frequencies in $Cl_3SnCo(CO)_4$ occur at 35.02 MHz ($\pm 5/2 \to \pm 7/2$), 23.37 MHz ($\pm 3/2 \to \pm 5/2$), and 11.68 MHz ($\pm 1/2 \to \pm 3/2$). Calculate η and e^2Qq. (Hint: What are the ratios of the frequencies when $\eta = 0$?)

9. A compound having the formula CH_3InI_2 is known. It is believed to be an ionic compound, $(CH_3)_2In^+InI_4^-$. Provided there are no crystallographically non-equivalent cations or anions, how many resonance lines would you expect for ^{115}In and ^{127}I in this compound? The cation has $\eta = 0.05$. Which structure of this cation is suggested by this small value?

10. In solid pyridine (C_5H_5N) at 77 K, ^{14}N lines are found at 3.90 and 2.95 MHz. What are e^2Qq/h and η for nitrogen in pyridine?

11. The ^{35}Cl lines in the spectrum of $HgCl_2$ lie at 22.05 and 22.25 MHz at 300 K.

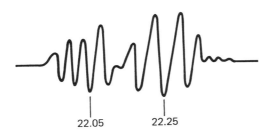

In $HgCl_2 \cdot$ dioxane, a single ^{35}Cl line at 20.50 MHz is found at 300 K. In the dioxanate complex, an $Hg \leftarrow O\!\!<$ donor-acceptor interaction occurs.

   a. What is the probable source of the line splitting in pure $HgCl_2$?

   b. The electric quadrupole moments of ^{35}Cl and ^{37}Cl are 0.079×10^{-24} cm^2 and 0.062×10^{-24} cm^2, respectively. At what frequency would you expect to find ^{37}Cl resonances in $HgCl_2$?

   c. In terms of the p orbital populations of equation (14–13), rationalize the decrease in the ^{35}Cl resonance frequency in the dioxanate adduct of $HgCl_2$ compared to that in pure $HgCl_2$.

15 Mössbauer Spectroscopy

15–1 INTRODUCTION

Mössbauer spectroscopy,[1] abbreviated as MB spectroscopy, involves nuclear transitions that result from the absorption of γ-rays by the sample. This transition is characterized by a change in the nuclear spin quantum number, I. The conditions for absorption depend upon the electron density about the nucleus, and the number of peaks obtained is related to the symmetry of the compound. As a result, structural information can be obtained. Many of the concepts and symbols used in this chapter have been previously discussed in Chapter 14.

To understand the principles of this method, first consider a gaseous system consisting of a radioactive source of γ-rays and the sample, which can absorb γ-rays. When a gamma ray is emitted by the source nucleus, it decays to the ground state. The energies of the emitted γ-rays, E_γ, have a range of 10 to 100 keV and are given by equation (15–1):

$$E_\gamma = E_r + D - R \tag{15–1}$$

where E_r is the difference in energy between the excited state and ground state of the source nucleus; D, the Doppler shift, is due to the translational motion of the nucleus, and R is the recoil energy of the nucleus. The recoil energy, similar to that occurring when a bullet leaves a gun, is generally 10^{-2} to 10^{-3} eV and is given by the equation:

$$R = E_\gamma^2/2mc^2 \tag{15–2}$$

where m is the mass of the nucleus and c is the velocity of light. The Doppler shift accounts for the fact that the energy of a γ-ray emitted from a nucleus in a gas molecule moving in the same direction as the emitted ray is different from the energy of a γ-ray from a nucleus in a gas molecule moving in the opposite direction. The distribution of energies resulting from the translational motion of the source nuclei in many directions is referred to as Doppler broadening. The left-hand curve of Fig. 15–1 represents the distribution of energies of emitted γ-rays, E_γ, from this effect. The breadth of the curve results from Doppler broadening. The dashed line in Fig. 15–1 is taken as E_r, the energy difference between the nuclear ground and excited states of the source. The energy difference,

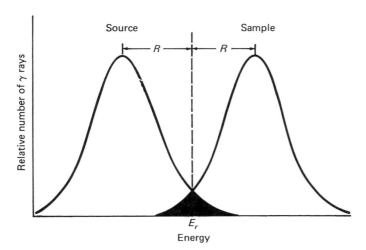

FIGURE 15–1 Distribution of energies of emitted and absorbed γ-rays.

R, between the dotted line and the average energy of the left-hand curve is the recoil energy transmitted to the source nucleus when a γ-ray is emitted.

In MB spectroscopy the energy of the γ-ray absorbed for a transition in the sample is given by:

$$E_\gamma = E_r + D + R \tag{15-3}$$

In this case, R is added because the exciting γ-ray must have energy necessary to bring about the transition and effect recoil of the absorbing nucleus. The quantity D has the same significance as before, and the value of E_r is assumed to be the same for the source and the sample. The curve in the right half of Fig. 15–1 shows the distribution of γ-ray energies necessary for absorption. The relationship of the sample and source energies can be seen from the entire figure. As indicated by the shaded region, there is only a very slight probability that the γ-ray energy from the source will match that required for absorption by the sample. Since the nuclear energy levels are quantized, there is accordingly a very low probability that the γ-ray from the source will be absorbed to give a nuclear transition in the sample. The main cause for nonmatching of γ-ray energies is the recoil energy, with the distribution for emission centered about $E_r - R$, whereas that for absorption is centered about $E_r + R$. The quantity R for a gaseous molecule ($\sim 10^{-1}$ eV) is very much larger than the typical Doppler energy. The source would have to move with a velocity of 2×10^4 cm sec^{-1} to obtain a Doppler effect large enough to make the source and sample peaks overlap, and these velocities are not readily obtainable. However, if the quantity R could be reduced, or if conditions for a recoilless transition could be found, the sample would have a higher probability of absorbing γ-rays from the source. As indicated by equation (15–2), R can be decreased by increasing m, the mass. By placing the nucleus of the sample and source in a solid, the mass is effectively that of the solid and the recoil energy will be small as indicated by equation (15–2). For this reason, MB spectra are almost always obtained on solid samples employing solid sources.

By placing the source and sample in solid lattices, we have not effected recoilless transitions for all nuclei, but we have increased the probability of a recoilless transition. The reason for this is that the energy of the γ-ray may cause excitation of lattice vibrational modes. This energy term would function in the same way as the recoil energy in the gas; *i.e.*, it would decrease the energy of the emitted particle and increase the energy required for absorption. Certain crystal properties and experimental conditions for emission or absorption will leave the lattice in its initial vibrational state; *i.e.*, conditions for a recoilless transition will be satisfied. It should be emphasized that these conditions simply determine the intensity of the peaks obtained, for it is only the number of particles with matching energy that is determined by this effect. We shall not be concerned with the absolute intensity of a band, so this aspect of MB spectroscopy will not be discussed. It should be mentioned, however, that for some materials (usually molecular solids), lattice and molecular vibrational modes are excited to such an extent that very few recoilless transitions occur at room temperature and no spectrum is obtained. Frequently, the spectrum can be obtained by lowering the temperature of the sample.

By going to the solid state we have very much reduced the widths of the resonance lines over that shown in Fig. 15–1. The Doppler broadening is now negligible, and R becomes $\sim 10^{-4}$ eV for a 100 keV gamma ray and an emitting mass number of 100. The full width of a resonance line at half height is given by the Heisenberg uncertainty principle as $\Delta E = h/\tau = 4.56 \times 10^{-16}/0.977 \times 10^{-7} = 4.67 \times 10^{-9}$ eV or 0.097 mm sec^{-1} (for ^{57}Fe). The line widths are infinitesimal compared to the source energy of 1.4×10^4 eV. The range of excited state lifetimes for Mössbauer nuclei is $\sim 10^{-5}$ sec to 10^{-10} sec, and this leads to line widths of 10^{-11} eV to 10^{-6} eV for most nuclei. This subject is treated in references 1 to 5, which contain a more detailed discussion of the entire subject of MB spectroscopy.

Our main concern will be with the factors affecting the energy required for γ-ray absorption by the sample. There are three main types of interaction of the nuclei with the chemical environment that result in small changes in the energy required for absorption: (1) resonance line shifts from changes in electron environment, (2) quadrupole interactions, and (3) magnetic interactions. These effects give us information of chemical significance and will be our prime concern.

Before discussing these factors, it is best to describe the procedure for obtaining spectra and to illustrate a typical MB spectrum. The electron environment about the nucleus influences the energy of the γ-ray necessary to cause the *nuclear* transition from the ground to excited state, *i.e.*, E_r in the sample. The energy of γ-rays from the source can be varied over the range of the energy differences arising from electron environments in different samples by moving the source relative to the sample. The higher the velocity at which the source is moved toward the sample, the higher the average energy of the emitted γ-ray (by the Doppler effect) and vice versa. The energy change ΔE_s of a photon associated with the source moving relative to the sample is given by:

$$\Delta E_s = \frac{v_0}{c} E_\gamma \cos \theta \qquad (15\text{–}4)$$

where E_γ is the stationary energy of the photon, v_0 is the velocity of the source, and θ is the angle between the velocity of the source and the line connecting the source and the sample. When the source is moving directly toward the sample, $\cos\theta = 1$. In order to obtain an MB spectrum, the source is moved relative to the sample, and the source velocity at which maximum absorption of γ-rays occurs is determined.

Consider, as a simple example, the MB spectrum of $Fe^{3+}Fe^{III}(CN)_6$ [where Fe^{3+} and Fe^{III} designate weak and strong field iron(III), respectively]. This substance contains iron in two different chemical environments, and γ-rays of two different energies are required to cause transitions in the different nuclei. To obtain the MB spectrum, the source is moved relative to the fixed sample, and the absorption of γ-rays is plotted as a function of source velocity as shown in Fig. 15–2. The peaks correspond to source velocities at which maximum γ-ray

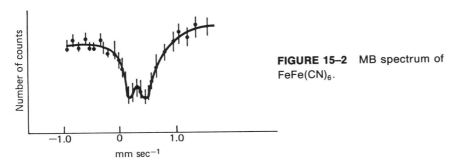

FIGURE 15–2 MB spectrum of $FeFe(CN)_6$.

absorption by the sample occurs. Negative relative velocities correspond to moving the source away from the sample, and positive relative velocities correspond to moving the source toward the sample. The relative velocity at which the source is being moved is plotted along the abscissa of Fig. 15–2, and this quantity is related to the energy of the γ-rays. For a ^{57}Fe source emitting a 14.4 keV γ-ray, the energy is changed by 4.8×10^{-8} eV or 0.0011 cal mole^{-1} for every mm sec^{-1} of velocity imposed upon the source. This result can be calculated from equation (15–4):

$$\Delta E_s = \frac{1 \text{ mm sec}^{-1}}{3.00 \times 10^{11} \text{ mm sec}^{-1}} \times 14.4 \times 10^3 \text{ eV} = 4.80 \times 10^{-8} \text{ eV}$$

This energy is equivalent to a frequency of 11.6 MHz ($v = E/h$, where $h = 4.14 \times 10^{-15}$ eV sec). For other nuclei having a γ-ray energy of E_γ (in keV),

$$1 \text{ mm sec}^{-1} = 11.6 \times \frac{E_\gamma}{14.4} \text{ MHz}$$

Referring again to the abscissa of Fig. 15–2, one sees that the energy difference between the nuclear transitions for Fe^{3+} and Fe^{III} in $FeFe(CN)_6$ is very small, corresponding to about 2×10^{-8} eV. The peak in the spectrum in Fig. 15–2 at 0.03 mm sec^{-1} is assigned[6] to Fe^{III} and that at 0.53 to the cation Fe^{3+} by comparison of this spectrum with those for a large number of cyanide complexes of iron. Different line positions that result from different chemical environments are indicated by the values for the source velocity in units of cm^{-1} or mm sec^{-1}, and are referred to as *isomer shifts*, *center shifts*, or *chemical shifts*. We shall now proceed with a discussion of the information contained in the parameters obtained from the spectrum.

15–2 INTERPRETATION OF ISOMER SHIFTS

The two different peaks in Fig. 15–2 arise from the isomer shift differences of the two different iron atoms in octahedral sites. The isomer shift results from the electrostatic interaction of the charge distribution in the nucleus with the electron density that has a finite probability of existing at the nucleus. Only s electrons have a finite probability of overlapping the nuclear density, so the isomer shift can be evaluated by considering this interaction. It should be remembered that p, d, and other electron densities can influence s electron density by screening the s density from the nuclear charge. Assuming the nucleus to be a uniformly charged sphere of radius R and the s electron density over the nucleus to be a constant given by $\psi_s^2(0)$, the difference between the electrostatic interaction of a spherical distribution of electron density with a point nucleus and that for a nucleus with radius R is given by

$$\delta E = K[\psi_s^2(0)]R^2 \qquad (15\text{–}5)$$

where K is a nuclear constant. Since R will have different values for the ground state and the excited state, the electron density at the nucleus will interact differently with the two states and thus will influence the energy of the transition; i.e.,

$$\delta E_e - \delta E_g = K[\psi_s^2(0)](R_e^2 - R_g^2) \qquad (15\text{–}6a)$$

where the subscript e refers to the excited state and g to the ground state. The influence of $\psi_s^2(0)$ on the energy of the transition is illustrated in Fig. 15–3 for ^{57}Fe, which has $I = \frac{1}{2}$ for the ground state and $I = \frac{3}{2}$ for the excited state. The energies of these two states are affected differently by $\psi_s^2(0)$, and the transition energy is changed.

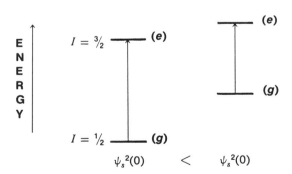

FIGURE 15–3 Changes in the energy of the Mössbauer transition for different values of $\psi_s^2(0)$. This is a graphical illustration of equation (15–16a) with two different values of $\psi_s^2(0)$ for an ^{57}Fe nucleus. The differences in $\psi_s^2(0)$ must result from a cubic or spherical distribution of bonded atoms in order for this diagram to apply.

The R values are constant for a given nucleus, but $\psi_s^2(0)$ varies from compound to compound. The center shift in the Mössbauer spectrum is the difference between the energy of this transition in the sample (or absorber) and the energy of the same transition in the source. This difference is given by the difference in equations of the form of (15–6a) for the source and sample, or:

$$\text{C.S.} = K(R_e^2 - R_g^2)\{[\psi_s^2(0)]_a - [\psi_s^2(0)]_b\} \qquad (15\text{–}6b)$$

where the subscripts a and b refer to absorber and source, respectively. Standard sources are usually employed (*e.g.*, ^{57}Co in Pd for iron Mössbauer spectra, or BaSnO$_3$ for tin spectra). The ^{57}Co decays to ^{57}Fe in an excited state *via* electron capture. The excited ^{57}Fe decays to stable ^{57}Fe by γ-ray emission. When standard sources are employed $[\psi_s^2(0)]_b$ is replaced by a constant, C. Furthermore, the change in radius $R_e - R_g$ is very small, leading to the following commonly employed expression for the center shift:

$$\text{C.S.} = K' \frac{\delta R}{R} [\psi_s^2(0)_a - C] \tag{15-6c}$$

where $\delta R = R_e - R_g$; C is a constant characteristic of the source; and K' is $2KR^2$. Both K' and $\delta R/R$ are constants for a given nucleus, so the center shift is directly proportional to the s electron density at the sample nucleus. The term *center shift* is used for the experimentally determined center of the peak; the term *isomer shift* is now used when the center shift has been corrected for the small Doppler contribution from the thermal motion of the Mössbauer atom. The sign of δR depends upon the difference between the effective nuclear charge radius, R, of the excited and ground states ($R_e^2 - R_g^2$). For the ^{57}Fe nucleus, the excited state is smaller than the ground state, and an increase in s-electron density produces a negative shift. In tin, the sign of δR is positive, so the opposite trend of shift with s electron density is observed.

As mentioned above, electron density in p or d orbitals can screen the electron density from the nuclear charge by virtue of the fact that the electron density in d and p orbitals penetrates the s orbital. Hartree-Fock calculations show [6,7] that a decrease in the number of d electrons causes a marked increase in the total s electron density at the iron nucleus. Accordingly, with comparable ligands and with negative $\delta R/R$, Fe^{2+} has an appreciably larger center shift than Fe^{3+}. When these ions are examined in a series of molecules, the interpretation becomes more difficult, for the d, s and p electron densities are modified by covalent bonding. For ^{57}Fe, for example, an increase in $4s$ density decreases the center shift, while an increase in $3d$ density increases the center shift. A series of high spin iron complexes have been interpreted on this basis.[7] In the case of ^{119}Sn, the center shift increases with an increased s electron density and decreases with an increase in p electron density.[8]

15-3 QUADRUPOLE INTERACTIONS

The discussion of the center shift in the previous section applies to systems with a spherical or cubic distribution of electron density. As discussed in Chapter 14, the degeneracy of nuclear energy levels for nuclei with $I > \frac{1}{2}$ is removed by a non-cubic electron or ligand distribution. For non-integral spins, the splitting does not remove the $+$ or $-$ degeneracy of the m_I levels, but we obtain a different level for each $\pm m_I$ set. Thus, the electric field gradient can lead to $I + \frac{1}{2}$ different levels for half-integer values of I (*e.g.*, two for $I = \frac{3}{2}$ corresponding to $\pm \frac{1}{2}$ and $\pm \frac{3}{2}$). For integer values of I we obtain $2I + 1$ levels (*e.g.*, five for $I = 2$ corresponding to 2, 1, 0, -1, -2). The influence of this splitting on the nuclear energy levels and the spectral appearance is illustrated[9] in Fig. 15-4 for ^{57}Fe. The ground state is not split but the excited state is split, leading to two peaks

632 Chapter 15 Mössbauer Spectroscopy

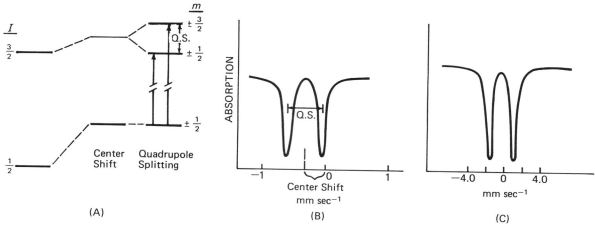

FIGURE 15–4 The influence of a non-cubic electronic environment on (A) the nuclear energy states of ^{57}Fe and (B) the Mössbauer spectrum. (C) The iron MB spectrum of Fe(CO)$_5$ at liquid N$_2$ temperature.[9]

in the spectrum. The center shift is determined from the center of the two resulting peaks. When both the ground and excited states have large values for I, complex Mössbauer spectra result.

The Hamiltonian for the quadrupole coupling is the same as that discussed for nqr.

$$\hat{H}_Q = \frac{e^2 Qq}{4I(2I-1)} [(3\hat{I}_z^2 - I(I+1) + (\eta/2)(\hat{I}_+^2 + \hat{I}_-^2)]$$

For the $I = 3/2$ case (^{57}Fe and ^{119}Sn), the quadrupole splitting Q.S. is given by

$$\text{Q.S.} = \frac{1}{2} e^2 Qq (1 + \eta^2/3)^{1/2} \qquad (15\text{-}7)$$

The symbols have all been defined in the nqr chapter. For ^{57}Fe, q and η cannot be determined from the quadrupole splitting. The sign of the quadrupole coupling constant is another quantity of interest. If $m_I = \pm 3/2$ is at high energy, the sign is positive; the sign is negative if $\pm 1/2$ from $I = 3/2$ is highest. From powder spectra, the intensities of the transitions to $\pm 1/2$ and $\pm 3/2$ are similar, and it becomes difficult to determine the sign. The sign can be obtained from spectra of ordered systems or from measurement of a polycrystalline sample in a magnetic field (*vide infra*). For systems in which the I values of the ground and excited states are larger than those for iron, the spectra are more complex and contain more information. The splitting of the excited state will not occur in a spherically symmetric or cubic field but will occur only when there is a field gradient at the nucleus caused by asymmetric p or d electron distribution in the compound. A field gradient exists in the trigonal bipyramidal molecule iron pentacarbonyl, so a splitting of the nuclear excited state is expected, giving rise to a doublet in the spectrum as indicated in Fig. 15–4(C).

If the t_{2g} set and the e_g set of orbitals in octahedral transition metal ion complexes have equal populations in the component orbitals, the quadrupole splitting will be zero. Low spin iron(II) complexes (t_{2g}^6) will not give rise to quadrupole splitting unless the degeneracy is removed, and these orbitals can interact differently with the ligand molecular orbitals. On the other hand, high spin iron(II) ($t_{2g}^4 e_g^2$) has an imbalance in the t_{2g} set, and a large quadrupole splitting is often seen. If the ligand environment about iron(II) were perfectly octahedral, d_{xy}, d_{yz}, and d_{xz} would be degenerate and no splitting would be detected. However, this system is subject to Jahn-Teller distortion, which can lead to a large field gradient. When the energy separaton of the t_{2g} orbitals from Jahn-Teller effects is of the order of magnitude of kT, a very temperature-dependent quadrupole splitting is observed. The ground state in the distorted complex can be obtained if the sign of q is known. The sign can be obtained from oriented systems or from studies in a large magnetic field. Similar considerations apply to high spin and low spin iron(III) compounds.

The factors contributing to the magnitude of the field gradient were discussed in Chapter 14. It was shown there that these data were of limited utility in providing further information about bond types.

15–4 PARAMAGNETIC MÖSSBAUER SPECTRA

As shown in Fig. 15–4, a non-cubic electronic environment in an iron complex splits the degenerate $I = ^3/_2$ excited state of ^{57}Fe into m_I states $\pm ^1/_2$ and $\pm ^3/_2$. The Mössbauer spectrum consists of a doublet corresponding to the two transitions shown in Fig. 15–4. When an effective magnetic field, H_{eff}, acts on this system, the degeneracies of the $\pm ^1/_2$ ground state as well as the $\pm ^3/_2$ excited states are removed[10] as shown in Fig. 15–5. The ordering of the levels reflects the fact that the ground-state moment is positive and the excited state moment is negative. The resulting six-line spectrum is shown in Fig. 15–5(B). Since the transitions are magnetic dipole in character, the selection rule is $\Delta m = 0$, ± 1 as shown in Fig. 15–5(A). The Hamiltonian is

$$\hat{H}_m = -g_N \mu_N \hat{I} \cdot H_{eff} \tag{15–8}$$

where $g_N = 0.18$ and $I = ^1/_2$ for the nuclear ground state and $g = -0.10$ and $I = ^3/_2$ for the excited state. The intensity of the lines changes with θ, the orientation of the H_{eff} relative to the direction of the Mössbauer radiation. For an ordered sample, the $\Delta m = 0$ transition has zero intensity for a field parallel to the radiation. When $\theta = 90°$, the $\Delta m = 0$ transitions are the most intense of the six.

The effective field of equation (15–8) consists of the internal field and the applied field:

$$H_{eff} = H + H_{int} \tag{15–9}$$

and is given by

$$H_{int} = -\langle S \rangle \cdot \mathbf{A}/g_N \mu_N \tag{15–10}$$

where $\langle S \rangle$ is the expectation value of the spin discussed in Chapters 11 and 12 and **A** the hyperfine coupling tensor. The internal magnetic field controls the magnetic hyperfine and it depends in a complex way on the zero field splitting

FIGURE 15-5
Magnetic and quadrupole splitting in a ferromagnetic ^{57}Fe compound. (A) Energy level diagram. (B) Expected Mössbauer spectrum. [Copyright © 1973 McGraw-Hill Book Co. (UK) Limited. From G. M. Bancroft, "Mössbauer Spectroscopy." Reproduced by permission.]

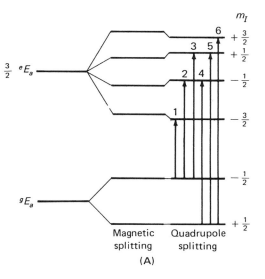
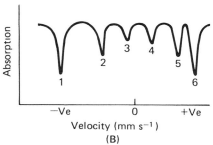

parameters D and E, on the electronic g tensors and $\mathbf{A}$ tensors, and on the orientation of the molecule relative to the applied field.

Internal fields exist in some systems in the absence of an applied magnetic field. The $\langle S \rangle$ of equation (15–10) is zero for diamagnetic systems and for systems with integer values of S in the absence of an applied field.[11] These spectra consist of quadrupole doublets. For compounds with half-integer spin (Kramers' doublets), two cases need be considered. With fast electron relaxation, the $\langle S \rangle$ averaged over all thermally accessible states is zero in zero magnetic field, leading to quadrupole doublets. In the slow relaxation limit, magnetic Mössbauer spectra can be obtained, generally at liquid helium temperatures. These spectra generally are studied at weak applied field to simplify them by decoupling hyperfine interactions with ligand nuclei.

In applied fields, magnetic Mössbauer splittings can be observed for both integer and half-integer spin systems. The applied field mixes the states producing a non-zero expectation value of the spin. The resulting spectra are interpreted using the spin Hamiltonian formalism[10,12] given in equation (15–11).

$$\hat{H} = D\left[\hat{S}_z^2 - \frac{1}{3}S(S+1)\right] + E(\hat{S}_x^2 - \hat{S}_y^2) + \beta\hat{S}\cdot\mathbf{g}\cdot\vec{H} + \hat{S}\cdot\mathbf{A}\cdot\hat{I}$$

$$- g_N\beta_N\hat{I}\cdot\vec{H} + \frac{e^2Qq}{4I(2I-1)}\{3\hat{I}_z^2 - I(I+1) + \eta(\hat{I}_x^2 - \hat{I}_y^2)\}_{\text{EFG}}$$

(15–11)

By standing the spectra as a function of temperature, field strength, and field direction, the various quantities in the spin Hamiltonian can be determined. For Kramers' doublets complimentary information can be obtained from the epr. However for integer S systems, *e.g.*, high spin iron(II), magnetic Mössbauer provides the sole means of obtaining electronic structural details.[13] Reference 14 presents applications of magnetic Mössbauer.

15-5 MÖSSBAUER EMISSION SPECTROSCOPY

^{57}Co decays by an electron capture to ^{57}Fe ($T_{1/2}$ for ^{57}Fe is 0.1 μsec), populating an excited state of the iron nucleus. The emitted γ-rays can be absorbed by a standard single-line absorber to investigate the energy levels of the ^{57}Fe nuclei produced when the source decays. The cobalt-57 analogue of the compound to be studied is prepared and used as the source. Information regarding the short-lived iron complex in the source is obtained[15] from this experiment. One must be sure that the desired iron complex remains intact when the high energy iron atoms are formed in the cobalt decay process. The results obtained from some oxygenated complexes of ^{57}Co protoporphyrin IX dimethyl ester as a function of the axial base attached are shown in Table 15-1. These are to be compared with values of ΔE_Q and δ for oxygenated hemoglobin, obtained by absorption measurements, of 2.23 and 0.27 respectively.

TABLE 15-1. Mössbauer Emission Studies.[12]
(Quadrupole splittings and isometric shifts for oxygenated complexes of ^{57}Co protoporphyrin IX dimethyl ester. The ligands coordinated trans to dioxygen are listed in the first column.)

Ligand	ΔE_Q (mm/sec)	δ_{Fe} (mm/sec)
1-methyl imidazole	2.17	0.29
1,2-dimethyl imidazole	2.32	0.30
pyridine	2.28	0.27
piperidine	2.25	0.30
ethylmethyl sulfide	2.27	0.30

Fig. 15-6(A) shows the emission spectrum of the five-coordinate 1-methyl imidazole complex before oxygenation; Fig. 15-6(B) shows the spectrum of the same complex after oxygenation.

This technique is particularly important when the parent iron compound is difficult to prepare and isolate.

15-6 APPLICATIONS

A few chemical applications of Mössbauer spectroscopy have been selected for discussion that are illustrative of the kind of information that can be obtained. Table 15-2 summarizes pertinent information about isotopes that have been studied by this technique.

Facsimiles of spectra obtained on some iron complexes are given in Fig. 15-7. As mentioned previously, for high spin iron complexes in which all six

636 Chapter 15 Mössbauer Spectroscopy

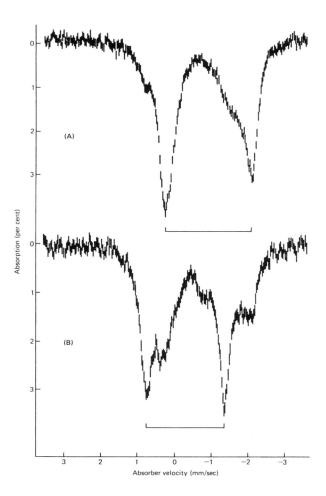

FIGURE 15–6 Emission Mössbauer spectra of (A) the 1-methyl imidazole adduct of cobalt protoporphyrin IX dimethyl ester and (B) its O_2 adduct.

ligands are equivalent, a virtually spherical electric field at the nucleus is expected for $Fe^{3+}(d^5)(t_{2g}{}^3 e_g{}^2)$ but not for $Fe^{2+}(d^6)(t_{2g}{}^4 e_g{}^2)$. As a result of the field gradients at the nucleus, quadrupole splitting should be detected in the spectra of high spin iron(II) complexes but not for high spin iron(III) complexes. This is borne out in spectra A and B of the complexes illustrated in Fig. 15–7. For low spin complexes, iron(II) has a configuration $t_{2g}{}^6$ and iron(III) has $t_{2g}{}^5$. As a result, quadrupole splitting is now expected for iron(III) but not iron(II) in the strong field complexes. This conclusion is confirmed experimentally by the spectra of ferrocyanide and ferricyanide ions. When the ligand arrangement in a strong field iron(II) complex does not consist of six equivalent ligands, e.g., $[Fe(CN)_5NH_3]^{3-}$, quadrupole splitting of the strong field iron(II) will result. The quadrupole splitting is roughly related to the differences in the d orbital populations (see Chapter 3) by

$$q_{\text{valence}} = K_d\left[-N_{d_{z^2}} + N_{d_{x^2-y^2}} + N_{d_{xy}} - \frac{1}{2}(N_{d_{xz}} + N_{d_{yz}})\right]$$

TABLE 15–2. Mössbauer Isotopes of Chemical Interest[a]

Isotope	Gamma energy (keV)	Half life[b] of precursor	Half width (mm sec^{-1})	Spin	Q[c] (Barns)	σ_0[d] (10^{-19} cm^2)	Natural abundance (%)	$10^4 \dfrac{\delta R}{R}$
^{57}Fe	14.4	270 d	0.19	$\frac{1}{2} \to \frac{3}{2}$	+0.3	23.5	2.19	-18 ± 4
^{61}Ni	67.4	1.7 h; 3.3 h	0.78	$\frac{3}{2} \to \frac{5}{2}$	0.13	7.21	1.19	
^{67}Zn	93.3	60 h; 78 h	3.13×10^{-4}	$\frac{5}{2} \to \frac{3}{2}$	+0.17	1.18	4.11	
^{73}Ge	67.0	76 d	2.19	$\frac{9}{2} \to \frac{7}{2}$	−0.26	3.54	7.76	
^{83}Kr	9.3	83 d, 2.4 h	0.20	$\frac{9}{2} \to \frac{7}{2}$	+0.44	18.9	11.55	$+4 \pm 2$
^{99}Ru	90	16.1 d	0.15	$\frac{5}{2} \to \frac{3}{2}$	>0.15	1.42	12.72	
^{107}Ag	93.1	6.6 h	6.68×10^{-11}	$\frac{1}{2} \to \frac{7}{2}$	—	0.54	51.35	
^{119}Sn	23.9	245 d	0.62	$\frac{1}{2} \to \frac{3}{2}$	−0.07	13.8	8.58	$+3.3 \pm 1$
^{121}Sb	37.2	76 y	2.10	$\frac{5}{2} \to \frac{7}{2}$	−0.29	2.04	57.25	-8.5 ± 3
^{125}Te	35.5	58 d	4.94	$\frac{1}{2} \to \frac{3}{2}$	±0.19	2.72	6.99	+1
^{129}I	27.8	33 d; 70 m	0.63	$\frac{7}{2} \to \frac{5}{2}$	−0.55	3.97	0	+3
^{129}Xe	39.6	1.6×10^7 y	6.84	$\frac{1}{2} \to \frac{3}{2}$	−0.41	1.95	26.44	0.3
^{133}Cs	81.0	7.2 y	0.54	$\frac{7}{2} \to \frac{5}{2}$	−0.003	1.02	100	
^{177}Hf	113.0	6.7 d, 56 h	4.66	$\frac{7}{2} \to \frac{9}{2}$	+3	1.20	18.50	
^{181}Ta	6.25	140 d, 45 d	6.48×10^{-3}	$\frac{7}{2} \to \frac{9}{2}$	+4.2	17.2	99.99	
^{182}W	100.1	115 d	2.00	$0 \to 2$	−1.87	2.46	26.41	+1.3
^{187}Re	134.2	23.8 h	203.8	$\frac{5}{2} \to \frac{7}{2}$	+2.6	0.54	62.93	
^{186}Os	137.2	90 h	2.37	$0 \to 2$	+1.54	2.89	1.64	
^{193}Ir	73.1	32 h	0.59	$\frac{3}{2} \to \frac{1}{2}$	+1.5	0.30	62.7	+0.6
^{195}Pt	98.7	183 d	17.30	$\frac{1}{2} \to \frac{3}{2}$	—	0.63	33.8	
^{197}Au	77.3	65 h, 20 h	1.85	$\frac{3}{2} \to \frac{1}{2}$	+0.58	0.44	100	±3

[a] Copyright © 1973 McGraw-Hill, (UK) Limited. From G. M. Bancroft, "Mössbauer Spectroscopy." Reproduced by permission.
[b] d = days, h = hours, y = years, m = minutes.
[c] The ground-state quadrupole moment, where both ground and excited states have $I > \frac{1}{2}$.
[d] Cross-section for absorption of a Mössbauer gamma ray.

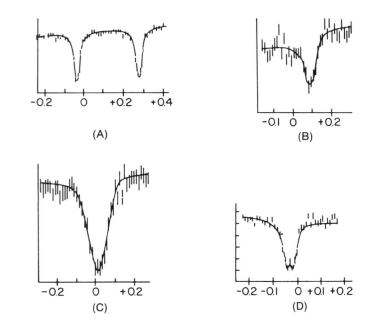

FIGURE 15–7 Mössbauer spectra of some iron(II) and iron(III) complexes.
(A) Spin-free iron(II)—FeSO$_4$·7H$_2$O.
(B) Spin-free iron(III)—FeCl$_3$.
(C) Spin-paired iron(II)—K$_4$Fe(CN)$_6$·3H$_2$O.
(D) Spin-paired iron(III)—K$_3$Fe(CN)$_6$. [From P. R. Brady, P. P. F. Wigley, and J. F. Duncan, Rev. Pure Appl. Chem., *12*, 181 (1962).]

Here, q_{valence} is the contribution to q from valence electrons in the d orbitals. For p-electrons we have

$$q_{\text{valence}} = K_p\left[-N_{p_z} + \frac{1}{2}(N_{p_y} + N_{p_x})\right]$$

TABLE 15–3. Quadrupole Splitting, ΔE_Q, and Isomer Shift, δ, for Some Iron Compounds (δ and ΔE_Q in mm sec^{-1})

Compound	ΔE_Q	δ	Compound	ΔE_Q	δ
High Spin Fe(II)			**Low Spin Fe(II)**		
FeSO$_4$·7H$_2$O	3.2	1.19	K$_4$[Fe(CN)$_6$]·3H$_2$O	—	−0.13
	3.15	1.3		—	−0.16
FeSO$_4$ (anhydrous)	2.7	1.2		<0.1	+0.05
Fe(NH$_4$)$_2$(SO$_4$)$_2$·6H$_2$O	1.75	1.19	Na$_4$[Fe(CN)$_6$]·10H$_2$O	<0.2	−1.01
	1.75	1.3	Na$_3$[Fe(CN)$_5$NH$_3$]	0.6	−0.05
FeCl$_2$·4H$_2$O	3.00	1.35	K$_2$[Fe(CN)$_5$NO]	1.85	−0.27
FeC$_4$H$_4$O$_6$	2.6	1.25		1.76	−0.28
FeF$_2$	2.68	—	Zn[Fe(CN)$_5$NO]	1.90	−0.27
FeC$_2$O$_4$·2H$_2$O	1.7	1.25			
High Spin Fe(III)			**Low Spin Fe(III)**		
			K$_3$[Fe(CN)$_6$]	—	−0.12
FeCl$_3$·6H$_2$O	0.2	0.85		—	−0.17
FeCl$_3$ (anhydrous)	0.2	0.5		0.26	−0.15
FeCl$_3$·2NH$_4$Cl·H$_2$O	0.3	0.45	Na$_3$[Fe(CN)$_6$]	0.60	−0.17
Fe(NO$_3$)$_3$·9H$_2$O	0.4	0.4			
Fe$_2$(C$_2$O$_4$)$_3$	0.5	0.45			
Fe$_2$(C$_4$H$_4$O$_6$)$_3$	0.77	0.43			
Fe$_2$O$_3$	0.12	0.47			

Values measured at room temperature for ΔE_Q and the isomer shift, δ, for a number of iron complexes have been collected[1] and are listed in Table 15–3. For iron complexes, isomer shifts in a positive direction correspond to a decrease in electron density in the region of the nucleus. For high spin complexes, a correlation exists between isomer shift and s electron density. An increase in δ of 0.2 mm sec^{-1} is equivalent to a decrease in charge density of 8% at the nucleus.[8] The negative values obtained for the low spin ferricyanides compared to high spin iron(III) complexes indicate more electron density at the nucleus in the ferricyanide ions. This has been explained as being due to extensive π bonding in the ferricyanides, which removes d electron density from the metal ion, which in turn decreases the shielding of the s electrons. This effect increases s electron density at the nucleus and decreases δ. Both strong σ donors and strong π acceptors decrease δ.

The MB spectrum of the material prepared from iron(III) sulfate and $K_4Fe(CN)_6$ is identical to the spectra for the compounds prepared either from iron(II) sulfate and $K_3Fe(CN)_6$ or by atmospheric oxidation of the compound from iron(II) sulfate and $K_4Fe(CN)_6$. The spectra of these materials indicate that the cation is high spin iron(III), while the anion is low spin iron(II).

The MB spectrum of sodium nitroprusside, $Na_2Fe(CN)_5NO$, has been investigated.[9] This material has been formulated earlier as iron(II) and NO^+ because the complex is diamagnetic. The MB spectrum consists of a doublet with a ΔE_Q value of 1.76 mm sec^{-1} and a δ value of -0.165 mm sec^{-1}. Comparison of this value with reported results[8] on a series of iron complexes led the authors to conclude that the iron δ value is close to that of iron(IV). The magnetism and MB spectrum are consistent with a structure in which there is extensive π bonding between the odd electron in the t_{2g} set of the orbitals of iron and the odd electron on nitrogen, as illustrated in Fig. 15–8. The filled π bonding orbital would need a large contribution from the nitrogen atomic orbital, and the empty π antibonding orbital would have a larger contribution from the iron atomic orbital to produce iron(IV). More of the π electron density would be localized on nitrogen and the δ value for iron would approach that of iron(IV) because of decreased shielding of the s electrons by the d electrons. Since electron density is being placed in what was previously a π antibonding orbital of nitric oxide, a decrease in the N—O infrared stretching frequency is observed. The very large quadrupole splitting is consistent with very extensive π bonding in the Fe—N—O link.

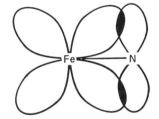

FIGURE 15–8 Orbitals involved in the Fe—N π bonding to the NO group in $[Fe(CN)_5(NO)]^{2-}$.

Mössbauer spectra are often useful in determining the oxidation state of atoms. It has been shown that the spectra originally assigned to some high spin iron(II) chelates of salicylaldoxime and other chelates were, in effect, those of an iron(III) oxidation product. The oxidized materials gave a center shift [relative to $Na_2Fe(CN)_5NO \cdot 2H_2O$] of ~ 0.6 mm sec^{-1} and a small quadrupole splitting. This smaller-than-expected quadrupole splitting for iron(II) was rationalized in terms of π-backbonding. The authentic material prepared under conditions that were rigorously air-free gave normal iron(II) center shifts and large quadrupole splittings, ~ 1.4. Values for the center shifts of various oxidation states of high spin iron compounds are shown in Table 15–4.

The effect of pressure on the Mössbauer spectra of a wide variety of iron compounds is reported.[16] Upon application of pressure (~ 165 kbar) to trisacetylacetonato iron(III), a new species forms that is attributed to iron(II). The change is reversed upon removal of the pressure.

The determination of the oxidation states of tin compounds from MB spectra

TABLE 15–4. Isomer Shifts for High Spin Iron Compounds [Shifts Relative to Na$_2$Fe(CN)$_5$NO · 2H$_2$O]

Oxidation state	+1	+2	+3	+4	+6
I.S.	~+2.2	~+1.4	~+0.7	~+0.2	~−0.6

is not as clear-cut as in the case of iron. Values of δ below 2.65 mm sec^{-1} are often due to tin(IV), and those above that value to tin(II). Exceptions are known. The isomer shifts of some four- and six-coordinate tin(IV) componds vary considerably as the attached anion is varied.[17]

The MB spectra of the iron pentacarbonyls Fe(CO)$_5$, Fe$_2$(CO)$_9$, and Fe$_3$(CO)$_{12}$ have been reported.[18,19] The results are as expected from the known structures for both Fe(CO)$_5$ and Fe$_2$(CO)$_9$. The structure of Fe$_3$(CO)$_{12}$ deduced from its MB spectrum was at odds with infrared results and a preliminary x-ray study. The MB indicated more than one type of iron, as shown in Fig. 15–9(A). The outer two lines are assigned to one type of iron and the inner two to a second type. In general, the areas are roughly proportional to the number of a particular type of iron present. A definitive crystal structure study has supported the structure shown in Fig. 15–9(B), which was predicted from the Mössbauer results. The spectra of Fe(II)X$_2$(CO)$_2$P$_2$ (where X = Cl, Br, and I, and P = phosphines and phosphites) have been interpreted in terms of the five different isomers that exist.

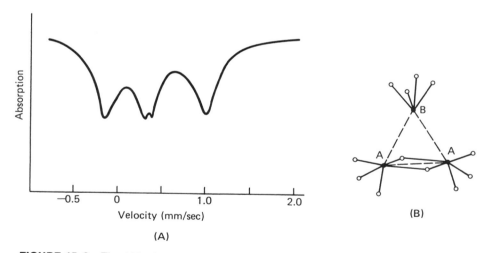

FIGURE 15–9 The Mössbauer spectrum (A) and structure (B) of Fe$_3$(CO)$_{12}$.

Several systems involving spin equilibrium between high spin and low spin iron(II) complexes have been studied by Mössbauer spectroscopy. A typical result[20] involves the hexadentate ligand {4-[(6-R)-2-pyridyl]-3-azabutenyl}$_3$ amine. The spectra obtained when two or three of the R groups are methyl are characteristic of low spin iron(II) ($^1A_1$) at 77 K, while at 294 K the large isomer shift and quadrupole splitting found are characteristic of high spin iron(II) ($^5T_2$). At intermediate temperatures, both forms are observed in the spectrum. This establishes that it is a true equilibrium ($^5T_2 \rightleftarrows {}^1A_1$) and that these states are long-lived on the Mössbauer time scale; *i.e.*, the lifetime must be 10^{-9} sec or greater.

There are several interesting biological applications of Mössbauer spectroscopy.[5,12,21,22] Horseradish peroxidase is an iron(III) heme protein. It can be oxidized in two one-electron steps, producing red and green compounds. The Mössbauer spectrum changes[23] upon oxidation to either form and in both cases is interpreted in terms of iron(IV). The removal of the second electron leading to the green form is believed to come from the protein or porphyrin.

Mössbauer studies have been of considerable utility in the study of the redox centers that exist in several classes of iron-sulfur proteins. A crystal structure investigation of the ferredoxin HP_{red} photosynthetic (high potential) protein from *Chromatium* has been carried out, and it has been shown to contain an $Fe_4S_4(SCys)_4$ (SCys refers to cystein) core with the cubane-like structure shown in Fig. 15–10. Similar units are present in other proteins. The compound HP_{red} is readily oxidized to form HP_{ox}. The crystal structure does not provide information about the oxidation states of the iron or the charge on the cluster. The Mössbauer spectra of several proteins containing this unit consist of a quadrupole split doublet.[24] By comparing the isomer shift in HP_{red} with those of other iron systems, it was concluded[24] that the core consists of an average of two Fe^{3+} and two Fe^{2+} ions. The irons are antiferromagnetically coupled, leading to a diamagnetic material, and the metal electrons in the system are delocalized so that only one kind of iron atom exists. The sensitivity of isomer shift to oxidation state is indicated in Table 15–5.

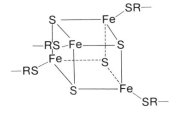

FIGURE 15–10 Cubane-like structure of the $Fe_4S_4(S—R—)_4$ core of some ferredoxins.

TABLE 15–5. Center Shifts for Various Iron-Sulfur Compounds

Species^a	δ
Fe^{3+} (rubredoxin)	0.25
$3Fe^{3+} + 1Fe^{2+}$ (*Chromatium* HP_{ox})	0.32
$2Fe^{3+} + 2Fe^{2+}$ (*Chromatium* HP_{red} and ox. ferredoxin)	0.42
$1Fe^{3+} + 3Fe^{2+}$ (red. ferredoxin)	0.57
Fe^{2+} (rubredoxin)	0.65

^{a} An entry such as $3Fe^{3+} + 1Fe^{2+}$ is meant to imply an average oxidation state for the iron atoms corresponding to the given combination.

The model compound $[(C_2H_5)_4N]_2[Fe_4S_4(SCH_2C_6H_5)_4]$ has been prepared.[25] The charge on this anion is known to be minus two, so it must be a $2Fe^{2+} + 2Fe^{3+}$ case. The Mössbauer spectrum[26] shown in Fig. 15–11 and the crystal structure are similar to those of HP_{red} or oxidized ferredoxin with equivalent iron atoms and an isomer shift of 0.36 mm sec^{-1}. This supports the assignment[24] of the corresponding ferredoxins as $2Fe^{2+} + 2Fe^{3+}$ systems.

The Mössbauer spectrum of $(Et_4N)_2[Fe_4S_4(SCH_2C_6H_5)_4]$ shown in Fig. 15–11(A) is a quadrupole split doublet arising from the low symmetry about the iron center. The lines are further split by a magnetic field, as shown in (B). The solid line in (A) is a least squares fit of the data, employing Lorentzian line shapes. The solid line in (B) is a computer-generated spectrum employing $H = 80$ kilogauss, $\eta = 0$, and $\Delta E_Q = 1.26$ mm sec^{-1}. The sign of the principal component of the field gradient is found to be positive from the spectral fit.

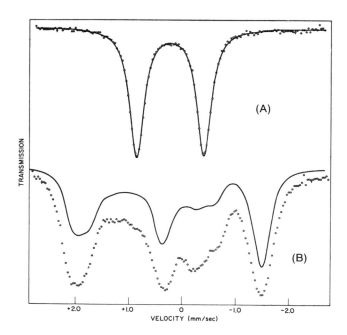

FIGURE 15–11 Mössbauer spectrum of $(Et_4N)_2[Fe_4S_4(SCH_2C_6H_4)_4]$ (A) at 1.5 K and $H_0 = 0$, and (B) at 4.7 K and $H_0 = 80$ kOersteds. [Reprinted with permission from R. H. Holm, et al., J. Amer. Chem. Soc., 96, 2644 (1974). Copyright by the American Chemical Society.]

As we have mentioned several times, it is difficult to interpret field gradients. However, it has been found possible to parametrize ions and groups attached to a central metal ion and to use these parameters, called *additive partial quadrupole splittings*, to predict the quadrupole coupling. The basic model is one of a point charge. In a diagonal electric field gradient coordinate system, the contributions to V_{xx}, V_{yy}, and V_{zz} from a charge Z are given by

$$V_{xx} = Zer^{-3}(3\sin^2\theta\cos^2\varphi - 1)$$
$$V_{yy} = Zer^{-3}(3\sin^2\theta\sin^2\varphi - 1)$$
$$V_{zz} = Zer^{-3}(3\cos^2\theta - 1)$$

where θ and φ have their usual definitions in the polar coordinate system. For a tetragonal complex (Fig. 15–12), for example, we must take each ligand (a point charge) and sum the individual contributions to V_{xx}, V_{yy}, and V_{zz}. Table 15–6 contains a summary of this calculation for the *cis* and *trans* complexes in Fig. 15–12.

Summing the contributions in Table 15–6, we obtain for the *trans* complex

$$V_{xx} = V_{yy} = (-2[A] + 2[B])e$$
$$V_{zz} = (4[A] - 4[B])e$$

where [A] and [B] are unspecified contributions from ligands A and B, respectively. For the *cis* complex we obtain

$$V_{xx} = V_{yy} = ([A] - [B])e$$
$$V_{zz} = (-2[A] + 2[B])e$$

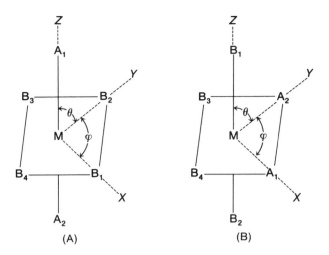

FIGURE 15–12 Geometries and coordinates for (A) *trans* and (B) *cis* MB$_4$A$_2$.

TABLE 15–6. Individual Point Charge Contributions to the EFG Tensor in *trans*- and *cis*-MA$_2$B$_4$.[a] The Quantity [A] Equals $Z_A e/r_A^3$.

			trans-MA$_2$B$_4$						
Ligand (Fig. 15–12)	θ	φ	sin θ	cos θ	sin φ	cos φ	V_{xx}/e	V_{yy}/e	V_{zz}/e
A$_1$	0	0	0	1	0	1	−[A]	−[A]	+2[A]
A$_2$	180	0	0	−1	0	1	−[A]	−[A]	+2[A]
B$_1$	90	0	1	0	0	1	+2[B]	−[B]	−[B]
B$_2$	90	90	1	0	1	0	−[B]	+2[B]	−[B]
B$_3$	90	180	1	0	0	−1	+2[B]	−[B]	−[B]
B$_4$	90	270	1	0	−1	0	−[B]	+2[B]	−[B]
			cis-MA$_2$B$_4$						
Ligand	θ	φ	sin θ	cos θ	sin φ	cos φ	V_{xx}/e	V_{yy}/e	V_{zz}/e
A$_1$	90	0	1	0	0	1	+2[A]	−[A]	−[A]
A$_2$	90	90	1	0	1	0	−[A]	+2[A]	−[A]
B$_1$	0	0	0	1	0	1	−[B]	−[B]	+2[B]
B$_2$	180	0	0	−1	0	1	−[B]	−[B]	+2[B]
B$_3$	90	180	1	0	0	−1	+2[B]	−[B]	−[B]
B$_4$	90	270	1	0	−1	0	−[B]	+2[B]	−[B]

[a] Copyright © 1973 McGraw-Hill (UK) Limited. From G. M. Bancroft, "Mössbauer Spectroscopy." Reproduced by permission.

The ratio of the *trans* to *cis* quadrupole splittings is

$$\frac{(4[A] - 4[B])e}{(-2[A] + 2[B])e} = -2$$

If both isomers are available, Mössbauer spectra provide a ready means of distinguishing them, with this ratio. In the partial quadrupole coupling approach, magnitudes are empirically assigned to [A] and [B], and the field gradients are predicted. A *cis* or *trans* geometry is readily determined with the above equations. For other geometries, the appropriate equations must be derived. The compound $Fe(CO)_2(PMe_3)_2I_2$ can exist as five different isomers whose calculated quadrupole splittings differ considerably. Two isomers can be prepared with observed quadrupole splittings of -0.90 and $+1.31$ mm sec^{-1}. These are readily identified as the (*trans* P, *cis* I, *cis* CO) and *all trans* isomers, respectively.[27] Partial quadrupole splitting parameters[27] for various ligands bonded to iron(II) are listed in Table 15–7.

When the values for the ligands in Table 15–7 are used with appropriate equations for the complexes in Table 15–8, the values listed under the column labeled "predicted" result.[27]

TABLE 15–7. Ligand Partial Quadrupole Splitting (PQS) Parameters[27] for Iron(II)[a]

Ligand	PQS value	Ligand	PQS value
NO^{+}[b]	$+0.01$	$P(OPh)_3$	-0.55
X^-	-0.30	CO	-0.55
N_2	-0.37	PPh_2Et	-0.58
N_3^-	-0.38	PPh_2Me	-0.58
CH_3CN	-0.43	depb/2	-0.59
$SnCl_3^-$	-0.43	$P(OEt)_3$	-0.63
H_2O[b]	-0.45	depe/2	-0.65
$SnPh_3$	-0.50	$P(OMe)_3$	-0.65
NCS^-	-0.51	PMe_3	-0.66
$AsPh_3$	-0.51	dmpe/2	-0.70
NH_3[b]	-0.52	ArNC	-0.70
NCO^-	-0.52	CN^-[b]	-0.84
PPh_3	-0.53	H^-	-1.04

[a] Copyright © 1973 McGraw-Hill (UK) Limited. From G. M. Bancroft, "Mössbauer Spectroscopy." Reproduced by permission.
[b] PQS values derived from room temperature data.

TABLE 15–8. Predicted and Observed[27,28] QS (mm sec^{-1}) at 295 K[a]

	Observed	Predicted
trans-$FeCl_2(ArNC)_4$	$+1.55$	
cis-$FeCl_2(ArNC)_4$	-0.78	-0.78
$[FeCl(ArNC)_5]ClO_4$	0.73	$+0.78$
trans-$Fe(SnCl_3)_2(ArNC)_4$	$+1.05$	
cis-$Fe(SnCl_3)_2(ArNC)_4$	0.50	-0.52
cis-$FeClSnCl_3(ArNC)_4$	0.61	-0.69 ($\eta = 0.60$)
trans-$FeH_2(depb)_2$	-1.84	
trans-$FeHCl(depe)_2$	<0.12	-0.20
trans-$Fe(EtNC)_4(CN)_2$	-0.60	
cis-$Fe(EtNC)_4(CN)_2$	0.29	$+0.30$
trans-$[FeH(ArNC)(depe)_2]BPh_4$	-1.14	-0.98
trans-$[FeH(CO)(depe)_2]BPh_4$	1.00	-0.46

[a] Copyright © 1973 McGraw-Hill (UK) Limited. From G. M. Bancroft, "Mössbauer Spectroscopy." Reproduced by permission.

Other parameters have been reported for use with tin compounds.[27] The approach has been applied to a large number of systems with a high degree of success.[27,28] There clearly are some shortcomings in the point charge model,[29] but more work is required to enable one to predict when it will break down in empirical type applications.

The sign and magnitude of the field gradient can be used to provide information about the electronic ground state of a transition metal ion complex. The *approximate* value of the field gradient for different ground states can be estimated by adding the d orbital contributions of the populated orbitals, employing Table 15–9.

TABLE 15–9. Magnitude of q and η for Various Atomic Orbitals

Orbital	q	η
p_z	$-\dfrac{4}{5}\langle r^{-3}\rangle$	0
p_x	$+\dfrac{2}{5}\langle r^{-3}\rangle$	-3
p_y	$+\dfrac{2}{5}\langle r^{-3}\rangle$	$+3$
$d_{x^2-y^2}$	$+\dfrac{4}{7}\langle r^{-3}\rangle$	0
d_{z^2}	$-\dfrac{4}{7}\langle t^{-3}\rangle$	0
d_{xy}	$+\dfrac{4}{7}\langle r^{-3}\rangle$	0
d_{xz}	$-\dfrac{2}{7}\langle r^{-3}\rangle$	$+3$
d_{yz}	$-\dfrac{2}{7}\langle r^{-3}\rangle$	-3

The quantity $\langle r^{-3}\rangle$ is the expectation value of $1/r^3$ for the appropriate orbital function. This table is constructed by using the various orbital functions to evaluate the matrix elements:

$$q_{zz} = q = -\left\langle \psi \left| \frac{3\cos^2\theta - 1}{r^3} \right| \psi \right\rangle$$

$$\eta q = \left\langle \psi \left| \frac{3\sin^2\theta \cos 2\varphi}{r^3} \right| \psi \right\rangle$$

In a typical application, the $a_{1g}^{\;2}(d_{z^2})$, $e_{2g}^{\;4}(d_{xy}, d_{x^2-y^2})$ ground state for ferrocene is predicted to have a q_{zz} value of $2(-4/7\langle r^{-3}\rangle) + 4(4/7\langle r^{-3}\rangle) = 8/7\langle r^{-3}\rangle$. A large positive field gradient is observed. In forming the ferricenium cation, a decrease in the quadrupole splitting is observed, which is consistent with the loss of an e_{2g} electron.[30]

REFERENCES CITED

1. P. G. Debrunner and H. Frauenfelder, in "Introduction to Mössbauer Spectroscopy," ed. L. May, Plenum Press, New York, 1971.
2. G. M. Bancroft, "Mössbauer Spectroscopy—An Introduction for Inorganic Chemists and Geochemists,' McGraw-Hill (UK), 1973.
3. M. Weissbluth, Structure and Bonding, *2*, 1 (1967).
4. A. J. Bearden and W. R. Dunham, Structure and Bonding, *8*, 1 (1970).
5. T. C. Gibb and N. N. Greenwood, "Mössbauer Spectroscopy," Chapman & Hall, 1971.
6. J. F. Duncan and P. W. R. Wigley, J. Chem. Soc., 1963, 1120.
7. R. E. Watson and A. J. Freeman, Phys. Rev., *120*, 1125 (1960).
8. L. R. Walker, G. K. Wertheim, and V. Jaccarino, Phys. Rev. Lett., *6*, 98 (1961).
9. L. M. Epstein, J. Chem. Phys., *36*, 2731 (1962).
10. a. E. Münck and B. H. Huynh in "ESR and NMR of Paramagnetic Species in Biological and Related Systems," eds. I. Bertini and R. S. Drago, D. Reidel, Dordrecht, 1980.
    b. E. Münck, Methods Enzymol., *54*, 346 (1978).
11. A. Abragam and B. Bleaney, "Electron Paramagnetic Resonance of Transition Ions," Chapter 15, Oxford University Press, New York, 1970.
12. a. R. L. Collins and J. C. Travis, Mössbauer Effect Methodol., *3*, 123 (1967).
    b. G. Lang, Quart. Rev. Biophys., *3*, 1 (1970).
13. a. E. Münck and P. M. Champion, J. Physique, Suppl. *12*/35, C6–33 (1974).
    b. "Mössbauer Spectroscopy," Topics in Applied Physics, Volume 5, ed. U. Gonser, Springer Verlag, New York, 1975.
14. a. R. L. Collins, J. Chem. Phys., *42*, 1072 (1965).
    b. E. Münck, in "The Porphyrins," ed. D. H. Dolphin, Academic Press, New York, 1975.
    c. W. T. Oosterhuis, Structure and Bonding, *20*, 59 (1974).
    d. G. Lang, Quart. Rev. Biophys., *3*, 1 (1970).
15. L. Marchant, M. Sharrock, B. M. Hoffman, and E. Münck, Proc. Natl. Acad., Sci. USA, *69*, 2396 (1972).
16. A. R. Champion, R. W. Vaughan, and H. G. Drickamer, J. Chem. Phys., *47*, 2583 (1967) and references therein.
17. R. V. Parish, Prog. Inorg. Chem., *15*, 101 (1972).
18. R. H. Herber, W. R. Kingston, and G. K. Wertheim, Inorg. Chem., *2*, 153 (1963).
19. R. H. Herber, R. B. Kingston, and G. K. Wertheim, Inorg. Chem., *3*, 101 (1964).
20. M. A. Hoselton, L. J. Wilson, and R. S. Drago, J. Amer. Chem. Soc., *97*, 1722 (1975).
21. W. Marshall and G. Lang, Proc. Phys. Soc. (London), *87*, 3 (1966).
22. T. H. Moss, A. Ehrenberg, and A. J. Bearden, Biochemistry, *8*, 4159 (1969).
23. P. Debrunner, in "Spectroscopic Approaches to Biomolecular Conformation," ed. D. W. Urry, American Medical Association Press, Chicago, 1969.
24. C. L. Thompson *et al.*, J. Biochem. *139*, 97 (1974) and references therein.
25. R. H. Holm, Endeavour, 1975, 38 and references therein.
26. R. H. Holm *et al.*, J. Amer. Chem. Soc., *96*, 2644 (1974) and references therein.
27. G. M. Bancroft, Coord. Chem. Rev., *11*, 247 (1973) and references therein.
28. G. M. Bancroft and K. D. Butler, Inorg. Chim. Acta, *15*, 57 (1975).
29. A. P. Marks, R. S. Drago, R. H. Herber, and M. J. Potasek, Inorg. Chem., *15*, 259 (1976).
30. W. H. Morrison, Jr., and D. N. Hendrickson, Inorg. Chem., *13*, 2279 (1974).

SERIES

"Mössbauer Effect Methodology," Volumes 1 to 9, Plenum Press, New York, 1965.

EXERCISES

1. What effect does increasing electron density at the nucleus have on the relative energies of the ground and excited states of ^{57}Fe and ^{119}Sn? Explain the expected isomer shift from this change in terms of effective nuclear charge radii of these states.

2. Suppose you read an article in which the author claimed that the two peaks in a MB spectrum of a low spin iron(III) complex were the result of Jahn-Teller distortion. Criticize this conclusion.

3. Draw the structure for SnF$_4$ and explain why quadrupole splitting is observed in this compound but not in SnCl$_4$.

4. Suppose you were interested in determining whether Sn—O or Sn—S π bonding were present, and which was greater, in the compounds (C$_6$H$_5$)$_3$SnOCH$_3$ and (C$_6$H$_5$)$_3$SnSCH$_3$. Describe experiments involving MB spectroscopy that might shed light on this problem.

5. Would you expect the ΔE_Q value to be greatest in SnCl$_2$Br$_2$, SnCl$_3$Br, or SnF$_3$I? Explain.

6. The product obtained from the reaction of ferrous sulfate and potassium ferricyanide gives rise to the spectrum below. Interpret this spectrum.

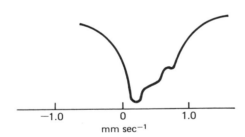

7. Using a point charge expression, indicate the three diagonal components of the electric field gradient for the two isomers of MA$_3$B$_3$. Predict the sign of the quadrupole splitting in the two isomers.

8. (CH$_3$)$_2$SnCl$_2$ is a chloro-bridged polymer with octahedral coordination about tin. The methyl groups are *trans*. With the CH$_3$—Sn bond as the z field gradient axis, predict the signs of q and e^2Qq.

9. Which componds would have the largest quadrupole splittings for the starred atom:

    a. cis or trans *Fe(CO)$_4$Cl$_2$?

    b. (CH$_3$)$_3$*SnCl (C$_{3v}$) or polymeric, chloro-bridged (CH$_3$)$_3$*SnCl (D$_{3h}$ about tin)?

    c. high spin cis or trans *Fe(NH$_3$)$_4$Cl$_2$?

10. The Mössbauer spectrum for Fe(CO)$_5$ at liquid nitrogen temperature is shown below.

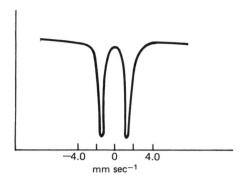

a. Why are the unusual units on the *x*-axis equivalent to energy?

b. What gives rise to the doublet observed? Show (and label) the energy levels involved.

11. The enzyme putidaredoxin has sites with two iron atoms. The oxidized form gives rise to the spectrum shown below:

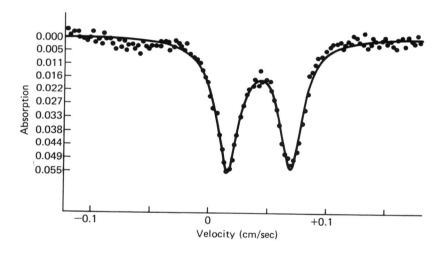

The *g*-tensors are anisotropic. Does the enzyme consist of a single type of iron site or two different iron sites?

12. One of the Mössbauer spectra below is $K_3Fe(CN)_6$. The other is $K_4Fe(CN)_6$. Which is which? Why?

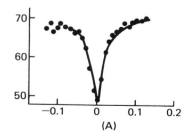

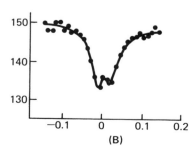

13. An article [J. Amer. Chem. Soc., 97, 6714 (1975)] dealt with a series of square-pyramidal iron(III) complexes having an N_4S (the sulfur donor is axial) set of donor atoms. These complexes varied considerably in electronic structure, depending upon the exact nature of the N_4 macrocyclic ligand and the axial ligand, as shown by various spectroscopic measurements.

   a. Using one-electron energy level diagrams, indicate which spin states might reasonably be expected for square-pyramidal iron(III). Remember that the relative energies of the *d* orbitals may vary slightly as a function of the donor set.

   b. Using spectra from the figure below, assign probable spin states to complexes A, B, and C. Explain what features of the Mössbauer spectra influenced your decision, and why.

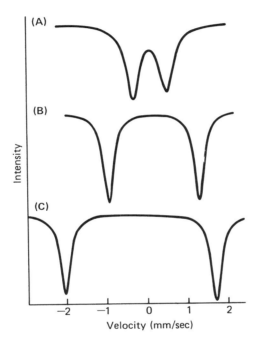

16 Ionization Methods: Mass Spectrometry, Ion Cyclotron Resonance, Photoelectron Spectroscopy

Mass Spectrometry

16-1 INSTRUMENT OPERATION AND PRESENTATION OF SPECTRA

There are many different types of mass spectrometers available, and the details of the construction and relative merits of the various types of instruments have been described.[1-6] Most of the basic principles of mass spectrometry can be illustrated with the aid of Fig. 16-1. The sample, contained in a reservoir, is added through the port, enters the ion source (a), and passes through the electron beam at (c). (The beam is indicated by a dotted line.) Interaction of the sample with an energetic electron produces a positive ion, which moves toward the accelerating plates (d) and (e) because of a small-potential difference between the back wall (inlet) and the front wall of this compartment. The back wall, which

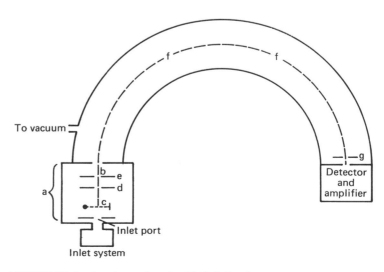

FIGURE 16-1 A schematic of a 180° deflection mass spectrometer.

is positively charged relative to the front, attracts and discharges the negative ions. The positive ions pass through (d) and (e), are accelerated by virtue of a large potential difference (a few thousand volts) between these plates, and leave the ion source through (b). Charged ions move in a circular path under the influence of a magnetic field. The semicircle indicated by (f) is the path traced by an accelerated ion moving in a magnetic field of strength H. The radius of the semicircle, r, depends upon the following: (1) the accelerating potential, V [*i.e.*, the potential difference between plates (d) and (e)]; (2) the mass, m, of the ion; (3) the charge, e, of the ion; and (4) the magnetic field strength, H. The relationship between these quantities is given* by equation (16–1):

$$\frac{m}{e} = \frac{H^2 r^2}{2V} \qquad (16\text{–}1)$$

When the ions pass through slit (g) into the detector, a signal is recorded. The source and the path through which the ions pass must be kept under 10^{-7} mm of mercury pressure in order to provide a long mean free path for the ion. The sample vapor pressure in the inlet should be at least 10^{-2} mm Hg at the temperature of the inlet system, although special techniques permit use of lower pressures. In general, only one or two micromoles of sample are required.

Since H, V, and r can be controlled experimentally, the ratio m/e can be determined. Note that a dipositive ion of mass 54 gives rise to the same m/e as a monopositive ion of mass 27. Under usual conditions for running a mass spectrometer, most of the ions are produced as singly charged species. Doubly charged ions are much less frequently encountered, whereas more highly charged ions are not present in significant concentrations.

Obtaining the mass spectrum consists in determining the m/e ratio for all fragments produced when a molecule is bombarded by a high intensity electron beam. To do this, the detector slit could be moved and the value of r measured for all particles continuously produced by electron bombardment in the ion source. This is not feasible experimentally. It is much simpler to vary H or V continuously [see equation (16–1)] so that all particles eventually travel in a semicircle of fixed radius. The signal intensity, which is directly related to the number of ions striking the detector, can be plotted as a function of H or V, whichever is being varied. In practice it is easiest to measure a varying potential, so the field is often held constant. If H and r are constant, V is inversely related to m/e [by equation (16–1)] and m/e can be plotted versus the signal intensity to produce a conventional mass spectrum. A narrow region of a typical spectrum is illustrated in Fig. 16–2.

On instruments in which the potential is varied, the results are readily presented in terms of m/e values. The accuracy varies with mass, as can be seen by comparing the potential difference corresponding to m/e values of 400

---

* This equation is simply derived: The potential energy of the ion, eV, is converted to kinetic energy $(^1/_2)mv^2$ (where m is the mass and v the velocity). At full acceleration,

$$eV = \tfrac{1}{2}mv^2 \qquad (16\text{–}2)$$

In the magnetic field the centrifugal force mv^2/r is balanced by centripetal force Hev. Solving $Hev = mv^2/r$ for r yields $r = mv/eH$. Using this equation to eliminate v from (16–2) produces (16–1).

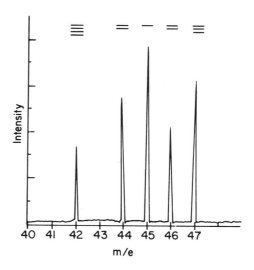

FIGURE 16-2 Mass spectrum of fragments with *m/e* in the range from 40 to 48. The peaks have been automatically attenuated, and the number of horizontal lines above each peak indicates the attenuation factor.

and 401 with the difference for *m/e* values of 20 and 21 [calculated by using equation (16–1)]. With modern low-resolution instruments, one can count on ± 0.4 to ± 1 *m/e* accuracy up to 1000 mass units. For high-resolution work, accuracy to 2 or 3 ppm can be obtained up to very high masses (>3000) if reference compounds are available that have accurately known peaks within 10 to 12% of the unknown mass peaks. Perfluorokerosine is a common reference compound, with many peaks that are accurately known up to about 900 mass units.

The mass spectrum obtained from the instrument may be a long and cumbersome record. As a result, the data are often centroided and summarized by a bar graph that plots intensity (relative abundance of each fragment) on the ordinate and the *m/e* ratio on the abscissa. The relative abundance is given as the percent intensity of a given peak relative to the most intense peak in the spectrum. Often it is informative to express the intensity in terms of *total ionization units*, S. This is an expression of the percentage that each peak contributes to the total ionization, and is obtained by dividing the intensity of a given peak by the sum of the intensities of all the peaks. When the spectrum has not been obtained over the range from 1 to the molecular weight of the material, the lower limit is indicated on the ordinate as a subscript to Σ (*e.g.*, Σ_{12} means that the spectrum was recorded from 12 to the molecular weight). A typical graph is illustrated in Fig. 16–3. For the comparison of the intensity of various peaks in the same spectrum, the relative abundance is adequate. For comparison of the intensity of peaks in different spectra, total ionization units should be employed. In general, a peak with an intensity equal to 0.5% of Σ is easily detected.

Most modern spectrometers contain computers to store and process the large amount of information from molecular fragmentation and to control the experimental variables involved in running the instrument. Accurate masses can be obtained with computer interfacing by introducing standards with the sample. The computer recognizes the standard peaks and employs them to determine the mass of the unknown peaks. Computer libraries[7] are available to match the sample spectra with that of known compounds in 70-eV electron impact ionization.

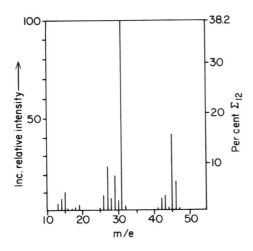

FIGURE 16–3
Representation of the mass spectrum of ethanol.

Our next concern is with the method of indicating the resolving power of the instrument. For many inorganic and organic applications, it is necessary to know the m/e ratio to an accuracy of one unit (*i.e.*, whether it is 249 or 250). The resolution of the instrument is sometimes expressed as $m/\Delta m$, where two peaks m and $m + \Delta m$ are separately resolved and the minimum intensity between the two peaks is only 10% of the total of m. For example, a resolution of 250 means that two peaks with m/e values of 251 and 250 are separated and that at the minimum between them the pen returns to the base line to within 10% of the total ion current (plotted as intensity in Fig. 16–2). Instruments with poor resolution will not do this with high mass peaks, and the magnitude of the m/e value for which peaks are resolved is the criterion for resolution.

Modern mass spectrometers provide a wide range of options in inlet systems, ion sources, mass analyzers, detectors, and signal processors. Batch inlet systems volatilize a sample externally and allow it to leak into the ionization region. The sample must have an appreciable vapor pressure below $\sim 500°$ C. The direct probe inlet introduces non-volatile and thermally unstable compounds with a sample probe that is inserted through a vacuum lock. The output from a chromatographic column can also be used as the inlet in GC-MS systems.

Most commercial mass spectrometers are equipped with accessories that allow use of several ion sources interchangeably. Some common ones are listed in Table 16–1. The desorption types use a sample probe from which energy directly transfers ions or molecules in the condensed phase to ions in the gas phase. Non-volatile and thermally fragile molecules can be studied. Electron impact, EI, uses a beam of energetic electrons to ionize the gaseous molecule. Field ionization and field desorption employ large voltages (10^{-20} kV) to produce large electric fields (10^8 V/cm) to ionize molecules. Gaseous samples are used in the former and non-volatile samples on a probe in the latter. Fragmentation of the molecule is minimized with the latter technique and the molecular ion, M^+, and one with a mass one unit larger, $(M + 1)^+$, are the major peaks in the spectrum. In chemical ionization, CI, the sample is bombarded by positive ions produced by electron bombardment of a gaseous atom or molecule (often CH_4) present with the sample and in 1000- to 10,000-fold

TABLE 16-1. Some Common Ion Sources

Name	Abbreviation	Type	Ionizing Agent
electron ionization	EI	gas phase	energetic electrons
field ionization	FI	gas phase	high-potential electrode
chemical ionization	CI	gas phase	reagent positive ions or electron capture
fast atom bombardment	FAB	desorption^a	energetic atoms
field desorption	FD	desorption^b	high-potential electrode
laser desorption	LD	desportion^b	laser beam
plasma desorption	PD	desorption^b	high-energy fission fragments from ^{252}Cf
secondary ion mass spectrometry	SIMS	desorption^b	1 to 20 keV ions
thermal desorption	TIMS	desorption^b	heat

^{a} Samples as solids or solutions
^{b} Samples as solids, gases, or solutions.

excess. We can illustrate the reactions in chemical ionization that lead to product ions with CH_4 as an example. The CH_4^+ ion formed by ionization reacts with another methane molecule forming CH_5^+. The CH_5^+ or CH_4^+ can react with substrate SH forming SH_2^+ and CH_4 or CH_3 forming the substrate $(M + 1)^+$ ion. With some substrates, CH_5^+ or CH_4^+ can abstract hydride forming H_2 and S^+, where S^+ is the $(M - 1)^+$ ion. In both instances, reactions of the ion with the sample lead to sample ions. Cluster ions can also form in CI, e.g., reactions occur to form $(M + C_2H_5)^+$ and $(M + C_3H_5)^+$ peaks in methane CI. Less fragmentation results with CI than with EI. In fast atom bombardment (FAB), the sample in a condensed state (often a glycerol matrix) is ionized by bombardment with energetic Xe or Ar *atoms*. Molecular ions are often formed even for large molecular-weight samples.

Analyzers are commonly radio-frequency filters or electromagnets in combination with electrostatic fields. The radial magnetic field shown in Fig. 16–1 employs single-focusing magnetic sector analysis. Double focusing analyzers first pass the beam through a radial electrostatic field and then the radial magnetic field. The electrostatic field focuses particles with the same kinetic energy on a second slit which introduces the beam into the magnetic field. Much better resolution is achieved. The quadrupole analyzer consists of four, short, parallel metal rods arranged symmetrically around the beam. One pair of opposite rods is attached to the positive terminal of a variable dc source and the other pair to the negative terminal. On top of this, variable radio-frequency ac potentials, that are 180° out of phase are applied to each pair of rods. The combined fields cause the charged particles in the beam to oscillate about a central axis describing their direction of travel. Only fields with a certain mass to charge ratio pass through the rods. The others collide with one of the rods. The m/e ratio is scanned by varying the frequency of the ac source or by varying the potential of the two sources keeping their ratio and frequency constant. The ion trap uses a field generated by a sandwich geometry with a ring electrode in the middle and caps on each end to trap and store ions of a selected mass-to-charge ratio. The mass spectrum is produced by varying the electric field to sequentially eject ions of different m/e ratios for detection.

In a time-of-flight instrument, the semicircular path and magnet are replaced by a straight path (no magnetic field). The ions are not produced continuously but in spurts and are allowed to diffuse toward the detector. Heavy ions will move more slowly than light ions. The spectrum is a plot of intensity versus the time of flight of the particle. The application of the Fourier transform technique to mass spectrometry has many of the advantages of applying this technique to other spectroscopies. This method will be discussed in the next section on ion cyclotron resonance.

16–2 PROCESSES THAT CAN OCCUR WHEN A MOLECULE AND A HIGH ENERGY ELECTRON COMBINE

It is important to emphasize the rather obvious point that a detector analyzes only those species impinging on it. We must, therefore, be concerned not only with the species produced in the ionization process but also with the reactions these species may undergo in the 10^{-5} sec required to travel from the accelerating plates to the detector. When a molecule, A, is bombarded with electrons of moderate energy, the initial processes that can occur are summarized by equations (16–3) to (16–5):

$$A + e^- \rightarrow A^+ + 2e^- \qquad (16\text{--}3)$$

$$A + e^- \rightarrow A^{n+} + (n+1)e^- \qquad (16\text{--}4)$$

$$A + e^- \rightarrow A^- \qquad (16\text{--}5)$$

The process represented by reaction (16–3) is the most common and the most important in mass spectrometry. It will occur if the energy of the bombarding electron is equal to or higher than the ionization energy of the molecule (7 to 15 eV). When the energy of the bombarding electron beam is just equal to the ionization potential, all of the electron's energy must be transferred to the molecule to remove an electron. The probability of this happening is low. As the energy of the bombarding electron increases, the probability that a collision will induce ionization increases, and a higher intensity peak results. As the energy of the bombarding electron is increased further, much of this excess energy can be given to the molecular ion that is formed. This excess energy can be high enough to break bonds in the ion, and fragmentation of the particle results. The acceleration potential of the bombarding electron that is just great enough to initiate fragmentation is referred to as the *appearance potential* of the fragment ion. When the electron energy is large enough, more than one bond in the molecule can be broken. The following sequence summarizes the processes that can occur when a hypothetical molecule B—C—D—E is bombarded with an electron:

(a) Ionization process

$$BCDE + e^- \rightarrow BCDE^+ + 2e^- \qquad (16\text{--}6)$$

(b) Fragmentation of the positive ion

$$BCDE^+ \rightarrow B^+ + CDE\cdot \qquad (16\text{–}7)$$

$$BCDE^+ \rightarrow BC^+ + DE\cdot \qquad (16\text{–}8)$$

$$BC^+ \rightarrow B^+ + C\cdot \qquad (16\text{–}9)$$

$$\text{or} \quad BC^+ \rightarrow B\cdot + C^+ \qquad (16\text{–}10)$$

$$BCDE^+ \rightarrow DE^+ + BC\cdot \qquad (16\text{–}11)$$

$$DE^+ \rightarrow D^+ + E\cdot \qquad (16\text{–}12)$$

$$\text{or} \quad DE^+ \rightarrow D\cdot + E^+ \qquad (16\text{–}13)$$

$$BCDE^+ \rightarrow BE^+ + CD\cdot \quad \text{etc.} \qquad (16\text{–}14)$$

$$BCDE^+ \rightarrow CD^+ + BE\cdot \quad \text{etc.} \qquad (16\text{–}15)$$

(c) Pair production

$$BCDE + e^- \rightarrow BC^+ + DE^- + e^- \qquad (16\text{–}16)$$

(d) Resonance capture

$$BCDE + e^- \rightarrow BCDE^- \qquad (16\text{–}17)$$

Other modes of cleavage are possible, but only positively charged species will travel to the detector and give rise to peaks in the mass spectrum. For the scheme above, peaks corresponding to B^+, BC^+, C^+, DE^+, D^+, E^+, BE^+, and CD^+ will occur in the spectrum if B, C, D, and E have different masses. More energy will have to be imparted to the $BCDE^+$ ion to get cleavage into B^+, $C\cdot$, and $DE\cdot$ [equation (16–9)] than for cleavage into $BC^+ + DE\cdot$ [equation (16–8)].

In equation (16–14), the ion has rearranged in the dissociation process, leading to fragments that contain bonds that are not originally present in the molecule BCDE. These rearrangement processes complicate the interpretation of a mass spectrum, and experience is necessary to be able to predict when rearrangements will occur e.g., alkenes commonly rearrange. In general, rearrangements are invoked to explain the occurrence of peaks of unexpected mass or unexpected intensity.

Ion-molecule reactions can give rise to mass peaks in the spectrum that are greater than the molecular weight of the sample. This process is represented by equation (16–18):

$$BCDE^+ + CDE\cdot \rightarrow BCDEC^+ + DE\cdot \qquad (16\text{–}18)$$

The reaction involves a collision of the molecular ion with a neutral molecule and as such is a second-order rate process, the rate of which is proportional to the product of the concentrations of the reactants. The intensity of peaks resulting from this process will depend upon the product of the partial pressures of $BCDE^+$ and CDE. Examination of the spectrum at different pressures causes variation in the relative intensities of these peaks, and as a result, the occurrence of this process can be easily detected.

The reactions discussed above are all unimolecular decay reactions. Production of ions by electron bombardment often involves loss of the least tightly held electron, and ions are often formed in vibrationally excited states that have an excess of internal energy. In some molecules of the sample, a low energy electron is removed, leaving an ion in an excited electronic state. The excited state ion can undergo internal conversion of energy, producing the electronic ground state of the ion having an excess of vibrational energy. The molecule could dissociate in any of the excited states involved in the internal conversions associated with the radiationless transfer of energy. In other molecules of the sample, ions are formed with energy in excess of the dissociation energy. In this case, the ion will fragment as soon as it starts to vibrate. Thus, in a given sample, ions with a wide distribution of energies are produced and many mechanisms are available for fragmentation processes. It is informative to compare the time scales for some of the processes we have been discussing. The time for a bond vibration is $\sim 10^{-13}$ sec, the maximum lifetime of an excited state ion is $\sim 10^{-8}$ sec, and the time an ion spends in the mass spectrometer ion chamber is 10^{-5} to 10^{-6} sec. There is ample time for the excess electronic energy in an ion to be converted into an excess of vibrational energy in a lower electronically excited state. Accordingly, we view the processes in the ionization chamber as producing molecular ions in different energy states, which undergo rapid internal energy conversion to produce individual ions with varying amounts of excess energy. Fragmentation takes place *via* a first order process at different rates, depending on the electronic state and excess vibrational energy of the individual ion. This is why all of the different processes represented above for the fragmentation of the ion $BCDE^+$ can occur and be reflected in the mass spectrum.

16–3 FINGERPRINT APPLICATION

The fingerprint application is immediately obvious. For this purpose an electron beam of 70 eV is usually employed to yield reproducible spectra, for this accelerating potential is above the appearance potential of most fragments. As indicated by equations (16–6) to (16–16), a large number of different fragmentation processes can occur, resulting in a large number of peaks in the spectra of simple molecules. Figure 16–3 contains the peaks with appreciable intensity that are found in the mass spectrum of ethanol. Counting the very weak peaks, which are not illustrated, a total of about 30 peaks are found. These weak peaks are valuable for a fingerprint application but generally are not accounted for in the interpretation of the spectrum (*i.e.*, assigning the fragmentation processes leading to the peaks).

For a fingerprint application, the more peaks the better. Accordingly, EI methods are preferred. Spectral libraries for EI at 70 eV are available in text[8] and computer form[7] for matching an unknown spectrum to a known one. The relative intensities are of less help in unknown identification for they vary with instrument and with experimental conditions. It is best to make a direct comparison of the spectrum of the unknown compound with that of a known sample on the same instrument if a decision is to be based on relative intensities.

An interesting fingerprint application is illustrated in Fig. 16–4, where the mass spectra of the three isomers of ethylpyridine are indicated. Pronounced differences occur in the spectra of these similar compounds that are of value in the fingerprint application. On the other hand, similarities in the spectra of hydrocarbons makes the fingerprint application of limited utility for this class of compounds. Optical isomers and racemates give rise to identical spectra. Impurities create a problem in fingerprint applications because the major fragments of these impurities give rise to several low intensity peaks in the

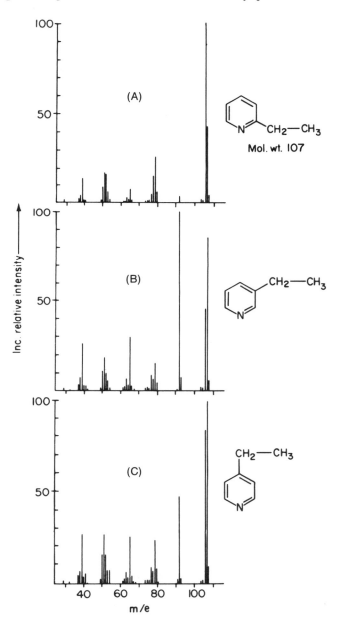

FIGURE 16–4 Mass spectra of three isomers of ethylpyridine. [Copyright © 1960 McGraw-Hill Book Company. From K. Biemann, "Mass Spectrometry." Reproduced by permission.]

spectrum. If the same material is prepared in two different solvents, the spectra may appear to be quite different if all solvent has not been removed. Contamination from hydrocarbon grease also gives rise to many lines.

16–4 INTERPRETATION OF MASS SPECTRA

The interpretation of a mass spectrum involves assigning each of the major peaks in a spectrum to a particular fragment. An intense peak corresponds to a high probability for the formation of this ion in the fragmentation process. In the absence of rearrangement [equation (16–14)], the arrangement of atoms in the molecule can often be deduced from the masses of the fragments that are produced. For example, a strong peak at $m/e = 30$ for the compound methyl hydroxylamine would favor the structure CH_3NHOH over H_2NOCH_3 because an $m/e = 30$ peak could result from cleavage of the O—N bond in the former case but cannot result by any simple cleavage mechanism from the latter compound. The higher mass fragments are usually more important than the smaller ones for structure determination.

It is often helpful in assigning the peaks in a spectrum to be able to predict probable fragmentation products for various molecular structures. The energy required to produce a fragment from the molecular ion depends upon the activation energy for bond cleavage, which is often related to the strength of the bond to be broken. The distribution of ions detected depends not only on this but also on the stability of the resulting positive ion. In most cases it is found that the stability of the positive ions is of greatest importance. This stability is related to the effectiveness with which the resulting fragment can delocalize the positive charge. Fragmentation of $HOCH_2CH_2NH_2^+$ can occur to produce $\cdot CH_2OH$ and $CH_2NH_2^+$ ($m/e = 30$) or $\cdot CH_2NH_2$ and CH_2OH^+ ($m/e = 31$). Since nitrogen is not as electronegative as oxygen, the resonance form $CH_2=NH_2^+$ contributes more to the stability of this ion than a similar form, $CH_2=OH^+$, does to its ion. As a result, charge is more effectively delocalized in the species $CH_2NH_2^+$ than in CH_2OH^+, and the $m/e = 30$ peak is about ten times more intense than the $m/e = 31$ peak. Charge is not stabilized as effectively by sulfur as it is by oxygen because carbon-sulfur π bonding is not as effective as carbon-oxygen π bonding. Thus, the $m/e = 31$ peak for CH_2OH^+ from $HSCH_2CH_2OH$ has about twice the intensity of the $m/e = 47$ peak that arises from CH_2SH^+.

Rearrangements of the positive ion will occur when a more stable species results. For example, the ion [structure] rearranges to [structure] and the benzyl cation rearranges to the tropylium ion, $C_7H_7^+$. An intense $m/e = 93$ peak from this fragment is seen in alkyl benzenes.

The production of many different fragments is often helpful in putting together the structure of the molecule. However, one must employ caution even in this application. The ion produced in the ion chamber undergoes many vibrations, during which rearrangement could occur to produce bonds that did not exist in the parent compound [see, *e.g.*, equation (16–14)]. The production of all these different ions makes it difficult to determine the chemical processes that lead to the various peaks. This in turn makes it difficult to infer the influence that bond strength or other properties of the molecule have on the relative abundances of the ion fragments formed. A quantitative treatment of mass spectrometric fragmentation has been attempted and is referred to as the *quasi-equilibrium theory*.[10] The internal energy is distributed over all the available oscillators and rotators in the molecule, and the rates of decomposition *via* different paths are calculated. A weighting factor or frequency factor (*i.e.*, an entropic term) is given to each vibration level. The full analysis is complex for a molecule of reasonable size. Approximations have been introduced, leading to a highly parametrized approach.[10,11]

The mass spectrometric shift rule[11] has been of considerable utility in elucidation of the structure of alkaloids and illustrates a basic idea of general utility. If there are low energy pathways for the breakdown of a complex molecule and this breakdown is not influenced by the addition of a substituent, the location of the substituent can often be determined. This is accomplished by finding an increase in the molecular weight of the fragment to which the substituent is bonded that corresponds to the weight of the substituent or a characteristic fragment of the substituent.

Mass spectrometry is used for routine sequencing of small peptides. The interested reader is referred to the review described in reference 12.

The low volatility of many substances hampers their analysis by mass spectroscopy. The volatility can often be increased by making derivatives of the polar groups in the molecule; *e.g.*, carboxyl groups can be converted to methyl esters or trimethylsilylesters. Field ionization techniques (*vide infra*) are also advantageous for this problem.

The combination of mass spectroscopy with GLC provides an excellent method for analysis of mixtures. Very small amounts of material are needed. The mass spectrometer may be used as the GLC detector, and numerous mass spectra can be accumulated as each component emerges from the column. A partially resolved GLC peak is readily detected by the change in mass spectra of the peak with time.

Many more examples and a thorough discussion of factors leading to stable ions produced from organic compounds are contained in the textbook references at the end of the chapter and reference 9. Generalizations for predicting when rearrangements are expected are also discussed. If, starting with a given structure, one can account for the principal fragments and assign the peaks in the mass spectrum by invoking a reasonable fragmentation pattern, this assignment amounts to considerable support for that structure.

16–5 EFFECT OF ISOTOPES ON THE APPEARANCE OF A MASS SPECTRUM

When the spectrum of a compound containing an element that has more than one stable, abundant isotope is examined, more than one peak will be found for each fragment containing this element. In the spectrum of CH_3Br, two peaks of nearly equal intensity will occur at m/e values of 94 and 96, corresponding mainly to $(CH_3{}^{79}Br)^+$ and $(CH_3{}^{81}Br)^+$. The abundances of ^{79}Br and ^{81}Br are almost the same (50.54 versus 49.46%), so two peaks of nearly equal intensity, separated by two mass units, will occur for all bromine-containing fragments. In a fragment containing two equivalent bromine atoms, a triplet with ratios 1:2:1 would result from different combinations of isotopes. In addition to these peaks, there will be small peaks resulting from the small natural abundances of D and ^{13}C, corresponding to all combinations of masses of ^{12}C, ^{13}C, D, H, ^{79}Br, and ^{81}Br. The resulting cluster of peaks for a given fragment is important in establishing the assignment of the peaks to a fragment. Their relative intensities will depend upon the relative abundances of the various naturally occurring isotopes of the atoms in the fragment; *e.g.*, CO$^+$ can consist of mass fragments at 28, 29, 30, and 31. The relative abundances of these fragments can be calculated from simple probability theory.[13,14] Computer programs have been reported to carry out these calculations.[15,16] These characteristic patterns are quite useful in assigning spectra of molecules that contain an atom with more than one abundant isotope. Molecules containing transition metals often give such isotope patterns. Use of the ^{13}C peaks enables one to determine the number of carbon atoms in a fragment.

The advantages of high resolution mass spectrometry can be illustrated[17] by the incorrect assignment of a peak at 56 in the low resolution spectrum of Fe[(C$_2$H$_5$)$_2$NCS$_2$]$_3$ to iron. At high resolution, a peak is expected at 55.9500, but none is found. Instead, one is obtained at 56.0350, which is assigned to the fragment C$_3$H$_6$N.

Another important application of the mass spectrometer involving isotopes is the study of exchange reactions involving nonradioactive isotopes. The product of the exchange from labeled starting material is examined for isotope content as a function of time to obtain the rate of exchange. The product or starting material can be degraded to a gaseous material containing the label, and the isotopic ratio is obtained from the mass spectrum. These materials may also be examined directly, and the location and amount of label incorporated can be deduced from an analysis of the change in spectrum of various fragments. By determining which peaks in the spectrum change on incorporation of the isotope, one can determine which parts of a molecule have undergone exchange. In the reaction of methanol with benzoic acid, it has been shown by a tracer study involving mass spectral analysis that the ester oxygen in the product comes from methanol:

$$C_6H_5C(=O)-OH + CH_3{}^{18}OH \rightarrow C_6H_5C(=O)-{}^{18}OCH_3 + H_2O$$

In another interesting application it was shown that the following exchange reactions occurred:

$$BF_3 + BX_3 \rightarrow BX_2F + BF_2X$$
$$3RBX_2 + 2BF_3 \rightarrow 3RBF_2 + 2BX_3 \text{ (also } BF_nX_m\text{)}$$
$$3R_2BX + BF_3 \rightarrow 3R_2BF + BX_3 \text{ (also } BF_nX_m\text{)}$$

where R is alkyl or vinyl and X is Cl or Br. Fragments corresponding to the products were obtained, although only starting materials were recovered on attempted separation.[18] A four-center intermediate of the type

$$\begin{array}{c} R \diagdown \quad Cl \diagdown \quad F \\ \quad B \quad \quad B \\ Cl \diagup \quad F \diagup \quad F \end{array}$$

was proposed for the exchange. In order to determine

whether or not alkyl groups were exchanged in the reaction:

$$3RBX_2 + 2BF_3 \rightarrow 3RBF_2 + 2BX_3 \text{ (and } BX_mF_n)$$

the boron trifluoride was enriched in ^{10}B. The absence of enrichment of ^{10}B in fragments in the mass spectrum containing alkyl or vinyl groups enabled the authors to conclude that neither alkyl nor vinyl groups were exchanged under conditions where RBX_2 species were stable.

It should be pointed out that in all of the above applications it is not necessary to label the compound completely. A slight enrichment will suffice.

16–6 MOLECULAR WEIGHT DETERMINATIONS; FIELD IONIZATION TECHNIQUES

Determination of the molecular weight requires methods that will produce the parent ion peak. The dramatic difference in EI and FI methods is illustrated with the mass spectra in Fig. 16–5. The M + 1 peak occurs from addition of a proton in the presence of the reagent ion. In some cases, proton abstraction leads to an M − 1 peak.

In Fig. 16–6, the spectra of glutamic acid[19a]

$$HOOCCH(NH_2)CH_2CH_2COOH$$

FIGURE 16–5 (A) Formula of D-ribose. (B) Electron bombardment mass spectrum. (C) Field ionization spectrum.

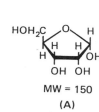

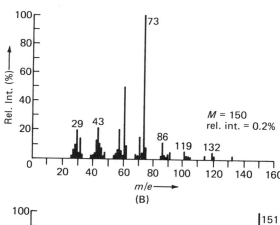

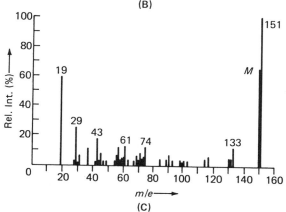

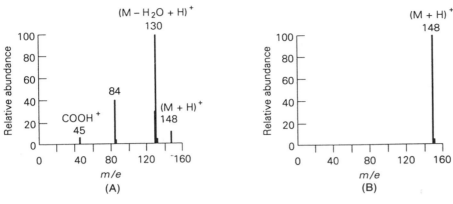

FIGURE 16-6 Mass spectra for glutamic acid: (A) field ionization; (B) field desorption.

obtained by field ionization and field desorption methods are compared. Very little fragmentation occurs with field desorption. Either method is adequate for molecular weight determination. FAB is the most frequently used method for molecular weight determination of non-volatile materials giving dominant $(M + H)^+$ and $(M - H)^+$ ions for organic materials. With organic materials and organometallic compounds, significant fragment ion information is obtained providing structural information. A comparison of the results on organometallic compounds from various ionization techniques is reported.[19b]

16-7 EVALUATION OF HEATS OF SUBLIMATION AND SPECIES IN THE VAPOR OVER HIGH MELTING SOLIDS

Evaluation of the heat of sublimation is based upon the fact that the intensity of the peaks in a spectrum is directly proportional to the pressure of the sample in the ion source. The sample is placed in a reservoir containing a very small pinhole (a Knudsen cell), which is connected to the ion source so that the only way that the sample can enter the source is by diffusion through the hole. If the cell is thermostated and enough sample is placed in the cell so that the solid phase is always present, the heat of sublimation of the solid can be obtained by studying the change in peak intensity (which is related to vapor pressure) as a function of sample temperature. The small amount of sample diffusing into the ion beam does not radically affect the equilibrium in the cell. Some interesting results concerning the nature of the species present in the vapor over some high melting solids have been obtained from this type of study. Monomers, dimers, and trimers were found over lithium chloride, while monomers and dimers were found in the vapor over sodium, potassium, and cesium chloride.[20]

The species Cr, CrO, CrO_2, O, and O_2 were found over solid Cr_2O_3. Appearance potentials and bond dissociation energies of these species are reported.[21] The vapors over MoO_3 were found to consist of trimer, tetramer, and pentamer. Vapor pressures, free energy changes, and enthalpies of sublimation were evaluated.[22]

16–8 APPEARANCE POTENTIALS AND IONIZATION POTENTIALS

As mentioned earlier, the molecular ion is produced whenever collision occurs with an electron with energy equal to or greater than the ionization energy of the molecule. A typical curve relating electron energy to the number of ion fragments of a particular type produced (*i.e.*, relative intensity of a given peak) is illustrated in Fig. 16–6. This is referred to as an ionization efficiency curve. At electron energies well below the ionization energy, no ions are produced. When the energy of the electron beam equals the ionization energy, a very low intensity peak results, for in the collision all of the energy of the electron will have to be imparted to the molecule, and this is not too probable. As the electron energy is increased, the probability that the electron will impart enough energy to the molecule to cause ionization is increased, and a more intense peak results until a plateau finally occurs in the curve. The tail of this curve at low energies results because of the variation in the energies of the electrons in the bombarding beam. Therefore, the curve has to be extrapolated (dotted line in Fig. 16–7) to produce the ionization energy. Various procedures for extrapolation and the error introduced by these procedures have been discussed in detail.[21] When the peak observed is that of the molecular ion, $e^- + RX \rightarrow RX^+ + 2e^-$, the ionization energy of the molecule can be obtained by extrapolation of the ionization efficiency curve. When the peak is that of a fragment, extrapolation of the ionization efficiency curve produces the appearance potential of that fragment. For example, if the peak being investigated is that of the fragment R^+ from the molecule R—X, the appearance potential, A_{R^+}, is obtained by extrapolation of the ion efficiency curve for this peak. The appearance potential is related to the following quantities:

$$A_{R^+} = D_{R-X} + I_R + E_k + E_e \qquad (16\text{–}19)$$

where D_{R-X} is the gas phase dissociation energy of the bond R—X; I_R is the ionization potential of R; E_k is the kinetic energy of the particles produced; and E_e is the excitation energy of the fragments (*i.e.*, the electronic, vibrational and

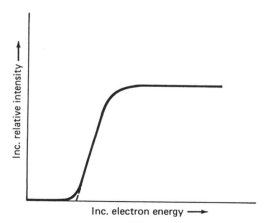

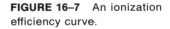
FIGURE 16–7 An ionization efficiency curve.

rotational energy if the fragments are produced in excited states). Generally, E_k and E_e are small and equation (16–19) is adequately approximated by:

$$A_{R^+} = D_{R-X} + I_R \qquad (16\text{–}20)$$

If D_{R-X} is known, I_R can be calculated from appearance potential data. Often, I_R is known, and D_{R-X} can be calculated. The value for I_R must be less than I_X for equation (16–20) to apply; otherwise, X^+ is dissociated or electronically excited. Experiments of this sort provide one of the best methods for evaluating bond dissociation energies but give less exact ionization energy data than can be obtained by other means.

An article on the mass spectrometric study of phosphine and diphosphine contains a nice summary of some of the information that can be obtained from these studies. Ionization energies and appearance potentials of the principal positive ions formed are reported. The energetics of the fragmentation processes are discussed and a mechanism is proposed.[22]

16–9 THE FOURIER TRANSFORM ION CYCLOTRON RESONANCE TECHNIQUE

FTICR/MS

In FTICR/MS, an ion is generated with mass, m, and charge, e, as previously discussed for mass spectroscopy. In a uniform magnetic field, the magnetic force acting on a particle with initial velocity causes the ion to follow a helical path along the axis of the magnetic field. The frequency of the circular motion in Hertz, *i.e.*, the cyclotron frequency is given by

$$v_c = eH/2\pi m \qquad (16\text{–}21)$$

The cyclotron frequency is independent of the velocity of the ion, v. The velocity distribution will give rise to a distribution of helical circular radii, r, for ions with the same cyclotron frequency according to

$$v_c = v/r$$

For magnetic fields of 1 T, the cyclotron frequency falls in the r.f. range (0.01 to 7.00 MHz).

In an ICR trapping cell, two trapping plates perpendicular to the magnetic field are used to prevent the ions from travelling along the field direction. The plates are maintained at a potential of about $+1$ V for positive ions and -1 V for negative ions. The magnetic field and trapping potential constrain the ions to a region of space in the center of the trap. Two parallel receiving plates and two parallel transmitter plates lying along the axis of the field and in between the trapping plates form a box shaped cell of six plates. The cell is mounted in a high vacuum in a strong magnetic field (typically ≥ 1 tesla).

Application of an external oscillating field with a frequency corresponding to the cyclotron frequency of an ion, v_c, across the transmitter plates causes ions of that mass [equation (16–21)] to move into resonance with the applied field and spiral out to orbits of larger radii. The kinetic energy of the particle is increased and its radius also increases according to

$$E_k = 2\pi^2 m v_c r^2 \qquad (16\text{-}22)$$

As the resonance condition is fulfilled, the random distribution of phases for the ions is changed to a "packet" of ions that all move simultaneously with the applied field. If the applied field is turned off or its frequency changed so it moves out of resonance with v_c, the ion packet persists long enough to induce an image current in the detector plate. The signal decays with time as collisions of the ion with neutral molecules restore the original random distribution of phases. The signal in the time domain contains information about the frequency of the ion and the ion concentration.[23]

A mass spectrum of the total ion population in the cell is detected by applying a field to the transmitter plates whose radio frequency changes rapidly over the range corresponding to the masses of interest [recall equation (16–21)]. As each ion undergoes resonance, a superposition of image currents is observed in the detector circuit and stored in a computer. The transmit-detect cycle is repeated many times and signal averaged. A Fourier transform of this data produces a plot of amplitude versus frequency or mass [equation (16–21)] of all the ions present, *i.e.*, the mass spectrum. The mass limit of detection depends on the magnetic field. At 3 T, good mass resolution up to 3000 amu results.

After the ions are formed in the cell and before detection, any one ion can be excited or, if they absorb sufficient energy, be caused to spiral to such large orbits that they are ejected from the cell. The excited ions collide with neutral molecules and fragment, a phenomenon called chemically induced dissociation (CID). This technique is advantageous for structure determination or for providing endothermicities of bond dissociative pathways. If the ejected ion is reformed with time by a chemical reaction, the rate of the reaction can be studied. By monitoring changes that occur in ion concentrations when a reactive species is introduced into the trapped ion cell, reaction rates and equilibria for gas phase reactions can be measured.[23,24]

Adiabatic electron affinities for a metal complex, M^-, can be studied by measuring the equilibrium constant for the reaction:

$$M^- + SO_2 \leftrightarrows SO_2^- + M$$

and knowing that the EA of SO_2 is 1.097 eV. If the reaction does not occur other reference compounds besides SO_2 can be employed. If the reaction goes to completion with a compound X and does not occur with SO_2, the electron affinity of M^- is bracketed between X and SO_2. About 100 organic compounds have been bracketed between known systems over a range of 3 eV by this technique.[25] Adiabatic ionization energies for a substance M are obtained in an analogous fashion by studying the reaction

$$M^+ + X \rightleftarrows M + X^+$$

where the IE of X is known.

Ligand exchange reactions have also been studied enabling one to obtain ligand basicity orders toward various cations.[26a]

$$ML^+ + L' \rightarrow ML'^+ + L$$

In some instances the equilibrium constant can be measured, and in other instances the relative basicities are estimated by laddering. Several excellent reviews provide details concerning the above applications.[26] Analytical applications of ICR have also been reviewed.[27]

Surface Science Techniques

16–10 INTRODUCTION

In this section we include techniques that, when applied to solids, provide information primarily about the surface of the material. Some of the methods can also be used to study gaseous and liquid samples. UPS is included in this section even though it is used primarily on gaseous samples because of its similarity to XPS. There is an immense literature background in this area. A brief introduction to selected methods will be presented that discusses the principles of operation and provides an illustration of the kind of information obtained. Most of the methods use an ion, electron, or x-ray source and monitor the energy or spatial resolution of the scattered (or resulting) ion, electron, or photon beam produced after the source impinges on the sample. Since the source can be very high in energy, induced reactions unfortunately can lead to the detection of materials that are not characteristic of the sample. One must also be concerned with the homogeneity of the surface when methods that produce information about a small area are employed. Methodology that permits examination of the surface lateral and transverse planes addresses this concern. Finally, one must be careful not to infer properties about the bulk sample from a surface measurement unless it is known that the surface actually represents the bulk property.

16–11 PHOTOELECTRON SPECTROSCOPY

Photoelectron spectroscopy has its basis in the photoelectric effect. There is a threshold frequency, v_t, required in order for radiation impinging on a solid to be able to eject an electron from the solid into a vacuum. The work function ϕ is defined as the corresponding minimum energy required,

$$e\phi = hv_t \qquad (16\text{–}23)$$

where e is the charge of the electron and h is Planck's constant. At higher frequencies of incident radiation, the emitted electrons have a kinetic energy corresponding to this excess frequency, *i.e.*,

$$E_{\text{kin}} = hv - e\phi \qquad (16\text{–}24)$$

The emitted electrons with maximum kinetic energy come from the conduction band. Electrons in orbitals below the conduction band are bound and emerge with less energy than those in the conduction band.

$$E_{kin} = h\nu - e\phi - E_B \qquad (16\text{–}25)$$

where E_B is the binding energy. Using photons whose energies are greater than the work function and measuring the kinetic energy of the ionized electrons produces the photoelectron spectrum. The spectrum is a plot of the number of electrons emitted versus the kinetic energy of the electron, (see Fig. 16–8). The peaks in the spectrum provide E_B for various electron states of the surface species. If a sample contains the same element in different chemical environments (*e.g.*, the nitrogen in N-N-O), different values for E_B will result. Only a single electron is ionized from a given molecule so the different signals arise from different molecules. The difference in energy for the different E_B values corresponding to different environments is called the *chemical shift*.

The source photon is monochromatic. Depending upon whether x-ray or ultraviolet radiation is employed, two different experiments result. The x-ray source[28a] experiment is called either x-ray photoelectron spectroscopy (XPS) or electron spectroscopy for chemical analysis (ESCA). The ultraviolet source[28b] is called ultraviolet photoelectron spectroscopy (UPS or PES). XPS is concerned with both valence and non-valence shell (core) electrons and UPS with valence shell electrons. We shall discuss the applications of these two spectroscopic methods separately.

XPS

Solids (including frozen solutions), gases, liquids, and solutions have been studied by XPS. Binding energies, E_B, of both core and valence electrons can be measured as seen in equation (16–25). In the XPS experiment, the binding energies are expressed relative to a reference level. Since the solid sample is in electrical contact with the spectrometer, the Fermi level, $e\phi$, of the sample and spectrometer are the same. As a result, only the spectrometer work function is needed to calculate E_B, referred to the Fermi level, from a measured kinetic energy. With an insulating sample a buildup of positive charge can occur near the surface resulting in a dipole layer and perturbation of the binding energies. An external standard with a known core energy, (*e.g.*, Au) can be deposited on the surface and binding energies measured relative to the standard. The energies can also be measured relative to the edge of the valence band.[29]

The binding energies of atoms vary with oxidation state and partial charges on the atoms. Compilations exist[30] that enable one to use binding energies to determine these quantities by a fingerprint type of application. Generally, the binding energies of cations increase with increasing oxidation state. Opposite behavior is observed for PbO in comparison to PbO_2 and this is attributed to differences in the Madelung potential in the two structures. Very small differences in binding energy are found for Ag_2O compared to Ag and for Cu_2O compared to Cu. On the other hand, the E_B (1s) for Cu and CuO differ by 4.4 eV.

It is interesting to note that linear correlations have been found between core-electron binding energies and Mössbauer isomer shifts for compounds of tin and iron.[31a] Also, a correlation of chlorine core binding energies with nuclear quadrupole resonance frequencies has been found.[31b]

Attempts have been made to correlate the binding energies to orbital energies from molecular orbital calculations. The basis for this approach is Koopmans' theorem, which states that the vertical ionization energy for removal of an electron from a molecular orbital corresponds to the negative of the eigenvalue obtained from the molecular orbital calculation. It is thus assumed that the molecular orbitals for the parent molecule and the ionized molecule are the same. Any electronic relaxation or change in correlation energies causes Koopmans' theorem to break down leading to limited quantitative success for this approach.[32]

The photoemission process comprises the following events:

1. Absorption of the photon by the bound electron, thereby increasing the kinetic energy of the electron.
2. Ejection of the electron from the atom using some of its kinetic energy to overcome the Coulomb attraction of the nucleus.
3. Simultaneous with 2, the electrons in the outer orbitals readjust (intra-atomic relaxation) to a lower-energy final state, transferring this energy to the outgoing electron.

Relaxation does not necessarily result in the formation of the ground state of the ion. Simultaneous with the photoionization of an electron, there can be excitation of one of the remaining electrons to an initially unoccupied orbital. This phenomenon is called a "shakeup" process. The higher energy, low intensity structure on XPS core-electron peaks can then be used to study various electronic transitions occurring simultaneously with the photoionization. These satellite features are found in the range from zero to 50 eV toward higher binding energy than the main peak. Obviously, an electronic absorption at 50 eV ($= 404,000$ $cm^{-1} = 25$ nm) is a very high energy absorption in the vacuum ultraviolet. Photoemission from states belonging to a shell with an empty or partially filled subshell will excite all states accessible by redistributing the electrons within that shell that have the same symmetry as the state produced by the one electron transition.

An example of such satellite XPS features appears in Fig. 16–8. Broad features are seen toward higher binding energy than the two O_{1s} peaks for O_2. The features marked A, B, and C are "shakeup" peaks appearing as satellites on the valence-electron peaks.

In ions with a partially filled outer shell, the coupling of the outer shell to the core hole (all spin and angular momenta) will result in a set of final states with distinct energies. The spectrum will exhibit fine structure (*vide infra*).

The "mixed-valent" species that was described in Chapter 10 $[(NH_3)_5Ru\text{-}(pyrazine)Ru(NH_3)_5]^{5+}$, has also been studied by XPS.[33] The XPS spectrum has been reported to contain ionization peaks from two non-equivalent transition metal ions. The carbon 1s peaks occur in the same spectral region as the metal ionization peaks, and the conclusion rests on the ability to subtract these peaks from the spectrum. The XPS technique has a very short time scale, on the order of 10^{-17} sec. However, when a core electron is removed from a mixed-valent compound, only discrete valences are obtained and fluctuations cannot persist in the final state. Depending on the spin of the core electron removed, two different spin states can result from the final ion. Thus, two "discrete peaks" in

670 Chapter 16 Ionization Methods

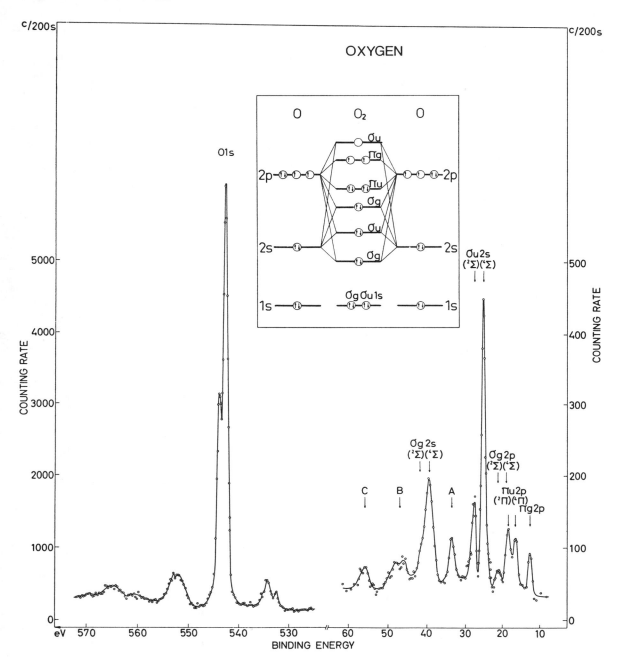

FIGURE 16–8 XPS spectrum of core and valence electrons in O_2 excited by Mg Kα X-radiation. Two peaks are obtained for each fully occupied orbital, including the oxygen 1s core orbital.[27] The peaks A, B, and C, and the peaks toward higher binding energy from the O_{1s} peak, are satellite peaks (*vide infra*). (From K. Siegbahn, *et al.*, "ESCA Applied to Free Molecules," North Holland/American Elsevier, New York, 1969.)

the XPS do not distinguish mixed valence from delocalized systems. The reader is referred to reference 34 for a critical discussion of the application of physical methods to mixed-valent systems.

UPS

In the UPS experiment, vacuum ultraviolet radiation is used as the ionizing radiation; usually this is provided by a helium [singly ionized, indicated as He(I)] resonance lamp with an energy of 21.21 eV. Other discharge lamps have been used. The energy of these lamps limits UPS to studies of valence electrons and, in general, measurements have been mostly confined to gaseous samples. There have been some reports of work on solutions[35] and solids.[36]

The UPS spectrum of a gaseous sample of N_2 is shown in Fig. 16–9. With the He(I) source determining an ionizing limit of 21.21 eV, three vibrationally structured photoionizations are seen (~ 15.6, ~ 17.0, and ~ 18.8 eV). These can be assigned to ionizations from the three highest-energy filled molecular orbitals for N_2, the $2\sigma_u$, π_u, and $3\sigma_g$ orbitals. The peaks have been assigned on the basis of the observed vibrational structure. As an aside, it should be noted that an XPS spectrum has the same three peaks (vibrational structure not seen because of lower resolution) in addition to a peak at 37.3 eV for ionization from the $2\sigma_g$ level as well as a single peak at 409.9 eV for both the $1\sigma_g$ and $1\sigma_u$ levels.[28a]

In Chapter 3 the symmetry and construction of the molecular orbitals of NH_3 were worked out. There it was found that the seven atomic orbitals in C_{3v} symmetry form a representation that is reduced to give *three* a_1 and *two* e irreducible species. The eight valence electrons of NH_3 fill two of the a_1 and one of the e molecular orbitals to give a ground state configuration of

$$\cdots (2a_1)^2(1e)^4(3a_1)^2$$

The only other filled orbital, the $1a_1$ orbital, is essentially the nitrogen $1s$ atomic orbital. The He(I) (21.21 eV) spectrum is shown in Fig. 16–10, where vertical ionizations are seen at 10.88 eV for the $3a_1$ level and at 16.0 eV (first maximum) for the $1e$ level. These assignments were made with the use of results from various m.o. calculations. The more energetic He(II) (42.42 eV) source has been used to observe the $2a_1$ vertical photoionization at 27.0 eV.[38] In passing, it is of interest to note the doubled or split character of the $1e$ peak at ~ 16 eV. This splitting has been assigned to Jahn-Teller splitting in the ion resulting from a $(2a_1)^2(1e)^3(3a_1)^2$ configuration. The splitting is 0.78 eV. Other examples of Jahn-Teller splitting have been reported.[28b]

The vibrational structure on a UPS band tells something about the bonding characteristics of the electron that is ionized. The ionization of a non-bonding electron will result in an ion with the same internuclear distance as the parent molecule. In this case the lowest energy ($v = 0, v' = 0$) vibrational peak will dominate the spectrum (see Fig. 16–11). Referring to the N_2 spectrum in Fig. 16–9, we see that ionization of electrons from either the $3\sigma_g$ level at 15.6 eV or the $2\sigma_u$ level at 18.8 eV in each case gives essentially only one strong peak. These levels are weakly antibonding and weakly bonding, respectively.

Photoionization of either a strongly bonding electron or a strongly antibonding electron will result in changes in equilibrium bond distances, as illustrated in Fig. 16–11. In these cases, the most intense peak (*i.e.*, vertical ionization)

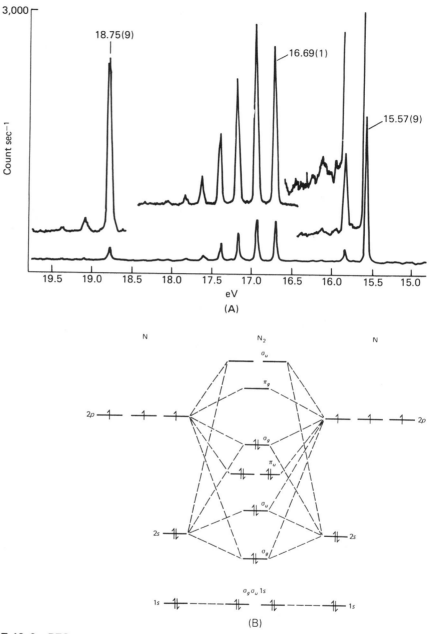

FIGURE 16–9 PES spectrum for a gaseous nitrogen sample. [From D. W. Turner and D. P. May, J. Chem. Phys., *45*, 471 (1966).]

will be at higher energy than the peak corresponding to adiabatic ionization. This is exemplified by the π_u peak for N_2. In fact, the relative intensities (called Franck-Condon factors) of the respective vibrational peaks can be theoretically calculated with reasonable success.[37] The vibrational frequencies observed for the ionized molecular state are also of further assistance in characterizing the molecular orbital from which the electron is ionized. The ground state stretching

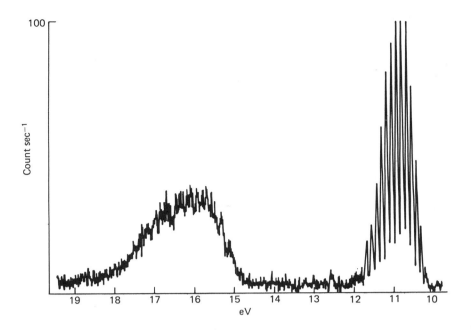

FIGURE 16–10 PES spectrum of a gaseous sample of ammonia. (From D. W. Turner, *et al.*, "Molecular photoelectron Spectroscopy," Wiley-Interscience, New York, 1970.)

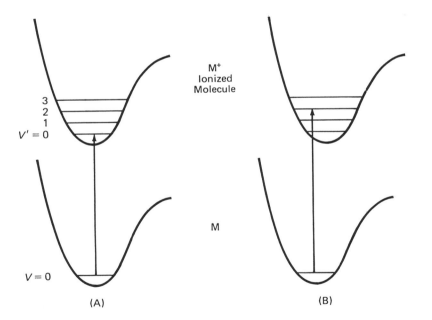

FIGURE 16–11 Vibrational states for the parent molecule M and the ionized molecule M^+. (A) In this case the electron ionized comes from a non-bonding orbital and the two electronic potential wells lie one above the other. The most intense transition (indicated with an arrow) is the $V' = 0 \leftarrow V = 0$ transition. (B) The wells are displaced (molecule dimensions change) when the electron ionized comes from either a bonding or an antibonding orbital. In this case the $V' = 2 \leftarrow V = 0$ transition is the most intense band seen.

frequency for N_2 is 2345 cm^{-1}. Table 16–2 lists the stretching frequencies seen for various N_2^+ ionized states. A comparison of each of these values with the 2345 cm^{-1} N_2 frequency leads to the implied m.o. characters of the lost electron given in Table 16–2.

In addition to vibrational structure, other fine structure is seen in UPS spectra. The first (*i.e.*, lowest binding energy) peak in the spectra of CH_3Cl, CH_3Br, and CH_3I is, obviously, from the highest occupied molecular orbital, and this is largely localized on the halogen atom (with some X—H anti-bonding

TABLE 16–2. Vibrational Structure in the N₂ UPS Spectrum

Peak	ν(N—N)	Implied M.O. Character
$\sigma_g(2p)$	2150 cm^{-1}	weakly bonding
$\pi_u(2p)$	1810 cm^{-1}	strongly bonding
$\sigma_u(2s)$	2390 cm^{-1}	weakly antibonding

and C—H bonding character.[38] Spin-orbit splitting is seen in this peak, where the separation varies in the series: CH_3I^+, 0.62 eV; CH_3Br^+, 0.31 eV; and CH_3Cl^+, ~0 eV.[28b] These spin-orbit interactions agree with Mulliken's predications of 0.625, 0.32, and 0.08 eV, respectively.[39] Many other such observations of spin-orbit interactions have been noted. In fact, recent work at the high resolution of 10 meV has detected spin-orbit splittings in NO of 0.012 to 0.016 eV.[40]

A considerable number of small gaseous molecules have now been studied with UPS and, for that matter, XPS. Some radical and excited state species have also been studied. The three expected states (^2P, ^4S, and ^2D) of O^+ have been detected.[41] An electrodeless microwave discharge was used to produce the excited state species $O_2(^1\Delta_g)$, and the photoelectron spectrum of this species shows a vibrationally structured peak owing to formation of $O_2^+(^2\Pi_g)$. Comparison with the peak corresponding to ionization of ground state $O_2(^3\Sigma_g^-)$ to the same ionic state gives a value of 11.09 ± 0.005 eV for the adiabatic ionization potential of $O_2(^1\Delta_g)$.[42] Several other such species (e.g., SO and NF_2) have been studied.

In comparison with diamagnetic compounds, paramagnetic species show additional complexity in the UPS (and XPS) spectra. The oxygen molecule has two unpaired π_g electrons. The UPS spectrum of O_2 is given in Fig. 16–12. The photoionization of an electron from the partly filled, antibonding $\pi_g(2p)$ molecular orbital appears as the first peak in the UPS spectrum. Only one ionic state is realized. On the other hand, photoionization of an electron from one of the other filled molecular orbitals leads, in each case, to *two* electronic states of the ion O_2^+. Thus, if an electron is ionized from the filled bonding π_u level, the *unpaired electron remaining* in the π_u orbital can be aligned either *with* or *against* the two unpaired electrons in the antibonding π_g level. When the electron is aligned with the two π_g electrons, there are three unpaired electrons, the total spin $S = ^3/_2$, and the electronic state of the O_2^+ molecule is $^4\Pi_u$. The other alignment gives a $^2\Pi_u$ electronic state for the O_2^+ molecule. The $^4\Pi_u$ and $^2\Pi_u$ states of O_2^+ are at different energies, and thus there is a splitting of the π_u orbital ionization peak. In Table 16–3, the observed features for the O_2 molecule from both UPS and XPS spectra are listed.

The XPS spectrum of gaseous O_2 is shown in Fig. 16–8. The oxygen 1s peak is also split, in this case by 1.1 eV. This splitting is *not* the difference between the O_2 molecular orbitals $1s\sigma_u$ and $1s\sigma_g$, which is calculated to be small. Recall that no such splitting is seen in the case of N_2. Even further, the intensity ratio of the two O_{1s} features is 2:1 and not the 1:1 ratio expected if the $1s\sigma_u/1s\sigma_g$ explanation applied. Again, as occurred in the valence orbitals, there is an interaction between the unpaired electron in the O_{1s} level and the two unpaired

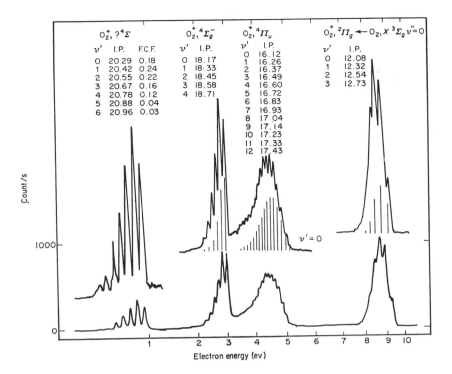

FIGURE 16–12 Oxygen PES spectrum. (From H. A. O. Hill and P. Day, "Physical Methods in Advanced Inorganic Chemistry," Interscience, New York, p. 88, 1968.)

TABLE 16–3. Vertical O_2 Ionization Data[a]

One-Electron Molecular Orbital	Ionic Electronic State	XPS, eV	PES, eV
$\pi_g 2p$	$^2\pi_g$	13.1	12.10
$\pi_u 2p$	$^4\pi_u$	17.0	16.26
	$^2\pi_u$	?	?
$\sigma_g 2p$	$^4\Sigma_g$	18.8	18.18
	$^2\Sigma_g$	21.1	20.31
$\sigma_u 2s$	$^4\Sigma_u$	25.3	24.5 [viaHe(II)]
	$^2\Sigma_u$	27.9	
$\sigma_g 2s$	$^4\Sigma_g$	39.6	
	$^2\Sigma_g$	41.6	
Oxygen 1s	$^4\Sigma$	543.1	
	$^2\Sigma$	544.2	

[a] See reference 27.

electrons in the $O_2^+ \pi_g$ antibonding molecular orbital. Ionization of even these oxygen core electrons leads to two O_2^+ states, $^2\Sigma$ and $^4\Sigma$, that differ appreciably in energy.

The observation of such splittings is very interesting. One-center exchange integrals dominate the energy difference between the quartet and doublet states. Thus, interelectron repulsions (e^2/r_{ij}, where r_{ij} is the distance between the ith and jth electrons and e is the charge on the electron) of the exchange type are appreciable even between valence unpaired electrons and the oxygen core electron. Such interelectron spin polarizations are found to be of importance in explaining the shifts in nmr peaks for paramagnetic molecules (see Chapter 12). Electron exchange splitting is also seen for core electrons of paramagnetic transition metal complexes, as is described in a later section.

It is clear from the above discussion that the UPS spectra of relatively large molecules are quite rich with information about ionization potentials, vibrational quanta for the ionized molecule, spin-orbit interactions, Jahn-Teller splittings, and electron exchange interactions. Unfortunately, there is frequently an overlapping of features, and broad peaks appear with no resolved vibrational structure. As an example of a small molecule with many photoionization peaks, Fig. 16–13 illustrates some of the spectral detail obtained[40] for gaseous NO at a resolution of 10 meV for a He(I) source and at 25 meV for a He(II) source. For

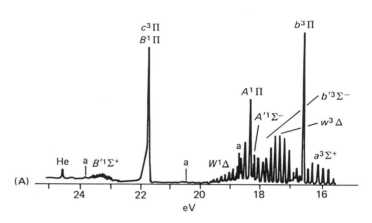

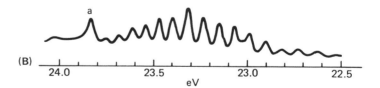

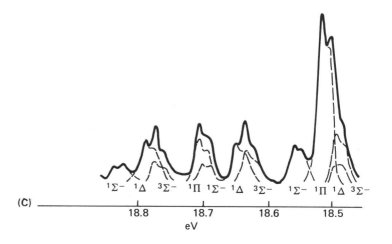

FIGURE 16–13 (A) Photoelectron spectrum of NO using the He 304-Å line; three small peaks marked "a" are due to the He 320 Å line; recording time was 75 h. (B) Photoelectron spectrum of the $B'\,^1\Sigma^+$ state of NO using the He 304 Å line; the peak at "a" is again due to the 320 Å line. (C) Photoelectron spectrum of NO using the 584 Å line. Resolution 10 meV. Deconvolution of the peaks is indicated by the dotted lines. [From O. Edqvist, L. Åsbrink, and E. Lindholm, Z. Naturforsch., *26a*, 1407 (1971).]

the fine points of assignment the reader is referred to reference 40. Figure 16–13 shows that both exchange and spin-orbit splittings have been resolved.

16–12 SIMS (SECONDARY ION MASS SPECTROMETRY)

Secondary ion mass spectrometry[43,44] employs an energetic (>4 keV) beam of primary ions (*e.g.*, $^{40}Ar^+$, $^{202}Hg^+$, $^{16}O^-$) to collide with surface species and dislodge them as secondary ions. Upon collision with the surface, the primary ion is implanted in the solid and its kinetic energy is transferred to atoms in the condensed phase. This energy causes desorption (sputtering) of a surface species producing gas phase ions. The mass spectra of these secondary ions are then detected. Both positive ion and negative ion spectra are detected. Figure 16–14 shows both types of spectra that result from Ar^+ bombardment of an aluminium surface.

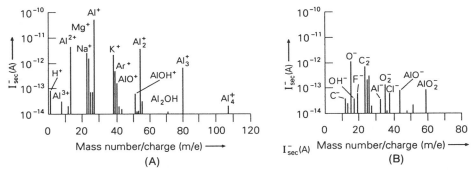

FIGURE 16–14 SIMS spectrum. Secondary ion intensity versus mass number/ion charge of an aluminum target under Ar bombardment. (A) Positive secondary ion spectrum. (B) Negative secondary ion spectrum.

The escape depth for sputtered particles ranges from the surface to greater than 20 Å, depending on the target material and energy of the primary ion. The predominant secondary ion species observed in SIMS are singly charged atomic and molecular ions. As in mass spectroscopy, the isotopic composition aids in identification of the fragments, which in turn leads to information about the chemical composition of atomic layers at or near the surface. This is referred to as static SIMS. By using ion etching (high primary ion current densities), underlying layers can be exposed and the SIMS determined as a function of depth. This is referred to as dynamic SIMS. In the dynamic SIMS mode, the chemical composition of the material in the deeper layers is changed by the prebombardment of the surface. In addition to chemical reactions, the more volatile components can be "boiled" off. Static SIMS minimizes this problem with negligible surface perturbation by the bombardment but provides information only about processes occurring on the surface. Even in static SIMS perturbation of the emission region before the secondary ion particle is emitted cannot be excluded.[44] SIMS instruments are available that provide lateral or *xy* information about the surface. Ion microprobes provide this information through probe imaging. Other instruments provide a direct image of the surface. Since the detection limits are very low (1 ppb for many elements), surface contamination by hydrocarbons, Cu, CO_2, H_2, N_2, O_2, and H_2O is common.

The surfaces close to the sample can also be contaminated by sputtered material. An additional complication is decomposition of the surface material by the high energy primary ion source, thereby producing materials that are not actually on the surface. The reader is referred to references 43 and 44 for a discussion of experimental problems.

SIMS is used to identify surface species and also to study the dynamics of surface processes (*e.g.*, diffusion and corrosion). The catalytic decomposition and synthesis of ammonia on iron metal has been studied.[45] The surface species present on the commercial silver catalyst used to convert ethylene to ethylene oxide have also been investigated.[46] The reader is referred to references 47, 48, and 49 for reviews of SIMS applications.

In molecular SIMS, the primary beam expels molecular secondary ions from samples of organic molecules supported on the surface. The molecular ions can dissociate to give fragments that contribute to the SIMS spectrum. The region just above the surface, called the selvedge, has a relatively high pressure where ion-neutral molecule association reactions can occur. For example, the SIMS spectrum of a silver film studied in the presence of benzene vapor produces $(AgC_6H_6)^+$ in SIMS. In the vacuum chamber, unimolecular dissociation reactions occur in the course of traveling to the analyzer.

Molecular species are converted into secondary ions via three distinct processes listed in order of efficiency: direct sputtering, cationization (or anionization), and electron loss or gain. In sputtering, momentum transfer from the primary ion dislodges organic cations and anions from the solid to the gas phase. Organic dications in solids have been vaporized as gaseous doubly charged ions. The limited degree of dissociation seen in SIMS spectra illustrates the low internal energy of the ions formed.[49] Cationization, in molecular SIMS, results from the ion-molecule reaction during or following the formation of support metal ions and the thermally induced evaporation of a supported organic molecule from the surface. One of the complications in the interpretation of SIMS involves distinguishing weak aggregates in a solid from aggregates formed in the selvedge under ion bombardment. The ion-molecule complex formed in the selvedge can fragment in the course of travelling to the analyzer much in the same way as the unimolecular dissociations mentioned previously. The final process to be discussed involves electron transfer from or to the surface molecule generating M^+ and M^-. These ions can undergo unimolecular decomposition leading to fragment ions. Examples of compounds that produce secondary ions by these mechanisms have been reported.[49]

Investigation of the chemistry that can occur when reactive primary ion beams are employed represents a recent growth area for SIMS. Studies show that N_2^+ is a reactive primary beam.[50]

16–13 LEED, AES, AND HREELS SPECTROSCOPY

LEED

In low energy electron diffraction (LEED), a monoenergetic beam of electrons (10 to 500 eV) is incident on one face of a single crystal. About half of the electrons are back scattered and the elastically scattered fraction hits a fluorescent screen. A well-ordered surface results in a diffraction pattern. Surface crystallography is carried out[51] by finding the surface structure that optimizes the fit of an estimated calculated pattern and the experimental one over a significantly large range of diffraction conditions (electron energy, beam angles and number of diffracted beams). In order to obtain information normal to the surface, the variation in the intensity of a given spot in the diffraction pattern is studied as a function of the incident beam energy. The theory involved in the calculation of the experimental observables is complex.[51] Studies have enabled the comparison of the surface structure with bulk structure and have shown the presence of a terrace geometry involving stepped surfaces[51] (often one atom in height). Oxidation or carbidization of the surface has been studied and the corresponding change in surface structure determined. Reactivity of the stepped surfaces are found to be different from those that are not stepped and different for various types of steps.[52]

A very interesting application of LEED involves determination of the structure of ordered adsorbate-substrate structures. Adsorbate-substrate distances accurate to 0.1 Å are claimed. A large number of surface structures for gases (*e.g.*, O_2, H_2S, C_2H_4, C_6H_6, CO, CH_4, C_5H_5N) adsorbed on substrates have been compiled.[51]

AES (AUGER ELECTRON SPECTROSCOPY)

In Auger spectroscopy (AES), the incident electron beam ionizes an atomic core level electron. The energy released when a higher-energy electron falls back to fill the core level is transferred to a valence electron, which is ejected into the gas phase where it is detected as an AUGER electron with an electron spectrometer.[53] The AUGER electrons appear as small peaks on a large background of backscattered electrons including the elastically scattered primary electrons, of LEED. The LEED optics can therefore be used to obtain the AUGER spectrum and the two methods are often used in combination. The energy of the AUGER electron varies with the surface element and its oxidation state. The AUGER electron escape depth varies from ~4 to 25 Å depending on the material, so this is clearly only a surface method.[53] Diffusion, corrosion, surface aggregation, and segregation can be monitored. The reaction of sulfur with O_2 on a Ni(111) surface was studied[54] by AES in conjunction with mass spectrometry. The rate of growth of the oxygen peak and decrease in the sulfur peak was monitored as a function of O_2 pressure. The mechanism involved O_2 adsorption, reaction of adsorbed O_2 with adsorbed sulfur to produce adsorbed SO_2, and desorption of SO_2. The reaction to form adsorbed SO_2 is rate controlling. Adhesion, friction, and wear have been investigated showing changes in surface segregation of various components.[51] Changes in the surface composition of heterogeneous catalysts have also been studied and correlated with catalyst poisoning.[55,56]

HREELS

High resolution electron energy loss spectroscopy (HREELS) is another form of electron beam spectroscopy. Vibrational excitation of surface atoms is induced by inelastic reflections of low energy electrons. The structure of the surface and adsorbed species is determined. The HREELS spectrum of CO coadsorbed with K on the Ru(111) surface is shown in Fig. 16–15. The stretching frequencies

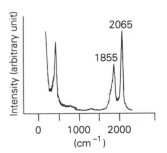

FIGURE 16–15 HREELS spectra of CO coadsorbed with K on a Rh(111) surface.

for two different types of bound CO are seen at 1855 and 2065 cm^{-1}. HREELS, in combination with the foregoing techniques, has provided considerable insight into heterogeneous catalysis. Studies of the hydrogenation of ethylene on platinum, the hydrodesulfurization of thiophene on sulfided molybdenum, the reforming of hexane, and the hydrogenation of carbon monoxide have been summarized.[57]

16–14 STM (SCANNING TUNNELING MICROSCOPY) AND AFM (ATOMIC FORCE MICROSCOPY)

The resolution of microscopic techniques is determined by the size of the measuring probe. With conventional microscopies, using light or electron beams, the area resolved encompasses more than 10^{11} atomic sites, producing an averaged macroscopic picture of the surface. In 1982, the Nobel Prize in physics was awarded to G. Binning and H. Rohrer[58] for developing the scanning tunneling microscope—a method that provides the topography of surfaces with atomic resolution.[58]

In STM[59,60] a sharp conducting metal tip is brought to within 5 to 10 Å of the sample. The tip traces the contours of the surface with atomic resolution so it must be very sharp, ideally terminating in a single atom. At 5 to 10 Å, the wave functions of the sample and the tip overlap. If a bias is applied to the sample, an electron tunneling current flows[61] between the tip and the surface. The electron flow can be in either direction depending on the sign of the bias imposed. The tip can be moved in the x, y, and z directions using three orthogonal x, y, and z piezoelectric translators. The tip moves in steps of ~1 nm (~three atoms) per volt applied to the translator. Moving the x and y translators maps the surface while the z translator varies the tip-surface distance. The tunneling current is very sensitive to the tip-surface distance, typically changing by a factor of 2 for a 0.1 nm change. Differences of 1/100 of an atomic diameter can be determined in the z direction. As one performs the scan, the voltage into the z translator is changed to change the tip-surface distance in order to maintain a constant current, typically 1 nA. Alternatively the tip-surface distance can be kept fixed and the tunneling current measured. With a tip the size of an atom, atomic resolution is achieved. Figure 16–16 is an STM image showing *carbon atoms* in a sample of highly oriented pyrolytic graphite. The dashed lines in the figure correspond to a scan in the x direction. The voltage along z required to vary the distance and keep the current constant is measured as a function of the voltage applied to move x. The image is made by multiple scans changing the y direction for each.

The STM has been used in technological applications to guide the manufacture of a spectral grating master[62] and in improving magnetic recording heads.[63] By removing and depositing atoms, images with atomic dimensions have been made. By exciting individual atoms, nanoscale surface chemistry can be induced.[64] The surface chemistry of silicon has been explored and changes that occur when chemical reactions take place monitored.[60] Reference 60 is highly recommended reading.

The application of STM is limited to conducting samples. Some success has been achieved by studying conducting images of non-metallic substances,[65] by direct ejection of electrons into the conduction band,[66] and by using

FIGURE 16–16 An STM image of carbon atoms in a highly oriented pyrolytic graphite sample. The image is a series of line scans, each displaced in the y direction from the previous one, displaying the path of the tip over the surface.

nonlinear alternating current STM.[67] The atomic force microscope, AFM, can image non-conducting surfaces.[68] A sharp tip, *e.g.*, a small fractured diamond fragment, attached to a spring in the form of a cantilever, traces over the sample and senses the repulsive forces between the tip and the sample. Repulsive forces usually in the range 10^{-6} to 10^{-9} N are recorded by measuring minute deflections of the cantilever. The spring deflection is measured by electron tunneling with an interferometer by deflection of a laser beam off a mirror on the spring or with a fiber-optic interferometer. Boron nitride has been imaged in this way.[69]

16–15 INTRODUCTION

EXAFS and XANES

Every element shows discontinuous jumps or "edges" in its X-ray absorption coefficient corresponding to the excitation of a core electron (Fig. 16–17(A)). The energy of the edge depends on the element and the core electron excited. The K edge (1s state) is observed for elements between sulfur and cadmium and the L edge (2s and 2p states) for elements of higher atomic numbers. Extended X-ray absorption fine structure (EXAFS) employs high intensity X-ray beams (energies to 40 keV; wavelength to 0.25 Å) from a synchnotron source. The brightness is as much as 10 orders of magnitude larger than that from conventional X-ray tubes. A typical plot of the absorption coefficient as a function of the synchnotron photon energy, E, is shown in Fig. 16–17(B). X-ray absorption spectroscopy[70] refers to the study of the structured absorption which is superimposed on the edge.

The EXAFS region is ~50 to 1000 eV above the edge, whereas the X-ray absorption near-edge structure (XANES) region refers to absorption within 50 eV of both sides of the edge. The same principles govern absorption in both EXAFS and XANES, but different information can be obtained.

The incident photon energy in EXAFS is much larger than the core-binding energy leading to an excited photoelectron with significant kinetic energy. The emitted photoelectron can be pictured as a wave propagating from the excited atom. Upon reaching a neighboring ligand, the wave front is scattered back toward the absorber. As the energy of the X-ray is increased, constructive and destructive interference of the outgoing and backscattered wave occurs near the origin giving rise to the sinusoidal variation of the amplitude versus E, (Fig. 16–17(B)), *i.e.*, EXAFS fine structure. The amplitude and frequency of this sinusoidal modulation depends on the type and bonding of the neighboring atoms and their distances away from the absorbing atom. The scattering is similar for atoms in the same row of the periodic table. Thus, C, N, and O are not distinguished but N and S are. After considerable mathematical manipulation, EXAFS provides the same compositional information as can be obtained from X-ray crystallography with the advantage of being applicable to non-crystalline systems. However, the mean free path damping factor limits information to ~4 Å from the absorbing atom and does not provide information regarding the angular arrangement of the surrounding atoms. It should be emphasized that EXAFS measures a bulk property and is not a surface technique. If the sample is not pure, a meaningless average EXAFS spectrum will result.

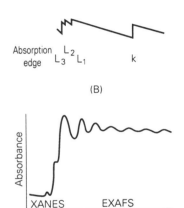

FIGURE 16–17 (A) Schematic representation of absorption edges in X-ray absorption spectroscopy. (B) Fine structure on a absorption edge.

16-16 APPLICATIONS

Many of the above limitations turn out to be an advantage in the elucidation of the coordination environment of metals and metal clusters in biological systems. The molybdenum site in a nitrogenase enzyme was shown[71] to be predominantly coordinated by sulfur ligands with another metal (inferred to be iron) ~2.7 Å from a molybdenum center. Several model compounds were synthesized with structures consistent with the EXAFS results. These models contain oxygen and nitrogen ligands that could not be seen in the EXAFS of the models in the presence of the sulfur ligands and thus are not expected to be seen in the EXAFS of the enzyme. EXAFS was also used[72] to show sulfur coordination to the iron heme of cytochrome P-450, which is retained throughout the catalytic oxidation cycle. EXAFS provides very accurate bond distances. It improved[73] considerably on the iron-sulfur distances from an x-ray crystal structure of the iron-sulfur protein, rubredoxin. The initial x-ray study reported non-equivalent metal-sulfur distances. EXAFS showed them to be the same to within 0.04 Å. Other applications illustrating the advantages and limitations of the technique have been summarized.[74]

REFERENCES CITED

1. R. A. W. Johnstone, "Mass Spectrometry for Organic Chemists," Cambridge University Press, London, 1972.
2. M. R. Litzow and T. R. Spalding, "Mass Spectrometry of Inorganic and Organometallic Compounds," Elsevier, New York, 1973; J. Lewis and B. F. G. Johnson, Acc. Chem. Res., *1*, 245 (1968); M. I. Bruce, Adv. Organomet. Chem., *6*, 273 (1968).
3. M. E. Rose and R. A. W. Johnstone, "Mass Spectrometry for Chemists and Biochemists," Cambridge University Press, Cambridge, MA; 1982.
4. I. Howe, D. H. Williams, and R. D. Bowen, "Mass Spectrometry: Principles and Applications," 2nd ed., McGraw-Hill, New York, 1981.
5. F. W. McLafferty, "Interpretation of Mass Spectra," 3rd ed., University Science Books, Mill Valley, CA, 1980.
6. For brief reviews of recent developments in the field, see the annual summaries in Analytical Chemistry.
7. a. G. M. Pesyna, R. Venkataraghavan, H. E. Dayringer, and F. W. McLafferty, Anal. Chem., *48*, 1362, (1976).
   b. D. Henneberg, Adv. Mass Spectrom., *8*, 1511 (1979); S. R. Heller and G. W. A. Milne, Amer. Lab., *12*(3), 33 (1980).
8. a. "Registry of Mass Spectral Data," E. Stenhagen, S. Abrahamsson, and F. W. McLafferty, Wiley, New York, 1974, 4 vols.
   b. "Eight Peak Index of Mass Spectra," 2nd ed., Mass Spectroscopy Data Centre, Aldermaston, Reading, U.K., 1974.
9. D. Williams, in "Determination of Organic Structures by Physical Methods," Vol. 3, eds. F. C. Nachod and J. J. Zuckerman, Academic Press, New York, 1971.
10. H. M. Rosenstock and M. Krauss, in "Mass Spectrometry of Organic Ions," ed. F. W. McLafferty, Academic press, New York, 1963.
11. K. Biemann and G. Spiteller, J. Amer. Chem. Soc., *84*, 4578 (1962).
12. J. H. Jones, Quart. Rev. Chem. Soc., *22*, 302 (1968).
13. E. Hugentobler and J. Löliger, J. Chem. Ed., *49*, 610 (1972).
14. J. Lederberg, J. Chem. Ed., *49*, 613 (1972).
15. D. Maurer et al., J. Chem. Ed., *51*, 463 (1974).
16. H. M. Bell, J. Chem. Ed., *51*, 548 (1974).

17. J. Collard, Ph.D. thesis, U. of Illinois, Urbana (1975).
18. F. E. Brinckman and F. G. A. Stone, J. Amer. Chem. Soc., *82*, 6235 (1960).
19. **a.** H. D. Beckey, A. Heindrichs, and V. Winkler, Int. J. Mass Spectrom. Ion Phys., *3*, 11 (1970).
    b. R. B. van Breeman, L. R. B. Martin, and A. F. Schreiner, Anal. Chem., *60*, 1314 (1988).
20. T. A. Milne and H. M. Klein, J. Chem. Phys., *33*, 1628 (1960).
21. R. T. Grimley, R. P. Burns, and M. G. Inghram, J. Chem. Phys., *34*, 664 (1961).
22. J. Berkowitz, M. G. Inghram, and W. A. Chupka, J. Chem. Phys., *26*, 842 (1957).
23. **a.** T. E. Sharp and J. R. Eyler, Int. J. Mass Spectrom. Ion Phys., *9*, 421 (1972).
    b. M. B. Comisarow, Anal. Chim. Acta, *178*, 1–15 (1985).
    c. J. D. Baldeschwieler, Science, *159*, 263 (1968).
24. **a.** T. A. Lehman and M. M. Bursey, "Ion Cyclotron Resonance Spectrometry," Wiley, New York, 1976.
    b. R. T. McIver, Jr. Amer. Lab, *12*(11), 26 (1980).
25. **a.** P. Kebarle and S. Chowdhury, Chem. Rev., *87*, 513 (1987).
    b. S. Chowdhury and P. Kebarle, J. Amer. Chem. Soc., *108*, 5453–5459 (1986).
26. **a.** R. G. Keesee and A. W. Castleman, Jr. J. Phys. Chem. Ref. Data, *15*, 1011 (1986).
    b. M. T. Bowers, ed. "Gas Phase Ion Chemistry," Academic Press, New York, 1979, Vols. 1 and 2.
    c. P. Sharpe and D. E. Richardson, Coord. Chem. Rev., *93*, 59 (1989).
    d. J. Allison, Prog. Inorg. Chem., *34*, 628 (1986).
    e. B. S. Freiser, Talanta, *32*, 697 (1985).
27. K. W. S. Chan and K. D. Cook, J. Amer. Chem. Soc., *104*, 5031 (1982) (see opening remarks).
28. **a.** K. Siegbahn, C. Nordling, G. Johansson, J. Hedman, P. F. Heden, K. Hamrin, U. Gelius, T. Bergmark, L. O. Werme, R. Manne, and Y. Baer, "ESCA Applied to Free Molecules," North-Holland/American Elsevier, New York, 1969.
    b. D. W. Turner, C. Baker, A. D. Baker, and C. R. Brundle, "Molecular Photoelectron Spectroscopy," Wiley-Interscience, New York 1970.
29. S. Evans, "Handbook of X-Ray and Photoelectron Spectroscopy," ed. D. Briggs, Heyden, London, 1977.
30. J. W. Robinson, "CRC Handbook of Spectroscopy," Volume 1, CRC Press, Cleveland, Ohio, 1974, pp. 517–752.
31. **a.** I. Adams, J. M. Thomas, G. M. Bancroft, K. D. Butler, and M. Barber, Chem. Commun., *751* (1972); W. E. Swartz, P. H. Watts, E. R. Lippincott, J. C. Watts, and J. E. Huheey, Inorg. Chem., *11*, 2632 (1972); M. Barber, P. Swift, D. Cunningham, and M. J. Frazer, Chem. Commun., 1338 (1970).
    b. D. T. Clark, D. Briggs and D. B. Adams, J. Chem. Soc., Dalton, 169 (1973).
32. D. A. Shirley, "Photoemission in Solids," Volume 1, Springer-Verlag, Berlin, 1978.
33. P. H. Citrin, J. Amer. Chem. Soc., *95*, 6472 (1973).
34. A. Ludi, J. Amer. Chem. Soc., *106*, 121 (1984).
35. H. Aulich, B. Baron, and P. Delahay, J. Chem. Phys., *58*, 603 (1973).
36. D. E. Eastman, "Photoemission Spectroscopy of Metals," in "Techniques in Metals Research VI," ed. E. Passaglia, Interscience, New York, 1971; T. E. Fischer, Surface Sci., *13*, 30 (1969); C. R. Brundle, "Surface and Defect Properties of Solids," Volume 1, eds. M. W. Roberts and J. M. Thomas, Specialist Periodical Reports, The Chemical Society, London, 1972.
37. D. W. Turner and D. P. May, J. Chem. Phys., *45*, 471 (1966).
38. H. Kato, K. Morokuma, T. Yonezawa, and K. Fukui, Bull. Chem. Soc. Japan, *38*, 1749 (1965).
39. R. S. Mulliken, Phys. Rev., *47*, 413 (1935).
40. O. Edquist, L. Åsbrink, and E. Lindholm, Z. Naturforsch, *26a*, 1407 (1971).
41. N. Jonathan, A. Morris, M. Okuda, D. J. Smith, and K. J. Ross, "Electron Spectroscopy," ed. D. A. Shirley, North-Holland, Amsterdam, 1972.

42. N. Jonathan, D. J. Smith, and K. J. Ross, J. Chem. Phys., *53*, 3758 (1970).
43. J. A. McHugh, "Methods of Surface Analysis," Elsevier, Amsterdam, 1975, Chapter 6.
44. A. Berringhoven et al. "Secondary Ion Mass Spectrometers—Basic Concepts, Instrumental Aspects, Applications and Trends," Wiley, New York, 1987.
45. V. I. Shvachko and Ya. M. Fogel, Kinet. Katal., *7*, 722 (1966).
46. A. Benninghoven, Surface Sci., *28*, 541 (1971).
47. Ya. M. Fogel, Int. J. Mass Spectrom. Ion Phys., *9*, 109 (1972).
48. A. Benninghoven, Surface Sci., *35*, 427 (1973).
49. R. S. Day, S. E. Unger, and R. G. Cooks, Anal. Chem., *52*, 557A (1980).
50. G. M. Lancaster, F. Honda, Y. Fukuda, and J. W. Rabalais, Chem. Phys. Lett., *59*, 356 (1978).
51. G. A. Somorjai and L. C. Kesmodel, "Surface Chemistry and Colloids," Chapter 1.
52. B. Lang, R. W. Joyner, and G. A. Somorjai, Surface Science, *30*, 454 (1972).
53. A. Joshi, L. E. Davis, and P. W. Palmberg, "Methods of Surface Analysis," Elsevier, Amsterdam, 1975, Chapter 5.
54. P. H. Holloway, and J. B. Hudson, Surface Sci., *33*, 56 (1972).
55. G. A. Somorjai, J. Catal., *23*, 453 (1972).
56. M. M. Bhasin, J. Catal., *34*, 356 (1974).
57. F. Zaera, A. J. Gellman, and G. A. Somorjai, Acc. Chem. Res., *19*, 24 (1986).
58. G. Binning and H. Rohrer, Rev. Mod. Phys., *59*, 615 (1987); contains the acceptance speech.
59. P. K. Hansma, V. B. Elings, O. Marti, and C. E. Bracker, Science, *242*, 157 (1988).
60. P. Avouris, J. Phys. Chem., *94*, 2246 (1990).
61. J. Bardeen, J. Phys. Rev. Lett., *6*, 57 (1961).
62. Perkin Elmer Corp., Irvine, CA in reference 59.
63. Censtor Corp., San Jose, CA in reference 59.
64. E. E. Erichs, S. Yoon, and A. L. de Lozanne, Appl. Phys. Lett., *53*, 2287 (1988).
65. M. Amrein et al., Science, *240*, 514 (1988).
66. P. Avouris and R. Wolkow, Appl. Phys. Lett., *55*, 1074 (1989).
67. G. P. Kochanski, Phys. Rev. Lett., *62*, 2285 (1969).
68. G. Binning, C. F. Quate, and Ch. Gerber, Phys. Rev. Lett., *56*, 930 (1986).
69. T. R. Albrecht and C. F. Quate, J. Appl. Phys., *62*, 2599 (1987).
70. B. K. Teo "EXAFS: Basic Principles and Data Analysis," Springer-Verlag, New York, 1986.
71. **a.** S. P. Cramer, K. O. Hodgson, W. O. Gillum, and L. E. Mortenson, J. Amer. Chem. Soc., *100*, 3398–3407 (1978). S. P. Cramer, W. O. Gillum, K. O. Hodgson, L. E. Mortenson, E. I. Stiefel, J. R. Chisnell, W. J. Brill, and V. K. Shaw, *ibid.*, 3814–3819.
    b. S. D. Conradson, B. K. Burgess, W. E. Newton, K. O. Hcdgson, J. W. McDonald, J. F. Rubinson, S. F. Gheller, L. E. Mortenson, M. W. W. Adams, P. K. Mascharak, W. A. Armstrong, and R. H. Holm, J. Amer. Chem. Soc., *107*, 7935–7940, (1985).
72. J. E. Hahn, K. O. Hodgson, L. A. Andersson, and J. H. Dawson, J. Biol. Chem., *257*, 10,934–10,941, (1982). S. P. Cramer, J. H. Dawson, K. O. Hodgson, and L. P. Hager, J. Amer. Chem. Soc., *100*, 7282–7290, (1978).
73. R. G. Shulman, P. Eisenberger, W. E. Blumnber, and N. A. Stombaugh, Proc. Natl. Acad. Sci. USA, *72*, 4002–4007 (1975); D. E. Sayers, E. A. Stern and J. R. Herriott, J. Chem. Phys., *64*, 427–428, (1976); B. Bunker and E. A. Stern, Biophys. J., *19*, 253–264 (1977). R. G. Shulman et al., J. Mol. Biol., *124*, 305–321, (1978).
74. **a.** J. E. Penner-Hahn, "Metal Clusters in Biological Systems," ACS Symposium Series, ed. L. Que, Washington, DC, 1988.
    b. J. C. J. Bart, Adv. Catal., *34*, 203 (1986).

EXERCISES

1. If the accelerating potential in a mass spectrometer is decreased in running a spectrum, will the large or small m/e ratios be recorded first?

2. Sulfur-carbon π bonding is not as effective as nitrogen-carbon π bonding. In the mass spectrum of $HSCH_2CH_2NH_2$, would the $m/e = 30$ or $m/e = 47$ peak be more intense?

3. Refer to Fig. 16–4 and recall the discussion on the relation of charge delocalization and stability of the positive ion. Explain why the mass 92 peak corresponding to

    ⌬—CH$_2$$^+$ is less intense for the isomer of ethyl pyridine with the ethyl group in the 3 position.

4. **a.** What m/e peak in the mass spectrum of CH_4 would you examine as a function of accelerating potential in order to determine the ionization potential of the methyl radical?
   b. Write an equation for the appearance potential of the fragment in part (a) in terms of ionization potential and dissociation energy.
   c. In evaluating the dissociation energy of part (b) from thermochemical data, what is wrong with using one-fourth value for the heat of formation (from gaseous carbon and hydrogen) of CH_4?

5. UPS spectra for even relatively small molecules like butadiene are complicated (see figures). Using Hückel calculations on butadiene, assign the peaks in the butadiene spectrum (the peaks at 8.6 and 10.95 eV are *impurity* peaks.) Three vibrational progressions (1520, 1180, and 500 cm^{-1}) have been tentatively identified in the first band at 9.08 eV. Discuss the vibrational structure on this band (both shape and magnitude) with respect to your assignment. (Bands are seen in the infrared spectrum of ground state butadiene at 1643, 1205, and 513 cm^{-1}.)

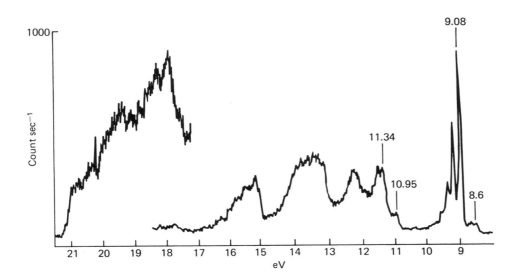

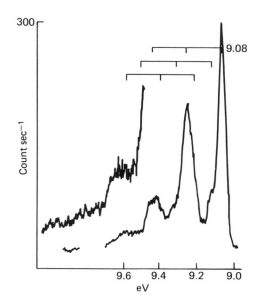

6. The effects of coordination of small molecules to transition metals can be studied with photoelectron spectroscopy. Spectra are given below for CO and $W(CO)_6$. Qualitatively explain what observations bearing on the bonding in this complex are possible. Take into account symmetry and energy considerations if applicable.

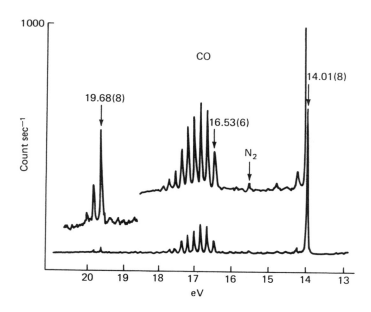

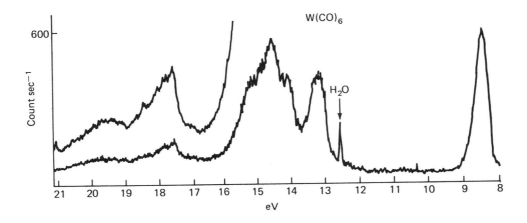

7. Given below are the UV photoelectron spectrum and m.o. diagram for ammonia. Bonding (b), non-bonding (n), and antibonding (*) levels are indicated.

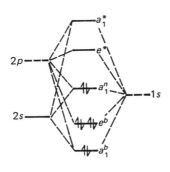

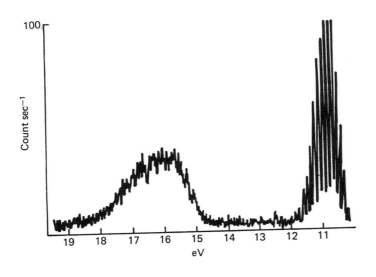

a. What phenomena give rise to the two bands centered at 11 and 16 eV.
b. What gives rise to the fine structure observed for each band? (Explain in one or two sentences.)
c. Is the fine structure what you would predict, assuming the m.o. diagram given? Why or why not? How might you justify any anomalies? Hint: consider the shapes of the "before" and "after" species.

8. Describe which of the techniques described in this chapter you would use to:
   a. determine the molecular weight of a non-volatile solid.
   b. measure the C—O stretching frequency of CO adsorbed on a surface.
   c. determine the oxidation state of a metal surface species after a metal film was exposed to Cl_2.
   d. determine if the entire film in (C) had reacted with Cl_2.
   e. determine the atoms in the coordination sphere of a metal ion in a metallo-protein.

9. Describe the differences and features in common for LEED, AUGER, and HREELs.

X-Ray Crystallography   17

JOSEPH W. ZILLER AND ARNOLD L. RHEINGOLD*

17-1 INTRODUCTION

X-ray crystallography is the most powerful and unambiguous method for the structure elucidation of solids available to modern scientists. X-ray diffraction has grown steadily since Max von Laue discovered in 1912 that a copper sulfate crystal could act as a three-dimensional diffraction grating upon irridation with X-rays. Early diffraction experiments were recorded on photographic plates or film. The labor involved in determining a three-dimensional structure from these early experiments could easily lead to one's thesis being based solely on one or two structural determinations. The advent of modern high speed computers, automated diffractometers, and powerful structure solution programs has allowed X-ray diffraction to become widely accepted as a necessary standard technique.

Progress in X-ray diffraction is probably best traced by the number and complexity of structures being reported each year. Information compiled in the Cambridge Crystallographic Database[1] illustrates how rapidly structural information is growing (Table 17–1). It is clear that researchers are increasingly relying on crystallography.

TABLE 17–1. Growth in the Cambridge Crystallographic Database

Year	Structures Added	Avg. Number of Atoms/Molecule	Total Structures
1960	224	15.1	583
1970	1260	27.9	5577
1980	4324	38.6	30475
1988	7162	51.7	71619
1989	8005	53.1	82130

---

* This chapter was written by Joseph W. Ziller, University of California – Irvine and Arnold L. Rheingold, University of Delaware.

The basic requirement to conduct an X-ray diffraction experiment is a single crystal that can withstand exposure to X-rays of a given frequency for many hours or days. Depending on the experimental conditions and the information desired from the diffraction experiment (*e.g.*, does one need to establish atom connectivity or the absolute configuration of a resolved chiral molecule?), an X-ray diffraction experiment may be carried out in as little as a few hours[2] or take several days.

It is the intent of this chapter to familiarize the novice with general X-ray diffraction principles, illustrate useful techniques for growing/handling/mounting crystals, explain briefly the diffraction experiment, and illustrate how to interpret and evaluate the results of a single-crystal X-ray diffraction study.

Principles

17–2 DIFFRACTION OF X-RAYS

In order to understand how the interaction of X-rays with single crystals yields a diffraction pattern and ultimately (in most cases) a three-dimensional crystal structure, it is necessary to know basic diffraction physics. A convenient and conceptually easy way to describe diffraction of X-rays from crystals is to consider that diffracted beams are "reflected" from planes in the crystal lattice analogous to the reflection of an object from a mirror. The mathematical expression used to describe how X-rays interact with crystals to produce a diffraction pattern is given by Bragg's law:

$$n\lambda = 2d \sin \theta \qquad (17\text{–}1)$$

where n is an integer, λ is the wavelength of the radiation, d is the perpendicular spacing between adjacent planes in the crystal lattice, and θ is the angle of incidence and "reflection" of the X-ray beam. Figure 17–1 shows how X-rays are "reflected" from planes in the crystal lattice. Electromagnetic waves 1 and 2 strike planes P1 and P2, respectively, at points A and B making the angle θ. For diffraction to occur, it is necessary for the waves generated (1' and 2') to be "in-phase" when "reflected" from P1 and P2. Constructive interference (in-phase) of the waves emanating from points A and B occurs only when the path lengths travelled are an integral multiple of the wavelength, $n\lambda$. From Fig. 17–1 it is seen that the path-length difference is $2d \sin \theta$. When this equals $n\lambda$, the Bragg equation is satisfied and the resulting constructive interference produces a diffraction maximum or "reflection."

Since the foregoing discussion considers diffraction of X-rays from lattice planes in a unit cell, these planes must be designated in a consistent manner. This is done by assigning Miller indices to the lattice planes. Miller indices are represented by (hkl) values, which are universally used to represent indices. Miller indices are also used to designate lattice points hkl that correspond to the (hkl) family of planes. Each "reflection" of an X-ray from a crystal is assigned a unique hkl value. (Note: when Miller indices are written in parentheses (hkl), they designate lattice planes or crystal faces and when simply written as hkl, they refer to a lattice point or "reflection"). Miller indices representing lattice planes and lattice points are shown in Figs. 17–2(A) and 17–2(B), respectively.

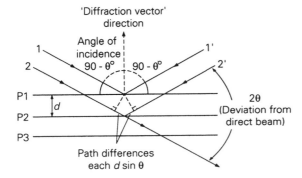

FIGURE 17–1 "Reflection" of X-rays from crystal lattice planes.

17–3 REFLECTION AND RECIPROCAL SPACE

It is now necessary to relate diffraction from crystal planes to Bragg's law and the direction of the diffracted beams. Rewriting Bragg's law in the form $\sin\theta = n\lambda/2(1/d)$, it is clear that $\sin\theta$ is inversely proportional to the d-spacing in the crystal lattice. As a result, structures having large d-spacings exhibit compressed diffraction patterns and structures with small d-spacings show expanded patterns. It would be more convenient to express a direct relationship between $\sin\theta$ and d. This is accomplished by construction of a reciprocal lattice in which $1/d$ is directly proportional to $\sin\theta$. A reciprocal lattice can be constructed from a direct lattice as shown in Figs. 17–2(A) and 17–2(B).

From the origin O, lines P, Q, and R are drawn normal to the direct lattice planes (hkl). Points along the normals are marked off at distances $1/d_{hkl}$ from O where d_{hkl} is the d-spacing between lattice planes. The resulting reciprocal lattice for a primitive (P) monoclinic cell is shown in Fig. 17–2(b). The mathematical relationships between direct and reciprocal lattices can be found in reference 3a. The relationship of the direct and reciprocal lattices is shown below for a monoclinic cell.

The reciprocal lattice parameters (a^*, b^*, c^*, β^*, and V^*) are determined from the direct lattice parameters according to Table 17–2. Figure 17–3 shows

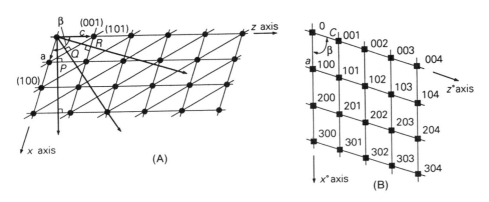

FIGURE 17–2 (A) Direct and (B) reciprocal lattices. (Reproduced with permission from M. F. C. Ladd and R. A. Palmer, "Structure Determination by X-Ray Crystallography." 2nd ed. Plenum Press, New York.)

TABLE 17–2. Relationships between Direct and Reciprocal Lattices for the Monoclinic Cell

$a^* = 1/(a \sin \beta)$	$a = 1/(a^* \sin \beta^*)$	$\alpha = \gamma = \alpha^* = \gamma^* = 90°$
$b^* = 1/b$	$b = 1/b^*$	$\beta^* = 180° - \beta$
$c^* = 1/(c \sin \beta)$	$c = 1/(c^* \sin b^*)$	$V^* = 1/V = a^*b^*c^* \sin \beta^*$
	$V = 1/V^* = abc \sin \beta$	

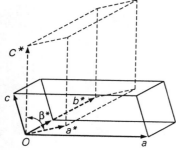

FIGURE 17–3 Direct (solid) and reciprocal (dashed) monoclinic cells. (Reproduced with permission from G. H. Stout and L. H. Jensen, "X-Ray Structure Determination. A Practical Guide." 2nd ed. John Wiley & Sons, New York, 1989.)

the reciprocal lattice outlined with dashes and the direct cell with solid lines. The reciprocal lattice may be used to illustrate how Bragg's law is satisfied.

Figure 17–4 shows the a^*c^* section of a reciprocal lattice of a crystal bathed in X-rays with wavelength λ. The X-ray beam (line XO) passes through the reciprocal lattice origin O. The circle with radius $1/\lambda$ is centered at point C such that the origin O is located on its circumference. Point P is a reciprocal lattice point at any convenient point on the circle and angle OPB is a right angle because it lies within the semicircle. The following equations are derived from Fig. 17–4(a):

$$\sin OBP = \sin \theta = \frac{OP}{OB} = \frac{OP}{2\lambda} \quad (17\text{–}2)$$

$$\sin \theta = (OP/2)\lambda \quad (17\text{–}3)$$

Because P is a reciprocal lattice point, $1/d_{hkl}$ is the length of OP. Substitution and rearrangement yields $\sin \theta = 1\lambda/2d_{hkl}$ or Bragg's law with $n = 1$, i.e., $1\lambda = 2d \sin \theta$. Since Bragg's law is satisfied, reflection occurs. Figure 17–4(B) shows that the reflecting plane is perpendicular to OP and therefore parallel to BP because OPB is a right angle. The angles θ that the incident and diffracted beams make with the reflecting plane must be equal for constructive interference and thus the diffracted beam OD is at the necessary 2θ angle to the incident beam. When Fig. 17–4(a) is constructed in three dimensions, a^*c^*, a^*b^*, and b^*c^*, the resulting sphere is called the "sphere of reflection." Whenever a point on this sphere satisfies the foregoing arguments, the Bragg condition for reflection is satisfied and a diffraction maximum can be measured.

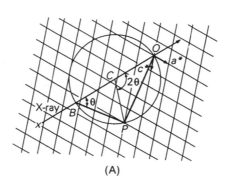
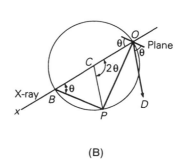

FIGURE 17–4 (A) Reciprocal lattice; (B) direct plane. (Reproduced with permission from G. H. Stout and L. H. Jensen, "X-Ray Structure Determination. A Practical Guide." 2nd ed. John Wiley & Sons, New York, 1989.)

17-4 THE DIFFRACTION PATTERN

The diffraction maximum or X-ray data collected from a single crystal yields an intensity, Miller indices, and the diffraction angle 2θ for each "reflection." The intensity of the reflection depends on the nature and arrangement of atoms in the unit cell and 2θ is dependent only on the dimensions of the crystal lattice. As previously stated, for reflection to occur, Bragg's law must be satisfied, but it must be recognized that atoms are not usually arranged in a unit cell such that all atoms lie directly on lattice planes. The "reflections" observed are composed of the sum of waves scattered from different atoms at different positions in the unit cell. Actually, all atoms (in fact, all electrons) contribute to the intensities of all reflections.

Superposition of waves is the method of combining different waves to generate the resultant wave observed as the "reflection" or hkl. This principle states that the resultant wave amplitude in a given direction is obtained by summing the individual waves scattered in that direction. A vector representation of this superposition is shown in Fig. 17-5. The magnitude $|F|$ and phase angle α may be calculated for the resultant F. The formulas may be found in reference 3a.

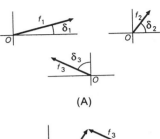

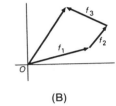

FIGURE 17-5 (A) Vector representation of the superposition of waves f_1, f_2, and f_3. (B) Resultant wave if F with phase angle α. (Reproduced with permission from G. H. Stout and L. H. Jensen, "X-Ray Structure Determination. A Practical Guide." 2nd ed. John Wiley & Sons, New York, 1989.)

17-5 X-RAY SCATTERING BY ATOMS AND STRUCTURES

The scattering of X-rays by the electrons in the crystal produces the diffraction pattern. The scattering power of a given atom is a function of the atom type, i.e., number of electrons and $(\sin \theta)/\lambda$. The scattering amplitude of an atom is designated as the "atomic scattering factor" or "atom form factor" (f_0). At $2\theta = 0$ ($\sin \theta/\lambda = 0$) the scattering factor is equal to the total number of electrons in the atom. As $(\sin \theta)/\lambda$ increases, the scattering of X-rays from different electrons in the atom will become more out of phase. Figure 17-6 illustrates the decrease in the scattering factor for several atoms at increasing $(\sin \theta)/\lambda$ values.

Since the frequency of X-radiation is high, the diffraction time scale is very short and all motion is frozen. However, since the structure determined will represent the frozen atomic positions averaged over 10^{15} to 10^{20} unit cells, the structure will show the effects of thermal activity, and any differences in atomic arrangement within the unit cell will be revealed as disorder.

The scattering power of a real atom is also affected by thermal motion or vibration of the atom. Thermal motion causes the electron cloud to be more diffuse. This phenomenon is usually associated with such factors as temperature, how tightly bound atoms are in a molecule, or how rigid the molecules are packed together in the crystal lattice. The correct expression of the structure factor for a spherical or isotropic atom is:

$$f = f_0 e^{-B(\sin^2 \theta)/\lambda^2} \qquad (17-4)$$

where B is a temperature parameter related to the mean-square amplitude of vibration $\overline{(\mu^2)}$ by $B = 8\pi^2 \overline{\mu^2}$.

Since we are interested in the scattering from a group of atoms that constitutes the structure being investigated, we must define another quantity, the "structure

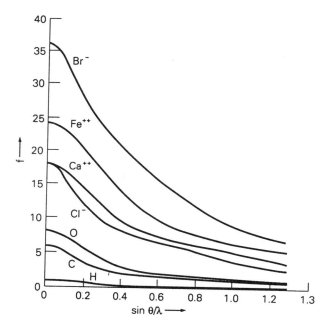

FIGURE 17–6 Scattering factor curves for atoms as a function of (sin θ/λ). (Reproduced with permission from J. P. Glusker and K. N. Trueblood, "Crystal Structure Analysis. A Primer." Oxford University Press, Oxford, U.K., 1985.)

factor" $|F|$ or $|F_{hkl}|$. The structure factor is proportional to the observed intensity taken according to equation (17–5). (See reference 4).

$$I = K|F|^2(L_p)(\text{Abs}) \qquad (17\text{–}5)$$

K is an initially unknown scale factor, L_p is the Lorentz and polarization correction factor and (Abs) is absorption. The observed intensities or "reflections" are converted into $|F_{obs}|$ (F observed). These quantities are used to calculate electron density maps from which atomic positions and ultimately the three-dimensional structure are determined.

The electron density at any point in the unit cell is given by the Fourier series in equation (17–6).

$$\rho(x, y, z) = \frac{1}{V} \sum_h \sum_k \sum_l F_{hkl}\, e^{-2\pi i(h_x + k_y + l_z)} \qquad (17\text{–}6)$$

Satisfying this equation results in a three-dimensional mapping of the electron density of the unit cell. An atom would be located at each value of $\rho(x, y, z)$ in the map with atomic coordinates x, y, z. Unfortunately, the foregoing equation cannot be solved straightforwardly because only the magnitudes and not the phases of the F_{hkl} values are contained in the X-ray data. The uncertainty of knowing the phases of the measured F-values is known as the phase problem in crystallography. Fortunately for the modern crystallographer (and especially for the amateur crystallographer), there exist excellent computer programs that have been very successful in cracking the phase problem (see reference 3a, p. 245 for a review of the phase problem and reference 4, p. 248 for a list of computer programs). Once the phase problem is solved, subsequent least-squares analyses and difference-Fourier syntheses are used to refine atomic positions and locate any missing atoms. (As a testament to the importance of the work that resulted in direct calculation of phases, in 1985 the Nobel Prize in chemistry was awarded to Jerome Karle and Herbert Hauptman.)

17–6 CRYSTAL GROWTH

The growing of high quality single crystals is often referred to as an "art." There are many different methodologies and schemes used to grow crystals. The answer to the question, "What is the best method for growing good crystals of this compound?" is "The best method is the one that works!" Some of the standard crystallization techniques are discussed here.

Most crystallization attempts begin with a saturated solution. It is usually best to begin with as much solute as possible since working with only a few milligrams of compound often results in very small or poor quality crystals.

Slow evaporation of the solvent is probably the most often used and simplest crystallization technique. During evaporation it is best to minimize agitation or vibration of the sample because this may cause crystals to "crash out" too quickly or cause developing crystals to go back into solution. A small vial or flask fitted with a septum into which a small needle is inserted makes an excellent and cheap crystallization device. If the solution requires heat to dissolve the solid, it is often best to allow the solution to cool slowly rather that immersing it in an ice bath or putting it in a freezer. This can be done by simply placing the crystallization vessel on a bench top. For very slow cooling, the vessel may be placed in a warm oil bath and both the bath and sample allowed to reach ambient temperature on their own. Alternatively, the temperature may be electronically decreased over a period of time.

Vapor diffusion is another method for obtaining high quality crystals. Figure 17–7 shows an "H-tube" apparatus. A solute/solvent solution (S1) is placed into side A and a second solvent (S2) is placed into side B. S2 should be a solvent in which the solute is less soluble. Upon mixing of the solvents by diffusion, crystals should form since they should be less soluble in the mixed solvent system than in the solvent (S1). An H-tube is quite versatile since it may be easily heated or cooled or used under an inert atmosphere such as nitrogen or argon for air-sensitive compounds.

Liquid diffusion produces crystals at the zone where two solvents mix. In another approach, a solute/solvent solution (S1) is placed into the vessel and a solvent of different density is "layered" on top of the mixture; Fig. 17–8 as the two solvents diffuse, crystals should appear at the interface and grow as mixing proceeds.

A rather elaborate method for growing crystals of low melting compounds directly on an X-ray diffractometer has been developed.[5] Focused heat radiation and simultaneous cooling in a capillary affords crystals that may be examined on the same machine on which they were grown.

A method for obtaining good quality crystals for low-yield compounds or for those that crystallize poorly is to complex the target molecule with something such as triphenylphosphine oxide (TTPO), which itself forms high quality crystals.[6] Basically, the approach involves either coupling or cocrystallizing a compound that forms high quality crystals with one that does not. TTPO is a good proton acceptor, and when it couples with the desired crystal, it often imparts its crystallizing characteristics to the product.

Crystals

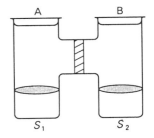

FIGURE 17–7 H-tube apparatus for growing crystals by vapor diffusion.

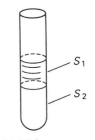

FIGURE 17–8 Liquid-diffusion vessel.

17-7 SELECTION OF CRYSTALS

A crystal suitable for a single-crystal X-ray diffraction study must possess internal order and be of the proper size and shape. Internal order means that a crystal should be "pure" on the atomic level. A "twinned" crystal is one that has more than one orientation of a crystal in the lattice, but is often confused with a "multiple" or "grown-together" crystal. Twinned crystals may sometimes be detected from the diffraction pattern. Observations to note are multiple spots (reflections) lying close together on a rotation photograph or "split" reflection profiles. It is often difficult to determine the unit cell of a twinned crystal and if it is decided that a crystal is twinned, the best and least frustrating way to deal with it is to throw it away and try another crystal. However, in some cases it is a property of all specimens.

Crystals should not contain cracks or have satellite crystals stuck to them and should be as equidimensional as possible; however, this is not a strict requirement. Many X-ray structures have been correctly determined and properly refined using plates, needles, or even "potato-shaped" crystals. If the nature of the crystal permits, it may be ground to a sphere using a device such as that depicted in Fig. 17–9. Air (or N_2, etc.) causes the crystal to tumble around in the abrasive cavity producing a spherical or nearly spherical shape. If grinding is not desired or successful (more likely), crystals can usually be cut and sized with a razor blade.

A polarizing microscope may be used to determine whether a crystal is single or made up of several fragments. Under polarized light, a good crystal should appear uniformly light and dark with every 90° rotation. Finally, the diffraction pattern of a good crystal should show sharp single spots with no "tails" or severe powder rings.

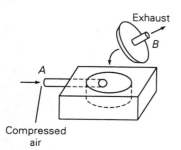

FIGURE 17–9 Air-driven crystal grinder. (Reproduced with permission from G. H. Stout and L. H. Jensen, "X-Ray Structure Determination. A Practical Guide." Macmillan, New York, 1968.)

The next concern we have is the size of the crystal. Several factors need to be considered. First, the crystal must be small enough to remain within the uniform intensity of the primary X-ray beam since the entire crystal should be irradiated by the same intensity radiation. This uniformity is usually about 0.50 mm × 0.50 mm. Crystals larger than 0.50 mm in any direction should be avoided or cut smaller. The optimum size depending on the diffracting power of the crystal and its absorption coefficient is about 0.10 to 0.40 mm on an edge.

As stated previously, absorption of X-rays by crystals must be considered when deciding how large a crystal to use or which radiation to use. The resultant intensity I of the X-ray beam after passing through an absorbing crystal of thickness (t) is given by the equation $I = I_0 e^{-\mu t}$, where I_0 is the incident beam intensity and (μ) is the linear absorption coefficient of the compound. Unfortunately, since the linear absorption coefficient is a function of wavelength of the radiation used and atomic number, it cannot be calculated until the makeup of the structure is known. This does not pose many difficulties since it is not usual for one to begin an X-ray study assuming a "light-atom-structure" and later determine that the crystal was a heavy-atom complex. The linear absorption coefficient for the complex is calculated by summing the mass absorption coefficients (μ/ρ) for the individual atoms present according to equation (17–7).

$$\mu = (\text{density}) \sum (\text{mass \%})(\mu/\rho) \tag{17-7}$$

Mass absorption coefficients for elements for several different types of radiation are listed in the International Tables, Vol. III, pp. 161–165.[7a]

Since the X-ray beam is only uniform to just over 0.50 mm, this should be used as an upper three-dimensional limit. For "light-atom," organic-type crystals with no heavy atoms, such as Br, I, or higher (in atomic number), crystals approaching the upper limit may be used. It is usually necessary to use large crystals for "light-atom" structures due to their low scattering power. With organometallic or inorganic compounds that contain higher percentages of transition-metals, lanthanides, actinides, or other elements that absorb X-rays, crystals should be as uniform as possible and not exceed about 0.20 to 0.40 mm on an edge.

17–8 MOUNTING CRYSTALS

Depending on the nature of the crystal and the apparatus to be used, there are many different methods of crystal mounting that may be utilized. Once good quality crystals are obtained and the size/shape requirements are met, the crystal must be mounted on the diffractometer. Unlike traditional film methods, modern diffractometers do not require that a crystal be oriented in any special direction relative to the X-ray beam. It is therefore not necessary to be concerned about the orientation of the crystal on the mounting device (goniometer head) but it is preferable to mount crystals so that their most nearly cylindrical axis is aligned with the z-axis of the goniometer head. A typical *xyz* style head is shown in Fig. 17–10. The goniometer head allows the crystal to be moved in three directions in order to center it in the X-ray beam. Typically, a crystal is secured to a glass fiber or wedged in a capillary tube that is inserted into a brass or aluminum pin, which in turn is inserted into the goniometer head. The assembly is then screwed onto the instrument.

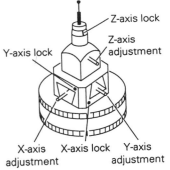

FIGURE 17–10 Goniometer head. (Reproduced with permission from Nicolet/Siemens, User's Guide. Siemens Analytical X-Ray Instruments, Inc., Madison, Wis.)

The method used to mount a crystal depends on the nature of the crystal and on the experimental conditions. The most often-used method to mount air/moisture-stable crystals is to glue them to a glass fiber. Five-minute epoxy is an excellent adhesive in most cases. A very small bead of epoxy will usually suffice and, in fact, large globs of glue should be avoided since some of the crystal may be obstructed making optical alignment of the crystal difficult and beam intensity will be diminished by random scatter. For crystals that are air sensitive, moisture sensitive, or lose solvent, a capillary tube may be used. Crystals should be placed into the capillary tube under an inert atmosphere either in a glove box or glove bag. The capillary is then flame sealed to prevent decomposition. Capillaries are suited for either room or low temperature data collection; however, they are somewhat awkward when using a low temperature device that utilizes a cold air or nitrogen stream to cool the crystal. A routine method of "oil mounting" crystals for low temperature studies has been developed.[8] The crystal is immersed in a thick hydrocarbon oil and attached to a glass fiber, which is mounted on the end of an elongated brass or copper pin. The oil acts as an adhesive and also as an insulator, keeping the crystal from decomposing until it is quick-frozen in the cold stream. This method is very useful for mounting crystals that decompose when separated from their mother liquor or solvent. Since mounting crystals in a glove bag or glove box can be an exercise in futility, this method allows one to even shape (with a razor blade) the desired crystal under a microscope on the benchtop. Although "oil mounting" may not be applicable for every crystalline sample, experience has shown that if one works quickly, even the most delicate and sensitive crystals stand a very good chance of surviving this procedure. (Note: at the University of California–Irvine, nearly all structural determinations, approximately 100/year, are done using this method).

Methodology

17-9 DIFFRACTION EQUIPMENT

The principal instrument used in modern X-ray laboratories is the four-circle diffractometer. Although more traditional film methods are still employed, these are becoming less standard practice. Modern diffractometers do not require alignment of crystals in anything but a random fashion. The orientation of the crystal relative to the diffractometer geometry is determined automatically by a computer.

The main advantages of the automated diffractometer are that the accuracy of results is far greater than that available from film, the instrument runs automatically via computer control, and the rate at which reflections are measured allows researchers to employ X-ray diffraction as a "routine," albeit, unparalleled, analytical tool.

There are four basic components of a four-circle diffractometer:

1. The X-ray source, which includes a high voltage generator and a sealed X-ray tube or a high intensity rotating anode
2. The goniometer or "crystal orientor"
3. The detector, which is usually a scintillation counter
4. The computer, which is used for diffractometer control and data manipulation.

17-10 DIFFRACTOMETER DATA COLLECTION

The heart of the diffractometer is the goniometer. The most widely used geometry is that of the four-circle or Eulerian cradle. The Eulerian cradle consists of four circles that together act to orient the crystal such that all possible reflecting planes may be examined automatically. These circles are designated 2-theta, omega, phi, and chi. Only three of the circles operate independently since 2-theta and omega are usually coupled. Figure 17–11 shows that 2-theta and omega are mounted about the same axis.

A structural determination is usually begun by mounting a crystal on the instrument and taking a photograph, known as a rotation photograph, using Polaroid film. The crystal is rotated on the spindle axis phi for several complete revolutions taking about 10 minutes; the diffraction pattern is recorded on the film. Film measurements are input to the computer which positions the goniometer to locate and center (determine the angular settings of 2-theta, omega, phi, and chi) on 10 to 15 reflections. An alternate method is to allow the computer to randomly search for reflections on which to center. This method works well for strongly diffracting crystals but does not allow one to visually inspect the diffraction pattern of the crystal. Visual inspection is advised since problem crystals may be discovered and dealt with by mounting a new crystal.

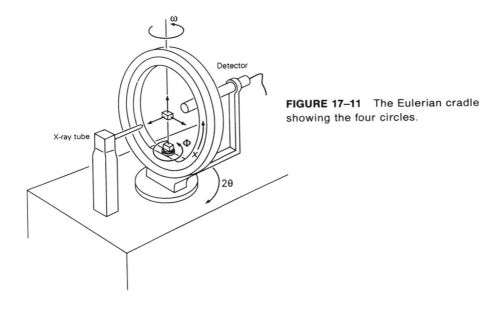

FIGURE 17–11 The Eulerian cradle showing the four circles.

Once several reflections have been centered, automatic computer routines are used with careful attention by the crystallographer to determine an orientation matrix that relates the crystal's geometry to that of the diffractometer and the crystal's unit-cell parameters (a, b, c, α, β, γ, and cell volume). Photographs along the cell axes should be taken to determine the presence or absence of symmetry. In most laboratories, this is also done on the diffractometer rather than by traditional film techniques.

Data collection should begin only after accurate cell parameters are obtained and after as much information about the crystal as possible is known. One should always try to determine the crystal's space group, possibly from a fast limited data set, prior to final data collection. The number of unique data points to be collected depends on the type of radiation used, the Laue symmetry of the crystal, and the quality of the crystal.

The total number of accessible reflections for a crystal is known as the "limiting sphere" and depends only on the wavelength and is given without further explanation as $N = 33.5$ (vol. direct cell/λ^3). The volume of a molecule containing a typical mix of light and heavy atoms may be approximated by the remarkably consistant observation that the average atom occupies 9–11 Å^3; when dominated by heavier atoms, often using 18 Å^3 for just the non-hydrogen atoms will prove more reliable. For a cell with volume = 3000 Å^3, the maximum theoretical number of reflections for Mo radiation ($\lambda = 0.71073$) is approximately 280,000, whereas for Cu ($\lambda = 1.5418$) the number is approximately 27,000. Fortunately, it is not necessary to collect all of these reflections. According to Friedel's law, $hkl \equiv \overline{hkl}$, therefore the maximum number of reflections necessary will be one-half of the preceding numbers. For higher-symmetry crystal systems, the number of unique data is reduced even further. For instance, if the crystal system is orthorhombic, the symmetry is *mmm* (three mutually perpendicular mirror planes) and only one octet of data needs to be collected since

$$hkl = \overline{h}kl = h\overline{k}l = hk\overline{l} = \overline{hk}l = \overline{h}k\overline{l} = h\overline{kl} = \overline{hkl}$$

Typical *hkl* ranges for triclinic, monoclinic, and orthorhombic crystal systems are:

$$0\bar{k}\bar{l} \to hkl, \text{ triclinic}$$

$$00\bar{l} \to hkl, \text{ monoclinic}$$

$$000 \to hkl, \text{ orthorhombic}$$

Another factor that determines how much data should be collected is the fact that the intensity of reflections decreases with increasing $(\sin \theta)/\lambda$. Data are collected between a min and max 2-theta (or theta) limit. Typical values of 2θ(max) for Mo radiation are 45 to 55° (or 0.54 to 0.65 $(\sin \theta)/\lambda$) and the total number of observable reflections is given by $N_0 = N_t \times \sin^3 \theta$. (For the foregoing, $N_0 = 280{,}000 \times \sin^3(25) = 21{,}135$ for 2θ-max = 50°.) Crystal quality, possibly due to thermal motion, may dictate that there will be no usable data beyond the 50° limit.

A method used to improve data quality that is quickly becoming routine in many laboratories is low temperature data collection. As already discussed, low temperature data collection allows for easier handling and mounting of crystals. Reflection profiles are sharpened and backgrounds are usually reduced. Improving the signal-to-noise ratio aids in resolution of weaker reflections and those at higher $(\sin \theta)/\lambda$ values. Thermal motion is also reduced, although not eliminated. Decomposition can be reduced or may be completely eliminated for crystals that are prone to decay in the X-ray beam. Scan speeds for data collection can be increased resulting in less X-ray exposure and ultimately in a greater number of data sets being collected.

17–11 COMPUTERS

Many crystallographic facilities are moving away from large mainframe computers to mini, micro, or even personal computers. Most crystallographic software can be run on a variety of computer platforms. The usual laboratory setup includes one computer that is used for both data collection and data reduction. Graphics terminals, color printers/plotters, and laser printers enable one to produce publication quality output without the necessity of employing specialized services such as a draftsperson.

Some Future Developments

17–12 AREA DETECTORS

The field of protein crystallography is growing steadily. Because of the necessity to determine the structures of large macromolecules, area detectors were developed. Standard diffractometer practice is to collect one reflection at a time and record the intensity using some type of scintillation counter. Area detectors are able to measure the intensities of many reflections simultaneously. This is of particular importance since it is necessary to measure a far greater number of reflections for macromolecules. Also, measuring many reflections at once results in less decay of the crystal. This is obviously a major concern when exposing biological samples to X-rays. Future advances in this area will result in the elucidation of larger and more complex crystal structures. As the resolution of area detectors improves, they may also provide advantages for small-molecule work as well.

17-13 X-RAY VERSUS NEUTRON DIFFRACTION

X-ray and neutron diffraction can be used as complimentary techniques because of both similarities and differences between them. Whereas X-ray diffraction relies on scattering by electrons, neutrons are scattered by nuclei. One of the most common uses of neutron diffraction is to locate hydrogen atoms with high accuracy because the degree of scattering by nuclei does not vary significantly with atomic number. In fact, because hydrogen has a negative scattering factor for neutrons, whereas that for deuterium is positive, they are easily distinguished from one another. A disadvantage of neutron diffraction is that much larger crystals are required than for an X-ray analysis.

17-14 SYNCHROTRON RADIATION

A highly specialized type of X-ray source is available through the use of particle accelerators. Synchrotron radiation is produced by high energy electrons in a storage ring when a magnetic field is applied, causing their path to bend. The energy emitted is 100 to 10^4 times as intense as a conventional X-ray tube. The wavelength of synchrotron radiation can be tuned for a particular experiment, typically in the 0.8 to 2.0-Å range desired by crystallographers. The main disadvantages of synchrotron radiation is that there are a limited number of installations and expenses are high.

17-15 CRYSTAL CLASSES

A unit cell is designated by the six lattice parameters a, b, c, α, β, and γ. These correspond to the axial lengths and the angles between them. α is the angle between b and c, β is the angle between a and c, and γ is the angle between a and b. (See Fig. 17-12.) The unit-cell volume, V may be obtained from

Symmetry and Related Concerns

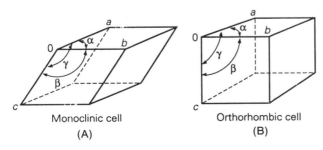

FIGURE 17-12 Monoclinic and orthorhombic unit cells showing the axial lengths and angles between them.

$$V = abc(1 - \cos^2 \alpha - \cos^2 \beta - \cos^2 \gamma + 2 \cos \alpha \cos \beta \cos \gamma)^{1/2}$$

The values of the six parameters may be either unique or they may be constrained due to the symmetry of one of the seven crystal classes listed in Table 17–3. The symbol $\neq$ should be read "not constrained to equal," since it is possible for

TABLE 17–3. The Seven Crystal Classes

Crystal System	No. Independent Parameters	Parameters	Lattice Symmetry
triclinic	6	$a \neq b \neq c; \alpha \neq \beta \neq \gamma$	$\bar{1}$
monoclinic	4	$a \neq b \neq c; \alpha = \gamma = 90°; \beta \neq 90°$	2/m
orthorhombic	3	$a \neq b \neq c; \alpha = \beta = \gamma + 90°$	mmm
tetragonal	2	$a = b \neq c; \alpha = \beta = \gamma = 90°$	4/mmm
rhombohedral	2	$a = b = c; \alpha = \beta = \gamma \neq 90°$	$\bar{3}$m
hexagonal	2	$a = b \neq c; \alpha = \beta = 90°; \gamma = 120°$	6/mmm
cubic	1	$a = b = c; \alpha = \beta = \gamma = 90°$	m3m

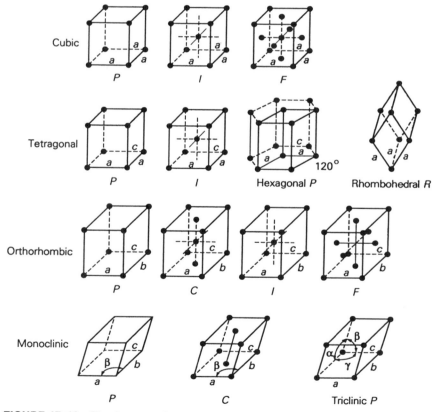

FIGURE 17–13 The fourteen Bravais lattices distributed among the seven crystal systems. The primitive lattice (P) has 1 equivalent point per unit cell at (000); the body centered lattice (C) has 2 equivalent points per unit cell at (000; $\frac{11}{22}$0); and the face-centered lattice has 4 equivalent points per unit cell at (000; $0\frac{11}{22}$; $\frac{1}{2}0\frac{1}{2}$; $\frac{11}{22}0$). All equivalent lattice points are related to each other by pure translation and have identical environments.

accidental equalities to occur, but this is rare. Addition of face and body centering to the seven crystal systems leads to the 14 Bravais lattices shown in Fig. 17–13.

17–16 SPACE GROUPS

The point-group operations of rotation, mirror reflection, and inversion discussed in Chapter 1 define the 32 unique point groups. Combining the 32 point groups with the 14 Bravais lattices result in 230 space groups that are a group of symmetry operations that describe how molecules are arranged, pack, interact, etc., in the unit cell of a crystal lattice. It is now necessary to define two additional "space-symmetry operators" that account for translational motion. These additional operators are the screw axis and the glide plane.

An n-fold screw axis is produced by combining a rotation axis and a translation parallel to this axis. The rotation is $(360/n)°$ and the translation is a fraction (m/n) of a unit translation. The integer n may take the values 1, 2, 3, 4, or 6. The screw axis defined by the symbol 3_1 means that a point is rotated around the axis by 120° (360/3) and then translated by one-third of a unit translation. This is shown graphically in Fig. 17–14. The points represented by 2 and 3 result from successive 3_1 operations. Point 1' is indistinguishable from 1 since it is the result of a 3/3 or unit translation.

A glide plane is the result of combining a translation parallel to a mirror plane with subsequent reflection across the mirror. The translation is either along a cell edge or a face diagonal. Glide planes are designated as a, b, c, n, or d depending on the translation. An a-glide is shown in Fig. 17–15. The translation for glides a, b, and c is one-half of the axial length, for n-glides it is one-half of the diagonal length. Diamond or d-glides occur for centered lattices and correspond to a one-fourth diagonal translation. Since all glide planes translate either by one-half or one-fourth, after either two or four glide operations, the resulting point corresponds to a unit translation. A glide operation inverts chirality; a screw does not.

In addition to screw axes and glide planes, centering operations may also be present. P or primitive lattices have no centered lattice points. A, B, and C lattices have lattice points centered on their respective cell faces. I lattices have a lattice point at the center of the cell and F lattices have lattice points on all faces. Centered lattices are shown in Fig. 17–13.

All 230 space groups are compiled in the International Tables, Vol I.[7b] Equivalent positions, special positions, and reflection conditions for each are listed. The reflection conditions observed for the diffraction data are used to determine the space group for a given crystal structure. Unfortunately, several space groups have identical reflection conditions (usually referred to as systematic absences) and are not distinguishable from one another solely on this basis. Table 17–4 lists symmetry elements and their associated systematic absences. This table may be used to determine the space group of a crystal structure based on the fact that symmetry elements give rise to specific systematically absent reflections. The equivalent positions (fractional coordinates where two identical molecules are located in the unit cell) for the monoclinic space group $P2_1$ are illustrated in Fig. 17–16. The 2_1 screw axis relates every point at xyz to an equivalent point at $-x, y + \frac{1}{2}, -z$. The screw axis is along b such that the equipoint $-x, y + \frac{1}{2}, -z$ results from rotation about b and then translation of one-half of a unit cell (the sign of y remains the same, whereas x and z become negative).

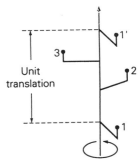

FIGURE 17–14 Three-fold screw axis. (Reproduced with permission from G. H. Stout and L. H. Jensen, "X-Ray Structure Determination. A Practical Guide." 2nd ed. John Wiley & Sons, New York, 1989.)

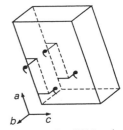

FIGURE 17–15 Glide plane a. (Reproduced with permission from G. H. Stout and L. H. Jensen, "X-Ray Structure Determination. A Practical Guide." 2nd ed. John Wiley & Sons, New York, 1989.)

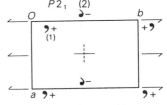

$P2_1$, equivalent positions
(1) x, y, z; (2) $\bar{x}, y + \frac{1}{2}, \bar{z}$

FIGURE 17–16 Equivalent position in space group $P2_1$. (Reproduced with permission from G. H. Stout and L. H. Jensen, "X-Ray Structure Determination. A Practical Guide." 2nd ed. John Wiley & Sons, New York, 1989.)

TABLE 17–4. Symmetry Elements and Reflection Conditions

Symmetry Element	Affected Reflection		Condition for Systematic Absence of Reflection
2-fold screw (2_1) ⎫	a	$h00$	$h = 2n + 1 =$ odd
4-fold screw (4_2) ⎬ along	b	$0k0$	$k = 2n + 1$
6-fold screw (6_3) ⎭	c	$00l$	$l = 2n + 1$
3-fold screw (3_1, 3_2) ⎫ along	c^a	$00l$	$l = 3n + 1, 3n + 2$,
6-fold screw (6_2, 6_4) ⎭			*i.e.*, not evenly divisible by 3
4-fold screw (4_1, 4_3) along	a	$h00$	$h = 4n + 1, 2,$ or 3
	b	$0k0$	$k = 4n + 1, 2,$ or 3
	c	$00l$	$l = 4n + 1, 2,$ or 3
6-fold screw (6_1, 6_5) along	c^a	$00l$	$l = 6n + 1, 2, 3, 4,$ or 5
Glide plane perpendicular to	a		
translation $b/2$ (b glide)		$0kl$	$k = 2n + 1$
$c/2$ (c glide)			$l = 2n + 1$
$b/2 + c/2$ (n glide)			$k + l = 2n + 1$
$b/4 + c/4$ (d glide)			$k + l = 4n + 1, 2,$ or 3
Glide plane perpendicular to	b		
translation $a/2$ (a glide)		$h0l$	$h = 2n + 1$
$c/2$ (c glide)			$l = 2n + 1$
$a/2 + c/2$ (n glide)			$h + l = 2n + 1$
$a/4 + c/4$ (d glide)			$h + l = 4n + 1, 2,$ or 3
Glide plane perpendicular to	c		
translation $a/2$ (a glide)		$hk0$	$h = 2n + 1$
$b/2$ (b glide)			$k = 2n + 1$
$a/2 + b/2$ (n glide)			$h + k = 2n + 1$
$a/4 + b/4$ (d glide)			$h + k = 4n + 1, 2,$ or 3
A-centered lattice (A)		hkl	$k + l = 2n + 1$
B-centered lattice (B)			$h + l = 2n + 1$
C-centered lattice (C)			$h + k = 2n + 1$
Face-centered lattice (F)			$h + k = 2n + 1$ ⎫ *i.e.*, h, k, l not
			$h + l = 2n + 1$ ⎬ all even or all
			$k + l = 2n + 1$ ⎭ odd
Body-centered lattice (I)			$h + k - l = 2n + 1$

^{a} Note that in the crystal classes in which 3- and 6-fold screws occur as cell axes, these are conventionally assigned to be c, so only the $00l$ reflections need be considered.

17–17 SPACE-GROUP DETERMINATION

Despite the possible need to choose from among 230 options, in fact, most molecular species crystallize in just a few different space groups: $P\bar{1}$, $P2_1/c$, $C2/c$, $P2_12_12_1$, *Pbca*, and *Pnma* account for about 85% of all structures.

The following two examples will illustrate how to determine the space group of a crystal from the diffraction symmetry and systematic absences. The diffraction symmetry may be determined using photographs and the systematically absent reflections may be determined from the intensity data. (Refer to Table 17–4 for the symmetry element/systematic absences relationships.)

Case 1. The diffraction symmetry is $2/m$ (a twofold rotation axis with a mirror plane perpendicular to it), indicating a monoclinic crystal system. There are no reflection conditions consistent with a centering operation so the lattice is primitive P. Reflections of the type $h0l$ are absent when l is odd and the $0k0$ reflections are absent when k is odd. According to

Table 17–4, the $h0l$ absence is a result of a c-glide plane perpendicular to the b axis and the $0k0$ absence is indicative of a 2_1 screw axis along b. The space group is uniquely defined as $P2_1/c$.

Case 2. The diffraction symmetry is $2/m$ indicating a monoclinic cell. The only systematic absences observed are for hkl reflections when the value of $h + k$ is odd. This information is consistent with a C-centered lattice. Since there is no screw axis or glide plane indicated, the only possible space groups are $C2$, Cm, or $C2/m$. These space groups are indistinguishable from one another and it may be necessary to examine each by trial structure solution/refinement to determine which is the correct space group.

Case 2 represents a situation where one must make a choice between centrosymmetric and non-centrosymmetric space groups ($C2/m$ is centric, Cm and $C2$ are acentric). If a structure contains an equal mix of left- and right-handed molecules, it must crystallize in a centrosymmetric space group. This is true even if the molecule is asymmetric. The handedness or chirality of a crystal means that a structure cannot be superimposed on its mirror image. If a crystal is known to be chiral and resolved, it must not be centrosymmetric. When more than one space group is possible, this information may assist one in determining whether a centric or non-centric space group is most likely the correct choice.

17–18 AVOIDING CRYSTALLOGRAPHIC MISTAKES

The aforementioned problem of determining whether a crystal is best described as centrosymmetric or non-centrosymmetric requires further discussion. Since centric/acentric space group "pairs" (*e.g.*, $C2/m$, Cm, $C2$) have the same systematic absences, these are of no help in sorting out the ambiguity.

Consider the monoclinic space groups $P2_1/m$ and $P2_1$ that are centrosymmetric and non-centrosymmetric, respectively. The only systematically absent reflections are the $0k0$ when k is odd (twofold screw axis perpendicular to b). If a structure is solved in $P2_1/m$, it may also be described in $P2_1$ with a doubling of the Z value (The Z value refers to the number of molecules in the unit cell. The Z value for $P2_1/m$ with one molecule in a general position is 4, whereas for $P2_1$ it is 2.) If a structure has molecular symmetry such that a mirror plane relates one-half of the molecule to the other, then it is possible (but not required) that the molecule possesses the same crystallographic symmetry and be located on the crystallographic mirror plane in $P2_1/m$. This structure could also be described in $P2_1$, but all atoms would have to be located. (The discussion of molecular versus crystallographic symmetry will illustrate the requirement to locate only the symmetry unique atoms for naphthalene.)

Frequently, a structure will be described as ordered in a non-centrosymmetric space group rather than as disordered in a centrosymmetric one.[9] When this situation occurs, it may be more correct to describe the structure as centrosymmetric and disordered, indicating that only the "average" structure has been determined. Several structures have been "republished" after investigation of the original work revealed that the space-group assignment was incorrect.[10]

An example of a structure that was originally described in a non-centrosymmetric space group and later shown to be more correctly described as centric is

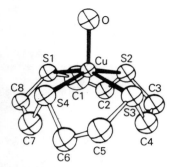

FIGURE 17–17 Isotropic thermal plot of [(1,4,7,10-tetrathiacyclododecane) copper(II) *bis*-perchlorate-monohydrate].

taken from reference 9. The structure of [(1,4,7,10-tetrathiacyclododecane) copper(II)*bis*-perchlorate-monohydrate] (Fig. 17–17) was described in the non-centric orthorhombic space group *Pbc2*. Potential problems that should have been carefully considered were:

1. A pseudomirror plane perpendicular to the *c*-axis that would be consistent with the mirror plane in *Pbcm* (centric).
2. Problems in least-squares refinement of heavy atom positions.
3. Refinement of S atoms with anisotropic temperature factors led to unrealistic bond distances requiring isotropic refinement of these atoms. (An anisotropic temperature factor is an exponential expression that accounts for the vibration of an atom. This vibration is described as an ellipsoid rather than spherically symmetric or isotropic.) See Fig. 17–19 for an example of a molecule refined with anisotropic thermal parameters and Fig. 17–17 for the isotropic plot of [(1,4,7,10-tetrathiacyclododecane)copper(II)*bis*-perchlorate-monohydrate].
4. A high *R* value of 12.7% (*R* values are quantities that are used to compare observed and calculated structure factors. See Section 17–20.
5. Large ranges for the Cu-S and S-C distances 2.30(1) to 2.37(1) and 1.73(4) to 1.9(4) Å.
6. Short C-C distances 1.41(3) Å.

Refinement in the centrosymmetric space group *Pbcm* yielded the following significant results.

1. Statistically equivalent Cu-S [2.325(3)] and S-C [1.815(9)] distances with esd's about one-half as large as the original.
2. A much lower *R* value of 6.8% with all atoms anisotropic.

There is considerable disorder in the *Pbcm* structure; the three S-C-C-S groupings are disordered across planar conformations and there is a high degree of disorder for the two perchlorate ions (a common problem). it is clear that the ordered *Pbc2* structure is just one of many possible orientations of the structure that is better described by the "average" *Pbcm* model.

The moral of the foregoing example is to examine each possible space group carefully when a choice must be made between a centrosymmetric or non-centrosymmetric model. Some of the following things to note are:

1. Does the molecule possess any molecular symmetry as a mirror plane, inversion center, or rotation axis that may also be crystallographic?
2. How many molecules are in the asymmetric unit? What is the expected and observed *Z*-value?
3. Are pairs of atom coordinates similar such that they may be related by some symmetry operator?
4. Are interatomic distances and angles consistent within accepted ranges?
5. Are the shapes of thermal ellipsoids reasonable?

The best rule to follow when one has a choice of centrosymmetric or non-centrosymmetric space groups is to test the centrosymmetric group first. If the structure cannot be solved or refined properly then attempt the solution with the non-centrosymmetric space group. If any of the foregoing problems are noted or if the structure fails to refine well, then investigate the centrosymmetric group again until it is clear that the structure can only be described in the non-centrosymmetric space group.

17-19 MOLECULAR VERSUS CRYSTALLOGRAPHIC SYMMETRY

There are often relationships between a structure's molecular and crystallographic symmetry. consider the case of naphthalene and pyrene (Figs. 17–18(A) and 17–18(B)). Both molecules are highly symmetric with planar structures. Viewing Figs. 17–17(A) and 17–17(B), it is clear that both molecules possess *mmm* molecular symmetry (three perpendicular mirror planes), inversion centers, and rotation axes. Naphthalene may be described as having mirror planes (1), the

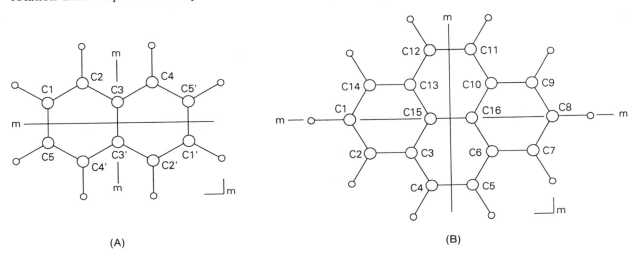

FIGURE 17-18 Ball-and-stick drawings of (A) naphthalene and (B) pyrene.

plane of the molecule, (2) a mirror along C3-C3′, which would relate each half of the molecule to itself, and (3) a plane bisecting the C1-C5, C3-C3′, and C1′-C5′ bonds. There are also twofold rotation axes perpendicular to these mirrors that relate halves of the molecule. In addition, there is an inversion center midway between C3-C3′ that interconverts the two rings. Whereas naphthalene exhibits a high degree of molecular symmetry, the only observed crystallographic symmetry is the inversion center between C3-C3′. Table 17–5 lists atom coordinates for the inversion-related atom pairs. The inversion center is located at (0, 0, 0). When solving the structure of naphthalene, only one-half of the total number of atoms needs to be located and refined since the other half is related by symmetry. All of the aforementioned noted molecular symmetry elements are also present for pyrene. This would lead one to expect that at least some of these elements are present in the crystal structure; however, pyrene has no crystallographic symmetry and all atoms must be located and included in the refinement. These two examples should illustrate the fact that molecular symmetry does not necessarily lead to crystallographic symmetry.

17-20 QUALITY ASSESSMENT

X-ray diffraction experiments produce results that are usually believed implicitly. After all, there is a picture of the molecule so it must be correct and well done. Although most X-ray studies are accurate, one should remember that the derived

TABLE 17–5. Atomic Coordinates and Their Inversion-related Pairs for Naphthalene

Atom	Axes		
	x	y	z
C1	0.08232	0.01856	0.32836
C2	0.11295	0.16382	0.22289
C3	0.04799	0.10518	0.03714
C4	0.07656	0.25183	−0.07582
C5	−0.01320	−0.19021	0.25460
H1	0.12420	0.05890	0.45540
H2	0.17870	0.30560	0.27110
H3	0.14180	0.39070	−0.02360
H4	−0.03330	−0.29520	0.33130
Inversion-related Atoms			
C1′	−0.08232	−0.01856	−0.32836
C2′	−0.11295	−0.16384	−0.22289
C3′	−0.04799	−0.10518	−0.03714
C4′	−0.07656	−0.25183	0.07582
C5′	0.01320	0.19021	−0.25460
H1′	−0.12420	−0.05890	−0.45540
H2′	−0.17870	−0.30560	−0.27110
H3′	−0.14180	−0.39070	0.02360
H4′	0.03330	0.29520	−0.33130

results were obtained experimentally and that errors may occur either in the diffraction experiment or in the interpretation of results. The following discussion focuses on what to look for in a crystallographic paper and how to decide whether or not the reported results are correct.

The two quantities that should be examined to determine the accuracy of a crystal structure are the R values and the estimated standard deviations (esd's) of derived parameters. The conventional R value is given by equation (17–8).

$$R = \sum |F_o - F_c| / \sum F_o \qquad (17\text{–}8)$$

where F_o are the observed structure factors derived from the measured reflection intensities and F_c are the structure factors calculated for the refined model. From equation (17–8) it would appear that the lower the R value, the better the structure since the observed and calculated F's will be in better agreement. Although this is generally true, the R value may be artificially lowered using a number of techniques. The most common way of reducing the R value of a structure is to omit weak reflections. This is an accepted practice by most crystallographers since all crystals exhibit weak reflections at higher diffraction angles and a cutoff intensity (I or F) should be included in the paper. An intensity cutoff of $I > 3\sigma(I)$ means that only reflections with intensities greater than 3 times their measured deviation (sigma) will be used in calculations ($3\sigma(I)$ corresponds to $6\sigma(F)$ since I is proportional to F^2). One should regard as suspect papers that report a data cutoff greater than the $3\sigma(I)$ or $6\sigma(F)$ threshold. In addition to the R (or R_F values), one should be aware of the weighted R (also listed as wR or R_{wF}). Weighted R values are based not only on F values but on the sigma of F as well. Reflections with higher σF will be downweighted from those with lower σF values. A typical weighting scheme is given by equation (17–9).

$$w^{-1} = \sigma^2(F) + gF^2 \qquad (17-9)$$

where g is a small value (0.01 to 0.001). The resulting weighted R value is shown in equation (17–10).

$$R_{wF} = \frac{\sum w^{1/2}|F_o - F_c|}{\sum w^{1/2} F_o} \qquad (17-10)$$

Since the weighted R value is based on more information (σF), it is the better value on which to base the accuracy of the structure determination. If it is reported that a structure was refined using "unit weights" (all reflections weighted equally), this structure should be reviewed carefully since a proper weighting scheme will usually lead to lower esd's for derived parameters.

The best criterion on which to judge the accuracy of a crystal structure determination is found in the derived parameters and their esd's. If interatomic distances and angles fall within expected ranges and their esd's are small, then it is reasonable to assume that the structural study was done properly. Omitting too many weak reflections will lead to increased esd's so this should be avoided. Failing to measure enough data will also result in higher esd's since the data-to-parameter ratio will be low. A data-to-parameter ratio of at least 8 to 10:1 should be considered a lower limit for a well-defined structure, although this is not always achievable.

Since all structural reports should include a plot of the molecule, the reliability of derived parameters may often be determined simply by looking at the picture. The molecular plot should be a thermal ellipsoid plot such as that shown in Fig. 17–19[11] rather than ball and stick plots as shown in Figs. 17–18(A) and 17–18(B). The thermal motion of an atom is represented by an ellipsoid within which the electrons of the atom have a probability of being. This probability, which is usually 50%, should be noted on the plot. The molecule in Fig. 17–19 shows a high degree of thermal motion for the carbon atoms (particularly the methyl carbons) of the pentamethylcyclopentadienyl ligands even though the data set was collected at 173 K. Since these ligands can rotate freely, it is understandable that the carbon atoms show this high degree of libration. If an atom is disordered (more than one possible position for an atom-static disorder), a large thermal motion may either alert one to this possibility or may serve to "smear out" the disordered components.

A few ions that are often disordered are Et_4N^+, ClO_4^-, BF_4^- and PF_6^-. Et_4N^+ usually exhibits static disorder between the alpha carbon atoms of the ethyl groups, whereas ClO_4^-, BF_4^-, and PF_6^- show spherical disorder due to their spherical nature. Figure 17–20 shows a disordered Et_4N^+ ion.[12] Whenever possible, these ions should be avoided when attempting to grow single crystals for a diffraction study since it may not be possible to completely resolve the resulting disorder problems. Some alternative counter ions that can be employed, and that are usually ordered, are triflate, $B(C_6H_5)_4^-$, $((C_6H_5)_3P)_2N^+$, and $(C_6H_5)_4As^+$.

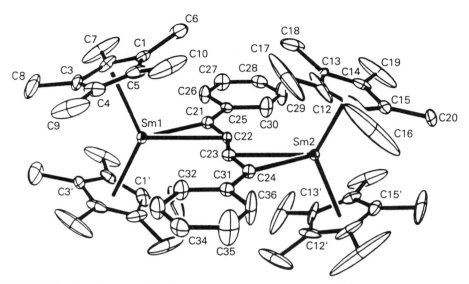

FIGURE 17–19 Thermal ellipsoid plot of $[(C_5Me_5)_2Sm]_2(u\text{-}n^2:n^2\text{-}PhC_4Ph)\cdot 2C_6H_5Me$.

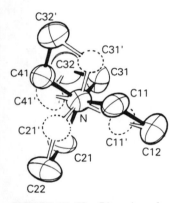

FIGURE 17–20 Disorder of the ethyl groups in Et_4N^+.

As the result of the co-crystallization of two closely related compounds, a crystal will not be pure. Often the contaminant will be present at only a few mole percent. This can give rise to another kind of disorder: compositional disorder. An example is found in the structure of *cis,fac*-$(R_3P)_3Mo(=O)Cl_2$, which is now known to contain small quantities of *fac*-$(R_3P)_3MoCl_3$. An earlier report of "anomolously" long Mo—O bond distances in the former complex has been reinterpreted as the superimposition of the Mo—O bond with a previously unrecognized small percentage of longer Mo—Cl bonds.[13]

17–21 CRYSTALLOGRAPHIC DATA

The advent of fast computers and automated data collection routines has resulted in an ever-increasing amount of structural data being generated and published. Fortunately, there are a number of computer databases and collections of printed material that one can access to utilize this structural information.

The largest compilation of structural data is the Cambridge Structural Database (CSD) operated by Cambridge University (England). All published or directly deposited organic and organometallic complexes characterized by either x-ray or neutron diffraction are included. The CSD is far more than a reference database. Included are atomic coordinates, unit cell parameters, space groups, *R* values, and esd's. Cambridge has also developed the computer programs necessary to access the database. A graphical user interface allows one easy access to all structural information as well as the ability to generate molecular geometries and plots. The database contained over 82,000 entries in 1989 with 8000 to 9000 new entries yearly (Table 17–1).

The Protein Data Bank of the Brookhaven National Laboratory contains atomic coordinates for protein and nucleic acid structures. The Inorganic Crystal Structure Data Base is composed of all compounds lacking carbon-carbon or carbon-hydrogen bonds but does include metal carbides. (See also reference 4.)

REFERENCES CITED

1. *The Newsletter of the Cambridge Structural Database System*, Cambridge Crystallographic Data Centre, Cambridge, U.K., April 1990.
2. H. Hope, and B. G. Nichols, Acta Cryst., *B37*, 158–161 (1981).
3. G. H. Stout, and L. H. Jensen, "X-Ray Structure Determination. A Practical Guide," 2nd ed., Wiley, New York, 1989.
4. J. P. Glusker and K. N. Trueblood, "Crystal Structure Analysis. A Primer," Oxford University Press, Oxford, U.K., 1985.
5. D. Brodalla, D. Mootz, R. Boese, and W. Osswald, J. Appl. Cryst., *18*, 316–319 (1985).
6. M. C. Etter, and P. W. Baures, J. Amer. Chem. Soc., *110*, 639–640 (1988).
7. a. "International Tables for X-Ray Crystallography," Vol. III, eds. C. H. Macgillavry and G. D. Rieck, Kynoch Press, Birmingham, England, 1962.
   b. "International Tables for X-Ray Crystallography," Vol. I, eds. N. F. M. Henry and K. Lonsdale, Kynoch Press, Birmingham, England, 1952.
8. H. Hope, University of California, Davis. Personal communication to J. Ziller, 1987.
9. R. E. Marsh, Acta. Cryst., *B42*, 193–198 (1986).
10. R. E. Marsh, and F. H. Herbstein, Acta. Cryst., *B44*, 77–88 (1988).
11. W. J. Evans, R. A. Keyer, and J. W. Ziller, Organometallics, *9*, 2628–2631 (1990).
12. M. R. Churchill, and S. W.-Y. Chang, Inorg. Chem., *13*, 2413–2419 (1974).
13. K. Yoon, G. Parkin, A. L. Rheingold, J. Am. Chem. Soc., *113*, 1437–1439 (1991).

EXERCISES

1. A crystal of [InIMes$_2$]$_2$ is monoclinic with reciprocal lattice parameters $a^* = 0.038930$, $b^* = 0.058389$, $c^* = 0.127765$ and $\beta^* = 75.298$. Calculate the direct cell parameters and unit-cell volume. (Mes is mesityl).

2. The density of the above complex is 1.792 mg/m^3. The number of molecules in the unit cell may be calculated using: $n = [(\text{density})(V)]/[(\text{Avogadro constant})(\text{MW})]$. What is n (Z value) for the complex?

3. The systematic absences observed for the above complex are hkl for $h + k = 2n + 1$ and $h0l$ for $l = 2n + 1$. What are the two possible monoclinic space groups? Given the calculated Z value, is it possible for the molecule to lie on a general position in both space groups? What point symmetry is possible in one of these space groups? (See International Tables.)

4. Determine the space group(s) consistent with the following systematic absences and crystal class.

   a. Monoclinic: $0k0$ for $k = 2n + 1$; $h0l$ for $h + l = 2n + 1$.

   b. Orthorhombic: hkl for $h + k$, $k + l$; $0kl$ for $k + l = 4n + 1$; $h0l$ for $n + l = 4n + 1$; $hk0$ for $h + k = 4n + 1$

   c. Orthorhombic: hkl for $h + k + l = 2n + 1$; $0kl$ for $k = 2n + 1$; $h0l$ for $h = 2n + 1$

   d. Orthorhombic: $0kl$ for $k = 2n + 1$; $h0l$ for $l = 2n + 1$; $hk0$ for $h = 2n + 1$

   e. Same as for (d) but also hkl for $h + k + l = 2n + 1$

   f. Triclinic: hkl for $h + k + l = 2n + 1$

5. What is the absorption coefficient (μ) for $(Me_5C_5)_3Sm$ given the density is 1.375 mg/m^3?

6. What is the number of accessible reflections for the complex in question (1) for both Mo and Cu radiation and how many octets of data will need to be collected? Why may the resulting number of reflections appear to be less than required? How many reflections will there be for Mo radiation with a 2-theta limit of 55 degrees?

7. If a molecule has a chiral center and is known to be resolved, what is a specific reason why it may not crystallize in space group $P2_1/c$?

8. What will be the maximum and minimum absorption effects on the reflections from the crystal in problem 5 if the crystal measurements are 0.10 × 0.40 × 0.80 mm? Should a correction be made? Should this crystal be used in the X-ray diffraction study?

Character Tables for Chemically Important Symmetry Groups

APPENDIX A

1. THE NONAXIAL GROUPS

C_1	E
A	1

C_s	E	σ_h		
A'	1	1	x, y, R_z	x^2, y^2, z^2, xy
A''	1	-1	z, R_x, R_y	yz, xz

C_i	E	i		
A_g	1	1	R_x, R_y, R_z	$x^2, y^2, z^2, xy, xz, yz$
A_u	1	-1	x, y, z	

2. THE C_n GROUPS

C_2	E	C_2		
A	1	1	z, R_z	x^2, y^2, z^2, xy
B	1	-1	x, y, R_x, R_y	yz, xz

C_3	E	C_3	C_3^2		$\varepsilon = \exp(2\pi i/3)$
A	1	1	1	z, R_z	$x^2 + y^2, z^2$
E	$\begin{cases}1 \\ 1\end{cases}$	$\begin{matrix}\varepsilon \\ \varepsilon^*\end{matrix}$	$\begin{matrix}\varepsilon^* \\ \varepsilon\end{matrix}$	$(x, y)(R_x, R_y)$	$(x^2 - y^2, xy)(yz, xz)$

C_4	E	C_4	C_2	C_4^3		
A	1	1	1	1	z, R_z	$x^2 + y^2, z^2$
B	1	-1	1	-1		$x^2 - y^2, xy$
E	$\begin{cases}1 \\ 1\end{cases}$	$\begin{matrix}i \\ -i\end{matrix}$	$\begin{matrix}-1 \\ -1\end{matrix}$	$\begin{matrix}-i \\ i\end{matrix}$	$(x, y)(R_x, R_y)$	(yz, xz)

2. THE C_n GROUPS (continued)

C_5	E	C_5	C_5^2	C_5^3	C_5^4			$\varepsilon = \exp(2\pi i/5)$
A	1	1	1	1	1	z, R_z		$x^2 + y^2, z^2$
E_1	$\begin{cases}1\\1\end{cases}$	$\begin{matrix}\varepsilon\\\varepsilon^*\end{matrix}$	$\begin{matrix}\varepsilon^2\\\varepsilon^{2*}\end{matrix}$	$\begin{matrix}\varepsilon^{2*}\\\varepsilon^2\end{matrix}$	$\begin{matrix}\varepsilon^*\\\varepsilon\end{matrix}$	$(x, y)(R_x, R_y)$		(yz, xz)
E_2	$\begin{cases}1\\1\end{cases}$	$\begin{matrix}\varepsilon^2\\\varepsilon^{2*}\end{matrix}$	$\begin{matrix}\varepsilon^*\\\varepsilon\end{matrix}$	$\begin{matrix}\varepsilon\\\varepsilon^*\end{matrix}$	$\begin{matrix}\varepsilon^{2*}\\\varepsilon^2\end{matrix}$			$(x^2 - y^2, xy)$

C_6	E	C_6	C_3	C_2	C_3^2	C_6^5			$\varepsilon = \exp(2\pi i/6)$
A	1	1	1	1	1	1	z, R_z		$x^2 + y^2, z^2$
B	1	-1	1	-1	1	-1			
E_1	$\begin{cases}1\\1\end{cases}$	$\begin{matrix}\varepsilon\\\varepsilon^*\end{matrix}$	$\begin{matrix}-\varepsilon^*\\-\varepsilon\end{matrix}$	$\begin{matrix}-1\\-1\end{matrix}$	$\begin{matrix}-\varepsilon\\-\varepsilon^*\end{matrix}$	$\begin{matrix}\varepsilon^*\\\varepsilon\end{matrix}$	(x, y) (R_x, R_y)		(xz, yz)
E_2	$\begin{cases}1\\1\end{cases}$	$\begin{matrix}-\varepsilon^*\\-\varepsilon\end{matrix}$	$\begin{matrix}-\varepsilon\\-\varepsilon^*\end{matrix}$	$\begin{matrix}1\\1\end{matrix}$	$\begin{matrix}-\varepsilon^*\\-\varepsilon\end{matrix}$	$\begin{matrix}-\varepsilon\\-\varepsilon^*\end{matrix}$			$(x^2 - y^2, xy)$

C_7	E	C_7	C_7^2	C_7^3	C_7^4	C_7^5	C_7^6			$\varepsilon = \exp(2\pi i/7)$
A	1	1	1	1	1	1	1	z, R_z		$x^2 + y^2, z^2$
E_1	$\begin{cases}1\\1\end{cases}$	$\begin{matrix}\varepsilon\\\varepsilon^*\end{matrix}$	$\begin{matrix}\varepsilon^2\\\varepsilon^{2*}\end{matrix}$	$\begin{matrix}\varepsilon^3\\\varepsilon^{3*}\end{matrix}$	$\begin{matrix}\varepsilon^{3*}\\\varepsilon^3\end{matrix}$	$\begin{matrix}\varepsilon^{2*}\\\varepsilon^2\end{matrix}$	$\begin{matrix}\varepsilon^*\\\varepsilon\end{matrix}$	(x, y) (R_x, R_y)		(xz, yz)
E_2	$\begin{cases}1\\1\end{cases}$	$\begin{matrix}\varepsilon^2\\\varepsilon^{2*}\end{matrix}$	$\begin{matrix}\varepsilon^{3*}\\\varepsilon^3\end{matrix}$	$\begin{matrix}\varepsilon^*\\\varepsilon\end{matrix}$	$\begin{matrix}\varepsilon\\\varepsilon^*\end{matrix}$	$\begin{matrix}\varepsilon^3\\\varepsilon^{3*}\end{matrix}$	$\begin{matrix}\varepsilon^{2*}\\\varepsilon^2\end{matrix}$			$(x^2 - y^2, xy)$
E_3	$\begin{cases}1\\1\end{cases}$	$\begin{matrix}\varepsilon^3\\\varepsilon^{3*}\end{matrix}$	$\begin{matrix}\varepsilon^*\\\varepsilon\end{matrix}$	$\begin{matrix}\varepsilon^2\\\varepsilon^{2*}\end{matrix}$	$\begin{matrix}\varepsilon^{2*}\\\varepsilon^2\end{matrix}$	$\begin{matrix}\varepsilon\\\varepsilon^*\end{matrix}$	$\begin{matrix}\varepsilon^{3*}\\\varepsilon^3\end{matrix}$			

C_8	E	C_8	C_4	C_2	C_4^3	C_8^3	C_8^5	C_8^7			$\varepsilon = \exp(2\pi i/8)$
A	1	1	1	1	1	1	1	1	z, R_z		$x^2 + y^2, z^2$
B	1	-1	1	1	1	-1	-1	-1			
E_1	$\begin{cases}1\\1\end{cases}$	$\begin{matrix}\varepsilon\\\varepsilon^*\end{matrix}$	$\begin{matrix}i\\-i\end{matrix}$	$\begin{matrix}-1\\-1\end{matrix}$	$\begin{matrix}-i\\i\end{matrix}$	$\begin{matrix}-\varepsilon^*\\-\varepsilon\end{matrix}$	$\begin{matrix}-\varepsilon\\-\varepsilon^*\end{matrix}$	$\begin{matrix}\varepsilon^*\\\varepsilon\end{matrix}$	(x, y) (R_x, R_y)		(xz, yz)
E_2	$\begin{cases}1\\1\end{cases}$	$\begin{matrix}i\\-i\end{matrix}$	$\begin{matrix}-1\\-1\end{matrix}$	$\begin{matrix}1\\1\end{matrix}$	$\begin{matrix}-1\\-1\end{matrix}$	$\begin{matrix}-i\\i\end{matrix}$	$\begin{matrix}i\\-i\end{matrix}$	$\begin{matrix}-i\\i\end{matrix}$			$(x^2 - y^2, xy)$
E_3	$\begin{cases}1\\1\end{cases}$	$\begin{matrix}-\varepsilon^*\\-\varepsilon^*\end{matrix}$	$\begin{matrix}i\\-i\end{matrix}$	$\begin{matrix}-1\\-1\end{matrix}$	$\begin{matrix}-i\\i\end{matrix}$	$\begin{matrix}\varepsilon^*\\\varepsilon\end{matrix}$	$\begin{matrix}\varepsilon\\\varepsilon^*\end{matrix}$	$\begin{matrix}-\varepsilon^*\\-\varepsilon\end{matrix}$			

3. THE D_n GROUPS

D_2	E	$C_3(z)$	$C_2(y)$	$C_2(x)$		
A	1	1	1	1		$x^2 + y^2, z^2$
B_1	1	1	-1	-1	z, R_z	xy
B_2	1	-1	1	-1	y, R_y	xz
B_3	1	-1	-1	1	x, R_x	yz

D_3	E	$2C_3$	$3C_2$		
A_1	1	1	1		$x^2 + y^2, z^2$
A_2	1	1	-1	z, R_z	
E	2	-1	0	$(x, y)(R_x, R_y)$	$(x^2 - y^2, xy)(xz, yz)$

Appendix A   Character Tables for Chemically Important Symmetry Groups

D_4	E	$2C_4$	$C_2(=C_4^2)$	$2C_2'$	$2C_2''$		
A_1	1	1	1	1	1		x^2+y^2, z^2
A_2	1	1	1	-1	-1	z, R_z	
B_1	1	-1	1	1	-1		x^2-y^2
B_2	1	-1	1	-1	1		xy
E	2	0	-2	0	0	$(x, y)(R_x, R_y)$	(xz, yz)

3. THE D_n GROUPS (continued)

D_5	E	$2C_5$	$2C_5^2$	$5C_2$		
A_1	1	1	1	1		x^2+y^2, z^2
A_2	1	1	1	-1	z, R_z	
E_1	2	2 cos 72°	2 cos 144°	0	$(x, y)(R_x, R_y)$	(xz, yz)
E_2	2	2 cos 144°	2 cos 72°	0		(x^2-y^2, xy)

D_6	E	$2C_6$	$2C_3$	C_2	$3C_2'$	$3C_2''$		
A_1	1	1	1	1	1	1		x^2+y^2, z^2
A_2	1	1	1	1	-1	-1	z, R_z	
B_1	1	-1	1	-1	1	-1		
B_2	1	-1	1	-1	-1	1		
E_1	2	1	-1	-2	0	0	$(x, y)(R_x, R_y)$	(xz, yz)
E_2	2	-1	-1	2	0	0		(x^2-y^2, xy)

4. THE C_{nv} GROUPS

$C_{2v}{}^a$	E	C_2	$\sigma_v(xz)$	$\sigma_v'(yz)$		
A_1	1	1	1	1	z	x^2, y^2, z^2
A_2	1	1	-1	-1	R_z	xy
B_1	1	-1	1	-1	x, R_y	xz
B_2	1	-1	-1	1	y, R_x	yz

[a] For a planar molecule the x-axis is taken perpendicular to the plane.

C_{3v}	E	$2C_3$	$3\sigma_v$		
A_1	1	1	1	z	x^2+y^2, z^2
A_2	1	1	-1	R_z	
E	2	-1	0	$(x, y)(R_x, R_y)$	$(x^2-y^2, xy)(xz, yz)$

$C_{4v}{}^a$	E	$2C_4$	C_2	$2\sigma_v$	$2\sigma_d$		
A_1	1	1	1	1	1	z	x^2+y^2, z^2
A_2	1	1	1	-1	-1	R_z	
B_1	1	-1	1	1	-1		x^2-y^2
B_2	1	-1	1	-1	1		xy
E	2	0	-2	0	0	$(x, y)(R_x, R_y)$	(xz, yz)

[a] If the C_{4v} molecule contains a square array of atoms, the σ_v planes should pass through the larger number of atoms of the square array or should intersect the largest possible number of bonds.

Appendix A Character Tables for Chemically Important Symmetry Groups

C_{5v}	E	$2C_5$	$2C_5^2$	$5\sigma_v$		
A_1	1	1	1	1	z	x^2+y^2, z^2
A_2	1	1	1	-1	R_z	
E_1	2	2 cos 72°	2 cos 144°	0	$(x, y)(R_x, R_y)$	(xz, yz)
E_2	2	2 cos 144°	2 cos 72°	0		(x^2-y^2, xy)

C_{6v}	E	$2C_6$	$2C_3$	C_2	$3\sigma_v$	$3\sigma_d$		
A_1	1	1	1	1	1	1	z	x^2+y^2, z^2
A_2	1	1	1	1	-1	-1	R_z	
B_1	1	-1	1	-1	1	-1		
B_2	1	-1	1	-1	-1	1		
E_1	2	1	-1	-2	0	0	$(x, y)(R_x, R_y)$	(xz, yz)
E_2	2	-1	-1	2	0	0		(x^2-y^2, xy)

5. THE C_{nh} GROUPS

C_{2h}	E	C_2	i	σ_h		
A_g	1	1	1	1	R_z	x^2, y^2, z^2, xy
B_g	1	-1	1	-1	R_x, R_y	xz, yz
A_u	1	1	-1	-1	z	
B_u	1	-1	-1	1	x, y	

C_{3h}	E	C_3	C_3^2	σ_h	S_3	S_3^5		$\varepsilon = \exp(2\pi i/3)$
A'	1	1	1	1	1	1	R_z	x^2+y^2, z^2
E'	$\begin{cases}1\\1\end{cases}$	$\begin{matrix}\varepsilon\\\varepsilon^*\end{matrix}$	$\begin{matrix}\varepsilon^*\\\varepsilon\end{matrix}$	$\begin{matrix}1\\1\end{matrix}$	$\begin{matrix}\varepsilon\\\varepsilon^*\end{matrix}$	$\begin{matrix}\varepsilon^*\\\varepsilon\end{matrix}$	(x, y)	(x^2-y^2, xy)
A''	1	1	1	-1	-1	-1	z	
E''	$\begin{cases}1\\1\end{cases}$	$\begin{matrix}\varepsilon\\\varepsilon^*\end{matrix}$	$\begin{matrix}\varepsilon^*\\\varepsilon\end{matrix}$	$\begin{matrix}-1\\-1\end{matrix}$	$\begin{matrix}-\varepsilon\\-\varepsilon^*\end{matrix}$	$\begin{matrix}-\varepsilon^*\\-\varepsilon\end{matrix}$	(R_x, R_y)	(xz, yz)

C_{4h}	E	C_4	C_2	C_4^3	i	S_4^3	σ_h	S_4		
A_g	1	1	1	1	1	1	1	1	R_z	x^2+y^2, z^2
B_g	1	-1	1	-1	1	-1	1	-1		x^2-y^2, xy
E_g	$\begin{cases}1\\1\end{cases}$	$\begin{matrix}i\\-i\end{matrix}$	$\begin{matrix}-1\\-1\end{matrix}$	$\begin{matrix}-i\\i\end{matrix}$	$\begin{matrix}1\\1\end{matrix}$	$\begin{matrix}i\\-i\end{matrix}$	$\begin{matrix}-1\\-1\end{matrix}$	$\begin{matrix}-i\\i\end{matrix}$	(R_x, R_y)	(xz, yz)
A_u	1	1	1	1	-1	-1	-1	-1	z	
B_u	1	-1	1	-1	-1	1	-1	1		
E_u	$\begin{cases}1\\1\end{cases}$	$\begin{matrix}i\\-i\end{matrix}$	$\begin{matrix}-1\\-1\end{matrix}$	$\begin{matrix}-i\\i\end{matrix}$	$\begin{matrix}-1\\-1\end{matrix}$	$\begin{matrix}-i\\i\end{matrix}$	$\begin{matrix}1\\1\end{matrix}$	$\begin{matrix}i\\-i\end{matrix}$	(x, y)	

Appendix A Character Tables for Chemically Important Symmetry Groups

C_{5h}	E	C_5	C_5^2	C_5^3	C_5^4	σ_h	S_5	S_5^7	S_5^3	S_5^9			$\varepsilon = \exp(2\pi i/5)$
A'	1	1	1	1	1	1	1	1	1	1	R_z		x^2+y^2, z^2
E_1'	$\begin{cases}1\\1\end{cases}$	$\begin{matrix}\varepsilon\\\varepsilon^*\end{matrix}$	$\begin{matrix}\varepsilon^2\\\varepsilon^{2*}\end{matrix}$	$\begin{matrix}\varepsilon^{2*}\\\varepsilon^2\end{matrix}$	$\begin{matrix}\varepsilon^*\\\varepsilon\end{matrix}$	$\begin{matrix}1\\1\end{matrix}$	$\begin{matrix}\varepsilon\\\varepsilon^*\end{matrix}$	$\begin{matrix}\varepsilon^2\\\varepsilon^{2*}\end{matrix}$	$\begin{matrix}\varepsilon^{2*}\\\varepsilon^2\end{matrix}$	$\begin{matrix}\varepsilon^*\\\varepsilon\end{matrix}$	(x, y)		
E_2'	$\begin{cases}1\\1\end{cases}$	$\begin{matrix}\varepsilon^2\\\varepsilon^{2*}\end{matrix}$	$\begin{matrix}\varepsilon^*\\\varepsilon\end{matrix}$	$\begin{matrix}\varepsilon\\\varepsilon^*\end{matrix}$	$\begin{matrix}\varepsilon^{2*}\\\varepsilon^2\end{matrix}$	$\begin{matrix}1\\1\end{matrix}$	$\begin{matrix}\varepsilon^2\\\varepsilon^{2*}\end{matrix}$	$\begin{matrix}\varepsilon^*\\\varepsilon\end{matrix}$	$\begin{matrix}\varepsilon\\\varepsilon^*\end{matrix}$	$\begin{matrix}\varepsilon^{2*}\\\varepsilon^2\end{matrix}$			(x^2-y^2, xy)
A''	1	1	1	1	1	-1	-1	-1	-1	-1	z		
E_1''	$\begin{cases}1\\1\end{cases}$	$\begin{matrix}\varepsilon\\\varepsilon^*\end{matrix}$	$\begin{matrix}\varepsilon^2\\\varepsilon^{2*}\end{matrix}$	$\begin{matrix}\varepsilon^{2*}\\\varepsilon^2\end{matrix}$	$\begin{matrix}\varepsilon^*\\\varepsilon\end{matrix}$	$\begin{matrix}-1\\-1\end{matrix}$	$\begin{matrix}-\varepsilon\\-\varepsilon^*\end{matrix}$	$\begin{matrix}-\varepsilon^2\\-\varepsilon^{2*}\end{matrix}$	$\begin{matrix}-\varepsilon^{2*}\\-\varepsilon^2\end{matrix}$	$\begin{matrix}-\varepsilon^*\\-\varepsilon\end{matrix}$	(R_x, R_y)		(xz, yz)
E_2''	$\begin{cases}1\\1\end{cases}$	$\begin{matrix}\varepsilon^2\\\varepsilon^{2*}\end{matrix}$	$\begin{matrix}\varepsilon^*\\\varepsilon\end{matrix}$	$\begin{matrix}\varepsilon\\\varepsilon^*\end{matrix}$	$\begin{matrix}\varepsilon^{2*}\\\varepsilon^2\end{matrix}$	$\begin{matrix}-1\\-1\end{matrix}$	$\begin{matrix}-\varepsilon^2\\-\varepsilon^{2*}\end{matrix}$	$\begin{matrix}-\varepsilon^*\\-\varepsilon\end{matrix}$	$\begin{matrix}-\varepsilon\\-\varepsilon^*\end{matrix}$	$\begin{matrix}-\varepsilon^{2*}\\-\varepsilon^2\end{matrix}$			

C_{6h}	E	C_6	C_3	C_2	C_3^2	C_6^5	i	S_3^5	S_6^5	σ_h	S_6	S_3			$\varepsilon = \exp(2\pi i/6)$
A_g	1	1	1	1	1	1	1	1	1	1	1	1	R_z		x^2+y^2, z^2
B_g	1	-1	1	-1	1	-1	1	-1	1	-1	1	-1			
E_{1g}	$\begin{cases}1\\1\end{cases}$	$\begin{matrix}\varepsilon\\\varepsilon^*\end{matrix}$	$\begin{matrix}-\varepsilon^*\\-\varepsilon\end{matrix}$	$\begin{matrix}-1\\-1\end{matrix}$	$\begin{matrix}-\varepsilon\\-\varepsilon^*\end{matrix}$	$\begin{matrix}\varepsilon^*\\\varepsilon\end{matrix}$	$\begin{matrix}1\\1\end{matrix}$	$\begin{matrix}\varepsilon\\\varepsilon^*\end{matrix}$	$\begin{matrix}-\varepsilon^*\\-\varepsilon\end{matrix}$	$\begin{matrix}-1\\-1\end{matrix}$	$\begin{matrix}-\varepsilon\\-\varepsilon^*\end{matrix}$	$\begin{matrix}\varepsilon^*\\\varepsilon\end{matrix}$	(R_x, R_y)		(xz, yz)
E_{2g}	$\begin{cases}1\\1\end{cases}$	$\begin{matrix}-\varepsilon^*\\-\varepsilon\end{matrix}$	$\begin{matrix}-\varepsilon\\-\varepsilon^*\end{matrix}$	$\begin{matrix}1\\1\end{matrix}$	$\begin{matrix}-\varepsilon^*\\-\varepsilon\end{matrix}$	$\begin{matrix}-\varepsilon\\-\varepsilon^*\end{matrix}$	$\begin{matrix}1\\1\end{matrix}$	$\begin{matrix}-\varepsilon^*\\-\varepsilon\end{matrix}$	$\begin{matrix}-\varepsilon\\-\varepsilon^*\end{matrix}$	$\begin{matrix}1\\1\end{matrix}$	$\begin{matrix}-\varepsilon^*\\-\varepsilon\end{matrix}$	$\begin{matrix}-\varepsilon\\-\varepsilon^*\end{matrix}$			(x^2-y^2, xy)
A_u	1	1	1	1	1	1	-1	-1	-1	-1	-1	-1	z		
B_u	1	-1	1	-1	1	-1	-1	1	-1	1	-1	1			
E_{1u}	$\begin{cases}1\\1\end{cases}$	$\begin{matrix}\varepsilon\\\varepsilon^*\end{matrix}$	$\begin{matrix}-\varepsilon^*\\-\varepsilon\end{matrix}$	$\begin{matrix}-1\\-1\end{matrix}$	$\begin{matrix}-\varepsilon\\-\varepsilon^*\end{matrix}$	$\begin{matrix}\varepsilon^*\\\varepsilon\end{matrix}$	$\begin{matrix}-1\\-1\end{matrix}$	$\begin{matrix}-\varepsilon\\-\varepsilon^*\end{matrix}$	$\begin{matrix}\varepsilon^*\\\varepsilon\end{matrix}$	$\begin{matrix}1\\1\end{matrix}$	$\begin{matrix}\varepsilon\\\varepsilon^*\end{matrix}$	$\begin{matrix}-\varepsilon^*\\-\varepsilon\end{matrix}$	(x, y)		
E_{2u}	$\begin{cases}1\\1\end{cases}$	$\begin{matrix}-\varepsilon^*\\-\varepsilon\end{matrix}$	$\begin{matrix}-\varepsilon\\-\varepsilon^*\end{matrix}$	$\begin{matrix}1\\1\end{matrix}$	$\begin{matrix}-\varepsilon^*\\-\varepsilon\end{matrix}$	$\begin{matrix}-\varepsilon\\-\varepsilon^*\end{matrix}$	$\begin{matrix}-1\\-1\end{matrix}$	$\begin{matrix}\varepsilon^*\\\varepsilon\end{matrix}$	$\begin{matrix}\varepsilon\\\varepsilon^*\end{matrix}$	$\begin{matrix}-1\\-1\end{matrix}$	$\begin{matrix}\varepsilon^*\\\varepsilon\end{matrix}$	$\begin{matrix}\varepsilon\\\varepsilon^*\end{matrix}$			

6. THE D_{nh} GROUPS

D_{2h}	E	$C_2(z)$	$C_2(y)$	$C_2(x)$	i	$\sigma(xy)$	$\sigma(xz)$	$\sigma(yz)$		
A_g	1	1	1	1	1	1	1	1		x^2, y^2, z^2
B_{1g}	1	1	-1	-1	1	1	-1	-1	R_z	xy
B_{2g}	1	-1	1	-1	1	-1	1	-1	R_y	xz
B_{3g}	1	-1	-1	1	1	-1	-1	1	R_x	yz
A_u	1	1	1	1	-1	-1	-1	-1		
B_{1u}	1	1	-1	-1	-1	-1	1	1	z	
B_{2u}	1	-1	1	-1	-1	1	-1	1	y	
B_{3u}	1	-1	-1	1	-1	1	1	-1	x	

D_{3h}	E	$2C_3$	$3C_2$	σ_h	$2S_3$	$3\sigma_v$		
A_1'	1	1	1	1	1	1		x^2+y^2, z^2
A_2'	1	1	-1	1	1	-1	R_z	
E'	2	-1	0	2	-1	0	(x, y)	(x^2-y^2, xy)
A_1''	1	1	1	-1	-1	-1		
A_2''	1	1	-1	-1	-1	1	z	
E''	2	-1	0	-2	1	0	(R_x, R_y)	(xz, yz)

$D_{4h}{}^a$	E	$2C_4$	C_2	$2C_2'$	$2C_2''$	i	$2S_4$	σ_h	$2\sigma_v$	$2\sigma_d$		
A_{1g}	1	1	1	1	1	1	1	1	1	1		x^2+y^2, z^2
A_{2g}	1	1	1	−1	−1	1	1	1	−1	−1	R_z	
B_{1g}	1	−1	1	1	−1	1	−1	1	1	−1		x^2-y^2
B_{2g}	1	−1	1	−1	1	1	−1	1	−1	1		xy
E_g	2	0	−2	0	0	2	0	−2	0	0	(R_x, R_y)	(xz, yz)
A_{1u}	1	1	1	1	1	−1	−1	−1	−1	−1		
A_{2u}	1	1	1	−1	−1	−1	−1	−1	1	1	z	
B_{1u}	1	−1	1	1	−1	−1	1	−1	−1	1		
B_{2u}	1	−1	1	−1	1	−1	1	−1	1	−1		
E_u	2	0	−2	0	0	−2	0	2	0	0	(x, y)	

^{a} σ_v passes through the atoms and σ_d bisects the bond angles.

D_{5h}	E	$2C_5$	$2C_5{}^2$	$5C_2$	σ_h	$2S_5$	$2S_5{}^3$	$5\sigma_v$		
A_1'	1	1	1	1	1	1	1	1		x^2+y^2, z^2
A_2'	1	1	1	−1	1	1	1	−1	R_z	
E_1'	2	2 cos 72°	2 cos 144°	0	2	2 cos 72°	2 cos 144°	0	(x, y)	
E_2'	2	2 cos 144°	2 cos 72°	0	2	2 cos 144°	2 cos 72°	0		(x^2-y^2, xy)
A_1''	1	1	1	1	−1	−1	−1	−1		
A_2''	1	1	1	−1	−1	−1	−1	1	z	
E_1''	2	2 cos 72°	2 cos 144°	0	−2	−2 cos 72°	−2 cos 144°	0	(R_x, R_y)	(xz, yz)
E_2''	2	2 cos 144°	2 cos 72°	0	−2	−2 cos 144°	−2 cos 72°	0		

D_{6h}	E	$2C_6$	$2C_3$	C_2	$3C_2'$	$3C_2''$	i	$2S_3$	$2S_6$	σ_h	$3\sigma_d$	$3\sigma_v$		
A_{1g}	1	1	1	1	1	1	1	1	1	1	1	1		x^2+y^2, z^2
A_{2g}	1	1	1	1	−1	−1	1	1	1	1	−1	−1	R_z	
B_{1g}	1	−1	1	−1	1	−1	1	−1	1	−1	1	−1		
B_{2g}	1	−1	1	−1	−1	1	1	−1	1	−1	−1	1		
E_{1g}	2	1	−1	−2	0	0	2	1	−1	−2	0	0	(R_x, R_y)	(xz, yz)
E_{2g}	2	−1	−1	2	0	0	2	−1	−1	2	0	0		(x^2-y^2, xy)
A_{1u}	1	1	1	1	1	1	−1	−1	−1	−1	−1	−1		
A_{2u}	1	1	1	1	−1	−1	−1	−1	−1	−1	1	1	z	
B_{1u}	1	−1	1	−1	1	−1	−1	1	−1	1	−1	1		
B_{2u}	1	−1	1	−1	−1	1	−1	1	−1	1	1	−1		
E_{1u}	2	1	−1	−2	0	0	−2	−1	1	2	0	0	(x, y)	
E_{2u}	2	−1	−1	2	0	0	−2	1	1	−2	0	0		

6. THE D_{nh} GROUPS (continued)

D_{8h}	E	$2C_8$	$2C_8^3$	$2C_4$	C_2	$4C_2'$	$4C_2''$	i	$2S_8$	$2S_8^3$	$2S_4$	σ_h	$4\sigma_d$	$4\sigma_v$		
A_{1g}	1	1	1	1	1	1	1	1	1	1	1	1	1	1		x^2+y^2, z^2
A_{2g}	1	1	1	1	1	−1	−1	1	1	1	1	1	−1	−1	R_z	
B_{1g}	1	−1	−1	1	1	1	−1	1	−1	−1	1	1	1	−1		
B_{2g}	1	−1	−1	1	1	−1	1	1	−1	−1	1	1	−1	1		
E_{1g}	2	$\sqrt{2}$	$-\sqrt{2}$	0	−2	0	0	2	$\sqrt{2}$	$-\sqrt{2}$	0	−2	0	0	(R_x, R_y)	(xz, yz)
E_{2g}	2	0	0	−2	2	0	0	2	0	0	−2	2	0	0		(x^2-y^2, xy)
E_{3g}	2	$-\sqrt{2}$	$\sqrt{2}$	0	−2	0	0	2	$-\sqrt{2}$	$\sqrt{2}$	0	−2	0	0		
A_{1u}	1	1	1	1	1	1	1	−1	−1	−1	−1	−1	−1	−1		
A_{2u}	1	1	1	1	1	−1	−1	−1	−1	−1	−1	−1	1	1	z	
B_{1u}	1	−1	−1	1	1	1	−1	−1	1	1	−1	−1	−1	1		
B_{2u}	1	−1	−1	1	1	−1	1	−1	1	1	−1	−1	1	−1		
E_{1u}	2	$\sqrt{2}$	$-\sqrt{2}$	0	−2	0	0	−2	$-\sqrt{2}$	$\sqrt{2}$	0	2	0	0	(x, y)	
E_{2u}	2	0	0	−2	2	0	0	−2	0	0	2	−2	0	0		
E_{3u}	2	$-\sqrt{2}$	$\sqrt{2}$	0	−2	0	0	−2	$\sqrt{2}$	$-\sqrt{2}$	0	2	0	0		

7. THE D_{nd} GROUPS

D_{2d}	E	$2S_4$	C_2	$2C_2'$	$2\sigma_d$		
A_1	1	1	1	1	1		x^2+y^2, z^2
A_2	1	1	1	-1	-1	R_z	
B_1	1	-1	1	1	-1		x^2-y^2
B_2	1	-1	1	-1	1	z	xy
E	2	0	-2	0	0	$(x, y), (R_x, R_y)$	(xz, yz)

D_{3d}	E	$2C_3$	$3C_2$	i	$2S_6$	$3\sigma_d$		
A_{1g}	1	1	1	1	1	1		x^2+y^2, z^2
A_{2g}	1	1	-1	1	1	-1	R_z	
E_g	2	-1	0	2	-1	0	(R_x, R_y)	$(x^2-y^2, xy), (xz, yz)$
A_{1u}	1	1	1	-1	-1	-1		
A_{2u}	1	1	-1	-1	-1	1	z	
E_u	2	-1	0	-2	1	0	(x, y)	

D_{4d}	E	$2S_8$	$2C_4$	$2S_8^3$	C_2	$4C_2'$	$4\sigma_d$		
A_1	1	1	1	1	1	1	1		x^2+y^2, z^2
A_2	1	1	1	1	1	-1	-1	R_z	
B_1	1	-1	1	-1	1	1	-1		
B_2	1	-1	1	-1	1	-1	1	z	
E_1	2	$\sqrt{2}$	0	$-\sqrt{2}$	-2	0	0	(x, y)	
E_2	2	0	-2	0	2	0	0		(x^2-y^2, xy)
E_3	2	$-\sqrt{2}$	0	$\sqrt{2}$	-2	0	0	(R_x, R_y)	(xz, yz)

D_{5d}	E	$2C_5$	$2C_5^2$	$5C_2$	i	$2S_{10}^3$	$2S_{10}$	$5\sigma_d$		
A_{1g}	1	1	1	1	1	1	1	1		x^2+y^2, z^2
A_{2g}	1	1	1	-1	1	1	1	-1	R_z	
E_{1g}	2	$2\cos 72°$	$2\cos 144°$	0	2	$2\cos 72°$	$2\cos 144°$	0	(R_x, R_y)	(xz, yz)
E_{2g}	2	$2\cos 144°$	$2\cos 72°$	0	2	$2\cos 144°$	$2\cos 72°$	0		(x^2-y^2, xy)
A_{1u}	1	1	1	1	-1	-1	-1	-1		
A_{2u}	1	1	1	-1	-1	-1	-1	1	z	
E_{1u}	2	$2\cos 72°$	$2\cos 144°$	0	-2	$-2\cos 72°$	$-2\cos 144°$	0	(x, y)	
E_{2u}	2	$2\cos 144°$	$2\cos 72°$	0	-2	$-2\cos 144°$	$-2\cos 72°$	0		

7. THE D_{nd} GROUPS (continued)

D_{6d}	E	$2S_{12}$	$2C_6$	$2S_4$	$2C_3$	$2S_{12}^{\,5}$	C_2	$6C_2'$	$6\sigma_d$		
A_1	1	1	1	1	1	1	1	1	1		x^2+y^2, z^2
A_2	1	1	1	1	1	1	1	−1	−1	R_z	
B_1	1	−1	1	−1	1	−1	1	1	−1		
B_2	1	−1	1	−1	1	−1	1	−1	1	z	
E_1	2	$\sqrt{3}$	1	0	−1	$-\sqrt{3}$	−2	0	0	(x,y)	
E_2	2	1	−1	−2	−1	1	2	0	0		(x^2-y^2, xy)
E_3	2	0	−2	0	2	0	−2	0	0		
E_4	2	−1	−1	2	−1	−1	2	0	0		
E_5	2	$-\sqrt{3}$	1	0	−1	$\sqrt{3}$	−2	0	0	(R_x, R_y)	(xz, yz)

8. THE S_n GROUPS

S_4	E	S_4	C_2	$S_4^{\,3}$		
A	1	1	1	1	R_z	x^2+y^2, z^2 ; x^2-y^2, xy
B	1	−1	1	−1	z	
E	$\begin{Bmatrix}1\\1\end{Bmatrix}$	$\begin{Bmatrix}i\\-i\end{Bmatrix}$	$\begin{Bmatrix}-1\\-1\end{Bmatrix}$	$\begin{Bmatrix}-i\\i\end{Bmatrix}$	$(x,y); (R_x, R_y)$	(xz, yz)

S_6	E	C_3	$C_3^{\,2}$	i	$S_6^{\,5}$	S_6		$\varepsilon = \exp(2\pi i/3)$
A_g	1	1	1	1	1	1	R_z	x^2+y^2, z^2
E_g	$\begin{Bmatrix}1\\1\end{Bmatrix}$	$\begin{Bmatrix}\varepsilon\\\varepsilon^*\end{Bmatrix}$	$\begin{Bmatrix}\varepsilon^*\\\varepsilon\end{Bmatrix}$	$\begin{Bmatrix}1\\1\end{Bmatrix}$	$\begin{Bmatrix}\varepsilon\\\varepsilon^*\end{Bmatrix}$	$\begin{Bmatrix}\varepsilon^*\\\varepsilon\end{Bmatrix}$	(R_x, R_y)	$(x^2-y^2, xy);$ (xz, yz)
A_u	1	1	1	−1	−1	−1	z	
E_u	$\begin{Bmatrix}1\\1\end{Bmatrix}$	$\begin{Bmatrix}\varepsilon\\\varepsilon^*\end{Bmatrix}$	$\begin{Bmatrix}\varepsilon^*\\\varepsilon\end{Bmatrix}$	$\begin{Bmatrix}-1\\-1\end{Bmatrix}$	$\begin{Bmatrix}-\varepsilon\\-\varepsilon^*\end{Bmatrix}$	$\begin{Bmatrix}-\varepsilon^*\\-\varepsilon\end{Bmatrix}$	(x, y)	

S_8	E	S_8	C_4	$S_8^{\,3}$	C_2	$S_8^{\,5}$	$C_4^{\,3}$	$S_8^{\,7}$		$\varepsilon = \exp(2\pi i/8)$
A	1	1	1	1	1	1	1	1	R_z	x^2+y^2, z^2
B	1	−1	1	−1	1	−1	1	−1	z	
E_1	$\begin{Bmatrix}1\\1\end{Bmatrix}$	$\begin{Bmatrix}\varepsilon\\\varepsilon^*\end{Bmatrix}$	$\begin{Bmatrix}i\\-i\end{Bmatrix}$	$\begin{Bmatrix}-\varepsilon^*\\-\varepsilon\end{Bmatrix}$	$\begin{Bmatrix}-1\\-1\end{Bmatrix}$	$\begin{Bmatrix}-\varepsilon\\-\varepsilon^*\end{Bmatrix}$	$\begin{Bmatrix}-i\\i\end{Bmatrix}$	$\begin{Bmatrix}\varepsilon^*\\\varepsilon\end{Bmatrix}$	$(x, y);$ (R_x, R_y)	
E_2	$\begin{Bmatrix}1\\1\end{Bmatrix}$	$\begin{Bmatrix}i\\-i\end{Bmatrix}$	$\begin{Bmatrix}-1\\-1\end{Bmatrix}$	$\begin{Bmatrix}-i\\i\end{Bmatrix}$	$\begin{Bmatrix}1\\1\end{Bmatrix}$	$\begin{Bmatrix}i\\-i\end{Bmatrix}$	$\begin{Bmatrix}-1\\-1\end{Bmatrix}$	$\begin{Bmatrix}-i\\i\end{Bmatrix}$		(x^2-y^2, xy)
E_3	$\begin{Bmatrix}1\\1\end{Bmatrix}$	$\begin{Bmatrix}-\varepsilon^*\\-\varepsilon\end{Bmatrix}$	$\begin{Bmatrix}-i\\i\end{Bmatrix}$	$\begin{Bmatrix}\varepsilon\\\varepsilon^*\end{Bmatrix}$	$\begin{Bmatrix}-1\\-1\end{Bmatrix}$	$\begin{Bmatrix}\varepsilon^*\\\varepsilon\end{Bmatrix}$	$\begin{Bmatrix}i\\-i\end{Bmatrix}$	$\begin{Bmatrix}-\varepsilon\\-\varepsilon^*\end{Bmatrix}$		(xz, yz)

9. THE CUBIC GROUPS

T	E	$4C_3$	$4C_3^{\,2}$	$3C_2$		$\varepsilon = \exp(2\pi i/3)$
A	1	1	1	1		$x^2+y^2+z^2$
E	$\begin{Bmatrix}1\\1\end{Bmatrix}$	$\begin{Bmatrix}\varepsilon\\\varepsilon^*\end{Bmatrix}$	$\begin{Bmatrix}\varepsilon^*\\\varepsilon\end{Bmatrix}$	$\begin{Bmatrix}1\\1\end{Bmatrix}$		$(2z^2-x^2-y^2,$ $x^2-y^2)$
T	3	0	0	−1	$(R_x, R_y, R_z); (x, y, z)$	(xy, xz, yz)

9. THE CUBIC GROUPS (continued)

T_h	E	$4C_3$	$4C_3^2$	$3C_2$	i	$4S_6$	$4S_6^5$	$3\sigma_h$		$\varepsilon = \exp(2\pi i/3)$
A_g	1	1	1	1	1	1	1	1		$x^2 + y^2 + z^2$
A_u	1	1	1	1	-1	-1	-1	-1		
E_g	$\begin{cases}1\\1\end{cases}$	ε ε^*	ε^* ε	1 1	1 1	ε ε^*	ε^* ε	$\begin{matrix}1\\1\end{matrix}$		$(2z^2 - x^2 - y^2,$ $x^2 - y^2)$
E_u	$\begin{cases}1\\1\end{cases}$	ε ε^*	ε^* ε	1 1	-1 -1	$-\varepsilon$ $-\varepsilon^*$	$-\varepsilon^*$ $-\varepsilon$	$\begin{matrix}-1\\-1\end{matrix}$		
T_g	3	0	0	-1	1	0	0	-1	(R_x, R_y, R_z)	(xz, yz, xy)
T_u	3	0	0	-1	-1	0	0	1	(x, y, z)	

T_d	E	$8C_3$	$3C_2$	$6S_4$	$6\sigma_d$		
A_1	1	1	1	1	1		$x^2 + y^2 + z^2$
A_2	1	1	1	-1	-1		
E	2	-1	2	0	0		$(2z^2 - x^2 - y^2,$ $x^2 - y^2)$
T_1	3	0	-1	1	-1	(R_x, R_y, R_z)	
T_2	3	0	-1	-1	1	(x, y, z)	(xy, xz, yz)

O	E	$6C_4$	$3C_2(=C_4^2)$	$8C_3$	$6C_2$		
A_1	1	1	1	1	1		$x^2 + y^2 + z^2$
A_2	1	-1	1	1	-1		
E	2	0	2	-1	0		$(2z^2 - x^2 - y^2,$ $x^2 - y^2)$
T_1	3	1	-1	0	-1	$(R_x, R_y, R_z); (x, y, z)$	
T_2	3	-1	-1	0	1		(xy, xz, yz)

O_h	E	$8C_3$	$6C_2$	$6C_4$	$3C_2(=C_4^2)$	i	$6S_4$	$8S_6$	$3\sigma_h$	$6\sigma_d$		
A_{1g}	1	1	1	1	1	1	1	1	1	1		$x^2 + y^2 + z^2$
A_{2g}	1	1	-1	-1	1	1	-1	1	1	-1		
E_g	2	-1	0	0	2	2	0	-1	2	0		$(2z^2 - x^2 - y^2,$ $x^2 - y^2)$
T_{1g}	3	0	-1	1	-1	3	1	0	-1	-1	(R_x, R_y, R_z)	
T_{2g}	3	0	1	-1	-1	3	-1	0	-1	1		(xz, yz, xy)
A_{1u}	1	1	1	1	1	-1	-1	-1	-1	-1		
A_{2u}	1	1	-1	-1	1	-1	1	-1	-1	1		
E_u	2	-1	0	0	2	-2	0	1	-2	0		
T_{1u}	3	0	-1	1	-1	-3	-1	0	1	1	(x, y, z)	
T_{2u}	3	0	1	-1	-1	-3	1	0	1	-1		

Appendix A Character Tables for Chemically Important Symmetry Groups   723

10. THE GROUPS $C_{\infty v}$ AND $D_{\infty h}$ FOR LINEAR MOLECULES

$C_{\infty v}$	E	$2C_\infty^\Phi$	...	$\infty \sigma_v$		
$A_1 \equiv \Sigma^+$	1	1	...	1	z	$x^2 + y^2, z^2$
$A_2 \equiv \Sigma^-$	1	1	...	-1	R_z	
$E_1 \equiv \Pi$	2	$2\cos\Phi$	...	0	$(x, y); (R_x, R_y)$	(xz, yz)
$E_2 \equiv \Delta$	2	$2\cos 2\Phi$	...	0		$(x^2 - y^2, xy)$
$E_3 \equiv \Phi$	2	$2\cos 3\Phi$	...	0		
...	...	...	...	...		

$D_{\infty h}$	E	$2C_\infty^\Phi$	...	$\infty \sigma_v$	i	$2S_\infty^\Phi$	...	∞C_2		
Σ_g^+	1	1	...	1	1	1	...	1		$x^2 + y^2, z^2$
Σ_g^-	1	1	...	-1	1	1	...	-1	R_z	
Π_g	2	$2\cos\Phi$	...	0	2	$-2\cos\Phi$	...	0	(R_x, R_y)	(xz, yz)
Δ_g	2	$2\cos 2\Phi$	...	0	2	$2\cos 2\Phi$	...	0		$(x^2 - y^2, xy)$
...	...	...	...	...	...	...	...	...		
Σ_u^+	1	1	...	1	-1	-1	...	-1	z	
Σ_u^-	1	1	...	-1	-1	-1	...	1		
Π_u	2	$2\cos\Phi$	...	0	-2	$2\cos\Phi$	...	0	(x, y)	
Δ_u	2	$2\cos 2\Phi$	...	0	-2	$-2\cos 2\Phi$	...	0		
...	...	...	...	...	...	...	...	...		

11. THE ICOSAHEDRAL GROUPS[a]

I_h	E	$12C_5$	$12C_5^2$	$20C_3$	$15C_2$	i	$12S_{10}$	$12S_{10}^3$	$20S_6$	15σ		
A_g	1	1	1	1	1	1	1	1	1	1		$x^2 + y^2 + z^2$
T_{1g}	3	$\frac{1}{2}(1+\sqrt{5})$	$\frac{1}{2}(1-\sqrt{5})$	0	-1	3	$\frac{1}{2}(1-\sqrt{5})$	$\frac{1}{2}(1+\sqrt{5})$	0	-1	(R_x, R_y, R_z)	
T_{2g}	3	$\frac{1}{2}(1-\sqrt{5})$	$\frac{1}{2}(1+\sqrt{5})$	0	-1	3	$\frac{1}{2}(1+\sqrt{5})$	$\frac{1}{2}(1-\sqrt{5})$	0	-1		
G_g	4	-1	-1	1	0	4	-1	-1	1	0		$(2z^2 - x^2 - y^2,$
H_g	5	0	0	-1	1	5	0	0	-1	1		$x^2 - y^2,$ $xy, yz, zx)$
A_u	1	1	1	1	1	-1	-1	-1	-1	-1		
T_{1u}	3	$\frac{1}{2}(1+\sqrt{5})$	$\frac{1}{2}(1-\sqrt{5})$	0	-1	-3	$-\frac{1}{2}(1-\sqrt{5})$	$-\frac{1}{2}(1+\sqrt{5})$	0	1	(x, y, z)	
T_{2u}	3	$\frac{1}{2}(1-\sqrt{5})$	$\frac{1}{2}(1+\sqrt{5})$	0	-1	-3	$-\frac{1}{2}(1+\sqrt{5})$	$-\frac{1}{2}(1-\sqrt{5})$	0	1		
G_u	4	-1	-1	1	0	-4	1	1	-1	0		
H_u	5	0	0	-1	1	-5	0	0	1	-1		

[a] For the pure rotation group I, the outlined section in the upper left is the character table; the g subscripts should, of course, be dropped and (x, y, z) assigned to the T_1 representation.

APPENDIX B

Character Tables for Some Double Groups

GROUP C_3'

C_3'		E	R	C_3	$C_3 R$	C_3^2	$C_3^2 R$	
A'	Γ_1	1	1	1	1	1	1	z, L_z
E'	$\{\Gamma_2$	1	1	ω^2	ω^2	$-\omega$	$-\omega$	$\}\, x, y, L_x, L_y$
	$\{\Gamma_3$	1	1	$-\omega$	$-\omega$	ω^2	ω^2	
$E_{1/2}'$	$\{\Gamma_4$	1	-1	ω	$-\omega$	ω^2	$-\omega^2$	
	$\{\Gamma_5$	1	-1	$-\omega^2$	ω^2	$-\omega$	ω	
$B_{3/2}'$	Γ_6	1	-1	-1	1	1	-1	

$$\omega = \exp(i\pi/3)$$

GROUPS D_3' AND C_{3v}'

D_3'		E	R	C_3, $C_3^2 R$	C_3^2, $C_3 R$	$3C_2'$	$3C_2' R$	D_3'	
C_{3v}'		E	R	C_3, $C_3^2 R$	C_3^2, $C_3 R$	$3\sigma_v$	$3\sigma_v R$		C_{3v}'
A_1'	Γ_1	1	1	1	1	1	1		z
A_2'	Γ_2	1	1	1	1	-1	-1	z, L_z	L_z
E'	Γ_3	2	2	-1	-1	0	0	x, y, L_x, L_y	x, y, L_x, L_y
$E_{3/2}'$	$\{(A_1')\, \Gamma_4$	1	-1	-1	1	i	$-i$		
	$\{(A_2')\, \Gamma_5$	1	-1	-1	1	$-i$	i		
$E_{1/2}$	$(E')\, \Gamma_6$	2	-2	1	-1	0	0		

Appendix B Character Tables for Some Double Groups

GROUPS C_4' AND S_4'

C_4'		E	R	C_4	$C_4 R$	C_4^2	$C_4^2 R$	C_4^3	$C_4^3 R$	C_4'	S_4'
S_4'		E	R	S_4	$S_4 R$	$C_2 R$	C_2	S_4^3	$S_4^3 R$		
A'	Γ_1	1	1	1	1	1	1	1	1	z	L_z
B'	Γ_2	1	1	-1	-1	1	1	-1	-1	L_z	z
E'	Γ_3	1	1	i	i	-1	-1	$-i$	$-i$	$\left\{ \begin{array}{c} x \pm iy \\ L_x \pm iL_y \end{array} \right.$	$\left\{ \begin{array}{c} x \pm iy \\ L_x \pm iL_y \end{array} \right.$
	Γ_4	1	1	$-i$	$-i$	-1	-1	i	i		
$E_{1/2}'$	Γ_5	1	-1	ω	$-\omega$	i	$-i$	ω^3	$-\omega^3$		
	Γ_6	1	-1	$-\omega^3$	ω^3	$-i$	i	$-\omega$	ω		
$E_{3/2}'$	Γ_7	1	-1	$-\omega$	ω	i	$-i$	$-\omega^3$	ω^3		
	Γ_8	1	-1	ω^3	$-\omega^3$	$-i$	i	ω	$-\omega$		

$$\omega = \exp(i\pi/4)$$

GROUP T'

T'		E	R	$3C_2$ $3C_2 R$	$4C_3$	$4C_3 R$	$4C_3^2$	$4C_3^2 R$	
A'	Γ_1	1	1	1	1	1	1	1	
E'	Γ_2	1	1	1	ω	ω	ω^2	ω^2	
	Γ_3	1	1	1	ω^2	ω^2	ω	ω	
T'	Γ_4	3	3	-1	0	0	0	0	x, y, z, L_x, L_y, L_z
$E_{1/2}'$	Γ_5	2	-2	0	1	-1	1	-1	
$G_{3/2}'$	Γ_6	2	-2	0	ω	$-\omega$	ω^2	$-\omega^2$	
	Γ_7	2	-2	0	ω^2	$-\omega^2$	ω	$-\omega$	

$$\omega = \exp(2\pi i/3)$$

GROUPS O' AND T_d'

O'		E	R	$4C_3$ $4C_3^2 R$	$4C_3^2$ $4C_3 R$	$3C_4^2$ $3C_4^2 R$	$3C_4$ $3C_4^3 R$	$3C_4^3$ $3C_4 R$	$3C_2'$ $3C_2' R$	O'	T_d'
T_d'		E	R	$4C_3$ $4C_3^2 R$	$4C_3^2$ $4C_3 R$	$3C_4^2$ $3C_4^2 R$	$3S_4$ $3S_4^3 R$	$3S_4^3$ $3S_4 R$	$6\sigma_d$ $6\sigma_d R$		
A_1' Γ_1		1	1	1	1	1	1	1	1		
A_2' Γ_2		1	1	1	1	1	-1	-1	-1		
E' Γ_3		2	2	-1	-1	2	0	0	0		
T_1' Γ_4		3	3	0	0	-1	1	1	-1	x, y, z, L_x, L_y, L_z	L_x, L_y, L_z x, y, z
T_2' Γ_5		3	3	0	0	-1	-1	-1	1		
(E_1') $E_{1/2}'$ Γ_6		2	-2	1	-1	0	$\sqrt{2}$	$-\sqrt{2}$	0		
(E_2') $E_{5/2}'$ Γ_7		2	-2	1	-1	0	$-\sqrt{2}$	$\sqrt{2}$	0		
(G') $G_{3/2}'$ Γ_8		4	-4	-1	1	0	0	0	0		

Appendix B Character Tables for Some Double Groups

GROUP D_4'

D_4'		E	R	C_4, $C_4^3 R$	C_4^3, $C_4 R$	C_2, $C_2 R$	$2C_2'$, $2C_2' R$	$2C_2''$, $2C_2'' R$
A_1'	Γ_1	1	1	1	1	1	1	1
A_2'	Γ_2	1	1	1	1	1	-1	-1
B_1'	Γ_3	1	1	-1	-1	1	1	-1
B_2'	Γ_4	1	1	-1	-1	1	-1	1
E_1'	Γ_5	2	2	0	0	-2	0	0
E_2'	Γ_6	2	-2	$\sqrt{2}$	$-\sqrt{2}$	0	0	0
E_3'	Γ_7	2	-2	$-\sqrt{2}$	$\sqrt{2}$	0	0	0

APPENDIX C

Normal Vibration Modes for Common Structures

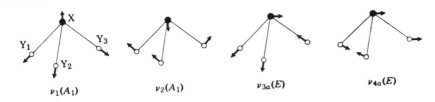

Pyramidal XY$_3$ Molecules.

Tetrahedral XY$_4$ Molecules.

Planar XY$_3$ Molecules.

Appendix C Normal Vibration Modes for Common Structures

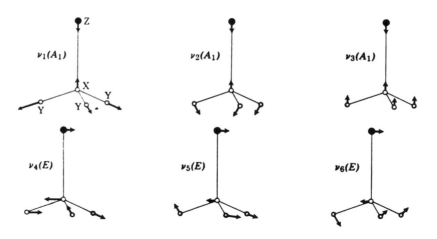

C_{3v} ZXY_3 Molecules.

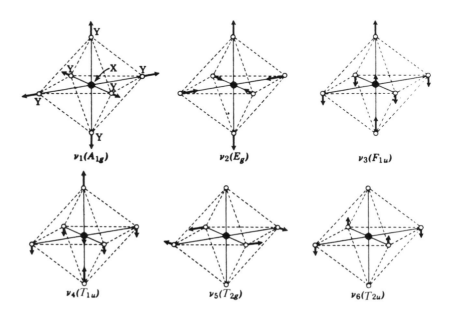

Octahedral XY_6 Molecules.

Appendix C Normal Vibration Modes for Common Structures   729

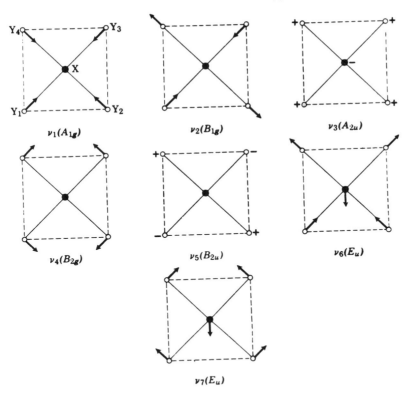

Square-Planar XY$_4$ Molecules.

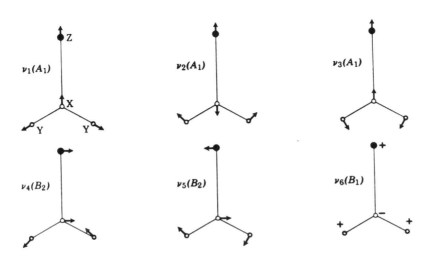

Planar ZXY$_2$ Molecules.

730 Appendix C Normal Vibration Modes for Common Structures

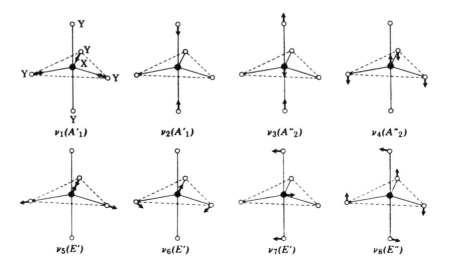

Trigonal Bipyramidal XY$_5$ Molecules.

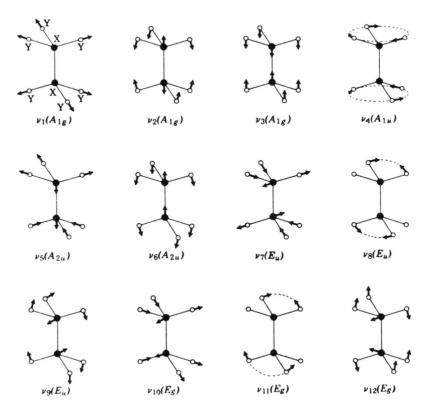

Ethane-Type X$_2$Y$_6$ Molecules.

Appendix C Normal Vibration Modes for Common Structures 731

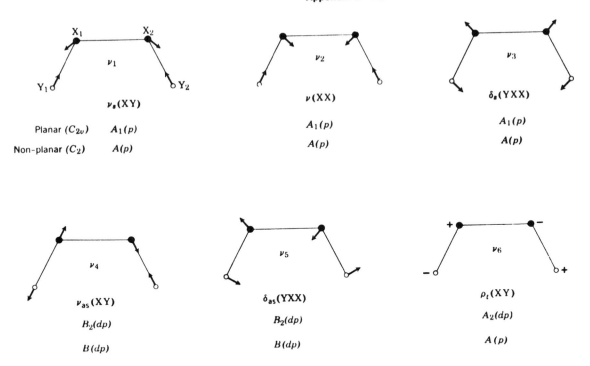

Nonlinear X_2Y_2 Molecules (*p*: Polarized; *dp*: Depolarized).

APPENDIX D

Tanabe and Sugano Diagrams for O_h Fields*

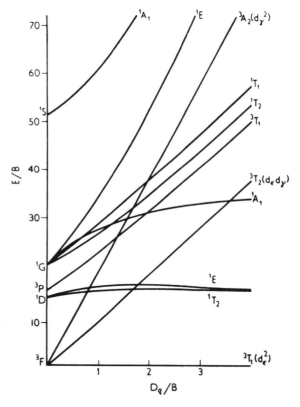

Energy diagram for the configuration d^2.

---

*These are complete energy diagrams for the configurations indicated, reproduced from Y. Tanabe and S. Sugano, J. Phys. Soc. Japan, 9, 753, 766 (1954).

γ refers to the ratio of the Racah parameters C/B. Heavy lines perpendicular to the Dq/B axis in d^4, d^5, d^6, and d^7 indicate transitions from weak to strong fields. The calculation of and the assumptions inherent in the calculations are contained in the original paper. The diagrams do not apply to any particular complex but give a qualitative indication of the energies of the various states as a function of Dq/B.

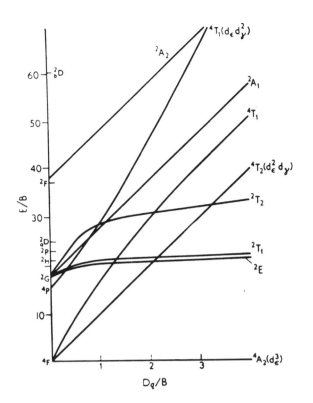

Energy diagram for the configuration d^3.

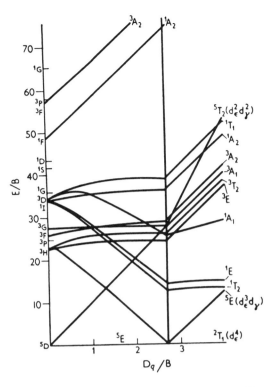

Energy diagram for the configuration d^4.

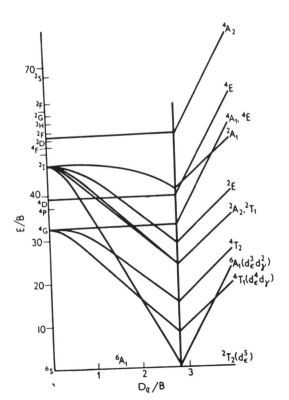

Energy diagram for the configuration d^5.

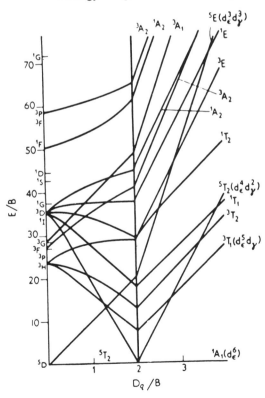

Energy diagram for the configuration d^6.

Appendix D Tanabe and Sugano Diagrams for O_h Fields

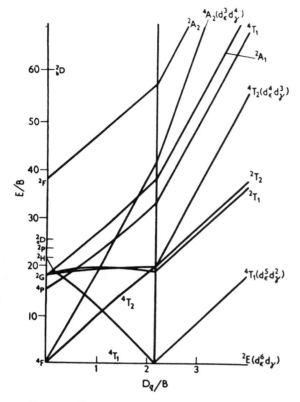

Energy diagram for the configuration d^7.

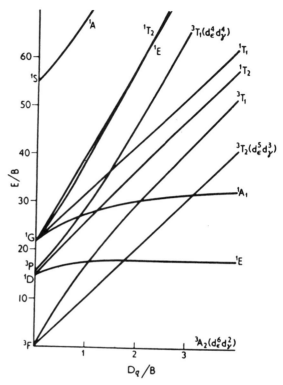

Energy diagram for the configuration d^8.

APPENDIX E

Calculation of Δ (10Dq) and β for O_h Ni^{II} and T_d Co^{II} Complexes

CALCULATION OF Δ AND β FOR OCTAHEDRAL Ni^{2+} COMPLEXES

The data in Table 10–6 for the Ni[(CH$_3$)$_2$SO]$_6$(ClO$_4$)$_2$ complex will be employed to illustrate the calculation of Δ, β, and the frequency for the $^3A_{2g} \rightarrow {}^3T_{1g}(F)$ band. The value for Δ, or 10Dq, is obtained directly from the lowest energy transition, $^3A_{2g} \rightarrow {}^3T_{2g}$, which occurs at 7728 cm^{-1}. Equation (10–12),

$$[6Dqp - 16(Dq)^2] + [-6Dq - p]E + E^2 = 0,$$

is employed to calculate the experimental ^3P energy value, that is, p of equation (10–12). The quantity p is equal to 15B for nickel(II), where B is a Racah parameter. Racah parameters indicate the magnitude of the interelectronic repulsion between various levels in the gaseous ion. The quantity B is a constant that enables one to express the energy difference between the levels of highest spin multiplicity in terms of some integer, n, times B; that is, nB. Both n and B vary for different ions; in the case of Ni^{2+}, the energy difference between ^3F and ^3P is 15B. The same term adjusted for the complex is 15B'. To use equation (10–12) it is necessary to employ the energy values for the $^3T_{1g}(P)$ state. This is the energy observed for the $^3T_{1g}(P)$ transition (24,038 cm^{-1}) plus the energy of the $^3A_{2g}$ level, because

E observed for transition (24,038 cm^{-1}) = energy of $^3T_{1g}(P)$ − energy of $^3A_{2g}$

so

energy of $^3T_{1g}(P) = E$ for transition + E of $^3A_{2g}$

Thus, combining this with equation (10–11) (E of $^3A_{2g} = -12Dq$; note that $Dq = 7728/10 \approx 773$), we obtain

$$E \text{ of } {}^3T_{1g}(P) = 24{,}038 \text{ cm}^{-1} - 12Dq = 14{,}762 \text{ cm}^{-1}$$

The value of $E = 14{,}762$ cm^{-1} is employed in equation (10–12) along with $Dq = 773$ to yield a value

$$p = 13{,}818 \text{ cm}^{-1} = 15B'$$

The gaseous ion $E(^3P)$ value for Ni^{2+} is $15B = 15{,}840$ cm^{-1} and β is [equation (10–14) or (10–13)]:

$$\beta = \frac{13{,}818}{15{,}840} = \frac{15B'}{15B} = 0.872$$

or

$$\beta^\circ = \frac{15{,}840 - 13{,}818}{15{,}840} \times 100 = 12.8\%$$

To calculate the energies for the $^3T_{1g}(F)$ and $^3T_{1g}(P)$ states, the values $p = 13{,}818$ cm^{-1} and $Dq = 773$ cm^{-1} are substituted into equation (10–12) and the equation is solved for E. Two roots, $E = 14{,}762$ cm^{-1} and $E = 3{,}694$ cm^{-1}, are obtained. Since the transition is $^3A_{2g} \to {}^3T_{1g}(F)$ or $^3T_{1g}(P)$, the absorption bands will correspond to the differences

$$E[^3T_{1g}(F)] - E[^3A_{2g}] \quad \text{and} \quad E[^3T_{1g}(P)] - E[^3A_{2g}]$$

or

$$[^3A_{2g} \to {}^3T_{1g}(F)] = 3{,}694 - [-12(773)] = 12{,}970 \text{ cm}^{-1}$$

and

$$[^3A_{2g} \to {}^3T_{1g}(P)] = 14{,}762 - [-12(773)] = 24{,}038 \text{ cm}^{-1}$$

The agreement of the calculated and experimental values for the 12,970 cm^{-1} band supports the β and Dq values reported above.

CALCULATION OF Δ AND β FOR T_d Co^{2+} COMPLEXES

In a field of tetrahedral symmetry, the ^4F ground state of Co^{2+} is split into $^4A_2$, $^4T_2$, and $^4T_1(F)$. The transitions $^4A_2 \to {}^4T_2$, $^4A_2 \to {}^4T_1(F)$, and $^4A_2 \to {}^4T_1(P)$ are designated as ν_1, ν_2, and ν_3, respectively. The following relationships are used to calculate Δ and β:

$$\nu_1 = \Delta \quad\quad\quad (E-1)$$

$$\nu_2 = 1.5\Delta + 7.5B' - Q \quad\quad\quad (E-2)$$

$$\nu_3 = 1.5\Delta + 7.5B' + Q \quad\quad\quad (E-3)$$

$$Q = \frac{1}{2}[(0.6\Delta - 15B')^2 + 0.64\Delta^2]^{1/2} \quad\quad\quad (E-4)$$

where B' is the effective value of the Racah interelectronic repulsion term in the complex. To repeat equations (10–14) and (10–13),

$$\beta = \frac{B'_{\text{complex}}}{B_{\text{free ion}}}$$

or

$$\beta° = \frac{B_{\text{free ion}} - B'_{\text{complex}}}{B_{\text{free ion}}} \times 100 \qquad \text{(E-5)}$$

To demonstrate the calculation, let us consider the spectrum of tetrahedral Co(TMG)$_4^{2+}$ (where TMG is tetramethylguanidine).[1] The band assigned to v_3 is a doublet with maxima at 530 mμ (18,867 cm^{-1}), $\varepsilon = 204$, and 590 mμ (16,949 cm^{-1}), $\varepsilon = 269$. The near infrared spectrum yields v_2 as a triplet: 1204 mμ (8306 cm^{-1}), $\varepsilon = 91.5$; 1320 mμ (7576 cm^{-1}), $\varepsilon = 85.0$; and 1540 mμ (6494 cm^{-1}), $\varepsilon = 23.5$. The T_1 states are split by spin-orbit coupling to the following extent[2]: $-9/4\lambda'$, $+6/4\lambda'$, and $+15/4\lambda'$. The energy of $^4A_2 \to {}^4T_1(F)$ is obtained by averaging the three peaks for the v_2 band, using the above weighting factors.

$$\frac{9}{4}(6494) = 14{,}612$$

$$\frac{6}{4}(7576) = 11{,}364$$

$$\frac{15}{4}(8306) = 31{,}148$$

Totals $\quad \frac{30}{4} \qquad 57{,}124$

The average energy of v_2 is thus $57{,}124 \div 30/4 = 7617$ cm^{-1}. The energy of the transition from $^4A_2$ to $^4T_1(P)$ (i.e., v_3) is obtained by averaging the two peaks to produce 17,908 cm^{-1}. The series of equations (E–1) to (E–4) are now solved to obtain Δ and β. Adding equations (E–2) and (E–3) produces:

$$\Delta = \frac{v_2 + v_3 - 15B'}{3}$$

and substituting v_2 and v_3 for Co(TMG)$_4^{2+}$ produces $\Delta = 5(1702 \text{ cm}^{-1} - B')$. Subtracting equation (E–2) from (E–3) produces

$$Q = \frac{1}{2}(v_3 - v_2) = 5146 \text{ cm}^{-1}$$

Squaring both sides of equation (E–4) and rearranging produces

$$4Q^2 = \Delta^2 - 18B'\Delta + 225(B')^2 \qquad \text{(E-6)}$$

Substituting $Q = 5146$ cm^{-1} and $\Delta = 5(1702$ cm$^{-1} - B')$ into equation (E–6) yields an equation that can be solved from B'. One root is 821 cm^{-1}, and the other root is negative. When the positive root is substituted into $\Delta = 5(1702$ cm$^{-1} - B')$, the value of $\Delta = 4405$ cm^{-1} is obtained; β is evaluated from equation (E–5).

REFERENCES CITED

1. R. S. Drago and R. L. Longhi, Inorg. Chem., *4*, 11 (1965).
2. R. Stahl-Broda and W. Low, Phys. Rev., *113*, 775 (1959).

Conversion of Chemical Shift Data

APPENDIX F

The chemical shifts, δ (in ppm), relative to $\delta = 0$ for tetramethylsilane for some compounds often employed as external standards are: cyclohexane, -1.6; dioxane, -3.8; H_2O, -5.2; CH_2Cl_2, -5.8; C_6H_6, -6.9; $CHCl_3$, -7.7; and H_2SO_4 (sp. gr. 1.857), -11.6 ppm. (The larger negative value indicates less shielding.) These shifts are obtained on the pure liquids relative to an external standard. As a result, they can be employed to convert data and allow comparison of results between the various materials as external standards. To convert δ obtained toward C_6H_6 as a reference to $Si(CH_3)_4$ (external standard), subtract 6.9 ppm from the C_6H_6 value. One should check to be sure that the sign convention for Δ is that described for protons ($\sigma_S - \sigma_R$).

The conversion of results obtained relative to an external standard to an internal standard is not quite as straightforward. If chemical shift values at infinite dilution in CCl_4 are converted to $Si(CH_3)_4$ as a reference by using the above data, τ values do not result. This procedure converts the data to the reference pure $Si(CH_3)_4$. The difference in δ for $Si(CH_3)_4$ in the pure liquid and at infinite dilution in CCl_4 is about 0.4 ppm. The pure liquid is more shielded.

For fluorine shifts, $F_2 = 0$ ppm is often taken as the standard. Shifts, in ppm, for other liquids relative to F_2 are SF_6, 375.6; $CFCl_3$, 414.3; CF_3Cl, 454.2; CF_4, 491.0; CF_3COOH, 507.6; C_6H_5F, 543.2; SiF_4, 598.9; and HF, 625, where the positive value indicates a more highly shielded fluorine.

Many phosphorus chemical shifts have been reported relative to 85% H_3PO_4 as the standard.

APPENDIX G

Solution of the Secular Determinant for the NMR Coupling of the AB Spin System

To evaluate these matrix elements for a second-order, AB system in Section 7–20 we need to know how the $\hat{I}_A \cdot \hat{I}_B$ operator works. This problem is simplified by defining the so-called *raising and lowering operators*. We shall have occasion to use these operators in other problems. Remember that

$$\hat{I}^2 = \hat{I}_X{}^2 + \hat{I}_Y{}^2 + \hat{I}_Z{}^2 \tag{G-1}$$

but that we can only find simultaneously $\hat{I}^2$ and the component in one direction. The raising and lowering operators, $\hat{I}_+$ and $\hat{I}_-$, are defined by taking linear combinations of $\hat{I}_X$ and $\hat{I}_Y$ such that:

$$\hat{I}_+ = \hat{I}_X + i\hat{I}_Y \tag{G-2}$$

$$\hat{I}_- = \hat{I}_X - i\hat{I}_Y \tag{G-3}$$

These operators have the property that when they operate on a wave function $|I, m_I\rangle$ (i.e., one defined by quantum numbers I and m_I) we get:

$$\hat{I}_+ |I, m_I\rangle = [I(I + 1) - m_I(m_I + 1)]^{1/2} |I, m_I + 1\rangle$$

$$\hat{I}_- |I, m_I\rangle = [(I(I + 1) - m_I(m_I - 1)]^{1/2} |I, m_I - 1\rangle$$

Then we find that these operators work on $|\alpha\rangle$ and $|\beta\rangle$ as follows:

$\hat{I}_+ |\alpha\rangle = 0$ (we cannot raise a $+\frac{1}{2}$ spin by 1 when $I = \frac{1}{2}$)

$\hat{I}_- |\alpha\rangle = |\beta\rangle$

$\hat{I}_+ |\beta\rangle = |\alpha\rangle$

$\hat{I}_- |\beta\rangle = 0$ (a $-\frac{1}{2}$ spin cannot be lowered by 1 when $I = \frac{1}{2}$)

The $\hat{I}_A \cdot \hat{I}_B$ operator for a two-spin AB system is given by:

$$\hat{I}_A \cdot \hat{I}_B = \hat{I}_{ZA} \cdot \hat{I}_{ZB} + \hat{I}_{XA} \cdot \hat{I}_{XB} + \hat{I}_{YA} \cdot \hat{I}_{YB}$$

$$= \hat{I}_{ZA} \cdot \hat{I}_{ZB} + (1/2)(\hat{I}_{+A}\hat{I}_{-B} + \hat{I}_{-A}\hat{I}_{+B}) \tag{G-4}$$

Appendix G Solution of the Secular Determinant for the NMR Coupling of the AB Spin System 741

Equation (8–18) in terms of the raising and lowering operators can be derived by solving equations (8–16) and (8–17) for $\hat{I}_X$ and $\hat{I}_Y$ and substituting this result into the equation for $\hat{I}_A \cdot \hat{I}_B$ in terms of X, Y, and Z components. In operating on the basis set ($\alpha\alpha$, etc.) with the $\hat{I}_A \cdot \hat{I}_B$ operator, the $\hat{I}_A$ spin operator acts only on the A nucleus (the first spin function listed) and the $\hat{I}_B$ operator acts only on the B nucleus (the second one listed). Accordingly,

$$\hat{I}_{ZA}|\alpha\beta\rangle = (1/2)|\alpha\beta\rangle$$

while

$$\hat{I}_{ZB}|\alpha\beta\rangle = -(1/2)|\alpha\beta\rangle$$

The general Hamiltonian for the coupled AB system, equation (8–15), becomes equation (G–5) when expressed in terms of the raising and lowering operators.

$$\hat{H}_{AB} = -v_0(1-\sigma_A)\hat{I}_{ZA} - v_0(1-\sigma_B)\hat{I}_{ZB} + J_{AB}\hat{I}_{ZA}\cdot\hat{I}_{ZB}$$
$$+ (1/2)J_{AB}(\hat{I}_{+A}\hat{I}_{-B} + \hat{I}_{-A}\hat{I}_{+B}) \quad (G-5)$$

Now we return to the evaluation of the matrix elements in the secular determinant given earlier. To evaluate $\langle\alpha\alpha|\hat{H}|\alpha\alpha\rangle$, we need to evaluate the effect of $\hat{H}$ on $|\alpha\alpha\rangle$ and then simply multiply by $\langle\alpha\alpha|$. Thus, we shall proceed by first evaluating $\hat{H}|\alpha\alpha\rangle$, $\hat{H}|\alpha\beta\rangle$, $\hat{H}|\beta\alpha\rangle$, and $\hat{H}|\beta\beta\rangle$.

$$\hat{H}|\alpha\alpha\rangle = [-v_0(1-\sigma_A)(1/2) - v_0(1-\sigma_B)(1/2) + (1/4)J]|\alpha\alpha\rangle$$

The $(1/4)J$ arises from $\hat{I}_{ZA}\cdot\hat{I}_{ZB}|\alpha\alpha\rangle$ for both $\hat{I}_{+A}\hat{I}_{-B}|\alpha\alpha\rangle$ and $\hat{I}_{-A}\hat{I}_{+B}|\alpha\alpha\rangle$ equal zero, since $\hat{I}_{+A}|\alpha\alpha\rangle = 0$ and $\hat{I}_{+B}|\alpha\alpha\rangle = 0$. Rearranging the above expression for $\hat{H}|\alpha\alpha\rangle$, we have

$$\hat{H}|\alpha\alpha\rangle = \left[v_0\left(-1 + \frac{1}{2}\sigma_A + \frac{1}{2}\sigma_B\right) + \frac{1}{4}J\right]|\alpha\alpha\rangle$$

$$\hat{H}|\alpha\beta\rangle = \left[-v_0(1-\sigma_A)\frac{1}{2} - v_0(1-\sigma_B)\left(-\frac{1}{2}\right) - \frac{1}{4}J\right]|\alpha\beta\rangle + \frac{1}{2}J[0 + |\beta\alpha\rangle]$$

$$= \left[+\frac{1}{2}v_0(\sigma_A - \sigma_B) - \frac{1}{4}J\right]|\alpha\beta\rangle + \frac{1}{2}J|\beta\alpha\rangle$$

$$\hat{H}|\beta\alpha\rangle = \left[-v_0(1-\sigma_A)\left(-\frac{1}{2}\right) - v_0(1-\sigma_B)\frac{1}{2} - \frac{1}{4}J\right]|\beta\alpha\rangle + \frac{1}{2}J[|\alpha\beta\rangle + 0]$$

$$= \left[-\frac{1}{2}v_0(\sigma_A - \sigma_B) - \frac{1}{4}J\right]|\beta\alpha\rangle + \frac{1}{2}J|\alpha\beta\rangle$$

$$\hat{H}|\beta\beta\rangle = \left[-v_0(1-\sigma_A)\left(-\frac{1}{2}\right) - v_0(1-\sigma_B)\left(-\frac{1}{2}\right) + \frac{1}{4}J\right]|\beta\beta\rangle$$

$$= \left[v_0\left(1 - \frac{1}{2}\sigma_A - \frac{1}{2}\sigma_B\right) + \frac{1}{4}J\right]|\beta\beta\rangle$$

With these quantities evaluated, it is a simple matter to multiply them by $\alpha\alpha$, $\alpha\beta$, etc., and thus to evaluate all the matrix elements in the secular determinant.

For example, $\langle\beta\alpha|\hat{H}|\alpha\beta\rangle = 0 + \frac{1}{2}J$. The results for all the matrix elements are contained in Table G–1. The result of $\hat{H}|\alpha\alpha\rangle$ is seen to yield $|\alpha\alpha\rangle$ back again, so this wave function is an eigenfunction of the Hamiltonian, as is $|\beta\beta\rangle$. However, $|\beta\alpha\rangle$ and $|\alpha\beta\rangle$ are not eigenfunctions. Accordingly, we find off-diagonal values of $(\frac{1}{2})J$ connecting $\alpha\beta$ and $\beta\alpha$. The determinant is block diagonalized, with two frequencies given as:

$$\frac{E_1}{h} = v_0\left[-1 + \frac{1}{2}\sigma_A + \frac{1}{2}\sigma_B\right] + \frac{J}{4} \tag{G-6}$$

$$\frac{E_4}{h} = v_0\left[1 - \frac{1}{2}\sigma_A - \frac{1}{2}\sigma_B\right] + \frac{J}{4} \tag{G-7}$$

The energies of E_2 and E_3 are obtained by solving the 2×2 block. This solution can be simplified by making a few definitions. The definitions to be made are not obvious *a priori* choices. They come with experience in solving determinantal equations and trying to relate the results to observed spectra. We will substitute:

TABLE G–1. Evaluation of the Matrix Elements for the Coupled AB System

	$\alpha\alpha$	$\alpha\beta$	$\beta\alpha$	$\beta\beta$
$\alpha\alpha$	$v_0\left[-1 + \frac{\sigma_A}{2} + \frac{\sigma_B}{2}\right] + \frac{J}{4} - \frac{E}{h}$	0	0	0
$\alpha\beta$	0	$+\frac{1}{2}v_0(\sigma_A - \sigma_B) - \frac{J}{4} - \frac{E}{h}$	$\frac{J}{2}$	0
$\beta\alpha$	0	$\frac{J}{2}$	$-\frac{1}{2}v_0(\sigma_A - \sigma_B) - \frac{J}{4} - \frac{E}{h}$	0
$\beta\beta$	0	0	0	$v_0\left[1 - \frac{\sigma_A}{2} - \frac{\sigma_B}{2}\right] + \frac{J}{4} - \frac{E}{h}$

$$\left(\frac{1}{2}\right)v_0(\sigma_A - \sigma_B) = \frac{\Delta}{2} = C\cos 2\theta$$

and

$$\frac{J}{2} = C\sin 2\theta$$

where $C = (\frac{1}{2})(J^2 + \Delta^2)^{1/2}$. The geometrical relation of these defined quantities is shown in Fig. G–1. With these definitions the 2×2 block of the secular determinant becomes:

$$\begin{vmatrix} -C\cos 2\theta - \frac{1}{4}J - \frac{E}{h} & C\sin 2\theta \\ C\sin 2\theta & C\cos 2\theta - \frac{1}{4}J - \frac{E}{h} \end{vmatrix} = 0$$

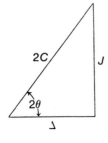

Appendix G Solution of the Secular determinant for the NMR Coupling of the AB Spin System

The solutions to this determinant are

$$\frac{E_2}{h} = -\frac{1}{4}J - C \qquad (G-8)$$

$$\frac{E_3}{h} = -\frac{1}{4}J + C \qquad (G-9)$$

To get the wave functions for these two states, we substitute the energies one at a time into the determinant given above. The matrix corresponding to this determinant (this matrix has the same form as the determinant) is multiplied by a column matrix (vector) of the coefficients. The product is zero. Expansion produces the two resulting, linearly dependent equations, which can be solved for C_1 and C_2 when the normalization requirement is imposed. In general, this is done for each energy. The procedure is similar to that used in Chapter 3 when Hückel theory was discussed. The wave functions ψ_2 and ψ_3 so obtained are listed below along with ψ_1 and ψ_4.

$$\psi_1 = |\alpha\alpha\rangle \qquad (G-10)$$

$$\psi_2 = \cos\theta|\alpha\beta\rangle - \sin\theta|\beta\alpha\rangle \qquad (G-11)$$

$$\psi_3 = \sin\theta|\alpha\beta\rangle + \cos\theta|\beta\alpha\rangle \qquad (G-12)$$

$$\psi_4 = |\beta\beta\rangle \qquad (G-13)$$

Index

A_1 symmetry, 169
AB molecule spectrum, relative magnitudes of J and delta, 263–264
AB spin system, nuclear magnetic resonance coupling, solution of secular determinant, 740–743
Absorption band, effects giving rise to, 154–156
Activation enthalpy, nuclear magnetic resonance, 290–295
Adiabatic electron affinity, M^-, 666
Allyl, intermediate neglect of differential overlap, 80–81
Allyl radical
  pi system, 69
  wave function, 76–78
Alternately double bonded hydrocarbon, molecular orbital theory, 83–84
Ammonia, ultraviolet photoelectron spectroscopy, 673
Angular momentum, 213
Angular overlap model, 452–458
Anharmonic vibration, 149–150
Anion, ambidentate, 198
Anionization, secondary ion mass spectrometry, 678
Anisotropy
  g value, 380–383
  hyperfine coupling, 383–390
A.O. basis set, nitrite ion, 61
Appearance potential, 655, 664–665
Asymmetric stretch, 154
Asymmetric top, 190–191
Asymmetry parameter, 269, 606
Atom, equivalent, 12–13
Atom r, q_r, electron density, 72–73
Atomic force microscopy, 680–681
Atomic transition, 92–95
Auger electron spectroscopy, 679

Band assignment, 156–160
  criteria, 118–119
Base-iodine addition compound, molecular orbital, 134
Base-iodine solution, spectra, 133
Basis set, 36

Bending vibration, 92
Benzene, molecular orbital, 116
Beta calculation, O_hNi(II) complex, 438–443
Bimetallic system, nuclear magnetic resonance, 543–552
Bloch equation, 218–220
Block diagonalized, 34
Blue shift, 118–119
Boltzmann constant, 95
Bond dissociation energy, force constant, 152
Bond order, 73, 74
Bond strength, frequency shift, 194–202
Bonding parameter, 450–458
  tetragonal complex, 450–452
Born-Oppenheimer approximation, 94
Bragg's law, 690
Bravais lattice, 703

^{13}C magnetic resonance, 319–323
C_{2v} point group, character table, 114
Cambridge Crystallographic Database, 689
Cambridge Structural Database, 710
Carbon dioxide, fundamental vibration modes, 155
Carbonyl compound, 131
Cartesian basis set, 36
Cartesian coordinate system, 30–43
Cationization, secondary ion mass spectrometry, 678
Character table, 35–36, 41–43
  C_{2v} point group, 114
  double group, 724–726
  symmetry group, 712–723
Charge-transfer spectrum, solvent polarity, 135–137
Charge-transfer transition, 127–128
Chemical exchange, spectral line width, 95–98
Chemical ionization, 654
Chemical shift
  data conversion, 739
  fluorine, 235
  interpretation, 232–240, 241–243

  local effects, 232–238
  measurement, 229–232
  remote effects, 238–240
Chromophore, heme, 171
Circular dichroism, 137–141
Closed shell system
  antishielding, 615
  shielding, 615–616
CO_2 molecule
  symmetric stretching vibration, polarizability change, 165
  vibrations, 155
Cobalt-substituted carbonic anhydrase, 543
Cobalt(II)
  diastereoisomerism, 535–537
  diastereotopism, 535–537
Co(Meacacen), electron paramagnetic resonance spectrum, 563
Complete neglect of differential overlap (CNDO), 80
Concentration, determination, 99–103
Configuration interaction, 117–118
Conjugate, 22
Constant-acid, constant-base frequency shift-enthalpy relations, 195
Contact shift, 507–508
  factoring, 512–514
  spin density, 514–518
Contrast agent, nuclear magnetic resonance, 530–531
Coordinate, normal, 93
Coordinate system, NH_3 location, 63
Coordination
  donor molecule spectra change, 196–198
  spectra and symmetry changes, 198–202
Copper (II) complex, 590
Coupling, quantum mechanical description, 259–263
Coupling combination band, symmetry, 176
Coupling constant, determining signs, 268–269
Crystal, 695–698
  classes, 701
  growth, 695

745

Crystal *(continued)*
    liquid-diffusion vessel, 695
    vapor diffusion, 695
    mounting, 697–698
    selection, 696–697
    size, 696
    X-ray absorption, 696–697
Crystal field, 416–432
Cu_2Co_2 superoxide dismutase, 545
Cu_2Co_2SOD derivative, ^1H nmr spectrum, 546
Curie relaxation, equations, 523–524
Curie temperature, 471
Cytochrome C peroxidase, X-band spectrum, 585

d orbital, 126–127
d orbital energy, ligand, 416–419
    symmetry, 419–427
d orbital energy level, distortion, 443–447
d^1, 578–579
d^2, 579
d^3, 579–581
d^4, 581
d^5, tetragonal field, 585
d^5 high spin, 583–586, 590
d^5 low spin, 581–583
d^9, 590–591
Decomposition formula, representation, 44–46
Degenerate representation, 38–40
Delta, formal charge, 73
Depolarization ratio, 169
Depolarized line, 169
Diamagnetic term, 234
Diamagnetism, 471–473
Diastereoisomerism, 535
    cobalt(II), 535–537
Diastereotopism, cobalt(II), 535–537
Diatomic molecule, 190
    electronic energy level, 109–110
    energy state, 94
    Morse energy curve, 109–110
    vibrational energy level, 110
Differential overlap, intermediate neglect, 78–83
Diffraction pattern, 693
Diffractometer, 697, 698
Dimeric complex, ^1H nmr, 544–545
Dimeric copper adenine complex, 592
1,2-Dimethyl imidazole, Mössbauer spectroscopy, 635
Dipolar hyperfine coupling, electrons in d orbitals, 575
Dipolar relaxation, equations, 522–523
Dipole moment, 13–14
Direct dipolar coupling, 340–341
Direct lattice, monoclinic cell, 690–692
Direct product, 46
Distance, polarizability, 167
Distortion, d orbital energy level, 443–447
Doppler energy, 627
Double group, 427–429
    character table, 724–726
Double resonance, nuclear quadrupole resonance spectroscopy, 620–622

Double resonance electron paramagnetic resonance, 594
Double resonance experiment, 267–268

e^2Qq, nuclear quadrupole resonance spectroscopy, 616–617
    crystal lattice effects, 617
    data interpretation, 616–617
Effective spin Hamiltonian, 363
Eigenvalue, 53, 81–82
Eigenvector, 81–82
Electric field gradient, 605
    molecular structure, 612–616
Electric quadrupole, intensity, 127
Electromagnetic radiation, plane-polarized, 91
Electron correlation, 80
Electron density, 72–73
    atom r, q_r, 72–73
Electron-electron interaction, 409–413
    term symbols, 409–410
Electron ionization, 654
Electron loss, secondary ion mass spectrometry, 678
Electron paramagnetic resonance (epr) spectroscopy, 360–399
    anisotropic effects, 380–399
    applications, 397–399
    Co(Meacacen), 563
    line width, 394–396
    metal cluster, 591–594
    principles, 360–363
    spectrum presentation, 368–370
    transition metal ion complexes, 559–594
        first-row survey, 578–591
    triplet state, 390–392
Electron relaxation, spin-orbit coupling, 578
Electron spin, 503–507
Electron spin state, zero-field splitting, 573
Electronic absorption spectroscopy, 109–143
    applications, 130–137
    nomenclature, 113–116
Electronic energy level, diatomic molecule, 109–110
Electronic relaxation time, 527–530
Electronic spectrum
    energy curve, 111–112
    oxo-bridged dinuclear iron center, 458–459
    structural evidence, 447–450
Electronic state, change, 92
Electronic transition, intensity, 120–130
Electrons in d orbitals, dipolar hyperfine coupling, 575
ENDOR experiment, 594
Energy curve, electronic spectrum, 111–112
Energy state, diatomic molecule, 94
Equivalent atom, 12
Et_4N^+, ethyl group disorder, 710
Ethylmethyl sulfide, Mössbauer spectroscopy, 635

Ethylpyridine, mass spectra of 3 isomers, 658
Eulerian cradle, 698–699
Expectation value, 53, 504–507
Extended X-ray absorption fine structure, 681

Far infrared vibrational frequency correlation chart, 188–189
Fast atom bombardment, 654
Fast chemical reaction, spectrum effect, 257–258
Fast nuclear relaxation, two-dimensional spectrum, 532–533
^{57}Fe
    magnetic splitting, 634
    non-cubic electronic environment, 632
    quadrupole splitting, 634
Fe_2S_2 ferrodoxin, ^1H nmr spectrum, 546–548
    oxidized, 547
    reduced, 547
$Fe_3(CO)_{12}$
    Mössbauer spectroscopy, 640
    structure, 640
$FeFe(CN)_6$, Mössbauer spectroscopy, 629
Fermi contact contribution, 576
Fermi contact coupling, 501
Fermi resonance, 155–156
    symmetry, 176
Ferrimyoglobin, X-band spectrum, 585
Field desorption, 654
    glutamic acid, 663
Field gradient
    pyridine, 619
    semiempirical approach, 615
Field gradient q, 269
Field gradient tensor, principal axis system, 605–606
Field ionization, 654
    glutamic acid, 663
Field ionization technique, 662–663
Fine splitting, 572
Fingerprinting, 106, 130–132, 184–186
    carbonyl compound, 131
    inorganic system, 132
    mass spectrometry, 657–659
    saturated molecule, 130–131
First overtone, 151
Fluorine, chemical shift, 235
Force constant, 149, 151–153
    bond dissociation energy, 152
Force constant matrix, compliance constants, 160
Formal charge, 73, 75–78
    delta, 73
Formaldehyde
    molecule shapes, 114
    spectrum, 123–124
Fourier transform epr, 594
Fourier transform ion cyclotron resonance technique, 665–667
Fourier transform nmr, 309–319
    experiment optimization, 314–315
    multipulse methods, 328–340
    other nuclei, 323–325

principles, 309–314
relaxation, 326–328
selective excitation and suppression, 333–334
sensitivity-enhancement methods, 331–332
spectral density, 326–328
T_1 measurement, 315–316
Franck-Condon factor, 672
Franck-Condon principle, 111
Free induction decay, 621
Free ion, spin-orbit coupling, 413–416
Free ion electronic state, 409–416
Frequency shift, bond strength, 194–202
Fundamental frequency, 151
Fundamental vibration frequency, 150

$g \rightarrow u$, 123
g value, anisotropy, 380–383
Gamma ray, 626, 627
Gas, spectra, 186–192
Geometric transformation, 30–34
Glide plane, 11
Glutamic acid, 663
Goniometer, 698
Goniometer head, 697
Gross atomic population, 75
Group H order, 20
Group multiplication table, 19–25
element classes, 23–25
properties, 19–22
Group theory, 18–35
element rules, 19
generation of symmetry combinations, 18
zero quantum mechanical integrals, 18
Group vibration, 160–162
limitations, 161

^{1}H nmr spectrum
Cu_2Co_2SOD derivative, 546
Fe_2S_2 ferrodoxin, 547
H_2, equivalent orientations, 1
H_2O molecule, internal coordinates, 157
Hahn spin-echo experiment, 328
Harmonic oscillator approximation, 156
Harmonic vibration, 149–150
Hückel procedure, 68–72
extended, 74–78
molecular orbital, 68–72
Heme, chromophore
ultraviolet (Soret), 171
visible, 171
Heme Raman spectrum, 170–171
Hemin-imidazole-cyanide, 538–539
Hemoglobin, schematic formula, 582
High melting solid, vapor over, 663
High resolution electron energy loss spectroscopy, 679–680
High resolution infrared spectrum, $Mo(CO)5Br$, 201
High resolution nmr, solids, 347–348
High spin-low spin equilibrium, magnetism, 489–490
Higher-state mixing, 117
Hooke's law constant, 149

Hybridization of ligand, 577
Hydrocarbon system, pi orbital, 68
Hydrogen atom, 363–368
Hydrogen-bonded phenol, infrared spectra, 194
Hydrogen bonding solvent, blue shift, 119
Hyperfine coupling, 571–576
anisotropy, 383–390
Hyperfine coupling constant, isotropic system, 374–379
Hyperfine splitting, nitroxide, 399
Hypsochromic shift, 118–119

$\hat{i}$, properties, 224–225
Identity, 2
Identity element, 19
Infrared group frequency range, 186–187
Infrared line, symmetry, 172–176
Infrared spectra, 194
Infrared spectroscopy
inorganic structure, 192–194
procedures, 179–184
Inorganic anion, characteristic absorption maxima, 132
Inorganic Crystal Structure Data Base, 710
Inorganic material, infrared group frequency range, 186–187
Inorganic structure, 192–194
Inorganic system, fingerprinting, 132
Integral zero, 121
Integrated intensity, 120
Intensity
electric quadrupole, 127
magnetic dipole, 127
spin-orbit, 124–126
spin-orbit coupling, 124–126
vibronic coupling, 124–126
Interaction force constant, 157
Interatomic ring current, 241
Intermediate neglect of differential overlap, allyl, 80–81
Intervalence electron transfer band, 459–461
Inversion center, 2
Iodine
molecular addition compound, 133–135
spectra, 133
Ion pair, N-methylpyridinium iodide, 136
Ion pairing, 538
Ionization, 650–682
Ionization efficiency curve, 664
Ionization potential, 664–665
Iron(II) complex
Mössbauer spectroscopy, 638
partial quadrupole splitting parameter, 644
Iron(III) complex, Mössbauer spectroscopy, 638
Iron (III) complex
spectrum, 584
splitting of energy levels, 584
Iron porphyrin, spin delocalization, 539–543
Iron-sulfur compound, center shift, 641

Irreducible representation, 34–35
wave function, 59–60
Isomer shift, Mössbauer spectroscopy, 630–631
Isomolar solution, Job's method, 106
Isosbestic point, 103–106
Isotope, mass spectrum, 660–662
Isotropic system, hyperfine coupling constant, 374–379

Jahn-Teller effect, 429–430
Job's method, isomolar solution, 106

Kramers' doublet, 559
rhombic symmetry, 586
Kramers' rule, 559

Lanthanides, 510–512
Laser desorption, 654
Lattice plane
Miller indices, 690–691
direct, 691
reflection, 691
Lattice vibration, 183
Ligand, d orbital energy, 416–419
symmetry, 419–427
Ligand hybridization, 577
Ligand hyperfine coupling, 576–578
Line width
chemical exchange processes, 561
electron paramagnetic resonance spectroscopy, 394–396
Linear combination of atomic orbitals-molecular orbital method, 54
Linear molecule
bending mode, 154
normal modes of vibration, 153
rotational mode, 154
Linear polyatomic molecule, 190
Local symmetry, 124
Longitudinal relaxation, 216
Low energy electron diffraction, 678–679

M^-, adiabatic electron affinity, 666
Magnetic behavior, types, 471–476
Magnetic coupling, metal ion cluster, 430–432
Magnetic dipole, intensity, 127
Magnetic field
magnet in, 213–214
nuclear quadrupole resonance spectroscopy, 611–612
T state, 570
Magnetic splitting, ^{57}Fe, 634
Magnetic susceptibility, measurement, 490–491
Magnetism, 469–494
high spin–low spin equilibrium, 489–490
intramolecular effects, 486–489
susceptibility measurement applications, 483–486
Magnetization vector, 215–216
Magnetocircular dichroism, 141–143
Marginal oscillator, peak, 609
Mass spectrometer, 653

Mass spectrometric shift rule, 660
Mass spectrometry, 650–665
 fingerprint, 657–659
 instrument operation, 650–665
 interpretation, 659–660
 isotope, 660–662
 molecule high energy electron combination, 655–657
 presentation, 650–665
Matrix, 27–30
Matrix formulation, molecular orbital calculation, 56–57
Matrix multiplication, 27–28
cis MB_4A_2
 coordinates, 643
 geometries, 643
trans MB_4A_2
 coordinates, 643
 geometries, 643
Metal ion cluster
 electron paramagnetic resonance spectrum, 591–594
 magnetic coupling, 430–432
1-Methyl imidazole, Mössbauer spectroscopy, 635
N-Methylpyridinium iodide, ion pair, 136
Microstate configuration, 409
Microwave spectroscopy, 177–179
Miller indices, lattice plane, 690–691
 direct, 691
Mirror plane, 4–5
$Mo(CO)_5Br$, high resolution infrared spectrum, 201
Molar absorptivity, 118–119
Molecular addition compound, iodine, 133–135
Molecular orbital
 base-iodine addition compound, 134
 benzene, 116
 calculations, 68–72
  Hückel procedure, 68–72
  projecting, 60–68
Molecular orbital calculation, matrix formulation, 56–57
Molecular orbital theory, 52–85
 alternately double bonded hydrocarbon, 83–84
Molecular structure
 electric field gradient, 612–616
 shift reagent determination, 550
Molecular transition, 92–95
Molecular vibration
 radiation absorption, 150–151
 symmetry, 172–179
Molecular weight determination, 662–663
Molecule
 asymmetric, 13
 dissymmetric, 13
 structures of excited states, 137
Monoclinic cell lattice, 690–692
Morse energy curve, diatomic molecule, 109–110
Morse potential, 109
Mössbauer emission spectroscopy, 635
Mössbauer isotope, 637

Mössbauer spectroscopy, 626–645
 applications, 635–645
 1,2-dimethyl imidazole, 635
 ethylmethyl sulfide, 635
 $Fe_3(CO)_{12}$, 640
 $FeFe(CN)_6$, 629
 iron(II) complex, 638
 iron(III) complex, 638
 isomer shift, 630–631
 1-methyl imidazole, 635
 non-cubic electronic environment, 632
 piperidine, 635
 principles, 626–629
 pyridine, 635
 quadrupole interaction, 631–633

$3N$–6(5) rule, 153–154
^{14}N quadrupole transition, pyridine, 619
Naphthalene, 707
 atomic coordinates, 708
Neighbor anisotropy, 238–239
Néel temperature, 471
Net atomic population, 74
Neutron diffraction, 701
NH_3 location, coordinate system, 63
Nickel, planar-tetrahedral equilibrium, 534–535
Nitrite ion
 A.O. basis set, 61
 pi-molecular orbital, 63
Nitrogen, ultraviolet photoelectron spectroscopy, 672
Nitroxide, hyperfine splitting, 399
NO, photoelectron spectrum, 676
NO_3, vibrational mode symmetry
 C_{2v} structure, 198
  correlation chart, 198
 C_s structure, 198
Non-diagonal representation, 36–40
Non-equivalent proton, 246
Non-linear polyatomic molecule, 190–192
Nonlinear molecule
 bending mode, 154
 normal modes of vibration, 153
 rotational mode, 154
Normal coordinate, 93
Normal coordinate analysis, 156–160
Normal mode, 92
Normal vibration, 92
NSF_3, vibration, fundamental frequencies, 193
Nuclear hyperfine splitting, 363–379
 more than one nucleus, 370–374
Nuclear magnetic resonance (nmr)
 activation enthalpy, 290–295
 bimetallic system, 543–552
 contrast agent, 530–531
 fluxional behavior, 300–305
 intramolecular rearrangements, 300–305
 liquid crystal solvent studies, 342–347
 paramagnetic substances in solution, 500–552
 quadrupolar nucleus, 319
 rate constant, 290–295
 reaction order determination, 295–297
 relaxation, 227–243

 solid studies, 341–342
 thermodynamic data evaluation, 290–291
NMR coupling, AB spin system, solution of secular determinant, 740–743
NMR kinetics, 201–309
 applications, 297–300
NMR spectroscopy, 211–270
 quantum mechanics, 224–227
 second-order systems, 265–266
 slow passage experiment, 220–224
 transition, 217–218
Nuclear Overhauser effect, 306–309, 531–532
Nuclear quadrupole interaction, 392–394
Nuclear quadrupole moment, 605
Nuclear quadrupole resonance (nqr) spectroscopy, 604–622
 applications, 616–620
 double resonance, 620–622
 e^2Qq, 616–617
  crystal lattice effects, 617
  data interpretation, 616–617
 magnetic field, 611–612
 structural information, 618–620
Nuclear relaxation, paramagnetic system, 519–524
Nuclear spin quantum number, 626

O-S-O bending mode, 154
Octahedral complex, vibration, 126
Octahedral iron (III) complex
 spectrum, 584
 splitting of energy levels, 584
Octahedral symmetry, deviations, 582
O_h field
 electronic spectra survey, 433–438
 Tanabe and Sugano diagram, 732–734
O_hNi(II) complex
 beta calculation, 438–443, 735–737
 D_q calculation, 438–443, 735–737
One electron molecular orbital, 54
Operator, 52–56
OPF_3, vibration, fundamental frequencies, 193
Optical activity, 13
Optical rotary dispersion, 137–141
Orbital, nitrite ion, 63
Order of group h, 20
Organic material, infrared group frequency range, 186–187
Orgel diagram, 427
Oscillator strength, 120
Overhauser effect, 306–309
Overlap population, 74–75
Oxo-bridged dinuclear iron center, electronic spectrum, 458–459
Oxygen
 ultraviolet photoelectron spectroscopy, 674
 vertical ionization data, 675
Oxyhemerythrin, resonance Raman analysis, 171

P-branch, 191
p orbital, 126–127

P_{AB}, bond order, 73–74
Parallel band, 190
Paramagnetic compound, properties, 500–503
Paramagnetic Mössbauer spectrum, 633–635
Paramagnetic substances in solution, nmr, 500–552
Paramagnetic system, nuclear relaxation, 519–524
Paramagnetic term, 233
Paramagnetism, simple system, 473–476
Partial quadrupole splitting parameter, iron(II), 644
Peak assignment, T_1, 317–319
Perturbation theory, 57–59
Phenol, infrared spectra, 194
Phonon mode, 183
Photoacoustic Fourier transform spectroscopy, 182
Photoelectron spectroscopy, 667
Photoemission, 669
Pi orbital, hydrocarbon system, 68
Pi system, allyl radical, 69
Piperidine, Mössbauer spectroscopy, 635
Planar-tetrahedral equilibrium, nickel, 534–535
Planck's constant, 150
Plasma desorption, 654
Point group, 8–11
  symmetry element, 9
Polarizability
  distance, 167
  vibration, 165–167
Polarized absorption spectra, 128–130
Polarized line, 169
Polyatomic molecule, vibration, 153–162
Product, 19
Product ground state wave function, 84–85
Projection operator, 60
Protein Data Bank, 710
Protonation, 119
Pseudocontact shift, 508–510
  factoring, 512–514
Pyrene, 707
Pyridine
  field gradient, 619
  Mössbauer spectroscopy, 635
  ^{14}N quadrupole transition, 619

Q-branch, 190, 191
Quadrupolar nucleus, nmr, 319
Quadrupole energy level
  axially symmetric field, 607
  spherical field, 607
Quadrupole interaction, Mössbauer spectroscopy, 631–633
Quadrupole moment nucleus, 269–270
Quadrupole splitting, ^{57}Fe, 634
Quadrupole transition, energies, 607–611
Quantum mechanics
  nmr spectroscopy, 224–227
  symmetry, 59–68
Quasi-equilibrium theory, 660

R-branch, 191
Radiation
  allowed transitions, 95
  corresponding energies, 91–92
  forbidden transitions, 95
  nature, 90–91
  selection rules, 95
  wave numbers, 92
Radiation absorption, molecular vibration, 150–151
Raman line
  depolarized, 168–169
  polarized, 168–169
  symmetry, 172–176
Raman spectroscopy, 162–171
  inorganic structure, 192–194
  procedures, 184
  selection rules, 164–168
Rapid nuclear quadrupole relaxation, 269
Rate constant, nmr, 290–295
Reciprocal lattice, monoclinic cell, 690–692
Reciprocal space, 691–692
Reflectance technique, 181
Reflection, 691–692
  lattice plane, 691
Relaxation, 215–216
  Fourier transform nmr, 326–328
  nmr, 227–243
  spectral line width, 95–98
Relaxometry, 524–527
Representation, 30–34, 43–44
  decomposition formula, 44–46
  degenerate, 38–40
  irreducible, 34–35
  non-diagonal, 36–40
Resonance Raman analysis, oxyhemerythrin, 171
Resonance Raman spectroscopy, 170–171
Rhombic symmetry, Kramers' doublet, 586
Rotating axis system, 214–215
Rotating frame, 214–215
Rotation axis, 2–4
  dissymmetric, 13
  improper, 5
  proper, 2
Rotation-reflection axis, 5–7
Rotational line, Zeeman splitting, 179
Rotational Raman spectrum, 179
Rotational state, 94

S=1/2 system, orbitally non-degenerate ground states, 565–569
Saturated molecule, fingerprinting, 130–131
Scanning tunneling microscopy, 680–681
Scattering factor curve, 694
SCF-INDO, 78–83
Screw axis, 11, 703
Second-order spectra, 259
Second overtone, 151
Secondary ion mass spectrometry, 654, 677–678
  aluminum target, 677
  anionization, 678
  cationization, 678
  electron loss, 678
  sputtering, 678
  vs. mass number/ion charge, 677
Self-consistent field, 79
Shift reagent, 549–552
Simple harmonic motion, 149
Single crystal study, polarized, 129–130
Site symmetry, 183
SO_2, infrared spectrum, 154
Solvent, Z-values, 136
Solvent polarity, charge-transfer spectrum, 135–137
Space symmetry, 11
Spectral density, Fourier transform nmr, 326–328
Spectral line width
  chemical exchange, 95–98
  relaxation, 95–98
Spectroscopy, 90–106
  applications, 98–106
  See also Electronic absorption spectroscopy
Spherical top, 190–191
Spin delocalization, iron porphyrin, 539–543
Spin density, contact shift, 514–518
Spin echo, 329–331
Spin Hamiltonian, 396–397, 573–574
Spin-lattice relaxation, 216
Spin-orbit coupling, 117, 483–486
  electron relaxation, 578
  free ion, 413–416
  intensity, 124–126
  large, 569–571
  T state, 571
Spin-polarization, 375
  results, 576
Spin saturation labeling, 305–306
Spin-spin coupling
  bond nature, 247–249
  bond number, 247–249
  scalar mechanisms, 249–251
  structure determination, 252–257
Spin-spin relaxation, 216
Spin-spin splitting, 243–270
  spectrum effect, 243–246
Spin temperature, 621
Spin-tickling experiment, 267–268
Sputtering, secondary ion mass spectrometry, 678
Square matrix, 27
Square planar Schiff base ligand, 590
Sternheimer effect, 615
Stokes line, 169
Stretching vibration, 92
Structure determination, spin-spin coupling, 252–257
Sublimation heat, 663
Sulfur dioxide, three fundamental vibrations, 155
Superparamagnetism, 491–494
Superregenerative oscillator, peak, 608–609
Surface science technique, 667–681

Susceptibility measurement applications, magnetism, 483–486
Symmetric stretch, 154
Symmetric stretching vibration, CO_2 molecule, polarizability change, 165
Symmetric top, 190–191
Symmetry
  center, 2
  coupling combination band, 176
  definition, 1
  Fermi resonance, 176
  infrared line, 172–176
  plane of, 4–5
  quantum mechanics, 59–68
  Raman line, 172–176
  X-ray crystallography, 701–710
    molecular vs. crystallographic, 707
  See also Space symmetry
Symmetry element, 2–7
  equivalent, 12–13
  point group, 9
Symmetry group, character table, 712–723
Symmetry operation, 12
Symmetry species of state, 115
Synchrotron radiation, 701

T state
  magnetic field, 570
  spin-orbit coupling, 571
T_1, peak assignment, 317–319
Tanabe and Sugano diagram, 426
  O_h field, 732–734
  $T_d Co^{II}$ complex, calculation of delta($10Dq$) and beta, 735–737
Tetragonal complex, bonding parameter, 450–452
  pi, 450–452
  sigma, 450–452
Tetragonal field, d^5
  strong, 585
  weak, 585
Thermal desorption, 654
Thermodynamic data evaluation, nmr, 290–291
Thiocarbonyl compound
  carbonyl compound, 131
  spectra, 131

Through-space dipolar coupling, 501
Total ionization unit, 652
Transition metal ion, 409–462
  g-values, 564–570
Transition metal ion complexes, epr spectrum, 559–594
  first-row survey, 578–591
Transition moment integral, 120–122
Transition probabilities, 225–227
Transmission infrared, 181
Transverse relaxation time, 216
Triplet state, epr spectroscopy, 390–392
Two-dimensional nmr, 334–340
Two-dimensional spectrum, fast nuclear relaxation, 532–533

$u \to g$, 123
Ultraviolet photoelectron spectroscopy, 671–677
  ammonia, 673
  nitrogen, 672
  oxygen, 674
Ultraviolet radiation, 94
Unit matrix, 27
Unpaired electron density, 375

Van Vleck's equation, 476–483
Variation energy, 55
Variational principle, 55
Vector, 25–26
Vibration, 92–94
  bending, 92
  harmonic, 149–150
  nomenclature, 172
  normal, 92
  normal modes for common structures, 727–731
  NSF_3, fundamental frequencies, 193
  octahedral complex, 126
  OPF_3, fundamental frequencies, 193
  polarizability, 165–167
  polyatomic molecule, 153–162
  stretching, 92
  symmetric, 165
  See also Specific type
Vibrational energy level, diatomic molecule, 110

Vibrational energy state, 92
Vibrational mode symmetry, NO_3^-
  C_{2v} structure, 198
  correlation chart, 198
  C_s structure, 198
Vibronic coupling, intensity, 124–126

Wave function, 52
  allyl radical, 76–78
  irreducible representation, 59–60
  mixing p-character, 127
  properties derived from, 72–74
  symmetrized, 84–85

X-ray
  atom and structure scattering, 693–695
  diffraction, 690
X-ray absorption near-edge structure, 681
X-ray crystallography, 689–710
  area detectors, 700–701
  avoiding crystallographic mistakes, 705–707
  computers, 700
  crystallographic data, 710
  diffraction equipment, 698
  diffractometer data collection, 698–700
  future developments, 700–701
  methodology, 698–700
  principles, 690–695
  quality assessment, 707–709
  reflection conditions, 704
  space-group determination, 704–705
  space groups, 702–703
  symmetry, 701–710
    molecular vs. crystallographic, 707
  symmetry elements, 704
  vs. neutron diffraction, 701
X-ray photoelectron spectroscopy, 668–671

Z-axis compression, in D_{4h}, 564
Zeeman Hamiltonian, 565
Zeeman splitting, rotational line, 179
Zero-field splitting, 391, 571–576
  anisotropic, 573
  electron spin state, 573